AF411432

Enhanced Geothermal Systems and other Deep Geothermal Applications throughout Europe: The MEET Project

Enhanced Geothermal Systems and other Deep Geothermal Applications throughout Europe: The MEET Project

Editors

Béatrice A. Ledésert
Ronan L. Hébert
Ghislain Trullenque
Albert Genter
Eléonore Dalmais
Jean Hérisson

MDPI • Basel • Beijing • Wuhan • Barcelona • Belgrade • Manchester • Tokyo • Cluj • Tianjin

Editors

Béatrice A. Ledésert
Geosciences and
Environment Cergy
CY Cergy Paris University
Cergy
France

Ronan L. Hébert
Geosciences and
Environment Cergy
CY Cergy Paris University
Cergy
France

Ghislain Trullenque
Bassins - Réservoirs -
Ressources URR 7511
UniLaSalle
Beauvais
France

Albert Genter
ES Geothermie
Electricité de Strasbourg
Strasbourg
France

Eléonore Dalmais
ES Géothermie
Electricité de Strasbourg
Strasbourg
France

Jean Hérisson
Benkei
Lyon
France

Editorial Office
MDPI
St. Alban-Anlage 66
4052 Basel, Switzerland

This is a reprint of articles from the Special Issue published online in the open access journal *Geosciences* (ISSN 2076-3263) (available at: www.mdpi.com/journal/geosciences/special_issues/meet_project).

For citation purposes, cite each article independently as indicated on the article page online and as indicated below:

LastName, A.A.; LastName, B.B.; LastName, C.C. Article Title. *Journal Name* **Year**, *Volume Number*, Page Range.

ISBN 978-3-0365-6054-0 (Hbk)
ISBN 978-3-0365-6053-3 (PDF)

Cover image courtesy of Alexandre Nachbauer and Electricité de Strasbourg

Contents

About the Editors . . . . **ix**

Preface to "Enhanced Geothermal Systems and other Deep Geothermal Applications throughout Europe: The MEET Project" . . . . **xi**

Béatrice A. Ledésert, Ronan L. Hébert, Ghislain Trullenque, Albert Genter, Eléonore Dalmais and Jean Herisson
Editorial of Special Issue "Enhanced Geothermal Systems and Other Deep Geothermal Applications throughout Europe: The MEET Project"
Reprinted from: *Geosciences* **2022**, *12*, 341, doi:10.3390/geosciences12090341 . . . . **1**

Béatrice A. Ledésert, Ronan L. Hébert, Justine Mouchot, Clio Bosia, Guillaume Ravier and Olivier Seibel et al.
Scaling in a Geothermal Heat Exchanger at Soultz-Sous-Forts (Upper Rhine Graben, France): A XRD and SEM-EDS Characterization of Sulfide Precipitates
Reprinted from: *Geosciences* **2021**, *11*, 271, doi:10.3390/geosciences11070271 . . . . **11**

Pierce Kunan, Guillaume Ravier, Eléonore Dalmais, Marion Ducousso and Pierre Cezac
Thermodynamic and Kinetic Modelling of Scales Formation at the Soultz-sous-Forts Geothermal Power Plant
Reprinted from: *Geosciences* **2021**, *11*, 483, doi:10.3390/geosciences11120483 . . . . **35**

Clément Baujard, Pauline Rolin, Éléonore Dalmais, Régis Hehn and Albert Genter
Soultz-sous-Forts Geothermal Reservoir: Structural Model Update and Thermo-Hydraulic Numerical Simulations Based on Three Years of Operation Data
Reprinted from: *Geosciences* **2021**, *11*, 502, doi:10.3390/geosciences11120502 . . . . **63**

Saeed Mahmoodpour, Mrityunjay Singh, Aysegul Turan, Kristian Bär and Ingo Sass
Hydro-Thermal Modeling for Geothermal Energy Extraction from Soultz-sous-Forts, France
Reprinted from: *Geosciences* **2021**, *11*, 464, doi:10.3390/geosciences11110464 . . . . **79**

Yassine Abdelfettah and Christophe Barnes
Sensitivity Analysis of FWI Applied to OVSP Synthetic Data for Fault Detection and Characterization in Crystalline Rocks
Reprinted from: *Geosciences* **2021**, *11*, 442, doi:10.3390/geosciences11110442 . . . . **99**

Johanne Klee, Sébastien Potel, Béatrice A. Ledésert, Ronan L. Hébert, Arezki Chabani and Pascal Barrier et al.
Fluid-Rock Interactions in a Paleo-Geothermal Reservoir (Noble Hills Granite, California, USA). Part 1: Granite Pervasive Alteration Processes away from Fracture Zones
Reprinted from: *Geosciences* **2021**, *11*, 325, doi:10.3390/geosciences11080325 . . . . **133**

Johanne Klee, Arezki Chabani, Béatrice A. Ledésert, Sébastien Potel, Ronan L. Hébert and Ghislain Trullenque
Fluid-Rock Interactions in a Paleo-Geothermal Reservoir (Noble Hills Granite, California, USA). Part 2: The Influence of Fracturing on Granite Alteration Processes and Fluid Circulation at Low to Moderate Regional Strain
Reprinted from: *Geosciences* **2021**, *11*, 433, doi:10.3390/geosciences11110433 . . . . **167**

Arezki Chabani, Ghislain Trullenque, Johanne Klee and Béatrice A. Ledésert
Fracture Spacing Variability and the Distribution of Fracture Patterns in Granitic Geothermal Reservoir: A Case Study in the Noble Hills Range (Death Valley, CA, USA)
Reprinted from: *Geosciences* **2021**, *11*, 520, doi:10.3390/geosciences11120520 . . . . **207**

Arezki Chabani, Ghislain Trullenque, Béatrice A. Ledésert and Johanne Klee
Multiscale Characterization of Fracture Patterns: A Case Study of the Noble Hills Range (Death
Valley, CA, USA), Application to Geothermal Reservoirs
Reprinted from: *Geosciences* **2021**, *11*, 280, doi:10.3390/geosciences11070280 . . . . **241**

Katja E. Schulz, Kristian Bär and Ingo Sass
Lab-Scale Permeability Enhancement by Chemical Treatment in Fractured Granite (Cornubian
Batholith) for the United Downs Deep Geothermal Power Project, Cornwall (UK)
Reprinted from: *Geosciences* **2022**, *12*, 35, doi:10.3390/geosciences12010035 . . . . **269**

**Saeed Mahmoodpour, Mrityunjay Singh, Christian Obaje, Sri Kalyan Tangirala, John
Reinecker and Kristian Bär et al.**
Hydrothermal Numerical Simulation of Injection Operations at United Downs, Cornwall, UK
Reprinted from: *Geosciences* **2022**, *12*, 296, doi:10.3390/geosciences12080296 . . . . **293**

D. C. P. Peacock, David J. Sanderson and Bernd Leiss
Use of Analogue Exposures of Fractured Rock for Enhanced Geothermal Systems
Reprinted from: *Geosciences* **2022**, *12*, 318, doi:10.3390/geosciences12090318 . . . . **311**

**Chaojie Cheng, Johannes Herrmann, Bianca Wagner, Bernd Leiss, Jessica A. Stammeier and
Erik Rybacki et al.**
Long-Term Evolution of Fracture Permeability in Slate: An Experimental Study with
Implications for Enhanced Geothermal Systems (EGS)
Reprinted from: *Geosciences* **2021**, *11*, 443, doi:10.3390/geosciences11110443 . . . . **327**

**Marina Cabidoche, Yves Vanbrabant, Serge Brouyère, Vinciane Stenmans, Bruno Meyvis
and Thomas Goovaerts et al.**
Spring Water Geochemistry: A Geothermal Exploration Tool in the Rhenohercynian
Fold-and-Thrust Belt in Belgium
Reprinted from: *Geosciences* **2021**, *11*, 437, doi:10.3390/geosciences11110437 . . . . **351**

D.C.P. Peacock, David J. Sanderson and Bernd Leiss
Use of Mohr Diagrams to Predict Fracturing in a Potential Geothermal Reservoir
Reprinted from: *Geosciences* **2021**, *11*, 501, doi:10.3390/geosciences11120501 . . . . **379**

Johannes Herrmann, Valerian Schuster, Chaojie Cheng, Harald Milsch and Erik Rybacki
Fracture Transmissivity in Prospective Host Rocks for Enhanced Geothermal Systems (EGS)
Reprinted from: *Geosciences* **2022**, *12*, 195, doi:10.3390/geosciences12050195 . . . . **407**

Saeed Mahmoodpour, Mrityunjay Singh, Kristian Bär and Ingo Sass
Impact of Well Placement in the Fractured Geothermal Reservoirs Based on Available Discrete
Fractured System
Reprinted from: *Geosciences* **2022**, *12*, 19, doi:10.3390/geosciences12010019 . . . . **441**

**Svava Davíðsdóttir, Baldur Geir Gunnarsson, Kjartan Björgvin Kristjánsson, Béatrice A.
Ledésert and Dagur Ingi Ólafsson**
Study of Corrosion Resistance Properties of Heat Exchanger Metals in Two Different
Geothermal Environments
Reprinted from: *Geosciences* **2021**, *11*, 498, doi:10.3390/geosciences11120498 . . . . **459**

Dmitry Romanov and Bernd Leiss
Analysis of Enhanced Geothermal System Development Scenarios for District Heating and
Cooling of the Göttingen University Campus
Reprinted from: *Geosciences* **2021**, *11*, 349, doi:10.3390/geosciences11080349 . . . . **483**

Josipa Hranić, Sara Raos, Eric Leoutre and Ivan Rajšl
Two-Stage Geothermal Well Clustering for Oil-to-Water Conversion on Mature Oil Fields
Reprinted from: *Geosciences* **2021**, *11*, 470, doi:10.3390/geosciences11110470 . . . . **503**

About the Editors

Béatrice A. Ledésert

Pr. Béatrice A. Ledésert is full professor at CY Cergy Paris Université (France) where she teaches applied geology, including geothermal exploration. Her research focuses on the exploration of deep geothermal reservoirs, based on mineralogy, petrography, and well data when available. The aim is to pinpoint the paths followed by the geothermal fluid within the reservoirs. She is also interested in processes that occur in surface installations (scaling and corrosion). Her research is supported by the French Agency for ecological transition, the European Research Council and the Eutopia European university alliance.

Ronan L. Hébert

Dr. Ronan L. Hebert is assistant-professor at CY Cergy Paris Université (France). He is a petrologist focusing on fluid–rock interactions in different contexts (geothermal, cultural heritage, building materials). His research focuses on hydrothermal alteration products, as well as clogging materials that may hinder the flow of fluid within geothermal reservoirs. These are crucial informations regarding reservoir stimulations. He has been working in the field of geothermal energy for more than 15 years.

Ghislain Trullenque

Dr. Ghislain Trullenque is a structural geologist with a solid fieldwork experience. His research is dedicated to rock petrophysical properties evolution during progressive deformation. He has expertise in the fields of experimental deformation, rock microfabric analysis and fluid–rock interaction processes. He is the scientific coordinator of the MEET project and leads several EU funded research and education projects.

Albert Genter

Dr. Albert Genter is a senior scientist and a worldwide expert in geothermal energy at Electricité de Strasbourg, France, a utility company providing energy on its territory in Eastern part of France. Specialized on structural geology and hydrothermal alteration of basement rocks, his research topics are related to fractured reservoirs in geothermal systems. By working for industry, he focused on environmental monitoring, non-technical barrier, and life cycle assessment of geothermal systems.

Eléonore Dalmais

M.Sc. Eléonore Dalmais is a geosciences engineer with a solid experience in resources exploration and estimation, 3D modelling and geophysical logging. She joined ES-Géothermie in 2013 where she contributes to the exploration and drilling of deep wells in Alsace (France) for Enhanced Geothermal Systems and is also involved in geothermal plant exploitation. She participated to several national and international R&D projects dealing with geothermal energy.

Jean Hérisson

Dr. Jean HERISSON studied chemistry, biology, environment, and material (culture heritage) building, making him a scientist capable of working in multidisciplinary environments. His experience with standardization committees, scientific networking through RILEM working groups and the different industrial projects he was responsible for, have oriented him toward consulting in the field of innovation. He is now a consultant for Benkei. He is currently managing 3 European projects and 3 French national ones.

Preface to "Enhanced Geothermal Systems and other Deep Geothermal Applications throughout Europe: The MEET Project"

The contribution of geothermal energy to the energetic mix of European countries has been steadily increasing in the last two decades. This resource, being virtually infinite and permanently available, with a negligible environmental impact, is to be seen as a pillar of the energy transition from fossil and nuclear fuels towards renewable sources. In addition, geothermal brines might also be an important source for the extraction of raw materials such as lithium for battery production in the near future.

Depending on the existing surface infrastructures and needs, geothermal energy can be used directly, in the form of heat, or converted into electricity, and related applications, such as cooling and heat storage, are also feasible.

Gains in geothermal energy can be achieved using a variety of techniques, depending on the geological setting of the underground. Among the list of exploitation concepts, enhanced geothermal systems (EGS) are particularly interesting, as their application is much less independent of the underground setting, allowing, in turn, a large geographical deployment and market penetration in Europe. The challenges of EGS are multiple in terms of investment costs, the testing of novel reservoir exploitation approaches with an inherent risk of induced seismicity, and the presence of aggressive geothermal brines, damaging infrastructures.

The multidisciplinary and multi-context demonstration of enhanced geothermal systems exploration and exploitation techniques and potentials (MEET) project has received funding from the European Union's Horizon 2020 research and innovation program under grant agreement No 792037. A European consortium of academic and industrial partners aims to analyze these challenges and propose a series of tools dedicated to potential users and investors in terms of developing EGS and other deep geothermal applications throughout Europe

In order to reach its goal, the MEET project mainly addresses the need to capitalize on the exploitation of the widest range of fluid temperature in EGS plants, use co-produced hot brines in mature oil fields and apply EGS in different geological settings.

The approach is based on a combination of research and demonstration activities in order to make EGS safe and sustainable. This Special Issue summarizes the output of the MEET project based on laboratory experiments, geological field works on high-quality analogues, advanced reservoir modeling, the development of a decision-maker tool for investors and specific demonstration activities, such as chemical stimulation or the innovative monitoring of deep geothermal wells, and the production of electrical power via small-scale binary technology tested in various geological contexts in Europe.

Béatrice A. Ledésert, Ronan L. Hébert, Ghislain Trullenque, Albert Genter, Eléonore Dalmais, and Jean Hérisson

Editors

 *geosciences*

Editorial

Editorial of Special Issue "Enhanced Geothermal Systems and Other Deep Geothermal Applications throughout Europe: The MEET Project"

Béatrice A. Ledésert [1,*], Ronan L. Hébert [1], Ghislain Trullenque [2], Albert Genter [3], Eléonore Dalmais [3] and Jean Herisson [4]

[1] Geosciences and Environment Cergy Laboratory, CY Cergy Paris Université, 1 Rue Descartes, F-95000 Neuville-sur-Oise, France
[2] UniLaSalle, 19 Rue Pierre Waguet BP30313, CEDEX, F-60026 Beauvais, France
[3] ÉS Géothermie, 26 bd Président Wilson, F-67000 Strasbourg, France
[4] Benkei, 97 Cours Gambetta, F-69003 Lyon, France
* Correspondence: beatrice.ledesert@cyu.fr; Tel.: +33-134-257-357

Citation: Ledésert, B.A.; Hébert, R.L.; Trullenque, G.; Genter, A.; Dalmais, E.; Herisson, J. Editorial of Special Issue "Enhanced Geothermal Systems and Other Deep Geothermal Applications throughout Europe: The MEET Project". *Geosciences* **2022**, *12*, 341. https://doi.org/10.3390/geosciences12090341

Received: 5 September 2022
Accepted: 5 September 2022
Published: 13 September 2022

Publisher's Note: MDPI stays neutral with regard to jurisdictional claims in published maps and institutional affiliations.

1. Introduction

The MEET project is a Multidisciplinary and multi-context demonstration of Enhanced Geothermal Systems exploration and Exploitation Techniques and potentials, which received funding from the European Commission in the framework of the Horizon 2020 program. During the four years of the project, two main types of exploitations were investigated: Enhanced Geothermal Systems (EGS) and oil-to-geothermal conversion or co-production. The following topics were addressed: the upscaling of thermal power production and optimized operation of EGS plants (papers [1–5]); variscan geothermal reservoirs in granitic and metamorphic rocks (papers [6–17]); and technical, economic and environmental assessment for oil-to-geothermal fields and EGS integration into energy systems (papers [18–20]). These 20 papers give an overview of some of the work performed in the MEET project, but they are not exhaustive of all the results obtained in this frame. Additional results are available at https://zenodo.org/communities/eu_project_meet/ (accessed on 1 September 2022). The MEET Project (Figure 1) received funding by the European Commission in the framework of the H2020 Program (Grant Agreement No. 792037) for a **M**ultidisciplinary and multi-context demonstration of **E**nhanced Geothermal Systems exploration and **E**xploitation **T**echniques and potentials.

Figure 1. Logo of the MEET project, URL (https://www.meet-h2020.com (accessed on 1 September 2022)) of its website and that of the platform (https://zenodo.org/communities/eu_project_meet/ (accessed on 1 September 2022)) on which all of the documents produced by the MEET project are stored.

The MEET project dealt with the gains in geothermal energy that can be achieved using a variety of techniques, depending on the geological setting of the underground. Among the list of exploitation concepts, **Enhanced Geothermal Systems (EGS)** are particularly interesting, as their application is little dependent on the underground setting, allowing, in turn, for a large geographical deployment and market penetration in Europe. The challenges of EGS are multiple in terms of investment costs, the testing of novel reservoir exploitation approaches with an inherent risk of induced seismicity and the presence of aggressive geothermal brines that can damage infrastructures due to scaling and corrosion. The **use of co-produced hot brines in mature oil fields** is another target of the project.

MEET aimed at 1- gathering knowledge of EGS heat and power production in various geological settings; 2- helping increase heat production from existing plants and convert oil wells into geothermal wells; 3- enhancing heat-to-power conversion at low flow (<10 l/s) and/or low temperature (60–90 °C) by using smart mobile Organic Rankine Cycle (ORC) units; and 4- replicating the technology by promoting the penetration of EGS power and/or heat plants.

In order to reach those objectives, the MEET project mainly addressed the need to capitalize on the exploitation of the widest range of fluid temperature in oil and geothermal fields (from 90 °C to 160 °C) and apply EGS in different geological settings (sedimentary basins and basements affected or not by post extensional tectonics). The approach was based on a combination of research and demonstration activities. It relied on the study of demonstration sites (Figure 2), either at the **exploration stage** (United Downs Deep Geothermal Project (UDDGP)—UK; Göttingen University campus—Germany; Havelange—Belgium) by studying **analogues** (e.g., Death Valley—CA, USA; Carnmenellis granite and Cornubian batholith—UK; Dinant synclinorium—Belgium; Rhenish massif and Harz mountains—Germany) or the **exploitation stage** (e.g., Soultz-sous-Forêts EGS plant—France; Chaunoy and Cazaux oil fields—France; Condorcet High School—France; Grásteinn farm and Krauma spa—Iceland).

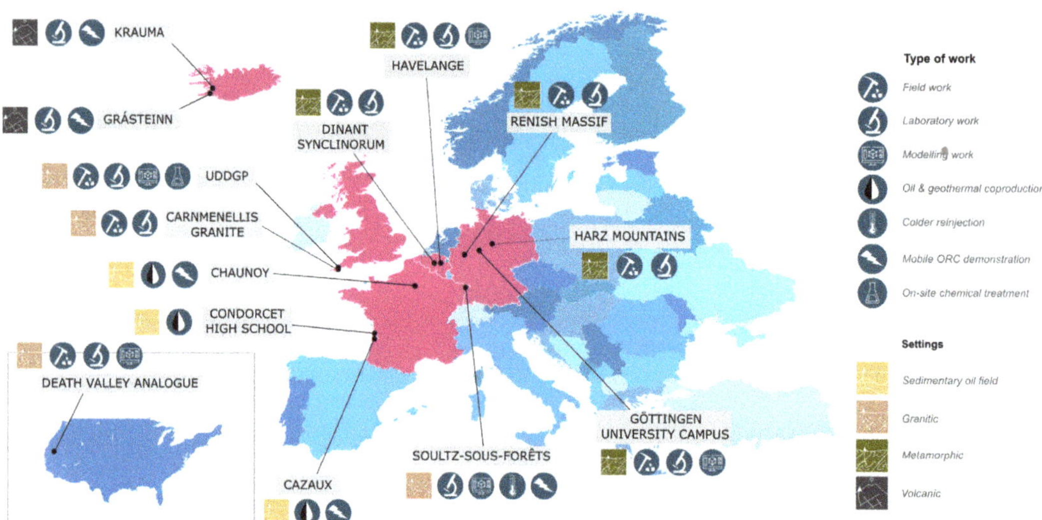

Figure 2. Location map of the sites and type of works performed in the MEET project.

Various geological settings are represented: volcanic areas (Krauma and Grásteinn), sedimentary basins (Chaunoy, Condorcet High School and Cazaux), granitic basements (Soultz-sous-Forêts; UDDGP; Death Valley analogue) and metamorphic basements (Havelange, Göttingen University campus, Harz Mountains and Rhenish Massif analogues).

After four years of transdisciplinary work, this Special Issue compiles some of the most recent geoscience results obtained by the 16 academic and industrial partners (ES-Géothermie (coordinator), UniLaSalle, GIM-Labs, CY Cergy Paris Université, Technische Universität Darmstadt, Universitätsenergie Göttingen GmbH, Georg-August-Universität Göttingen, Vermilion, Enogia, GFZ, Febus-Optics, University of Zagreb—Faculty of Electrical Engineering and Computing, ICETEC, Geological Survey of Belgium, GeoThermal Engineering and Benkei) coming from 5 countries (Belgium, Croatia, France, Germany and Iceland) that joined their efforts for the development of geothermal energy across Europe. The MEET project was organised in dedicated work-packages (WP) through the technical themes that were developed during the project: upscaling of thermal power production and optimized operation of EGS plants (WP3); enhancing and/or converting petroleum sedimentary basins for geothermal electricity and thermal power production (WP4); variscan geothermal reservoirs (granitic and metamorphic rocks; WP5); demonstration of electricity and thermal power generation (WP6); and economic and environmental assessment for EGS integration into energy systems (WP7). This *Geosciences* Special Issue is mostly presenting the geoscientific work performed within MEET. WP3 and WP5 have a dedicated section, whereas WP4, 6 and 7 are grouped in the section "Technical, economic and environmental assessment for oil to geothermal fields and EGS integration into energy systems". Other works performed within the MEET project can be found on the Zenodo platform: https://zenodo.org/communities/eu_project_meet/ (accessed on 1 September 2022).

The geoscience work performed in the MEET project was based on geological field work on high-quality analogues; laboratory experiments; advanced reservoir modelling; specific demonstration activities such as chemical stimulation, colder reinjection or innovative monitoring of deep geothermal wells; and production of electrical power via small-scale binary technology tested in the various geological contexts in Europe shown in Figure 2.

2. Upscaling of Thermal Power Production and Optimized Operation of EGS Plants

The Soultz-sous-Forêts demonstration site (called Soultz in the following) has been operated since 2016 for the production of electricity from a granitic basement. The temperature of the brine produced at the well head of GPK-2 drill hole is around 160 °C. It is currently reinjected in the ground at a temperature close to 70 °C.

In those operation conditions, minerals precipitate during the lowering of the temperature into the surface installations, producing deposits called scales. Scales have a negative impact on the power plant in lowering the electricity production and inducing specific waste management issues. Thus, scales have to be thoroughly characterized [1] and their deposition process modelled [2] to control scaling processes in surface installations. In order to increase energy output, a small-scale heat exchanger prototype, called SHEx in the following, has been tested for 3 months in order to lower the reinjection temperature down to 40 °C. The SHEx is of tubular type, made of an entrance, an exit, water boxes at each end, and tubes made of several alloys and metals in between. The lowering of the temperature of reinjection might have an impact on the geothermal reservoir ([3,4]) while the structure of the reservoir ([3,5]) is another key parameter to be considered for the sustainability of the EGS.

Ledésert et al. [1], in "Scaling in a Geothermal Heat Exchanger at Soultz-Sous-Forêts (Upper Rhine Graben, France): A XRD and SEM-EDS Characterization of Sulfide Precipitates", studied the sulfide scales that deposited in the SHEx by using X-Ray Diffraction (XRD) and a Scanning Electron Microscope (SEM) coupled with an Energy-Dispersive Spectrometer (EDS). The effect of the flow regime on the shape of sulfide crystals was questioned, as well as that of the lowering of temperature on the thickness of the deposit. The scales deposited in the SHEx were compared to scales deposited in normal industrial conditions.

In "Thermodynamic and Kinetic Modelling of Scales Formation at the Soultz-sous-Forêts Geothermal Power Plant", Kunan et al. [2] performed the thermodynamic and kinetic modelling of scale deposition by using Phreeqc and the Thermoddem database thanks to the data on the chemical elements, minerals, and gas it contains. The model generated a rough prediction of the scale formation when operating the plant with sulfate scales inhibitors at the Soultz geothermal plant and showed only a small deviation between simulated results and the actual case.

"Soultz-sous-Forêts Geothermal Reservoir: Structural Model Update and Thermo-Hydraulic Numerical Simulations Based on Three Years of Operation Data" by Baujard et al. [3] presents a thermo-hydraulic numerical simulation to better constrain the parameters that govern the functioning of the Soultz granitic exchanger at depth when colder reinjection is performed. In this article, a 3D hydrothermal study was performed in order to evaluate the spreading of the thermal front during colder reinjection and its impact on production temperature. The fault scale was investigated first, integrating pre-existing models from seismic profiles, seismic cloud structure and borehole image logs calibrated with well data. Secondly, this geometrical model was adapted to be able to run hydrothermal simulations. In a third step, a 3D hydrothermal model was built based on the structural model. After calibration, the effect of colder reinjection on the production temperature was calculated. Finally, the accuracy of the structural model on which the simulations are based is discussed and an update of the structural model is proposed in order to better reproduce the observations.

Mahmoodpour et al., [4] in "Hydro-Thermal Modeling for Geothermal Energy Extraction from Soultz-sous-Forêts, France", propose another hydro-thermal modelling for geothermal energy extraction from Soultz, based on structures identified in [3], at temperature lower than the current 70 °C fluid reinjection temperature. Two injection-production rate scenarios were modelled, and the drop in the production wellhead temperature for 100 years of operation was quantified. For each scenario, reinjection temperatures of 40°C, 50°C and 60°C were chosen and compared with the 70 °C current reinjection condition.

In "Sensitivity Analysis of FWI Applied to OVSP Synthetic Data for Fault Detection and Characterization in Crystalline Rocks", Abdelfettah and Barnes [5] used the Full Wave Inversion (FWI) method to detect, delineate and better characterize faults in the granitic geothermal reservoir, from Multi-Offset Vertical Seismic Profile (OVSP) data in order to further characterize the geothermal resource at Soultz. They made several sensitivity studies to show the dip and thickness of a fault that can be imaged by FWI, even in the presence of additive Gaussian noise. Their work was applied to the Soultz site to help a better characterization of the fracture network.

3. Variscan Geothermal Reservoirs in Granitic and Metamorphic Rocks

Given the depth of EGS reservoirs and the difficulty to obtain data to characterize them, surface analogues were studied in metamorphic and granitic environments. Analogues are surface sites that are easier to access and have similarities with the geothermal reservoirs in terms of rock nature and geological context. They allow researchers to better characterize the different kinds of EGS reservoirs.

3.1. Granitic Rocks

The granitic reservoir topic is developed through two sites: the Death Valley granitic surface analogue and the Cornish granites around and within the wells of the United Downs Deep Geothermal Project (UDDGP, Carnmenellis batholith). In the framework of MEET, reservoir improvements by using soft stimulation are planned in Cornwall at EDEN geothermal site where a 5 km deep well has been drilled in a fractured granite close to the UDDGP site.

3.1.1. Death Valley Analogue

The Death Valley granitic analogue is located in the Noble Hills (Noble Hills Granite, NHG, CA, USA). Granitic rocks affected by present or past hydrothermal fluid circulation, typically undergo a variety of alteration processes called hydrothermal alteration. This is due to the instability of the primary mineralogical assemblages under the new physico-chemical conditions, which leads to the formation of new mineral phases. Hydrothermal alteration was described for outcrops far from the faults [6] and close to them [7]. The structural characterization of the fracture network that conducted the fluids responsible for the hydrothermal alterations is presented in [8,9].

Far from the faults, Klee et al. [6] present "Fluid-Rock Interactions in a Paleo-Geothermal Reservoir (Noble Hills Granite, CA, USA). Part 1: Granite Pervasive Alteration Processes away from Fracture Zones", aimed at studying the impact of the regional geological context on the rock facies and alteration types. The NHG was first characterized from a geochemical and petrographical point of view. Once alteration parageneses were identified, an attempt was made to correlate the rock hydration rate given by the loss on ignition obtained during Inductively Coupled Plasma spectrometry to porosity, calcite content and chemical composition. Illite crystallinity was used to pinpoint a likely regional temperature gradient in the NHG.

In "Fluid-Rock Interactions in a Paleo-Geothermal Reservoir (Noble Hills Granite, CA, USA). Part 2: The Influence of Fracturing on Granite Alteration Processes and Fluid Circulation at Low to Moderate Regional Strain", Klee et al. [7] decipher the role of fractures in the hydrothermal alteration of the rock they crosscut. Several generations of fluids have percolated through the granitic reservoir. The alteration degree, the porosity and the calcite content were evaluated approaching fracture zones. A correlation between the degree of alteration and the fracture density and the amount of strain is proposed.

Chabani et al. [8] proposed a geometrical description and a quantification of the multiscale network organization and its effect on connectivity using a wide-ranging scale analysis. The statistical analyses were performed from regional maps to thin sections. The aim of "Fracture Spacing Variability and the Distribution of Fracture Patterns in Granitic Geothermal Reservoir: A Case Study in the Noble Hills Range (Death Valley, CA, USA)" was to show which class of fractures (small, medium, large) and which orientations ruled the connectivity and hence the ability of fractures to conduct hydrothermal fluids.

In "Multiscale Characterization of Fracture Patterns: A Case Study of the Noble Hills Range (Death Valley, CA, USA), Application to Geothermal Reservoirs", Chabani et al. [9] further characterized the fracture network in the NHG by proposing geometric description and quantifying the multiscale network organization and its effect on connectivity, using a wide-ranging scale analysis and scale order classification. The statistical analyses were performed on real (measured in the field) and virtual (established from photographs) scanlines intersecting fracture networks.

3.1.2. United Downs Deep Geothermal Project (UDDGP) Demonstration Site

In the United Downs Deep Geothermal Project (UDDGP) demonstration site settled in another granite body in Cornwall (UK), both field and borehole samples were collected and analyzed, and numerical simulations were performed. A hydrothermal doublet system was drilled in a fault-related granitic reservoir. It targets the Porthtowan Fault Zone (PTF), which transects the Carnmenellis granite, one of the onshore plutons of the Cornubian Batholith in SW England. At 5058 m depth (TVD; 5275 m MD) up to 190 °C were reached in the dedicated production well. The injection well, UD-2, is aligned vertically above the production well, UD-1, and reaches a depth of 2393 m MD.

Schulz et al. [10] propose a "Lab-Scale Permeability Enhancement by Chemical Treatment in Fractured Granite (Cornubian Batholith) for the United Downs Deep Geothermal Power Project, Cornwall (UK)". Lab-scale acidification experiments were performed on outcrop analogue samples from the Cornubian Batholith, which include mineralized veins. The experimental setup is based on autoclave experiments on sample powder and plugs and core flooding tests on sample plugs. These tests were designed to investigate to what degree the permeability of natural and artificial (saw-cut) fractures can be enhanced. Petrological and petrophysical analysis of the samples was performed before and after the acidification experiments to track changes resulting from the acidification for the prediction of likely chemical stimulation.

In "Hydrothermal Numerical Simulation of Injection Operations at United Downs, Cornwall, UK", Mahmoodpour et al. [11] present numerical simulations to analyze the hydraulic stimulation results and evaluate the increase in permeability of the reservoir. Experimental and field data were used to characterize the initial reservoir static model. Based on experimental and field data, stochastic discrete fracture networks (DFN) were developed to mimic the reservoir permeability behavior. Equivalent permeability fields were calculated to create a computationally feasible model. Hydraulic testing and stimulation data from the UD-1 borehole were used together with hydraulic testing and stimulation data from the UD-2 borehole used for validation.

3.2. Metamorphic Rocks

The geothermal potential of metamorphic basement was also investigated through the study of the Harz Mountains and Rhenish Massif as analogues for the EGS projects in Göttingen and Havelange. The inputs and difficulties of studying surface analogues were investigated [12]. Fluid flow laboratory experiments were conducted within slate samples from the Harz Mountains [13] and water samples collected from springs around a very deep borehole in Belgium were analyzed [14]. The fluid flow pathways were investigated through the use of the Mohr diagram [15] and the experimental determination of hydraulic properties of fractures [16]. The placement of wells for geothermal projects in rock basements was also investigated [17].

In "Use of Analogue Exposures of Fractured Rock for Enhanced Geothermal Systems", Peacock et al. [12] present an overview of the input and difficulties of the study of field exposures as analogues for EGS sub-surface reservoirs. This contribution discusses general lessons learnt about the use of deformed Devonian and Carboniferous meta-sedimentary rocks in the Harz Mountains (Germany), as analogues for a proposed EGS at Göttingen University campus (Germany). It indicates that the objectives of analogue studies must be clarified in order to explain to people from other disciplines the information that can and cannot be obtained from surface exposures. The parameters that have to govern the choice of an analogue are also highlighted.

Cheng et al. [13], in "Long-Term Evolution of Fracture Permeability in Slate: An Experimental Study with Implications for Enhanced Geothermal Systems (EGS)", present an experimental study for the evaluation of long-term evolution of fracture permeability in saw-cut slate samples from the Harz Mountains (Germany). The purpose was to investigate fracture permeability evolution at temperatures up to 90 °C using both deionized water (DI) and a NaCl solution as the pore fluid.

Cabidoche et al. [14] characterized water samples collected in 50 springs to evaluate the geothermal potential around the Havelange (Belgium) deep borehole, within the Rhenohercynian fold and thrust belt. They based their work on a heat map as well as on hydrogeochemistry and geothermometry analyses to define the main water types, and produced the paper entitled "Spring Water Geochemistry: A Geothermal Exploration Tool in the Rhenohercynian Fold-and-Thrust Belt in Belgium".

As regards fluid flow pathways, Peacock et al. [15] propose a "Use of Mohr Diagrams to Predict Fracturing in a Potential Geothermal Reservoir". Inferences have to be made about likely structures and their effects on fluid flow in a geothermal reservoir at the pre-drilling stage. Simple mechanical modelling was used here to predict the range of possible structures that are likely to exist in the sub-surface and that may be generated during the stimulation of a potential geothermal reservoir. In particular, Mohr diagrams are used to show under what fluid pressure and stress different types and orientations of fractures are likely to be reactivated or generated. The approach enables the effects of parameters to be modelled individually and defines the type and orientation of fractures to be considered. This modelling is useful for helping geoscientists to consider, model and predict the ranges of mechanical properties of rock, stresses, fluid pressures and the resultant fractures that are likely to occur in the sub-surface. Here, the modelling was applied to folded and thrusted greywackes and slates, which are planned to be developed as an Enhanced Geothermal System beneath Göttingen (Germany).

In "Fracture Transmissivity in Prospective Host Rocks for Enhanced Geothermal Systems (EGS)", Herrmann et al. [16] experimentally determined the hydraulic properties of fractures within various rock types, focusing on a variety of Variscan rocks. Flow-through experiments were performed on slate, graywacke, quartzite, granite, natural fault gouge, and claystone samples containing an artificial fracture with a given roughness. For slate samples, the hydraulic transmissivity of the fractures was measured at confining pressures up to 50 MPa, temperatures between 25 and 100 °C and differential stress perpendicular to the fracture surface of up to 45 MPa.

In their paper "Impact of Well Placement in the Fractured Geothermal Reservoirs Based on Available Discrete Fractured System", Mahmoodpour et al. [17] show how necessary well placement in a given geological setting for a fractured geothermal reservoir is for enhanced geothermal operations. Fully coupled thermo-hydromechanical (THM) processes are simulated in 2D in the fractured reservoir to estimate maximum geothermal energy extraction potential by optimizing well placement. To enhance the knowledge of well placement for different working fluids, the authors examine different injection–production well doublet positions in a given fracture network using coupled THM numerical simulations. Thermal breakthrough time, mass flux, and the energy extraction potential are examined to assess the impact of well position in a two-dimensional reservoir framework.

4. Technical, Economic and Environmental Assessment for Oil to Geothermal Fields and EGS Integration into Energy Systems

Numerous oil fields are approaching the end of their lifetime and have great geothermal potential considering temperature and water cut. EGS is also a promising source of energy. However, electricity and thermal power generation is threatened by technical issues [18], and a proper economic evaluation of different scenarios is crucial for further implementation of these solution at larger scale [19,20].

In "Study of Corrosion Resistance Properties of Heat Exchanger Metals in Two Different Geothermal Environments", Davíðsdóttir et al. [18] investigated the corrosion resistance of different alloy candidates for heat exchangers. They exposed in situ corrosion-resistant alloy coupon samples 316L, 254SMO, Inconel 625 and titanium grade 2 at two locations and geological settings (Triassic clastic sediments, Paris Basin, France; volcanic setting, Iceland). Coupons were exposed for four months at the Chaunoy oil field in France and one month at the Reykjanes powerplant in Iceland. After exposition, the tested alloys were analysed regarding corrosion with macro- and microscopic techniques using optical and electron microscopes.

Romanov and Leiss [19] focused their study entitled "Analysis of Enhanced Geothermal System Development Scenarios for District Heating and Cooling of the Göttingen University Campus" on potential scenarios of EGS development in the poorly known Variscan basement below Göttingen, for district heating and cooling of the University campus. On average, they demonstrated that a single EGS doublet could cover about 20% of the heat demand and 6% of the cooling demand of the campus. The levelized cost of heat (LCOH), net present value (NPV) and CO_2 abatement cost were evaluated. Based on a sensitivity analysis, the EGS heat output was estimated for potential profitability. The most influential parameters on the outcome were identified and are presented in this paper. Key prerequisites for launching EGS project in Göttingen are also given.

Hranić et al. [20] worked on oil fields that are approaching the end of their lifetime and have great geothermal potential considering temperature and water cut, for which oil companies consider switching from oil business to investments into geothermal projects on existing oil wells. The used methodology presents the evaluation of the existing geothermal potential for several oil fields in terms of water temperature, flow rate and spatial distribution of existing oil wells. This paper entitled "Two-Stage Geothermal Well Clustering for Oil-to-Water Conversion on Mature Oil Fields" proposes a two-stage clustering approach for grouping similar wells in terms of temperature and then spatial arrangement to optimize the location of production facilities. The outputs regarding the production quantities and economic and environmental aspects provide insight into the optimal scenario for oil-to-water conversion. A case study has also been developed.

This Special Issue hence compiles 20 scientific contributions resulting from some of the work performed during the H2020 MEET project. It shows that a multidisciplinary approach including geology, material science, petrophysics, geophysics, reservoir modeling and the collaboration between academics and the industry is essential. This Special Issue brings some tangible scientific content to convince the readers about the opportunity to generalize EGS in Europe in different geological contexts and cogenerate hot water and oil in order to tackle the challenge of energy transition.

Funding: This work has received funding from the European Union's Horizon 2020 research and innovation program (Grant agreement No. 792037—MEET Project).

Acknowledgments: The partners of the MEET project wish to thank GEIE Exploitation Minière de la Chaleur, UDDGP, EDEN, Vermilion, Krauma spa and Grásteinn farm for site access and data sharing. The Fonds Régional d'Aide aux Porteurs de Projets Européens (FRAPPE) of Hauts-de-France is acknowledged for its support.

Conflicts of Interest: The authors declare no conflict of interest.

References

1. Ledésert, B.A.; Hébert, R.L.; Mouchot, J.; Bosia, C.; Ravier, G.; Seibel, O.; Dalmais, É.; Ledésert, M.; Trullenque, G.; Sengelen, X. Scaling in a geothermal heat exchanger at soultz-sous-forêts (Upper Rhine Graben, France): A XRD and SEM-EDS characterization of sulfide precipitates. *Geosciences* **2021**, *11*, 271. [CrossRef]
2. Kunan, P.; Ravier, G.; Dalmais, E.; Ducousso, M.; Cezac, P. Thermodynamic and Kinetic Modelling of Scales Formation at the Soultz-sous-Forêts Geothermal Power Plant. *Geosciences* **2021**, *11*, 483. [CrossRef]
3. Baujard, C.; Rolin, P.; Dalmais, É.; Hehn, R.; Genter, A. Soultz-sous-Forêts Geothermal Reservoir: Structural Model Update and Thermo-Hydraulic Numerical Simulations Based on Three Years of Operation Data. *Geosciences* **2021**, *11*, 502. [CrossRef]
4. Mahmoodpour, S.; Singh, M.; Turan, A.; Bär, K.; Sass, I. Hydro-Thermal Modeling for Geothermal Energy Extraction from Soultz-sous-Forêts, France. *Geosciences* **2021**, *11*, 464. [CrossRef]
5. Abdelfettah, Y.; Barnes, C. Sensitivity Analysis of FWI Applied to OVSP Synthetic Data for Fault Detection and Characterization in Crystalline Rocks. *Geosciences* **2021**, *11*, 442. [CrossRef]
6. Klee, J.; Potel, S.; Ledésert, B.A.; Hébert, R.L.; Chabani, A.; Barrier, P.; Trullenque, G. Fluid-Rock Interactions in a Paleo-Geothermal Reservoir (Noble Hills Granite, California, USA). Part 1: Granite Pervasive Alteration Processes away from Fracture Zones. *Geosciences* **2021**, *11*, 325. [CrossRef]
7. Klee, J.; Chabani, A.; Ledésert, B.A.; Potel, S.; Hébert, R.L.; Trullenque, G. Fluid-Rock Interactions in a Paleo-Geothermal Reservoir (Noble Hills Granite, California, USA). Part 2: The Influence of Fracturing on Granite Alteration Processes and Fluid Circulation at Low to Moderate Regional Strain. *Geosciences* **2021**, *11*, 433. [CrossRef]
8. Chabani, A.; Trullenque, G.; Klee, J.; Ledésert, B.A. Fracture Spacing Variability and the Distribution of Fracture Patterns in Granitic Geothermal Reservoir: A Case Study in the Noble Hills Range (Death Valley, CA, USA). *Geosciences* **2021**, *11*, 520. [CrossRef]
9. Chabani, A.; Trullenque, G.; Ledésert, B.A.; Klee, J. Multiscale Characterization of fracture patterns: A case study of the Noble Hills Range (Death Valley, CA, USA), application to geothermal reservoirs. *Geosciences* **2021**, *11*, 280. [CrossRef]
10. Schulz, K.E.; Bär, K.; Sass, I. Lab-Scale Permeability Enhancement by Chemical Treatment in Fractured Granite (Cornubian Batholith) for the United Downs Deep Geothermal Power Project, Cornwall (UK). *Geosciences* **2022**, *12*, 35. [CrossRef]
11. Mahmoodpour, S.; Singh, M.; Obaje, C.; Tangirala, S.K.; Reinecker, J.; Bär, K.; Sass, I. Hydrothermal Numerical Simulation of Injection Operations at United Downs, Cornwall, UK. *Geosciences* **2022**, *12*, 296. [CrossRef]
12. Peacock, D.C.P.; Sanderson, D.J.; Leiss, B. Use of Analogue Exposures of Fractured Rock for Enhanced Geothermal Systems. *Geosciences* **2022**, *12*, 318. [CrossRef]
13. Cheng, C.; Herrmann, J.; Wagner, B.; Leiss, B.; Stammeier, J.A.; Rybacki, E.; Milsch, H. Long-Term Evolution of Fracture Permeability in Slate: An Experimental Study with Implications for Enhanced Geothermal Systems (EGS). *Geosciences* **2021**, *11*, 443. [CrossRef]
14. Cabidoche, M.; Vanbrabant, Y.; Brouyère, S.; Stenmans, V.; Meyvis, B.; Goovaerts, T.; Petitclerc, E.; Burlet, C. Spring Water Geochemistry: A Geothermal Exploration Tool in the Rhenohercynian Fold-and-Thrust Belt in Belgium. *Geosciences* **2021**, *11*, 437. [CrossRef]
15. Peacock, D.C.P.; Sanderson, D.J.; Leiss, B. Use of Mohr Diagrams to Predict Fracturing in a Potential Geothermal Reservoir. *Geosciences* **2021**, *11*, 501. [CrossRef]
16. Herrmann, J.; Schuster, V.; Cheng, C.; Milsch, H.; Rybacki, E. Fracture Transmissivity in Prospective Host Rocks for Enhanced Geothermal Systems (EGS). *Geosciences* **2022**, *12*, 195. [CrossRef]

17. Mahmoodpour, S.; Singh, M.; Bär, K.; Sass, I. Impact of Well Placement in the Fractured Geothermal Reservoirs Based on Available Discrete Fractured System. *Geosciences* **2022**, *12*, 19. [CrossRef]
18. Davíðsdóttir, S.; Gunnarsson, B.G.; Kristjánsson, K.B.; Ledésert, B.A.; Ólafsson, D.I. Study of Corrosion Resistance Properties of Heat Exchanger Metals in Two Different Geothermal Environments. *Geosciences* **2021**, *11*, 498. [CrossRef]
19. Romanov, D.; Leiss, B. Analysis of Enhanced Geothermal System Development Scenarios for District Heating and Cooling of the Göttingen University Campus. *Geosciences* **2021**, *11*, 349. [CrossRef]
20. Hranić, J.; Raos, S.; Leoutre, E.; Rajšl, I. Two-Stage Geothermal Well Clustering for Oil-to-Water Conversion on Mature Oil Fields. *Geosciences* **2021**, *11*, 470. [CrossRef]

 *geosciences*

Article

Scaling in a Geothermal Heat Exchanger at Soultz-Sous-Forêts (Upper Rhine Graben, France): A XRD and SEM-EDS Characterization of Sulfide Precipitates

Béatrice A. Ledésert [1,*], Ronan L. Hébert [1], Justine Mouchot [2], Clio Bosia [2], Guillaume Ravier [2], Olivier Seibel [2], Éléonore Dalmais [2], Mariannick Ledésert [3,†], Ghislain Trullenque [4], Xavier Sengelen [1] and Albert Genter [2]

[1] CY Cergy Paris Université, Géosciences et Environnement Cergy, 1 Rue Descartes, 95000 Neuville-sur-Oise, France; ronan.hebert@cyu.fr (R.L.H.); xavier.sengelen@cyu.fr (X.S.)
[2] Électricité de Strasbourg Géothermie, 26 Boulevard du Président Wilson, 67000 Strasbourg, France; justine.mouchot@arverne.earth (J.M.); clio.bosia@es.fr (C.B.); guillaume.ravier@es.fr (G.R.); olivier.seibel@es.fr (O.S.); eleonore.dalmais@es.fr (É.D.); albert.genter@es.fr (A.G.)
[3] Cristallography Laboratory, 14000 Caen, France; ledesert@cyu.fr
[4] UniLaSalle, Collège Géosciences, équipe B2R, 19 Rue Pierre Waguet, 60000 Beauvais, France; ghislain.trullenque@unilasalle.fr
* Correspondence: beatrice.ledesert@cyu.fr; Tel.: +33-134257357
† Retired.

Citation: Ledésert, B.A.; Hébert, R.L.; Mouchot, J.; Bosia, C.; Ravier, G.; Seibel, O.; Dalmais, É.; Ledésert, M.; Trullenque, G.; Sengelen, X.; et al. Scaling in a Geothermal Heat Exchanger at Soultz-Sous-Forêts (Upper Rhine Graben, France): A XRD and SEM-EDS Characterization of Sulfide Precipitates. *Geosciences* **2021**, *11*, 271. https://doi.org/10.3390/geosciences11070271

Academic Editors: Jesus Martinez-Frias and Matteo Alvaro

Received: 5 May 2021
Accepted: 23 June 2021
Published: 28 June 2021

Abstract: The Soultz-Sous-Forêts geothermal site (France) operates three deep wells for electricity production. During operation, scales precipitate within the surface installation as (Ba, Sr) sulfate and (Pb, As, Sb) sulfide types. Scales have an impact on lowering energy production and inducing specific waste management issues. Thus scaling needs to be reduced for which a thorough characterization of the scales has to be performed. The geothermal brine is produced at 160 °C and reinjected at 70 °C during normal operation. In the frame of the H2020 MEET project, a small heat exchanger was tested in order to allow higher energy production, by reinjecting the geothermal fluid at 40 °C. Samples of scales were analyzed by XRD and SEM-EDS, highlighting that mostly galena precipitates and shows various crystal shapes. These shapes can be related to the turbulence of the flow and the speed of crystal growth. Where the flow is turbulent (entrance, water box, exit), crystals grow quickly and mainly show dendritic shape. In the tubes, where the flow is laminar, crystals grow more slowly and some of them are characterized by well-developed faces leading to cubes and derived shapes. The major consequence of the temperature decrease is the increased scaling phenomenon.

Keywords: Soultz-Sous-Forêts; geothermal site; heat exchanger; scales; sulfates; sulfides; As and Sb-bearing galena; crystal growth; crystal shapes

1. Introduction

Geothermal power production is a very attractive resource with characteristics such as low cost, little environmental pollution and worldwide distribution [1–3]. In addition, it is available all the time, whatever climatic conditions and day/night alternation, as opposed to wind- or solar-derived energy. Geothermal power plants take their energy from deep underground water that is pumped to the surface. During its residence time in the ground at sometimes high temperature (150–300 °C), the water acquires specific properties by interaction with the rock reservoir, generally resulting in high salinity and acidity. These characteristics are responsible for corrosion and scaling (deposition) issues in wells and in surface installations. Those phenomena are known from the very beginning of industrial high-temperature geothermal operations and are identified to be responsible for economic issues [4,5]. Scaling is encountered in both low enthalpy [6] and high enthalpy [5,7] geothermal systems. According to [8], among the most abundant scales are silica, carbonates, sulfates, sulfides, and native metals such as antimony (Sb). It is also known that metal sulfide scaling frequently occurs in volcanic geological context or high Cl environments [9].

Among all the geothermal geological contexts, grabens present specific characteristics. In the Upper Rhine Graben (URG; at the border between France and Germany) hydrothermal fluids percolate within fault zones and are brought relatively close to the surface, favoring the development of high-energy geothermal plants. These are dedicated to the production of either electricity (e.g., Soultz-Sous-Forêts, called Soultz in the following) or heat (e.g., Rittershoffen). The URG deep ground water system is characterized by a brine with high salinity (99–107 g/L at Soultz) and moderate low pH (around 5) [10] responsible for strong corrosion of geothermal surface installations (pipes, heat exchangers) and also deposition of minerals within these installations [11,12].

The Soultz geothermal production plant is based on three 5000 m-deep wells, GPK-2, GPK-3 and GPK-4 (Figure 1) penetrating the granitic basement. GPK-2 is the production well while the total reinjection of the geothermal fluid is performed through GPK-3 and GPK-4. In industrial operation conditions, the brine is produced at 160 °C and is reinjected at 70 °C [13], showing no difference in its chemical composition when the temperature decreases. In the framework of the H2020 MEET European program [14,15], an additional small heat exchanger (SHEx) has been installed temporarily in order to assess the optimization of energy production by lowering the temperature of the reinjected fluid from 70 °C to 40 °C [16]. This SHEx received 10% of the total flow and was tested over 3 months. It was designed with six alloys in order to test their reaction to corrosion and scaling. Only scaling phenomena are described here and several points are addressed: (1) determination of the chemical and mineralogical composition of the scales, (2) impact of temperature lowering on the scaling processes (composition/morphology/thickening of the deposits), (3) influence of alloys on scaling development. An X-ray diffraction (XRD) and scanning electron microscopy coupled with energy dispersive spectrometry (SEM-EDS) survey was performed to answer those questions.

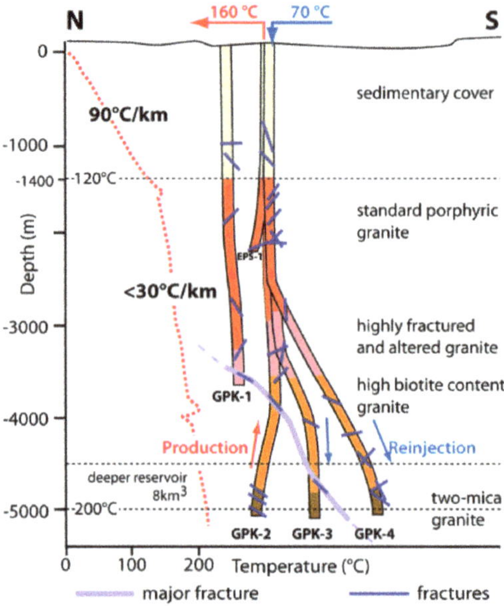

Figure 1. The Soultz Enhanced Geothermal System with its 5 wells among which the 3 deepest ones (GPK-2, GPK-3 and GPK-4) are used for electricity production. The brine is produced at 160 °C in GPK-2 well and reinjected at 70 °C in GPK-3 and GPK-4 in normal operation. The temperature gradient recorded at Soultz (red dashed line) and key figures are indicated on the left. Figure modified after [17,18]; size of the deep reservoir from [19].

2. Technical Context

2.1. Scaling Phenomenon in Geothermal Power Plants Worldwide

Scaling is a common phenomenon in geothermal power plants worldwide. Specific site analyses on scaling issues occurring in various geothermal plant types are reported in abundant literature [12,20–30], showing the importance of this topic for the plant operation. Scales act as insulators and thus lower the thermal exchanges. They also reduce the diameter of the pipes and inside volume of the exchangers which lowers the overall productivity of the geothermal power plants. Scales may also trigger the accumulation of toxic chemical elements, inducing additional risk and cost issues during the operational phase. In addition, scale formation is generally linked to degassing process, which has a major impact in terms of corrosion on the surface installation of geothermal power plants. Reducing the formation of scales is a challenge for operators who thus use inhibitors in their industrial process in order to prevent their formation [2,21,31].

2.2. Scaling Phenomeon at Soultz Geothermal Power Plant

At Soultz, surface installations are composed of several parts among which pipes and heat exchangers in which the natural geothermal fluid provides its thermal energy to an industrial fluid. An Organic Rankine Cycle (ORC) allows the production of electricity. The temperature of the brine lowers within the exchangers and the chemical equilibrium changes, resulting in the precipitation of minerals. The geochemistry of the brine provided by [10,13] shows a high salinity due to a great abundance of Na, K and Ca as major cations and Cl and SO_4 as major anions. It is summarized in Table 1, for the elements found in the scales. Li, Zn, Ba and other minor elements are also present in significant abundance. In addition, the geothermal fluid is characterized by a high CO_2 content, and natural anoxic conditions [10,13].

Table 1. Chemical composition of the brine after [10,13] from 22 brine samples collected in GPK-1, and the three deep wells at Soultz. Sb is not given in [13] (ng: not given), the only data about Sb comes from [10]. MRCC: most representative chemical composition of the native geothermal brine, in [13].

	Na (g/L)	K (g/L)	Ca (g/L)	Mg (mg/L)	Cl (g/L)	SO_4 (mg/L)	SiO_2 (mg/L)	As (mg/L)	Sb (µg/L) in [10]	Pb (µg/L)
Min-max values [13]	21–28.2	2.38–3.38	3.46–7.30	75–411	32.6–61	150–255	63–409	0.6–11	ng	181–782
MRCC [13] except for Sb	27.5	3.25	6.90	125	59	159	427	6	57.4	300

From this brine, sulfates of barite type ($(Sr, Ba)SO_4$) and minor sulfides of galena type ($(Pb, As, Sb)S$) precipitate during the lowering of the temperature in surface installations when no antiscalants are used [27,30]. The same phenomenon is also encountered in German geothermal plants located in the URG [12]. In the URG, and at Soultz in particular where the geothermal brine circulates within a granitic basement, those scales are known to accumulate radionuclides, ^{226}Ra for sulfates and ^{210}Pb for sulfides [30], and are thus to be disposed of as Naturally Occurring Radioactive Material waste (NORM classification, [32]. In such conditions when no inhibitors were used, the Soultz power plant needed to be stopped and cleaned three times a year, inducing high maintenance cost, loss of energy production and waste management issues [33].

For safety reasons and power plant healthy operation, the formation of barite needs to be inhibited continuously [31,33]. Antiscalants are well known from the oil and gas industry and mainly consist of phosphonates and polycarboxylates when preventing barite formation [34]. The scaling phenomenon being closely linked with corrosion phenomena described by [35–37], both an antiscalant and a corrosion inhibitor are currently used at Soultz. The corrosion inhibitor agent is based on amines. Each type of mineral scale has an antiscalant which is more suitable for lowering its deposited amount. Antiscalants are either inhibitors of crystallization known to be very powerful to control barite deposition or dispersants that better control metal sulfide scaling [33], or even a mixture of both. When sulfate production is made impossible thanks to antiscalants, sulfides precipitate as observed by [12] in geothermal power plants of the URG. In the following, chemicals used to prevent the deposition of scales will be simply called inhibitors.

2.3. The Tested Small Heat Exchanger (SHEx; Soultz)

In a geothermal exchanger, the natural hot brine provides its thermal energy to a working fluid and then is reinjected. Both flows are totally independent of one another and never mix. The SHEx was installed as bypass on the reinjection line [17]. It was tested over three months (late January to April 2019) in the presence of inhibitors, after which it was dismantled to allow scaling and corrosion studies. The SHEx consists of a tubular heat exchanger made with tubes of six different alloys (Figure 2), an entrance, an exit, and one water box at each end with different designs (Figure 2A) made of a seventh alloy. The west water box is separated into three compartments, while the east one is made of only two parts (Figure 2B). The cross-section of the shirt and the included tubes with their alloy is shown in Figure 2C. The tubes are organized in three parallel layers. The hot fluid comes into the SHEx through the entrance and flows through the three layers of tube with a constant decrease of the temperature: around 65–70 °C at the entrance, ~60 °C in the upper layer of tubes, 50 °C in the intermediate layer, then 40 °C in the lower layer and the exit (Figure 2B). Each water box is closed by a flange (Figure 2D).

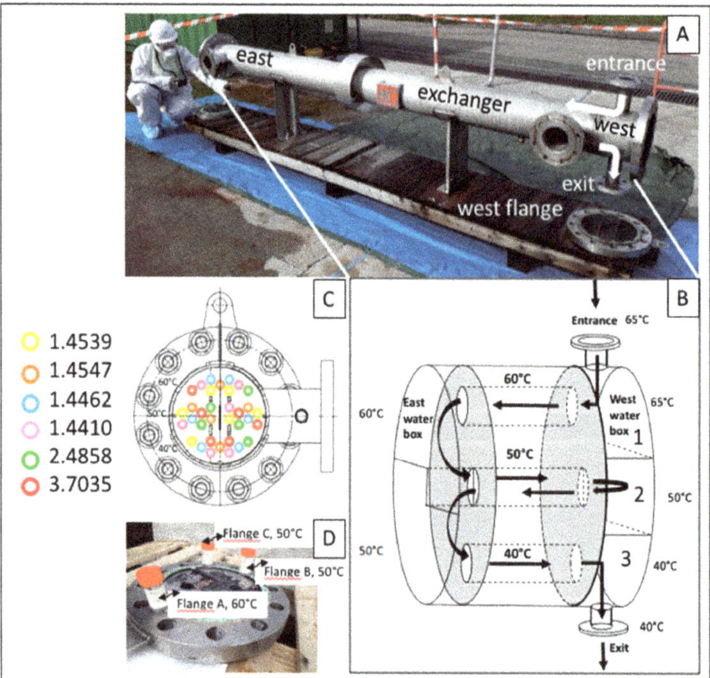

Figure 2. (**A**) overview of the SHEx; (**B**) schematic section of the cooling-down loop with 4 passes; (**C**): schematic front view with the tested alloys after [17]; (**D**): location of samples on flange closing the east water box. Note that only one tube is represented in (**B**) for each temperature (see Figure 2C for exact front representation of the location of tubes, after [17]). The intermediate layer of tubes is separated vertically in two parts (Figures 2B,C and 3) as can be seen in the east water box. Only the circulation of the geothermal brine is schematized.

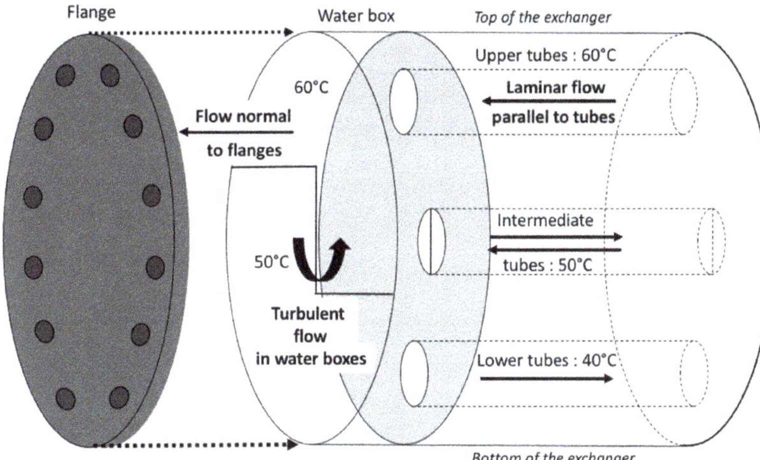

Figure 3. Temperature and flow inside the exchanger. Note that only one tube is represented for each temperature (see Figure 2C for exact cross section representation of the location and alloy of tubes). The intermediate layer of tubes is separated in two parts by a vertical panel. Thus, the flow occurs in both directions but in separate tubes. The west water box is not represented.

The path followed by the geothermal fluid within the SHEx and the flow regime is shown in Figure 3: turbulent in the entrance and in water boxes, laminar into the tubes, and perpendicular to the flanges, hence allowing to examine the likely influence of the flow regime on the scales. The three layers of tubes allow the examination of the likely impact of temperature on the scaling phenomenon. The six tested alloys are 1.4539 (904 L), 1.4547 (254 SMO), 1.4462 (DX 2205), 1.4410 (SDX 2507), 2.4858 (Alloy 825) and 3.7035 (TiGr2) [17] as visible in Figure 2C. The potential influence of the alloys on the scales will also be discussed. Mundhenk (2012) [26] proposed a ranking of metals as regards corrosion in geothermal brine conditions of the URG. The industrial exchangers currently operated at Soultz and Rittershoffen are made of 1.4410 (SDX 2507).

3. Material and Methods

3.1. Scales

In this fluid circulation test performed with the use of inhibitors, scales occur as black deposits, either as a powder (for example in water boxes, Table 2), or as a continuous plating forming a thin solid layer (like in tubes, Figure 4).

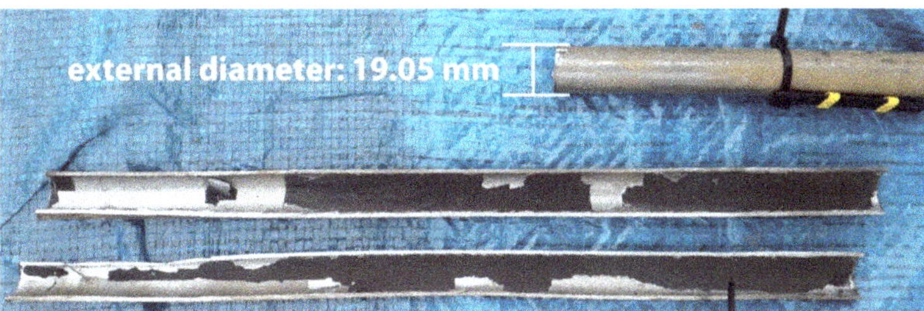

Figure 4. Macroscopic view of continuous scales while still in contact with a metal tube during dismantling of the SHEx in April 2019. The tube has been cut all along for the recovery of scales.

3.2. Preparation of Samples

The SHEx was only drained and not rinsed before dismantling. Thirty-five samples were collected in the SHEx and one in the operated industrial plant for SEM-EDS analyses (Table 2). They were collected in order to be representative of hydrodynamic and thermal conditions for each alloy. They were neither rinsed with clear water nor ground. They were simply dried at ambient temperature. The industrial sample was collected in 2017 in the operated power plant at the exit of the industrial heat exchanger, just before the reinjection line, where the geothermal fluid is circulating at a temperature of 75–65 °C in contact with a 1.4410 steel.

Millimeter-size fragments of continuous solid scales were collected and glued on metal stubs with carbon lacquer (Figure 5) for observation by reflection optical microscopy (ROM) and scanning electron microscopy (SEM) coupled with energy dispersive spectrometry (EDS) for local chemical analyses. For each sample, the side in contact with the metal (called metal side, MS in the following), the side in contact with the fluid (called fluid side, FS in the following) and the cross-section were prepared systematically (Figure 5). The MS surface looks bright and smooth while FS surface is velvety and rough (Figures 4 and 5). The cross-section was prepared in order to study the thickness of the deposits, but the preparation frequently failed. It is to be noted that on alloy 1.4462, the scales separated systematically into two layers (MS and FS) during sampling, which is the reason why three samples were collected for each temperature (MS, FS and total). Scales from industrial sample, entrance and water boxes of the SHEx occur as a powder (Figure 6) and they were just spread over carbon lacquer (Figure 6).

Figure 5. ROM view of tiny samples of continuous scales with smooth MS (tube side), velvety rough FS and section of samples collected in a tube made of 1.4410 super duplex steel, at 60 °C, glued with carbon lacquer on a metal stub for SEM-EDS analysis.

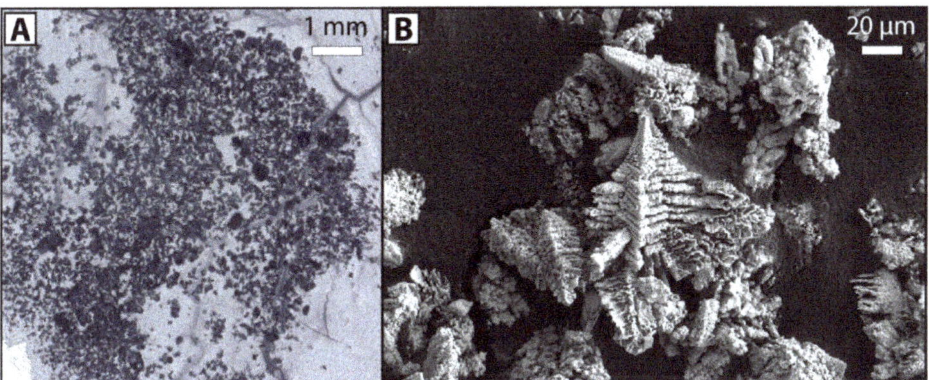

Figure 6. Powder collected in the entrance of the SHEX made of 1.4307, (**A**) as viewed by ROM and (**B**) by SEM.

3.3. X-ray Diffraction

More than 1 g of scales being necessary for X-ray diffraction (XRD), the amount of scales was insufficient in any tube of the SHEx. Only entrance, exit and water boxes 1 and 3, all of them made of 1.4307, provided 4 samples. XRD was performed by ORANO company with a Panalytical diffractometer using Co Kα radiation (λ = 1791 Å) in order to avoid potential Fe fluorescence. Neither internal nor external standards were used. The results were compared to JCPDS files for the determination of the mineral phases present in the samples.

3.4. Scanning Electron Microscopy Coupled with Energy Dispersive Spectrometry (SEM-EDS)

The SEM used in this study, a Zeiss GeminiSEM 300 coupled with a Bruker EDS, belongs to the IMAT analysis facility of CY Cergy Paris Université. No metal coating was necessary prior to observation. Observation was performed with a secondary electron detector by using a chosen acceleration voltage (between 10 and 15 kV) at a working distance between 6 and 8 mm that allowed EDS analyses in high vacuum mode. A low acceleration voltage was deliberately chosen in order to lower the beam/sample interaction volume size and to ascertain that the X-ray pulses came exclusively from the scale particles. This modest acceleration voltage presents a second advantage as it reduces significantly charging phenomena. However, this induces a poor quality of analyses enhanced by the marked topography of samples which is not ideal in terms of quantification as the beam/sample interaction volume might be truncated or shadowed. Thus, the analyses can only be used for relative abundance of the elements within and between the different scale samples.

4. Results

Results obtained by XRD and SEM-EDS are presented below and discussed in Section 5.

4.1. X-ray Diffraction (XRD)

XRD patterns that were obtained for the four samples are provided in Figure 7. Entrance and exit of the SHEx show the same diffractograms, indicating the presence of galena (PbS), and likely minor dufrénoysite (Pb$_2$As$_2$S$_5$) as regards the weak intensity of the peaks, over the whole range of temperatures (65 °C and 40 °C). The water boxes also show the presence of galena and likely dufrénoysite, together with that of halite (NaCl) for the two temperatures under concern (65 °C and 40 °C). The samples not being reduced into powder it was possible to determine hkl diffraction planes for galena, namely 111, 200, 220, 311 and 222 (Figure 7).

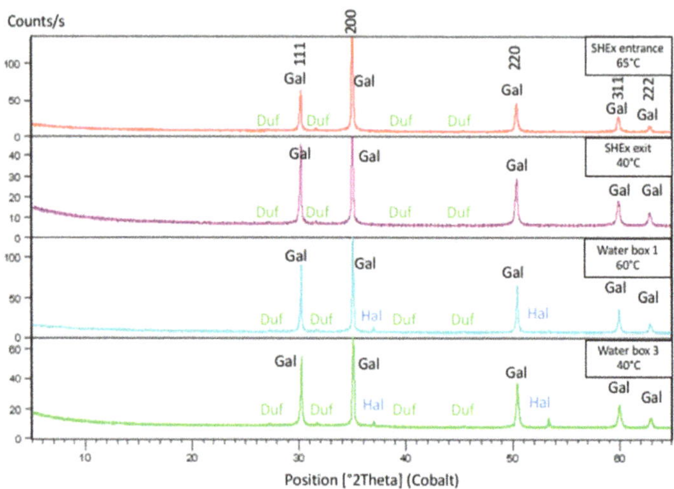

Figure 7. X-ray diffraction diagrams obtained for scales collected in the entrance, the exit and water boxes 1 and 3 of the SHEx. They show galena (Gal; As and Sb bearing PbS), additional halite (Hal; NaCl) in the water boxes, and likely traces of dufrénoysite (Duf; $Pb_2As_2S_5$), all of these sites being made of 1.4307 alloy. The numbers indicated vertically represent the hkl diffraction planes of galena crystals.

4.2. Scanning Electron Microscopy coupled with Energy Dispersive Spectrometry (SEM-EDS)

The observation by SEM allows to distinguish several features.

4.2.1. Structure of the Scales

As indicated before, scales occur as a continuous solid deposit (Figure 5) or as a powder (Figure 6, Table 2). In that first case, the deposit shows a rough FS in contact with the geothermal brine (Figure 8A,B) and a smooth MS in contact with the metal (Figure 8C,D).

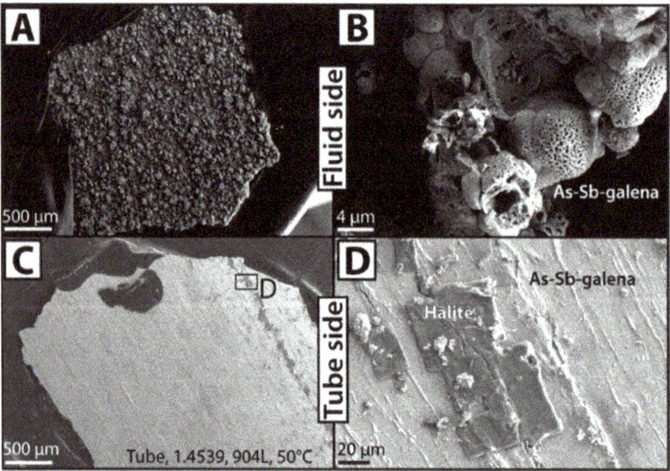

Figure 8. Sample 1.4539, 50 °C, surface of the scales: rough in contact with the brine (**A** and zoom in **B**), smooth in contact with the metal (**C** and zoom in **D**), showing As-Sb-galena and halite on MS and only galena on FS in this example.

Table 2. Summary of sampling points, performed analyses and observation by SEM. Industrial: industrial heat exchanger. All other sampling points are within the SHEx. FS: fluid side, MS: metal side, Total: FS+MS sampled at the same time when they separate. The alloys are from [17]. The analyses performed on each sample are indicated. Wwb: West water box. (co): cuboctahedron. (oct): octahedron. Underlined figures are for 1.4410 alloy for comparison between three temperatures.

Sampling Points	Alloy	Temperature (°C)	Structure of scales	Analyses	Whole Thickness of Scales (µm)	Thickness of Smooth Zone (µm)	Dendrite	Needle	Coral	Cube	Fibro-Radiated
Industrial	1.4410	65–75	Powder	SEM			x				
Entrance	1.4307	65	Powder	XRD, SEM			x				
Exit	1.4307	40	Powder	XRD							
Wwb 1 (top)	1.4307	65	Powder	XRD, SEM			x	x		x	x
Wwb 2 (Middle)	1.4307	55	Powder	SEM					x		
Wwb 3 (Bottom)	1.4307	40	Powder	XRD, SEM					x		x
Flange A (East)	1.4307	60	Continuous	SEM	-	15	x		x		x
Flange B (East)	1.4307	50	Continuous	SEM	30	12		x	x	x	
Flange C (East)	1.4307	50	Continuous	SEM	110	15					
Tube	1.4539	60	Continuous	SEM				x	x	x (co)	
Tube	1.4547	60	Continuous	SEM					x		
Tube (FS)	1.4462	60	Continuous	SEM					-		
Tube (MS)	1.4462	60	Continuous	SEM					-		
Tube (Total)	1.4462	60	Continuous	SEM	22	5			x?		
Tube	1.4410	60	Continuous	SEM	50	10			x		
Tube	2.4858	60	Continuous	SEM					x		
Tube	3.7035	60	Continuous	SEM	15	6			-		
Tube	1.4539	50	Continuous	SEM				x	x		x
Tube	1.4547	50	Continuous	SEM	270	-			x		x
Tube (FS)	1.4462	50	Continuous	SEM					-		
Tube (MS)	1.4462	50	Continuous	SEM					-		
Tube (Total)	1.4462	50	Continuous	SEM					x?		
Tube	1.4410	50	Continuous	SEM	80	10			x		x
Tube	2.4858	50	Continuous	SEM	50	30		x	x		x
Tube	3.7035	50	Continuous	SEM					x		
Tube	1.4539	40	Continuous	SEM	200	10			x		
Tube	1.4547	40	Continuous	SEM	270	25			x		
Tube (FS)	1.4462	40	Continuous	SEM							
Tube (MS)	1.4462	40	Continuous	SEM							
Tube (Total)	1.4462	40	Continuous	SEM				x	x	x (oct)	x
Tube	1.4410	40	Continuous	SEM	220	10			x		
Tube	2.4858	40	Continuous	SEM	60	30			x		x
Tube	3.7035	40	Continuous	SEM							

The scales are composed of several superimposed layers (Figure 9A) and the smooth layer in contact with the metal is generally divided in several parallel sub-layers (Figure 9B).

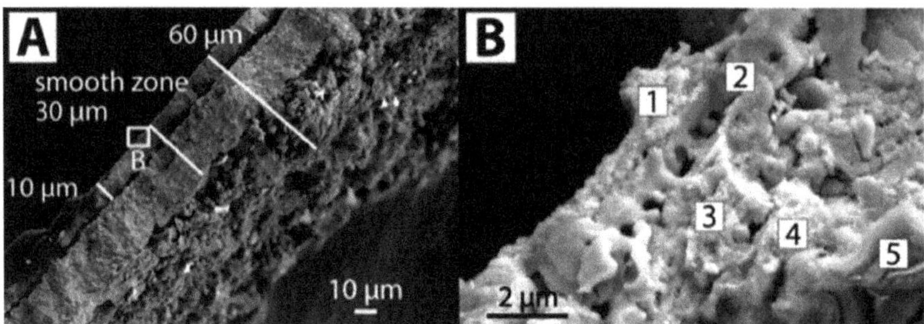

Figure 9. Sample of scales collected in a tube made of 2.4858 at 40 °C. (**A**) Cross-section of the scales divided into several layers, (**B**) the very first one (in contact with the metal) being itself composed of several micro-porous sub-layers. B is a focus on the white rectangle in A and shows five successive sub-layers that deposited on the top of one another. Sub-layer 1 is the oldest, in direct contact with the metal.

4.2.2. Thickness of the Scales

The thickness of scales was measured whenever it was possible, which was not very frequent. It could be measured systematically only for scales deposited in tubes made of 1.4410 (Figure 10) at decreasing temperatures. One has to note that measuring the thickness by SEM provides values with a non-negligible uncertainty as the measurement was sometimes not exactly normal to the deposit. However, the magnitude of the measurement remains true. The thickness observed in tubes of 1.4410 after the 3-month test was around 50 μm at 60 °C, 80 μm at 50 °C and locally up to 220 μm at 40 °C (Table 2, Figure 10), thus representing a deposition rate of about 17 μm/month at 60 °C to 73 μm/month at 40 °C, considering a constant deposition rate. The same trend of increasing thickness with decreasing temperatures tends to be seen on other alloys (Table 2): small at 60 °C (22, 50, and 15 μm, in tubes made of 1.4462, 1.4410, 3.7035 respectively), generally greater at 50 °C (270, 80, and 50μm, in tubes made of 1.4547, 1.4410, 2.4858 respectively) and in general the biggest at 40 °C (200, 270, 220, 60 μm, in tubes made of 1.4539, 1.4547, 1.4410, 2.4858 respectively).

Wherever it could be measured, the smooth zone always shows a thickness smaller than or equal to 30 μm, while the whole thickness of the deposit reaches 270 μm. Nothing much can be said about the thickness of scales deposited on the east flange as it could not be measured as a whole at 60 °C and varies from 30 to 110 μm at 50 °C. No sample is available at 40 °C as the flange is separated in only two zones: 60 °C and 50 °C.

4.2.3. SEM-EDS Chemistry of the Samples

Whatever the location (water box, entrance, exit and tubes), either on MS or FS of the scales when they are continuous, or in powder, and whatever the alloy on which they deposited, SEM-EDS spectra and maps show the presence of Pb, S $\pm$ As and Sb compounds of galena type (Figures 11 and 12). Halite (NaCl) is also frequently observed. Both of these two phases were encountered on the XRD patterns (Figure 7). Because of the analytical limitations exposed in the Methods section and their likely small size, crystals of sulfosalts such as dufrénoysite were not identified by SEM-EDS.

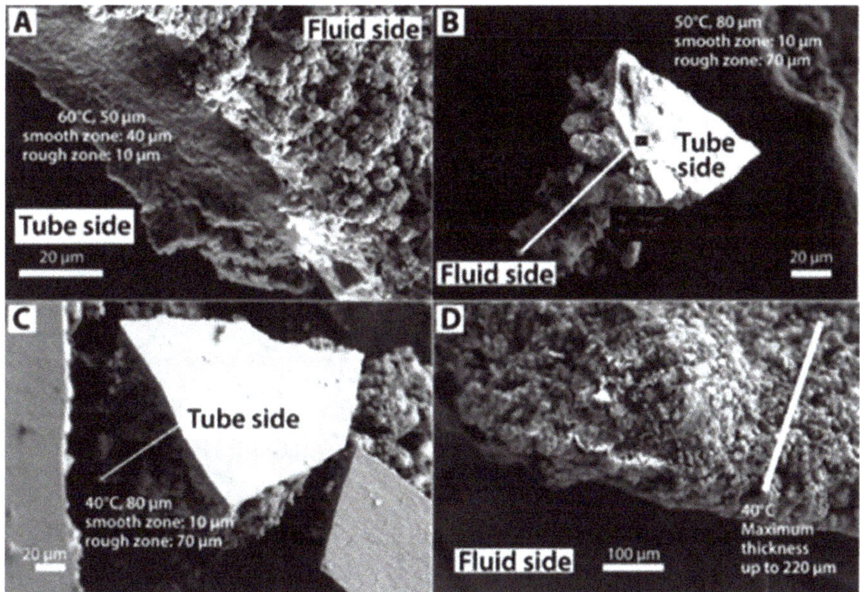

Figure 10. Tubes made of 1.4410, thickness as a function of temperature. The deposit in contact with the metal is smooth while it is rough when in contact with the brine. The thickness of the scales tends to increase when the temperature decreases within the SHEx, from 50 μm at 60 °C to 220 μm at 40 °C. (**A**) 60 °C, (**B**) 50 °C, (**C,D**) 40 °C.

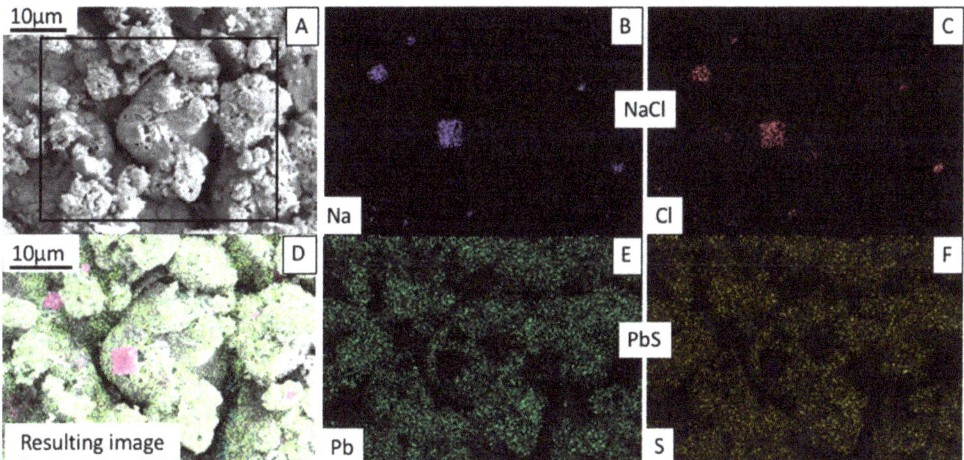

Figure 11. EDS map of FS of scales collected in a tube made of 2.4858, at 40 °C (**A–F**) showing (**A**) the SEM image and the black rectangle in which the elementary maps were performed, (**B,C**) maps of elementary concentration for Na and Cl characteristic of halite (NaCl) with its common cubic shape, (**D**) resulting map with all elements, (**E,F**) maps of elementary concentration for Pb and S characteristic of galena (PbS). As and Sb were also encountered together with Pb and S but in such a small amount that the images are not contrasted enough to be included.

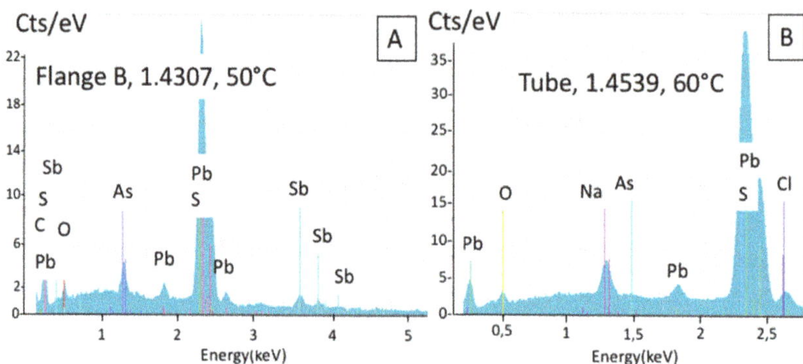

Figure 12. EDS spectra obtained on FS of scales collected (**A**) on a flange at 50 °C and (**B**) in a tube made of 1.4539, at 60 °C.

Table 3 shows examples of semi-quantitative analyses. The same elements (Pb, As, Sb, S) are present in each of the studied samples (analyses 1, 2 and 3), but in varying relative abundance. Sb is sometimes quite abundant (analyses 2 and 3) but no Sb-bearing sulfosalts were discovered either on XRD diagrams (Figure 7) or by SEM. Na and Cl are also locally detected in the samples.

Table 3. Three normalized semi-quantitative analyses obtained by SEM-EDS. Note the limitations of accuracy due to the low acceleration voltage, the topography of samples and their low thickness. The analyses can only be used for relative abundance of the elements within and between the different scale samples and mostly indicate the elements present in the samples.

	1	2	3
	Tube	Tube	Tube
	3.7035	2.4855	2.4855
	60 °C	60 °C	40 °C
Elements	Content (wt.%)	Content (wt.%)	Content (wt.%)
S	11.75	16.28	15.00
As	3.89	8.57	3.70
Sb	11.46	25.08	17.82
Pb	72.90	50.07	63.48
Total	100	100	100

From these semi-quantitative results, no difference appears in the chemistry of the scales, whatever the alloy on which they deposited or the temperature of the brine from which they precipitated.

4.2.4. Shapes of Galena Crystals

Various galena crystal shapes were observed thanks to SEM on the tiny fragments described in Section 3 (Materials and Methods). The crystals are of micrometer size in millimeter-sized samples. Thus, these observations might not be exhaustive but give an overview of the crystal shapes of galena.

1. Dendrites

Dendritic crystal shape was observed in the scales collected in the industrial installation (reference sample), as well as in the entrance of the SHEx, and on the upper part of the eastern flange (Figure 13). The industrial sample and that collected at the entrance of the SHEx are made of only dendrites (Figure 13A,B). The only dendrite observed on the flange (Figure 13C) seems to have been deposited by the flow as it is free and not embedded in the matrix. No dendrites were found in any of the tubes, whatever the temperature.

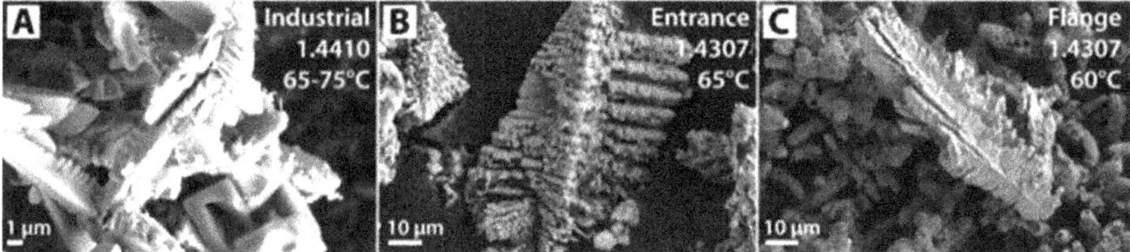

Figure 13. Dendritic shapes of galena crystals (**A**) in the industrial sample (1.4410 alloy), (**B**) in the entrance of the SHEx and (**C**) on the upper part of the eastern flange (both made of 1.4307). No dendrites were found in any of the tubes.

2.   Needles

Needle shape is found in tubes of 1.4462 (50 °C and 40 °C), 1.4539 (60 °C and 50 °C), and 2.4858 (50 °C), in water boxes (1.4307 alloy, at 65 °C, 55 °C and 40 °C), and on a flange (1.4307 alloy, 50 °C), thus at various temperatures (from 65 °C to 40 °C, Figure 14) and alloys. Needles were not observed on other samples. Needles can be parallel to each other (A) or perpendicular (B to H) and sometimes in three orthogonal directions (C). Needles have a square section as visible mostly in B and F. They were observed in zones where the flow is rather laminar (tubes) or turbulent (water boxes and flanges).

Figure 14. Needle shapes of galena crystals (**A,B**) on a flange and (**C**) in water box and (**D–F**) in tubes, at different temperatures and on three different alloys (1.4307, 1.4462, 1.4539). (**A**) needles parallel to each other, (**B–F**): needles grew in perpendicular directions. (**B,F**): Note that the needles are monocrystals with square section.

3.   Coral Shapes

Coral-like shapes are of various types (Figure 15) that all show an important internal porosity. They are the most common shapes encountered in the exchanger (Table 4). They were observed in different zones of the SHEx (entrance, 1.4307 alloy; water box, 1.4307 alloy; flange, 1.4307 alloy) and in tubes of all alloys, and at temperatures varying from 65 °C (entrance and water box) to 40 °C (water box and tubes).

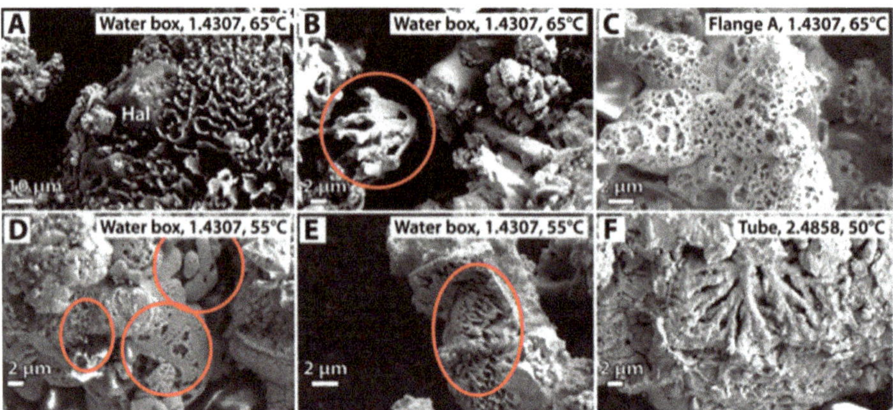

Figure 15. PbS coral shapes of different kinds. All of them show an important internal porosity. The zones to be observed are highlighted by red ellipses. Hal: halite. (**A**) halite crystal embedded in coral shape galena, water box, 1.4307, 65 °C, (**B**) 3D view of coral shape galena, water box, 1.4307, 65 °C, (**C**) coral shape made of botryoids with abundant porosity, flange A, 1.4307, 65 °C, (**D**) coral shape made of botryoids with abundant porosity, water box, 1.4307, 55 °C, (**E**) coral shape made of numerous contiguous needles, water box, 1.4307, 55 °C, (**F**) cross section of coral shape galena.

4. Cube and Cubic-Derived Shapes

Several cubic or cubic-derived shapes were observed in the samples. Cubes were found in a tube at 60 °C (1.4539) where it shows exactly the same hollow shape as on a flange (1.4307, 40 °C; Figure 16A). Cubes were also observed in a water box (1.4307, 40 °C; Figure 16B,C) where they occur either as massive structures (Figure 16B) or as a kind of skeleton made of needles oriented in the three directions of space (Figure 16C), those two features being in close contact in the same sample. Cubes are thus found on at least two different alloys (1.4307 and 1.4539) and at temperatures from 60 °C to 40 °C.

Flange, 1.4307, 65°C Water box, 1.4307, 65°C Water box, 1.4307, 65°C

Figure 16. Cubic-derived shape of PbS crystals, occurring as (**A**) hollow cubes, (**B**) massive cubes, or (**C**) skeleton made of needles in three orthogonal directions inside the orange ellipse. The massive cube (**B**) is seen in the lower left-hand corner of (**C**).

Other PbS shapes derived from the cube were locally observed in tubes, as shown in Figure 17, such as a cuboctahedron (Figure 17A) and an octahedron (Figure 17B), on two different alloys and at 60 °C and 40 °C respectively. The octahedron (Figure 17B) in found in the vicinity of orthogonal needles not visible on the photograph.

Figure 17. PbS cuboctahedron (**A**) and PbS octahedron (**B**) with their indexed faces, observed in two tubes made of 1.4539, at 60 °C (**A**) and 1.4462, at 40 °C (**B**).

Thus, cubes and cubic-derived shapes were observed on three different alloys and at temperatures from 65 to 40 °C (Table 4).

5. Fibro-Radiated Botryoidal Shape

The fibro-radiated botryoidal type (Figure 18A) is made of needles organized in 3D fan shape (Figure 18B,C) with several superimposed layers (Figure 18B,C). No to minor porosity is observed as opposed to coral shape. These three examples were observed occurred at a 50 °C temperature, on three different alloys. Table 2 shows all the locations where botryoids were observed, from 65 °C to 40 °C.

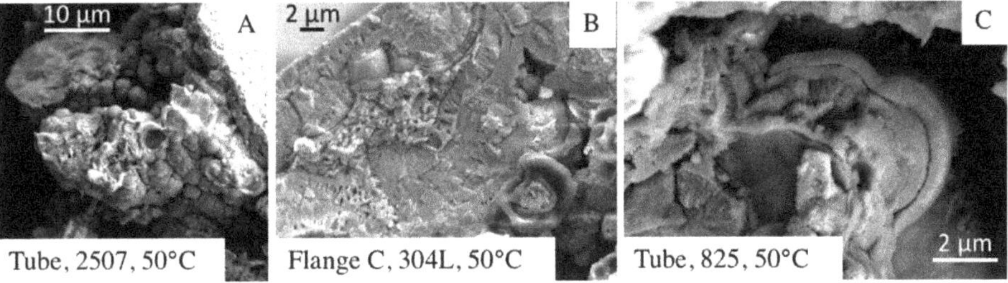

Figure 18. Fibro-radiated botryoidal shape from three different samples collected on a flange, and in two tubes of different alloys. All of the examples presented here were observed for a 50 °C temperature, but they also occurred at 65 °C and 40 °C (see Table 2). (**A**) general view, on a tube made of 2507 at 50 °C, (**B,C**) close view of a cross-section of botryoid on a flange and on a tube, at 50 °C.

Table 2 recapitulates all of the shapes that were observed by SEM, as a function of the location inside the SHEx, the alloy type and the temperature. Some samples do not show any characteristic crystal shapes, because of their poor quality (tubes of 1.4462 and 3.7035 alloys, Table 4). In the other samples, the coral shape is the most widely observed, whatever the alloy and the temperature. Other crystal shapes are frequently found in association with it in samples collected in tubes. Cubes are associated with coral shape in the west water box and on a flange, and in association with needles in both water boxes and in some tubes. Dendrites were observed in great abundance and not in association with other shapes in the industrial installation and in the entrance. The only dendrite found on a flange appears to be free on the surface of the scale and not embedded in the deposit. The various shapes were observed whatever the temperature and the alloy, except for dendrites which were observed only at the highest temperature.

5. Discussion

Galena is studied in ore deposits for scientific and economical purposes [38–40], and because of its toxicity in mining environments [41–43]. Galena is also well known in the industry, in particular for its semi-conductor properties and is thus thoroughly studied [44–46]. Natural galena is rarely a pure PbS component and frequently contains arsenic [47] and antimony [40,48]. Various galena shapes and chemical compositions are reported in natural environments [49]. Laboratory growth experiments show that the shape and chemistry of galena crystals can be controlled by several factors among which time, temperature and concentration of elements in the solvent [44–46,50]. All these previous studies might be useful for understanding the growth process in the SHEx at Soultz, even though the chemistry of the solution and other parameters are different. As regards shapes of PbS crystals, the literature reports laboratory growth of hopper (skeletal) crystals [51], dendrites, nanocubes, and truncated nanocubes [50], dendrites with different shapes [52], nanocoral [53], and many others.

5.1. Structure and Chemistry of Scales

In the SHEx, scales occur either as a powder or as layered deposits (Table 4).

Table 4. Structure of the scales sampled in the SHEx.

Structure	Location	Flow
Powder	Industrial Entrance Exit	Turbulent
Layers	Water boxes Tubes Flange	Laminar Perpendicular

Where the flow is turbulent, the scales deposit as a powder. Layered deposits are structured into sub-layers likely related to the operation of the power plant. In tubes made of 1.4462, the scales occurred as two major layers which were difficult to extract together and fell into small pieces during sampling, inducing a poor quality of samples. Scales deposited in 3.7035 tubes were strongly attached to the metal and were difficult to collect, resulting also in a bad quality of samples. This explains the lack of information about crystal shapes for those two alloys (Table 2). The micro-porosity observed in the superimposed thin layers, as well as between and inside the crystals (coral-shape for example) might be due to local turbulence of the flow.

The deposits that formed in the SHEx indeed contain galena, whatever the occurrence (entrance, water boxes or exit), as indicated by XRD diagrams (Figure 7) when compared to [50] who also used Cobalt anticathod. Their characteristics are summarized in Table 5.

Table 5. Characteristics of galena crystals determined from XRD.

Sharp Peaks	Location of Peaks	Preferential Growth
Well-crystallized [52]	Face-centered cubic structure, Fm3 m space group [50]	Strong intensity of (200) reflection peak [50,51], thus, preferential growth in the <100> direction
	Based on JCPDS, 5-592 [51] or ASTM file card No. 030660020 [53]	

XRD analyses of the scales indeed show the presence of galena but give no information about its chemistry which was surveyed by SEM-EDS. It is homogeneous with the systematic presence of As and Sb in varying abundance, in addition to Pb and S, which is a well-known phenomenon in natural ore systems [40,47,48]. The chemistry of scale surveyed by SEM-EDS is summarized in Table 6.

Table 6. Summary of the chemistry of scales as surveyed by SEM-EDS.

Location	Elements	Phases	Shapes
All samples	Pb, As, Sb, S	Galena As, Sb sulfosalts (e.g., dufrénoysite, Figure 7)	Various (Table 2) Undetermined
All samples	Na, Cl	Halite	Cubes (Figures 8, 11 and 14)

Other sulfosalts might also occur given the high relative amount of As and Sb given by EDS analyses, but their small abundance did not allow us to see them on the diffractograms. In addition, the semi-quantification performed thanks to SEM-EDS did not allow to determine either their presence or their amount in the samples.

Halite was detected by XRD neither in the SHEx entrance, nor in its exit, probably because of its too small abundance, but it was seen by SEM-EDS. Indeed, the intensity of the peaks related to halite in the water boxes is weak on the XRD patterns (Figure 7).

To summarize, the composition of scales is homogeneous whatever the metal on which they formed and whatever the temperature of deposition (from 65 °C to 40 °C). Thus, it appears that these two parameters (alloy and temperature) do not influence the chemistry of scales.

5.2. Thickness of Scales

It is to be noted that the measurement of the deposit thickness (Table 2) might be considered as only semi-quantitative. Indeed, it could not always be performed strictly perpendicular to the deposit, which induced an uncertainty in the obtained value. However, a general tendency is observed: the thickness increases (e.g., from 50 to 220 µm for 1.4410 tubes) when the temperature decreases (from 60 °C to 40 °C). The deposition of scales in an exchanger has several effects, some of them positive, such as protection against corrosion, others negative, such as insulation reducing the heat exchange and hence energy production, or the decrease of the diameter of tubes which reduces the fluid flow. The thicker the deposit, the better the protection against corrosion but the lower the energy production. The thickness of scales has thus to be carefully monitored and controlled by addition of inhibitors to the process and maintenance when necessary, in order to allow optimal energy production.

5.3. Conditions of Scale Formation

It is likely that halite crystals developed when the SHEx was dismantled, during its draining and drying, as they are not embedded in the scales and sometimes grew on the MS of the scales (Figure 8A).

The conditions for scale formation during the geothermal process are summarized in Table 7 which shows the changes undergone by the deposit through time, with a decreasing influence of metal.

Table 7. Summary of conditions for scale formation.

Order of Layer Formation	Layer Structure	Location	Influence of Metal
1st	Smooth (Figures 8D, 9A and 10)	Contact with metal	Strong
2nd	Smooth to rough (Figures 9 and 10)	Contact with 1st layer	Low to none
3rd	Smooth to rough	Contact with 2nd layer	Low to none
4th and more	Rough (Figures 8A,B and 10)	Contact with previous layer	None

The influence of alloy nature on the shape of galena crystals being eliminated, the likely influencing parameters that remain to explain the crystal shapes are the chemical composition of the brine, its temperature and the flow regime.

Dendrites were observed exclusively at the exit of the industrial installation and in the entrance of the SHEx. Indeed, dendrites are known to crystallize quickly [31], which is permitted by the highest temperature (75–65 °C) and the turbulence in these locations.

Abundant recent literature presents the conditions of galena synthesis in laboratory and the shapes of crystals obtained [44,45,51–56]. Shapes identical to those found in the SHEx are encountered in conditions described as hydrothermal, meaning with water as a solvent and maintained at a temperature of, for example, 80 °C [50] to 200 °C [45], with durations of 2 h [45] to 48 h [50]. Song et al. (2012) [44] report that the concentration of the solution and reaction time (24 h at 170 °C) are key parameters for obtaining controlled PbS crystal shapes. In those conditions, they report cubes, dendrites, stars, and wires. These various shapes are required for specific industrial uses where the optical, magnetic and electronic properties of semi-conductors are of high importance [45]. Other authors [52] conducted solvothermal syntheses meaning with organic solvents and imposed temperature conditions leading to PbS dendrites. Nanocoral shape was obtained by [53], by vapor-solid deposition at high temperature (1050 °C) and thus at conditions drastically different from ours. [51] provide examples of various shapes obtained at a constant temperature (120 °C) but for various synthesis durations (3 to 24 h). [57] report various shapes of PbS nanoparticles (cubic, needle-like, spherical) due to the use of a number of capping agents. Wang et al. (2003) [52] who conducted their syntheses at the constant 120 °C temperature with various starting agents and several solvents, including water, obtained various types of PbS dendrites and other shapes. Hence, all these experiments show that there is no clear relationship between the parameters of the synthesis and the shapes obtained.

At Soultz, the solvent of the brine is water, but the inhibitors that are injected in the process are composed of organic molecules which play a role in the crystallization of galena. In fact, when no such agent is used, mostly sulfates (barite group $(Ba,Sr,Ca)SO_4$ solid-solution) are produced in the URG [12] and at Soultz in particular [27,31,33].

In the case of galena crystallization (use of inhibitors), when it occurs as cubes or derived shapes it results from a preferential growth along <111> direction inducing {100} faces to develop, which is not consistent with the major growth in <100> direction deduced from XRD. However, this is not abnormal since cubes and cubic-derived shapes are rarely found. Indeed, [54] proposed relations between crystal structure and crystal morphology on an energy basis. According to [54] the morphology of a crystal is governed by chains of strong bonds running through the structure, called periodic bond chain (P.B.C.) vectors. Crystal faces are divided into three classes. Flat faces (F) contain two or more coplanar P.B.C. vectors and are the most important faces. Stepped faces (S) are parallel to only one P.B.C vector and are of medium importance. Kinked (K) faces are parallel to no P.B.C. vector and are very rare or do not occur at all. For PbS crystals that belong to the fcc structure, F faces are {100}, S faces are {110} and K faces are {111}. According to Hartmann's and Perdok's theory [54] only F faces should appear at equilibrium, giving cubes as in Figure 16. In fact, during crystal growth, impurities such as the inhibitors used in the geothermal industrial process at Soultz, or As and Sb ions present in the geothermal brine, are adsorbed on K faces ({111} in this case) which promotes their development, together with that of cube (faces {100}), inducing the formation of cuboctahedrons (Figure 17A). In other cases (Figure 17B), only faces {111} develop, leading to octahedral crystals. Crystals with such planar faces (cubes, cuboctahedron, octahedron) appear in conditions of small growth rate, here in laminar flow, as opposed to dendrites which are obtained by a high growth rate in a single direction, here in the hottest and most turbulent flow. As seen in Figure 19, when a crystal grows, the faces which are kept at the end are those where the setting up of atoms is the slowest (faces {100} in the case of galena cubes). Indeed [55] indicates that the faster the growth in a given direction, the smaller the area of the face developed perpendicular to that direction (Figure 19, face {111}).

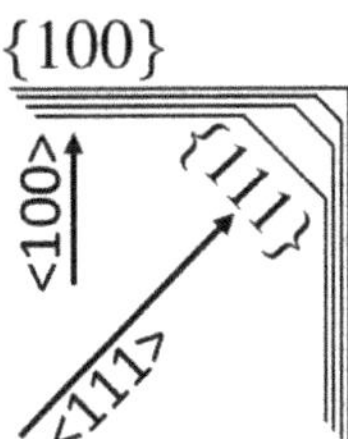

Figure 19. Schematic growth of a crystal face {xxx} as a function of the atom setting up rate in the direction perpendicular to that face <xxx>.

Hence, F faces are visible in the final crystal. Changes in morphology are related to F faces showing a different (higher) growth rate. As a consequence, dendrites and needles that develop in a preferential direction grow very quickly (turbulence and/or high temperature) while cubes and derived shapes grow slowly, in laminar flow and at temperatures which can be low (down to 40 °C). In Figure 14C, needles developed in three perpendicular directions grow on a preexisting 30 µm wide galena cube. This succession of shapes might be controlled by very local changes in the parameters of the surrounding medium.

Table 8 summarizes the occurrence of galena crystal shapes.

Table 8. Occurrence of galena crystal shapes as a function of their location, the temperature and the flow.

Crystal Shape		Abundance among Location Samples	Temperature (°C)	Flow
Dendrite	+	Industrial, Entrance(flange)	75–65	Turbulent
Needle	++	Water box, flange, tubes	65–40	Turbulent, laminar
Coral	+++	Water box, flange, tubes	65–40	Turbulent, laminar
Cube	+	Water box, flange, tubes	65–40	Turbulent, laminar
Fibro-radiated	++	Water box, flange, tubes	65–40	Turbulent, laminar

Thus, except for dendrites, the location and hence turbulence degree of the place, nature of alloy and temperature are not controlling the shape of PbS crystals that formed in the SHEx, which can be found mixed at given places (Table 2) as opposed to what is described in the literature for syntheses in the laboratory. This might be due to the fact that in the SHEx, the parameters are not controlled as in the laboratory, which allows various shapes to crystallize at the same place. In addition, the temperature range is low, from 65 °C to 40 °C and is probably not discriminating for promoting specific crystal shapes.

The presence of As sulfosalts such as dufrénoysite, and maybe others containing Sb as indicated by the EDS semi-quantitative analyses (Table 5), might also be responsible for some of the shapes that were encountered during this study. However, it was not possible to identify them during SEM-EDS survey.

6. Conclusions and Outlook

One way to improve the energy production of geothermal power plants in the URG is to decrease the reinjection temperature. This might induce several problems in the power plants including cooling of the rock reservoir, promotion of a chemical disequilibrium into it, and increase of scaling phenomenon as observed for the samples collected in the SHEx at Soultz. When inhibitors are used, those scales are mostly composed of lead sulfide (galena, PbS) together with minor sulfosalts The galena crystals collected at the interface between the metals from which the SHEx was made and the geothermal fluid, after three months of operation, show a homogeneous chemical composition including As and Sb, whatever the alloy on which the scales deposited and whatever the temperature (65 °C to

40 °C). Thus, there is no influence of the alloy on the scaling phenomenon, as opposed to what is observed for corrosion. Those galena crystals show several shapes that cannot be evidently connected to the alloy, the temperature or the large-scale flow regime, except for dendrites. Indeed, the alloy is insulated from the fluid by the very first layers of deposit, the temperature does not vary drastically from the entrance to the exit of the SHEx (25 °C gradient only, at rather low temperatures), the chemical composition of the brine is constant during the industrial process (Ravier, personal communication), and after three months of operation, the scales are rough at the contact with the fluid and the flow is certainly very slow because of this rugosity, allowing the slow growth of crystals with various shapes including cubes and derived shapes. Dendrites are the only shape to be found exclusively at the highest temperature (65–75 °C) and in a turbulent environment. To go further, investigation could be performed with Raman to characterize the sulfosalts likely present in the samples, and with XANES to assess the oxidation state of As, Pb and Sb. In addition, statistics of the various shapes encountered at the different locations ought to be performed to pinpoint likely influencing parameters at the micro-scale. Furthermore, one could also conduct laboratory experiments with the brine produced from the geothermal reservoir, and with varying parameters such as temperature, alloy, speed of flow rate, type and amounts of inhibitors, etc. Finally, at present, scales produced at Soultz have to be disposed of as waste due to their toxicity. One can rather imagine an industrial valorization, especially for those deposited at the entrance and exit of the exchangers where mostly dendrites are formed, which is a sought-after shape for the industry [52].

Author Contributions: Conceptualization, B.A.L. and M.L.; methodology, G.R., O.S., B.A.L., É.D., A.G., X.S.; validation, J.M., C.B., É.D., A.G., Ayming and H2020 MEET partners; formal analysis, B.A.L.; investigation, B.A.L.; resources, G.R., O.S., J.M., C.B., É.D.; data curation, B.A.L. and M.L.; writing—original draft preparation, B.A.L. and M.L.; writing—review and editing, R.L.H., J.M., C.B., G.R., O.S., É.D., G.T., X.S., A.G.; supervision, B.A.L., A.G., G.T; project administration, A.G., É.D., G.T.; funding acquisition, all the co-authors and H2020 MEET consortium. All authors have read and agreed to the published version of the manuscript. Authorship is limited to those who have contributed substantially to the work reported.

Funding: This project has received funding from the European Union's Horizon 2020 research and innovation program under grant agreement No 792037 (MEET project).

Acknowledgments: The authors thank the Soultz-Sous-Forêts site owner, GEIE Exploitation Minière de la Chaleur, for giving access to their geothermal installation. Valérie Granger (Orano) is acknowledged for XRD analysis and its interpretation. We gratefully thank Jean Hérisson (Ayming) for reviewing this paper as well as relationship with European Commission and overview of the H2020 MEET project. We gratefully acknowledge the two reviewers for their helpful review which greatly helped us to improve the manuscript.

Conflicts of Interest: The authors declare no conflict of interest. The funders had no role in the design of the study; in the collection, analyses, or interpretation of data; in the writing of the manuscript, or in the decision to publish the results.

References

1. Assad, M.E.H.; Bani-Hani, E.; Khalil, M. Performance of Geothermal Power Plants (Single, Dual, and Binary) to Compensate for LHC-CERN Power Consumption: Comparative Study. *Geotherm. Energy* **2017**, *5*, 17. [CrossRef]
2. Spinthaki, A.; Kamaratou, M.; Skordalou, G.; Petratos, G.; Tramaux, A.; David, G.; Demadis, K.D. A Universal Scale Inhibitor: A Dual Inhibition/Dispersion Performance Evaluation under Difficult Brine Stresses. *Geothermics* **2021**, *89*, 101972. [CrossRef]
3. van der Zwaan, B.; Dalla Longa, F. Integrated Assessment Projections for Global Geothermal Energy Use. *Geothermics* **2019**, *82*, 203–211. [CrossRef]
4. Shannon, D.W. Economic Impact of Corrosion and Scaling Problems in Geothermal Energy Systems. Master's Thesis, Battelle Pacific Northwest Labs, Richland, WA, USA, 1975.
5. Tardiff, G.E. Using Salton Sea Geothermal Brines for Electrical Power: A Review of Progress in Chemistry and Materials Technology, 1976 Status. In Proceedings of the Twelfth Intersociety Energy Conversion Engineering Conference, Washington, DC, USA, 28 August–2 September 1977.

6. Criaud, A.; Fouillac, C. Sulfide Scaling in Low Enthalpy Geothermal Environments: A Survey. *Geothermics* **1989**, *18*, 73–81. [CrossRef]
7. Scheiber, J.; Nitschke, F.; Seibt, A.; Genter, A. Geochemical and Mineralogical Monitoring of the Geothermal Power Plant in Soultz-Sous-Forêts (France). In Proceedings of the 37th Workshop on Geothermal Reservoir Engineering, Stanford, CA, USA, 30 January–1 February 2012; pp. 1033–1042.
8. Ngothai, Y.; Yanagisawa, N.; Pring, A.; Rose, P.; O'Neill, B.; Brugger, J. Mineral Scaling in Geothermal Fields: A Review. In Proceedings of the 2010 Australian Geothermal Energy Conference, Adelaide, Autralia, 17–19 November 2010; pp. 405–409.
9. Skinner, B.J.; White, D.E.; Rose, H.J.; Mays, R.E. Sulfides Associated with the Salton Sea Geothermal Brine. *Econ. Geol.* **1967**, *62*, 316–330. [CrossRef]
10. Sanjuan, B.; Millot, R.; Innocent, C.; Dezayes, C.; Scheiber, J.; Brach, M. Major Geochemical Characteristics of Geothermal Brines from the Upper Rhine Graben Granitic Basement with Constraints on Temperature and Circulation. *Chem. Geol.* **2016**, *428*, 27–47. [CrossRef]
11. Baticci, F.; Genter, A.; Huttenloch, P.; Zorn, R. Corrosion and Scaling Detection in the Soultz EGS Power Plant, Upper Rhine Graben, France. In Proceedings of the World Geothermal Congress, WGC2010, Bali, Indonesia, 25–30 April 2010.
12. Haas-Nüesch, R.; Heberling, F.; Schild, D.; Rothe, J.; Dardenne, K.; Jähnichen, S.; Eiche, E.; Marquardt, C.; Metz, V.; Schäfer, T. Mineralogical Characterization of Scalings Formed in Geothermal Sites in the Upper Rhine Graben before and after the Application of Sulfate Inhibitors. *Geothermics* **2018**, *71*, 264–273. [CrossRef]
13. Mouchot, J.; Genter, A.; Cuenot, N.; Scheiber, J.; Seibel, O.; Bosia, C.; Ravier, G. First Year of Operation from EGS Geothermal Plants in Alsace, France: Scaling Issues. In Proceedings of the 43rd Workshop on Geothermal Reservoir Engeneering, Stanford, CA, USA, 12–14 February 2018.
14. Dalmais, E.; Genter, A.; Trullenque, G.; Leoutre, E.; Leiss, B.; Wagner, B.; Mintsa, A.C.; Bär, K.; Rajsl, I. MEET Project: Toward the Spreading of EGS across Europe. In Proceedings of the European Geothermal Congress, Den Haag, The Netherlands, 11–14 June 2019.
15. Trullenque, G.; Genter, A.; Leiss, B.; Wagner, B.; Bouchet, R.; Léoutre, E.; Malnar, B.; Bär, K.; Rajšl, I. Upscaling of EGS in Different Geological Conditions: A European Perspective. In Proceedings of the 43rd Workshop on Geothermal Reservoir Engineering, Stanford, CA, USA, 12–14 February 2018.
16. Ravier, G.; Seibel, O.; Pratiwi, A.S.; Mouchot, J.; Genter, A.; Ragnarsdóttir, K.R.; Sengelen, X. Towards an Optimized Operation of the EGS Soultz-Sous-Forêts Power Plant (Upper Rhine Graben, France). In Proceedings of the European Geothermal Congress, Den Haag, The Netherlands, 11–14 June 2019.
17. Ledésert, B.A.; Hébert, R.L. How Can Deep Geothermal Projects Provide Information on the Temperature Distribution in the Upper Rhine Graben? The Example of the Soultz-Sous-Forêts-Enhanced Geothermal System. *Geosciences* **2020**, *10*, 459. [CrossRef]
18. Cuenot, N.; Charléty, J.; Dorbath, L.; Haessler, H. Faulting Mechanisms and Stress Regime at the European HDR Site of Soultz-Sous-Forêts, France. *Geothermics* **2006**, *35*, 561–575. [CrossRef]
19. Gunnarsson, I.; Arnórsson, S. Impact of Silica Scaling on the Efficiency of Heat Extraction from High-Temperature Geothermal Fluids. *Geothermics* **2005**, *34*, 320–329. [CrossRef]
20. Haklıdır, F.S.T.; Balaban, T.Ö. A Review of Mineral Precipitation and Effective Scale Inhibition Methods at Geothermal Power Plants in West Anatolia (Turkey). *Geothermics* **2019**, *80*, 103–118. [CrossRef]
21. Hardardóttir, V.; Ármannsson, H.; Þórhallsson, S. Characterization of Sulfide-Rich Scales in Brine at Reykjanes. In Proceedings of the World Geothermal Congress 2005, Antalya, Turkey, 24–29 April 2005.
22. Jamero, J.; Zarrouk, S.J.; Mroczek, E. Mineral Scaling in Two-Phase Geothermal Pipelines: Two Case Studies. *Geothermics* **2018**, *72*, 1–14. [CrossRef]
23. Köhl, B.; Grundy, J.; Baumann, T. Rippled Scales in a Geothermal Facility in the Bavarian Molasse Basin: A Key to Understand the Calcite Scaling Process. *Geotherm. Energy* **2020**, *8*, 1–27. [CrossRef]
24. Kristmannsdóttir, H. Types of Scaling Occurring by Geothermal Utilization in Iceland. *Geothermics* **1989**, *18*, 183–190. [CrossRef]
25. Mundhenk, N. Corrosion and Scaling in Utilization of Geothermal Energy in the Upper Rhine Graben. Ph.D. Thesis, Karlsruher Instituts für Technologie (KIT), Karlsruhe, Germany, 2013. [CrossRef]
26. Nitschke, F.; Scheiber, J.; Kramar, U.; Neumann, T. Formation of Alternating Layered Ba-Sr-Sulfate and Pb-Sulfide Scaling in the Geothermal Plant of Soultz-Sous-Forêts. *Neues Jahrb. Für Mineral. Abh. J. Mineral. Geochem.* **2014**, *191*, 145–156. [CrossRef]
27. Raymond, J.; Williams-Jones, A.E.; Clark, J.R. Mineralization Associated with Scale and Altered Rock and Pipe Fragments from the Berlin Geothermal Field, El Salvador; Implications for Metal Transport in Natural Systems. *J. Volcanol. Geotherm. Res.* **2005**, *145*, 81–96. [CrossRef]
28. Reyes, A.G.; Trompetter, W.J.; Britten, K.; Searle, J. Mineral Deposits in the Rotokawa Geothermal Pipelines, New Zealand. *J. Volcanol. Geotherm. Res.* **2003**, *119*, 215–239. [CrossRef]
29. Genter, A.; Cuenot, N.; Melchert, B.; Moeckes, W.; Ravier, G.; Sanjuan, B.; Sanjuan, R.; Scheiber, J.; Schill, E.; Schmittbuhl, J. Main achievements from the multi-well EGS Soultz project during geothermal exploitation from 2010 and 2012. In Proceedings of the European Geothermal Congress, Pisa, Italy, 3–7 June 2013.
30. Scheiber, J.; Seibt, A.; Birner, J.; Genter, A.; Cuenot, N.; Moeckes, W. Scale Inhibition at the Soultz-Sous-Forêts (France) EGS Site: Laboratory and on-Site Studies. In Proceedings of the World Geothermal Congress 2015, Melbourne, Australia, 16–24 April 2015.

31. Sanjuan, B.; Millot, R.; Dezayes, C.; Brach, M. Main Characteristics of the Deep Geothermal Brine (5 Km) at Soultz-Sous-Forêts (France) Determined Using Geochemical and Tracer Test Data. *Comptes Rendus Geosci.* **2010**, *342*, 546–559. [CrossRef]
32. Cuenot, N.; Goerke, X.; Guery, B.; Bruzac, S.; Sontot, O.; Meneust, P.; Maquet, J.; Vidal, J. Evolution of the Natural Radioactivity within the Soultz Geothermal Installation. In Proceedings of the Soultz Geothermal Conference 2011, Soultz-Sous-Forêts, France, 5–6 October 2011; Volume 5, p. 19.
33. Mouchot, J.; Scheiber, J.; Florencio, J.; Seibt, A.; Jähnichen, S. Scale and Corrosion Control Program, Example of Two Geothermal Plants in Operation in the Upper Rhine Graben. In Proceedings of the European Geothermal Congress 2019, Den Haag, The Netherlands, 11–14 June 2019.
34. He, S.; Oddo, J.E.; Tomson, M.B. The Inhibition of Gypsum and Barite Nucleation in NaCl Brines at Temperatures from 25 to 90 °C. *Appl. Geochem.* **1994**, *9*, 561–567. [CrossRef]
35. Karlsdottir, S.N.; Ragnarsdottir, K.R.; Thorbjornsson, I.O.; Einarsson, A. Corrosion Testing in Superheated Geothermal Steam in Iceland. *Geothermics* **2015**, *53*, 281–290. [CrossRef]
36. Karlsdottir, S.N.; Thorbjornsson, I.O.; Ragnarsdottir, K.R.; Einarsson, A. Corrosion Testing of Heat Exchanger Tubes in Steam from the IDDP-1 Exploratory Geothermal Well in Krafla, Iceland. In Proceedings of the CORROSION 2014, San Antonio, TX, USA, 9–13 March 2014.
37. Ravier, G.; Huttenloch, P.; Scheiber, J.; Perrot, V.; Sioly, J.L. Design, Manufacturing and Commissioning of the ECOGI's Heat Exchangers at Rittershoffen (France): A Case Study. In Proceedings of the European Geothermal Conference, Strasbourg, France, 19–24 September 2016.
38. Apopei, A.I.; Damian, G.; Buzgar, N.; Buzatu, A. Mineralogy and Geochemistry of Pb–Sb/As-Sulfosalts from Coranda-Hondol Ore Deposit (Romania)—Conditions of Telluride Deposition. *Ore Geol. Rev.* **2016**, *72*, 857–873. [CrossRef]
39. Martínez-Abad, I.; Cepedal, A.; Arias, D.; Fuertes-Fuente, M. The Au–As (Ag–Pb–Zn–Cu–Sb) Vein-Disseminated Deposit of Arcos (Lugo, NW Spain): Mineral Paragenesis, Hydrothermal Alteration and Implications in Invisible Gold Deposition. *J. Geochem. Explor.* **2015**, *151*, 1–16. [CrossRef]
40. Pitcairn, I.K.; Olivo, G.R.; Teagle, D.A.; Craw, D. Sulfide Evolution during Prograde Metamorphism of the Otago and Alpine Schists, New Zealand. *Can. Mineral.* **2010**, *48*, 1267–1295. [CrossRef]
41. Medunić, G.; Bucković, D.; Crnić, A.P.; Bituh, T.; Srček, V.G.; Radošević, K.; Bajramovic, M.; Zgorelec, Z. Sulfur, Metal (Loid) s, Radioactivity, and Cytotoxicity in Abandoned Karstic Raša Coal-Mine Discharges (the North Adriatic Sea). *Rud. Geološko Naft. Zb.* **2020**, *35*. [CrossRef]
42. Majzlan, J.; Drahota, P.; Filippi, M. Parageneses and Crystal Chemistry of Arsenic Minerals. *Rev. Mineral. Geochem.* **2014**, *79*, 17–184. [CrossRef]
43. Karakaya, N.; Karakaya, M.C. Toxic Element Contamination in Waters from the Massive Sulfide Deposits and Wastes around Giresun, Turkey. *Turk. J. Earth Sci.* **2014**, *23*, 113–128. [CrossRef]
44. Song, C.; Jiang, L.; Zhang, Y.; Pang, L.; Wang, D. Shape Controllable Growth of PbS Polyhedral Crystals. *Cryst. Res. Technol.* **2012**, *47*, 1008–1013. [CrossRef]
45. Ding, B.; Shi, M.; Chen, F.; Zhou, R.; Deng, M.; Wang, M.; Chen, H. Shape-Controlled Syntheses of PbS Submicro-/Nano-Crystals via Hydrothermal Method. *J. Cryst. Growth* **2009**, *311*, 1533–1538. [CrossRef]
46. Ni, Y.; Wei, X.; Hong, J.; Ma, X. Hydrothermal Preparation of PbS Crystals and Shape Evolution. *Mater. Res. Bull.* **2007**, *42*, 17–26. [CrossRef]
47. Radosavljević-Mihajlović, A.S.; Stojanović, J.N.; Radosavljević, S.A.; Pačevski, A.M.; Vuković, N.S.; Tošović, R.D. Mineralogy and Genetic Features of the Cu-As-Ni-Sb-Pb Mineralization from the Mlakva Polymetallic Deposit (Serbia)—New Occurrence of (Ni-Sb)-Bearing Cu-Arsenides. *Ore Geol. Rev.* **2017**, *80*, 1245–1258. [CrossRef]
48. Adamczyk, Z.; Nowińska, K.; Melaniuk-Wolny, E.; Szewczenko, J. Variation of the Content of Accompanying Elements in Galena in Pyrometallurgical Process of Zinc and Lead Production. *Acta Montan. Slovaca* **2013**, *18*, 158–163.
49. Hagni, R.D. Platy Galena from the Viburnum Trend, Southeast Missouri: Character, Mine Distribution, Paragenetic Position, Trace Element Content, Nature of Twinning, and Conditions of Formation. *Minerals* **2018**, *8*, 93. [CrossRef]
50. Singh, K.; McLachlan, A.A.; Marangoni, D.G. Effect of Morphology and Concentration on Capping Ability of Surfactant in Shape Controlled Synthesis of PbS Nano-and Micro-Crystals. *Colloids Surf. A Physicochem. Eng. Asp.* **2009**, *345*, 82–87. [CrossRef]
51. Zhu, J.; Duan, W.; Sheng, Y. Uniform PbS Hopper (Skeletal) Crystals Grown by a Solution Approach. *J. Cryst. Growth* **2009**, *311*, 355–357. [CrossRef]
52. Wang, D.; Yu, D.; Shao, M.; Liu, X.; Yu, W.; Qian, Y. Dendritic Growth of PbS Crystals with Different Morphologies. *J. Cryst. Growth* **2003**, *257*, 384–389. [CrossRef]
53. Obaid, A.S.; Mahdi, M.A.; Hassan, Z. Nanocoral PbS Thin Film Growth by Solid-Vapor Deposition. *Optoelectron. Adv. Mater. Rapid Commun.* **2012**, *6*, 422–426.
54. Hartman, P.; Perdok, W.G. IUCr On the Relations between Structure and Morphology of Crystals. Available online: https://scripts.iucr.org/cgi-bin/paper?a01327 (accessed on 11 December 2020).
55. Pandey, G.; Shrivastav, S.; Sharma, H.K. Role of Solution PH and SDS on Shape Evolution of PbS Hexagonal Disk and Star/Flower Shaped Nanocrystals in Aqueous Media. *Phys. E Low Dimens. Syst. Nanostruct.* **2014**, *56*, 386–392. [CrossRef]

56. Esmaeili, E.; Sabet, M.; Salavati-Niasari, M.; Saberyan, K. Synthesis and Characterization of Lead Sulfide Nanostructures with Different Morphologies via Simple Hydrothermal Method. *High. Temp. Mater. Process.* **2016**, *35*, 559–566. [CrossRef]
57. Patel, A.A.; Wu, F.; Zhang, J.Z.; Torres-Martinez, C.L.; Mehra, R.K.; Yang, Y.; Risbud, S.H. Synthesis, Optical Spectroscopy and Ultrafast Electron Dynamics of PbS Nanoparticles with Different Surface Capping. *J. Phys. Chem. B* **2000**, *104*, 11598–11605. [CrossRef]

 *geosciences*

Article

Thermodynamic and Kinetic Modelling of Scales Formation at the Soultz-sous-Forêts Geothermal Power Plant

Pierce Kunan [1], Guillaume Ravier [1,*], Eléonore Dalmais [1], Marion Ducousso [2] and Pierre Cezac [2]

[1] ES-Géothermie, 5 Rue Ampère, 67450 Mundolsheim, France; piercegkunan@gmail.com (P.K.); eleonore.dalmais@es.fr (E.D.)

[2] École Nationale Supérieure en Génie des Technologies Industrielles, Bat D'Alembert, Rue Jules Ferry, BP 7511, CEDEX, 64075 Pau, France; marion.ducousso@univ-pau.fr (M.D.); pierre.cezac@univ-pau.fr (P.C.)

* Correspondence: guillaume.ravier@es.fr

Abstract: Geothermal energy has been a subject of great interest since the 1990s in the Upper Rhine Graben (URG), where the first European Enhanced Geothermal System (EGS) pilot site has been developed, in Soultz-sous-Forêts (SsF), France. Several studies have already been conducted on scales occurring at the reinjection side at the geothermal plants located in the URG. It has been observed that the composition of the scales changes as chemical treatment is applied to inhibit metal sulfate. The purpose of this study was to model the scaling phenomenon occurring in the surface pipes and the heat exchangers at the SsF geothermal plant. PhreeqC, a geochemical modelling software, was used to reproduce the scaling observations in the geothermal plant during exploitation. A suitable database was chosen based on the availability of chemical elements, minerals, and gas. A thermodynamic model and a kinetic model were proposed for modelling the scaling phenomenon. The thermodynamic model gave insight on possible minerals precipitated while the kinetic model, after modifying the initial rates equation, produced results that were close to the expected scale composition at the SsF geothermal plant. Additional laboratory studies on the kinetics of the scales are proposed to complement the current model.

Keywords: Upper Rhine Graben; Soultz-sous-Forêts; geothermal brine; scaling; metal sulfides; thermodynamic; kinetics

Citation: Kunan, P.; Ravier, G.; Dalmais, E.; Ducousso, M.; Cezac, P. Thermodynamic and Kinetic Modelling of Scales Formation at the Soultz-sous-Forêts Geothermal Power Plant. *Geosciences* **2021**, *11*, 483. https://doi.org/10.3390/geosciences11120483

Academic Editors: Antonio Paonita and Jesus Martinez-Frias

Received: 22 September 2021
Accepted: 18 November 2021
Published: 23 November 2021

Publisher's Note: MDPI stays neutral with regard to jurisdictional claims in published maps and institutional affiliations.

1. Introduction

1.1. Geothermal Energy in the Upper Rhine Graben

The Upper Rhine Graben (URG) is a rifting formation, oriented NNE, part of the European Cenozoic rift system. It extends for 300 km of length, from Basel (Switzerland) in the south to Mainz (Germany) in the north. Important thermal anomalies have been identified in the URG thanks to a rich geological exploration (Figure 1, [1]). These anomalies delineate thermal gradient locally over 100 °C/km in the first km of sediments and controlled with normal faults parallel to the graben direction. The first European Geothermal research project of Soultz-sous-Forêts (SsF) was conducted initially in the early 1990s. This project was based on the Hot Dry Rock (HDR) concept, where the goal was to create an artificial heat exchanger in the basement rocks by hydraulic fracturing [2]. However, the results obtained after the drilling of the first well at SsF showed the presence of natural fluid circulation through the existing fracture network of the reservoir [3]. Since then, the Enhanced Geothermal System (EGS) technology was incorporated into future development of the URG geothermal project. This approach consists of exploiting the natural thermal brine circulation by improving, if necessary, the connection between the geothermal wells and the reservoir with various chemical, hydraulic, and thermal treatments [4].

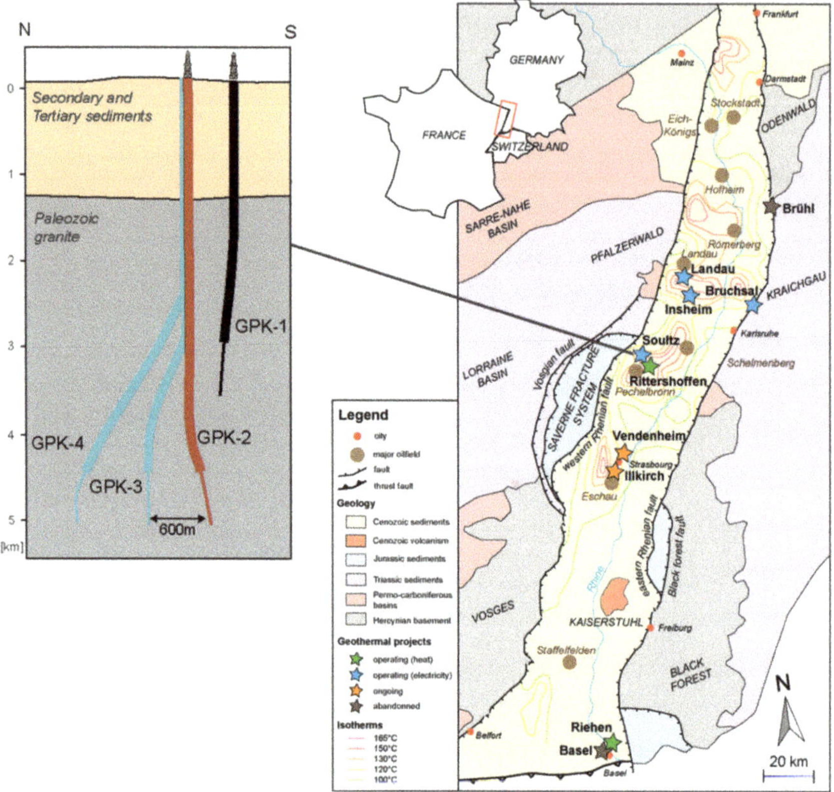

Figure 1. Map of the Upper Rhine Graben and schematic drawing of the deep wells at SsF geothermal plant [1].

There are several geothermal projects that have been developed in the French, German, and Swiss URG region over the past years. In France, two notable geothermal plants are in operation at SsF and Rittershoffen, respectively, for power and heat production while in Germany, three geothermal plants are in operation for power generation.

1.2. SsF Geothermal Power Plant

The Soultz-sous-Forêts geothermal project started in 1987 and is the cradle of the geothermal energy European research in granitic and fractured systems. Over 30 years of research, the geothermal site at SsF continues to exploit commercially the fractured basement for the EEIG Heat Mining. The actual geothermal system consists of three wells: one production well named GPK-2 and two injection wells named GPK-3 and GPK-4 which are drilled 5 km into the granitic basement. The geothermal brine is produced at a temperature of 150 °C, reaching the wellhead with a nominal flow rate of 30 kg/s provided by a downhole production Line Shaft Pump [5]. The installed gross capacity of the binary plant is around 1.7 MWe (Figure 2).

Figure 2. The SsF geothermal power plant (Source: EEIG Heat Mining).

The geothermal brine is flowed through a system that consists of three consecutive double pass tubular heat exchangers which supply heat to an Organic Rankine Cycle (ORC) to produce electricity. The geothermal brine is then fully reinjected into the granitic basement at around 65–70 °C. The volume of reinjected brine is split between the two injection wells without the need of reinjection pumps. The well-head overpressure in the surface infrastructure is regulated by using production pump which reaches about 23 bars to keep the gas dissolved in the brine. The reinjection temperature is linked to the conversion process. The geothermal plant has been successfully producing electricity commercially since September 2016, with an availability rate of about 90% for the past four years [6]. The granite reservoir is made of a porphyritic monzogranite rich in K-feldspar megacrysts. Primary silicate minerals are quartz, plagioclase, biotite, and hornblende. A chemical analysis on the composition of the brine was taken in February 2020 (Table 1, [7]), while an analysis on the gas dissolved in the brine was taken in April 2019 (Table 2, [7]).

Table 1. Composition of brine at the production well of the SsF geothermal plant [7].

GPK-2 (Production Well)												
Composition of brine (mg/L)	Na	Ca	K	Cl	Mg	Sr	Li	SiO_2	SO_4	Br	Mn	NH_4
	26,400	7020	3360	55,940	123	422	160	179	108	240	17	23.2
Composition of brine (mg/L)	As	Ba	Cs	Rb	B	Fe	Zn	F	I	Cu	Pb	Cd
	10	26	14	23	38	26.3	2.8	1.3	1.6	0.001	0.11	0.01
Composition of brine (mg/L)	Sb	Al	U	Ni	HCO_3	COT						
	0.06	0.05	0.001	0.0011	197	0.9						

Table 2. Composition of gas in brine at the production well of the SsF geothermal plant [7].

GPK-2 (Production Well)		
Gas dissolved in brine	%vol	Partial pressure (atm)
CO_2	0.882	0.882
N_2	0.0908	0.0908
CH_4	0.0239	0.0239

1.3. Geochemical Characterization of the Scale during Operation

In the Upper Rhine Graben region, scaling commonly occurs at the cold side of the SsF geothermal plant [8]. Therefore, in the Upper Rhine Graben, scale formation before the application of sulfate scale inhibitors was dominated by (Ba, Sr, Ca)SO_4 solid–solution scaling containing minor amounts of galena, pyrite, or poly-metallic sulfides phases [8–10]. The main scales observed related to deep geothermal activity have been studied not only because when represented at a significant amount of secondary precipitations they could plug the geothermal infrastructures (pipe, heat exchanger, well-head), but also because the scales have the properties to trap radiogenic elements such as ^{226}Ra and ^{210}Pb in their crystalline lattices [8,9].

By using sulfate inhibitors in the Upper Rhine Graben region, barite precipitation was strongly reduced [8,11]. However, brittle grey–dark scales are still precipitating on the pipe walls consisting of PbS, and elemental Pb, As, Sb are precipitating in the geothermal infrastructures [11]. Traces of halite are present on some samples, but it corresponds to a drying residue from the geothermal brine [11]. Based on Raman spectrum of the sulfide phase, a hydrothermal Pb-Sb-Cu-sulfide ($Pb_{13}CuSb_7S_{24}$) has been characterized as well as an amorphous phase [11].

Several studies at SsF geothermal plant [6,12] report on the effects of the chemical treatment used to inhibit the formation of sulfate scales at SsF geothermal plant. Complementary studies have been carried out in the framework of the MEET research project at temperature below 65 °C with a test heat exchanger [13]. A typical black scale deposit at the wall of a tube pipe of this heat exchanger is shown in Figure 3. CY Cergy Paris Université conducted a study on different scales found in the test heat exchanger with a Zeiss GeminiSEM 300 Scanning Electron Microscopy, coupled with a Bruker Energy Dispersive Spectrometry. Figure 4 details this typical scale, a (Pb,As,Sb)S fibro-radiated hilly scale found at 50 °C on 1.4410 stainless steel tube [14].

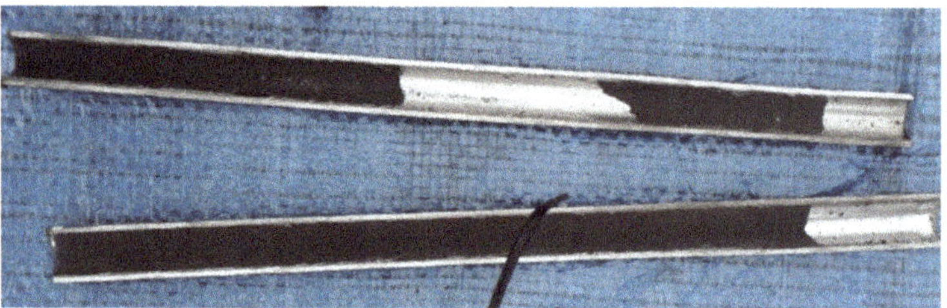

Figure 3. PbS scales deposited in tubes from the test heat exchanger.

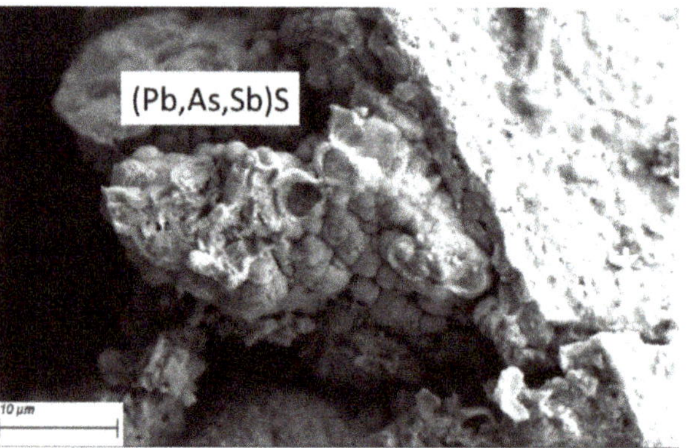

Figure 4. Microscopic photo of (Pb,As,Sb)S scale found at SsF plant [14].

Scales in the range between 150 °C and 65 °C have been sampled in June 2018 before cleaning operation in the ORC evaporator and preheaters after nearly one year of operation. Figure 5 presents a schematic drawing of the geothermal loop at SsF and the temperature gradient in the heat exchangers between the production well GPK-2 and injection wells GPK-3 and GPK4. Chemical composition of these scales has been determined using ICP MS method which is a type of mass spectrometry that uses an inductively couple plasma to ionize the sample. Scales in the range between 60 °C and 40 °C have been sampled in April 2019 in a test heat exchanger (HEX) designed with different metallurgy and installed at the SsF geothermal plant during three months in the framework of the MEET research project [13]. The latest chemical composition of scales observed at SsF geothermal plant within a range of temperature between 150 °C to 40 °C are presented in Table 3. Table 3 considers only scaling samples from tubes with 1.4410 metallurgy like the ORC heat exchanges to have a good comparison. A detail description of these scales is given by Ledésert et al. (2021) [14], and chemical composition was also determined using ICP MS method. Chemical treatment of the brine was almost the same for the two sets of scales. These scales consist of S, Pb, Sr, Ba, Sb, As, Fe, Si, and Cu elements.

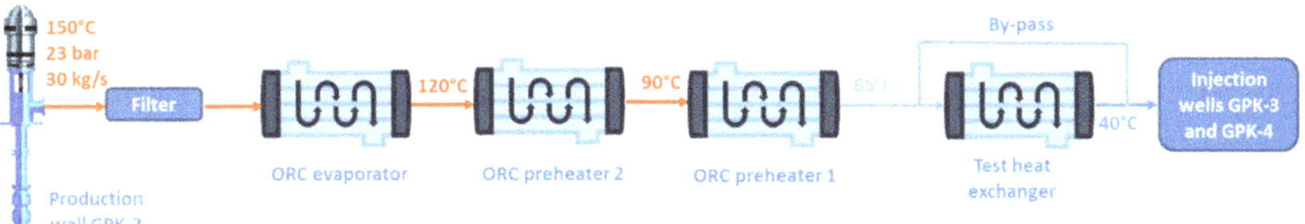

Figure 5. Schematic drawing of the geothermal loop at SsF.

Table 3. The mass composition of scales formed in the heat exchangers at the geothermal plant and in the test heat exchangers in percentage.

Temperature	S	Pb	Sr	Ba	Sb	As	Fe	Si	Cu	Exchanger
150	2.9%	2.0%	2.9%	0.94%	0.11%	0.53%	1.7%	3.8%	0.40%	ORC Inlet Evaporator
120	11.8%	26.5%	0.65%	1.9%	3.3%	6.6%	7.5%	8.0%	16.6%	ORC Inlet Preheater 2
90	11.2%	36.1%	0.86%	3.6%	3.1%	5.2%	8.0%	16.9%	5.1%	ORC Inlet Preheater 1
65	13.1%	46.3%	0.51%	2.2%	6.3%	7.3%	4.6%	8.4%	4.5%	ORC Outlet Preheater 1
60	13.1%	74.6%	0.01%	0.00%	6.4%	3.2%	0.07%	1.4%	0.40%	Test HEX
50	14.4%	66.5%	0.01%	0.01%	11.4%	4.3%	0.55%	1.0%	0.43%	Test HEX
40	16.7%	64.2%	0.01%	0.01%	10.9%	4.5%	0.48%	1.6%	0.36%	Test HEX

The presentation of the mass percentage of scales is based on the total elements found in the scales. Certain compounds, mainly carbonates, were omitted from Table 3 because they are not the main focus of this study which is dedicated to low temperature scale formation. There are also lesser amounts of the scales deposited in the higher temperature heat exchangers (ORC heat exchangers), while more scales are deposited in the lower temperature heat exchangers (Test HEX).

Lead is found primarily at lower temperatures notably at temperatures below 120 °C. Sulfur, arsenic, silicon, and antimony are also deposited at large quantities after lead. The rest of the elements are found in smaller traces (less than 5%). The test heat exchanger has a different concentration of scales compared to the ORC heat exchangers at the geothermal plant due to the difference in temperature. In the test heat exchanger, lead has a higher concentration than those in the main exchangers. The chemical treatment on the sulfate scales proved to be effective as the quantity of barium sulfate (barite) and strontium sulfate (celestite) are found in very small quantities which are less than 4% for any point of temperature, while before the application of such treatment (Ba, Sr, Ca)SO_4 solid–solution was dominating [8].

The main objective of this study was to model the scaling phenomenon occurring in the surface pipes and heat exchangers at the SsF geothermal plant. Scaling formation was firstly modelled according to thermodynamic perspective and the results are compared to the geochemical analyses presented in Table 3 and used as references. A previous investigation was conducted on available thermodynamical databases to find the most suitable one regarding geochemical elements and possible scaling minerals. Thermodynamical modeling was then completed with kinetic modeling to better represent real operational conditions in heat exchangers. The results of both modeling are later discussed.

2. Methods

The modelling of the geochemical fluids is done through the software, PhreeqC 3.6.4 which is a computer program that is written in C++ programming language. It is designed to perform numerous aqueous geochemical calculations. PhreeqC implements several types of aqueous models depending on the database used. This program was created by the U.S. Geological Survey (USGS). PhreeqC is freely distributed by the USGS and is currently an open source software.

PhreeqC uses a pre-established thermodynamic database to perform the calculations during modelling of a fluid. Each database has different sets of elements and aqueous species as well as different thermodynamic data which are taken from different references sources. There are several databases found within the installation of the PhreeqC program. Supplementary databases were also found in the PhreeqC Users forum. There are databases taken from studies such as e THERMOCHIMIE [15] and THEREDA [16]. The PhreeqC manual [17] was referred to when performing the modelling of formation of scales with PhreeqC. Table 4 shows the list of databases gathered which are listed from D1 to D19:

Table 4. PhreeqC databases and allocated nomenclature.

Databases	Nomenclature
Phreeqc	D1
Pitzer	D2
ColdChem	D3
Core10	D4
Frezchem	D5
Iso	D6
LLNL	D7
MINTEQ	D8
Minteq v4	D9
Pitzer_Old	D10
sit	D11
T_H	D12
WATEQ4F	D13
Thermoddem_06_2017	D14
PHREEQC_ThermoddemV1.10_15Dec2020	D15
ThermoChimie_PHREEQC_eDH_v9b0	D16
THEREDA_2020_PHRQ	D17
CEMDATA18.1-16-01-2019-phaseVol	D18
ThermoChimie_PhreeqC_SIT_oxygen_v10a	D19

2.1. Verification: Elements

In order to verify the validity of the databases to be used in the modelling process, the sets of elements available within the databases were compared to the elements found in the geothermal fluid at the SsF plant. The latest chemical analysis (taken in February 2020) on the composition of the brine at the SsF plant was used to cross-reference with the sets of elements found in the databases to narrow down the list of valid databases. This analysis showed that there was high concentration of Na and Cl ions in the brine. The recent study by Bosia et al. (2021) [7] provides further details on the geochemical dataset used. Databases with more supplementary elements were taken more into consideration due to the likelihood of simulating the actual fluid. Thus, the presence of the elements in the databases are compared to the elements found in the geothermal fluid at the SsF plant (Table 5)

Table 5. Geochemical elements in the databases.

	D1	D2	D3	D4	D5	D6	D7	D8	D9	D10	D11	D12	D13	D14	D15	D16	D17	D18	D19
S	x	x	x	x	x	x	x	x	x	x	x	x	x	x	x	x	x	x	x
Pb	x						x	x	x		x	x	x	x	x	x			x
Sr	x	x					x	x	x	x	x	x	x	x	x	x	x	x	x
Ba	x	x					x	x	x	x	x	x	x	x	x	x			x
Sb							x	x	x		x			x	x	x			x
As							x	x	x		x	x	x	x	x	x			x
Fe	x	x		x		x	x	x	x	x	x	x	x	x	x	x		x	x
Si	x	x		x		x	x	x	x		x	x	x	x	x	x	x	x	x
Cu	x			x			x	x	x		x	x	x	x	x	x			x
Al	x			x		x	x	x	x		x	x	x	x	x	x	x	x	x
B	x	x		x			x	x	x	x	x	x	x	x	x	x			x
Be							x	x	x					x	x				
Br	x	x				x	x	x	x	x	x	x	x	x	x	x			x
Ca	x		x	x	x	x	x	x	x	x	x	x	x	x	x	x	x	x	x
Cd	x						x	x	x		x	x	x	x	x	x			x
Ce							x							x	x				
Cl	x	x	x	x	x	x	x	x	x		x	x	x	x	x	x	x	x	x
Co				x			x		x		x			x	x	x			x
Cs							x				x	x	x	x	x	x	x		x
Dy							x							x	x				
Er							x							x	x				
Eu				x			x				x			x	x	x			x
F	x					x	x	x	x		x	x	x	x	x	x			x
Gd				x			x							x	x				
Ge														x	x				
Hg							x	x	x		x			x	x	x			
Ho							x				x			x	x	x			x
I							x	x	x		x	x	x	x	x	x			x
In							x							x	x				
K	x	x	x	x	x	x	x	x	x	x	x	x	x	x	x	x	x	x	x
La							x							x	x				
Li	x	x		x			x	x	x	x	x	x	x	x	x	x			x
Lu							x							x	x				
Mg	x	x	x	x	x	x	x	x	x	x	x	x	x	x	x	x	x	x	x
Mn	x	x		x			x	x	x	x	x	x	x	x	x	x			x
Mo				x			x		x		x			x	x	x			x
Na	x	x	x	x	x	x	x	x	x	x	x	x	x	x	x	x	x	x	x
Nd							x							x	x		x		
Ni				x			x	x	x		x		x	x	x	x			x
P	x			x		x	x	x	x		x	x	x	x	x	x	x		x
Pd							x					x		x	x	x			x
Pr							x							x	x				
Rb							x	x			x	x	x	x	x	x			x
Re							x							x	x				
Rh														x	x				
Sc				x			x							x	x				
Sm				x			x				x			x	x	x			x
Tb							x							x	x				
Tm							x							x	x				
W		x					x							x	x				
Y							x							x	x				
Yb							x							x	x				
Zn	x	x		x			x	x	x		x	x	x	x	x	x			x
HCO$_3$	x	x		x	x	x	x	x	x	x	x	x	x	x	x	x *	x	x	x
NH$_4$				x			x	x	x		x	x	x	x	x	x		x	x
SO$_3$				x			x	x	x		x	x *	x *	x	x	x		x	x
SO$_4$	x	x	x	x	x	x	x	x	x	x	x	x	x	x	x	x	x	x	x
Total	23	17	7	26	8	14	55	32	33	14	38	29	30	57	57	38	14	14	37

* = limited.

The geochemical elements from the Table 5 are represented in their aqueous state. From this study, the Thermoddem (D14 and D15) [18] and LLNL (D7) [19] databases, having respectively 57 and 55 elements of the 57 SsF brine chemical composition, are observed to be suitable for the purpose of this study as they possess the most amount elements found in the brine at the SsF plant. Further reference to the Thermoddem database will be the Thermoddem (D15) database instead of the Thermoddem (D14) database, because D15 is the latest version for the Thermoddem database.

2.2. Verification: Minerals

Another criterion set for the validation of the databases is the formation of probable minerals in the geothermal fluid at the SsF plant. A list of known minerals precipitated was made to compare to the minerals found in the databases. Furthermore, a list of probable minerals precipitated was made for minerals that have not been identified before in previous studies. These minerals that are susceptible to precipitation are identified by listing out minerals from the databases that consist of at least two of nine elements that are the majority in the analysis of scales conducted at the site. The nine principal elements are sulfur, lead, strontium, barium, antimony, arsenic, iron, silicon, and copper. A similar approach to the verification of elements was used in the verification of minerals in which a table with the list of minerals was cross-referenced with the database. The occurrences of known minerals and minerals susceptible to precipitation in the databases are tabulated (Table 6).

Table 6. Minerals in the databases.

		Databases																		
Known Minerals		D1	D2	D3	D4	D5	D6	D7	D8	D9	D10	D11	D12	D13	D14	D15	D16	D17	D18	D19
Galena	PbS							x	x	x		x	x	x	x	x	x			x
Quartz	SiO_2	x	x		x		x	x	x	x		x	x	x	x	x	x	x	x	x
Calcite	$CaCO_3$	x	x		x	x	x	x	x	x	x		x	x	x	x	x		x	x
Anhydrite	$CaSO_4$	x	x	x	x	x	x	x	x	x	x		x	x	x	x	x		x	x
Gypsum	$CaSO_4{:}2H_2O$	x	x	x	x	x	x	x	x	x	x	x	x	x	x	x	x		x	x
Barite	$BaSO_4$	x	x					x	x	x	x		x	x	x	x	x			x
Halite	NaCl	x	x	x	x	x	x	x	x	x	x	x	x	x	x	x	x	x		x
Goethite	FeOOH	x			x		x	x	x	x		x	x	x	x	x	x		x	x
Celestite	$SrSO_4$	x	x					x	x	x	x	x	x	x	x	x	x		x	x
Arsenopyrite	FeAsS							x							x	x				
Stibnite	Sb_2S_3							x	x	x		x			x	x	x			x
Possible Other Minerals																				
Hematite	Fe_2O_3	x			x		x	x	x	x		x	x	x	x	x	x		x	x
Strontianite	$SrCO_3$	x						x	x	x		x	x	x	x	x	x		x	x
Svanbergite	$SrAl_3(PO_4)(SO_4)(OH)_6$														x	x				
$Sr_3(AsO_4)_2$	$Sr_3(AsO_4)_2$							x				x			x	x	x			x
SrS	SrS							x				x			x	x	x			x
Anglesite	$PbSO_4$	x						x	x	x		x	x	x	x	x	x			x
Cerussite	$PbCO_3$	x						x	x	x		x	x	x	x	x	x			x
Alamosite	$PbSiO_3$							x	x			x	x	x	x	x	x			x
Beudantite	$PbFe_3(AsO_4)_2(OH)_5{:}H_2O$														x	x				
Corkite	$PbFe_3(PO_4)(OH)_6SO_4$							x							x	x				
Cotunnite	$PbCl_2$							x	x	x		x	x	x	x	x	x			x
Duftite	$PbCuAsO_4(OH)$														x	x				
Hinsdalite	$PbAl_3(PO_4)(SO_4)(OH)_6$							x	x	x			x	x	x	x				
Hydrocerussite	$Pb_3(CO_3)_2(OH)_2$							x		x		x	x	x	x	x	x			x
Jarosite(Pb)	$Pb_{0.5}Fe_3(SO_4)_2(OH)_6$														x	x				
Lanarkite	Pb_2SO_5							x	x	x		x	x	x	x	x	x			x
Mimetite	$Pb_5(AsO_4)_3Cl$														x	x				

Table 6. *Cont.*

		Databases																		
$Pb_3(AsO_4)_2$	$Pb_3(AsO_4)_2$								x	x		x	x	x			x			x
Pb_3SO_6	Pb_3SO_6							x	x	x			x	x						
$Pb_4(OH)_6SO_4$	$Pb_4(OH)_6SO_4$								x	x			x	x						
Pb_4SO_7	Pb_4SO_7							x	x	x			x	x						
$PbSO_4(NH_3)_2$	$PbSO_4(NH_3)_2$							x												
$PbSO_4(NH_3)_4$	$PbSO_4(NH_3)_4$							x												
Pb(Thiocyanate)$_2$	$Pb(SCN)_2$							x												
Philipsbornite	$PbAl_3(AsO_4)_2(OH)_5{:}H_2O$														x	x				
Tsumebite	$Pb_2Cu(PO_4)(SO_4)OH$							x							x	x				
Realgar	AsS							x	x	x		x	x	x	x	x	x			x
Orpiment	As_2S_3							x	x	x		x	x	x	x	x	x			x
Bornite	Cu_5FeS_4				x			x							x	x				
Chalcocite	Cu_2S				x			x	x	x			x	x	x	x				
Berthierite	$FeSb_2S_4$														x	x				
	Total	12	7	3	9	4	7	33	25	25	6	21	25	25	35	35	23	2	8	23

The similar conclusion as before can be drawn from this verification in which the two databases, Thermoddem (D15) and LLNL (D7) are suitable for the modelling of the geothermal fluids at the SsF plant due to possessing an extensive amount of thermodynamic data on known mineral found as deposits in the plant as well as possible minerals precipitated. The LLNL database has 33 mineral datasets out of the 42 possible minerals deposited, while the Thermoddem database has 35 out of the 42 possible minerals deposited.

Another step was carried out to verify the domain of validity for the minerals in the LLNL and Thermoddem databases. The range of temperature valid for each mineral was verified to ensure that it corresponds with the maximum modelling temperature of 200 °C. For the Thermoddem database, the thermodynamic data of all the minerals are valid within 0 °C to 300 °C. On the other hand, the LLNL database has different limits for each mineral. Fortunately, the minerals that were identified in Table 6 are well within the limits proposed in the LLNL database, as the lowest maximum temperature for the minerals found is at 200 °C.

2.3. Verification: B-Dot Model Database

The two databases of interest, the Thermoddem database and the LNLL database, utilize the B-Dot equation for the calculation of activity of the elements. The B-dot model is also known as the Truesdell–Jones model (TJ model). The ionic strength of the fluid was calculated from the major elements mentioned in the most recent published geochemical datasets in Bosia et al. (2021) [7] and found to be at 1.79 mol/kg for GPK-2 and at 1.8 mol/kg for GPK-3 (Table 7). The unit for the ionic strength can be represented as mol/L or mol/kg since the fluid is primarily composed of water while the effects of the ions in the conversion can be ignored due to their miniscule presence in the fluid. The validity of the B-dot model is verified in Figure 6 [20] as the ionic strength is well within the limit of the TJ model for both wells. The higher the ionic strength, the less accurate the results produced. When the ionic strength of the brine exceeds the limits of the TJ model (2.2 mol/kg), the results obtained from using the B-dot databases will no longer be valid.

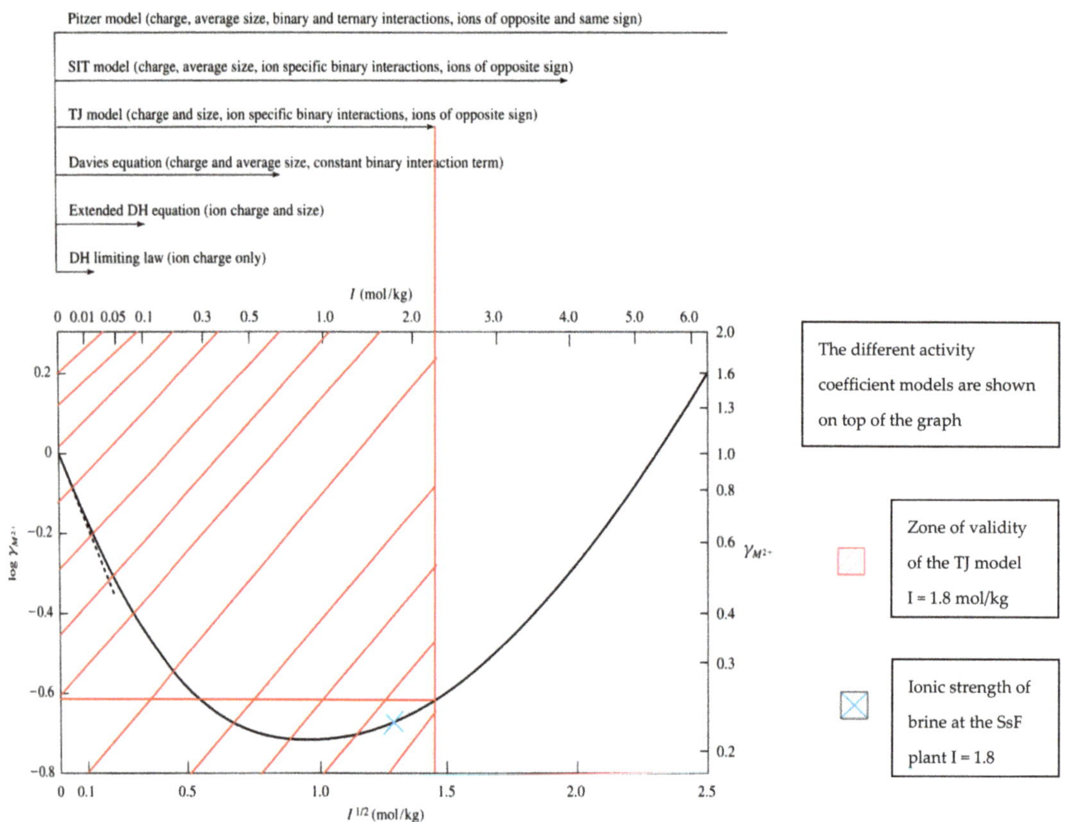

Figure 6. Schematic plot showing the general applicability of different activity coefficient models as a function of ionic strength for a divalent cation. The dashed tangent to the curve at its origin is a plot of the Debye–Hückel limiting law for the ion [18].

Table 7. Ionic strength calculations of the geothermal fluid sampled at GPK-2 and GPK-3.

	Molar Mass	GPK-2	GPK-3	GPK-2	GPK-3	GPK-2	GPK-3
	M (mg/mol)	mg/L		mol/L		Ionic Strength, I (mol/L or mol/kg)	
Na	23,000	26,400	26,700	1.148	1.161	0.574	0.580
Cl	35,500	57,490	57,490	1.619	1.619	0.810	0.810
K	39,100	3350	3350	0.086	0.086	0.043	0.043
Ca	40,100	7020	7030	0.175	0.175	0.350	0.351
Sr	87,620	422	434	0.005	0.005	0.010	0.010
Br	79,904	240	234	0.003	0.003	0.002	0.001
Li	6940	160	163	0.023	0.023	0.012	0.012
SiO$_2$	40,100	179	180	0.004	0.004		
Total		95,261	95,581	3.063	3.077	1.799	1.807

Since the ionic strength of the fluids at the SsF geothermal plant are well within the limits of the zone of validity, the two databases are thus used for the modelling of the fluids. Alsemgeest et al. (2021) [21] suggest being cautious when applying B-dot equation to SsF high saline geothermal brine. Nevertheless, they are also the most documented in terms of the geochemical elements and minerals.

2.4. Verification: Gas

The data available on the gases in the databases are compared to those required for modelling the geothermal fluid. The databases are then analyzed by initiating a preliminary modelling of the fluids to compare the results of the modelling with the results at the plant. For this preliminary modelling, the mixture of the gas dissolved in the brine (Table 2) was used. The conditions of the preliminary modelling are done at pH 5.2 and at two different temperatures, 80 °C and 150 °C. The saturation pressure of each database is compared and analyzed. For this analysis, the Thermoddem database, the LLNL database, and the Pitzer database were used. For the Thermoddem database and the LLNL database, as they were deemed suitable for the modelling of scales through the verification of elements and minerals, they are thus analyzed for the verification of gases. Even though the Pitzer database lacks several data on the elements and minerals, it is still considered for modelling of dissolved gases in the geothermal fluid because this database uses a different model for the calculation of activity of the elements. This may then give a more accurate result in the modelling of dissolved gases in the geothermal fluid. The results of the preliminary modelling at two different temperatures steps in terms of saturation pressure with the three databases are recorded in Table 8.

Table 8. Results of the saturation pressure of SsF gas for each database at two temperature steps.

Temperature (°C)	Pitzer	LLNL	Thermoddem
80	14 atm	10 atm	10 atm
150	18 atm	15 atm	16 atm

The LLNL and Thermoddem databases give out a similar result at both tested temperature while the Pitzer database shows a higher pressure compared to the two previous databases (Table 8). The saturation pressure obtained from modelling at 150 °C with the Pitzer database (18 atm = 18.2 bar) is closer to the actual case observed at the SsF plant [22] at the same temperature which ranges between 18.0 and 18.5 bar at relative pressure. The Thermoddem and LLNL databases provided results outside the range of saturation pressure observed at the SsF plant. Thus, the Pitzer database is found to be more suitable than the Thermoddem and LLNL databases for the gas modelling of the SsF plant.

Overall, the Thermoddem database was selected for the modelling of the formation of scales in the geothermal fluids as this database has more data than the LLNL database on the geochemical elements and possible minerals precipitated. Furthermore, the Thermoddem database has been compiled by a French geological survey company, BRGM which is specifically designed for waste derived from natural fluid precipitation [18]. As for modelling of the dissolved gas in the fluid, the Pitzer database was observed to have given a more satisfactory result as mentioned in the previous paragraph. Thus, the Pitzer database should be used for the modelling of the solubility of gas in the geothermal fluid.

2.5. Scale Modelling

When modelling the formation of scales with PhreeqC, the physical properties of the fluids such as the temperature, pressure and pH of the fluid are inputted into the software. The initial temperature, pressure and pH of the fluid are 25 °C, 1 bar, and pH 5.2 respectively representative of the laboratory conditions for brine analysis. The temperature and pressure were later changed to the production conditions of the brine at the SsF geothermal plant which are at 150 °C and 20 bars respectively. The pH of the fluid is also adjusted by the software to reflect the temperature and the composition of the fluid, thus there was no need to modify it. The unit for the concentration of each component in the fluids is also user-defined. In the case of this study, the unit used is in mg/kgw where kgw stands for a kilogram of water. Thus, the unit mg/kgw is the mass in milligrams of the element for each kilogram of water.

The formation of scales at the SsF plant is initially modeled by using thermodynamic modelling. This method uses the thermodynamic database researched in the previous section. The saturation index of each mineral is studied in this modelling process. For any minerals with a saturation index equal or higher than zero for the conditions of the fluid at the geothermal plant, that mineral can potentially precipitate. The amount of minerals precipitated was then calculated. This method provided insight on the potential minerals that could precipitate aside from the minerals already observed in previous studies such as those mentioned in Scheiber et al. (2012) [8], Sanjuan et al. (2011) [9], and Nitschke (2012) [10]. However, this method is limited to cases where thermodynamic equilibrium is reached.

Kinetic modelling was also considered to represent accurately the situation of the formation of scales at the geothermal plant. For this method, the amount of time that the fluids pass through the plant's exchangers is needed. It takes around 3 min for the fluid to circulate from the entrance of the first ORC heat exchanger to the exit of the final ORC heat exchanger. In these conditions, the kinetics of the reaction is also a crucial factor for the kinetic modelling. The kinetic data for chalcopyrite, galena, orpiment, and pyrite was taken from the database made by Zhang et al. (2019) [23]. The kinetic constant for stibnite was taken from Biver et al. (2011) [24] and adjusted into a modified kinetic equation for galena. For other minerals without any kinetic data, a modified kinetic equation of a similar mineral was used. The amount of minerals precipitated is calculated using its kinetic equation. This method refers to the saturation index of the mineral before calculating with the kinetic information available. As stated before, when the saturation index of the mineral is below zero, the kinetic calculation is skipped as the mineral does not precipitate. The duration for the kinetic modelling at each temperature was set to one minute because the velocity of the brine is estimated to be slightly less than 1 m/s and the length of the tubes of heat exchanger (30 m). This gives a duration of about 30 s to pass through a heat exchanger. Another 30 s was added to take into account the head cover and the pipes between each heat exchanger.

3. Results

As mentioned in the previous section, the modelling of scales in the geothermal fluids was done in Phreeqc with the Thermoddem database. For this modelling sequence, the range of temperature and pressure were set. The temperature starts from 150 °C which is the highest observable temperature at the SsF plant. The temperature then reduces until the lowest temperature found in the test heat exchanger which is at 40 °C. Additionally, two fictional temperatures were added which are at 175 °C and 200 °C in order to simulate the influence of such high temperatures on the formation of scales. These two temperatures are representative of temperatures found in the geothermal reservoir that is four to five kilometers deep under. The pressure was then fixed at 20 bars to simulate the exact conditions at the SsF geothermal plant.

3.1. Thermodynamic Modelling

The precipitation of the minerals was first studied through the observation made on the saturation index of each mineral. For the minerals with a saturation index equal or higher than zero, they are minerals that could possibly be present in the scales at thermodynamic equilibrium (Appendix A, Table A1). A list of potential minerals present within the set range of temperature was constructed from the observation of the saturation index of each mineral (Table 9).

Table 9. Presence of potential minerals at the set range of temperature according to saturation index.

	Pressure (bar)					**20**				
	Temperature (°C)	40	50	60	65	90	120	150	175	200
	Known Minerals									
SiO_2	Amorphous_silica	x	x	x	x	x	x			
$CaSO_4$	Anhydrite								x	x
Sb_2S_3	Stibnite	x	x	x	x	x	x	x		
$FeAsS$	Arsenopyrite				x	x	x			
$BaSO_4$	Barite	x	x	x	x	x	x	x	x	x
$CuFeS_2$	Chalcopyrite (alpha)	x	x	x	x	x	x	x	x	x
PbS	Galena	x	x	x	x	x	x	x	x	x
SiO_2	Quartz (alpha)	x	x	x	x	x	x	x	x	x
SiO_2	Quartz (beta)	x	x	x	x	x	x	x	x	x
	Possible Other Minerals									
$Cu_{1.75}S$	Anilite	x	x	x	x					
$FeSb_2S_4$	Berthierite	x	x	x	x	x	x			
Cu_5FeS_4	Bornite (alpha)	x	x	x	x	x	x	x		
SiO_2	Chalcedony	x	x	x	x	x	x	x	x	x
Cu_2S	Chalcocite (alpha)	x	x	x	x					
SiO_2	Coesite (alpha)	x	x	x	x	x	x			
CuS	Covellite	x	x	x	x					
SiO_2	Cristobalite (alpha)	x	x	x	x	x	x	x	x	x
SiO_2	Cristobalite (beta)	x	x	x	x	x	x	x	x	
$Cu_{1.934}S$	Djurleite	x	x	x	x					
$Fe_{10}S_{11}$	$Fe_{10}S_{11}$						x	x	x	x
$Fe_{11}S_{12}$	$Fe_{11}S_{12}$						x	x	x	x
$Fe_{7.016}S_8$	$Fe_{7.016}S_8$					x	x	x	x	x
Fe_9S_{10}	Fe_9S_{10}						x	x	x	x
FeS_2	Marcassite	x	x	x	x	x	x	x	x	x
As_2S_3	Orpiment	x	x	x	x	x				
FeS_2	Pyrite	x	x	x	x	x	x	x	x	x
$Na_2(Fe_3Fe_2)Si_8O_{22}(OH)_2$	Riebeckite									x

The next step for the modelling of scales formation at the SsF geothermal plant is to calculate the quantity of minerals precipitating in the given temperature range. An initial modelling based on the present minerals (Table 9) was done and the results showed that not all minerals with a positive saturation index precipitated (Table 10, left side). This is explained by the higher saturation index of several minerals which have higher priority to precipitate. The results of the thermodynamic modelling (Table 11) from using the minerals of the left side of Table 10 showed that majority of the minerals consist of silicates because of the high concentration of O and Si. At the range of temperature between 40 °C to 150 °C, silicate scales are not usually found at high amounts at the SsF geothermal plant.

Table 10. Mineral precipitated for thermodynamic modelling (For 40–200 °C).

Known Minerals			
Minerals precipitated according to saturation index		**Minerals considered for thermodynamic modelling**	
SiO_2	Amorphous silica	$CuFeS_2$	Chalcopyrite (alpha)
$CaSO_4$	Anhydrite	PbS	Galena
$BaSO_4$	Barite	Sb_2S_3	Stibnite
$CuFeS_2$	Chalcopyrite (alpha)		
PbS	Galena		
SiO_2	Quartz (alpha)		
SiO_2	Quartz (beta)		
Sb_2S_3	Stibnite		

Table 10. Cont.

Possible Other Minerals			
Minerals precipitated according to saturation index		**Minerals considered for thermodynamic modelling**	
$Cu_{1.75}S$	Anilite	$Cu_{1.75}S$	Anilite
$FeSb_2S_4$	Berthierite	$FeSb_2S_4$	Berthierite
Cu_5FeS_4	Bornite (alpha)	Cu_5FeS_4	Bornite (alpha)
SiO_2	Coesite (alpha)	CuS	Covellite
CuS	Covellite	FeS_2	Marcasite
SiO_2	Cristobalite (beta)	As_2S_3	Orpiment
FeS_2	Marcasite	FeS_2	Pyrite
As_2S_3	Orpiment		
FeS_2	Pyrite		

Table 11. Results of first thermodynamic modelling in weight percentage.

		Temperature (°C)								
	M (g/mol)	40	50	60	65	90	120	150	175	200
As	74.922	0.00%	0.00%	0.00%	0.00%	0.00%	0.00%	0.00%	0.00%	0.00%
Ba	137.33	2.3%	2.1%	4.4%	4.8%	5.1%	5.5%	4.9%	0.00%	0.00%
Ca	40.08	0.00%	0.00%	0.00%	0.00%	0.00%	0.00%	0.00%	5.0%	9.7%
Cu	63.546	0.00%	0.00%	0.00%	0.00%	0.00%	0.00%	0.00%	0.00%	0.00%
Fe	55.847	0.60%	0.52%	0.45%	0.65%	1.3%	1.4%	1.4%	1.2%	1.1%
O	15.999	51.6%	51.7%	50.8%	50.4%	49.5%	49.3%	49.5%	50.8%	49.9%
Pb	207.2	0.01%	0.01%	0.02%	0.00%	0.02%	0.00%	0.01%	0.00%	0.01%
S	32.066	1.2%	1.1%	1.5%	1.9%	2.7%	2.9%	2.7%	5.4%	9.0%
Sb	121.75	0.01%	0.01%	0.02%	0.01%	0.02%	0.02%	0.01%	0.00%	0.00%
Si	28.086	44.3%	44.5%	42.8%	42.3%	41.3%	41.0%	41.4%	37.6%	30.3%

To have a better focus on the modelling of scales at the SsF geothermal plant, the minerals considered for the thermodynamic model were then identified (Table 10, right side). Barite and celestite were excluded from future modelling sequence, because inhibitors are used by the operator to prevent the formation of these scales. For silicates, it is suspected that kinetic reaction prevents their deposition. That is why they were excluded to focus on the primary elements found in the scales found at the SsF geothermal plant as mentioned before. The results of the calculation are done at the different temperatures (Table 12).

Table 12. Results of refined thermodynamic modelling in weight percentage.

		Temperature (°C)								
	M (g/mol)	40	50	60	65	90	120	150	175	200
As	74.922	0.00%	0.00%	0.00%	0.00%	0.00%	0.00%	0.00%	0.00%	0.00%
Cu	63.546	0.04%	0.00%	0.00%	0.00%	0.00%	0.00%	0.00%	0.00%	0.00%
Fe	55.847	45.6%	45.6%	44.8%	45.9%	45.8%	46.2%	46.2%	46.5%	46.4%
Pb	207.2	0.72%	0.50%	1.5%	0.25%	0.52%	0.07%	0.43%	0.04%	0.30%
S	32.066	52.8%	52.9%	52.3%	53.1%	53.0%	53.2%	53.2%	53.4%	53.3%
Sb	121.75	0.77%	1.0%	1.5%	0.76%	0.67%	0.48%	0.17%	0.00%	0.00%

For each step of temperature, the modelling results show that sulfur and iron are the major elements with concentrations of 45% and 53% respectively (Table 12). On the other hand, the total amount of the other elements represents less than 3% of the total. Copper is only found at 40 °C and in extremely small quantities. Antimony and lead are also found in small quantities (less than 1.5%) at any given step of temperature.

3.2. Kinetic Modelling

The results given out by the calculation of the thermodynamic model gives insight on the precipitation of the minerals at thermodynamic equilibrium which may not necessarily be respected in the conditions studied. Modelling done from a kinetics aspect was proposed and the results from the thermodynamic model were compared and complimented with literature review and field knowledge to select the proper minerals which could precipitate. The kinetic information was mainly obtained from Zhang et al. (2019) [23] as mentioned in the Section 2. Initially, the model had little modification to the kinetic information used from the source with exception for minerals lacking their kinetic information. Rates equations for metal sulfides including the concentration of oxygen into the calculation are removed, because they serve no purpose due to the little to no oxygen content in the brine at the SsF geothermal plant.

For the initial model, two different sets of minerals were considered. The first set of minerals are galena (PbS), orpiment (As_2S_3), pyrite (FeS_2), amorphous silica (SiO_2), quartz (alpha) (SiO_2), and stibnite (Sb_2S_3). Galena and stibnite are known minerals already observed at the SsF plant [14]. Pyrite was considered over arsenopyrite (AsFeS) and chalcopyrite (CuFeS), because pyrite has a higher saturation index than arsenopyrite (Appendix A, Table A1); thus pyrite is more susceptible to precipitate than arsenopyrite. Chalcopyrite was dismissed as the principal provider of Fe precipitation because there is only a small amount of copper found in the analysis done at the SsF plant (Table 3) which is negligible compared to the quantity of Fe found. As for orpiment, this mineral is the only representative for presence of the element As. For amorphous silica and quartz (alpha), they were considered as they had a major influence in the thermodynamic modelling. Unfortunately, the desired modelling conditions do not fall within the domain of validity for the initial kinetic model created. For the formation of galena, this model is only valid for a temperature between 25 °C to 70 °C and a pH between one and three. For the formation of pyrite, this model is only valid for a temperature between 20 °C to 40 °C and a pH between one and four. For both cases, the range of pH is too acidic compared to the actual case. The model for the formation of orpiment is only valid for a temperature between 25 °C to 40 °C and a pH between 7.3 and 9.4 which is too alkaline. For the formation of amorphous silica, the model is only valid for a pH around 5.7, which is a bit too alkaline compared to the pH of the fluid at the SsF geothermal plant. For the formation of quartz (alpha), the model is within the proper zone of validity. Regardless, this model was used as an initial approach to modelling the minerals precipitated. For stibnite, no source for its kinetic information aside from its kinetic constant is found [24]. Thus, the kinetic equation of galena was taken and modified to suit the kinetic rate of stibnite. Minerals such as barite and celestite were not added, because their exclusion serves as a proxy to their inhibition by chemical treatment.

The second set of minerals consists of the same minerals from the first set, but excluding amorphous silica and quartz (alpha). These two minerals were excluded to better focus on the main minerals identified in the scales at the SsF geothermal plant. The modelling with both set of minerals was only done from 200 °C to 65 °C as it is complicated to model the circulation of fluids in the pipes between the ORC heat exchangers and the test heat exchangers. Furthermore, the residence time and the surface area of the heat exchangers in contact with the brine are different in both cases which will thus further complexify the model. To simplify the model, the ORC heat exchangers were chosen as the standard for the temperature to be modelled.

The first results showed that for the temperatures between 65 °C and 150 °C, S and Fe are the major elements in the simulated scales (Table 13). From 175 °C onwards, Si and O are the major elements while Pb, As, and Sb are found in negligible amounts.

Table 13. Results for initial kinetics model with first set of minerals in weight percentage.

Temperature	Pb	Fe	As	Sb	S	Si	O	Majority
65	8.88%	40.77%	0.14%	1.13%	48.73%	0.16%	0.18%	S
90	2.14%	45.11%	0.01%	0.04%	52.16%	0.24%	0.28%	S
120	0.95%	44.67%	0.00%	0.01%	51.45%	1.4%	1.6%	S
150	0.60%	35.72%	0.00%	0.00%	41.12%	10.5%	12.0%	S
175	0.22%	15.66%	0.00%	0.00%	18.01%	30.9%	35.2%	O
200	0.01%	3.80%	0.00%	0.00%	4.36%	42.9%	48.9%	O

The results show that sulfur is the majority for every step of temperature taking up to 53.4% of the composition of scales (Table 14). Iron is shown to be in second largest mass quantity with a weight percentage of around 46% except at 65 °C which is at 40.9%. Lead is shown to be in smaller quantity such as 8.9% at 65 °C and 2.2% at 90 °C, respectively. Between 200 °C and 120 °C, the quantity of lead is less than 1%. As for antimony and arsenic, both are found in extremely small quantities where antimony is at 1.1% and arsenic is at 0.14% for the temperature of 65 °C. Antimony and arsenic are not found at higher temperatures (above 150 °C).

Table 14. Results for initial kinetics model with second set of minerals in weight percentage.

Temperature	Pb	Fe	As	Sb	S	Majority
65	8.9%	40.9%	0.14%	1.1%	48.9%	S
90	2.2%	45.4%	0.01%	0.04%	52.4%	S
120	0.98%	46.0%	0.00%	0.01%	53.0%	S
150	0.78%	46.1%	0.00%	0.00%	53.1%	S
175	0.64%	46.2%	0.00%	0.00%	53.2%	S
200	0.11%	46.5%	0.00%	0.00%	53.4%	S

4. Discussion

4.1. Introduction

In this discussion, an analysis is done on the thermodynamic modelling and the kinetic modelling to identify the utility and shortcomings of each method. The factors that affect the results of each method are also discussed. Modifications were done on the kinetic model to better fit with the chemistry of scale observed at the SsF plant. Finally, new perspectives are proposed and discussed to improve further the proposed predictive kinetic model.

4.2. Thermodynamic Modelling Analysis

The thermodynamic modelling provides insight on possible precipitation of minerals at each temperature step. It can be observed that minerals containing strontium such as celestite were not listed as minerals precipitated by the modelling software (Table 9). In the analysis made on the scales at the SsF plant, traces of strontium were found and were identified to be celestite [8,9]. This discrepancy can be explained by the fact that the supposed mineral found at the plant, celestite, dissolves in favor of the precipitation of barite [25]. Since the results are calculated at thermodynamic equilibrium, the total consumption of celestite was already considered during the calculations made by PhreeqC. Another explanation is that the PhreeqC software does not consider the existence of solid solutions like barium/strontium sulfates. Hence, the software considers barite over celestite for their precipitation. Thus, strontium was excluded from the comparison of the weight percentage of the elements between the Ssf plant analyses, the thermodynamic models, and the kinetic models. Barite is shown to potentially precipitate at the given range of temperature (Table 9). However, as the temperature decreases, the saturation index of barite increases thus increasing its potential to precipitate (Appendix A, Table A1). A similar situation is observed in the formation of galena, albeit with a higher saturation index. For pyrite, it can also potentially precipitate at the given range of temperature. Its saturation index increases from 200 °C to 90 °C in which it starts to decrease thereafter. Precipitation of native metals could not be observed in neither thermodynamics modelling nor kinetics modelling, because the modelling software cannot take into account their formation.

When silicates were considered for the thermodynamic model, the results (Table 11) showed that Si and O take up the majority of the elements until it rendered the rest of the elements negligible in the simulated scales. This is not the case at the SsF geothermal plant as there were tiny amounts of silicate in the actual analyses. A second model was constructed by excluding the silicates to have a better focus on the known minerals found at the geothermal plant.

The amount of galena formed in the thermodynamic models is greatly inferior to the actual scaling at the SsF geothermal plant (Table 15). There is an unusually high amount of iron and sulfur in the thermodynamic modelling. Furthermore, the quantity of lead is still in the minority. Another problem is that the thermodynamic modelling simulates the precipitation of the minerals over a great amount of time which is until the fluid reaches thermodynamic equilibrium. At the SsF geothermal plant, the precipitation of the minerals is not necessarily at thermodynamic equilibrium since the residence time of the brine in the exchanger is only around three minutes. Furthermore, the initial amount of lead (Pb) (Table 2) is smaller than the rest of elements in the brine. This could explain the low amount of lead found in simulated scales compared to the other elements in this modelling method. Thus, the thermodynamic model proved to be not sufficient for the prediction of formation of scales at the SsF geothermal plant and kinetic effect must be considered.

Table 15. Comparison between Soultz-sous-Forêts, thermodynamic model, and kinetic model results in relative percentage by weight.

	Temperature	65	90	120	150
	SsF plant analyses	59.7%	56.8%	39.9%	27.3%
	Thermodynamic model 1	0.00%	0.02%	0.00%	0.01%
Pb	Thermodynamic model 2	0.25%	0.52%	0.07%	0.43%
	Kinetic Model 1	8.9%	2.1%	0.95%	0.60%
	Kinetic Model 2	8.9%	2.2%	0.98%	0.78%
	SsF plant analyses	5.9%	12.6%	12.1%	23.3%
	Thermodynamic model 1	0.65%	1.34%	1.37%	1.36%
Fe	Thermodynamic model 2	45.9%	45.8%	46.2%	46.2%
	Kinetic Model 1	40.8%	45.1%	44.7%	35.7%
	Kinetic Model 2	40.9%	45.4%	46.0%	46.1%
	SsF plant analyses	9%	8%	13%	7%
	Thermodynamic model 1	0%	0%	0%	0%
As	Thermodynamic model 2	0.00%	0.00%	0.00%	0.00%
	Kinetic Model 1	0.14%	0.01%	0.00%	0.00%
	Kinetic Model 2	0.14%	0.01%	0.00%	0.00%
	SsF plant analyses	8%	5%	3%	2%
	Thermodynamic model 1	0.01%	0.02%	0.02%	0.01%
Sb	Thermodynamic model 2	0.76%	0.67%	0.48%	0.17%
	Kinetic Model 1	1.13%	0.04%	0.01%	0.00%
	Kinetic Model 2	1.1%	0.04%	0.01%	0.00%
	SsF plant analyses	17%	18%	32%	41%
	Thermodynamic model 1	1.9%	2.7%	2.9%	2.7%
S	Thermodynamic model 2	53.1%	53.0%	53.2%	53.2%
	Kinetic Model 1	48.73%	52.2%	51.5%	41.1%
	Kinetic Model 2	48.9%	52.4%	53.0%	53.1%

4.3. Kinetic Modelling Analysis

The kinetics model with the first set of minerals (Table 13) showed improvements in the results when compared to the first thermodynamic model (Table 11). The kinetic model with the first set of minerals (Table 13) has significantly reduced the Si and O contents for the temperatures between 65 °C and 150 °C. This confirms that the kinetic effect controls the absence of silicates in the SsF scales.

However, for this range of temperature, sulfur (S) and iron (Fe) have the highest concentration with the highest percentage being 52.2% and 45.1% respectively (Table 15). Regardless, the concentration of each element for the kinetic model 1 does not reflect the actual concentration found in the SsF plant analyses.

As for the kinetic model 2, it showed similar improvements in the results to the results of kinetic model 1. At 65 °C, the quantity of lead has increased from 0.25% (thermodynamic model 2) to 8.9% (kinetic model 2) in the composition of elements found in the modelled scales (Table 14). However, iron and sulfur are still the major elements in the modelled scales. The lack of kinetic information on the formation of stibnite could also lead to inaccuracies in the results such as the low amount of antimony. In addition, the amount of sulfur present at each temperature is larger than the actual case. The discrepancies can be explained by the conditions of the modelled scales being outside the domain of validity for temperature and pH of the kinetic information used.

Therefore, to better simulate the scale formation at the SsF geothermal plant, a modified version of the initial model was created. In this second model, the kinetic information of the minerals was modified to reflect closely to the analyses done at the geothermal plant. The kinetic information was purposely modified until the model produces a result similar to the ones obtained at SsF geothermal plant at one temperature step. The modification was done iteratively until the results were in an approximate range of the actual case. Thus, the modified kinetic information is not indicative of any actual kinetic values. The two minerals (arsenopyrite and chalcopyrite) were added to compensate for the low amount of arsenic and the high amount of sulfur and iron. The kinetic information of chalcopyrite is taken from Zhang et al. (2019), whereas no kinetic data was found on arsenopyrite. Thus, the kinetic data of chalcopyrite was taken and modified for arsenopyrite. Next, the kinetic rate of pyrite was slowed down as this mineral has the greatest influence on the increases of percentage of iron and sulfur (Table 16). Overall, the kinetic information of all the minerals except galena and chalcopyrite was modified to obtain a general model for the formation of scales.

Table 16. Modification of the first kinetic model. n_x: representing the index used in the rates equation (Appendix B).

	Initial Model	**Modified Model**
Arsenopyrite	$n = 1.68$	$n = 0.8$
Orpiment	$n_2 = -1.26$	$n_2 = -1.48$
Stibnite	$n = 0.5$	$n = 0.475$
Pyrite	$n_1 = -0.5$	$n_1 = -0.25$
	$n_3 = 0.5$	$n_3 = 0.55$

The modified model presented a result that is closer to the analyses of scales at the geothermal plant (Tables 15 and 17). The percentage of sulfur is still higher than the actual case, but the increase in quantity of sulfur scales better than the unmodified kinetic information models. The quantity of iron is higher than the actual case for the temperature between 90 °C and 150 °C. In addition, there are no other minerals that contain antimony and arsenic that has a positive saturation index for temperatures above 120 °C. This leads to having small and negligible quantities of both elements at the mentioned temperature. All things considered, this model allows a rough prediction on the scale formation when operating the plant with sulfate scales inhibitors at the SsF geothermal plant as there is only a small deviation between simulated results and the actual case. The model becomes less accurate at higher temperatures such as at 150 °C, because of the lack of antimony and arsenic at this temperature (Table 15).

Table 17. Mass of elements in percentage for modified kinetics model.

Temperature	Pb	Fe	As	Sb	S	Cu	Majority
65	52.2%	5.0%	9.0%	8.5%	22.5%	2.8%	Lead
90	45.6%	16.0%	9.2%	1.2%	25.2%	2.8%	Lead
120	34.9%	23.5%	7.0%	0.42%	29.6%	4.6%	Lead
150	40.1%	22.7%	0.00%	0.01%	32.3%	4.9%	Lead
175	41.7%	23.0%	0.00%	0.00%	32.8%	2.5%	Lead
200	14.4%	38.0%	0.00%	0.00%	45.8%	1.8%	Sulfur

4.4. New Perspectives

For the modelling of scales for the SsF geothermal plant, a lot of information was lacking such as the kinetic information that is suited for the operating conditions of the plant. Future studies and analyses on the precipitation of the minerals are to be arranged to obtain the missing kinetic information and challenge the modified kinetic model. A laboratory study is necessary to investigate the precipitation of minerals at conditions of the SsF geothermal plant which is at around pH 5.2 and the temperature range of the ORC heat exchangers. The kinetic model for pyrite might also not be suitable for modelling the scales at the pH, pressure, and temperature of SsF geothermal plant which led to inaccuracies in the results pertaining to the amount of Fe and S. Therefore, the kinetic information of the precipitation of pyrite as well as galena, arsenopyrite, chalcopyrite, arsenides, sulfosalts, selenides, and other base metal sulfides are needed to be determined through this laboratory study so that a proper kinetic model can be constructed.

Furthermore, the inhibition of sulfates such as barium and celestite was just excluded from the calculation due to lack of information on their kinetics. Therefore, the inhibition process should also be analyzed and studied to obtain its kinetic information that can be integrated into the kinetic model. With a proper kinetic model, a more precise result can be obtained through the simulation on the formation of scales in the pipes and exchanger at the geothermal plant. Besides that, other reactions aside from precipitation should also be studied and integrated into the model such as the possibility of heavy metal corrosion in the pipes and heat exchanger, as mentioned in Lichti and Brown (2013) [26] and Lichti et al. (2016) [27]. This phenomenon should be studied at the SsF geothermal plant and be verified whether it affects the amount of scales formed at the plant. A study should also be conducted on the possibility of a chemical interaction between FeS and PbS. The results from the laboratory studies on this chemical interaction at the SsF operational condition could be integrated into the current prediction model for a more accurate result.

5. Conclusions

From the geochemical analyses done on the SsF geothermal plant, lead is found to be the major element in the composition of scales formed when operating the plant with sulfate anti-scales. The principal mineral formed was identified to be galena. This could change when additional chemical treatment is added to the process. To have an accurate prediction on the mineral and elements formed during the scaling phenomenon, a prediction model needs to be created.

The main goal of this study was to better characterize the scales formed at the SsF geothermal plant by establishing a geochemical model that allows the prediction of the formation of scales. Intensive bibliographic research was done to obtain the necessary thermodynamic and kinetic information used in the modelling of the formation of scales at the SsF geothermal plant. The two methods of modelling present their own set of challenges to reflect accurately the actual case.

For the thermodynamic modelling, this method is done over a great amount of time which is impractical for predicting the formation of scales in an actual case. The saturation index obtained from thermodynamic modelling however is a good indication on which mineral can precipitate in function of the temperature. Minerals such as silicate scales could potentially precipitate at the right conditions.

For the kinetic modelling, specific kinetic information such as the rates equation and the kinetic constant for the precipitation of the mineral are lacking for the desired range of temperature. Nevertheless, the modelling shows that silicate precipitation is strongly controlled by kinetic. Additionally, this method allows a more accurate prediction for the formation of scales with the caveat of having the proper kinetic information.

The results obtained in this study open up to new perspectives on the issue of lack of kinetic information. The proposed steps from the new perspectives can improve the current prediction model for future uses.

Author Contributions: Conceptualization, P.K., G.R. and M.D.; methodology, P.K., G.R. and M.D.; software, P.K.; validation, G.R., E.D., M.D. and P.C.; formal analysis, P.K., G.R. and M.D.; investigation, P.K. and G.R.; resources, P.K. and G.R.; data curation, G.R.; writing—original draft preparation, P.K. and G.R.; writing—review and editing, P.K., G.R., E.D., M.D. and P.C.; visualization, P.K. and G.R.; supervision, G.R. and M.D.; project administration, G.R. and E.D.; funding acquisition, G.R. and E.D. All authors have read and agreed to the published version of the manuscript.

Funding: This research was funded by the European Union's Horizon 2020 research and innovation program under grant agreement No 792037 (MEET project), as well as the Agence Nationale de la Recherche under grant agreement No 10-IEED-0811-05 (THERMA'LI project).

Data Availability Statement: All data supporting this paper can be found in the references.

Acknowledgments: The authors thank the Soultz-sous-Forêts site owner, GEIE Exploitation Minière de la Chaleur, for giving access to their geothermal installation. The authors would also thank Laurent André (BRGM) for the technical support, guidance, and insight on the modelling process. Finally, the author would like to thank Beatrice Ledésert (CY Cergy Paris Université) for providing the microscopic image of the (Pb,As,Sb)S scale found at SsF plant.

Appendix A

Table A1. Saturation Index of minerals with potential to precipitate.

Pressure (bar)						20				
Temperature (°C)		40	50	60	65	90	120	150	175	200
SiO_2	Amorphous_silica	0.45	0.38	0.32	0.29	0.16	0.03	−0.08	−0.16	−0.24
$CaSO_4$	Anhydrite	−0.98	−0.88	−0.78	−0.73	−0.53	−0.32	−0.1	0.08	0.25
$Cu_{1.75}S$	Anilite	2.61	1.97	1.31	0.97	−0.55	−1.95	−3.05	−3.86	−4.65
FeAsS	Arsenopyrite	−0.52	−0.28	−0.04	0.06	0.3	0.13	−0.24	−0.58	−0.9
$BaSO_4$	Barite	1.24	1.11	0.99	0.93	0.67	0.4	0.21	0.1	0.01
$FeSb_2S_4$	Berthierite	1.33	1.25	1.17	1.13	0.92	0.63	−0.02	−1.48	−3.15
Cu_5FeS_4	Bornite (alpha)	17.03	15.3	13.5	12.59	8.22	3.83	0.17	−2.56	−5.19
SiO_2	Chalcedony	1.16	1.05	0.96	0.91	0.72	0.52	0.34	0.21	0.1
Cu_2S	Chalcocite (alpha)	2.75	2.01	1.24	0.85	−0.89	−2.45	−3.66	−4.53	−5.38
$CuFeS_2$	Chalcopyrite (alpha)	6.21	6.07	5.91	5.8	5.12	4.08	3.01	2.17	1.36
SiO_2	Coesite (alpha)	0.64	0.55	0.47	0.43	0.26	0.09	−0.05	−0.16	−0.26
CuS	Covellite	1.42	1.07	0.71	0.53	−0.36	−1.28	−2.09	−2.7	−3.28
SiO_2	Cristobalite (alpha)	0.89	0.8	0.72	0.68	0.52	0.35	0.21	0.1	0
SiO_2	Cristobalite (beta)	0.83	0.74	0.66	0.62	0.47	0.31	0.17	0.07	−0.02
$Cu_{1.934}S$	Djurleite	2.76	2.05	1.3	0.93	−0.76	−2.29	−3.47	−4.34	−5.18

Table A1. Cont.

Pressure (bar)		20								
$Fe_{10}S_{11}$	$Fe_{10}S_{11}$	−19.75	−15.47	−11.34	−9.49	−2.77	0.56	1.65	1.98	2.14
$Fe_{11}S_{12}$	$Fe_{11}S_{12}$	−21.56	−16.83	−12.28	−10.23	−2.81	0.92	2.19	2.61	2.84
$Fe_{7.016}S_8$	$Fe_{7.016}S_8$	−11.61	−8.67	−5.83	−4.55	0.01	2.12	2.67	2.74	2.72
Fe_9S_{10}	Fe_9S_{10}	−16.97	−13.14	−9.45	−7.79	−1.8	1.11	2	2.23	2.31
PbS	Galena	2.57	2.54	2.51	2.49	2.18	1.6	1	0.53	0.05
FeS_2	Marcassite	4.15	4.39	4.63	4.72	4.87	4.42	3.73	3.13	2.56
As_2S_3	Orpiment	0.92	0.97	1.04	1.04	0.58	−0.82	−2.6	−4.1	−5.52
FeS_2	Pyrite	4.84	5.06	5.28	5.36	5.45	4.96	4.23	3.6	3
SiO_2	Quartz (alpha)	1.43	1.31	1.21	1.16	0.95	0.73	0.54	0.4	0.28
SiO_2	Quartz (beta)	1.21	1.11	1.02	0.97	0.78	0.59	0.42	0.29	0.18
$Na_2(Fe_3Fe_2)Si_8O_{22}(OH)_2$	Riebeckite	−7.54	−6.95	−6.34	−6.05	−4.66	−3.17	−1.68	−0.44	0.8
Sb_2S_3	Stibnite	3.25	2.76	2.29	2.08	1.25	0.7	0.02	−1.4	−3.01

Appendix B

The PhreeqC program code can be divided into several parts which signify different simulation iterations. Every part is ended with the line "End" to carry on to the next simulation. Each part is divided into several sections that carry out the different calculations for the modelling of fluids. Certain sections are not mandatory for the simulation as each of them serves different purposes. The first section is the "Database" in which we define the database to be use as a reference for the calculations. The next section is "Solution" in which the properties of the fluid are defined. Examples of the properties of the fluids which are added in this section are the temperature, pressure, and pH of the fluid. Furthermore, the composition of the fluid is also added in this section. The unit for the concentration of each component in the fluids is defined by the user. In the case of this study, the unit that was used is in mg/kgw (milligrams per kilogram of water).

The next section is the "Gas_Phase". For this section, it functions similarly as the "Solution" section in which the properties of the gas are defined and the composition in percentage of the gas is declared. The properties of the gas can be modified for the different simulation iterations by using the line "Gas Phase Modify". This enables the modification of volume, pressure, and the concentration of each component of the gas. In the case of this study, this line is only used to modify the pressure of the gas.

The line "Reaction_Temperature" is used to modify the temperature of the solution after the first simulation iteration. This section allows the modification of the initial temperature of the fluid to another designated temperature or to a range of temperature. The line "Equilibrium_Phases" is used to model and simulate the precipitation of minerals in the brine. This line allows the user to obtain the number of moles of the minerals precipitated or dissolved at thermodynamic equilibrium. The user is required to provide the saturation index of each the corresponding minerals at the desired temperature.

The fluid can also be simulated from a kinetic aspect by using the lines "Rates" and "Kinetics". In the "Rates" section, the user is required to provide the rate equation for the given mineral as well as the kinetics constant of the rate equation. The "Kinetics" section uses the information from the "Rates" section to properly calculate the number of moles of minerals precipitated for a given duration. In this section, the user is required to provide information on the number of moles of minerals present initially in the fluid, the desired duration of the precipitation of the minerals, the number of intervals between the given duration and the type of Runge Kutta equation used. The Runge Kutta method is a family of implicit and explicit iterative methods that includes the Euler method. This method is used in temporal discretization for the approximate solution of differential equations.

The final command line used is the "Selected_Output" command line. This section allows the user to output the certain parts of the results of the simulation into a text file or a csv file.

DATABASE C:\phreeqc\database\PHREEQC_ThermoddemV1.10_15Dec2020.dat
SOLUTION 1
Units mg/L

```
Temperature 25.0
Pressure 1.0
pH 5.2

Cl 55942
Na 26412
Ca 7018
K 3357
S 64.4
Pb 0.113
Sr 422.415
Ba 25.55
Sb 0.0645
As 9.676
Fe 26.3
Si 179
Cu 0.001

GAS_PHASE 1
   -Pressure 1.0
   -Fixed_Pressure
   -Temperature 25
   -Volume 1.03
CO2(g) 0.882
   N2(g)        0.0908
   CH4(g)       0.0239
END
USE SOLUTION 1
USE GAS_PHASE 1

GAS_PHASE_MODIFY 1
Pressure    19.7385

RATES

################
#arsenopyrite
################

-start
1 rem assuming Fe(III)>1e-4M is the switch point for Fe-promoted mechanism
10 R=8.31451
20 if TOT("Fe(3)")<=1e-4 then J=(10^-1.52)*EXP(-28200/(R*TK))*ACT("H+")^0.8
30 if (parm(1)>0) then SA0=parm(1) else SA0=1
40 if (M0<=0) then SA=SA0 else SA=SA0* (M/M0)^0.67
70 SR_mineral=SR("Arsenopyrite")
80 if (M<0) then goto 150
90 if (M=0 and SR_mineral<1) then goto 150
100 rate=J*SA*(1-SR_mineral) *parm(2)
120 moles=rate*Time
150 Save moles
-end

################
```

```
# chalcopyrite (Kimball et al. 2010)
###############
chalcopyrite(alpha)

# experimental condition range T=4-100C, pH=0-5, log C(Fe+++)=-5-0

-start
1 rem assuming Fe(III)>1e-4M is the switch point for Fe-promoted mechanism
10 R=8.31451
20 if TOT("Fe(3)")<=1e-4 then J=(10^-1.52)*EXP(-28200/(R*TK))*ACT("H+")^1.68
else J=(10^1.88)*EXP(-
48100/(R*TK))*ACT("H+")^0.8*TOT("Fe(3)")^0.42
30 if (parm(1)>0) then SA0=parm(1) else SA0=1
40 if (M0<=0) then SA=SA0 else SA=SA0* (M/M0)^0.67
70 SR_mineral=SR("Chalcopyrite(alpha)")
80 if (M<0) then goto 150
90 if (M=0 and SR_mineral<1) then goto 150
100 rate=J*SA*(1-SR_mineral)*parm(2)
120 moles=rate*Time
150 Save moles
-end


###############
# Galena (Acero et al., 2007)
###############
Galena

# experimental condition range T=25-70C, pH=1-3

-start
1 rem unit should be mol, kgw-1 and second-1
2 rem parm(1) is surface area in the unit of m2/kgw
3 rem calculation of surface area can be found in the note
4 rem M is current moles of minerals
5 rem M0 is the initial moles of minerals
6 rem parm(2) is a correction factor
40 SR_mineral=SR("Galena")
41 if (M<0) then goto 200
42 if (M=0 and SR_mineral<1) then goto 200
43 if (M0<=0) then SA=PARM(1) else SA=PARM(1)*(M/M0)^0.67
50 if (SA<=0) then SA=1
60 R=8.31451
70 J=10^-5.7*exp(-23000/R/TK)*ACT("H+")^0.43
90 Rate=J*(1-Sr_mineral)*SA*parm(2)
100 moles=Rate*Time
200 save moles
-end


###########
#As2S3(a)
###########
Orpiment

# from Palandri and Kharaka 2004
```

```
# experimental condition range T=25-40C, pH=7.3-9.4

-start
1 rem unit should be mol,kgw-1 and second-1
2 rem parm(1) is surface area in the unit of m2/kgw
3 rem calculation of surface area can be found in the note
4 rem M is current moles of minerals. M0 is the initial moles of minerals
5 rem parm(2) is a correction factor
10 rem acid solution parameters
11 a1=0
12 E1=0
13 n1=0
20 rem neutral solution parameters
21 a2=4.95E-09
22 E2=8700
23 n3=0.180
30 rem base solution parameters
31 a3=1.36E-16
32 E3=8700
33 n2=-1.48
36 rem rate=0 if no minerals and undersaturated
40 SR_mineral=SR("ORPIMENT")
41 if (M<0) then goto 200
42 if (M=0 and SR_mineral<1) then goto 200
43 if (M0<=0) then SA=PARM(1) else SA=PARM(1)*(M/M0)^0.67
50 if (SA<=0) then SA=1
60 R=8.31451
75 Rate1=a1*EXP(-E1/R/TK)*ACT("H+")^n1 #acid rate expression
80 Rate2=a2*EXP(-E2/R/TK)*ACT("O2")^n3 #neutral rate expression
85 Rate3=a3*EXP(-E3/R/TK)*ACT("H+")^n2 #base rate expression
90 Rate=(Rate1+Rate3)*(1-Sr_mineral)*SA*parm(2)
100 moles= rate*Time
200 save moles
-end


##############
#pyrite
###########
pyrite
# from Palandri and Kharaka 2004
# experimental condition range T=20-40C, pH=1-4

-start
1 rem unit should be mol,kgw-1 and second-1
2 rem parm(1) is surface area in the unit of m2/kgw
3 rem calculation of surface area can be found in the note
4 rem M is current moles of minerals. M0 is the initial moles of minerals
5 rem parm(2) is a correction factor
10 rem acid solution parameters
11 a1=2.82E+02
12 E1=56900
13 n1=-0.25
14 n3=0.55
30 rem neutral solution parameters
```

```
31 a3=2.64E+05
32 E3=56900
33 n2=0.500
36 rem rate=0 if no minerals and undersaturated
40 SR_mineral=SR("pyrite")
41 if (M<0) then goto 200
42 if (M=0 and SR_mineral<1) then goto 200
43 if (M0<=0) then SA=PARM(1) else SA=PARM(1)*(M/M0)^0.67
50 if (SA<=0) then SA=1
60 R=8.31451
75 Rate1=a1*EXP(-E1/R/TK)*ACT("H+")^n1*ACT("Fe+3")^n3 #acid rate expression
80 Rate2=a2*EXP(-E2/R/TK)*ACT("O2") #neutral rate expression
90 Rate=(Rate1)*(1-Sr_mineral)*SA*parm(2)
100 moles= rate*Time
200 save moles
-end

Stibnite
-start

1 rem unit should be mol,kgw-1 and second-1
2 rem parm(1) is surface area in the unit of m2/kgw
3 rem calculation of surface area can be found in the note
4 rem M is current moles of minerals
5 rem M0 is the initial moles of minerals
6 rem parm(2) is a correction factor
40 SR_mineral= SR("Stibnite")
41 if (M<0) then goto 200
42 if (M=0 and SR_mineral<1) then goto 200
43 if (M0<=0) then SA=PARM(1) else SA=PARM(1)*(M/M0)^0.67
50 if (SA<=0) then SA=1
60 k=1.25E-10*EXP(298.2/TK)
70 J=k*ACT("H+")^0.475
90 Rate=J*(1-SR_mineral)*SA*parm(2)
100 moles=Rate*Time
200 save moles
-end

KINETICS

    Arsenopyrite
      -M   0.0
      -M0   0.0
      -parms   1.0   1.0
      -tol   1e-8
    -steps 1 min
    -step_divide 10
    -runge_kutta 3

    Chalcopyrite(alpha)
      -M   0.0
      -M0   0.0
      -parms   1.0   1.0
      -tol   1e-8
```

```
  -steps 1 min
  -step_divide 10
  -runge_kutta 3

Galena
  -M    0.0
  -M0    0.0
  -parms   1.0   1.0
  -tol   1e-8
-steps 1 min
-step_divide 10
-runge_kutta 3

Orpiment
  -M    0.0
  -M0    0.0
  -parms   1.0   1.0
  -tol   1e-8
-steps 1 min
-step_divide 10
-runge_kutta 3

Pyrite
  -M    0.0
  -M0    0.0
  -parms   1.0   1.0
  -tol   1e-8
-steps 1 min
-step_divide 10
-runge_kutta 3

Stibnite
  -M    0.0
  -M0    0.0
  -parms   1.0   1.0
  -tol   1e-8
-steps 1 min
-step_divide 10
-runge_kutta 3

REACTION_TEMPERATURE 1
  40.0 50.0 60.0 65.0 90.0 120.0 150.0 175.0 200.0

EQUILIBRIUM_PHASES
40 °C
Amorphous_silica   0.45 0.0
Barite          1.24 0.0
Chalcedony        1.16 0.0
Coesite(alpha)       0.64 0.0
Cristobalite(alpha)    0.89 0.0
Cristobalite(beta)    0.83 0.0
Quartz(alpha)       1.43 0.0
Quartz(beta)       1.21 0.0
Anilite          2.61 0.0
```

Berthierite   1.33 0.0
Bornite(alpha)   17.03 0.0
Chalcocite(alpha)   2.75 0.0
Chalcopyrite(alpha)   6.21 0.0
Covellite   1.42 0.0
Djurleite   2.76 0.0
Galena   2.57 0.0
Marcassite   4.15 0.0
Orpiment   0.92 0.0
Pyrite   4.84 0.0
Stibnite   3.25 0.0

References

1. Glaas, C. Mineralogical and Structural Controls on Permeability of Deep Naturally Fractured Crystalline Reservoirs. Ph.D. Thesis, Univerité de Strasbourg, Strasbourg, France, 2021.
2. Genter, A.; Evans, K.; Cuenot, N.; Fritsch, D.; Sanjuan, B. Contribution of the exploration of deep crystalline fractured reservoir of Soultz to the knowledge of enhanced geothermal systems (EGS). *Comptes Rendus Geosci.* **2010**, *342*, 502–516. [CrossRef]
3. Vidal, J.; Genter, A. Overview of naturally permeable fractured reservoirs in the Upper Rhine Graben: Insights from geothermal wells. *Geothermics* **2018**, *74*, 57–73. [CrossRef]
4. Schill, E.; Genter, A.; Cuenot, N.; Kohl, T. Hydraulic performance history at the Soultz EGS reservoirs from stimulation and long-term circulation tests. *Geothermics* **2017**, *70*, 110–124. [CrossRef]
5. Baujard, C.; Genter, A.; Cuenot, N.; Mouchot, J.; Maurer, V.; Hehn, R.; Ravier, G.; Seibel, O.; Vidal, J. Experience Learnt from a Successful Soft Stimulation and Operational Feedback after 2 Years of Geothermal Power and Heat Production in Rittershoffen and Soultz-sous-Forêts Plants (Alsace, France). *GRC Trans.* **2018**, *42*. [CrossRef]
6. Mouchot, J.; Ravier, G.; Seibel, O.; Pratiwi, A. Deep Geothermal Plants Operation in Upper Rhine Graben: Lessons Learned. In Proceedings of the European Geothermal Congress, The Hague, The Netherlands, 11–14 June 2019.
7. Bosia, C.; Mouchot, J.; Ravier, G.; Seibt, A.; Jähnichen, S.; Degering, D.; Scheiber, J.; Dalmais, E.; Baujard, C.; Genter, A. Evolution of Brine Geochemical Composition during Operation of EGS Geothermal Plants (Alsace, France). In Proceedings of the 46th Workshop on Geothermal Reservoir Engineering, Stanford, CA, USA, 15–17 February 2021.
8. Scheiber, J.; Nitschke, F.; Seibt, A.; Genter, A. Geochemical and Mineralogical Monitoring of the Geothermal Power Plant in Soultz-sous-Forêts (France). In Proceedings of the 37th Workshop on Geothermal Reservoir Engineering, Stanford, CA, USA, 30 January–1 February 2012.
9. Sanjuan, B. *Soultz EGS Pilot Plant Exploitation—Phase III: Scientific Program about On-Site Operations of Geochemical Monitoring and Tracing (2010–2013)*, First Yearly Progress Report BRGM/RP-59902-FR. 2011; Volume 16, p. 92.
10. Nitschke, F. Geochemische Charakterisierung des Geothermalen Fluids und der Damit Verbundenen Scalings in der Geothermieanlage Soultz sous Forêts. Master's Thesis, Institut für Mineralogie und Geochemie (IMG) at Karlsruher Institut of Technology (KIT), Singapore, 2012; p. 126.
11. Haas-Nüesch, R.; Heberling, F.; Schild, D.; Rothe, J.; Dardenne, K.; Jähnichen, S.; Eiche, E.; Marquardt, C.; Metz, V.; Schäfer, T. Mineralogical characterization of scalings formed in geothermal sites in the Upper Rhine Graben before and after the application of sulfate inhibitors. *Geothermics* **2018**, *71*, 264–273. [CrossRef]
12. Mouchot, J.; Cuenot, N.; Bosia, C.; Genter, A.; Seibel, O.; Ravier, G.; Scheiber, J. First year of operation from EGS geothermal plants in Alsace, France: Scaling issues. In Proceedings of the 43rd Workshop on Geothermal Reservoir Engineering, Stanford, CA, USA, 12–14 February 2018.
13. Seibel, O.; Mouchot, J.; Ravier, G.; Ledésert, B.; Sengelen, X.; Hebert, R.; Ragnarsdótti, K.R.; Ólafsson, D.I.; Haraldsdóttir, H.Ó. Optimised valorisation of the geothermal resources for EGS plants in the Upper Rhine Graben. In Proceedings of the World Geothermal Congress 2020+1, Reykjavik, Iceland, 24–27 October 2021.
14. Ledésert, B.A.; Hébert, R.L.; Mouchot, J.; Bosia, C.; Ravier, G.; Seibel, O.; Dalmais, E.; Ledésert, M.; Trullenque, G.; Sengelen, X.; et al. Scaling in a Geothermal Heat Exchanger at Soultz-sous-Forêts (Upper Rhine Graben, France): A XRD and SEM-EDS Characterization of Sulfide Precipitates. *Geosciences* **2021**, *11*, 271. [CrossRef]
15. Gifaut, E.; Grivé, M.; Blanc, P.; Vieillard, P.; Colàs, E.; Gailhanou, H.; Gaboreau, S.; Marty, N.; Madé, B.; Duro, L. Andra thermodynamic database for performance assessment: Thermochimie. *Appl. Geochem.* **2014**, *49*, 225–236. [CrossRef]
16. Moog, H.C.; Bok, F.; Marquardt, C.M.; Brendler, V. Disposal of nuclear waste in host rock formations featuring high-saline solutions—Implementation of a thermodynamic reference database (THEREDA). *Appl. Geochem.* **2015**, *55*, 72–84. [CrossRef]
17. Parkhurst, D.L.; Appelo, C.A.J. Description of Input and Examples for PHREEQC Version 3—A Computer Program for Speciation, Batch-Reaction, One-Dimensional Transport, and Inverse Geochemical Calculations: U.S. Geological Survey Techniques and Methods 2013, Book 6, Chapter A43, p. 497. Available online: http://pubs.usgs.gov/tm/06/a43/ (accessed on 12 April 2021).
18. Blanc, P.; Lassin, A.; Piantone, P.; Azaroual, M.; Jacquemet, N.; Fabbri, A.; Gaucher, E.C. Thermoddem: A geochemical database focused on low temperature water/rock interactions and waste materials. *Appl. Geochem.* **2012**, *27*, 2107–2116. [CrossRef]

19. Delany, J.M.; Lundeen, S.R. *The LLNL Thermochemical Data Base–Revised Data and File Format for the EQ3/6 Package*; Lawrence Livermore National Lab.: Livermore, CA, USA, 1991.
20. Langmuir, D. *Aqueous Environmental Geochemistry*; Prentice Hall: Hoboken, NJ, USA, 1997.
21. Alsemgeest, J.; Auqué, L.F.; Gimeno, M.J. Verification and comparison of two thermodynamic databases through conversion to PHREEQC and multicomponent geothermometrical calculations. *Geothermics* **2021**, *91*, 102036. [CrossRef]
22. Hettkamp, T.; Baumgaertner, J.; Paredes, R.; Ravier, G.; Seibel, O. Industrial Experiences with Downhole Geothermal Line-Shaft Production Pumps in Hostile Environment in the Upper Rhine Valley. In Proceedings of the World Geothermal Congress 2020+1, Reykjavik, Iceland, 24–27 October 2021.
23. Zhang, Y.; Hu, B.; Teng, Y.; Tu, K.; Zhu, C. A library of BASIC scripts of reaction rates for geochemical modeling using PHREEQC. *Comput. Geosci.* **2019**, *133*, 104316. Available online: https://github.com/HydrogeoIU/PHREEQC-Kinetic-Library/blob/master/PHREEQC%20script.txt (accessed on 14 June 2021). [CrossRef]
24. Biver, M.; Shotyk, W. Stibnite (Sb2S3) oxidative dissolution kinetics from pH 1 to 11. *Geochim. Cosmochim. Acta* **2012**, *79*, 127–139. [CrossRef]
25. Poonoosamya, J.; Curti, E.; Kosakowski, G.; Grolimund, D.; Van Loon, L.R.; Mäder, U. Barite precipitation following celestite dissolution in a porous medium: A SEM/BSE and μ-XRD/XRF study. *Geochim. Cosmochim. Acta* **2016**, *182*, 131–144. [CrossRef]
26. Lichti, K.A.; Brown, K.L. Prediction and Monitoring of Scaling and Corrosion in pH Adjusted Geothermal Brine Solutions. In Proceedings of the NACE International Corrosion Conference & Expo 2013, Orlando, FL, USA, 17–21 March 2013.
27. Lichti, K.A.; Ko, M.; Kennedy, J. Heavy Metal Galvanic Corrosion of Carbon Steel in Geothermal Brines. In Proceedings of the NACE International Corrosion Conference & Expo 2016, Vancouver, BC, Canada, 6–10 March 2016.

Article

Soultz-sous-Forêts Geothermal Reservoir: Structural Model Update and Thermo-Hydraulic Numerical Simulations Based on Three Years of Operation Data

Clément Baujard [1,*], **Pauline Rolin** [2], **Éléonore Dalmais** [1], **Régis Hehn** [3] **and Albert Genter** [1]

[1] Électricité de Strasbourg Géothermie, 26 Boulevard du Président Wilson, F-67000 Strasbourg, France; eleonore.dalmais@es.fr (É.D.); albert.genter@es.fr (A.G.)

[2] ENSG, 2 rue du Doyen Marcel Roubault, BP 10162 CEDEX, F-54505 Vandoeuvre-lès-Nancy, France; pauline.rolin1996@gmail.com

[3] GéoPlusEnvironnement, 1 175 Route de Margès, F-26380 Peyrins, France; r.hehn.geoplus@orange.fr

* Correspondence: clement.baujard@es.fr

Abstract: The geothermal powerplant of Soultz-sous-Forêts (France) is investigating the possibility of producing more energy with the same infrastructure by reinjecting the geothermal fluid at lower temperatures. Indeed, during the operation of the powerplant, the geothermal fluid is currently reinjected at 60–70 °C in a deep fractured granite reservoir, and the MEET project aims to test its reinjection at 40 °C. A 3D hydrothermal study was performed in order to evaluate the spreading of the thermal front during colder reinjection and its impact on the production temperature. In the first step, a 3D structural model at fault scale was created, integrating pre-existing models from 2D vintage seismic profiles, vertical seismic profiles, seismic cloud structure and borehole image logs calibrated with well data. This geometrical model was then adapted to be able to run hydrothermal simulation. In the third step, a 3D hydrothermal model was built based on the structural model. After calibration, the effect of colder reinjection on the production temperature was calculated. The results show that a decrease of 10 °C in the injection temperature leads to a drop in the production temperature of 2 °C after 2 years, reaching 3 °C after 25 years of operation. Lastly, the accuracy of the structural model on which the simulations are based is discussed and an update of the structural model is proposed in order to better reproduce the observations.

Keywords: Soultz-sous-Forêts; deep geothermal reservoir; structural model; thermo-hydraulic simulations; MEET H2020 project

Citation: Baujard, C.; Rolin, P.; Dalmais, É.; Hehn, R.; Genter, A. Soultz-sous-Forêts Geothermal Reservoir: Structural Model Update and Thermo-Hydraulic Numerical Simulations Based on Three Years of Operation Data. *Geosciences* **2021**, *11*, 502. https://doi.org/10.3390/geosciences11120502

Academic Editors: Gianluca Groppelli and Jesus Martinez-Frias

Received: 28 September 2021
Accepted: 6 December 2021
Published: 9 December 2021

Publisher's Note: MDPI stays neutral with regard to jurisdictional claims in published maps and institutional affiliations.

1. Introduction

The Upper Rhine Graben is known for its great potential for the exploitation of geothermal energy at high temperatures. Indeed, it is characterized by strong local geothermal anomalies. Usually, the geothermal gradient in continental crust is approximately 30 °C/km. However, it can reach 100 °C/km in the Upper Rhine Graben (URG) thanks to large convection loops in the granitic basement and the Triassic sandstone [1], up to the Muschelkalk in some parts of the graben.

Soultz-sous-Forêts is located at around 50 km north of Strasbourg in the URG. The geothermal project began in 1984 and the first drilling began in 1987 [2]. The initial goal was to use the heat in the deep crystalline rocks to produce electricity by fracturing the granite to create an artificial heat exchanger as part of a hot dry rock (HDR) project. For this, an initial phase of drilling, stimulation, circulation tests and observation was carried out until 2007 to study the crystalline rock and the feasibility of future operations. This showed that hydrothermal fluid circulation was occurring in the natural fracture system. The geothermal fluid is a 100 g/L NaCl type brine. Hydraulic, thermal and chemical stimulations were performed to increase the permeability and the connections between the reservoir and the wells [3]. The term Enhanced Geothermal System (EGS) was defined from the research work on Soultz-sous-Forêts. The site gradually shifted from a research to industrial facility. Commercial electricity production began in June 2016 [4].

Three wells are currently operated, GPK-2, GPK-3 and GPK-4, reaching a depth of more than 5000 m. Their trajectories are distributed in the north–south direction, following the maximum horizontal stress direction. GPK-2 is the production well and GPK-3 and GPK-4 are used as injection wells. The powerplant uses an organic Rankine cycle (ORC) to convert the heat into electricity to produce a gross power of 1.7 MWe. Currently, the powerplant produces, in a sustainable manner, fluid at more than 150 °C and the injection temperature is approximately 60–70 °C [5].

The MEET H2020 project (Multidisciplinary and multi-context demonstration of EGS exploration and Exploitation Techniques and potentials) aims at improving deep geothermal energy development in Europe in different ways [6]. In existing EGS plants, it plans to demonstrate the feasibility of reinjecting at lower temperatures, down to 40 °C, thus increasing the potential heat valorization of 30%. For this purpose, various investigations have been carried out in the framework of the MEET project, regarding the on-site feasibility of colder reinjection [7] and its chemical effects [8]. The objective of the works presented here was to evaluate the potential consequences of a decrease in the injection temperature on the production temperature on a long-term basis.

2. Materials and Methods

2.1. Pre-Existing Soultz Structural Models

The European Cenozoic Rift System groups several rifts formed in response to Alpine and Pyrenean orogens at the beginning of the Cenozoic era. The URG is one of these Tertiary rifts. The rifting began during the Lower Eocene due the Pyrenean compressive tectonic phase, which created or reactivated N–S and NE–SW Variscan faults. It was followed by an E–W extension tectonic phase during the Oligocene, which was characterized by strong subsidence [9]. This crustal thinning induced the rise of the Moho, which formed regional thermal anomalies at URG scale.

The basement has been reached at 1400 m at Soultz-sous-Forêts. It is composed of two different granites: a porphyritic monzogranite and a fined-grained two-mica granite. The latter is found at the bottom of wells, at around 4500 mTVD [10]. The first one can be very hydrothermally altered, especially around nearly vertical fault zones, which are very abundant within the first km at the top basement. The small-scale fractures associated with fault zones are mostly sealed with secondary quartz, calcite and illite.

The monzogranite is covered by sedimentary layers from Permo-Trias to Quaternary. The Permian is poorly represented. The Trias sequence is the most important in this sedimentary cover, from 1350 m to 750 m, characterized by sandstones, alternation of limestone, marls and dolomites (from marine to fluvial environment). Then, Jurassic layers with alternation of limestones and marls appear from 750 m to 600 m. The Tertiary era marked the opening of the rift with the presence of evaporites and deposits of thin lacustrine sediments.

Three pre-existing structural models of the nearby Soultz reservoir have been used in order to build the complete structural model presented in Section 4.1:

- The first 3D structural model of Soultz-sous-Forêts was built in 1994 [10] from well data and seismic profiles interpreted by the BRGM [11]. This model contains 5 horizons from Jurassic to granite layers and faults in the sedimentary part.
- Later, a new interpretation of the PN84J seismic line [12] allowed the achievement of another 3D geological model. It considers 3 horizons: the granitic basement, the Buntsandstein unit and the Mélettes layer. It also contains faults in the sedimentary part.
- In another model [13], the two-mica granite interface was considered as a layer.

2.2. Operation Data and Calibration Dataset

The powerplant is operated with GPK-2 as a producer and GPK-3 and GPK-4 as injectors. GPK-2, GPK-3 and GPK-4 are cemented to 1431, 1447 and 1446 mMDGL (Measured Depth from Ground Level), respectively. At these depths, GPK-2 has been drilled in 8"1/2 until 5057 mMDGL and cased in 7" until 4440 mMDGL. GPK-3 and GPK-4 have been drilled in 9"5/8 until 4592 and 4767 mMDGL and in 8"1/2 until 5111 and 5270 mMDGL. In these wells, a zone between 4000 m and 4500 m TVD is cemented. GPK-2 is cemented between 4200 m and 4500 m.

Feeding zones behind the casing have been reported [14] (in the upper granite reservoir section) for GPK-2 and GPK-3, connecting the well to its annulus from, respectively, 1431 and 1447 mMDGL to 4170 and 3988 mMDGL. As GPK-4 does not present any significant leaks in its casing between 1400 m and 4500 m, only its open hole below 4767 mMDGL is considered. The trajectory of GPK-2 and GPK-3 was taken from the top of Buntsandstein the bottom to take into account the known feed zones. Indeed, the cased section is cemented at the top and bottom only, and nearly the entire granite section can be considered an open hole. The Buntsandstein is then cemented but has been taken into account for its hydrothermal connection with upper granite via nearly vertical fault zones.

Operation data used for model calibration range from 25 June 2016 to 24 June 2019 (3 years). The average production rate is 25–30 L/s. Injection is 100% in GPK-3 until the beginning of March 2017 (almost 7 months) and is then split into GPK-3 and GPK-4 (see Figure 1).

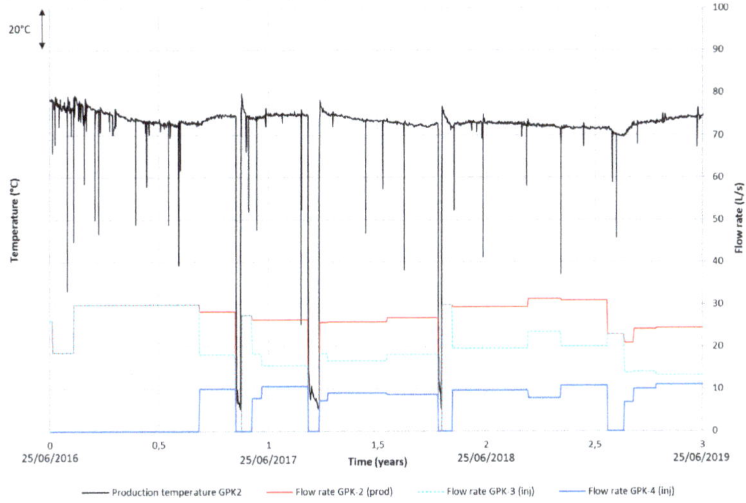

Figure 1. Soultz-sous-Forêts operation data used for calibration. Temperature absolute values are hidden for confidentiality reasons. Source: GEIE EMC.

2.3. Software and Codes Used

The works presented in this paper could be carried out using two commercial software licenses:

- The reservoir modeling software PETREL 2019, developed by Schlumberger, for structural modeling, using the Structural Framework workflow.
- The Finite Element simulation software FEFLOW 7, edited by DHI, for hydrothermal calculation.

The meshing of the simplified structural model was performed with MeshIt, a software program for the three-dimensional volumetric meshing of faulted reservoirs, developed by GFZ [15].

3. Results

3.1. Structural Model

3.1.1. Complete Structural Model

For this study, the model of Renard and Courrioux 1994 [10] was used as the basis. The horizons and regional faults provided in this model are based on vintage 2D seismic line interpretations. The horizons and regional faults of this model were loaded and reconstructed in Petrel. The well trajectories of GPK-1, GPK-2, GPK-3, GPK-4, EPS-1 and 4550 were also imported into the model. The two-mica granite layer [11] was also imported as a layer according to the depth in GPK-3 and GPK-4 where it was observed.

The major faults from this model, identified in the sedimentary units (Kutzenhausen, Soultz and Hermerswiller faults), were extended into the basement.

For the lower part of the structural model, information about local faults was collected thanks to different data [13,16]:

- Cuttings;
- Well logs including oriented borehole logs (caliper, gamma ray, Ultrasonic Borehole Imager);
- Vertical seismic profiles (VSPs);
- Microseismicity studies.

As a result, the constructed model contains six geological layers. A total of 50 structures representing local faults could be integrated into the Petrel model (Figure 2). Each fault was added by informing its orientation and its dip in the table of the well that is intersected. Then, in the 3D model, each fault identified in several wells was linked to create surfaces.

3.1.2. Simplified Structural Model for Meshing

The sedimentary faults are considered far enough from the wells to not have any influence on the simulations, so it was decided to not keep them in the 3D hydrothermal model. Indeed, these faults are located at a distance greater than the downhole distance between the wells, which is generally considered the well radius of influence in the reservoir for geothermal systems.

To study the hydrothermal circulation in the granite and between wells, faults were selected based on the following criteria:

- Permeable faults: they must present flow or thermal anomalies, or have been detected by a microseismicity cloud.
- Extension: they must intersect several wells to respect the connections between them.
- Contribution of the flow: recent flow logs and precedent studies [17] allowed the estimation of the flow produced and injected in the different sections of the wells. In order to respect these contributions in the hydraulic calibration, it was necessary to keep faults crossing the well in specific sections (Table 1, Figure 3).

Table 1. Estimated flow contributions of the faults, derived from flow log data.

Wells	FZ1800	FZ2120	FZ4770	FZ4760	FZ4925
GPK-2	-	65% (FZ2120 + FZ4770)		35%	-
GPK-3	65% (FZ1800 + FZ2120)		35%	-	-
GPK-4	-	-	-	-	100%

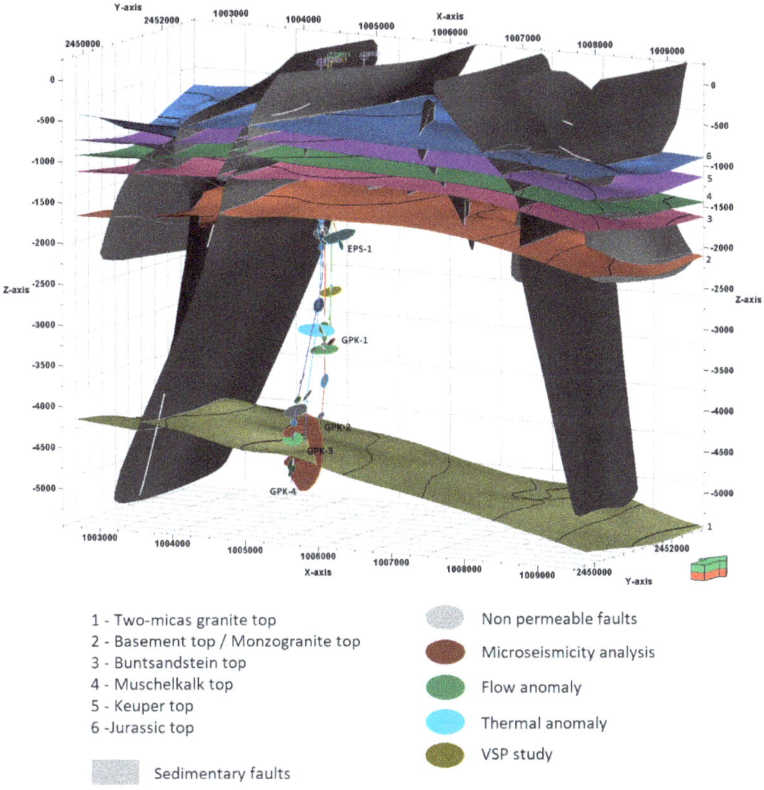

Figure 2. Complete 3D structural model of Soultz-sous-Forêts (in gray: regional fault zones from 2D seismic data interpretation; in color: local fault zones). The arrow on the bottom right of the figure indicates north.

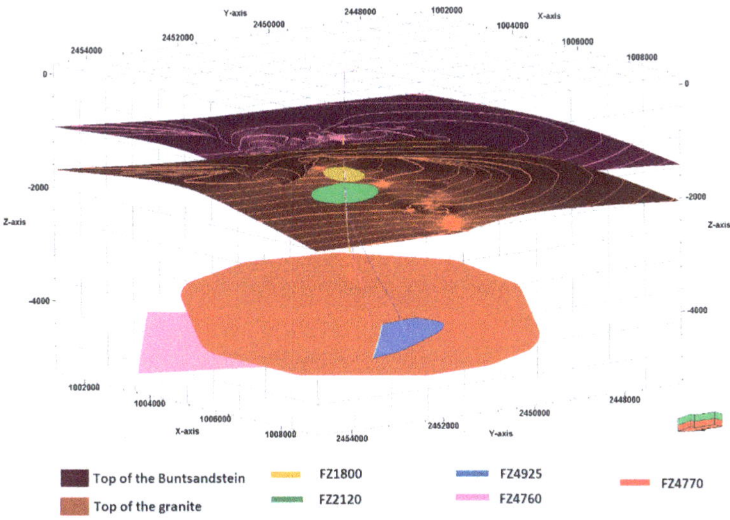

Figure 3. Simplified 3D structural model of Soultz-sous-Forêts. The arrow on the bottom right of the figure indicates north.

Finally, five faults were chosen (Figure 3):

- GPK-3-FZ4770 intersects GPK-3 at 4770 mMDGL (Measured Depth from Ground Level). It is the largest observed in the UBI. It controlled 70% of fluid losses during the hydraulic test and it matches with the microseismic structure MS-GPK-3-2003a. It also intersects the leakage of GPK-2 at 3900 mMDGL.
- Then, the microseismicity structure MS-GPK3-2003c fits with the FZ4925 fracture that intersects GPK-4 at 4924 mMDGL. It also intersects the fracture GPK-3-FZ4770, creating a needed connection with the other wells, so it is important to take it into account in the hydrothermal circulation.
- At a shallower level, GPK2-FZ2120 was selected because of its extent. Indeed, it intersects GPK-2 at 2123 mMDGL but also GPK-3 and GPK-4 at similar depths, connecting the three annuluses together. Moreover, it induces total mud losses and took 95% of the flow during drilling.
- Tracer tests demonstrated that around 60% of the flow comes from the far field through the open hole of GPK-2 but, recently, this contribution has decreased and is now estimated at around 35%. To represent this entry, the fault GPK2-FZ4760 was selected. Even if there were no geophysical measurements in the deepest part of GPK-2, an orientation of N170° and a dip of 65°W at 4760 m were inferred based on gamma ray, caliper logs and cutting observations [16]. This fault was extended to one of the boundaries of the model to simulate the far-field influx.
- GPK4-FZ1800 was the last fault selected because of its intersection with GPK-4 at 1801 mMDGL and GPK-3, the thickness of the damage zone (observed on UBI) and mud losses during drilling. A sensibility study revealed that this fault has not much impact on the model and it was chosen before the hypothesis of keeping only the open hole of GPK-4. Thus, it was kept in the structural model but not in the hydrothermal simulations.

3.2. Mesh

The horizons and faults were extracted from Petrel as surfaces and imported in MeshIt, intermediate software that allows the building of units. Once the model was structured, it was imported into FEFLOW. The entire model was refined by the Tetgen algorithm, especially near the surfaces of faults where the distance between two points was 20 m and the well trajectory where the refinement varied between 5 and 1 m near the intersection between faults and wells (Table 2). This created an unstructured 3D tetrahedral mesh (Figure 4). Faults were assigned to their corresponding surface as 2D discrete feature elements.

Table 2. Meshing parameters.

Parameter	Value
Mesh type	Tetrahedron
Number of units	3
Number of tetrahedrons	257,491
Number of nodes	43,013
Number of discrete features	5
Number of elements in discrete features	19,491
Volume total	411.6 km^3
North–south extent	7 km
East–west extent	7 km
Depth	−8 km above sea level
Top surface	400 m above sea level

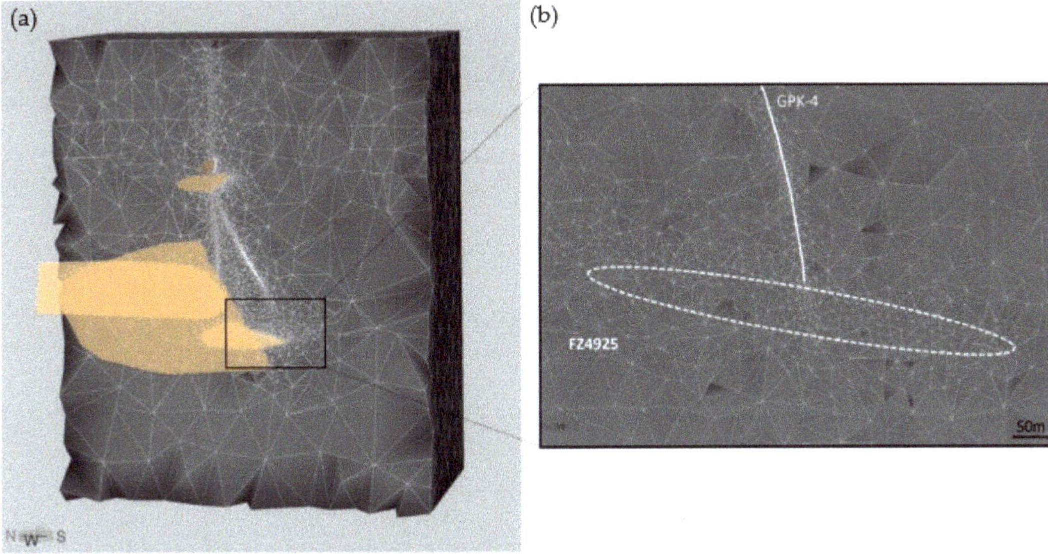

Figure 4. (**a**) Meshing of the 3D structural model and (**b**) magnification of the refinement meshing near GPK-4 and the fault FZ4925 (**b**).

3.3. Boundary Conditions and Initial State

The boundary conditions are defined as follows:

- Constant head of 10 m (=hydrostatic pressure gradient) boundary conditions on the lateral sides and on top of the model.
- A heat flux of 0.072 W/m^2 coming from the bottom of the model. This value was adjusted in order to reproduce the observed temperature gradient in the deepest part of the reservoir (see green line in Figure 5), where a conductive-only heat transport regime is assumed.
- The temperature is 10 °C at the surface (0 m).

The initial state is defined by:

- A hydrostatic pressure gradient in the model.
- A non-linear temperature distribution, varying with depth (Figure 5). As the model does not simulate convection and does not aim at reproducing long-term fluid flow, which would explain the observed temperature profile in Soultz (suggesting up-welling of hot fluid), the measured temperature profile at the wells was imported as a temperature distribution for the initial state. A simple transient hydrothermal simulation assuming no well use showed that this temperature profile remained almost unchanged over 50 years (see red points in Figure 5). Thus, within the timeframe of the simulation, the temperature distribution is representative of the real temperature profile of GPK-3 during the 30 years of simulations discussed in this study.

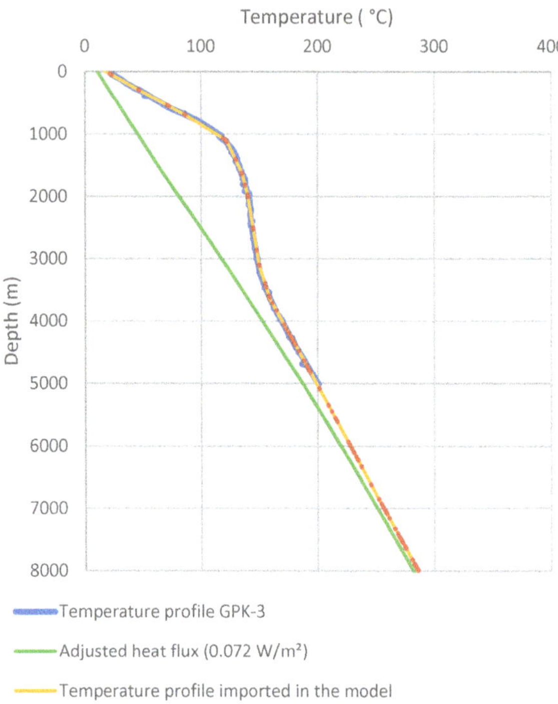

Figure 5. Comparison between GPK-3 temperature profile measured at equilibrium (in blue) and initial condition of the model (in yellow). The red points show the temperature extracted from the model after 50 years without pumping. The green line shows the result of a 0.072 W/m2 conductive-only thermal regime.

3.4. Calibration Results

The calibration must respect:

- The estimated contribution of the faults in each section of the wells (Table 1);
- The measured wellhead pressure during the past 3 years of operation—around 0 bar for GPK-2, 1 bar for GPK-3 and 18 bar for GPK-4 according to the flow rate attributed earlier;
- The production temperature observed in GPK-2, which is very dependent on the preferential fluid flow paths governed by the relative hydraulic properties of the different fault zones.

The calculations were run in a steady-state hydraulic regime and transient thermal regime. Fluid parameters were derived from the analysis realized at the nearby Rittershoffen geothermal site (thermal conductivity of 0.681 W/m/K and thermal capacity of 3.86 MJ/m^3/K). The calibration was done manually by varying the hydraulic parameters of faults and matrix as the hydraulic conductivity and the thickness of faults. Initial geological layers and faults' hydraulic parameters were derived from the literature [13,14,18,19]. Fault thickness was derived from geological data (cutting analysis), also reported in these publications.

A sensitivity analysis was manually performed in order to calibrate the model. It highlighted the high impact of the fault hydraulic properties (hydraulic conductivity and thickness). The calibration was carried out in two steps. At first, the flow contributions of the faults and well pressures were fitted at best (see Table 3) using realistic parameters, in order to calibrate the fault transmissivity. Then, the hydraulic conductivity and thickness of each fault were adjusted, keeping the transmissivity of each fault constant. As a result, GPK-

2′s production temperature could also be acceptably fitted (see Figure 6). It is important to mention that the output GPK-2 production temperature from FEFLOW is a mean temperature from the different contributions. Therefore, it was necessary to correct the FEFLOW output to take into account temperature losses in the well when the fluid circulates upwards over a few kilometers in the well in order to be able to compare the model results with observed temperature data (wellhead measurements). The temperature losses were calculated using a polynomial function of the flowrate and reservoir temperature, calculated using wellbore simulator HEX-B for the GPK-2 well [20]. The same work was realized on the input injection temperature for GPK-3 and GPK-4 (i.e., the input temperature given to FEFLOW is higher than operation values as it takes into account the heating of the fluid when going down in the well). It must be pointed out that numerical instabilities (oscillations) were observed in the calculated temperature after the shutdown/restart of the plant, inducing rapid fluid velocity changes in the model.

Table 3. Fault contributions obtained at the end of the calibration (in bold). Estimated real contributions from Table 1 are recalled in italic.

Wells	FZ1800	FZ2120	FZ4770	FZ4760	FZ4925
GPK-2	-	**60%** (FZ2120 + FZ4770)—*65%*		**37**–*35%*	-
GPK-3	**62%** (FZ1800 + FZ2120)—*65%*		**31**–*35%*	-	-
GPK-4	-	-	-	-	**99**–*100%*

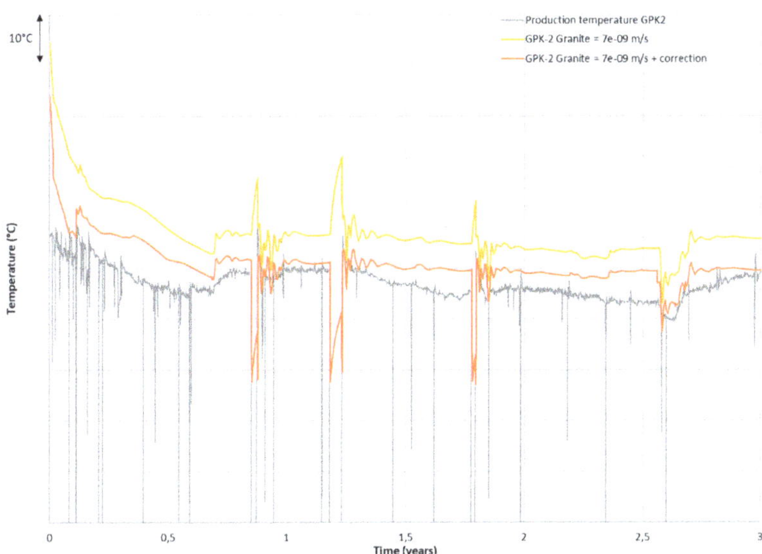

Figure 6. Comparison between simulated and operation data. The corrected value (orange line) takes into account temperature losses in the well in order to be comparable to wellhead values.

At the end of the calibration, the faults had transmissivities between 6.30×10^{-5} and 3×10^{-4} m^2/s, except the FZ4760 one, which showed significantly higher transmissivity. Indeed, this fault is known as a main contributor to GPK-2 from logging data. However, many structures contribute to the production of the GPK-2 open-hole with an important connection to the far-field [17]. In order to reproduce the high contribution of this zone to the system, the best solution was to artificially extend this fault to the boundary of the model. As a result, this fault represents the entire contribution of the far field to the production well. The final geological layers and fault parameters are shown in Tables 4 and 5.

Table 4. Geological layer parameters after calibration.

Parameter	Units	Upper Sediments	Buntsandstein	Granite
Hydraulic Conductivity	[m/s]	5e-08	1e-08	7e-09
Specific Storage	[1/m]	8e-07	5e-07	1.75e-08
Porosity	[-]	0.1	0.03	0.03
Thermal Conductivity	[W/m/K]	2.8	2.5	2.5
Thermal Capacity	[J/m^3/K]	2e06	3.2e06	2.9e06
Heat Production	[W/m^3]	5e-07	5e-07	3e-06

Table 5. Fault parameters after calibration.

Parameter	Units	FZ1800	FZ2120	FZ4760	FZ4770	FZ4925
Hydraulic Conductivity	[m/s]	6.08e-06	1.7e-05	0.05	2e-05	6.3e-05
Specific Storage	[1/m]	2e-06	2e-06	2e-06	2e-06	2e-06
Porosity	[-]	0.1	0.1	0.1	0.1	0.1
Thermal Conductivity	[W/m/K]	2.5	2.5	2.5	2.5	2.5
Thermal Capacity	[J/m^3/K]	2.9e06	2.9e06	2.9e06	2.9e06	2.9e06
Thickness	[m]	12	15	8	15	1
Heat Production	[W/m^3]	3e-06	3e-06	3e-06	3e-06	3e-06
Transmissivity	[m^2/s]	7.3e-05	2.55e-04	0.4	3e-04	6.3e-05

3.5. Long-Term Simulation Results

In order to calculate the reinjection temperature effect over a long period, the simulation was run over 30 years, including the first 3 years of calibration. Four runs were carried out, simulating reinjection temperatures of, respectively, 70, 60, 50 and 40 °C (see Figure 7).

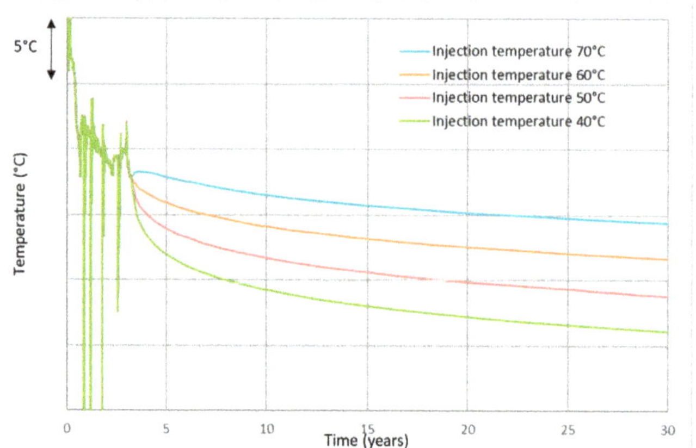

Figure 7. Variation in calculated production temperature in GPK-2 with injection temperatures, calculated over 30 years. The figure shows corrected values (comparable to wellhead values).

The results show that a decrease of 10 °C in the injection temperature in GPK-3 and GPK-4 is expected to produce a drop of approximately 2.8 °C in the production temperature at GPK-2 over 30 years. It is interesting to note that this drop does not increase significantly with time, as the drop is already significant (2 °C) after 2 years of operation (i.e., after 5 years of simulation).

3.6. Short-Term Simulation Results

Additional simulations were run using this numerical model, assuming a lower reinjection temperature in the reservoir for 4 months only, and then returning to the actual (70 °C) reinjection temperature (Figure 8). This showed that the effect of a lower injection temperature over a few months would have a short-term impact on the production temperature, which would drop from one to two degrees one year after the injection temperature changes. This effect disappears approximately 2 years after returning to the initial injection temperature of 70 °C.

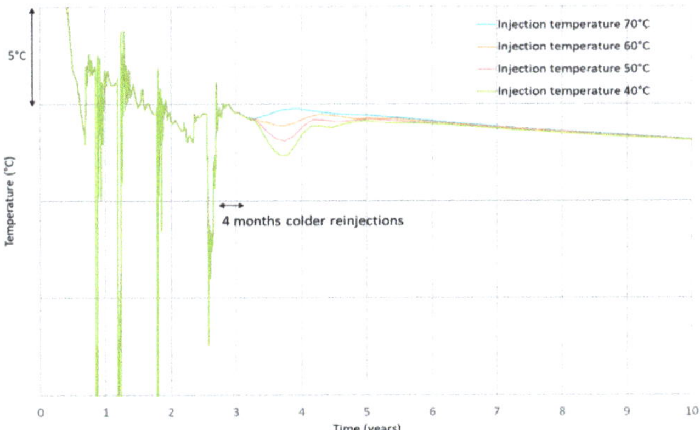

Figure 8. Variation in calculated production temperature in GPK-2 using colder reinjections over 4 months only, shown over 10 years. The figure shows corrected values (comparable to wellhead values).

4. Discussion

4.1. Outcomes of the Simulations

The works presented here are part of a wider simulation effort aiming at understanding the Soultz-sous-Forêts geothermal site's behavior. Thus, the results presented here are a work in progress, but the calculations to come aim at understanding how the far-field contribution integrates in the Soultz system (by which fault connections). Therefore, the remaining work will not change or update the results/conclusions of the works presented here. The simulation results show that:

- A temporary decrease (over a few months) in the injection temperature does not affect the production temperature over a long time, as the production temperature returns to the initial production temperature after a maximum of 2 years.
- A permanent decrease in the injection temperature will have a limited effect on the production temperature. According to the models, a 10 °C injection temperature decrease in GPK-3 and GPK-4 is expected to produce a drop of up to 3 °C in the production temperature at GPK-2 over 30 years.
- Moreover, colder reinjection does not significantly affect the temperature distribution in the reservoir. The cold front does not propagate faster, and the "cooled-down" volume reaches a lower temperature at its core but the temperature impact at the fringe remains limited, as the cold fluid will finally drain more energy from the surrounding rock. This is the reason that the temperature impact at the production well remains limited.

To summarize, a significant decrease in the injection temperature (from 70 °C down to 40 °C) in order to produce more energy appears to have a clear but limited impact on the reservoir and on the production temperature. This impact is mainly due to fast connections, likely to occur in the upper part of the reservoir, due to production casing integrity issues. Thus, from a reservoir point of view, this work confirms that using colder injections to increase the total produced energy is perfectly feasible.

4.2. Simulation Results vs. Observations

In the framework of the MEET project, a low-temperature ORC unit was tested in Soultz-sous-Forêts. It was installed after the existing high-temperature ORC unit, thus taking fluid at 70 °C as an input (output temperature of the existing ORC unit). This small additional ORC unit has been running for 3 months, from mid-March to mid-June 2021 [6]. The objective of this test was (1) to quantify the efficiency of such a low-temperature ORC unit and (2) to reinject fluid at a lower temperature in Soultz during a few months and observe the effect on the reservoir. This would have allowed comparison of the model results with observation data. Unfortunately, the thermal power of the ORC unit was too low to create a significant change in the reinjection temperature.

A recent observation of the production temperature suggests that the injection temperature has a slightly stronger effect than that inferred by the model (which shows that a decrease of 10 °C in the injection temperature is expected to produce a drop of approximately 2–3 °C in the production temperature after a few years). Indeed, it is suspected that the permeability of the main fault zones (which are the largest flow contributors to the wells) has been increasing since the system begun operation. This leads to a faster connection between wells. This seems to be confirmed by the pressure decrease observed at the injection wells.

The process behind the possible permeability increase in the most permeable faults is unclear. Indeed, mineral crystallization during brine circulation could lead to a permeability decrease in the circulation zones, as mineralization could progressively close the fractures [21]. Nevertheless, this effect might be compensated by other processes, either purely thermal (fracture aperture increasing as the surrounding rock is being cooled down) or mechanical (fracture aperture increasing due to failure processes). The cooling of the reservoir and subsequent stress change around the injection well could also positively impact the fracture's mechanical behavior, allowing fracture opening by jacking at lower pressures. Interestingly, this possible permeability increase in the most permeable faults leads to the conclusion that the most permeable fault network located in the first kilometer of the Soultz granitic basement is a promising geothermal target. The successful geothermal Rittershoffen project targeting such local faults located in the first kilometer of the top basement confirmed this observation [22].

4.3. Structural Model Update

The calibration effort led the model to reproduce realistic values of the wellhead pressures, to fit accurately the contribution of the faults at the wells and leading to a good fit of the production temperature over the 3 year calibration period. This certainly shows that the structural model is representative of the reservoir.

Nevertheless, the calibration also showed the limitations of the structural model behind the simulations. Indeed, it appeared necessary to extend one of the faults (FZ4760) to the boundary of the model and to attribute to this structure higher transmissivity values than inferred during the hydraulic tests of the well, in order to obtain the far-field contribution from the reservoir to the system. It is possible that the hydraulic properties of this fault increase with time since testing, as mentioned previously, but this possibly shows that some of the faults included in the model could be connected to great extension structures such as the Soultz horst border faults, which have at least 500 m of vertical offset that connects the system to the regional geothermal reservoir.

In 2018, Electricité de Strasbourg conducted a 3D seismic acquisition survey [23]. This seismic acquisition covered the Soultz-sous-Forêts wells and reservoir. Processing was on-going when the structural model presented here was built and the interpretation was finalized recently; thus, detailed structural results of the 3D seismic acquisition were not available for this work.

Thus, it was decided to update the local Soultz structural model by considering the 3D seismic interpretation results. In light of the seismic results, some faults in the granite included in the model, derived from imaging well data, are connected to regional-scale structures derived from the 3D seismic survey (see Figure 9), especially in the upper part of the reservoir (1500–2500 m—for example, FZ2120). This could possibly validate the contribution of the far field to the Soultz hydraulic system through the upper part of the reservoir. However, this update could also prove to be insufficient to reproduce the far-field contribution in the deep reservoir and show that connections to the far field exist in the deep reservoir but could not be seen in the 3D seismic data.

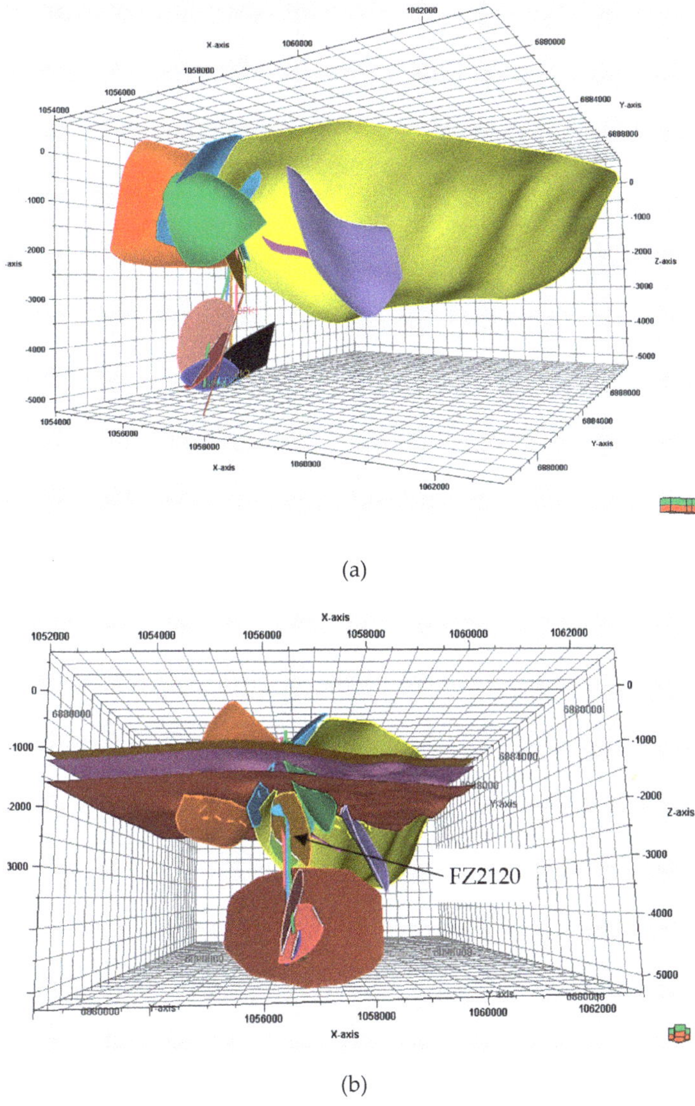

(a)

(b)

Figure 9. (**a**) the image shows the new regional faults on the upper part of the reservoir (yellow, orange, blue, green and violet structures) and faults in the deep reservoir used in the model presented in this paper. (**b**) the image shows the same new regional faults, the new horizons top Muschelkalk (brown), top Buntsandstein (pink) and top Granite (red) and the new structural fault model in the granite.

4.4. Future Work

New numerical simulations will be carried out for the testing of various fault contribution calibrated with operation data. In order to allow more flexibility during the gridding/meshing process (from the structural model to the calculator), it was decided to run these numerical simulations with ECLIPSE, which uses a regular grid, as it is a finite difference code, in contrast to FEFLOW, which needs a mesh as it is based on finite elements. The gridding process used by ECLIPSE should allow more flexibility at the interface between the structural model and the numerical simulation results. These new simulations should help to validate the hypothesis of a far-field reservoir connection through regional faults in the upper part of the reservoir.

Author Contributions: Materials and methods, C.B.; structural model, P.R., É.D. and A.G.; meshing, P.R. and R.H.; calibration, P.R. and R.H.; long-term simulation, P.R. and R.H.; discussion, C.B. Authorship is limited to those who have contributed substantially to the work reported. All authors have read and agreed to the published version of the manuscript.

Funding: This project has received funding from the European Union's Horizon 2020 research and innovation program under grant agreement no. 792037 (MEET project).

Data Availability Statement: Relies on mostly confidential data owned by us.

Acknowledgments: The authors would like to warmly thank the Soultz-sous-Forêts site owner, EEIG Exploitation Minière de la Chaleur, for data access and GFZ Potsdam, especially Mauro Cacace and Guido Blöcher, for their support and the use of MeshIt.

Conflicts of Interest: The authors declare no conflict of interest.

References

1. Vidal, J.; Genter, A.; Schmittbuhl, J. How do permeable fractures in the Triassic sediments of Northern Alsace characterize the top of hydrothermal convective cells? Evidence from Soultz geothermal boreholes (France). *Geotherm. Energy* **2015**, *3*, 8. [CrossRef]
2. Gérard, A.; Genter, A.; Kohl, T.; Lutz, P.; Rose, P.; Rummel, F. The deep EGS (Enhanced Geothermal System) project at Soultz-sous-Forêts (Alsace, France). *Geothermics* **2006**, *35*, 502–516. [CrossRef]
3. Schill, E.; Genter, A.; Cuenot, N.; Kohl, T. Hydraulic performance history at the Soultz EGS reservoirs from stimulation and long-term circulation tests. *Geothermics* **2017**, *70*, 110–124. [CrossRef]
4. Ravier, G.; Seibel, O.; Pratiwi, A.S.; Mouchot, J.; Genter, A.; Ragnarsdóttir, K.; Sengelen, X. Towards an optimized operation of the EGS Soultz-sous-Forets power plant (Upper Rhine Graben, France). In Proceedings of the European Geothermal Congress, The Hague, The Netherlands, 11–14 June 2019.
5. Mouchot, J.; Ravier, G.; Seibel, O.; Bosia, C.; Pratiwi, A. Deep geothermal plants operation in Upper Rhine Graben: Lessons learned, first year of operation from EGS geothermal plants in Alsace. In Proceedings of the European Geothermal Congress, The Hague, The Netherlands, 11–14 June 2019.
6. Dalmais, E.; Genter, A.; Trullenque, G.; Leoutre, E.; Leiss, B.; Bär, K.; Mintsa, A.-C.; Ólafsson, D.I.; Rajsl, I.; Wagner, B. MEET Project: Toward large scale deployment of deep geothermal energy in Europe. In Proceedings of the World Geothermal Congress 2020, Reykjavik, Iceland, 26 April–2 May 2020.
7. Seibel, O.; Mouchot, J.; Ravier, G.; Ledesert, B.; Sengelen, X.; Hebert, R.; Ragnarsdottir, K.; Olafsson, D.I.; Haraldsdottir, H.O. Optimised valorisation of the geothermal resources for EGS plants in the Upper Rhine Graben. In Proceedings of the World Geothermal Congress 2020, Reykjavik, Iceland, 26 April–2 May 2020.
8. Ledésert, B.; Hébert, R.; Mouchot, J.; Bosia, C.; Ravier, G.; Seibel, O.; Dalmais, É.; Ledésert, M.; Trullenque, G.; Sengelen, X.; et al. Scaling in a geothermal heat exchanger at soultz-sous-forêts (Upper Rhine Graben, France): A XRD and SEM-EDS characterization of sulfide precipitates. *Geosciences* **2021**, *11*, 271. [CrossRef]
9. Briais, J.; Lasseur, E.; Homberg, C.; Beccaletto, L.; Couëffé, R.; Bellahsen, N.; Chateauneuf, J.J. Sedimentary record and structural analysis of the opening of the European Cenozoic Rift System: The case of the Upper Rhine Graben. In Proceedings of the EGU General Assembly Conference Abstracts, Vienna, Austria, 23–28 April 2017; p. 17227.
10. Renard, P.; Courrioux, G. Three-dimensional geometric modeling of a faulted domain: The Soultz Horst example (Alsace, France). *Comput. Geosci.* **1994**, *20*, 1379–1390. [CrossRef]
11. Menjoz, A.; Cautru, J.P.; Criaud, A.; Genter, A. Stimulation des réservoirs géothermiques en milieu cristallin–Caractérisation des réservoirs fractures. *Rapp. Annu. D'activités* **1988**, 35–42.
12. Place, J. Caractérisation des Chemins de Circulations de Fluides Dans le Réseau Poreux d'un Batholite Granitique, Application au Site Géothermique de soultz-sous-Forêts. Ph.D. Thesis, Université de Strasbourg, Strasbourg, France, 2010.
13. Dezayes, C.; Genter, A.; Valley, B. Structure of the low permeable naturally fractured geothermal reservoir at Soultz. *C. R. Geosci.* **2010**, *342*, 517–530. [CrossRef]

14. Held, S.; Genter, A.; Kohl, T.; Kölbel, T.; Sausse, J.; Schoenball, M. Economic evaluation of geothermal reservoir performance through modeling the complexity of the operating EGS in Soultz-sous-Forêts. *Geothermics* **2014**, *51*, 270–280. [CrossRef]
15. Cacace, M.; Blöcher, G. MeshIt—a software for three dimensional volumetric meshing of complex faulted reservoirs. *Environ. Earth Sci.* **2015**, *74*, 5191–5209. [CrossRef]
16. Sausse, J.; Dezayes, C.; Dorbath, L.; Genter, A.; Place, J. 3D model of fracture zones at Soultz-sous-Forêts based on geological data, image logs, induced microseismicity and vertical seismic profiles. *C. R. Geosci.* **2010**, *342*, 531–545. [CrossRef]
17. Sanjuan, B.; Pinault, J.-L.; Rose, P.; Gérard, A.; Brach, M.; Braibant, G.; Crouzet, C.; Foucher, J.-C.; Gautier, A.; Touzelet, S. Tracer testing of the geothermal heat exchanger at Soultz-sous-Forêts (France) between 2000 and 2005. *Geothermics* **2006**, *35*, 622–653. [CrossRef]
18. Gentier, S.; Rachez, X.; Ngoc, T.D.T.; Peter-Borie, M.; Souque, C. 3D Flow modelling of the medium-term circulation test performed in the deep geothermal site of Soultz-sous-Forêts (France). In Proceedings of the World Geothermal Congress 2010, Bali, Indonesia, 25 April 2010.
19. Dezayes, C.H.; Chevremont, P.; Tourlière, B.; Homeier, G.; Genter, A. Geological Study of the GPK4 HFR Borehole and Correlation with the GPK3 Borehole (Soultz-sous-Forêts, France). Open File Report BRGM/RP-53697-FR. 2005. 85p. Available online: http://infoterre.brgm.fr/rapports/RP-53697-FR.pdf (accessed on 3 December 2021).
20. Megel, T.; Kohl, T.; Gérard, A.; Rybach, L.; Hopkirk, R. Downhole pressures derived from wellhead measurements during hydraulic experiments. In Proceedings of the World Geothermal Congress 2005, Antalya, Turkey, 24–29 April 2005.
21. Audigane, P.; Royer, J.; Kaieda, H. Permeability characterization of the Soultz and Ogachi large-scale reservoir using induced microseismicity. *Geophysics* **2002**, *67*, 204–211. [CrossRef]
22. Baujard, C.; Genter, A.; Dalmais, E.; Maurer, V.; Hehn, R.; Rosillette, R.; Vidal, J.; Schmittbuhl, J. Hydrothermal characterization of wells GRT-1 and GRT-2 in Rittershoffen, France: Implications on the understanding of natural flow systems in the rhine graben. *Geothermics* **2017**, *65*, 255–268. [CrossRef]
23. Richard, A.; Maurer, V.; Toubiana, H.; Carriere, X.; Genter, A. How to upscale geothermal energy from deep fractured basement in the Upper Rhine Graben? The impact of a new 3D seismic dataset. In Proceedings of the World Geothermal Congress 2020+1, Reykjavik, Iceland, 30 March–27 October 2021.

 *geosciences*

Article

Hydro-Thermal Modeling for Geothermal Energy Extraction from Soultz-sous-Forêts, France

Saeed Mahmoodpour [1,*], Mrityunjay Singh [1,*], Aysegul Turan [1], Kristian Bär [1] and Ingo Sass [1,2]

[1] Group of Geothermal Science and Technology, Institute of Applied Geosciences, Technische Universität Darmstadt, 64287 Darmstadt, Germany; turan@geo.tu-darmstadt.de (A.T.); baer@geo.tu-darmstadt.de (K.B.); sass@geo.tu-darmstadt.de (I.S.)

[2] Darmstadt Graduate School of Excellence Energy Science and Engineering, Technische Universität Darmstadt, 64287 Darmstadt, Germany

* Correspondence: saeed.mahmoodpour@tu-darmstadt.de (S.M.); mrityunjay.singh@tu-darmstadt.de (M.S.)

Abstract: The deep geothermal energy project at Soultz-sous-Forêts is located in the Upper Rhine Graben, France. As part of the Multidisciplinary and multi-contact demonstration of EGS exploration and Exploitation Techniques and potentials (MEET) project, this study aimed to evaluate the possibility of extracting higher amounts of energy from the existing industrial infrastructure. To achieve this objective, the effect of reinjecting fluid at lower temperature than the current fluid injection temperature of 70 °C was modeled and the drop in the production wellhead temperature for 100 years of operation was quantified. Two injection-production rate scenarios were considered and compared for their effect on overall production wellhead temperature. For each scenario, reinjection temperatures of 40, 50, and 60 °C were chosen and compared with the 70 °C injection case. For the lower production rate scenario, the results show that the production wellhead temperature is approximately 1–1.5 °C higher than for the higher production rate scenario after 100 years of operation. In conclusion, no significant thermal breakthrough was observed with the applied flow rates and lowered injection temperatures even after 100 years of operation.

Keywords: Soultz-sous-Forêts; EGS; hydro-thermal modeling

Citation: Mahmoodpour, S.; Singh, M.; Turan, A.; Bär, K.; Sass, I. Hydro-Thermal Modeling for Geothermal Energy Extraction from Soultz-sous-Forêts, France. *Geosciences* **2021**, *11*, 464. https://doi.org/10.3390/geosciences11110464

Academic Editors: Béatrice A. Ledésert, Ronan L. Hébert, Ghislain Trullenque, Albert Genter, Eléonore Dalmais, Jean Hérisson and Jesus Martinez-Frias

Received: 23 September 2021
Accepted: 7 November 2021
Published: 9 November 2021

Publisher's Note: MDPI stays neutral with regard to jurisdictional claims in published maps and institutional affiliations.

1. Introduction

Geothermal energy is a clean, renewable and low-cost solution for heating and power generation. One of the most challenging problems that humanity is facing is how to mitigate climate change and the anthropogenic emission of carbon dioxide, in order to achieve the target of the Paris agreement, which limits the atmospheric temperature rise to 2 °C or less [1]. Carbon geosequestration is the most desirable solution to this problem [2–5]. However associated cost and underdeveloped technology limits the industry from its implementation. Therefore, use of geothermal energy to replace the carbon-based energy sources is gaining momentum [6]. A milestone of the installation of 2 million heat pumps by the European geothermal heat pump market was achieved in 2019 [7]. The geothermal heat usage and electricity production in Europe is expected to grow up to 880–1050 TWh/year and 100–210 TWh/year in 2050 respectively. This contribution is equivalent to 4–7% of European power generation in the year 2050 [8]. As part of the Multidisciplinary and multi-contact demonstration of EGS exploration and Exploitation Techniques and potentials [9] project, a numerical hydrothermal model was developed to critically validate the flow behavior of the Soultz-sous-Forêts geothermal power plant from existing operational data. Furthermore, our model was enhanced by including discrete fault structures and validated with operational data to allow for a realistic prediction of the future operational behavior.

Soultz-sous-Forêts is located in the central Upper Rhine Graben, France and has a great potential for geothermal energy exploitation. Soultz-sous-Forêts is the most investigated site in terms of geoscientific studies. The top 1.5 km of the geological succession is made of thick Quaternary and Tertiary sediments, Mesozoic to Paleozoic sedimentary rocks above the crystalline basement, which is represented by naturally fractured granite. The Mesozoic to Paleozoic sedimentary rocks can be subdivided into two layers: Buntsandstein and Permian. The Buntsandstein is approximately 350 m thick and comprised of fluvial deposits whereas the Permian represents more alluvial continental deposit filling the paleo-basin of the variscan orogeny [10]. The basement is composed of monzogranite with K-feldspar mega crystals with localized concentration of biotite (depth between 1420 and 4700 m) and a two-mica granite containing muscovite (depth between 4700 and 5000 m) [11,12]. In Table 1, the rock properties for the two sandstone layers and granite are listed [13,14]. It must be noted that the data presented in Tables 1 and 2 are based on the calibration through the field data and discussed in the unpublished works of the MEET project. The sedimentary section has a maximum geothermal gradient of up to >100 K km^{-1} making the Soultz-sous-Forêts site ideal for geothermal energy extraction [15]. Figure 1 shows that the temperature around the wellbores of Soultz-sous-Forêts is higher than that of the surrounding region. Free convection along the major faults [16–18] is the primary reason causing the increased thermal gradients. For depths greater than 3700 m, the geothermal gradient becomes 10 K/km.

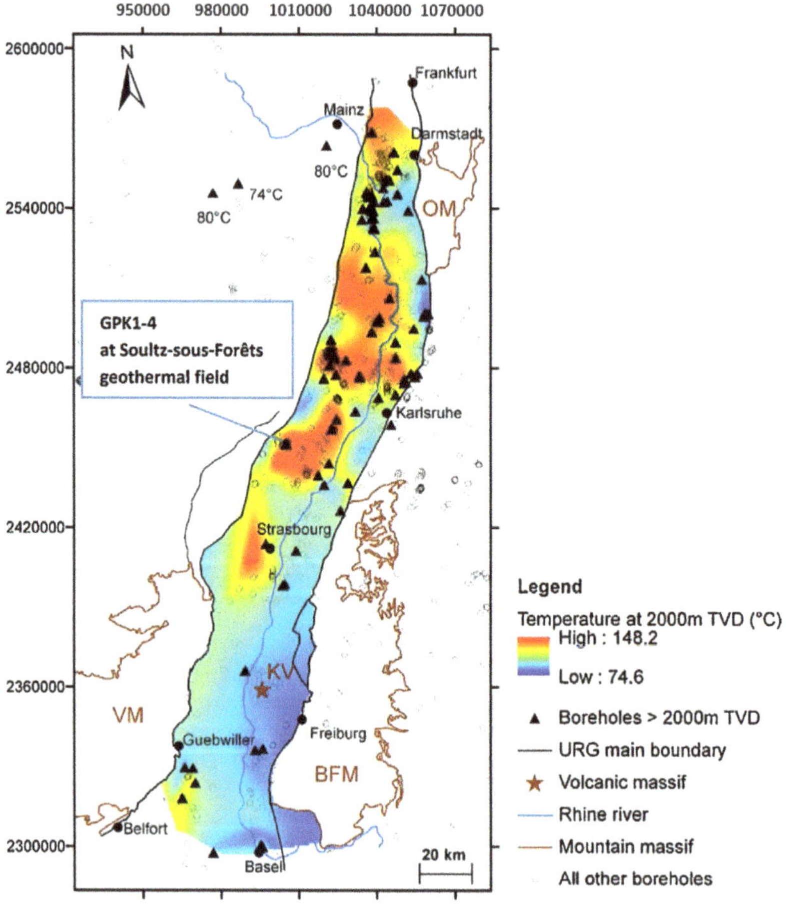

Figure 1. Temperature distribution at 2 km depth TVD in the Upper Rhine Graben [19].

Table 1. Rock matrix parameters [13,20–22].

Parameter	Unit	Upper Sediment	Buntsandstein	Granite
Hydraulic conductivity	$m·s^{-1}$	5×10^{-8}	1×10^{-8}	9×10^{-9}
Specific storage	$1·m^{-1}$	8×10^{-7}	5×10^{-7}	1.75×10^{-8}
Porosity	-	0.1	0.03	0.03
Thermal conductivity	$W·m^{-1}·K^{-1}$	2.8	2.5	2.5
Thermal capacity	$J·m^{-3}\,K^{-1}.$	2×10^{6}	3.2×10^{6}	2.9×10^{6}
Heat production	$W·m^{-3}$	5×10^{-7}	5×10^{-7}	3×10^{-7}

Table 2. Fault parameters [13].

Parameter	Unit	FZ1800	FZ2120	FZ4760	FZ4770	FZ4925
Hydraulic conductivity ($K_{f,0}$)	$m·s^{-1}$	6.08×10^{-6}	1.7×10^{-5}	0.05	2×10^{-5}	6.3×10^{-5}
Specific storage	$1·m^{-1}$	2×10^{-6}	2×10^{-6}	2×10^{-6}	2×10^{-6}	2×10^{-6}
Porosity	-	0.1	0.1	0.1	0.1	0.1
Thermal conductivity	$W·m^{-1}·K^{-1}$	2.5	2.5	2.5	2.5	2.5
Thermal capacity	$J·m^{-3}\,K^{-1}$	2.9×10^{-6}	2.9×10^{-6}	2.9×10^{-6}	2.9×10^{-6}	2.9×10^{-6}
Thickness (F_0)	m	12	15	8	15	1
Heat production	$W·m^{-3}$	3×10^{6}	3×10^{6}	3×10^{6}	3×10^{6}	3×10^{6}
Transmissivity	$m^{2}·s^{-1}$	7.3×10^{-5}	2.55×10^{-4}	0.4	3×10^{-4}	6.3×10^{-5}

Figure 2 shows the geothermal gradient at the Soultz-sous-Forêts site. Sausse et al. [23] and Dezayes et al. [24] used borehole image logs and core studies to characterize 3D realistic and static fractures of Soultz granite. Sausse et al. [23] found 53 structures including 39 fracture zones, seven microseismic structures and six vertical seismic profiles (VSP) at the Soultz-sous-Forêts site. In addition, Dezayes et al. [24] also identified 39 fractures aligned with a general strike of N160°E at the Soultz site. The sedimentary layer above 1400 m is considered for geothermal activity in the literature due to its remoteness from the main fluid circulation, and it is considered as a caprock.

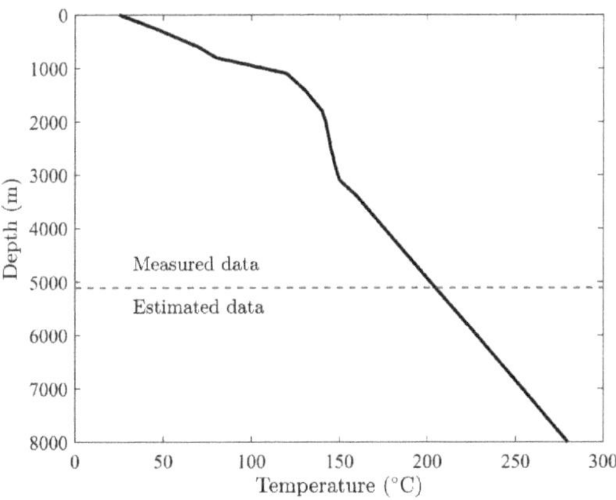

Figure 2. Geothermal gradient at the Soultz-sous-Forêts site. Here, an anomaly in temperature is observable in the top 3 km section or in the sedimentary layer. We assumed 10 °C temperature at the surface to calculate this geothermal gradient. The initial data up to the depth of 5.1 km is measured alongside GPK-2 by Pribnow and Schellschmidt [25] and further modified by Rolin et al. [13].

The geothermal project was commenced at Soultz-sous-Forêts in 1984 and the drilling started in 1987 [26]. The earliest plan was to create a fractured granite reservoir in the deep crystalline rock at a depth of 5 km to generate electricity. The industrial electricity production at this site started in June 2016. Presently, the Soultz-sous-Forêts site operates three wells with a maximum depth of up to 5000 m (GPK-2, GPK-3 and GPK-4, see Figure 3). These wells follow the main fault along the NNW–SSE direction. The binary geothermal power plant is working on an Organic Rankine Cycle (ORC) for the heat to electricity conversion. The production well is GPK-2 whereas two wells, GPK-3 and GPK-4, are re-injection wells. The hot fluid produced from GPK-2 is fed into the heat exchanger where the heat is transferred to the isobutane of the ORC cycle and reinjected after being cooled. The fluid production temperature at the Soultz plant is >150 °C and the injection temperature is 70 °C. The production well (GPK-2) and one injection well (GPK-3) indicate fluid leakage in the respective depth intervals at 1431–4170 m measured depth from ground level (MDGL) and 1447–3988 m MDGL, respectively [27]. There is not enough precise data available for the leakage zone and, therefore, it is assumed to be homogeneous over the depth. Both injection wells are cased only at the top, whereas the granitic reservoir section is not completed and in an open-hole condition.

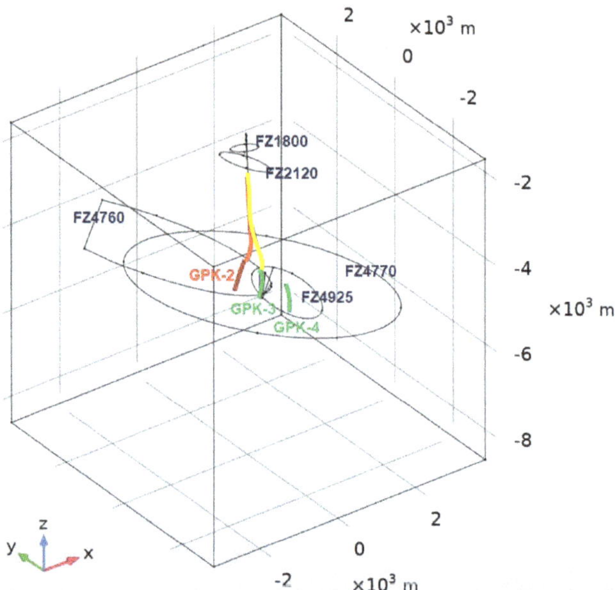

Figure 3. Geometry for numerical modeling of Soultz-sous-Forêts geothermal site. Elliptic geometries are faults listed in Table 1 (blue color). Open hole sections of the injection wells are denoted by green colors (GPK-3 and GPK-4) whereas open hole section of the production well is denoted by the dark red color (GPK-2). The leakage zone of the production well is denoted by light red whereas the leakage zone of the GPK-3 is shown by the yellow color.

For the model geometry, only the hydraulically active fractures with high permeability, as proven by thermal anomalies, detected microseismicity during stimulation and operation [23], and which are intersecting multiple wells, were included. The model is thus limited to only five major fractures out of 39 faults or fault zones as shown in Figure 3. The properties of these fractures (fault zones) are listed in Table 2.

Although the Soultz-sous-Forêts site has been the focus of more than 60 PhD theses and 300 peer-reviewed articles [19], only a few hydrothermal modeling studies have been conducted to understand the hydro-thermal behavior of the reservoir in detail. These studies were coupled with and validated by field operational data specifically with tracer tests to understand the flow path within the fractured granite [14].

The flow circulation between GPK-3 and GPK-2 wells was addressed by Sanjuan et al. [27] through an analytical dispersive transfer model, whereas Blumenthal et al. [28], Gessner et al. [29] and Egert et al. [30] also used dispersive transport models for the Soultz-sous-Forêts site. They investigated the hydraulic connectivity between the injection well (GPK3) and production wells (GPK2 and GPK4) using a multi-well tracer test. Gentier et al. [31] developed the first discrete fracture network (DFN) model while employing a particle tracking method to consider the hydraulically active parts and fracture sets for both wells. More recent modeling studies include Magnenet et al. [32], where a 2D THM model was developed based on a finite element grid (FEM); Aliyu and Chen [33], where finite element method (FEM) was used to model hydro-thermal (HT) processes of Soultz while using different working fluids; and most recently Vallier et al. [14], where a THM model based on FEM was developed at reservoir scale coupled with gravity measurements.

Previous studies showed that a single-fracture approach is not sufficient to represent the hydraulic flow existing at Soultz and 2D models are limited to represent the site in terms of the complex geometry and interconnection of dominating faults. Thus, this study takes its roots from the developed 3D THM model based on FEM while hosting five fractures (FZ1800, FZ2120, FZ4760, FZ4770 and FZ4925; also see Table 2) [13].

From the above literature, it is clear that cold water is injected at 70 °C through both the injection wells. Therefore, injection of cold water below this temperature may enable much higher geothermal energy extraction. However, no numerical studies have been conducted thus far to support this idea. In the presented study, the energy extraction potential from Soultz-Sous-Forêts for 100 years was investigated, allowing the thermal drawdown at the production well to be quantified. The major simplification of this study is neglecting the mechanical behavior. For the short term, as the temperature and pressure development are limited in the wellbore regions, this simplification is relevant and we can use the modeling hydro-thermal simulation result matching with the operational data to better characterize the wellbore effect and reservoir properties. In the ongoing study, we are trying to examine THM behavior of this system for a better prediction for the long term. Another simplification considered here is scaling in the reservoir. Possible scaling effects on the pipelines and heat exchanger devices are beyond the scope of this study. The reservoir size considered for the numerical simulation is large and computational modeling of kinetic controlled reactive fluid flow in such a reservoir requires significantly high computational resources. The possible incompatibility is insignificant because of the reinjection of the same fluid for the entire operation. However, the effect of temperature reduction on the chemical reactions requires experimental work to update the permeability variation.

The manuscript outline is as follows: First, we present a brief geological setting of Soultz-Sous-Forêts, followed by numerical modeling studies for the site. Furthermore, the mathematical and computational technique to model hydro-thermal processes during heat mining from a fractured reservoir is discussed. Next, the wellbore–reservoir coupling is demonstrated and its impact on wellhead temperature is quantified. In the following section, model results and their discussion are followed by final conclusions.

2. Methodology

In this section, the mathematical modeling is discussed in two stages. In the first part, governing equations for cold water dynamics in the porous media are presented, and in the second part a mathematical model for fluid leakage from the wellbore is discussed.

2.1. Reservoir Flow Modeling

A constant heat flux of 0.07 W/m^2 [17] was assigned at the bottom boundary of the domain. All other exterior boundaries of the modeled domain are defined as no flow for both fluid and heat transmission. Because the weather conditions of Soultz are not available, the monthly averaged daily weather fluctuation of Strasbourg, France was used for this study. Strasbourg is approximately 40 km SSE from the Soultz geothermal site. All fractures within the domain are regarded as internal boundaries, implicitly considering the mass and energy exchange between porous media and fractures or fault zones. In the injection well, the diameter of the well is small and can, as a simplification, be represented by a line.

The coupled heat and mass transfer in a fractured rock matrix can be modeled using the mass balance equation integrated with heat transport. The governing equation for heat and mass flow in porous media can be written as [34]:

$$\rho_1(\phi_m S_1 + (1 - \phi_m)S_m)\frac{\partial p}{\partial t} - \rho_1(\alpha_m(\phi_m \beta_1 + (1 - \phi_m)\beta_m))\frac{\partial T}{\partial t} = \nabla.(\frac{\rho_1 k_m}{\mu}\nabla p) \quad (1)$$

In the above equation, fluid pressure and temperature in the rock matrix are denoted by p and T, respectively. Here, rock porosity is ϕ_m, and storage coefficients for rock and fluid are S_1 and S_m. The thermal expansion coefficient of the fluid and rock matrix is denoted by β_1 and β_m, respectively. The fluid density and dynamic viscosity are indicated using ρ_1 and μ, whereas the reservoir permeability is denoted by k_m.

The fractures are assumed as internal boundaries and the flow along the internal fractures can be denoted by:

$$\rho_1(\phi_f S_1 + (1 - \phi_f)S_{mf})e_h\frac{\partial p}{\partial t} - \rho_1(\alpha_f(\phi_f \beta_1 + (1 - \phi_f)\beta_f))e_h\frac{\partial T}{\partial t} = \nabla_T.\left(\frac{e_h\rho_1 k_f}{\mu}\nabla_T p\right) + n.Q_m \quad (2)$$

Here, fluid pressure and temperature in the fracture are indicated by p and T respectively. Additionally, ϕ_f, S_f, β_f, e_h and k_f denote the fracture porosity, storage coefficients of the fracture, thermal expansion coefficient of the fracture, hydraulic aperture between the two fracture surfaces, and fracture permeability, respectively. The mass flux exchange between the fracture and matrix are denoted by $n.Q_m = n.(-\frac{\rho k_m}{\mu \nabla p})$, whereas the gradient operator applicable along the fracture tangential plane is indicated by ∇_T.

The local thermal non-equilibrium (LTNE) approach to model heat exchange between the rock matrix and water is implemented in this study. The conductive heat transfer between rock matrix and pore fluid is the dominant heat exchange mechanism. For the rock matrix, the heat transfer equation can be written as:

$$(1 - \phi_m)\rho_m C_{p,m}\frac{\partial T_m}{\partial t} = \nabla.((1 - \phi_m)\lambda_m\nabla T_m) + q_{ml}(T_l - T_m) \quad (3)$$

In the above equation, rock matrix and fluid temperatures are denoted by T_m and T_l, respectively. Here, rock density, rock-specific heat capacity, rock thermal conductivity and the rock–fluid heat transfer coefficient are denoted by ρ_m, $C_{p,m}$, λ_m and q_{ml}, respectively. The heat flux leaving the domain and received by the adjacent fracture can be written as:

$$(1 - \phi_f)e_h\rho_f C_{p,f}\frac{\partial T_m}{\partial t} = \nabla_T.((1 - \phi_f)e_h\lambda_f\nabla_T T_m) + e_h q_{fl}(T_l - T_m) + n.(-(1 - \phi_m)\lambda_m\nabla T_m \quad (4)$$

where T_m and T_l are the matrix and fluid temperatures in the fracture, respectively; ρ_f is the density of the fracture; $C_{p,f}$ is the specific heat capacity of the fracture; λ_f is the thermal conductivity of the fracture; and q_{fl} represents the rock fracture–fluid interface heat transfer coefficient, related to the fracture aperture. The last term on the right-hand side of Equation (4) represents the heat flux exchange between the rock matrix and the fracture.

The heat convection equation for the pore fluid can be written as:

$$\phi_m\rho_l C_{p,l}\frac{\partial T_l}{\partial t} + \phi_m\rho_l C_{p,l}(-\frac{k_m\nabla p}{\mu}).\nabla T_l = \nabla.(\phi_m\lambda_l\nabla T_l) + q_{ml}(T_m - T_l) \quad (5)$$

Here $C_{p,l}$ is the heat capacity of the fluid at a constant pressure and λ_l is the thermal conductivity of the fluid.

The heat flux coupling relationship of the fluid between the domain and the fracture is satisfied by:

$$\phi_f e_h \rho_l C_{p,l} \frac{\partial T_l}{\partial t} + \phi_f e_h \rho_l C_{p,l} \left(-\frac{k_f \nabla_T p}{\mu}\right).\nabla_T T_l = \nabla_T.(\phi_f e_h \lambda_l \nabla_T T_l) + e_h q_{fl}(T_m - T_l) + n.q_l \tag{6}$$

where the heat flux $n.q_l = n.(-\phi_l \lambda_l \nabla T_l)$ denotes the heat exchange of the fluid between porous media and the fracture.

Temperature-dependent fluid thermodynamic properties are implemented into the coupled hydrothermal mass and energy balance equations. The thermophysical properties of water as a function of temperature, including dynamic viscosity (μ), specific heat capacity (C_p), density (ρ) and thermal diffusivity (κ), are listed below [34]:

$$\begin{aligned}
\mu = {}& 1.38 - 2.12 \times 10^{-2} \times T^1 + 1.36 \times 10^{-4} \times T^2 - 4.65 \times 10^{-7} \times T^3 + 8.90 \times 10^{-10} \times T^4 \\
& -9.08 \times 10^{-13} \times T^5 + 3.85 \times 10^{-16} \times T^6 \qquad (273.15 - 413.15\ K)
\end{aligned} \tag{7}$$

$$\begin{aligned}
\mu = {}& 4.01 \times 10^{-3} - 2.11 \times 10^{-5} \times T^1 + 3.86 \times 10^{-8} \times T^2 - 2.40 \times 10^{-11} \times T^3 \\
& (413.15 - 553.15\ K)
\end{aligned} \tag{8}$$

$$C_p = 1.20 \times 10^4 - 8.04 \times 10^1 \times T^1 + 3.10 \times 10^{-1} \times T^2 - 5.38 \times 10^{-4} \times T^3 + 3.63 \times 10^{-7} \times T^4 \tag{9}$$

$$\rho = 1.03 \times 10^{-5} \times T^3 - 1.34 \times 10^{-2} \times T^2 + 4.97 \times T + 4.32 \times 10^2 \tag{10}$$

$$\kappa = -8.69 \times 10^{-1} + 8.95 \times 10^{-3} \times T^1 - 1.58 \times 10^{-5} \times T^2 + 7.98 \times 10^{-9} \times T^3 \tag{11}$$

We used the commercial software COMSOL Multiphysics, version 5.6 [34] for numerically solving the coupled mass and energy conservation equations listed above. COMSOL Multiphysics solves general-purpose partial differential equations using the finite element method.

2.2. Wellbore Leakage Modeling

Understanding the fluid flowing temperature along the wellbore can be useful for an accurate estimation of the overall heat production at the production wellhead temperature, and for estimating any possible leakage caused by heat loss along the wellbore. Several reliable analytical techniques are reported in the literature to calculate the flowing temperature distribution along a wellbore [35–37].

We integrated our reservoir simulation with a wellbore flow model as developed by Hasan et al. [36]. The model constitutes an analytical approach to estimate wellbore-fluid temperature distribution for steady state flow. The analytical equations are solved sequentially for each section. Figure 4 shows a simplification of a typical geothermal well with one deviation angle. The well is inclined at an angle α with the horizontal plane. The heat transfer between the wellbore fluid and the rock matrix occurs due to the temperature difference between them. A general energy balance equation for single phase fluid flow can be expressed as:

$$\frac{dH}{dz} - g\sin\alpha + v\frac{dv}{dz} = \pm\frac{Q}{w} \tag{12}$$

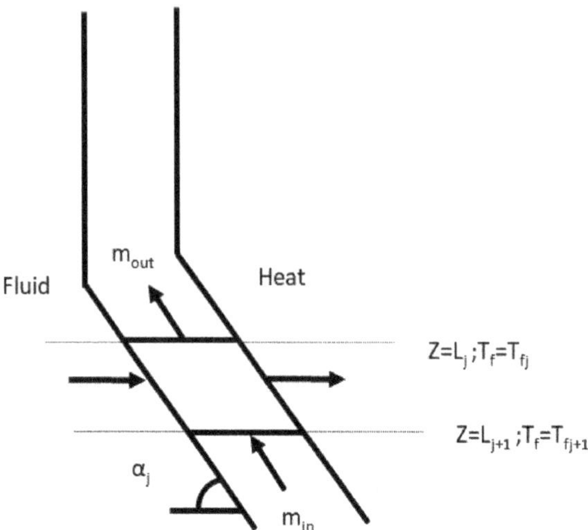

Figure 4. Wellbore heat loss modeling schematic.

Here, H is the fluid enthalpy, g is the gravitational constant, z is the variable well depth from the surface, v is the flow velocity, Q is the heat flux per unit of well length and w is the mass rate. When assuming no-phase change conditions, enthalpy will become:

$$dH = (\frac{\partial H}{\partial T})_p dT + (\frac{\partial H}{\partial p})_T dp = c_p dT - C_J c_p dp \tag{13}$$

In the above equation, T is the fluid temperature and p is the pressure, c_p is the specific heat capacity of fluid and C_J is the Joule–Thomson coefficient. If T_f is the fluid temperature, the energy balance equation will be:

$$\frac{dT_f}{dz} = C_J \frac{dp}{dz} + \frac{1}{c_p}(\pm \frac{Q}{W} + g \sin\alpha - v \frac{dv}{dz}) \tag{14}$$

The heat flux per unit wellbore length can be expressed as:

$$Q \equiv -L_R w c_p (T_f - T_{ei}) \tag{15}$$

Here, T_{ei} is the rock temperature, and L_R is the relaxation parameter defined as:

$$L_R = \frac{2\pi}{c_p w}[\frac{r_{to} U_{to} k_e}{\lambda_m + (r_{to} U_{to} T_D)}] \tag{16}$$

$$T_f = T_{ei} + \frac{1 - e^{(z-L)L_R}}{L_R}[g_G \sin\alpha + \Phi - \frac{g \sin\alpha}{c_p}] \tag{17}$$

In Equations (16) and (17), r_{to} is the tubing outside radius, U_{to} is the overall heat transfer coefficient, k_e is rock thermal conductivity, T_D is the nondimensional temperature, L is the measured depth of the wellbore, g_G is the geothermal gradient and Φ is the lumped parameter, which lumps the kinetic energy term and the Joule–Thomson coefficient term.

If V is the fluid specific volume and S is fluid entropy then from Maxwell identities, we can write:

$$(\frac{\partial H}{\partial p})_T = V + T(\frac{\partial S}{\partial p})_T \ \& \ (\frac{\partial S}{\partial p})_T = -(\frac{\partial V}{\partial T})_p \tag{18}$$

$$dH = c_p dT + [V - T(\frac{\partial V}{\partial T})_p]dp \tag{19}$$

$$c_p C_J = -[V - T\left(\frac{\partial V}{\partial T}\right)_p]$$ (20)

For liquids where ρ is the liquid density, volume expansivity (β) can be calculated as:

$$\beta \equiv \left(\frac{1}{V}\right)\left(\frac{\partial V}{\partial T}\right)_p \equiv \left(-\frac{1}{\rho}\right)\left(\frac{\partial \rho}{\partial T}\right)_p$$ (21)

$$dH = c_p dT + V(1 - \beta T)dp$$ (22)

$$c_p C_J = -V(1 - \beta T)$$ (23)

Therefore, the final output temperature from the wellhead will be:

$$T_{out} = \frac{\int m\, c_p\, T\, dz}{\int m\, c_p\, dz}$$ (24)

In this text, we considered three wells: GPK-3 and GPK-4 as two injection wells and GPK-2 as a production well. In GPK-3, the wellbore leakage was assumed between 1282 and 4852 m depth measured from the surface. In the case of GPK-2, the wellbore leakage was modeled between 1264 m to 4244 m depth measured from the surface. The fluid is single phase water flow and the model parameters are constant specific heat capacity of water as 4200 J·kg^{-1}K^{-1}, $L_R = 0.00001$ m^{-1}, and $\Phi = 0.00345$ Km^{-1}, respectively. Here, L_R and Φ accounts for the casing properties, cement properties and their thicknesses.

The coupling between the reservoir and the wellbore model is achieved through a sequential approach. First, the temperature drop due to heat exchange between the injection wellbore and the rock matrix is calculated through the analytical model. From this, the final wellbore bottom temperature is obtained, which is used as an input for the first iteration of the numerical reservoir model (heat exchange between the rock and the fluid). In the next stage, the wellbore heat exchange effect is implemented through the updated values for the reservoir temperature measured at the production wellbore bottom. The wellbore heat exchange effect is defined analytically and the temperature alongside the wellbore is obtained. Wellbore radius is very small compared to the reservoir size, and is considered as a line with the calculated temperature profile through the analytical model inside the reservoir simulation. The total number of elements in the geometry is 142,051, whereas boundary elements number 8305 and edge elements number 666.

3. Results and Discussions

In this section, first we present the benchmark for our numerical model. Then, the hydrothermal numerical modeling results are compared with the operational data measured at Soultz-sous-Forêts for three years of operation. Furthermore, new injection scenarios are proposed that can be adopted with the existing industrial setup to enhance the energy extraction capability. Finally, we perform sensitivity analysis on ten governing parameters and estimate their impact on the production temperature.

3.1. Benchmarking

For benchmarking the numerical model, we used the approach adopted by Cheng et al. [38] and Bongole et al. [39] by using a simplified 1D heat transfer problem for a single fracture system. This approach is used in the previous studies to benchmark the models. The analytical equation for heat transfer considers that the geometry is infinitely extended in both directions (see Figure 5), there is no flow boundary conditions for heat exchange, steady state fluid flow occurs only through the fracture and the rock permeability is zero, and the thermophysical properties of water are constant throughout the simulation. The temperature distribution for the fluid is identical to that of the rock matrix due to the local thermal equilibrium assumption. The analytical solution for the fluid temperature distribution is [38,39]:

$$T_{fD}(x,t) = \frac{T_0 - T_f(x,t)}{T_0 - T_{in}} = erfc\left[\left(\frac{2\lambda_m x}{e_h u_f \rho C_{p,l}}\right)\sqrt{\frac{u_f \rho_m C_{p,m}}{4\lambda_m\left(u_f t - x\right)}}\right]$$

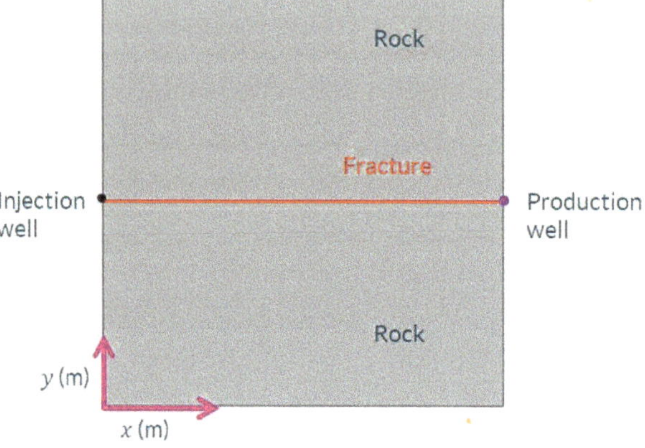

Figure 5. Geometry for the benchmarking problem.

In the above equation, T_{fD} is the nondimensional fracture temperature and u_f is the fluid velocity.

We observe a good agreement between the numerical and analytical solutions, as demonstrated in Figure 6.

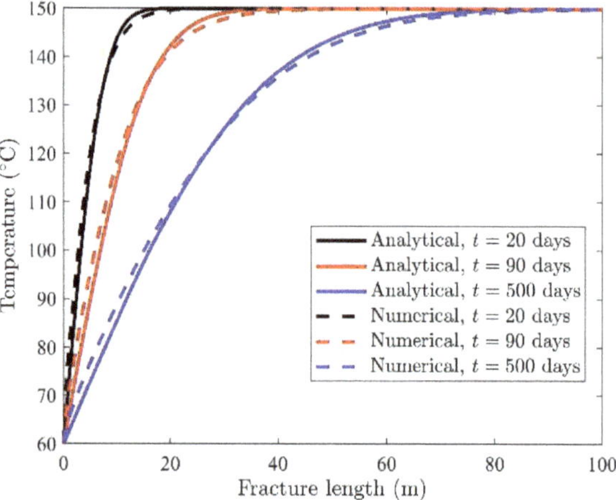

Figure 6. Comparison of the numerical solution with the analytical temperature distribution along the fracture length.

The operational data for three years was made available for Soultz-sous-Forêts site by the site operators and is used here to calibrate the coupled unsteady hydro-thermal model. Figure 7 shows the injection and production rates at the wellhead for 1163 days from June 2016 to September 2019. The fluid injection temperature is 70 °C for both the injection wells.

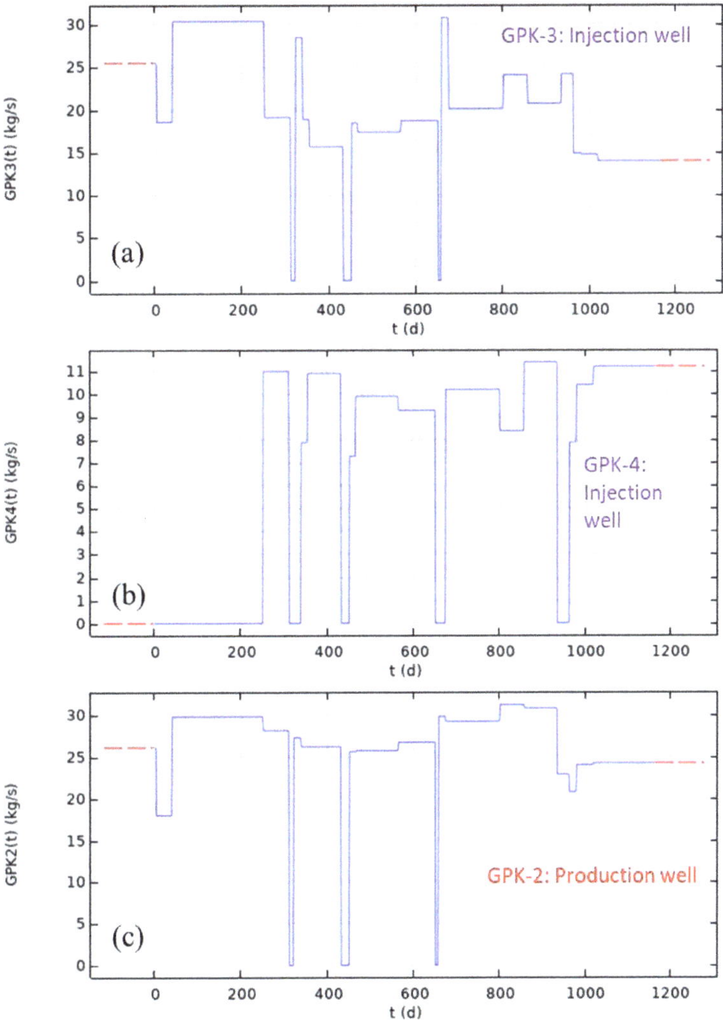

Figure 7. Injection schedule at (**a**) GPK-3 and (**b**) GPK-4 and (**c**) production schedule at production well GPK-2 for 1163 days of operation from June 2016 to September 2019. Here, the blue lines are the actual injection and production rates. The red dash lines indicate no operation period.

3.2. Validation with Operational Data

In Figure 8, the numerical model data is validated with operational data for the time period as described above. Unfortunately, it is not possible to publish the exact values of operational data due to concerns of our industrial partners and we can only show the amount of change. These are the actual temperature values shown by different colors for operational and simulations (not the differences). The measured temperature data is the operational data for 1163 days at the production wellhead. The temperature at the production well based on the hydro-thermal model is significantly different compared to the operational data. For most of the operational period, the predicted production well temperature is 15 °C higher than the measured temperature. Only operation onset and termination stages display smaller deviation in predicted temperature than observed temperature. Because the wellhead temperature measuring device may be affected by the local ambient temperature and the monthly average temperature near the geothermal site is almost the same for the corresponding months in each operational year, a correction factor to account for the weather impact on the measuring device based on the numerical model is introduced. Two scenarios of seasonal impact on the production fluid temperature are considered: (a) 20% impact of ambient temperature ($T_{effective} = T_{simulation} + 0.2 \times$ ambient temperature) and (b) 50% impact of ambient temperature ($T_{effective} = T_{simulation} + 0.5 \times$ ambient temperature).

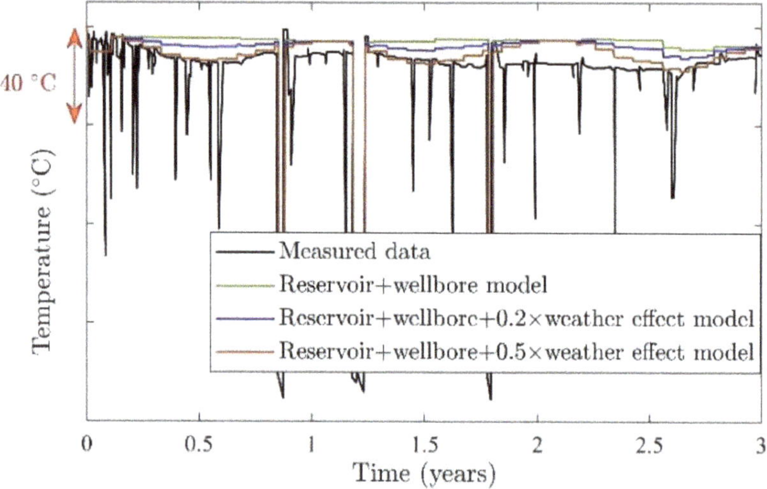

Figure 8. Difference between operational data from June 2016 to September 2019 and the data obtained from the numerical model.

Figure 8 shows the comparison of the operational data with the coupled reservoir-wellbore model and the weather-influenced production fluid temperature. The integrated wellbore–reservoir model has the highest overestimation of production temperature. However, when daily weather fluctuations in the integrated wellbore–reservoir model are considered, the prediction matches very well for most of the operation, as shown in Figure 8. The temperature differences are more relevant to understand its deviation from the actual value rather than the original temperature. The difference between operational and numerical data while considering 50% of the ambient temperature on the production temperature of the coupled wellbore–reservoir model has the best matching among all models. However, the model deviates by more than 15 °C from the operation data during the periods of 1.8 and 2.4 years. Because no other reasons for these deviations are provided with the operational data set, different measurement procedures or false measurements at the wellhead are assumed as reasons for these deviations.

3.3. Long-Term Operational Behavior

In the next study, the model was extended to a simulation period of 100 years of operation to predict the wellhead temperature development at the production well. In this section, different initial temperatures at the bottom hole section than the operationally measured data were used. The main objective of this study was to estimate the temperature at the production well (GPK-2) for different injection temperatures for long-term operational periods. In both scenarios, the injection rates for the first 1163 days are the same as in the provided operational data set. The recently designed heat exchanger at Soultz-sous-Forêts is capable of cooling the water from 70 to 40 °C using a cooling loop at 15 °C and 40 m^3/h [40]. Therefore, two scenarios were considered, A and B, for different injection temperatures. For the remaining operational period, scenario A considers four different fluid injection temperatures at the injection wellhead (70, 60, 50 and 40 °C).

The fluid injection rates are 13.3 and 11 L/s for GPK-3 and GPK-4, respectively, and the production fluid rate of GPK-2 is 24.3 L/s for the remaining operational period. In Scenario B, the injection rates after 1163 days are 19.6 and 9.7 L/s for GPK-3 and GPK-4, respectively, and the production rate of GPK-2 is 29.3 L/s; the same four injection wellhead temperatures as for scenario A were considered: 70, 60, 50 and 40 °C. These values of the injection and production rates are the operational requirements requested by our industrial partner.

Figure 9a–d shows the temperature along the wellbore for scenarios A and B, respectively, for both injection wells. The wellbore GPK-3 has an open hole section that causes a linear temperature drop along the wellbore instead of a nonlinear temperature drop, as shown in Figure 9a,b. It is interesting to note that instead of having different injection-production rates in all three wells, the fluid production temperature at the GPK-2 wellhead is almost similar for both of the scenarios A and B, as shown in Figure 9e,f. The small increase in temperature at the production wellhead is due to the sudden drop in the production wellhead pressure. The contribution in the fluid flow is due to the first pressure shock of the injection that comes from the faulted zones which are located at the bottom of the system with a higher temperature. As time proceeds, the contribution from the matrix and the leakage zone increases and reduces the temperature a few days after the beginning of the injection. To calculate the initial temperature at the wellhead, it is assumed that there is a steady state flow from the combination of the matrix and the fault zones. This initial temperature is slightly lower than that of the unsteady condition at the early time period. The fluid with the lower viscosity shows a delay in the development of the pressure shock resulting from the cold fluid injection. Therefore, the contribution from the matrix and the leakage zone for the fluid with the temperature 40 °C happens later and the main fluid flow from the faulted zone in the bottom of the system lasts for a longer time. Moreover, the temperature increase in scenario B is higher compared to that of scenario A due to the fact that scenario B has a higher production rate than scenario A which reduces the time for exchanging heat in the wellbore.

Figure 10 shows the comparison of temperature distribution in the fractures and along the wellbore for scenarios A and B. The higher production rate results in slightly faster thermal drawdown at the production well bottom for scenario B than scenario A. No thermal breakthrough was observed at the production well bottom even after 100 years of operation, as shown in Figure 10e,f.

3.4. Uncertainties

There are several uncertainties in this model. We considered the wellbore as a line source for the heat flow. The faulted zone is formulated using a fracture. Both of these assumptions are reliable because the size of the wellbore and the fault zone is negligible in comparison to the overall size of the reservoir. Data validation for the short operational period for production confirms this behavior. The matrix zone is considered as homogeneous and isotropic. As the permeability of matrix is lower than faulted zone, its contribution to the heat and mass flux is small. Therefore, this assumption holds true. As we do not know the exact point of the leakage zone alongside the casing area, we considered a homogeneous leakage and tried to compensate for the possible errors by performing a trial-and-error method to find an appropriate lumped parameter that defines the wellbore heat exchange effect. Furthermore, due to the unavailability of the geomechanical and geochemical data, we mainly focused on the hydrothermal behavior of the geothermal system. Short-term validation of this TH model gives an insight regarding the accurate system characterization, including the permeability and porosity distribution, fault placement and its contribution to the overall flow, the wellbore effect on the overall heat exchange, and fluid and rock properties. Therefore, it builds a basis for future THM or THMC (thermo-hydro-mechanical-chemical) simulations. Our expectation regarding the THM behavior is that permeability around the injection would increase (resulting from the localized thermoelastic stress reorientation and increased pore pressure). Therefore, the enhanced permeability will be favorable in the energy extraction. Based on this understanding, TH provides a preliminary basis for the thermoelastic and poroelastic stress development in this geothermal site.

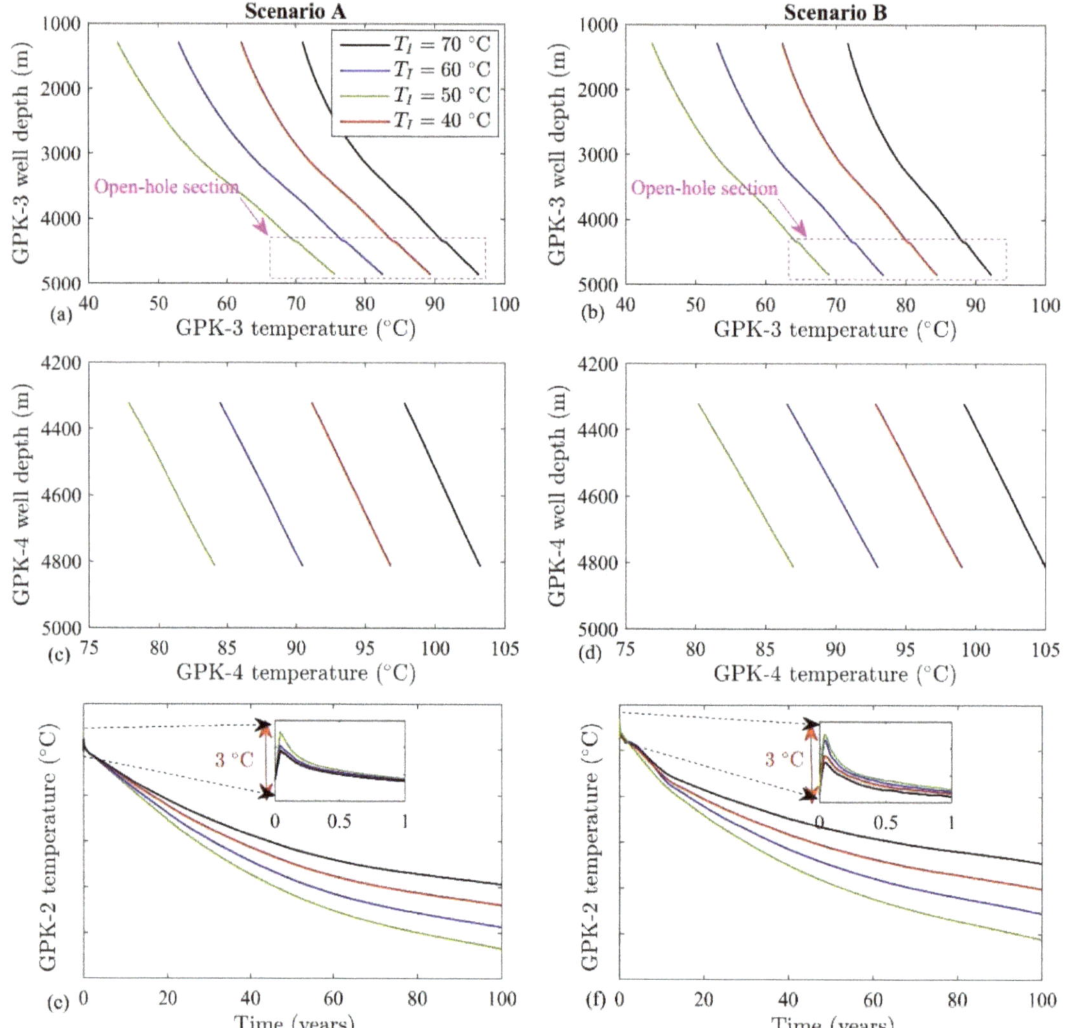

Figure 9. Comparison between scenarios A and B for (**a**,**b**) temperature along the injection well GPK-3, (**c**,**d**) temperature along the injection well GPK-4, and (**e**,**f**) wellhead temperature at the production well GPK-2.

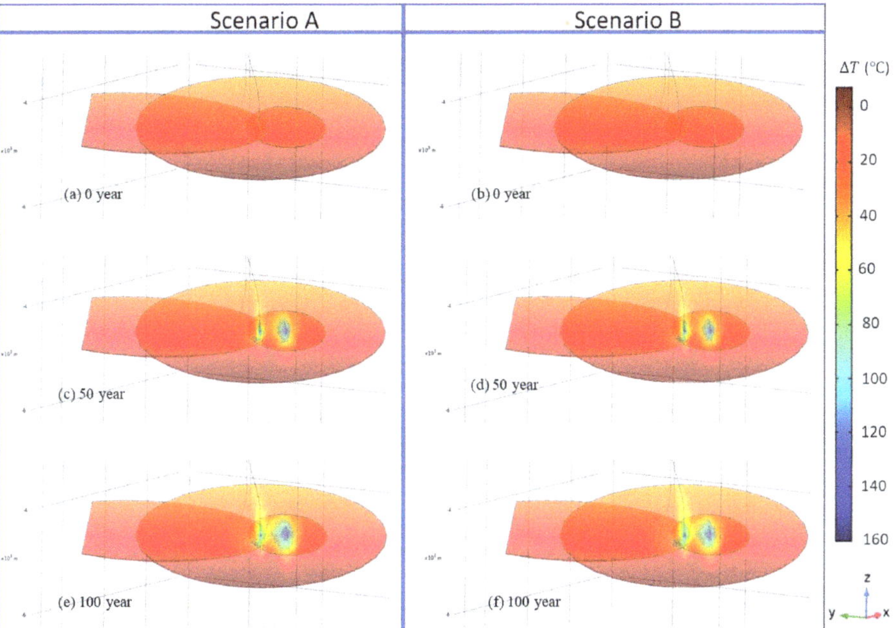

Figure 10. Comparison of temperature distribution (in SI units) in the fractures for scenarios A and B at time (**a,b**) 0 year, (**c,d**) 50 years and (**e,f**) 100 years. Here, ΔT is the temperature drop in the reservoir from the initial state.

3.5. Sensitivity Analysis of Hydrothermal Uncertainties

To examine the effect of the uncertainty of the involved parameters in hydro-thermal simulations for the Soultz-sous-Forêts geothermal reservoir, a detailed sensitivity analysis was performed, and its results are shown in Figure 11. The base case was selected as scenario A with an injection temperature of 40 °C. The range of parameters are mentioned in Table 3. Results show that the hydraulic conductivity of the matrix and fault zone and the wellbore leakage have a considerable effect on the production temperature. This finding is well aligned with the sensitivity analysis of the THM process in the fracture reservoir system [41,42]. However, these values are one order of magnitude lower than the wellbore heat exchange effect, as shown in Figure 9e. The other three parameters—fault thickness, matrix thermal conductivity and the matrix specific heat capacity—have approximately 1 °C variation over 100 years of operation. The porosity of the matrix and the fault zone, in addition to the fault zone thermal conductivity, have no impact on the temperature variation. Interestingly, heat flux has no effect on the production temperature at the surface due to the conductive heat flow in the reservoir and because the fault zones are farther away from the bottom boundary considered for the simulation. Important parameters in this sensitivity analysis show a monotonic effect on the production temperature behavior and they cannot explain the sinusoidal temperature fluctuation of more than 10 °C in each year. To check our assumption regarding the weather fluctuation impact on the production temperature, we tried to estimate the wellbore heat exchange effect. We found that the wellbore heat exchange is mainly flow-rate dependent parameter and the flow rates for the production data are constant from 41 to 250 days, whereas we can see a fluctuation in the recorded temperature. This indicates that the cyclic variability in the production temperature cannot be supported by the wellbore heat exchange argument. Therefore, the most suitable reason of the periodic production temperature variability is weather fluctuation on the measuring device.

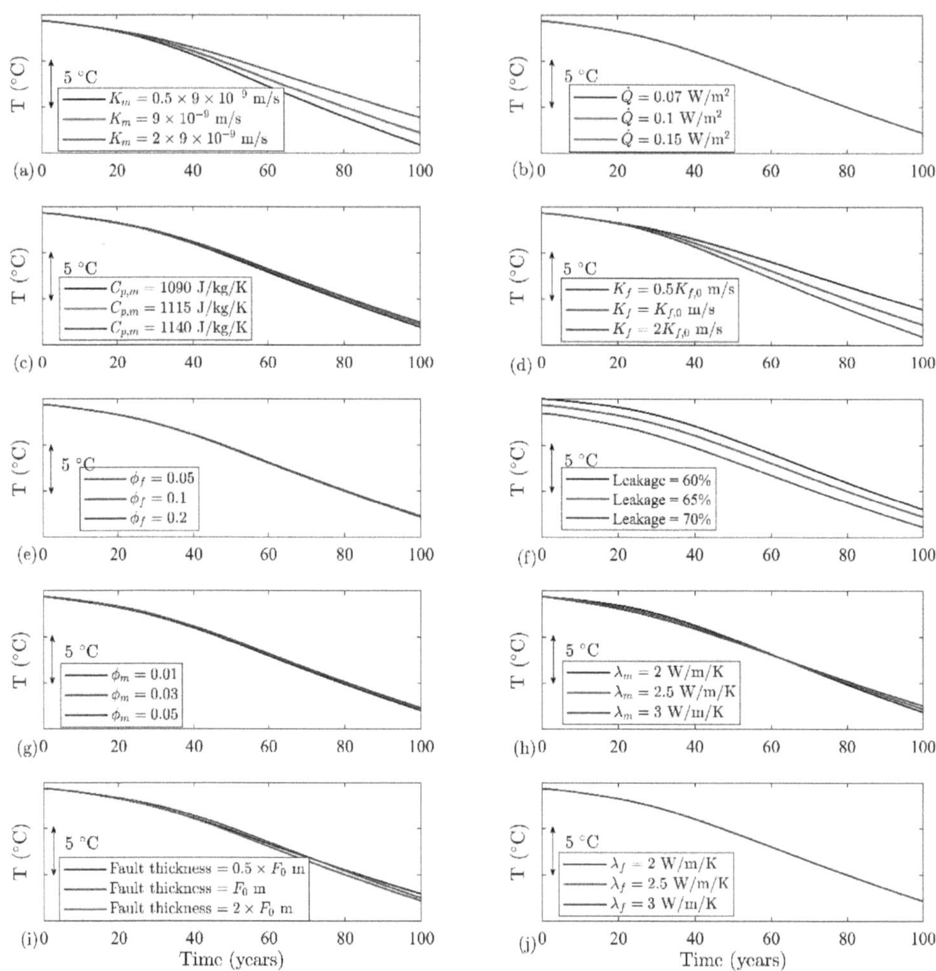

Figure 11. Sensitivity analysis for 10 parameters affecting the hydro-thermal processes at Soultz-sous-Forêts for (**a**) matrix hydraulic conductivity, (**b**) heat flux from the bottom boundary, (**c**) matrix specific heat capacity, (**d**) hydraulic conductivity of faults (here $K_{f,0}$ is the fault zone hydraulic conductivity as given in Table 2), (**e**) porosity of fault zone, (**f**) leakage contribution to the total fluid flow, (**g**) matrix porosity, (**h**) matrix thermal conductivity, (**i**) fault thickness (here F_0 is the fault thickness as given in Table 2), and (**j**) thermal conductivity of the fault zone.

Table 3. Range of parameters for sensitivity analysis.

Parameter	Base Case Value	1st Assumed Value	2nd Assumed Value
Matrix hydraulic conductivity	9×10^{-9} m/s	$0.5 \times 9 \times 10^{-9}$ m/s	$2 \times 9 \times 10^{-9}$ m/s
Heat flux from the bottom boundary	0.07 W/m^2	0.1 W/m^2	0.15 W/m^2
Matrix specific heat capacity	1115 J/kg/K	1090 J/kg/K	1140 J/kg/K
Hydraulic conductivity of fault zone (see Table 2)	$K_{f,0}$ m/s	$0.5 K_{f,0}$ m/s	$2 K_{f,0}$ m/s
Porosity of the fault zone	0.1	0.05	0.2

Table 3. Cont.

Parameter	Base Case Value	1st Assumed Value	2nd Assumed Value
Wellbore leakage fraction	65%	60%	70%
Matrix porosity	0.03	0.01	0.05
Thermal conductivity of the matrix	2.5 W/m/K	2 W/m/K	3 W/m/K
Fault thickness (see Table 2)	F_0 m	$0.5F_0$ m	$2F_0$ m
Thermal conductivity of the fault zone	2.5 W/m/K	2 W/m/K	3 W/m/K

4. Conclusions

As part of the MEET project, a coupled reservoir and wellbore model for hydraulic and thermal processes involved during the geothermal energy extraction operation at Soultz Sous Forêts was developed. Operational data from a period of 1163 days of operation was used to validate the numerical model. The validated hydro-thermal numerical model precisely simulates the geothermal energy extraction operation for 3 years. Furthermore, two operational scenarios for 100 years with four different injection wellhead temperatures—70, 60, 50 and 40 °C—were analyzed. It can be observed that even after 100 years of operation, the thermal breakthrough at the production well is only in the range of 10 to 20 °C. After 100 years of cold fluid injection and hot fluid production, the observed temperature drop at the production wellhead is less than 20 °C. Therefore, our numerical model predicts that 100 years of geothermal energy extraction operation at Soultz-sous-Forêts is feasible and will have a sufficiently high production temperature throughout the operation duration.

Author Contributions: Conceptualization, S.M. and K.B.; methodology, S.M. and M.S.; software, S.M. and M.S.; validation, S.M. and M.S. writing—original draft preparation, S.M. and M.S.; writing—review and editing, K.B., A.T.; visualization, S.M. and M.S.; supervision, K.B. and I.S.; project administration, K.B.; funding acquisition, K.B. All authors have read and agreed to the published version of the manuscript.

Funding: The work is conducted as part of the MEET project that has received funding from the European Union's Horizon 2020 research and innovation programme under grant agreement No 792037. Authors have received support from the Group of Geothermal Science and Technology, Institute of Applied Geosciences, Technische Universität Darmstadt.

Institutional Review Board Statement: Not applicable.

Informed Consent Statement: Not applicable.

Data Availability Statement: Restrictions apply to the availability of these data. Data was obtained from [ÉS-Géothermie] and are available [https://geothermie.es.fr/en/contact/, accessed on 9 November 2021] with the permission of [ÉS-Géothermie].

Acknowledgments: Authors would like to thank ÉS-Géothermie for providing Soultz-sous-Forêts operational data.

Conflicts of Interest: The authors declare no conflict of interest.

References

1. COP-21. Paris Agreement, United Nations Framework Convention on Climate Change, Conference of the Parties 21. Available online: https://unfccc.int/process-and-meetings/the-paris-agreement/the-paris-agreement (accessed on 15 June 2021).
2. Mahmoodpour, S.; Amooie, M.A.; Rostami, B.; Bahrami, F. Effect of gas impurity on the convective dissolution of CO_2 in porous media. *Energy* **2020**, *199*, 117397. [CrossRef]
3. Mahmoodpour, S.; Rostami, B.; Soltanian, M.R.; Amooie, A. Convective dissolution of carbon dioxide in deep saline aquifers: Insights from engineering a high-pressure porous cell. *Phys. Rev. Appl.* **2019**, *12*, 034016. [CrossRef]
4. Singh, M.; Chaudhuri, A.; Stauffer, P.H.; Pawar, R.J. Simulation of gravitational instability and thermo-solutal convection during the dissolution of CO_2 in deep storage reservoirs. *Wat. Resour. Resear.* **2020**, *56*, e2019WR026126.
5. Singh, M.; Tangirala, S.K.; Chaudhuri, A. Potential of CO_2 based geothermal energy extraction from hot sedimentary and dry rock reservoirs, and enabling carbon geo-sequestration. *Géoméch. Geophys. Geo. Energy Geo. Resour.* **2020**, *6*, 16. [CrossRef]

6. Singh, M.; Chaudhuri, A.; Soltanian, M.R.; Stauffer, P.H. Coupled multiphase flow and transport simulation to model CO_2 dissolution and local capillary trapping in permeability and capillary heterogeneous reservoir. *Int. J. Greenh. Gas Control.* **2021**, *108*, 103329. [CrossRef]
7. EGEC Geothermal. The Geothermal Energy Market Grows Exponentially, but Needs the Right Market Conditions to Thrive. Available online: https://www.egec.org/the-geothermal-energy-market-grows-exponentially-but-needs-the-right-market-conditions-to-thrive/ (accessed on 15 June 2021).
8. Longa, F.D.; Nogueira, L.P.; Limberger, J.; van Wees, J.-D.; van der Zwaan, B. Scenarios for geothermal energy deployment in Europe. *Energy* **2020**, *206*, 118060. [CrossRef]
9. MEET. (N.D.). Multidisciplinary and Multi-Context Demonstration of Enhanced Geothermal Systems Exploration and Exploitation Techniques and Potentials, European Union's Horizon 2020 Research and Innovation Program under Grant Agreement No. 792037. Available online: https://www.meet-h2020.com/ (accessed on 15 October 2021).
10. Duringer, P.; Aichholzer, C.; Orciani, S.; Genter, A. The complete lithostratigraphic section of the geothermal wells in Rittershoffen (Upper Rhine Graben, eastern France): A key for future geothermal wells. *BSGF Earth Sci. Bull.* **2019**, *190*, 13. [CrossRef]
11. Cocherie, A.; Guerrot, C.; Fanning, C.M.; Genter, A. Datation U–Pb Des Deux Faciès Du Granite de Soultz (Fossé Rhénan, France). *Comptes Rendus Geosci.* **2004**, *336*, 775–787. [CrossRef]
12. Schill, E.; Genter, A.; Cuenot, N.; Kohl, T. Hydraulic performance history at the Sultz EGS reservoirs from stimulation and long term circulation tests. *Geothermics* **2017**, *70*, 110–124. [CrossRef]
13. Rolin, P.; Hehn, R.; Dalmais, E.; Genter, A. D3.3 Hydrothermal Model Matching Colder Reinjection Design, WP3: Upscaling of Thermal Power Production and Optimized Operation of EGS Plants. 2018, H2020 Grant Agreement No-792037, Report. Available online: https://cordis.europa.eu/project/id/792037/reporting (accessed on 10 April 2021).
14. Vallier, B.; Magnenet, V.; Schmittbuhl, J.; Fond, C. THM modeling of gravity anamolies related to deep hydrothermal circulation at Soultz-sous-Forêts (France). *Geotherm. Energy* **2020**, *8*, 13. [CrossRef]
15. Genter, A.; Guillou-Frottier, L.; Feybesse, J.-L.; Nicol, N.; Dezayes, C.; Schwartz, S. Typology of potential Hot Fractured Rock resources in Europe. *Geothermics* **2003**, *32*, 701–710. [CrossRef]
16. Bächler, D.; Kohl, T.; Rybach, L. Impact of graben-parallel faults on hydrothermal convection—Rhine Graben case study. *Phys. Chem. Earth Parts A B C* **2003**, *28*, 431–441. [CrossRef]
17. Kohl, T.; Bächler, D.; Rybach, L. Steps towards a comprehensive thermo-hydraulic analysis of the HDR test site Soultz-sous-Forêts. In Proceedings of the World Geothermal Congress, Beppu/Morioka, Japan, 28 May–10 June 2000; pp. 2671–2676.
18. Pribnow, D.; Schellschmidt, R. Thermal tracking of upper crustal fluid flow in the Rhine graben. *Geophys. Res. Lett.* **2000**, *27*, 1957–1960. [CrossRef]
19. Baillieux, O.; Schill, E.; Edel, J.-B.; Mauri, G. Localization of temperature anomalies in the Upper Rhine Graben: Insights from geophysics and neotectonics activity. *Int. Geol. Rev.* **2013**, *55*, 1744–1762. [CrossRef]
20. GeORG. Potentiel Géologique Profond du Fossé Rhénan Supérieur. Parties 1 à 4. 2013. Available online: http://www.geopotenziale.eu (accessed on 10 April 2021).
21. Surma, F.; Geraud, Y. Porosity and Thermal Conductivity of the Soultz-sous-Forêts Granite. *Pure Appl. Geophys.* **2003**, *160*, 1125–1136. [CrossRef]
22. Vallier, B.; Magnenet, V.; Schmittbuhl, J.; Fond, C. Large scale hydro-thermal circulation in the deep geothermal reservoir of Soultz-sous-Forêts (France). *Geothermics* **2019**, *78*, 154–169. [CrossRef]
23. Sausse, J.; Dezayes, C.; Dorbath, L.; Genter, A.; Place, J. 3D model of fracture zones at Soultz-sous-Forêts based on geological data, image logs, induced microseismicity and vertical seismic profiles. *Comptes Rendus Geosci.* **2010**, *342*, 531–545. [CrossRef]
24. Dezayes, C.; Genter, A.; Valley, B. Structure of the low permeable naturally fractured geothermal reservoir at Soultz. *Comptes Rendus Geosci.* **2010**, *342*, 517–530. [CrossRef]
25. Guillou-Frottier, L.; Carré, C.; Bourgine, B.; Bouchot, V.; Genter, A. Structure of hydrothermal convection in the Upper Rhine Graben as inferred from corrected temperature data and basin-scale numerical models. *J. Volcanol. Geotherm. Res.* **2013**, *256*, 29–49. [CrossRef]
26. Gérard, A.; Genter, A.; Kohl, T.; Lutz, P.; Rose, P.; Rummel, F. The deep EGS (enhanced geothermal system) project at soultz-sous-Forêts (Alsace, France). *Geothermics* **2006**, *35*, 473–483. [CrossRef]
27. Sanjuan, B.; Pinault, J.L.; Rose, P.; Gerard, A.; Brach, M.; Braibant, G.; Crouzet, C.; Foucher, J.C.; Gautier, A. Geochemical fluid characteristics and main achievements about traces tests at Soultz-sous-Forêts (France). In Proceedings of the EHDRA Scientific Conference 2006, Soultz-sous-Forêts, France, 15–16 June 2006.
28. Blumenthal, M.; Kuhn, M.; Pape, H.; Rath, V.; Clauser, C. Hydraulic model of the deep reservoir quantifying the multi-well tracer test. In Proceedings of the EHDRA Scientific Conference, Soultz-sous-Forêts, France, 28–29 June 2007.
29. Gessner, K.; Kühn, M.; Rath, V.; Kosack, C.; Blumenthal, M.; Clauser, C. Coupled Process Models as a Tool for Analysing Hydrothermal Systems. *Surv. Geophys.* **2009**, *30*, 133–162. [CrossRef]
30. Egert, R.; Maziar, G.K.; Held, S.; Kohl, T. Implications on large-scale flow of the fractured EGS reservoir Soultz inferred from hydraulic data and tracer experiments. *Geothermics* **2020**, *84*, 101749. [CrossRef]
31. Gentier, S.; Rachez, X.; Ngoc, T.D.T.; Peter-Borie, M.; Souque, C. 3D flow modelling of the medium-term circulation test performed in the deep geothermal site of Soultz-sous-Forêts (France). In Proceedings of the World Geothermal Congress, Bali, Indonesia, 25–30 April 2010; p. 13.

32. Magnenet, V.; Fond, C.; Genter, A.; Schmittbuhl, J. Two-dimensional THM modelling of the large scale natural hydrothermal circulation at Soultz-sous-Forêts. *Geotherm. Energy* **2014**, *2*, 533. [CrossRef]
33. Aliyu, M.D.; Chen, H. Numerical modelling of coupled hydro-thermal processes of the Soultz heterogeneous geothermal system. In Proceedings of the Seventh European Congress on Computational Methods in Applied Sciences and Engineering ECCOMAS Congress Stanford University, Crete Island, Greece, 5–10 June 2016.
34. COMSOL Multiphysics® v. 5.6. COMSOL AB: Stockholm, Sweden. Available online: www.comsol.com (accessed on 11 November 2020).
35. Alves, I.; Alhanati, F.; Shoham, O. A Unified Model for Predicting Flowing Temperature Distribution in Wellbores and Pipelines. *SPE Prod. Eng.* **1992**, *7*, 363–367. [CrossRef]
36. Hasan, A.R.; Kabir, C.S.; Wang, X. A Robust Steady-State Model for Flowing-Fluid Temperature in Complex Wells. *SPE Prod. Oper.* **2009**, *24*, 269–276. [CrossRef]
37. Moradi, B.; Ayoub, M.; Bataee, M.; Mohammadian, E. Calculation of temperature profile in injection wells. *J. Pet. Explor. Prod. Technol.* **2020**, *10*, 687–697. [CrossRef]
38. Bongole, K.; Sun, Z.; Yao, J.; Mehmood, A.; Yueying, W.; Mboje, J.; Xin, Y. Multifracture response to supercritical CO_2-EGS and water-EGS based on thermo-hydro-mechanical coupling method. *Int. J. Energy Res.* **2019**, *43*, 7173–7196.
39. Cheng, A.H.-D.; Ghassemi, A.; Detournay, E. Integral equation solution of heat extraction from a fracture in hot dry rock. *Int. J. Numer. Anal. Methods Géoméch.* **2001**, *25*, 1327–1338. [CrossRef]
40. Ravier, G. Deliverable, D.3.1, Design of a Heat Exchanger for Cold Injection, WP3: Upscaling of Thermal Power Production and Optimized Operation of EGS Plants, Report Submitted to European Union. 2019. Available online: https://www.meet-h2020.com/project-results/deliverables/ (accessed on 11 November 2020).
41. Mahmoodpour, S.; Singh, M.; Turan, A.; Bär, K.; Sass, I. Key parameters affecting the performance of fractured geothermal reservoirs: A sensitivity analysis by thermo-hydraulic-mechanical simulation. *arXiv* **2021**, arXiv:2107.02277.
42. Mahmoodpour, S.; Singh, M.; Bär, K.; Sass, I. Thermo-hydro-mechanical modeling of an enhanced geothermal system in a fractured reservoir using CO_2 as heat transmission fluid—A sensitivity investigation. *arXiv* **2021**, arXiv:2108.05243.

Article

Sensitivity Analysis of FWI Applied to OVSP Synthetic Data for Fault Detection and Characterization in Crystalline Rocks

Yassine Abdelfettah [1,*] and Christophe Barnes [1,2]

[1]   Geophysical Inversion & Modeling Labs, 57 Allée de L'Albatros, 95800 Courdimanche, France; christophe.barnes@gim-labs.com
[2]   Geosciences and Environment Laboratory (GEC), Cergy Paris Université, 1 Rue Descartes, 95000 Neuville-sur-Oise, France
*   Correspondence: yassine.abdelfettah@gim-labs.com

Abstract: We have performed several sensitivity studies to assess the ability of the Full Wave Inversion method to detect, delineate and characterize faults in a crystalline geothermal reservoir from OVSP data. The distant goal is to apply the method to the Soultz-sous-Forêts site (France). Our approach consists of performing synthetic Full Wave 2D Inversion experiments using offset vertical seismic and comparing the estimated fields provided by the inversion, i.e., the estimated underground images, to the initial reference model including the fault target. We first tuned the inversion algorithmic parameters in order to adapt the FWI software, originally dedicated to a sedimentary context, to a crystalline context. In a second step, we studied the sensitivity of the FWI fault imaging results as a function of the acquisition geometry parameters, namely, the number of shots, the intershot distance, the maximum offset and also the antenna length and well deviation. From this study, we suggest rules to design the acquisition geometry in order to improve the fault detection, delineation and characterization. In a third step, we studied the sensitivity of the FWI fault imaging results as a function of the fault or the fault zone characteristics, namely, the fault dip, thickness and the contrast of physical parameters between the fault materials and the surrounding fresh rocks. We have shown that a fault with high dip, between 60 and 90° as thin as 10 m (i.e. lower than a tenth of the seismic wavelength of 120 m for Vp and 70 m for Vs) can be imaged by FWI, even in the presence of additive gaussian noise. In summary, for a crystalline geological context, and dealing with acceptable S/N ratio data, the FWI show a high potential for accurately detecting, delineating and characterizing the fault zones.

Keywords: geothermal; OVSP; well seismic data; fault; fracture; EGS; geothermal derisking; FWI; numerical modelling; inversion; imaging

Citation: Abdelfettah, Y.; Barnes, C. Sensitivity Analysis of FWI Applied to OVSP Synthetic Data for Fault Detection and Characterization in Crystalline Rocks. *Geosciences* **2021**, *11*, 442. https://doi.org/10.3390/geosciences11110442

Academic Editors: Béatrice A. Ledésert, Ronan L. Hébert, Ghislain Trullenque, Albert Genter, Eléonore Dalmais, Jean Hérisson and Jesus Martinez-Frias

Received: 23 August 2021
Accepted: 24 October 2021
Published: 27 October 2021

Publisher's Note: MDPI stays neutral with regard to jurisdictional claims in published maps and institutional affiliations.

1. Introduction

For any deep georesources exploration project, finding the location of underground resources is an important step. Deep geothermal exploration projects do not go against this rule. The location of the underground thermal anomalies as well as the different physical parameters of the subsurface are key to the success of the project. For instance, the porosity, permeability, heat flux and geomechanical stress field of the reservoir are determining parameters. For the case of an Enhanced Geothermal System (EGS), undoubtedly, the fractures and the faults play a major role for the reservoir quality and increase the geothermal fluid production (e.g., [1]. The surface of rock-to-hot fluid interaction is increasing with the interconnected fractures network. This improves heat exchange by driving the deep hot fluid to shallower and exploitable depths. The presence of an adequate fault network in geothermal fields is crucial. This was observed during the drilling and different hydraulic tests performed in Soultz-sous-Forêts geothermal boreholes (e.g., [2] and included references) that the geothermal brine inflows to the wells at depths where faults are imaged and clearly identified.

For geothermal reservoirs deeper than 1500 m, the main fault identification and characterization remains a difficult task, even using efficient geophysical exploration techniques. Using a dataset acquired from the surface, we can merely identify a major fault crossing our study area presenting an important vertical displacement. However, identification and characterization of faults related to the reservoir quality, at the reservoir scale, remains a challenge. These large faults could be identified, from surface, by using, for instance, surface seismic [3] or gravity and magnetotelluric surveys [4,5] However, hectometric faults as well as faults at the reservoir scale (i.e., metric) are more challenging to identify from the surface and their characterization, at depth, remains a difficult task that becomes impossible when their size is lower than the seismic resolution.

Once the first well is drilled, it is possible to identify and characterize the fractures and faults in the reservoir by analyzing the cuttings and the borehole data. We can characterize faults in the vicinity of the well, but not in the whole reservoir. These faults play a key role in the productivity and longevity of the project. One can also use the microseismicity generated during well stimulation or well cleaning to better understand the structure of the fault network located in the reservoir [6] but not to provide their physical parameters.

A Multi-Offset Vertical Seismic Profile (OVSP) or a 3D-VSP, which provide a better data redundancy, could be appropriate geophysical techniques to better identify and characterize these faults in a neighborhood of a few hundred of meters around the well (e.g., [7]. As the seismic waves cross the surface weathered zone (i.e., the near-surface zones presenting velocity variations) only once, and as the receivers located in the reservoir are close to the target, the seismic signal is generally more informative than the one from surface seismic, the frequency bandwidth is higher and the signal to noise ratio is often higher. Another advantage is that the area illuminated by the well seismic is better in the vicinity of the well due to a higher data redundancy. More precisely, the downgoing waves interact with the heterogeneities or faults around the well and therefore generate a complex scattered wavefield. The incident P-wave generates reflected P-waves and converted P-to-S waves (both reflected and transmitted), and the incident S-wave (often P-to-S downgoing conversion from shallower interfaces) generates reflected S-waves and converted S-to-P waves (both reflected and transmitted). This complex wavefield, recorded by the downhole receivers, is very informative for imaging and characterization purposes in the vicinity of the well.

Considering the standard processing techniques for VSP data (e.g., [8], as the receivers record both up and downgoing waves but also laterally incident waves and wave conversions, e.g., P-to-S and S-to-P, the generated and the recorded wavefield is more complex compared to the surface seismic. The crystalline body we consider here is also more complex than the OVSP in a more horizontally structured sedimentary body. In such a case, the classical and standard processing sequence, by separating up and downgoing waves, show some limitations due to several reasons, particularly when increasing the offset of the source. Due to the global velocity trend, increasing with depth, the long to very long offset (from offset equal to target depth up to several times the target depth), the direct P-wave, downgoing in the shallow depths, can propagate laterally when arriving at the well and sometimes even turn into an upgoing wave (diving waves, or refracted waves). For small or zero offset, wave separation still stands (except when lateral velocity variations are high above the target zone). For objects that are not nearly horizontal, such as faults, reflected waves may arrive laterally or with uncommon apparent velocity. The second reason is that when increasing the offset in sedimentary bodies, the scattered field for shear waves generally shows increasing energy, making it difficult to separate P and S waves. In addition, another issue may arise: the presence of P-to-S-to-P converted waves, for example, or other scattered waves of second order which can be energetic under certain conditions. In sedimentary context, is not uncommon to find 3rd order scattered phases, even 4th order in OVSP seismograms as soon as the offset is sufficient. Moreover, when imaging faults, diffracted waves due to point or linear scatters can be observed. The standard processing technique (wave separation, deconvolution and migration) does not take into account such phases, because the processing has been designed for primaries, i.e., for the first order scattered field. Of course, the 2nd order scattered field is less energetic than the first order, but if the S/N ratio is sufficient, the 2nd order scattered wavefield provides information and one can take advantage of it rather than trying to remove it or neglect it. The third reason is that the deconvolution of the upgoing waves by the downgoing waves is not an accurate process when increasing the offset as the downgoing raypath becomes different from the upgoing raypath. This issue can be important for target not very close to the well and in presence of lateral variations of velocity. As a comparison, the diffractions are considered as a noise for a classical processing approach whereas it carries useful information for FWI (e.g., [9], helping to accurately build an edge or discontinuous surface. In other words, the classical VSP data processing approach reduces the complexity of the recorded complex wavefield by separating the wave field to up and downgoing waves and often removes part of the information contained in the data. This approach provides a good result, but it is subject to limitations for specific and complex cases.

The recent sensitivity study performed on the granite where the data were processed by a classical approach [10] have showed the capability of this technique and pointed out some issues. During their synthetic sensitivity study [10] a systematic analysis optimizing the number of sources was proposed, and the authors showed that the VSP data have a high potential to detect faults in crystalline bodies. They also noted that, for some fault characteristics, even with an increased number of sources, some artefacts remain strong in the final migrated images, given interpretation ambiguities. They also pointed out that the faults with high dips (e.g., 70°) were more complicated to recover than the shallow dipping faults (e.g., 30°). In deep geothermal projects, especially in the granite context, the probability that the well crosses faults with dips around 30° is exceptionally low compared to highly dipping faults, and this defines an important limit of the classical processing approach. Other interesting works have also been conducted on the OVSP data of Soultz-sous-Forêts recorded in GPK1 and EPS1 wells using the standard approach (e.g., [11]. They used an adapted classical approach for crystalline rocks. The authors of [11] introduced a 3D parametric separation which treated the seismic wavelength in 3D, improving the 3D structural model, especially the faults, and the reservoir knowledge. This study was conducted for VSP 0-offset (<200 m of source-offset) and small sources-offset distance (<600 m), and for shallower targets, located at 2000 m for the EPS1 well and 3000 m for GPK1. A specific data processing queue should be adapted for the crystalline studies (e.g., see also [12]).

Full Wave Inversion (FWI) can deal with primaries, secondary or even higher order scattered phases, including diffractions and multiples issues (water layer or internal multiples). Its informative wavefield can be extracted to constrain the imaging purpose [13,14] We can therefore detect and even characterize faults located hundreds of meters from the well and not only the faults crossing the well. As the OVSP data are acquired in different azimuths, and by a different offset according to the receivers, the target is observed in different and complementary incident angles. The FWI technique applied to well seismic data (OVSP) should provide a better underground image and therefore better identify, delineate and characterize the faults in the well neighborhood. This approach has also been assessed using OVSP data in difficult contexts as subsalt imaging (e.g., [15]. Nevertheless, this technique remains underused, even in the sedimentary context for well seismic data, because the FWI requires building physical models with an adequate rheology (elastic including anisotropy and attenuation) before applying the inversion. Effectively, for well seismic, data redundancy is weak, and the starting model should be sufficiently close to the true one to start the FWI process. In other words, the starting model should reproduce qualitatively the recorded wavefield and the travel times of the major phases should be correct (to avoid the cycle skipping issue). Additionally, it requires a good S/N ratio because the technique is sensitive to noise. All these additional necessary preprocessing tasks are time consuming, which remains the major constraint, especially in industrial applications. As a consequence, data-driven techniques (data processing) are often preferred. Nevertheless, in the industry, FWI is used classically when working with the lower frequencies in order to improve the velocity model used for 3D migration, particularly in difficult geological contexts (e.g., [16]).

Previous works have demonstrated the added value of the FWI method for OVSP data [13,17–21] These studies were performed for oil and gas targets with high challenges, for instance, for subsalt imaging, velocity estimation for pore pressure, sub-basalt imaging, CO_2 sequestration and others, and authors have obtained better results compared to the surface seismic and compared to VSP data processed with a classical approach (e.g., [13]. In Barnes and Charara [13] FWI was applied to North Sea VSP data and recovered accurately the gas reservoir, which is confirmed by the well, whereas a standard VSP processing gave a wrong imaging result, despite the sedimentary context.

More recently, for geothermal purposes and in the Soultz-sous-Forêts geological context, where a granitic basement is located at 1.4 km depth, OVSP synthetic data have been successfully inverted and faults were well identified and characterized around GPK's wells using the FWI and Full Wave Modeling (FWM) [22,23] The obtained results confirm the high imaging potential for this inversion technique and open a new perspective for its geothermal use.

In this paper, we perform a synthetic study for fault delineation and characterization in the crystalline basement using FWM and FWI. No real data will be shown. The study was done in 2D assumption because we need to perform several experiments of different physical models, including different fault geometries and features. A 3D sensitivity analysis needs much more CPU time. The main objective of this paper is to test the ability of the FWI to detect and delineate faults and further characterize them in terms of velocities and density. We aimed first to check the applicability of the FWI in the hard rock. To provide an accurate sensitivity analysis, we have used synthetic experiments, because we can easily control the affecting parameters, e.g., faults characteristics and acquisition geometry, frequency and the receiver's location, etc. We need to know our data exactly to analyze the FWI results accurately and quantitatively. We added noise to the synthetic data to assess the robustness of the method in noisy environments. Nevertheless, we decided to perform the entire experiment without adding noise, except for experiments related to the noisy test where noises are added to the data, to ensure separation of the effects and assessing the FWI capabilities. Important issues such as the effects of a wrong starting model, of anisotropy or attenuation in the data, are not addressed in this paper.

With these purposes in mind, a complete sensitivity study is shown and discussed, including both noise-free data and noisy data. Different fault features as well as acquisition geometry were tested, and the results summarized. We have first introduced the method and tuned the inversion with the adequate parameters (Section 3), examining for instance the effect of the polarization, of the correlation lengths, etc. In a second part (Section 4), we have studied the effect of the acquisition geometry on the inversion results, including the intershot distance, number of shots used in the same run, the maximum offset, etc. Finally, in a third part (Section 5), we have studied the effect of the fault characteristics on the inversion results: fault thickness, dip and distance to the receivers, as well as the P- and S-waves velocities and density contrasts, including the multi-faults experiment and the robustness of the method with respect to the seismic noise.

2. Geological Setting

We performed this study inspired by the geological context of the geothermal site of Soultz-sous-Forêts. This site belongs to the Upper Rhine Graben (URG) and is characterized by an important thermal anomal [24] It is intensively documented by several papers and works, which in the early days studied the site for its oil and gas potential [25] but more recently for its geothermal potential for both electricity and heat co-generation [26,27].

A thick sediment pile overlies the crystalline basement which is located at 1400 m depth beneath GPK1 headwell. Sediment ages filling the graben range from Quaternary to Permia [28] the bedding structures dips shallowly to the south and east. The graben strikes N–S, to NNE–SSW in the northern part of the studied area. Several faults and fractures analyses have been performed on the continuous sample borehole core and or from borehole data in the different geothermal deep well in Soultz, for instance EPS1 and GPK1 to 4. They showed that the fractures are mainly dominated by NNW–SSE strike orientation [29,30] The current stress regime is strike-slip with a NW–SE compression direction in accordance with the general trend in Western Europe, and with the observed faults and fracture [31].

3. Methodology

3.1. Seismic Imaging Techniques for Well Data

Well seismic data are used for different purposes. For instance, VSP data can be used to estimate a depth-time relationship and calibrate the velocity model for surface seismic data in order that seismic horizons are located at the correct depth. Walkaway or walkaround seismic data are used for anisotropy estimation and walkaway for AVO analysis (Amplitude Versus Offset). Imaging using well seismic data is less common because the conventional imaging tool, the migration, has to overcome the problem of lack of data redundancy compared to surface seismic. It is an important issue even for dense 3D-VSP. Moreover, for the fault imaging goal, the conventional technique of downgoing and upgoing waves suffer from some limitations when increasing the offset. In principle, the FWI method does not suffer from the above limitations except the weak redundancy which can be overcome by using additional constraints in the inversion process. These constraints can derive from additional observations such as polarization (e.g., [32]) in the data space or for example, the spatial correlation (e.g., [13,33,34]), or the inter parameter correlation in the model space.

3.2. The Inverse Problem Applied to Seismic Data

The inverse problem can be expressed as the minimization of a misfit function as stated by Tarantola [35] for least squares, this function is a scalar function defined over the model space as:

$$S(\mathbf{m}) = \Delta\mathbf{d}^{\mathrm{T}}\mathbf{C}_{\mathrm{D}}^{-1}\Delta\mathbf{d} \qquad (1)$$

where $\mathbf{m}$ is a model, $\Delta\mathbf{d} = \mathbf{g}(\mathbf{m}) - \mathbf{d}_{obs}$ are the residuals with $\mathbf{d}_{obs}$ the observed data, and $\mathbf{g}(\mathbf{m})$ the synthetic data obtained by resolution of the wave equation, and where $\mathbf{C}_{\mathrm{D}}$ denotes the covariance matrix over the data space. This matrix can be not diagonal as when using the polarization constraint [34] i.e., the constraint given by the azimuth and incidence of the seismic waves. The misfit function measures the discrepancy between observed and calculated seismic data, i.e., between amplitudes of the signals for each trace. Because of the non-linearity and the complexity of the forward modelling, the misfit is reduced iteratively using a local method based on derivatives [35] The descent algorithm is based on the conjugate gradient method proposed by Polack and Ribière [36] The step optimization is defined along the conjugate gradient direction as in Crase [37] FWI software following these procedures has been developed continuously by GIM-labs for more than 15 years.

3.3. Modelling the Seismic Wave Propagation

The direct problem associated to the present fullwave inverse problem is the propagation of (visco) elastic waves in an (an) isotropic medium. As the imaging target, i.e., the fault network, is one or several 3D objects in a 3D nearly homogeneous medium, we aim at achieving a 3D inversion. However, the sensitivity analysis requires numerous inversions to check all the parameters separately and a 3D inversion cost tens of hours of CPU time (using our resources) even using a well-designed parallelized seismic modelling code. As a consequence, we will consider 2D wave propagation for the synthetic experiments used in the sensitivity analysis (see Section 5).

This assumption allows us to check a major part of the parameters of the sensitivity analysis. However, the main drawback of running 2D seismic modelling rather than 3D modelling is that the fault is always perpendicular to the propagation plane. In other word, the azimuth of the fault (the dip direction) corresponds to the propagation plane. In a 3D world, we need more shots with different azimuths to constrain the 3D FWI. A 3D sensitivity analysis of the dependency to the azimuth of the fault dip has not been performed due to the high cost in CPU time.

The finite differences method is based on the displacement formulation of the wave equation for the viscoelastic rheology (isotropic, VTI and HTI anisotropy). The spatial discretization is using a staggered grid, a 2nd order Taylor explicit scheme in time and 4th order differential operators in space. The FWM corresponding software has been developed continuously by GIM-labs for more than 20 years. It is parallelized (using OpenMP for domain decomposition and MPI for shots) and can be run on large clusters.

3.4. The Rheology Issue

The actual rheology is elastic for the synthetic FWI experiments in Section 5.1 and viscoelastic for the case of real seismic data FWI. When the data redundancy is weak as for borehole seismic, we need to extract the most information from the data. In order to achieve this goal, we must reduce all the additional noises: numerical, experimental and physical [14] The numerical noise is a trade-off between the calculation cost in time and the accuracy of results, reducing this noise is then easy but has a cost; moreover, this noise is not strongly structured [14] The experimental noise concerns the accuracy of the source and receiver location, the modelling of the source and receiver radiation pattern or the source time function. These parameters can be either better controlled or inverted during the FWI (for instance, the source time function). The physical noise is the most challenging one as it is strongly structured and may produce strong artefacts in results when the physics is not adequate [14] As a rule of thumb, all these noises should be less than the data noise in order to extract information from the data and reduce artefacts [14] Non additive structured noises have the largest impact on the results [14] and reducing the noise due to an inadequate rheology is thus the main issue. For example, using a simplified rheology as acoustic, even for marine data, implies a large artefact in the solution due to the wrong modelled AVO, as shown by Barnes and Charara [19].

Borehole seismic data generally exhibits a highly informative wavefield containing energetic 2nd or even 3rd order scattered waves [13] For example, downgoing P-to-S converted waves can convert back to P-wave after reflection on an interface. P-to-S conversions increase in amplitude according to the offset. Consequently, in most cases for borehole seismic with offset (OVSP, walkaway, 3D-VSP) and in a sedimentary context, once removing the downgoing direct P-wave, 80% of the energy is provided by the S-waves [13] For OVSP data in a crystalline context, S-waves are also present (as for instance the downgoing P-to-S wave converted at the top basement), sometimes more attenuated. The elastic rheology is then required.

Moreover, Equation (1) is based on amplitude, and thus seismic attenuation, often present in the data, should be considered as well, at least in the FW modeling (without inverting for attenuation parameters). Anelastic phenomena can be modeled by viscoelasticity at the scale of the seismic wavelength. One often defines a global Q-factor using a constant Q or a nearly constant Q (NCQ) model [38,39] or even a more general standard linear solid model (SLS). Charara and Barnes [40] have proposed modeling the attenuation by using two Q-factors, the Q_κ factor related to the bulk incompressibility modulus κ and the Q_μ factor related to the shear modulus μ. The first one is affected by fluids, in particular, liquid-gas mixing while the latter one is related to microstructures of the solid. In FWI, the viscoelastic parameter fields provide a poor information on the spatial resolution as the seismic attenuation is an integrative phenomenon but, when taking into account the seismic attenuation, we obtain a better resolution on other parameters (e.g., [41]).

The seismic anisotropy is another issue which is important in a sedimentary context. For crystalline rocks with a sedimentary cover, the anisotropy could be taken into account.

We did not address seismic attenuation or seismic anisotropy in the sensitivity study for the sake of simplicity. Of course, for real data, these seismic rheologies have to be taken into account (e.g., [41,42]).

3.5. Aim of the Sensitivity Studies

The first objective of the conducted sensitivity study is to evaluate the ability of the FWI to detect, delineate and characterize the fault zones in the granite. Some precisions about these goals:

- The fault detection allows to obtain qualitative information as in medical ultrasound technique;
- The delineation with correct localization is a quantitative goal aiming at understanding the fault network geometry;
- The characterization provides information about the physical properties of the various materials inside the fault zone (crushed, deposit, etc.) and the possible hydrothermal alteration of the granite in the fault vicinity. As such, after some calibration processes, the P- and S-wave velocities or other parameter fields could indirectly provide information about porosity, lithology or gas saturation for instance through a rock physic model (see [43–45] for CO_2 monitoring examples).

Two major sensitivity analyses have been carried out. The first one concerns the acquisition geometry and the second one concerns the fault characteristics.

We carefully investigated several acquisition parameters affecting the results of fault detection and characterization. The studied parameters are:

1. the shot number used in the same inversion run,
2. the intershot distance,
3. and the maximum offset.

We have investigated first the single shot problem for the understanding of the resolution power of the FWI. From that, we investigate the effect of this parameter on the final underground image reconstruction. In the first set of synthetic FWI experiments, the number of sources is varying from 1 to 17, while increasing the maximum offset and decreasing the intershot distance (as we are in 2D, this can apparently appear as a walkaway VSP experiment). For a second set of experiments the intershot distance is varying while the maximum offset is constant, quantifying the intershot distance effect alone.

After having defined the optimal acquisition parameters, we can perform the sensitivity analysis of the fault characteristics. We have considered constant velocity values (P and S) and density contrasts, according to the background, and we mainly investigated the effect of:

1. the fault thickness, from 5 to 50 m,
2. the dip, from 0 to 90°,
3. and the fault location according to the receivers, the fault crossing, or not, the well.

3.6. Tuning of the Inversion Algorithm and Definition of the Inversion Parameters

Our FWI software parameters are those usually used in the sedimentary geological domain, so we spent large time to define the right inversion parameters. Before using them in the granite context, we need to first choose adequate inversion parameters to maximize the likelihood of faults' detection, delineation and characterization. This step was done using several inversion experiments where their results were compared. The best inversion parameters are those which provide better results in term of fault imaging, and then minimize the difference with the true models.

Several inversion algorithmic parameters have been tested and validated. We have tested the data polarization, the physical parameter crosscorrelation, the physical parameter spatial correlation, the frequency content, and the antennae length (illuminated zone according to the receivers) effects. The three first are constraints in the FWI through the covariance matrices in the data space for the polarization and in the model space for both the parameter crosscorrelation and the spatial correlation (see [32,34]). Basically, we have tested the following parameter values:

Spatial correlation range: 20, 50 and 100 m, using the Laplace correlation function and stationary random field assumption. This means that, for each iteration, the perturbation of the parameter field values should be close in these ranges.

- Using the crosscorrelation between P- and S- velocities and density, or not. If used, the statistical relationship is applied between the inverted parameters; P- S-waves velocities and density. We consider a positive crosscorrelation, i.e., the physical parameter, e.g., the P-wave velocity and the density are varying together; as the P velocity increases, so does the density.
- Using data polarization constraint during inversion, or not. It is the same principle as the above crosscorrelation but in the data space, i.e., introduce through the covariance matrix on the data space. This corresponds to a crosscorrelation between the geophone components.
- Central frequencies of the source function: from 20 to 50 Hz for the same fault thickness. The frequency content of the data is an important issue when processing real data. In the present synthetic sensitivity analysis, the performed tests show that the frequency is not an issue when using synthetic data and the results are not impacted according to the dominant frequency. We therefore consider only the latter case with a central frequency of 50 Hz.

We also performed other experiments combining and mixing these parameters. The best results are obtained when used (1) the correlation length of 20 m, corresponding approximately to the fault thickness (or a bit smaller, depending on the considered fault thickness), (2) without using the polarization and (3) using the crosscorrelation between velocities and density (with positive coefficients of 0.9). We decided to use (1) and (2) and do not use (3) in order to evaluate independently the resolving power of FWI for the different physical parameter. Concerning the polarization, it is demonstrated from previous works (e.g., [32] that it provides better results in a sedimentary context, but this is not the case in the granite.

3.7. The Workflow Used for the Sensitivity Analysis

We run several synthetic FWI experiments with controlled parameters and part of the parameters are varying for the corresponding parametric study. All the used parameters including physical models, source location, receiver locations, the dominant frequency, the noise level, etc., are well known and accurately controlled. The general workflow is: (i) we chose the acquisition geometry parameters, for instance source and receiver locations, etc, (ii) define the target, i.e., the fault characteristics, (iii) generate synthetic data, and then (iv) invert these synthetic data and analyse the results.

The first step is choosing the geometry data acquisition where we define the parameters controlling the data acquisition experiment, for instance the number of sources, shot locations, deviated or vertical well, the receiver locations and their number, and so on.

In the second step, we define the characteristics of the target, i.e., the fault. We define its geometry, shape and area, its depth, thickness and dip. We should also define, for an elastic modelling, its physical parameters, for instance its P- and S- wave velocities and density values. These values depend on the contrast considered between the fault and the background, for instance 5, 10 or 15% (see Section 3.8 for more details on how these values are defined). From the fault definition and the background model, we build the 2D model named "true model" or considered model used in the next step (the FWM).

The last step before inversion, is to model the synthetic data according to the geometry acquisition and the source function (type, frequency), and the true model (including the fault) using the FWM (Figure 1). These synthetic data are calculated according to the maximum recording time. This time should be chosen large enough to record the data in the entire receivers including the time where the waves interact with fault. Once the synthetic data are calculated, we can use them as the observed data in the inversion process, directly for free noise experiments, or after adding noise for noisy data experiments.

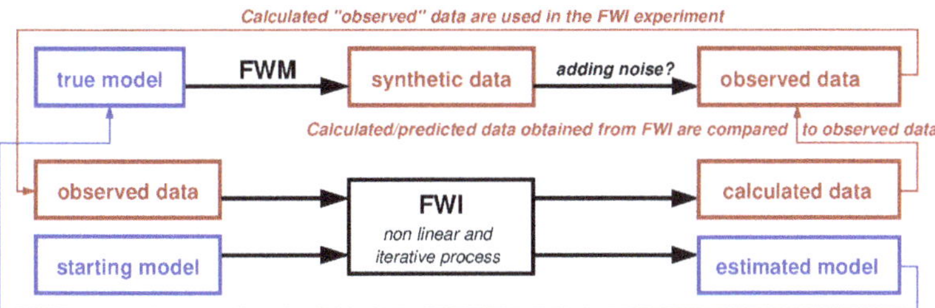

Figure 1. Workflow of the method used in the feasibility study. (1) From the true model (including the fault), we generate the synthetic data by FWM. (2) Add or not the noise according to the objective of the experiment. This data is then used in the inversion as observed data. (3) Start the inversion process from the starting model (which is the true model without the fault) and the observed data, and (4) once the nonlinear iterative FWI process has converged, we obtain (i) the estimated data that will be directly compared to the observed data, and (ii) the estimated models, which are directly compared to the true models. Here, the term "observed data" should not be confused with the term "real data" because we are dealing only with synthetic data.

Finally, we reach the inversion step and so the fault imaging goal. The objective in this step is to use the observed data obtained from the true model and the FWM in the previous step and invert these data to quantify the capability of the FWI for fault detection, delineation and characterization.

The workflow presented in Figure 1 summarizes the main steps during the FWM/FWI process.

3.8. Geological and Physical Models Building and Faults Modelling

In the frame of EGS Alsace project and ANR Cantare programs, a regional 3D structural model was built for Northern Alsace. It is built mainly by reprocessing the old vintage 2D seismic lines [46] which are focused on oil and gas exploration of the Tertiary layers. This geological model could show high uncertainties because at this time, and as the target is shallower than the geothermal reservoir, the used seismic parameters were not adequate to produce images of the underlying crystalline rocks. These uncertainties could mainly be found at the sediments – basement interface which can reach hundreds of meters, especially in the deeper part of the graben, for instance the eastern part. A total of five geological interfaces were modelled and included in this model: (1) "schistes à Poissons", (2) Tertiary unconformity, (3) top of Trias, (4) top of Muschelkalk and (5) top of Buntsandstein.

In order to build the P- and S-wave velocities as well as the density models from the structural model, we need initial values. For this purpose, we used existing physical, generally 1D, models. These models were obtained from previous studies and from borehole data. These models are good enough for our model building. Once the physical models were chosen, we used them jointly with the stratigraphical 3D model to generate a 3D P-wave, S-wave velocity models and a density model. We defined these parameters in the top and at the bottom of our stratigraphical model, and applied a linear gradient in-between. We show on Figure 2 cross-sections of P-wave and S-wave models and density model extracted from the complete 3D models.

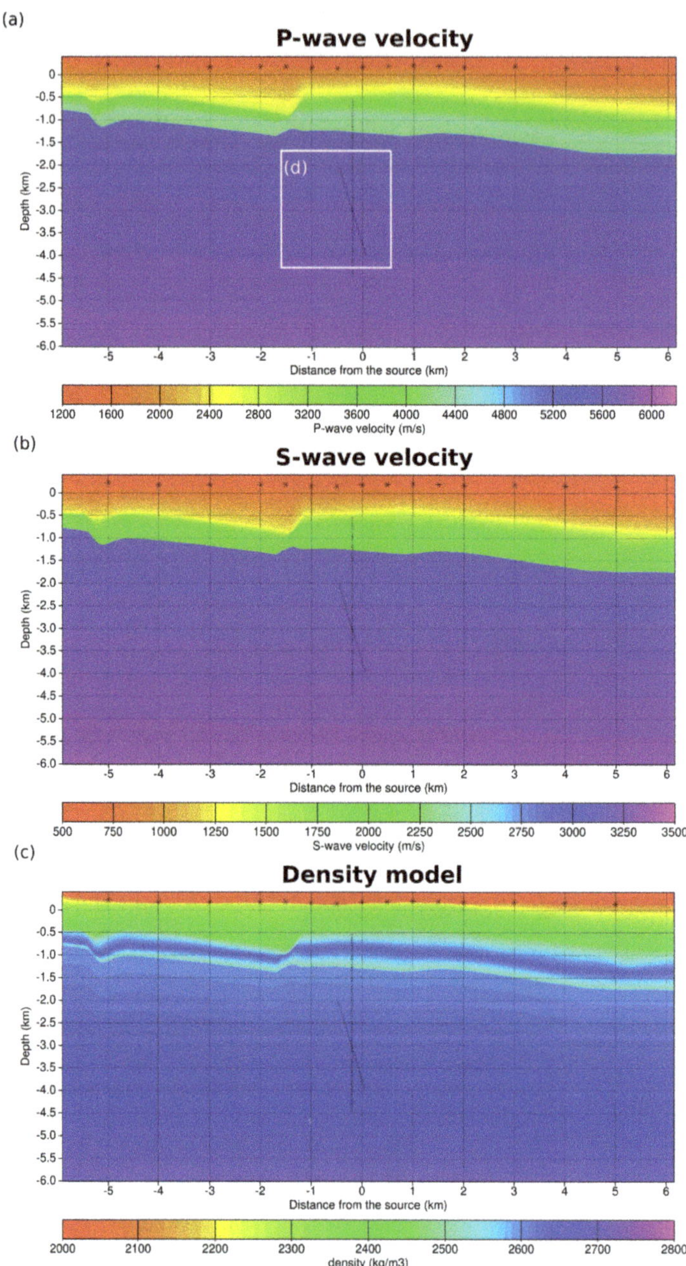

Figure 2. Physical models used to perform a sensitivity study of experiments. (**a**,**b**) the P-wave and S-wave velocity models, and (**c**) the density model. (**d**) shows the target area around the receivers where the results will be shown. The inverted triangles are the receivers located in the well and the stars are the source locations on the earth's surface. Note that the inter-shot is variable, 500 m around the well head to 1000 m far away the well head.

The sources are placed on the topographic surface. The waves propagate down and no reflected and or transmitted waves occur above the surface. As a complete wave field is modelled and the full seismic propagation equation used, several seismic waves have been considered, for instance up and downgoing waves, converted waves and even multiples.

To define the characteristics of the faulted zone, we carefully analysed the available borehole data for the GPK1, 2, 3 and 4 wells. The objective is to define their average thickness and their physical values, i.e., P- and S-wave velocities and density. We showed in Figure 3 the middle part of the GPK1 where we noted the interpreted faulted zones by the red arrows.

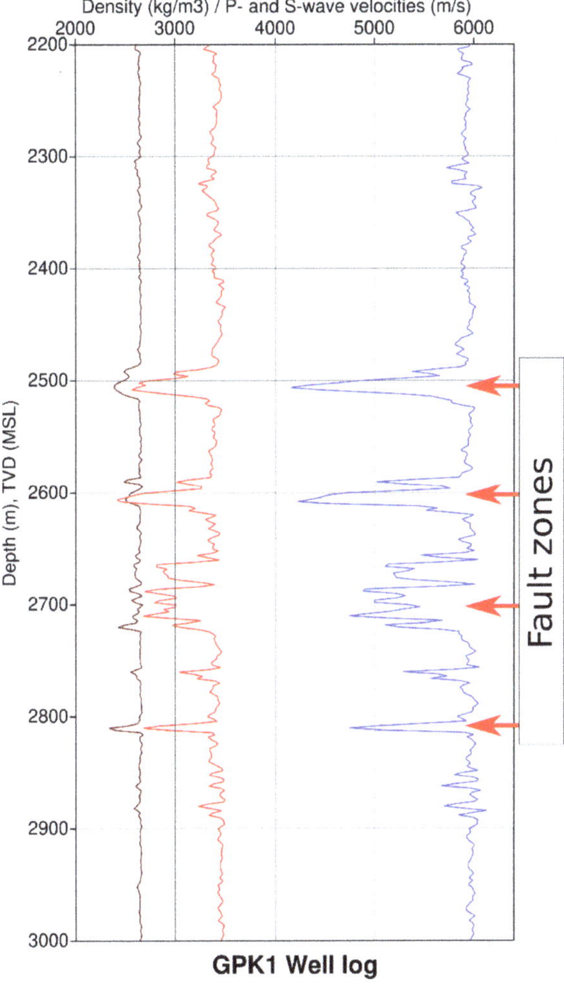

Figure 3. Middle part of the borehole logs for GPK1 (granite) where the blue profile is the P-wave velocity, the red profile is the S-wave velocity, and the brown profile is the density values. The red arrows show the location of some interpreted fault zones. A good correlation is observed between P- and S-wave velocities and density (especially in the granites). These borehole data are measured on a well crossing the dipping faults. The apparent vertical thickness of these fault zones varies between 15 and 80 m. If we consider a fault dip of 75°, we found that their real thickness varies between 5 and 25 m.

During our fault analysis performed on the GPK1 and GPK2 well log data (e.g., Figure 3) we noted that the apparent fault thickness varies between 15 and 80 m. If we consider a dip fault of 75°, which is a realistic value for GPK1 and the closer boreholes [47] we recompute the real fault thickness which varies between 5 and 25 m. More detailed fault and fractures data analysis can be found in Dezayes et al. [29] and Sausse et al. [30].

From the amplitudes, we also computed the amplitude decreases of the P- and S-wave velocities and density values according to the background, i.e., according to the mean values. From the well data of GPK1 and GPK2 (e.g., sonic and neutron density), the values decreased by 10% to 25% for P-wave velocity, between 10% to 20% for S-wave velocity and between 5% to 10% for densities (Figure 3).

Consequently, for our synthetic experiments, we considered an average fault characteristics value of the real values. We consider a fault zone with the following characteristics: (i) the faulted zone thickness is 30 m and (ii) the decrease of the physical values is 20% for the P- and S-wave velocities and 7% for density. We show on Figure 2 the reference physical models from which the fault contrast is computed. These models are the background physical models.

4. Sensitivity Analysis of the Parameters Describing the Acquisition Geometry

We consider an experimental setup which fits as much as possible the real OVSP setup. The sources are located on the Earth's surface, and the locations of the well-head as well as the well trajectory (where geophones are located) are realistic. In addition, we have used the seismic velocities and the density models extracted from the 3D models.

4.1. Single Shot Problem

We have started our analysis from the FWI results using only one source, located at zero-offset. This numerical experiment helps to understand the resolution power of the FWI. The initial P-wave velocity and S-wave velocity models used as well as the location of the fault are shown in Figure 2 (the density model is not shown). We have used the following acquisition geometry parameters:

- The well is vertical;
- The source is located at 0 m offset, on the topographic surface (at 175 m MSL), a pressure, Gaussian first derivative and centred at 50 Hz;
- The antenna is made of 198 2C geophones located from 550 m depth down to 4500 m and spaced every 20 m;
- The observed data are free of noise.

The fault characteristics are:

- Vertical fault extension is 2000 m;
- Fault cross the well at 3000 m MSL;
- Fault dip is 75°;
- Fault thickness is 30 m;
- Contrasts are −20% for P- and S-wave velocities and −7% for density.

We have also used the algorithmic parameters tuned in Section 3.6. These parameters have been set as follows: (i) FWI performed without polarization, (ii) spatial correlation range is 20 m, (iii) the main frequency is 50 Hz and (iv) without using the inter-correlation between the physical parameters in order to better understand the sensitivity of each parameter separately.

Besides the interest of the single shot problem for the understanding of the FWI method resolution power, we think that it is also remarkably interesting to study this case on a practical point of view. Indeed, due to restricted financial support compared to the oil and gas industry, geothermal project supervisors tend to choose this single-source data acquisition layout. Thus, it is particularly important to quantify the recovered physical fault images especially in the granite geological context.

We compare on Figure 4 seismograms obtained by FW modelling for the model with and without the fault. Recall that this experiment is done without adding noise. If we consider the standard processing, the reflected phase coming from the fault has a slope corresponding to the downgoing field, which leads us to a wrong interpretation. The apparent velocity between the downgoing waves and the reflections arising from the fault are very close, creating an ambiguity in their origin. The FWI method overcomes this problem by dealing with the full wavefield without wave separation.

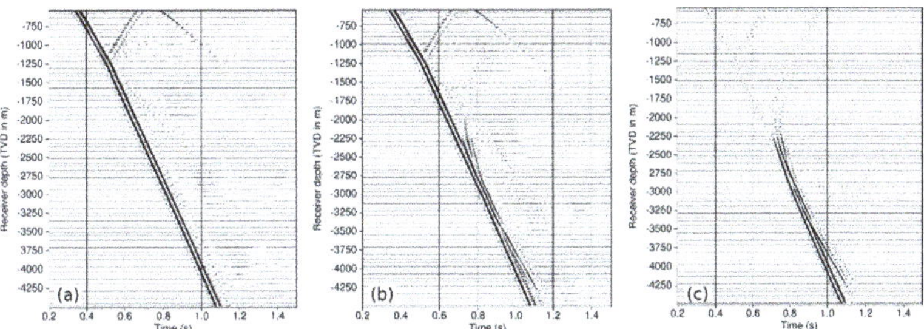

Figure 4. Seismograms showing the seismic effect of the fault. (**a**) synthetic VSP data of the model without the fault. (**b**) synthetic VSP data including the fault. We can mainly identify the scattered fields coming from the fault and crossing the first arrivals at ~−2400 m depth. (**c**) difference between (**a**) and (**b**). The acquisition geometry is identical to that showed in Figure 2, where we show only the data for source located at 0 m offset.

We show on Figure 5 the inversion result for a zero-offset single-source VSP experiment after 60 iterations. The result is poor. FWI has recovered a small part of the P-wave velocity field but with artefacts mainly in the upper part of the antenna leading to interpretation difficulties. For the S-wave velocity field, FWI has recovered a small section in the lower part of the antenna, but several artefacts can also be observed. For density, as expected, the result is bad, showing that this parameter is weakly sensitive to the fault and cannot resolve it when reflection redundancy is weak. From the one-source shot experiment, we conclude that in the granites and using the described acquisition geometry and the fault characteristics, we cannot image the modelled fault accurately enough for a reliable interpretation.

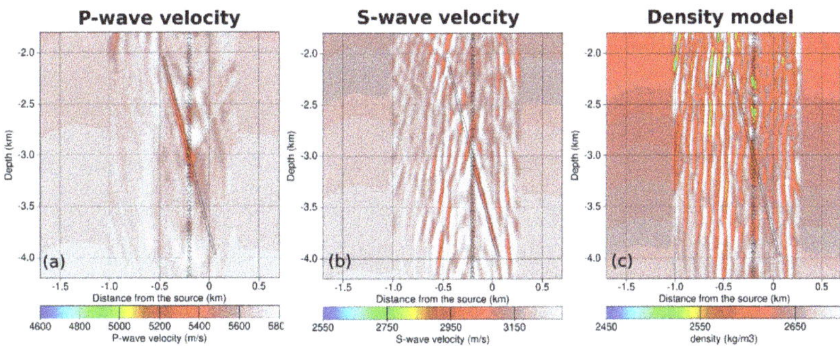

Figure 5. The physical models retrieved from FWI for one shot at zero offset. (**a**) P-wave velocity model shown for the interest area around the receivers (see the white box in Figure 2c). (**b**,**c**) recovered S-wave velocity and density models, respectively. The trace of the simulated fault is shown with black polygons.

We understand from this experiment that in the granite geological context, one zero–offset shot is not sufficient to detect, delineate and characterize faults having the described features.

4.2. Multi-Shots Problem

Several synthetic experiments have been performed for different data acquisition geometries (Table 1). The goal is to quantify the effects of (i) the maximum offset, (ii) the number of sources and iii) the intershot distance effect. Experiment 1 is that discussed in the previous section for a one-source problem. The idea is to compare the experiment results, for instance, to compare the results of experiments 5, 6 and 7 to quantify only the effect of the inter-shots, experiments 6 to 9 to quantify only the effect of the maximum offset and experiments 7 to 10 to understand and quantify the effect of the maximum offset.

Table 1. Geometry data acquisition parameters used in the sensitivity study for fault detection, delineation and characterization.

Expe.	Shots	Shot Offsets [km]	Max Offset [km]
1	1	0	0
2	3	−0.5, 0, 0.5	±0.5
3	5	−1, −0.5, 0, 0.5, 1	±1.0
4	7	−1.5, −1, −0.5, 0, 0.5, 1, 1.5	±1.5
5	5	−2, −1, 0, 1, 2	±2.0
6	9	−2, −1.5, −0.5, 0, 0.5, 1, 1.5, 2	±2.0
7	17	−2, −1.75, 1.5, −1.25, −1, −0.75, −0.5, −0.25, 0, 0.25, 0.5, 0.75, 1, 1.25, 1.5, 1.75, 2	±2.0
8	11	−3, −2, −1.5, −1, −0.5, 0, 0.5, 1, 1.5, 2, 3	±3.0
9	13	−4, −3, −2, −1.5, −1, −0.5, 0, 0.5, 1, 1.5, 2, 3, 4	±4.0
10	15	−5, −4, −3, −2, −1.5, −1, −0.5, 0, 0.5, 1, 1.5, 2, 3, 4, 5	±5.0

The objective is to change the number of shots used in the same inversion run as well as the intershot and the maximum offset to quantify the FWI ability for faults' detection and characterization. For instance, comparing the results of experiments 5, 6 and 7 to quantify only the effect of the inter-shot spacing, which is 1000 m for experiment 5, and 500 and 250 m for 6 and 7, respectively. The maximum offset of 2000 m remains unchanged for these three experiments. We can also compare the results of experiment 6 and 9 to quantify the effect of the maximum offset alone, which is 2000 m for the experiment 6 and 4000 m for experiment 9, where the inter-shot spacing of 500 m remains unchanged around the well.

Experiment 2 used only three shots distant by 500 m, which is the smallest maximum offset used after the single-shot zero-offset experiment. From experiments 2 to 7, we changed and increased the number of shots, the intershot distance as well as the maximum offset to reach a maximum offset of 2000 m. Experiments 7 to 10 will help us to understand and quantify the effect of the maximum offset. In all these experiments, we used the starting physical models shown in Figure 2 and the fault features are those used in the single-shot problem (Section 4.1).

We show in Figure 6 the results summary of experiments 2, 4, 6, 8 and 10, showing the recovered P- and S-wave velocity fields as well as the recovered density model. In Figure 6a, we show the retrieved models for experiment 2 (Table 1) where only three shots are used with a maximum offset of 500 m. The FWI has partly recovered the upper part of the fault in the P-wave field and density but less in the S-wave field. We note that in the P-wave field, the shape of the fault is nearly completely recovered. The physical values remain lower than those of the real model, however. The lower part of the fault is not well resolved for P velocity because beside the direct P, the only phase is the downgoing P-to-S converted wave. We note also that several significant artefacts are present that would lead to wrong geological interpretations.

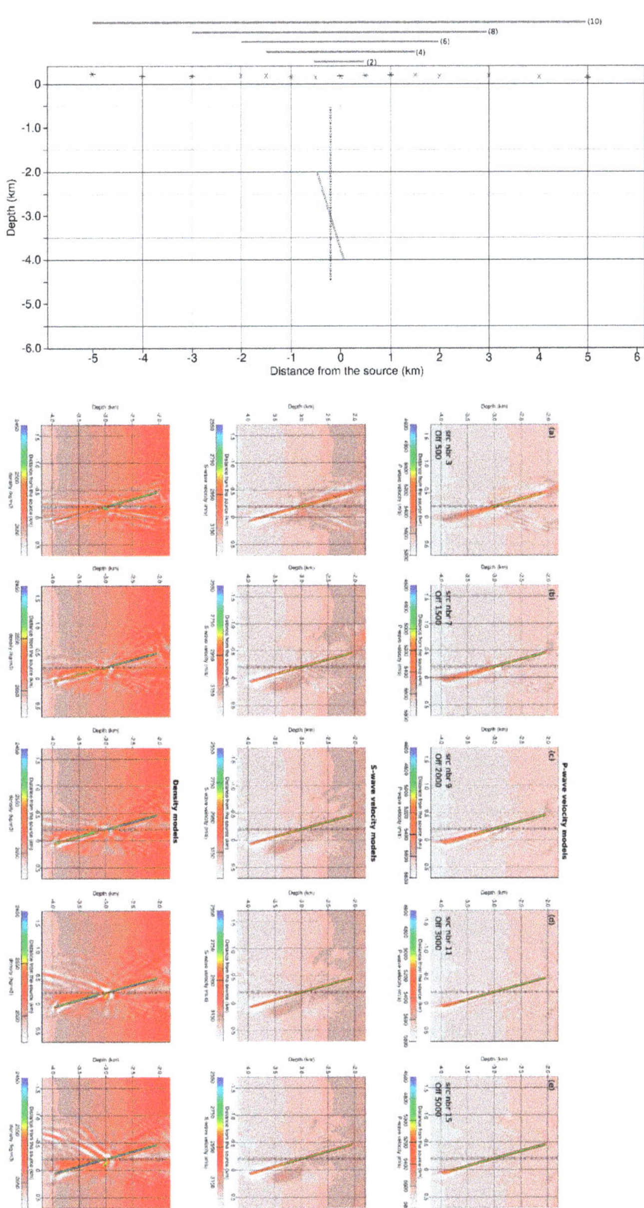

Figure 6. In the top panel, a simplified sketch showing the acquisition geometry used for each experiment is indicated by the number in-brackets. The modeled physical models remain unchanged throughout the different experiments whereas the number of sources changes. In the bottom panel, the physical models recovered from the sensitivity study of the acquisition geometry. At the top, we show P-wave field, in the middle S-wave field and at the bottom the density model. According to the Table 1, we show the results obtained from experiments 2 (**a**), 4 (**b**), 6 (**c**), 8 (**d**) and 10 (**e**). The inverted triangles are the receivers, and the black polygons follow the fault trace. The results are shown only in the target area.

When we increase the number of shots up to seven and the maximum offset to 1500 m (Table 1 and Figure 6b), the fault is better resolved. The upper part of the fault is better delineated for the three physical parameters. The lower part of the fault remains however not well resolved (better for the S-wave field due to the downgoing P-to-S converted phase). The artefacts remain relatively significant for the three parameters (Figure 6b).

In Figure 6c, we show the retrieved physical models for experiment 6 (Table 1) where nine shots and a maximum offset of 2000 m were used. The retrieved models, compared to experiments 4 and 2, are better resolved, especially for the lower part of the fault. This lower part is less revolved compared to the upper part, and the inverted physical values for the three parameters show underestimated contrasts. We also note the significant improvement of these results, for instance the artefacts present in experiments 1, 2 and 4 are attenuated, except for the density model as expected, because the density is always less resolved then P- and S-wave fields for borehole FWI context.

When increasing the maximum offset to 3000 m (experiment 8) and the number of shots to 11 (Figure 6d), the estimated physical models are improved. These improvements are higher for the P velocity field and the density field than for the S velocity field. Nevertheless, the retrieved density model still exhibits important artefacts.

The last results (Figure 6e) were obtained from experiment 10 (Table 1). The number of shots is 15 and the maximum offset is 5000 m. The fault is accurately retrieved, and the obtained values are compared to those of the real model, except for the lower fault extremity where the values are underestimated. This is mainly true for the P- and S-wave velocity fields, but not for the density. The density model is better recovered except for the area crossing the well, where some artefacts remain visible.

To conclude, the five experiments studied (e.g., experiments 2, 4, 6, 8 and 10) and documented in Table 1 reveal that in the crystalline basement and for a 2D model and seismic modelling):

- Three shots allow the detection of the fault but are not enough to delineate the fault and characterisation is not reliable. The delineation is not complete and not precise, and its physical values are different than the real modelled values.
- Even when using seven shots, fault delineation is not completely successful if the maximum offset is less than 1500 m.
- The fault is well delineated using nine shots or more with the maximum offset of 2000 m or more (to be compared to the target depth, in fact, the rule of thumbs is: maximum offset should be around the target depth).
- Increasing the number of shots in the same inversion is not necessarily the best way to go. For instance, the results are improved from experiment 5 to 6 but not between experiments 6 and 7, where the maximum offset of 2 km is the same for both. This means that the included model-part during the inversion which affects really the data is the same for both experiments. Even though the model complexity is the same between these experiments, we improved only the results of experiment 5. We also remember that parameters affecting the data, for instance attenuation, noise level (here free noise), diffractions, etc., are the same.
- It is important to make a balance between the number of shots and the maximum offset which should be used to accurately delineate the fault.
- The lower part of the fault is always less resolved as the downgoing scattered field has only the converted P-to-S converted wave while the upgoing scattered field is made of two phases, the reflected P-wave and the reflected P-to-S converted wave.

Note that the inter-shot spacing of the experiments discussed above is 500 m, while the number of shots and the maximum offset is varying. The shots at large offsets clearly provide more information through the scattered field. This is not the case with real data as the shots at large offsets exhibit low S/N ratios.

We would like also to quantify the effects of the number of sources for a constant maximum offset. We set the maximum offset to 2000 m and consider various numbers of shots. The number of shots of experiments 5, 6 and 7 (Table 1) are respectively 5, 9 and 17.

This means that the inter-shot spacing is different between these experiments, from 1000 m for experiment 5, to 500 m for experiment 6 and down to 250 m for experiment 7.

Slight or non-negligible improvement can be observed between experiments 5 and 6 in all recovered physical parameters. We recovered better the lower part of the fault in experiment 6. This is the direct effect of the number of shots and the intershot distance. Where a non-negligible effect is observed between the results of experiments 5 and 6, no visible effect was observed between the results of experiments 6 and 7. The reason is that for our target, the minimum inter-shot spacing is already reached in experiment 6 and even if we increase the shots number, and decrease the inter-shot distance, the imaging results are not improved. For real noisy data, a smaller inter-shot spacing distance and more sources should be needed, depending strongly on the noise level and the depth of the target. For a real case, a sensitivity analysis could be performed including different noise types and amplitudes to quantify their effect and then choose the optimizing acquisition geometry.

We conclude from these experiments that the number of shots and the inter-shot spacing are both important. It is not appropriate to use several sources with kilometric inter-shot spacing or inter-shot spacings smaller than 250 m (for a 3 km depth target fault and data main frequency of 50 Hz). We observed that using a small number of shots does not allow us to recover accurately our target fault, especially its lower part (e.g., Figure 6a,b), but also that using more shots with non-adequate inter-shot spacing do not improve the results (e.g., experiment 7, Table 1). We should define an appropriate distance from a sensitivity analysis, which could change according to the depth, thickness, dip, azimuth, and so on, of the target. According to the faults which could be met in the granite of Soultz-sous-Forêts, the best parameters deduced from the sensitivity study are: (i) the optimized number of shots is 9, (ii) the intershot distance is 800 m, and (iii) the maximum offset is 4000 m.

We recall also that these conclusions hold for noise-free data and for a 2D seismic modelling. For noisy data, the number of required shots would increase. And for a 3D model with azimuth coverage, the number of necessary shots N should be around $\pi n^2 / 4$, where n is the number of shots needed for 2D.

5. Sensitivity Analysis of the Fault Parameters

We now use the best acquisition geometry parameters determined in the previous section and study the seismic effect of the fault features: (i) the fault geometry (length, dip, distance from the receiver) and (ii) the fault physical characteristics (P- and S-wave velocities, density contrasts and thickness). The thickness of the fault zone is a geometrical parameter but as this parameter is related to the ability of the FWI to retrieve the true values of the physical parameters inside the fault zone, we consider it as a physical parameter. The ability to retrieve the true values of the physical parameters (fault characterisation) is also related to the main frequency of the seismic data.

5.1. Fault Geometry Parameters

We consider a fault with the same features as that described in the single-shot problem (Section 4.1). We vary the fault dip ($0°$, $30°$, $60°$, $75°$, $-75°$ and $90°$)) and the fault distance from the well: fault crossing the well, or horizontally shifted by 600 m or 1000 m from (Figure 7). The same inversion parameters were used: 20 m for the spatial correlation length, and without polarization. Concerning the correlation between the physical parameters contrasts, Vp, Vs and density, we consider two cases: (i) a strong correlation, i.e., crosscorrelation coefficients of 0.9 for the three couples of parameters) and (ii) a moderate to strong correlation, i.e., 0.8 for the (Vp, Vs) couple and 0.6 for the two other couples (Vp, density) and (Vs, density). We used the best geometrical data acquisition recovered in the previous sensitivity study (Section 5.2), namely (i) 800 m for the inter-shot spacing, (ii) maximum offset of 4000 m, and (iii) eleven shots located at the following offsets: -4000, -3200, -2400, -1600, -800, 0, 800, 1600, 2400, 3200 and 4000 m (Figure 7).

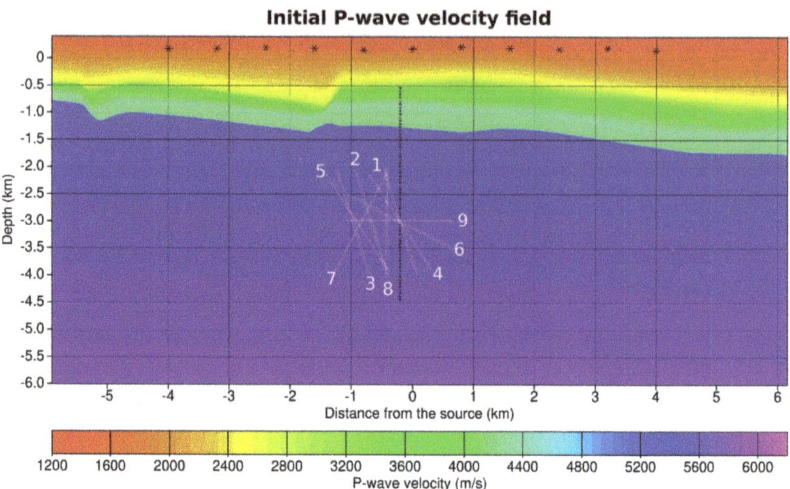

Figure 7. Different faults used in the sensitivity analysis with different dips shown on the P-wave velocity model. The center of the fault is at 3000 m depth. The faults are: (1) 75° dip, (2) 75° dip, shifted 600 m to the left (the lower part of the fault is about 200 m from the receivers), (3) 75° dip, shifted 800 m to the left, (4) 60° dip, (5) 60° dip, shifted 700 m to the left, (6) 30° dip, (7) −75° dip, shifted 650 m to the left, (8) vertical, shifted 200 m to the left and (9) horizontal, crossing the well at 3000 m. The stars show source locations and the reverted triangles, the receivers.

We have summarized in Figure 8 the sensitivity to fault dips for the faults crossing the well and the receivers. We can observe that the fault dip is not an issue even at high dips (>75°). Generally, the higher the fault dips the better the FWI retrieves the whole fault shape and contrasts. This is true except for a horizontal fault, where the fault parts close to the receivers are accurately recovered, but not the faraway parts (Figure 8d). Nevertheless, the physical parameters for the parts closest to the receivers are accurately recovered. We note that for high dip faults, i.e., >60° (Figure 8a,b), we recovered accurately the entire upper part of the fault, but not completely the lower part, as in the previous experiments (see Section 5). This is due to the scattered field, as already explained. The worst cases are for shallow dipping faults, probably between 10° to 40° more or less (e.g., Figure 8c). We note also the presence of some artefacts in the recovered density field, but again we can observe that for a horizontal fault (Figure 8d) and for high dip faults (e.g., Figure 8a), the retrieved density field is better than the retrieved P- and S-wave velocity field.

In a second step, we have considered the case of the distance of the fault from the well. What is the maximum distance where a fault could be accurately imaged or even detected? The results of the FWI for the faults 2, 3 and 5 are shown in Figure 9 and for the faults 7 and 8, in Figure 10. The FWI can accurately detect and delineate the fault even when it is far away from the receivers. The maximum distance varies from 500 m for the lower part of the fault to 1000 m for the upper part (e.g., Figure 9). This maximum distance decreases when the dip decreases. When the fault dip decreases from 75° to 60°, the farthest part of the fault is not detected (compare Figure 9a,c). When the fault dip decreases, we delineate with more accuracy the closest part than the furthest part, this can be noticed when comparing faults 2 and 5. This is expected, because when increasing the fault dip, we increase the distance between the upper part of the fault and the receivers, for the same fault length. We note also that the density field is recovered with some artefacts, and its physical values were recovered accurately. We recall that the inverted data are free of noise, and we expect these maximum distances to be smaller with real noisy data.

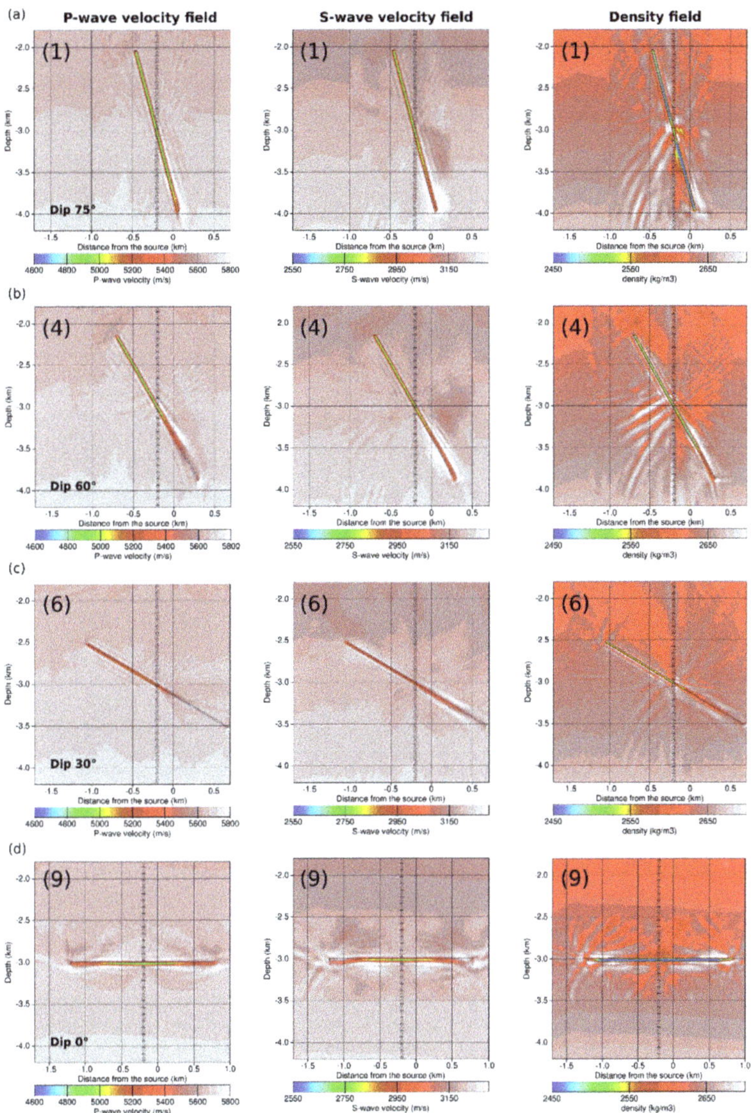

Figure 8. Sensitivity analysis results obtained for dip faults crossing the well at 3000 km depth. We show P-wave velocity (left), S-wave velocity (center) and density (right) fields for fault dips of (**a**) 75°, (**b**) 60°, (**c**) 30° and (**d**) 0°. These results correspond to faults 1, 4, 6 and 9 showed in Figure 7. We show only the interest area showed by the white rectangle in Figure 2c.

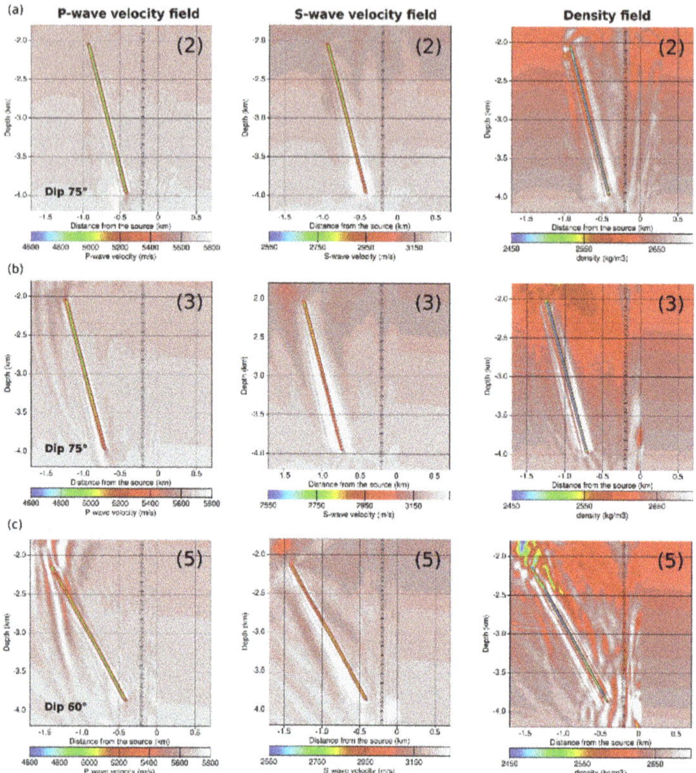

Figure 9. The retrieved P-wave velocity field (left), S-wave velocity field (center) and density field (right) for faults (**a**) 2, (**b**) 3 and (**c**) 5 (Figure 7). The faults 2 and 5 are 200 m far from the closest receivers whereas fault 3 is 500 m far from the closest receivers in the lower part of the fault. We show only the target area showed by the white rectangle in Figure 2c.

The conclusions of the previous section concerning the fault dip remain true. We observed that the higher the fault dips, the better we can delineate the fault and recover the contrasts in the physical parameters. This is the case even for a vertical fault (Figure 10b). In this case, the fault is accurately delineated, and the P-wave velocity and the density fields are recovered with a good precision. The S-wave velocity field show a correct delineation, but the contrasts are not well estimated in the deeper part of the fault. The reason is that in this part, the P-to-S reflected (from right side sources), or the transmitted converted S waves (from left side sources) provide information only at a few receivers (the deepest). Even for a vertical fault, the FWI shows a high potential to accurately recover its shape and the physical contrasts, which is not the case to the classical VSP processing approach (see discussion below).

The challenging fault shape is for a dip of $-75°$ when the fault is far away from the receivers and when reflected waves cannot reach the receivers. We can notice that for the P-wave velocity field (Figure 10a), only a small upper part is recovered, whereas for both the S-wave velocity and density fields, the whole fault is correctly delineated, but with underestimated contrasts. The S-wave velocity field is better estimated due to the P-to-S converted waves of the scattered field (Figure 11). Notice that in the well vicinity (distance less than 300 m) the FWI has accurately recovered the shape and the physical contrasts of the fault. The tendency is that the physical parameter contrasts are more underestimated when moving down and away from the well.

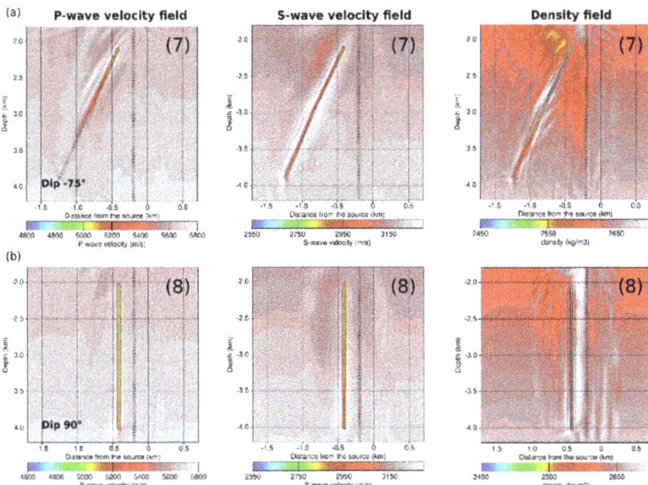

Figure 10. The retrieved P-wave velocity field, S-wave velocity field and density field for faults (**a**) 7 and (**b**) 8 (Figure 7). The fault 7 is 200 m far to the closest receivers in its upper part. We show only the interest area showed by the white rectangle in Figure 2c. Even for a vertical fault, the FWI shows a high potential to accurately recover its shape and the physical contrasts. In this case, the fault is accurately delineated, and the P-wave velocity and the density fields are recovered with a high precision. The S-wave velocity model show a correct delineation, but the contrasts are not well estimated in the deeper part of the fault, which could be explained by a low P-to-S reflected or less transmitted converted S-waves.

An interesting comparison could be made between the FWI technique and the standard OVSP data processing technique for their ability to detect and characterize faults in the granites. This sensitivity study shows us that we can fairly to accurately detect, delineate, and characterize fault zones in the granite context (in the well vicinity). Experiments show also that the higher the fault dip, the better is the fault delineation and also the better we characterize the physical parameter contrasts. A standard approach, where OVSP data are processed classically by separating the recorded field to up- and down- going fields and where only the first seismic order (primaries) of the upgoing fields is used, would fail for faults with high dips. Reiser et al. [10] obtained accurate fault imaging for faults, especially those with a shallow dip, and demonstrated that the Kirchhoff migration cannot provide accurate fault imaging for high fault dips: the faults dipping at 30° are, for instance, better imaged than those at 70° (also in a granite context). The authors have also noted the presence of important artefacts that cannot be removed even when they increase the source number due to the diffraction hyperbola. For the FWI method, the most challenging fault dip could be found for dip angles of 10–40°. For fault dips between 45 and 90°, which are the most frequently found in deep geothermal crystalline contexts, the FWI show a fully accurate fault imaging result. Very few faults may present this maximum dip (i.e., 90°), but the idea here is to check the ability of the applied method in very steepest contrasts. Additionally, the observed artefacts are removed when increasing the number of shots to seven, and completely removed with nine shots, because the multiple diffractions or high order scattered field constitutes useful information. In other words, as the full wave equation reproduces these phases, we can invert this part of the wavefield, introducing constraining information and helping the inversion to provide an accurate and realistic underground image. However, we recall that these FWI synthetic experiments have been performed for perfect data (noise free) and for a 2D world (i.e., the fault is always perpendicular to the propagation plane). We recall also that the aim of this study is to demonstrate the ability of the FWI to recover accurately faults with realistic parameters and characteristics, besides the real data complications (for instance attenuation, 3D effects, alterations and anthropogenic noise) which could not affect considerably the results of this study, because their effects (e.g., attenuation) do not play a major role in the FWI.

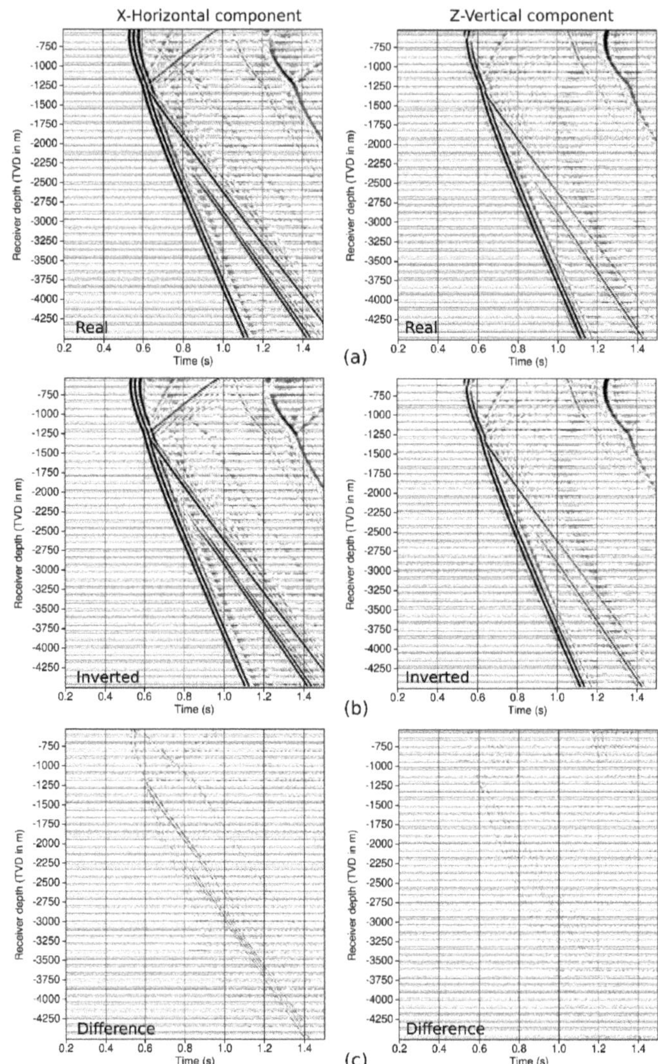

Figure 11. Seismograms recovered from the FWI of fault number 7 (for location, see Figure 7) for X (horizontal) component (left) and Z (vertical) component (right) obtained for zero offset source location. (**a**) observed X and Z components, (**b**) inverted X and Z components and (**c**) X and Z differences between (**a**) and (**b**).

We have performed an additional sensitivity study regarding antennae length. We have found as expected that we recover only the fault part located at the receiver depths. This means that when a fault is detected and delineated, its length could not be the complete length if the recovered fault length and the antennae are comparable.

The sensitivity analyses described and discussed above have been performed for a fault thickness of 30 m. What is the minimum fault thickness which can be recovered in the granites using the FWI technique? Even if the thickness is a geometrical parameter of the fault, its estimation is related to the physical parameter contrasts and the frequency content of the seismic signal (interaction between the wavelength of the P- and the S-wave and the fault thickness); this is why we have considered this parameter separately.

5.2. The Fault Thickness Issue

We start by analysing the quantitative parameters recovered in the previous experiments. For instance, we focus on the horizontal profile of the estimated physical parameters for fault no. 4 (see Table 1, Figure 8b), which represent a fault crossing the well at 3000 m, showing a dip of 60°, a thickness of 30 m and 20, 20 and 7% for P- and S-wave velocity and density contrasts, respectively. Qualitatively, remember that the upper part of this fault was accurately imaged by the three parameters (i.e., P-, S-waves and density), but its lower section was partly imaged. The contrast analysis (Figure 12) provides a quantitative view. The fault contrasts in its upper part before crossing the well is fully recovered where at least 97% of the maximum contrast is retrieved. Some oscillations can be observed for the density.

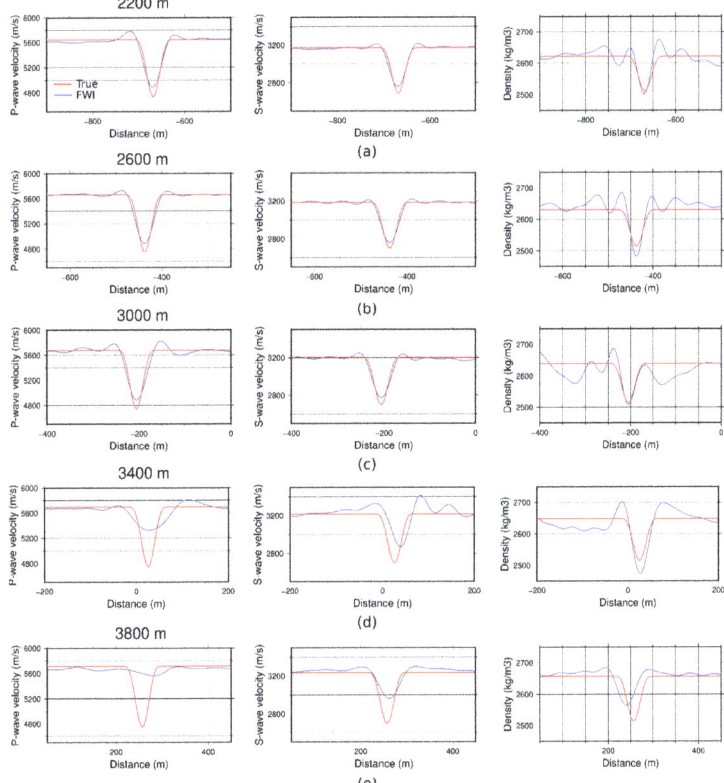

Figure 12. The estimated P- (right) and S-wave (middle) velocities and density (right) contrasts for five representative depths; (**a**) at the upper extremity (2200 m), (**b**) at 2600 m, (**c**) crossing the well at 3000 m, (**d**) at 3400 m and (**e**) the lower fault extremity at 3800 m. The fault no. 4 has been modelled and inverted (Figure 7), where the FWI image is shown in Figure 8b. Note that the true curves has no rectangular shape, because we smoothed the model to reduce diffractions arising from the model pixels. A filter smoothing 4 × 4 pixels (i.e., 20 × 20 m) have been applied to smooth the final physical, P-, S- waves and density models.

The physical parameter contrasts for the lower part of the fault are poorly recovered in the P-wave velocity field where 400 m below the receivers (i.e., at 3400 m depth), only 40% of the maximum contrast has been recovered (Figure 12d), whereas 70% of the S–wave contrast has been obtained, and an overestimation of 120% for the density. Considering the lower fault extremity (Figure 12e), less than 10% of the P-wave velocity contrast has been retrieved and around 45% for the S-wave velocity and 75% for the density. Note that the density provides better results for free noise data, and it is more sensitive to noise as the obtained results quality decreases when increasing the noise (Figure 12).

To quantitatively address the fault thickness, we consider the geometry data acquisition as that shown in Figure 7, and we consider the fault (1), dip of 75° with different thicknesses. We have studied the following fault thicknesses: 5, 10, 15, 20, 30 and 50 m. Both acquisition geometry and inversion parameters remain unchanged, for comparison with the previous images obtained for a fault thickness of 30 m.

We can notice that the fault contrasts have been accurately estimated by the FWI (Figure 13). Qualitatively, the fault has been accurately retrieved in all experiments. The fault location was rightly estimated, even for the very narrow fault of 5 m thickness. It is also important to quantitatively assess the FWI estimated contrasts especially for narrow faults, i.e., fault thickness less than 20 m (see discussion below).

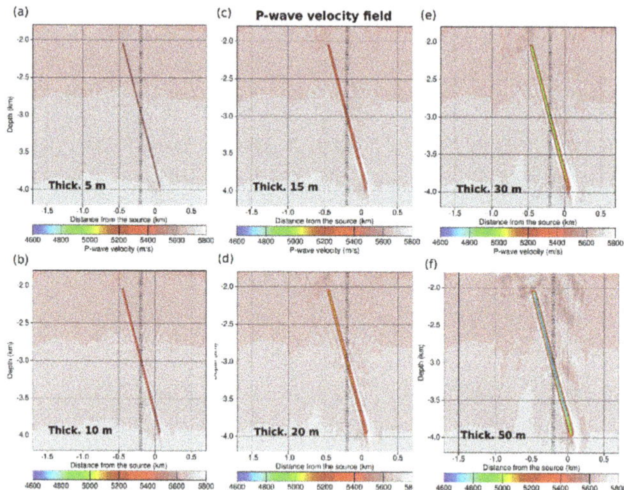

Figure 13. P-wave velocity estimated field for faults with different thicknesses: (**a**) 5 m, (**b**) 10 m, (**c**) 15 m, (**d**) 20 m, (**e**) 30 m and (**f**) 50 m. We show only the P-wave velocity in the target area limited by white block in Figure 2c. The quality results of the S-wave velocity and density are comparable to the P-wave velocity results.

In the granite, the P-wave velocity value is around 6000 m/s as showed by sonic logs acquired in GPK1 borehole (e.g., [48]. For a central frequency of 50 Hz, as is the case for our experiments, the seismic wavelength for P-waves is 120 m (i.e., $\lambda = 120$ m) and about half for the S-wave (70 m). Barnes and Charara [13] show that for OVSP data, when the scattered S-wave field is energetic (as often when offset is sufficient), the S-wave velocity estimated field provides a better spatial resolution than the P-wave velocity field, even when S-wave suffer from seismic attenuation because the scattered S-wave field is mainly generated in the well vicinity implying small propagation distances. Moreover, and because it deals with the complete wave equation, the FWI can detect and delineate faults even for thicknesses of $\lambda/10$ (i.e., for $\lambda p = 12$ m, and $\lambda s = 7$ m). In a sedimentary context, thin beds can also be detected with thicknesses of $\lambda/1$ [13] However, as the scattered energy decreases with the fault thickness, in the real world, the scattered energy from a thin fault can easily be covered by ambient noise. Therefore, for real data, the detection, delineation, and characterization of thin faults depend strongly on the data quality and on the S/N ratio.

As the seismic wavelength is around 120 m (70 m for S-waves), we observe also that either for faults of 5, 10 and 15 m thick (Figure 13a–c), we accurately recovered their locations and dips, but what about the recovered contrasts?

The obtained contrasts compared to the true contrasts are shown on Figure 14 for several experiments, including different fault thicknesses (5, 10, 15, 30 and 50 m) and at a particular depth of 3000 m. It was clear that the FWI will provide precise faults geometries and their characteristics, i.e., velocities and density contrasts, for thick faults (>30 m). Contrasts of these thick faults have been accurately retrieved, especially for P- and S-wave fields but less resolved for density fields, with some artefacts (Figure 14e,f). Surprisingly, precise contrasts have been recovered for very narrow faults of 5 and 10 m (Figure 14a,b). When comparing the FWI estimated contrasts to the true contrasts (Figure 14), precise values have been obtained for all thicknesses (Table 2), where the worst recovered contrast is 97% of the maximum true contrast, which is very satisfactory for fault thickness characterization, especially for narrow faults (i.e., fault thickness less than 15 m).

Table 2. The recovered contrast according to the maximum true contrasts, where True means the physical parameter contrast for the fault in the reference or True model and FWI means the physical parameter contrast for the fault in the models estimated by the FWI. Note that these contrasts are relative to the background values.

Fault Thickness [m]	P-Wave Velocity Contrast [m·s^{-1}]		S-Wave Velocity Contrast [m·s^{-1}]		Density Contrast [kg·m^{-3}]	
	True	FWI	True	FWI	True	FWI
50	−1031	−1089	−557	−492	−135	−189
30	−972	−839	−525	−456	−128	−188
20	−778	−603	−421	−327	−102	−141
15	−635	−501	−343	−279	−83	−64
10	−443	−373	−240	−203	−58	−78
5	−238	−198	−129	−111	−31	−34

5.3. The Multi-Faults Experiment

We now consider four faults with different lengths and thicknesses, and a deviated well. We thus mix different challenges in the same inversion experiment to explore the ability of the FWI method to separate and interpret the scattered fields of the different faults. The modelled fault features are shown in Table 3. The faults a, c and d cross the well, whereas fault b is located in the vicinity of the well at the right side, under the receivers.

Table 3. Fault features of the multi-fault experiment. The VC. Stand the Velocity Contrast (idem to C). The fault locations are shown in Figure 15.

Features	Faults			
	a	b	c	d
Thickness [m]	20	50	25	40
Dip [°]	−60	−60	75	80
P-wave VC. [%]	25	20	25	20
S-wave VC. [%]	25	20	25	20
Density C. [%]	10	7	10	7

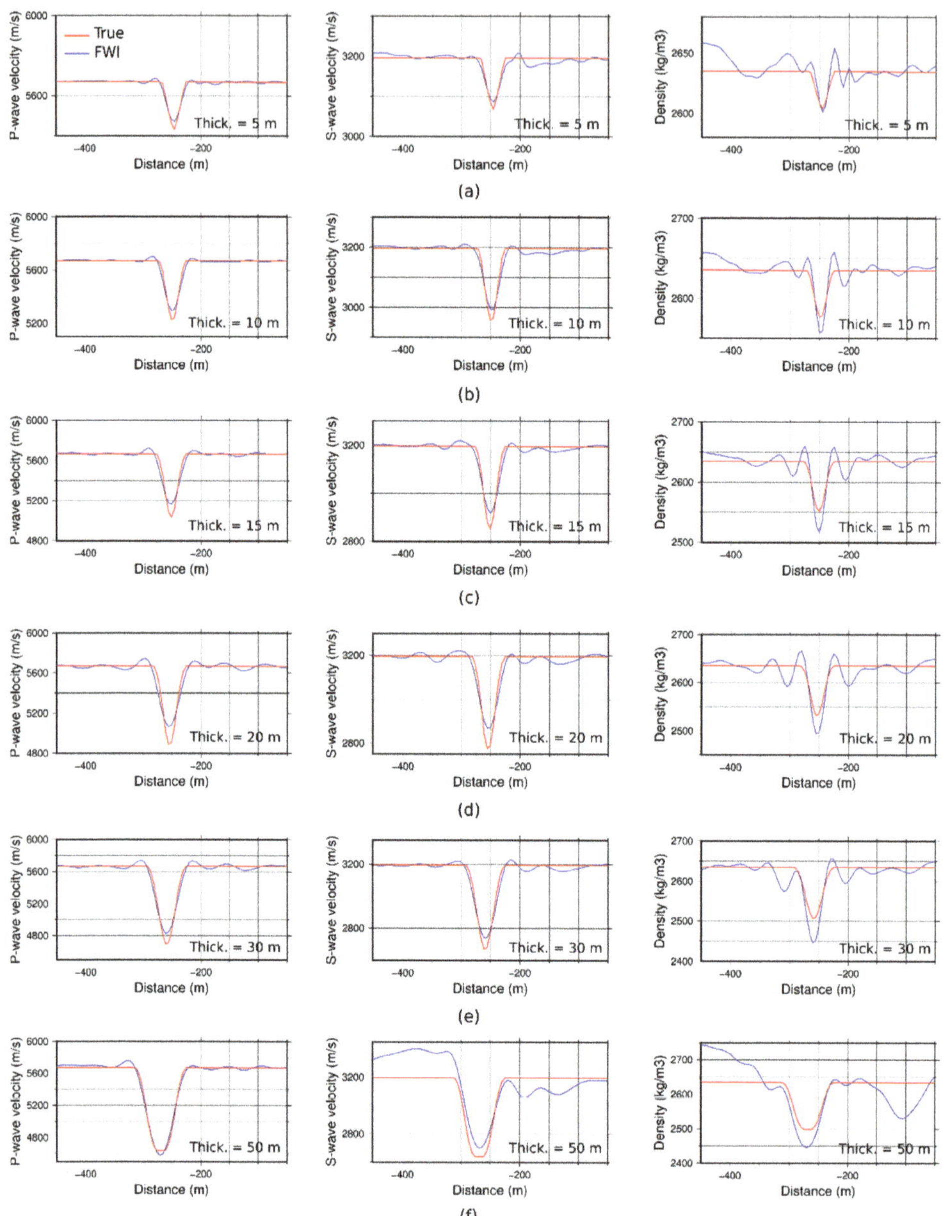

Figure 14. Estimated physical parameter values as a function of the distance for several fault thickness at 2800 m depth. The estimated P-wave (left), S-wave (middle) and density (right) from FWI for fault thicknesses; (**a**) 5 m, (**b**) 10 m, (**c**) 15 m, (**d**) 20 m, (**e**) 30 m and (**f**) 50 m. The estimations are good to very good but some artefacts are present in density curves, for instance, panels c-e. Surprisingly, the FWI provides accurate and precise fault characterization even for faults as thin as 5 or 10 m (a and b).

For acquisition geometry parameters, a total of 21 sources have been used with increasing inter-shot spacing from 250 m, around the well, to 500 m at intermediate distance (between 1 and 3 km away) to reach 1000 m at large offsets. The maxim offset used is 5000 m. The deviated well is inspired by GPK4 at Soultz-sous-Forêts. We do not add noise. The obtained P- and S-wave velocity and density estimated fields are shown in Figure 15 after 100 iterations. The FWI accurately recovers the fault locations and their shapes except for the deepest part of the faults c and d, as was observed in the previous experiments. These deep parts are better recovered for the S-wave estimated field than for the P-wave field and density as in previous experiments.

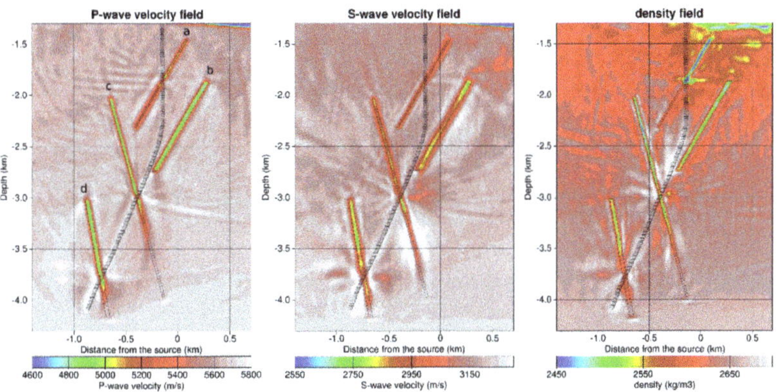

Figure 15. P- and S-wave velocity and density estimated fields from experiment with several faults having different features. The features of the modelled faults are shown in Table 3. The inverted triangles showed the receivers in the deviated well. We show the results only around the faults. The black polygon shows the location of the true modelled faults.

From the estimated fields, we can observe that the most important parameters affecting the results are the dip and location of the faults according to the receivers (Figure 15). We also notice that even if the fault does not cross the well, the FWI can retrieve accurately the fault characteristics. From the P-wave velocity estimated field, we can notice that the fault b is better resolved, then d, then c and finally a. The major difference between the resulted fault features for fault a and b, which have the same dip, is mainly their thickness, where the fault b thickness is 50 m, and only 20 m for fault a (Table 3).

The main result from the multi-fault experiment is that (i) the overall conclusions are the same as for previous synthetic FWI experiments and (ii) the presence of several faults does not affect the accuracy of results for each fault independently, at least in 2D and using noise-free data. In addition, we have understood from this experiment, that the FWI can provide a clear underground image of faults network, at least for fault zones having average characteristic values as those modelled here. For instance, fault a, which is only 20 m thick, is recovered in the right location and shape, but its thickness is underestimated. Undoubtedly, the narrow faults, typically with a thickness of less than 20 m, will be more challenging.

During the FWM, we can record the waves' propagations through the entire experiment. Figure 16 shows snapshots at times 2 ms, 660 ms, 850 ms and 1050 ms. The P- and S-wave front can be followed and their interactions with the faults are visible. P-to-S and S-to-P wave conversions are also visible. Combining these screen shots, we built a movie (see the Supplementary Material to watch it). Analysing continuously the seismograms and the movie, we better understand the different phases in the seismograms and their origins.

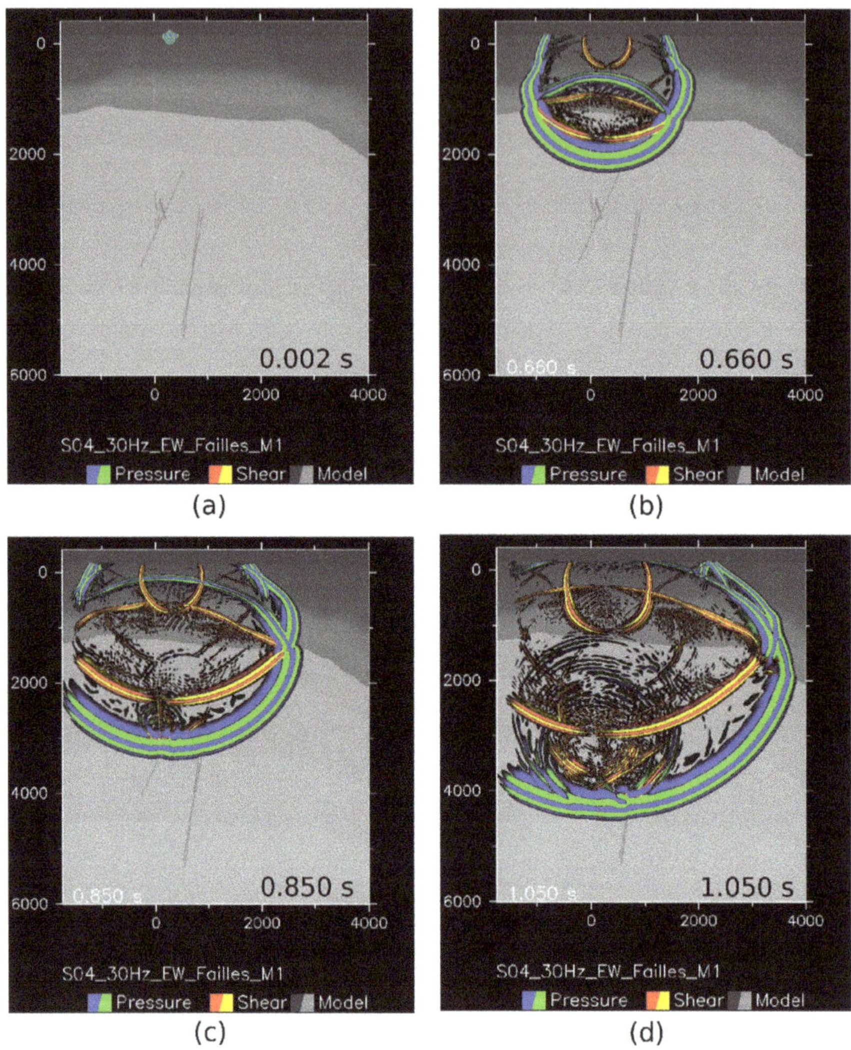

Figure 16. Representative snapshots for the multi-faults experiment at different times; (**a**) 2 ms, (**b**) 660 ms, (**c**) 850 ms and (**d**) 1050 ms. The divergence of displacements (P-waves) are shown in bleu (negative) and green (positive), and the curl of displacements (S-waves) are shown in red (negative) and yellow (positive). Notice the S-wave generated from the faults (i.e., panels c and d), the reflected and the converted waves from the top basement (e.g., panel b) and the multiples in the sedimentary layers (panel c and d), and several other waves.

6. Results and Discussion on the FWI Robustness for Noisy Data

In order to demonstrate the FWI ability for fault delineation and characterization, it is important to separate the data quality issue (information content of the data) and the starting model issue (non-linearity) from the imaging issue. This is why we have performed sensitivity analyses using noise-free data in the above sections (even the numerical noise is negligible as it is the same in the observed data and in the calculated data). We now test the FWI performance with moderate and high noise in the data using experiment 5 as a reference (Table 1). We considered the noise with the following characteristics:

- Additive Gaussian noise,
- Noise has the same f-k amplitude spectrum than the data,
- Ambient noise is unlocalized,
- Coda noise is localized a few periods after energetic phases.

We show in Figure 17 the Z-components for the three experiments namely Ref, MN and HN (Table 4). We can observe that for free noise data, we identify clearly the arrivals crossing the first arrival from -2000 m, whereas in MN these arrivals are more difficult to identify. For HN, these arrivals cannot be identified. We focus on these arrivals, because they are generated from our target (i.e., the fault).

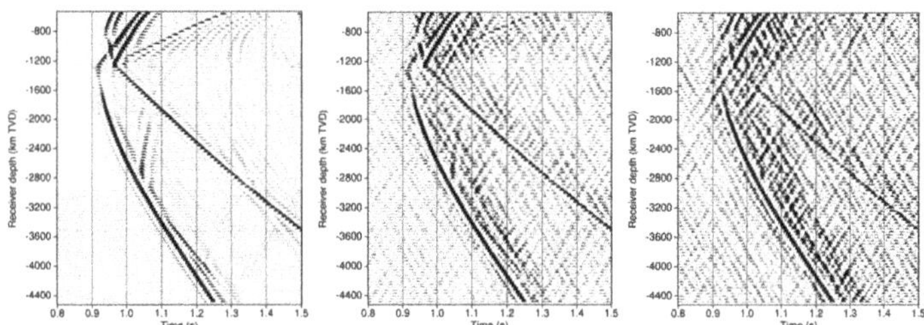

Figure 17. Vertical component (Z) for the three experiments showing the free noise data (**left**), MN data (**center**) and HN (**right**). We can notice that for the HN case, the fault scattered field is completely covered by the coda noise.

Table 4. The two noise parameters, ambient and coda, used in the noise study. The reference experiment is experiment 5 (Table 1).

N/S Amplitude Radio	Ambient [%]	Coda [%]
Noise free (Ref)	0	0
Moderate noise (MN)	15	25
High noise (HN)	30	50

We have run the FWI for these three experiments, as shown in Figure 2. The outcomes are shown in Figure 18. As expected, the estimated fields are noisy. The FWI detects the fault and retrieve quite accurately the shape and the location of the fault as well as its dip, even for noisy data (e.g., Figure 18b). The contrasts in the physical parameter are not well estimated. As previously observed, the lower part of the fault is partly recovered for both the P- and the S-wave velocity fields. This is due to the short maximum offset used in these experiments (2000 m) which is less than the optimal one (4000 m). Several artefacts can also be observed implying ambiguities in the interpretation (e.g., Figure 18c).

For the case of the moderate noise experiment (MN), the FWI is able to provide an accurate fault detection, delineation, and characterization (Figure 18b).

In the high-noise experiment (HN, Figure 18c), the upper part of the fault in the P-wave velocity field is recovered, but it is less well defined in the S-wave velocity field (and not well retrieved in density field, not shown here). The lower part of the fault is recovered partially in the P- and S-wave velocity fields, the delineation can be obtained but not the physical parameter contrasts. This HN experiment illustrates the capability of the FWI to detect and delineate faults, even for noisy data in the granite context. We recall here that this noise, even high, is structured, additive and Gaussian, which is theoretically compatible with the least square method used in the FWI. The application to real data is not as straightforward as in a sedimentary context. Effectively, the FWI needs a precise starting model including a complex rheology in order to account for the main heterogeneities (structures), the travel times of the main downgoing waves, and potentially, anisotropy or attenuation effects in the data. Some experiment parameters have to be inverted (as the source function for instance). Finally, the noise characteristics, depending on the dataset, should be carefully studied as the seismic data quality is always a key issue when using the FWI technique. All these issues have been already addressed with success in sedimentary contexts for oil and gas applications. The present results prepare the application of the FWI method to real borehole data in a crystalline basement and we hope that they will ease the interpretation of its outcomes.

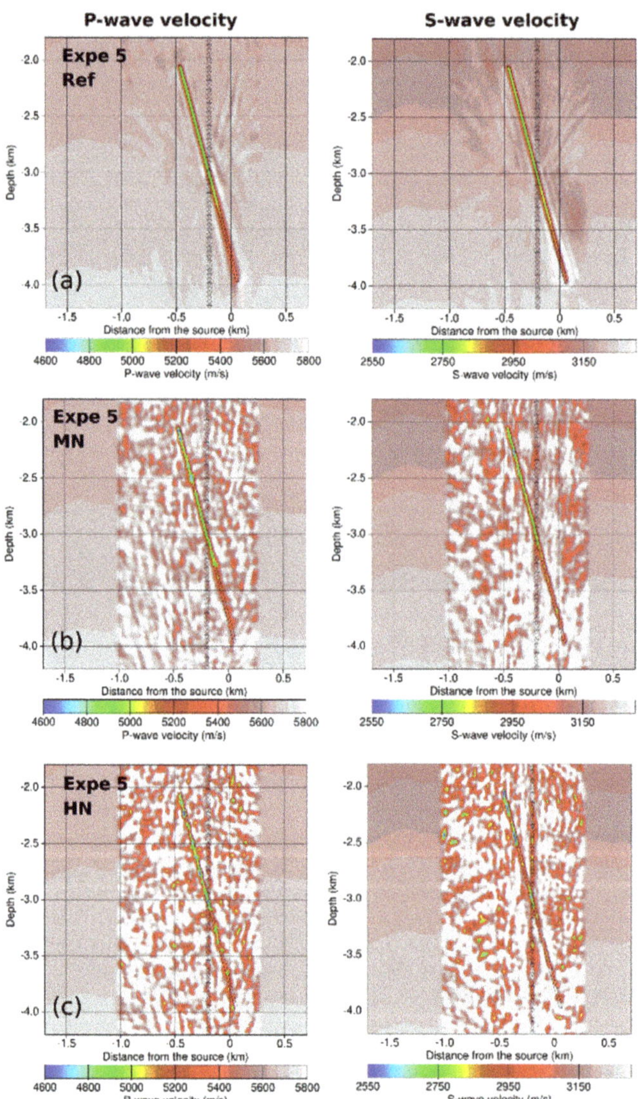

Figure 18. P- and S-wave velocity models recovered for; (**a**) the reference experiment (Ref), i.e., data free noise, (**b**) moderately noisy data (MN), and (**c**) highly noisy data (HN). We show only the target area around the fault for P-wave velocity (left) and S-wave velocity (right). The inverted triangles show the receivers and the black polygon show the shape of the fault.

7. Conclusions

We have performed several sensitivity analyses of the full wave inversion (FWI) method using numerical full wave modelling (FWM) in order to assess the capabilities of the FWI method for the purpose of detection, delineation, and characterization of faults in a crystalline basement. By adopting the following simplifications, we are somehow assessing the maximum capabilities of the FWI:

- Seismic modelling is performed in 2D using an elastic isotropic rheology. We do not address the attenuation and anisotropy issues. 2D modelling implies that the propagation plane is parallel to the dip direction of faults, therefore, the azimuthal dependency of the inversion results is not addressed (this would require 3D modelling).
- The starting model issue is not addressed.
- The acquisition geometry and experimental parameters for the source and receivers are perfectly known.

Most of the experiments are performed using noise-free data, i.e., perfect data. This allows us to check the method independently of the noise characteristics, i.e., the capabilities of the FWI methods when data conditions are perfect.

We have studied several sets of parameters: (i) the inversion algorithmic parameters for the crystalline basement context, (ii) the acquisition geometry parameters, and (iii) the fault characteristics parameters. For algorithmic parameters, the goal was to tune the inversion process in the FWI in order to optimize the results, and improve the final quality of the recovered underground images (i.e., the physical estimated fields). For instance, and contrary to sedimentary contexts, considering the data polarization increases the artefacts in the estimated fields. We have also tested several spatial correlation ranges. We have found that a range of 20 m, i.e., a little less than the fault thickness, yields the best results according to the thickness of the fault target. We have also tested the crosscorrelation between the P-, the S-wave velocities and the density parameters as strong positive correlations are noticeable in well logs, and we have found that the obtained images are improved, and the artefacts are clearly attenuated.

In the second step, we have studied the effect of the acquisition geometry: (i) the number of shots, (ii) the inter-shot distance and (iii) the maximum offset, which could be used to improve the final FWI results. Using a single shot in the granite is more challenging for the FWI to accurately characterize the fault. The presence of important artefacts in the estimated fields creates ambiguities in the interpretation of the fault images. As expected, these artefacts are attenuated when increasing the number of shots. With three shots, the quality of the retrieved images is improved but the estimated fields remain perturbed by artefacts. With five shots, we obtain better estimated fields and from seven shots and up, we accurately retrieve the fault with negligible artefacts. For a given number of shots, small inter-shot spacing (around 250 m) and large inter-shot spacing (around 1000 m) do not give suitable results. A reasonable intershot distance should be defined according to the depth of the target and to the seismic main frequency. Shots at far offsets (once to twice the target depth) provide constraining information to the FWI method. However, the quality of real, noisy data at far offsets is often not sufficient. The optimal acquisition geometry parameters depend on the S/N ratio. Of course, these conclusions have to be adapted when considering a 3D domain.

Considering the fault delineation and characterization goals, we studied the effect of the fault thickness and its dip in both configurations: a fault crossing the receivers in the well, and a fault far away from the well. The obtained spatial resolution is good even for a very thin fault of 5 m, where about 98% of the physical parameter contrasts have been recovered. This result stands for noise-free data but, as the energy of the scattered field decreases with the fault thickness, the S/N ratio is critical for thin fault zones. We can accurately detect, delineate and characterize faults with high dips (60° to 90°). Horizontal faults and their features were also retrieved accurately. The dips ranging between 10° and 40° remain a challenge. This may not be a critical problem because this fault dip range seems uncommon in granite, judging by the Soultz-sous-Forêts and Rittershoffen geothermal sites (e.g., [26,27,29]. The multi-faults case was also studied and the FWI showed a noticeable robustness and ability for delineation and characterization objectives. The delineation and characterization of a fault network could be considered in future applications. This capability of the FWI method has to be investigated further, especially in 3D, by studying the azimuthal effect for noisy OVSP data.

For a moderate noisy data, the FWI showed a high potential to detect, delineate and even characterize faults in the crystalline context. This opens new perspective for its future application on real data.

Supplementary Materials: Supplementary Materials: Video: Seismic wave propagation in a simplified elastic model of Soultz-sous-Forêts. Available online at https://www.mdpi.com/article/10.3390/geosciences11110442/s1. Description: Seismic wave propagation in a simplified elastic model of Soultz-sous-Forêts: Seismic wave propagation in a 2D earth model extracted from a 3D simplified elastic model of Soultz-sous-Forêts. The divergence of the displacement (P-waves) is indicated in blue/green while the curl of the displacement (S-waves) are in red/yellow. We can notice the complexity of the wavefield and the clear interaction with the faults. The calculation is performed using an elastic fullwave modelling code based on the finite difference method.

Author Contributions: Conceptualization, Y.A. and C.B.; methodology, C.B. and Y.A.; validation, Y.A. and C.B.; writing—original draft preparation, Y.A.; writing—review and editing, Y.A. and C.B. All authors have read and agreed to the published version of the manuscript.

Funding: This research was funded by MEET project, which has received funding from the European Union's Horizon 2020 research and innovation program under grant agreement No. 792037.

Acknowledgments: The authors thank EEIG "Heat Mining" for sharing geoscientific and technical data from Soultz-sous-Forêts geothermal site. This work was realized within the MEET project which has received funding from the European Union's Horizon 2020 research and innovation program under grant agreement No. 792037. We also thank Albert Genter, Eléonore Dalmais, Vincent Maurer and Nicolas Cuenot from Electricité de Strasbourg company for their fruitful discussions on various issues. We would like to thank Albert Genter for the first reading of this paper. We thanks Amine Ourabah and Bertrand Maillot for helping us to improve the english language.

Conflicts of Interest: The authors declare no conflict of interest.

References

1. Gentier, S.; Rachez, X.; Dezayes, C.; Blaisonneau, A.; Genter, A. How to understand the effect of the hydraulic stimultation in term of hydro-mechanical behavior at Soultz-sous-Forêts (France)? *GRC Trans.* **2005**, *29*, 159–166.
2. Schill, E.; Genter, A.; Cuenot, N.; Kohl, T. Hydraulic performance history at the Soultz EGS reservoirs from stimulation and long-term circulation tests. *Geothermics* **2017**, *70*, 110–124. [CrossRef]
3. Richard, A.; Gillot, E.; Maurer, V.; Klee, J. Northern Alsace (France): The Largest Geothermal Exploration by 3D Seismic Reflection. In European Geothermal Congress, The Hague-The Netherlands. 2019. Available online: http://europeangeothermalcongress.eu/wp-content/uploads/2019/07/proceedings-V2-2.pdf (accessed on 25 October 2021).
4. Abdelfettah, Y.; Sailhac, P.; Girard, J.-F.; Dalmais, E.; Maurer, V.; Genter, A. Resistivity Image under GRT1-2 Geothermal Doublet of the Rittershoffen EGS Project as Revealed by Magnetotelluric. In Eauropean Geothermal Congres, The Hagues-The Netherlands. 2019. Available online: http://europeangeothermalcongress.eu/wp-content/uploads/2019/07/352.pdf (accessed on 25 October 2021).
5. Abdelfettah, Y.; Hinderer, J.; Calvo, M.; Dalmais, E.; Maurer, V.; Genter, A. Using highly accurate land gravity and 3D geologic modeling to discriminate potential geothermal areas. *Geophysics* **2020**, *85*, G35–G56. [CrossRef]
6. Dorbath, L.; Cuenot, N.; Genter, A.; Frogneux, M. Seismic response of the fractured and faulted granite of Soultz-sous-Forêts (France) to 5 km deep massive water injection. *Geophys. J. Int.* **2009**, *177*, 653–675. [CrossRef]
7. Marie, J.-L.; Vergniault, C. Well Seismic Surveying and Acoustic Logging. *Sci. Ed.* **2018**, 1–140. Available online: https://www.edp-open.org/images/stories/books/fulldl/SeismicSurveying/9782759822638-SeismicSurveying_ebook.pdf (accessed on 25 October 2021).
8. Hardage, B.A. Vertical seismic profiling. *Lead. Edge* **1985**, *4*, 59. [CrossRef]
9. Dell, S.; Abakumov, I.; Znak, P.; Gajewski, D.; Kashtan, B.; Ponomarenko, A. On the role of diffractions in velocity model building: A full-waveform inversion example. *Studia Geophys. Et Geod.* **2019**, *63*, 538–553. [CrossRef]
10. Reiser, F.; Schmelzbach, C.; Maurer, H.; Greenhalgh, S.; Hellwig, O. Optimizing the design of vertical seismic profiling (VSP) for imaging fracture zones over hardrock basement geothermal environments. *J. Appl. Geophys.* **2017**, *139*, 25–35. [CrossRef]

11. Place, J.; Sausse, J.; Marthelot, J.-M.; Diraison, M.; Géraud, Y.; Naville, C. 3-D mapping of permeable structures affecting a deep granite basement using isotropic 3C VSP data. *Geophys. J. Int.* **2011**, *186*, 245–263. [CrossRef]

12. Lubrano, P. Traitement des Données de Sismique de Puits Acquises en 2007 sur le Site de Soultz-sous-Forêts Pour la Caractérisation de la Fracturation du Réservoir Géothermique. Ph.D. Thesis, Université de Strasbourg, Strasbourg, France, 2013.

13. Barnes, C.; Charara, M. Anisotropic anelastic full waveform inversion: Application to North Sea offset VSP data. In Proceedings of the SEG 80th Conference & Exhibition, Expanded Abstracts, Denver, CO, USA, 17–22 October 2010.

14. Barnes, C.; Charara, M. The choice of the physical law and constraints in seismic FWI. In Proceedings of the 2nd SEG Workshop on Full Waveform Inversion: Filling the Gaps, Abu Dhabi, United Arab Emirates, 30 March–1 April 2015.

15. Roberts, M.; Singh, S.; Hornby, B.E. Investigation into the use of 2D elastic wave inversion from look-ahead walk-away VSP surveys. *Geophys. Prospect.* **2008**, *56*, 883–895. [CrossRef]

16. Ben-Hadj-Ali, H.; Operto, S.; Virieux, J. Velocity model building by 3D frequency-domain, full-waveform inversion of wide-aperture seismic data. *Geophysics* **2008**, *73*, VE101–VE117. [CrossRef]

17. Charara, M.; Barnes, C.; Tarantola, A. The state of affairs in inversion of seismic data: An OVSP example. In Proceedings of the 66th SEG Conference & Exhibition, Houston, TX, USA, November 1996. [CrossRef]

18. Hirabayashi, N.; Leaney, W. Velocity profile prediction using walkaway VSP data. In Proceedings of the SEG 76th Conference & Exhibition, New Orleans, LA, USA, 1–6 October 2006. [CrossRef]

19. Barnes, C.; Charara, M. Viscoelastic full waveform inversion of North Sea offset VSP data. In Proceedings of the SEG 79th Conference & Exhibition, Expanded Abstracts, Houston, TX, USA, October 2009.

20. Podgornova, O.; Charara, M. Multiscale time-domain full-waveform inversion for anisotropic elastic media. In Proceedings of the DEG 81st Conference & Exhibition, Proceedings, Houston, TX, USA, August 2011. [CrossRef]

21. Owusu, J.; Podgornova, O.; Charara, M.; Leaney, S.; Campbell, A.; Ali, A.; Borodin, I.; Nutt, L.; Menkiti, H. Anisotropic elastic full-waveform inversion of walkaway VSP data from the arabian Gulf. *Geophys. Prospect.* **2016**, *64*, 38–53. [CrossRef]

22. Abdelfettah, Y.; Barnes, C.; Dalmais, E.; Maurer, V.; Genter, A. Full Wave Inversion of OVSP Seismic Data for Faults Delineation and Characterization in Granite Context. In Proceedings of the 1st Geoscience & Engineering in Energy Transition Conference, Strasbourg, France, 16–18 November 2020.

23. Barnes, C.; Abdelfettah, Y.; Cuenot, N.; Dalmais, E.; Genter, A. Reservoir seismic imaging using an elastic Full-wave Inversion approach applied to multi-OVSP Data in the context of Soultz-sous-Forêts. In Proceedings of the World Geothermal Congress 2020, Reykjavik, Iceland, 24–27 October 2020.

24. Pribnow, D.; Schellschmidt, R. Thermal tracking of upper crustal fluid flow in the Rhine graben. *Geophys. Res. Lett.* **2000**, *27*, 1957–1960. [CrossRef]

25. Sittler, C. Le pétrole dans le département du Haut-Rhin. Bilan d'un siècle et demi de recherches et d'exploitation. *Sci. Géol. Bulltins Mémoires* **1972**, *25*, 151–161. [CrossRef]

26. Genter, A.; Evans, K.; Cuenot, N.; Fritsch, D.; Sanjuan, B. Contribution of the exploration of deep crystalline fractured reservoir of Soultz to the knowledge of Enhanced Geothermal Systems (EGS). *Geoscience* **2010**, *342*, 502–516. [CrossRef]

27. Baujard, C.; Genter, A.; Dalmais, E.; Maurer, V.; Hehn, R.; Rosillette, R.; Schmittbuhl, J. Hydrothermal characterization of wells GRT-1 and GRT-2 in Rittershoffen, France: Implications on the understanding of natural flow systems in the Rhine Graben. *Geothermics* **2017**, *65*, 255–268. [CrossRef]

28. Aichholzer, C.; Duringer, P.; Orciani, S.; Genter, A. New stratigraphic interpretation of the Soultz-sous-Forêts 30-year-old geothermal wells calibrated on the recent one from Rittershoffen (Upper Rhine Graben, France). *Geotherm. Energy* **2016**, *4*, 1–26. [CrossRef]

29. Dezayes, C.; Genter, A.; Valley, B. Structure of the low permeable naturally fractured geothermal reservoir at Soultz. *Geosciences* **2010**, *342*, 517–530. [CrossRef]

30. Sausse, J.; Dezayes, C.; Dorbath, L.; Genter, A.; Place, J. 3D model of fracture zones at Soultz-sous-Forets based on geological data, image logs, induced microseismicity and vertical seismic profiles. *Comptes Rendus Geosci.* **2010**, *342*, 531–545. [CrossRef]

31. Valley, B.; Evans, K. Stress state at Soultz-sous-Forêts to 5 km depth from well-bore failure and hydraulic observations. In Proceedings of the Thirty-Second Workshop on Geothermal Reservoir Engineering, Stanford, CA, USA, 22–24 January 2007.

32. Barnes, C.; Charara, M. Constrained full-waveform inversion of multi-component seismic data. In Proceedings of the 72nd EAGE Conference and Exhibition Incorporating SPE EUROPEC, Barcelona, Spain, 14–17 June 2010.

33. Charara, M.; Barnes, C.; Tarantola, A. *Inverse Methods: Constrained Waveform Inversion of Seismic Well Data*; Springer: Berlin/Heidelberg, Germany, 1996.

34. Charara, M.; Barnes, C. Constrained full waveform inversion for borehole multicomponent seismic data. *Geosciences* **2019**, *9*, 45. [CrossRef]

35. Tarantola, A. *Inverse Problem Theory: Methods for Data Fitting and Model Parameter Estimation*; Elsevier: Amsterdam, The Netherlands, 1987.

36. Polack, E.; Ribière, G. Note sur la convergence de méthodes de directions conjuguées. *Rev. Française D'informatique Et De Rech. Opérationnelle* **1969**, *16*, 35–43. [CrossRef]

37. Crase, E. Robust Elastic Nonlinear Inversion of Seismic Waveform Data. Ph.D. Thesis, University of Houston, Houston, TX, USA, 1989.

38. Blanch, J.; Robertsson, J.; Symes, W. Modeling of a constant Q: Methodology and algorithm for an efficient and optimally inexpensive viscoelastic technique. *Geophysics* **1995**, *60*, 176–184. [CrossRef]
39. Christensen, R. *Theory of Viscoelasticity*; Academic Press: Cambridge, MA, USA, 1982.
40. Charara, M.; Barnes, C.; Tarantola, A. Full waveform inversion of seismic data for a viscoelastic medium. *Inverse Methods Interdiscip. Elem. Methodol. Comput. Appl.* **2000**, *92*, 68–81.
41. Barnes, C.; Charara, M.; Williamson, P. P & S wave attenuation effects on full-waveform inversion for marine seismic data. In Proceedings of the 2014 SEG Annual Meeting, Denver, Colorado, USA, October 2014.
42. Barnes, C.; Charara, M. The domain of applicability of acoustic full-waveform inversion for marine seismic data. *Geophysics* **2009**, *74*, WCC91–WCC103. [CrossRef]
43. Tsuchiya, T.; Yamada, N.; Iskandar, U.; Kurihara, M.; Barnes, B.; Charara, M. CO_2 Monitoring by Using VSP-FWI Synthetic Study on CO2-saturation and Pressure-buildup Differentiation. *Eur. Assoc. Geosci. Eng.* **2017**. [CrossRef]
44. Barnes, B.; Charara, M.; Tsuchiya, T.; Yamada, N. CO_2 Monitoring by Using VSP-FWI-Timelapse Elastic Parameters Estimation. *Eur. Assoc. Geosci. Eng.* **2017**. [CrossRef]
45. Charara, M.; Barnes, C.; Tsuchiya, T.; Yamada, N. Time-lapse VSP viscoelastic full-waveform inversion for CO_2 monitoring. In Proceedings of the SEG International Exposition and Annual Meeting, Houston, TX, USA, September 2017.
46. Maurer, V.; Perrinel, N.; Dalmais, E.; Richard, A.; Plevy, L.; Genter, A. Towards a 3D velocity model deduced from 2D seismic processing and interpretation in Northern Alsace. In Proceedings of the European Geothermal Congress, Strasbourg, France, September 2016.
47. Genter, A.; Traineau, H.; Bourgine, B.; Ledesert, B.; Gentier, S. Over 10 years of geological investigations within the European Soultz HDR project, France. In Proceedings of the World Geothermal Congress, Kyushu-Tohoku, Japan, 28 May–10 June 2000.
48. Kappelmeyer, O.; Gerard, A.; Schloemer, W.; Ferrandes, R.; Rummel, F.; Benderitter, Y. *European HDR Project at Soultz-sous-Forêts, General Presentation*; Geothermal Energy in Europe, The Soultz Hot Dry Rock Project; James, C., Ed.; Bresee Gordon and Breach Science Publishers: London, UK, 1992.

geosciences

 MDPI

Article

Fluid-Rock Interactions in a Paleo-Geothermal Reservoir (Noble Hills Granite, California, USA). Part 1: Granite Pervasive Alteration Processes away from Fracture Zones

Johanne Klee [1,*], Sébastien Potel [1], Béatrice A. Ledésert [2], Ronan L. Hébert [2], Arezki Chabani [1], Pascal Barrier [1] and Ghislain Trullenque [1]

[1] B2R, Geosciences Department, Institut Polytechnique UniLaSalle Beauvais, 19 Rue Pierre Waguet, F-60026 Beauvais, France; sebastien.potel@unilasalle.fr (S.P.); arezki.chabani@unilasalle.fr (A.C.); pascal.barrier@unilasalle.fr (P.B.); ghislain.trullenque@unilasalle.fr (G.T.)

[2] Geosciences and Environment Cergy, CY Cergy Paris Université, 1 Rue Descartes, F-95000 Neuville-sur-Oise, France; beatrice.ledesert@cyu.fr (B.A.L.); ronan.hebert@cyu.fr (R.L.H.)

* Correspondence: johanne.klee@unilasalle.fr; Tel.: +33-6-06-93-90-07

Citation: Klee, J.; Potel, S.; Ledésert, B.A.; Hébert, R.L.; Chabani, A.; Barrier, P.; Trullenque, G. Fluid-Rock Interactions in a Paleo-Geothermal Reservoir (Noble Hills Granite, California, USA). Part 1: Granite Pervasive Alteration Processes away from Fracture Zones. *Geosciences* **2021**, *11*, 325. https://doi.org/10.3390/geosciences11080325

Academic Editors: Chris Clark and Jesus Martinez-Frias

Received: 18 June 2021
Accepted: 27 July 2021
Published: 31 July 2021

Publisher's Note: MDPI stays neutral with regard to jurisdictional claims in published maps and institutional affiliations.

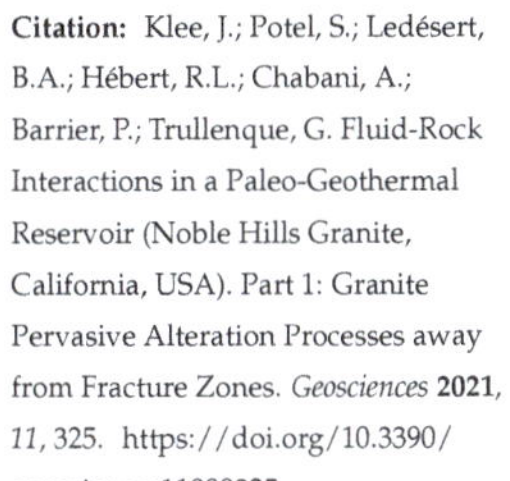

Abstract: Only few data from geothermal exploited reservoirs are available due to the restricted accessibility by drilling, which limits the understanding of the entire reservoir. Thus, analogue investigations are needed and were performed in the framework of the H2020 MEET project. The Noble Hills range, located along the southern branch of the Death Valley pull-apart (CA, USA), has been selected as a possible granitic paleo-reservoir. The aim is to characterize the pervasive alteration processes affecting this granite, away from the influence of the faults, in terms of mineralogical, petrophysical and chemical changes. Various methods were used as petrographic, geochemical and petrophysical analyses. Mineral changes, clay mineralogy, bulk rock chemical composition, calcite content and porosity were determined on different granite samples, collected in the Noble Hills granite, far from the faults and in the Owlshead Mountains, north of the Noble Hills, considered as its protolith. In order to complete the granite characterization, the metamorphic grade has been studied through the Noble Hills granite body. This complete characterization has allowed distinguishing the occurrence of three stages of alteration: (1) a pervasive propylitic alteration characterized by calcite-corrensite-epidote-K-white mica assemblage, (2) a more local one, only present in the Noble Hills granite, producing illite, kaolinite, illite/smectite, calcite and oxides, characteristic of the argillic alteration, which overprints the propylitic alteration and (3) weathering evidenced by the presence of montmorillonite in the Owlshead Mountains, which is considered as negligible in both granites. Alteration was also outlined by the correlation of the loss on ignition, representing the hydration rate, to porosity, calcite content and chemical composition. Moreover, the Kübler Index calculated from illite crystals allowed to identify a NW-SE temperature gradient in the Noble Hills.

Keywords: Noble Hills granite; Owlshead Mountains granite; metamorphic grade; fluid/rock interactions; newly formed minerals; element variations; geothermal reservoir

1. Introduction

Geothermal systems occur in different geological settings such as active volcanic fields, plutonic provinces, extensional domains, intracratonic basins and orogenic belts, i.e., anywhere with heat and fluids that are able to flow through the rocks [1]. The exploitation of geothermal energy is expanding worldwide due to the abundant resources and the progress of the technology that lead to Enhanced Geothermal Systems (EGS). EGS (1) defines a reservoir where the natural permeability of the rocks needs to be enhanced through stimulations in order to obtain a sufficient temperature/flow rate ratio [2] and (2) aims at transforming efficiently the geothermal resource into heat and electricity for human consumption [3].

The present study is part of the MEET H2020 project (Multidisciplinary and multi-context demonstration of EGS exploration and Exploitation Techniques and potentials) [4]. One aim of this project is to provide a characterization method of a geothermal granitic reservoir in a geological extensive context, such as the Great Basin region (USA), where normal fault zones also act as the most favorable structural setting for geothermal fluid flow [5]. In the Great Basin, other notable structural settings control fluid flow such as the intersection between normal faults and other structures like strike-slip faults (~22%), as well as pull-apart structures (4%) [5]. The fluid-rock interaction along and inside fracture zones results in hydrothermal alterations. They lead to geochemical, mineralogical and petrophysical (porosity and permeability) modifications of the rocks [6]. In granitic rocks, fluid circulations usually occur through the fracture network at different scales [6,7] involving a pervasive alteration which may influence up to cubic kilometers of rock [8]. Thus, an intense fluid/rock interaction [9] can significantly change the mineralogy, the chemistry and the texture of the bulk-rock [7] among which the common formation of clay minerals, including illite. Exploited geothermal reservoirs are located at depth, and the only and limited available data come from drillings (cores, cuttings) and seismic surveys. Studying exhumed geothermal reservoirs allows better understanding of the 3D features of the reservoir. To do so, the Noble Hills range (NH, Death Valley, CA, USA) has been selected as a possible granitic geothermal reservoir in a trans-tensional context. The NH are located in the southwestern part of the Great Basin region along the Southern Death Valley Fault Zone (SDVFZ), which constitutes the southern branch of the Death Valley pull-apart [10]. They extend over an area of 7 km long and 2 km wide and contain a part of the Cretaceous granitic pluton (~95 Ma) forming the Owlshead Mountains (OM) [11]. The arid climate prevents from a thick vegetal cover and the deep canyons that crosscut the range allow a thorough 3D investigation in order to characterize the evolution of the granite at the kilometer scale.

This paper aims at characterizing the pervasive alteration processes affecting the NH granite. This massif is considered as a possible paleo-geothermal reservoir. The study is based on the analysis of (1) rock mineralogical and related petrophysical properties changes and (2) associated chemical transfers between host rock and percolating fluids. All samples have been collected away from fractures described by [12] (this issue) in order to avoid the influence of strain and massive fluid flow. This sampling strategy ascertains the preservation of the protolith initial magmatic texture and mineralogy. A second paper (Klee et al., this issue) focusses on the role of deformation on hydrothermal alteration close to fractures. Data collected from the NH granite samples are directly compared to those obtained for the granite samples collected in the OM pluton, which is considered as the protolith. Analyses performed on targeted areas include macroscopic and microscopic petrographic studies, mineralogical characterization by X-ray diffraction (XRD) on whole rock and clay minerals, bulk rock chemical characterization by Inductively Coupled Plasma (ICP)—Mass Spectrometry (MS) and—Atomic Emission Spectrometry (AES), Scanning Electron Microscopy coupled with Energy Dispersive Spectrometry (SEM-EDS) for structural observation and local chemical analyses, as well as calcimetry and porosimetry analyses. To help at the granite characterization, the metamorphic evolution can be estimated through the Kübler Index (KI). Indeed, the temperature range of illite formation can be estimated thanks to the KI [13,14] based on illite "crystallinity" (IC). Temperature is thought to be the main factor controlling the IC evolution, but the lithology also has important effects [15–18]. Working at constant lithology, here granitic rock, allows us to avoid this effect. During the NH range formation, minerals could be transformed similarly as diagenetic reactions observed in feldspathic sandstones [19]. Inoue (1995) [9] has also shown that the rock alteration resulting from hot fluid storage during a long geological period, heated in-situ and in equilibrium with the surrounding rock is usually considered as diagenesis or metamorphism. Therefore, the terms defined by [13] for each diagenetic and metamorphic zones are considered in this study as eligible for granitic context.

2. Geological Setting

2.1. Death Valley

The area of interest for this study is the southern part of the Death Valley (DV) region, which extends for about 200 km. It is located southwest of the Basin and Range province [20], in the Eastern California Shear Zone/Walker Lane Belt (ECSZ/WLB) [21–23] (Figure 1a). DV is considered as one of the youngest regions where strike-slip deformation contemporaneously occurs with large-scale crustal extension within the Basin and Range province [24–26]. This extensional basin formation accompanying normal and associated strike-slip faulting would be active since 15 Ma according to [24]. It corresponds to a structural depression between the Panamit Range and the Black and Funeral Mountains [27] (Figure 1b), generally NNW-SSE oriented. This depression is related to tension along a segment of two strike-slip faults. Those two strike-slip faults consisting in the Northern Death Valley Fault Zone (NDVFZ) and the Southern Death Valley Fault Zone (SDVFZ), are characterized by en echelon traces [28] (Figure 1b). They have a general right-lateral movement, NW-SE oriented, from which results a "pull-apart" structure forming a N-S oriented basin [10] (Figure 1b).

The SDVFZ is composed of several branches. The SDVFZ formed the NH, at its southernmost part, by vertical displacement [29,30] (Figure 2a). It intersects the east-trending left lateral Garlock Fault Zone (GFZ), which ends at its western termination by the northwest-striking San Andreas fault zone [28] (Figure 1a). Recently, [11] suggested a net dextral slip along the SDVFZ of 40–41 km based on the offset positions of the granite-basement contact from the OM to the Avawatz Mountains. Much of the dextral slip, occurring before the deposition of the 6–8 Ma Neogene cover, is indicated by stratigraphic overlaps on fault rocks. This suggests an occurrence of the dextral slip during the main extension. The authors challenge the young feature of the DV pull-apart [25], by opting to a long-lived pull-apart which is consistent with regional evidence of the initiation in the middle Miocene of the dextral trans-tension in the ECSZ/WLB [31].

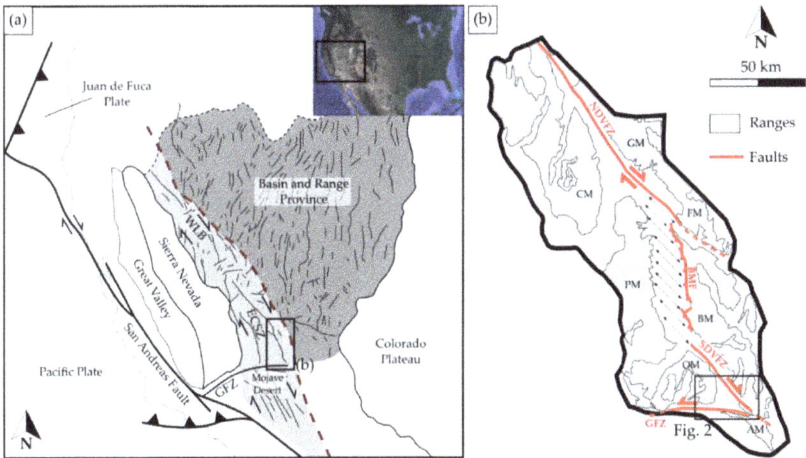

Figure 1. (**a**) Simplified tectonic map of the Great Basin region (western U.S. Cordillera) showing the tectonic provinces (modified after [32]). WLB—Walker Lane Belt; ECSZ—Eastern California Shear Zone; GFZ—Garlock Fault Zone. The Basin and Range Province is represented in dark grey and the WLB/ECSZ in light grey. The red dashed line marks the limit between these both domains. The dark lines within the Basin and Range Province and in the WLB-ECSZ zone represent the main faults. (**b**) Structural setting of the Death Valley region (modified after [33,34]. AM—Avawatz Mountains; BM—Black Mountains; BMF—Black Mountains Fault; CM—Cottonwood Mountain; FM—Funeral Mountains; GM—Grapevine Mountains; OM—Owlshead Mountains; PM—Panamint Mountains; GFZ—Garlock Fault Zone; NDVFZ—Northern Death Valley Fault Zone; SDVFZ—Southern Death Valley Fault Zone.

2.2. Noble Hills

The NH extend 14 km northwest of the northern Avawatz Mountains [30] (Figure 2a,b). A first general geological map of the NH was provided by [35], and later completed by [29,36–38]. They show that Precambrian gneiss covered by Crystal Spring Formation (CSF), a siliclastic-carbonate unit [39] of the Pahrump Group, were first intruded by 1.1 Ga diabase sills then by Mesozoic granitic rocks. All of these facies form the axial crystalline ridge defined by [29].

The NH granite was poorly studied previously. It is part of the calc-alkaline granitoid intrusion related to the emplacement of the Sierra Nevada batholith, which was formed due to the eastward dipping subduction of the Pacific plate under the North American continent [40]. It was only defined as both quartz monzonite [35] and leucocratic adamellite [41]. Brady (1986) [29] completed the description as a medium to coarse equigranular grained to slightly porphyritic leucocratic rock containing few biotite and little or no hornblende. The presence of sporadic mineralization due to hydrothermal alteration was also raised by [38]. After the emplacement of the granite and its exhumation, ~3.34 Ma Pliocene sediments of the Noble Hills Formation (NHF) were deposited [38]. The NHF consists in interbedded fine-grained clastic and evaporitic rocks, alluvial conglomerates, minor limestone and megabreccia. Recent work done by [42] in the NH has shown a more complex geometry of the axial ridge than described until now (Figure 2c). The undifferentiated Precambrian and Paleozoic rocks facies were described by [42] as a stacking of different CSF series, intruded by the Mesozoic granite. They seem to be dragged and stretched southeastward against the granite following the SDVFZ trend. Tertiary volcanism was also highlighted at the back of the range.

Based on the geology along the SDVFZ trace, Pavlis and Trullenque (2021) [11] suggests that the NH axial crystalline ridge must be a transported part of the Cretaceous granitic pluton (~95 Ma), forming the OM [34], which is relatively weakly deformed internally at its southern part [26]. This piece would have moved a minimum of 8 km according to [29,37] and around 28 km according to [43] along the SDVFZ (Figure 2b). Brittle shearing and large-scale *boudinage* characterize a brittle deformation, showing generally subhorizontal axes, which is prevalent within the Cretaceous granitic intrusion and Precambrian roof pendants (Figure 2c). Contractional deformation involving all members of the NHF is characterized along the NH and increases in intensity southeast toward the intersection with the GFZ [38].

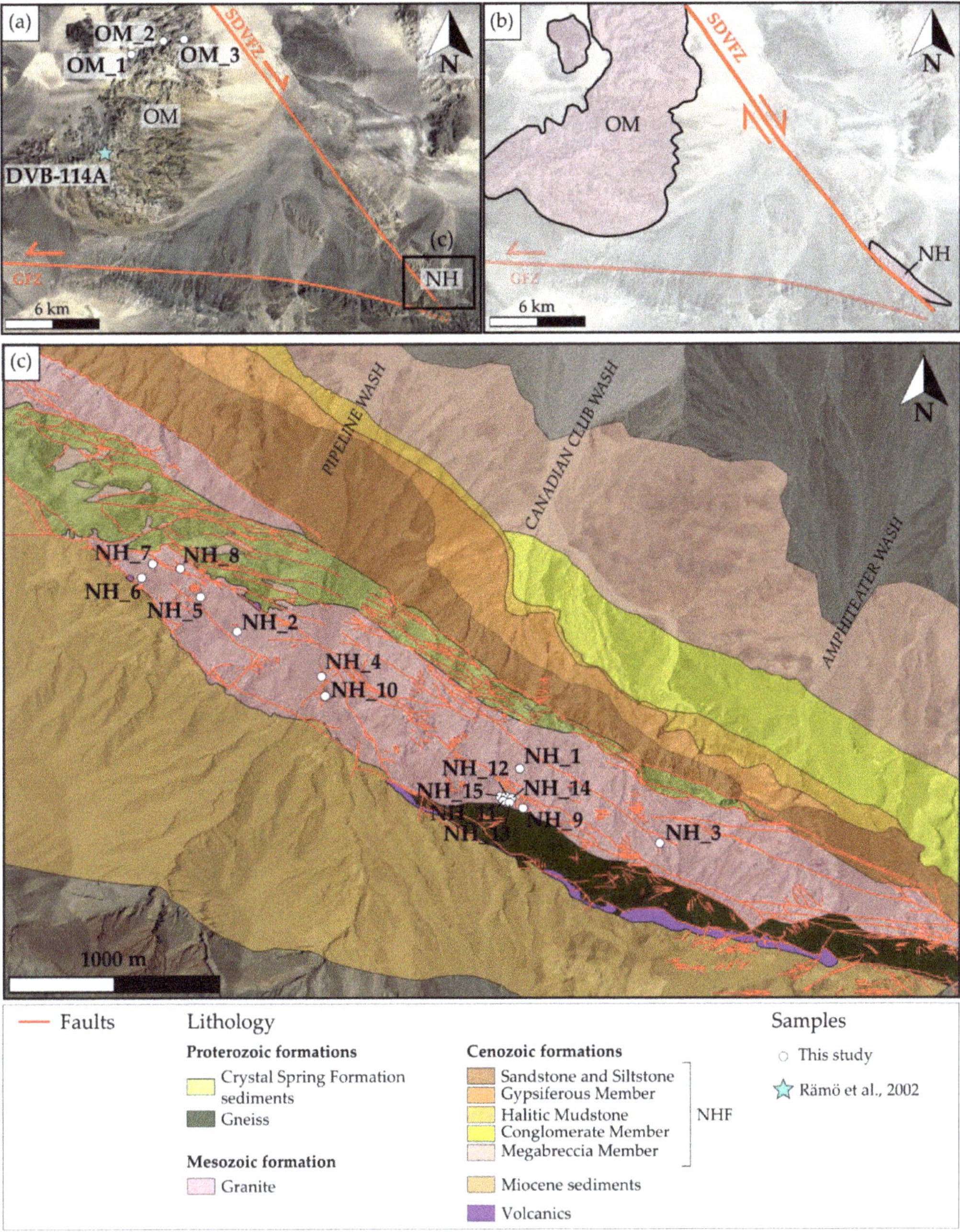

Figure 2. (**a**) Map presenting the structural setting of the southern part of the Death Valley region and the location of the Noble Hills range (NH) and the location of samples in the Owlshead Mountains (OM) represented by the white dots. (**b**) A scheme showing the displacement of the NH granite from the OM along the SDVFZ described by [11]. (**c**) The geological map of the Noble Hills range, modified after [38,42]. The quaternary is not displayed but available on [38] map. In (**a**,**b**), samples location for this study are represented by the white dots; the blue star corresponds to a sample from [34]. NHF—Noble Hills Formation.

3. Material and Methods

3.1. Material and Sampling Strategy

Fifteen samples from the NH granite and three samples from the OM granite (Figure 2) were collected in order to perform petrographical, mineralogical and geochemical characterizations. Hand specimens of centimetric size were collected in the two granite bodies (OM and NH) in order to characterize their mineralogical changes.

The selection of those samples through the NH was done under one scope, consisting of target zones away from the faults in order to avoid their influence. In that case, the magmatic texture is preserved. These faults are striking mainly NW/SE (SDVFZ direction) ([12], this issue). In order to be more precise in the sample selection, this latter was also based on the degrees of microfracturing defined by [44], from microscopical observations. This scale was improved by attributing a value of fracture density, based on scanlines realized on thin section mosaics, for each degree of microfracturing, which is described in the Methods section.

All the samples are georeferenced for database supply as well as located precisely (Figure 2). The OM being considered as the same batholith as the NH granite [11], samples were collected in an area unaffected by the SDVFZ activity in order to have a reference protolith of the studied area. Thin sections as well as powders were prepared to perform the following analyses.

3.2. Methods

3.2.1. Microscopic Observations

A petrographical study performed on covered and polished thin-sections of 15 samples from the NH and two samples from the OM (OM_1 and OM_3), was realized at Institut Polytechnique UniLaSalle in Beauvais, using a Leica DM4500-P optical microscope equipped with a Leica DFC450C camera. Images were acquired thanks to the software Leica Application Suite (LAS) v4.11.0. The system is also equipped with a multistep acquisition program to perform thin section mosaics. It consists in the acquisition of several photos that follow each other in order to scan the whole thin section. At the end, the pictures are merged to obtain the mosaic. All the mineral abbreviations used in this paper refer to mineral symbols defined by [45]. The degree of mineral transformation into secondary minerals was defined based on optical observations.

3.2.2. Fractures Density

The studied samples were selected away from fracture zones, but also according to a certain degree of microfracturing based on the scale defined by [44]. For this study, fractures density values were calculated for each degree of microfracturing by using thin section mosaics. For each mosaic, two scanlines [46,47] perpendicular to the main fractures were realized by digitalizing fractures along both lines. The P_{10} [46] has been calculated for each scanline. It consists in the number of fracture intersects per line length of scanline. The average was calculated to obtain a value of fracture density attributed to each thin section mosaic and corresponding to each microfracturing degree defined by [44]:

$F_d0 < 1687$ fracs/m—no to very low microfracturing

$F_d1 = 1687$ fracs/m—microfracturing of order less than the grain size

$F_d2 = 2694$ fracs/m, with a multiplicator factor of 1.6—microfracturing of grain size order with interconnections

$F_d3 = 3549$ fracs/m, with a multiplicator factor of 1.3—abundant microfracturing

$F_d4 \geq 5140$ fracs/m ([12] this issue), with a multiplicator factor of 1.4—very abundant microfracturing

Samples selected for this study have so a fracture density lower or equal to F_d2.

3.2.3. SEM-EDS

Scanning Electron Microscope (Hitachi S-3400N SEM) equipped with a Thermo Ultradry Energy Dispersive X-ray Spectrometer (EDS) probe was used on two polished thin sections and one hand specimen at Institut Polytechnique UniLaSalle in Beauvais. It aimed at analyzing the microstructure characteristics and to perform qualitative and semi-quantitative chemical analyses of various selected mineral phases.

A NORAN-type correction (©Thermo Fisher Scientific, Waltham, MA, USA) procedure was used for all data and all Fe was assumed to be ferrous for simplification. Polished thin sections were analyzed using a 50 µA beam current, an accelerating voltage of 20 kV and an acquisition time of 30 s.

3.2.4. X-ray Diffraction (XRD)

Experimental Conditions

XRD analyses were performed at Institut UniLaSalle Beauvais on the 15 samples from the NH and one sample from the OM (OM_2) using a D8-Advance Bruker-AXS (Siemens, Munich, Germany) diffractometer with a Ni-filtered CuKα radiation at 40 kV and 40 mA, a primary soller slit of 2.5°, divergence slit of 0.6 mm and a secondary soller slit of 2.5°, with a detector slit of 0.1 mm and an antiscattering slit of 0.6 mm. Samples were crushed with an agate pestle and mortar. Quantitative phase analysis based on Reference Intensity Ratio values were performed on randomly oriented bulk rock powders with a step length of 0.5° and a scan speed of 0.014°/s over the range 3°–70°2θ for bulk rocks composition. The uncertainty is estimated to be around 5%.

Determination of Illite Crystallinity and Kübler Index

XRD investigations were carried out on bulk rock powder specimens and clay fractions of 16 samples in order to identify and determine the relative abundance of mineral phases (semi-quantitative, around 3–5%) observed upstream under the optical microscope and the SEM. Clay mineral separation was conducted using techniques described by [48], following the recommendations of [49], and according to the standard techniques suggested by [50]. It consists into the collection of the <2 µm and 2–6 µm fractions from the sample powders put in suspension into water in decantation tubes. Oriented slides were then prepared by pipetting suspension onto glass slides (5 mg/cm^2) and air-drying. XRD measurements were then performed at air dried, solvated with ethylene glycol, and after heating (550 °C) conditions with a step length of 0.5° and a speed of 0.01°/s per step over the range 3°–35°2θ. The clay minerals identification, which is based on d-values and the relative intensity of their 00*l* peak reflections, was undertaken referring to [51,52]. The illite "crystallinity" (IC), defined as the full width at half maximum (FWHM) of the 10 Å (001) peak of illite, was calculated using the software DIFFRAC EVA v 4.2 (by ©Bruker AXS, Billerica, MA, USA). The obtained values were standardized using the crystallinity index-standard (CIS) samples of [53] in order to calculate the Kübler Index (KI). The KI values of raw data expressed in Δ°2θ, were measured into three slots, corresponding to different campaigns, inducing the three following standardizations:

$$KI_{(ULS1)} = 1.6987 \times IC_{measured} - 0.0842 \ (R^2 = 0.9724) \tag{1}$$

$$KI_{(ULS2)} = 1.5501 \times IC_{measured} - 0.0512 \ (R^2 = 0.9944) \tag{2}$$

$$KI_{(ULS3)} = 1.5337 \times IC_{measured} - 0.0498 \ (R^2 = 0.9975) \tag{3}$$

KI was used to define the limits of metamorphic zones [54], following the recommendations for Kübler-Index calibration of [50] and the CIS-KI transformation formalism of [55]. In siliciclastic rocks, the transitions from non-metamorphic to low-grade (referring to the term greenschist facies [56]) and from the very low-grade (chlorite zone [57]) to low-grade metamorphic zone (biotite zone [58]) take place through three zones defined by [13]: the diagenetic zone, the anchizone and the epizone. The zone boundary values are presented in Table 1. The smallest FWHM able to be measured by our diffractometer (limit detection) is 0.08Δ°2θ.

Table 1. Metamorphic zone boundaries [59] for Kübler Index (KI) values [55] and temperatures [60].

Metamorphic Zone	KI ($\Delta°2\theta$)	Temperature (°C)
Low Diagenesis	>1	~100
High Diagenesis	0.42–1	~200
Low Anchizone	0.30–0.42	
High Anchizone	0.25–0.30	~300
Epizone	<0.25	

3.2.5. ICP-MS—ICP-AES

The analyses of major, trace and rare earth elements were completed on five selected samples from the NH (NH_1, NH_2, NH_3, NH_4 and NH_12) and one from the OM (OM_3) by Bureau Veritas Minerals (Vancouver, Canada) using ICP-ES and ICP-MS.

Samples were crushed and mixed with $LiBO_2/LiB_4O_7$ flux. Crucibles were fused in a furnace at 980 °C. Then, the cooled bead was dissolved in ACS grade nitric acid and analyzed by ICP-AES and/or ICP-MS. Loss on ignition (LOI) was determined by igniting the samples split then measuring the weight loss.

3.2.6. Manocalcimetry

Calcite contents were determined using an OFITE 152-95 manocalcimeter. The analyses were performed on 15 samples from the NH and one sample from the OM at CY Cergy Paris University in the Geosciences and Environment Cergy (GEC) laboratory. It provides an indication of the total carbonate content in a sample and allows to assess the influence of calcite on permeability of the fluid pathways. This is achieved by measuring the rate of response of 10% hydrochloric acid on the samples. Calcimetry has also proved to be an efficient, easy and low-cost method to better understand the hydrothermal sealing of a reservoir [61].

Manocalcimetry consists into the measurements of CO_2 partial pressure when calcite is dissolved by HCl. The reaction that occurs is:

$$CaCO_{3(s)} + 2HCl_{(l)} \rightleftarrows CO_{2(g)} + H_2O_{(l)} + CaCl_2 \tag{4}$$

The calcimeter, composed of a glass flask and a high precision manometer calibrated with pure calcite reduced into fine powder. This allows us to determine the calcimeter coefficient. Variation of temperature and pressure can cause uncertainty on measurements, which was considered to obtain high quality results with a precision around 0.5 wt.%.

Prior to measurements, sample preparation was done according to [61]. The samples were reduced into powder in an agate mortar and put in the oven at 60 °C for 24 h. Then, 1.000 g of each sample was weighed and put in a sample holder, itself put in the calcimeter with a little glass filled with HCl. The amount of HCl is in excess in order to dissolve all the calcite present in the sample. The maximum value reached during the measurement was read on the manometer and the $CaCO_3$ percentage was determined as follows:

$$\%CaCO_3 = (\text{Measured value} \times 100)/\text{Calcimeter coefficient} \tag{5}$$

Two replicates were performed for each sample in order to check the reproducibility of the results, which is considered as good when the difference between the two results is lower than 0.5 wt.%, corresponding to the precision interval mentioned above.

3.2.7. Ethanol Saturation Porosimetry

The estimation of available volume for fluid storage is fundamental and can be quantified by porosity measurements [62]. The connected porosity was measured on 6 samples from the NH and one from the OM by the triple weighing method [63] defined by the RILEM standard (test n°I.1, 1978). It consists in the saturation of the samples after vacuum degassing. The measurements were carried out at CY Cergy Paris University in the Laboratory of Mechanics and Materials for Civil Engineering (L2MGC).

In this study, ethanol has been chosen instead of water, as used in classical methods, in order to avoid possible clay swelling [64], which could lead to the destruction of the sample and biased results. Even though ethanol (0.469 nm) is a molecule larger than that of water (0.343 nm), the pore volume is not estimated to be under-evaluated, as the pore size is much larger than that of ethanol molecules. The samples were first oven-dried at approximately 40 °C until obtaining a constant weight (W1). Samples were then soaked by capillary action with ethanol after staying for 12 h under a vaccum. When the samples were completely immersed, the vacuum was stopped and the samples left in the ethanol for 24 h. Finally, they were weighted twice: (1) weighing of saturated samples, W2; and (2) weighing of samples under "ethanostatic" conditions (suspended into ethanol under the balance), W3. The connected porosity ϕ is then calculated:

$$\phi(\%) = (W2 - W1/W2 - W3) \times 100 \tag{6}$$

4. Results

4.1. Petrographic Description

4.1.1. The Owlshead Granite

Samples of the OM granite appear rather fresh, meaning that primary minerals seem not transformed into secondary ones and present no sign of deformation. They show a light grey/whitish and yellowish granite (Figure 3a) with equant medium-size grains (0.1–1 cm). Primary assemblage is composed of plagioclase, quartz, K-feldspar and biotite. Biotite is generally surrounded by a yellowish oxide halo, showing its incipient alteration.

OM granite microscopic observations (Figure 3b,c, Table 2) confirm the freshness of the rock as well as the very low to absent microfracturing. Figure 3b shows a microfracturing around F_d1 (Table 2). Most of the fractures are open and a few of them are filled by calcite. Calcite is also present at the grain boundaries (Figure 4a,b).

Regarding mineralogical composition, plagioclase (35%, modal composition) occurs subautomorphous crystals, up to 5 mm, of oligoclase composition with an oscillatory zonation [65] (Figure 3b). Quartz (35%) forms up to 2 cm polycrystalline clusters of anhedral crystals which range up to 1 cm and present a slight rolling extinction. K-feldspar (25%) occurs as centimetric subautomorphous crystals of microcline and orthoclase with a Carlsbad twin, rich in perthites (albite vein or braid/patch shape [65] (Figure 4a,b). K-feldspar crystals can contain plagioclase and biotite inclusions. Biotite (5%) of around 2 mm in length is euhedral. Primary opaque minerals (<1%) are also found in this granite.

The core of plagioclase is slightly transformed into K-white mica (Figures 3c and 4c,d). They will be represented as Wm. Quartz and K-feldspar are not affected by the alteration and are only little undeformed as seen on quartz showing undulatory extinction (Figure 4a). Biotite can be slightly altered. Few epidotes are present nearby the biotite (Figure 4e,f).

Microscopic observations of samples OM_3 and OM_1 are consistent with whole rock XRD analyses performed on the sample OM_2 from the OM granite. The three samples have a similar mineralogical composition and are grouped together in Table 2.

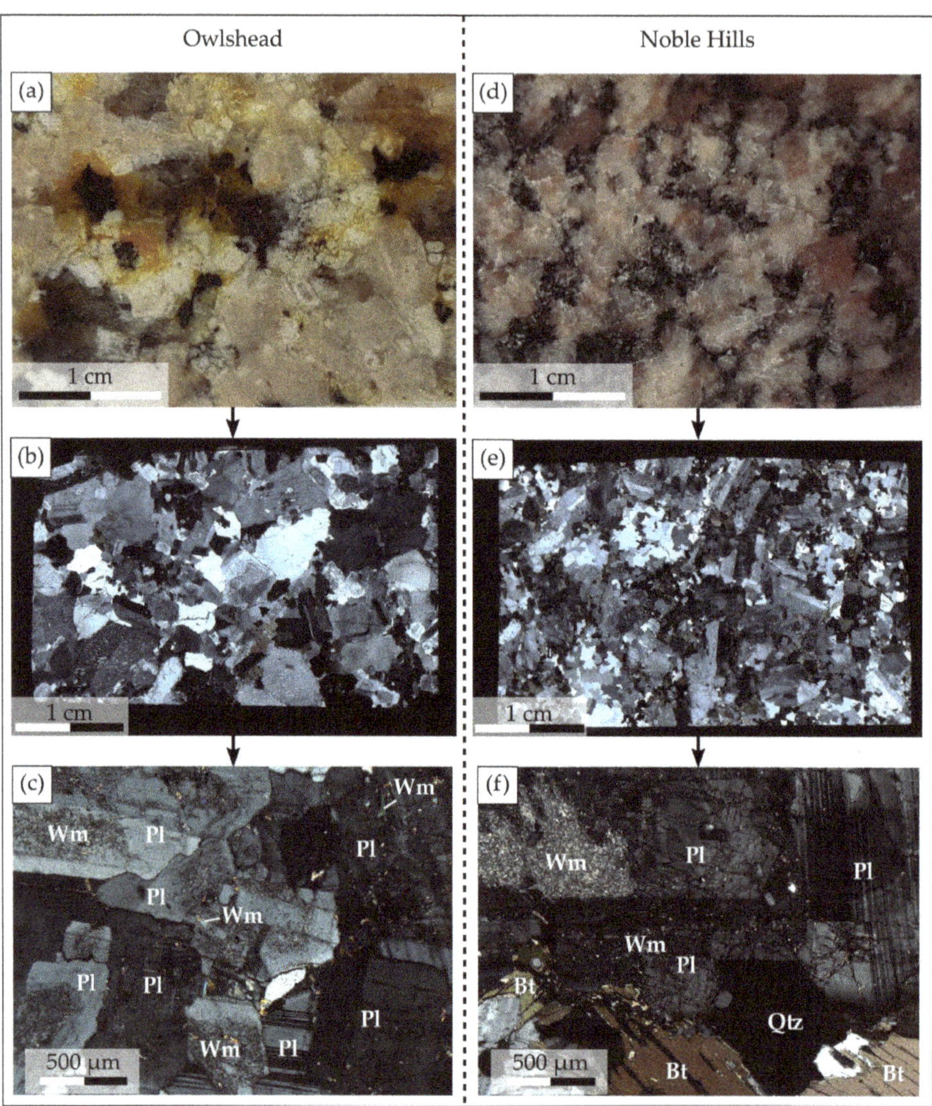

Figure 3. Macroscale to microscale photographs illustrating (**a–c**) the Owlshead and (**d–f**) the Noble Hills granites. (**a,d**) Hand specimens for each granite showing their different aspects. (**b,e**) Thin section mosaics realized for each granite under optical microscope in polarized—analyzed light and showing the difference in terms of grain size and microfracturing. (**c,f**) Thin sections zooms of each granite mosaic in polarized—analyzed light showing the difference in terms of alteration degree. Abbreviations (except for "Wm") after [45]: Bt—Biotite, Kfs—K-feldspar, Wm—K-white mica, Pl—Plagioclase, Qtz—Quartz.

Table 2. List of samples collected in the field with their mineralogical composition determined after microscopic observations and XRD analysis (primary minerals, alteration phases for plagioclase, biotite and K-feldspar, other secondary minerals and microfissuring after Castaing and Rabu scale). Abbreviations (except for "Ox and Wm") after [45]: Qtz—Quartz, Pl—Plagioclase, Mc—Microcline, Or—Orthoclase, Bt—Biotite, Kln—Kaolinite, Ep—Epidote, Ox—Oxide (when it could not be determined precisely), Wm—K-white mica, Cal—Calcite, Dol—Dolomite, Hem-Hematite, Gp—Gypsum, Fd—Fracture density.

Sample ID	Primary Minerals	Major Secondary Phases Within						Other Secondary Minerals	Microfracturing
		Pl			Bt		Kfs		
		Wm	Kln	Cal	Wm	Ox	Wm		
OM_1_2_3	Qtz, Pl, Mc, Or, Bt, Ox	√						Ep, Ox, Wm, Cal	Fd1
NH_1	Qtz, Pl, Mc, Or, Bt, Ox	√	√		√	√		Cal, Dol, Wm	Fd1
NH_2	Qtz, Pl, Or, Bt, Ox	√	√			√		Dol, Wm	Fd1
NH_3	Qtz, Pl, Or, Bt, Ox	√	√		√	√	√	Cal, Dol, Wm	Fd1-2
NH_4	Qtz, Pl, Or, Bt, Ox	√	√	√	√	√		Cal, Hem	Fd0-1
NH_5	Qtz, Pl, Or, Bt, Ox	√	√			√		Hem, Wm	Fd0-1
NH_6	Qtz, Pl, Or, Bt, Ox	√	√	√	√	√		Cal, Hem, Wm	Fd1-2
NH_7	Qtz, Pl, Or, Bt, Ox	√	√		√	√		Cal, Ox, Wm	Fd1
NH_8	Qtz, Pl, Or, Bt, Ox	√	√	√	√	√		Cal, Ox, Wm	Fd1-2
NH_9	Qtz, Pl, Or, Bt, Ox	√	√			√		Cal, Ox, Wm	Fd0-1
NH_10	Qtz, Pl, Or, Bt, Ox	√	√	√	√	√		Cal, Hem, Wm, Ep	Fd2
NH_11	Qtz, Pl, Or, Bt, Ox	√	√	√	√	√		Cal, Hem, Wm	Fd1-2
NH_12	Qtz, Pl, Mc, Or, Bt, Ox	√	√	√	√	√		Cal, Hem, Gp, Wm	Fd0-1
NH_13	Qtz, Pl, Mc, Or, Bt, Ox	√	√	√	√	√		Cal, Hem, Gp, Wm	Fd0-1
NH_14	Qtz, Pl, Mc, Or, Bt, Ox	√	√	√	√	√		Cal, Hem, Gp, Wm	Fd0-1
NH_15	Qtz, Pl, Or, Bt, Ox	√	√		√	√		Cal, Mag, Wm	Fd1-2

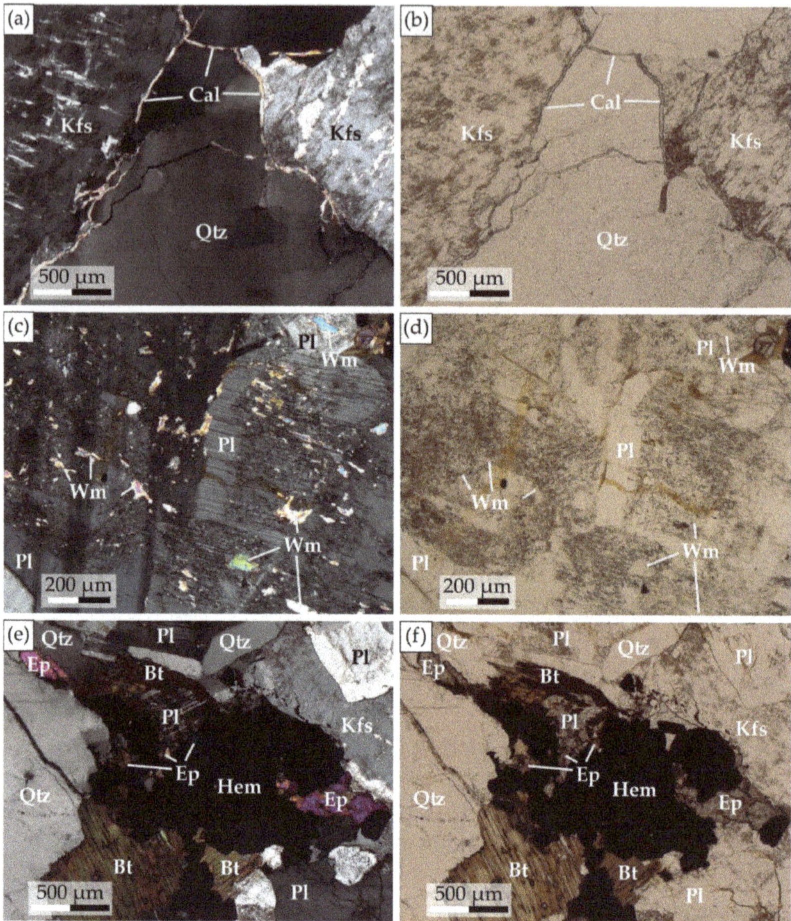

Figure 4. Photomicrographs of the Owlshead granite in polarized-analyzed and polarized—non-analyzed light showing (**a,b**) perthitic and unaltered orthoclase, unaltered quartz presenting a slight undulatory extinction and calcite veinlets at grain boundaries. (**c,d**) Plagioclase alteration with the formation of K-white mica. (**e,f**) Incipient biotite alteration associated to hematite and epidote formation. Abbreviations (except for "Wm") after [45]: Bt—Biotite, Cal—Calcite, Ep—Epidote, Hem—Hematite, Kfs—K-feldspar, Wm—K-white mica, Pl—Plagioclase, Qtz—Quartz.

4.1.2. The Noble Hills Granite

Far from the major faults, the NH granite appears, as a whole, microfractured and altered. with preserved "fresh" zones. The collected hand specimens (Figure 2b) are equant to slightly porphyritic and display a pinkish color (Figure 3d) or a whitish color for the samples with evidences of alteration. Primary assemblage, as for the OM granite, is made of plagioclase, quartz, K-feldspar and biotite.

At the microscale, the degree of microfracturing ranges from F_d0–1 to F_d2 (Table 2). As for the OM granite, the microfractures can be opened or filled by carbonates. As regards mineralogical composition, plagioclase (~35%, oligoclase in composition) occurs as sub-anhedral to euhedral crystals (up to 5 mm in length) showing growth zonation (see Figure 3f and [65]). Quartz (~30%) forms polycrystalline euhedral clusters of around 1 cm in size made of 3-mm-wide crystals with sometimes a slight undulatory extinction linked to low deformation. K-feldspar (~25%) is made of sub-anhedral to euhedral orthoclase (up to 1 cm in length) with vein shaped perthites [65]. Microcline is only present in the samples NH_12 and NH_14, in the southern part of the range (Figure 2 and Table 2). Some K-feldspar crystals contain inclusions of plagioclase or/and biotite. Depending on the samples, the amount of plagioclase can be equivalent to that of K-feldspar (~30% each), but most of the time, plagioclase dominate. Myrmekite can be observed at the interface between plagioclase and K-feldspar. Biotite (~10%) appears euhedral with crystals ~2 mm in length. Accessory minerals (<1%) as apatite are also observed. More rarely, primary muscovite, as well as primary opaque minerals (oxides) are observed in the granite. A comparison between the NH and the OM granites highlights a grain size difference. Quartz and K-feldspar crystals are smaller in the NH granite (up to 3 mm and 1 cm respectively, Figure 3e) than in the OM granite (up to 1 cm and centimetric crystals respectively, Figure 3b).

Primary minerals, function of their sensitivity to alteration [65], recrystallized into secondary minerals which are shown in Table 2. This is related to differences into chemical properties [66]. In the plagioclase, which is the most altered mineral (Figure 5), recrystallization propagate from the core of the crystal to the more albitic rim (Figure 5a) [67]. When alteration is intense, plagioclase is entirely replaced by newly formed minerals leaving only the pseudomorph of the plagioclase to subsist (Figure 5d). The newly formed minerals are the following:

- Illite is the most frequent and occurs as tiny flakes or needles 0.5 to 8 µm width and up to 40 µm long [68] (Figure 5b,c).
- Kaolinite is present as fan shape (Figure 5e,f) of 25 µm to 40 µm in diameter. Under SEM (Figure 5f), well crystallized kaolinite presents a porous structure which can contribute to the porosity of the rock. It is only present in the NH granite (Table 3) indicating that the NH granite has undergone a different alteration from that of the OM granite.
- Calcite, which occurs as small spots, is mainly associated with kaolinite (Figure 5d,e). It crystallizes in the porosity created by plagioclase dissolution.

  Illite and kaolinite can be present together in the same sample (Figure 5d).

  The K-feldspar remains always unaffected (Figure 5d), but when the alteration is relatively pronounced, perthites can be altered as well as the mineral inclusions like in sample NH_3 Table 2). Biotite is progressively replaced by K-white mica (illite) as compared to the biotite in the OM granite (Table 2). K-white mica crystallizes along the cleavages and can be associated to the crystallization of oxides, as hematite, also along the cleavages (Figure 6).

Table 3. List of samples with their respective clay minerals composition, FWHM and Kübler Index (KI) for the fractions < 2 µm and 2–6 µm. Abbreviations (except for "Cor") after [45]: Ill—Illite, Kln—Kaolinite, Cor—Corrensite, I/S—Illite/Smectite, Vrm—Dioctahedral Vermiculite, Bt—Biotite, Mnt—Montmorillonite, AD—Air-dried.

Sample ID	<2 µm					2–6 µm						<2 µm (AD)		2–6 µm (AD)	
	Ill	Kln	Cor	I/S	Bt	Ill	Kln	Cor	I/S	Mnt	Bt	FWHM	KI	FWHM	KI
OM_2	no material					—				+				0.18	0.22
NH_1	+	—			++	—	+				++	0.48	0.69	0.34	0.46
NH_2	——	—			++	——	—				++	0.31	0.43	0.32	0.44
NH_3	+	+	++			+	+	++				0.63	0.92	0.59	0.87
NH_4	++	+	—			++	+	—				0.81	1.29	0.73	1.16
NH_5	+	—	—			—	—	—			+	0.69	1.09	0.37	0.55
NH_6	+	++		—		+	++		—			1.37	2.25	1.15	1.87
NH_7	+	+		—		+	+		—		+	1.01	1.63	0.71	1.13
NH_8	+	+	+			+	++	+				1.10	1.79	1.18	1.93
NH_9	+	+	++			+	+	++				0.85	1.36	0.56	0.87
NH_10	+	—			++	+	—				++	0.46	0.65	0.51	0.73
NH_11	++	—	+			++	+	+				0.67	1.05	0.62	0.97
NH_12	++	+		—		++	+		—			0.57	0.89	0.46	0.69
NH_13	+	—	+			+	—	+				0.46	0.69	0.38	0.56
NH_14	+	——	—		+	+	——	—			+	0.59	0.92	0.44	0.66
NH_15	+	—	+			+	—	—				0.69	1.08	0.50	0.77

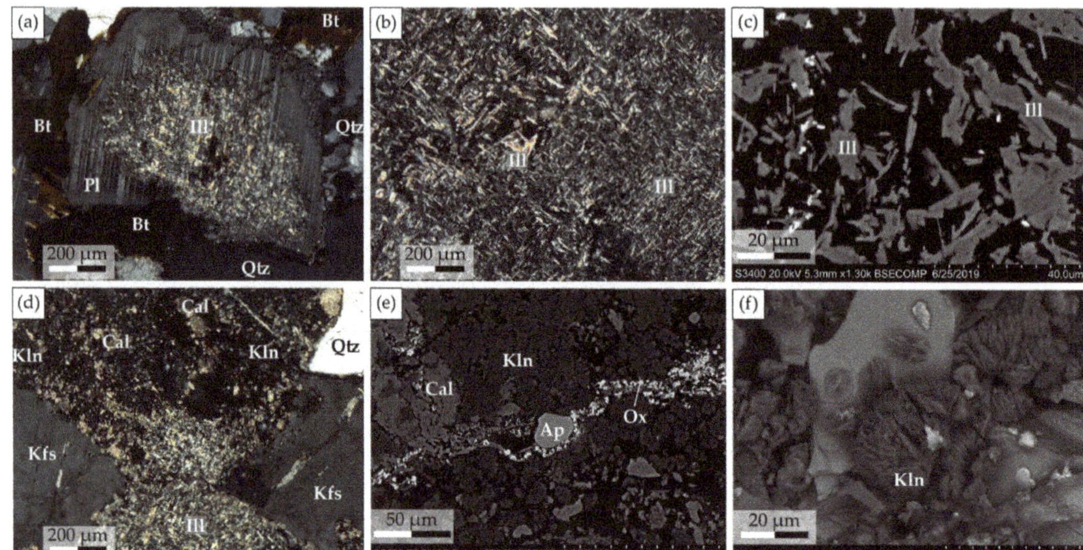

Figure 5. Photomicrographs showing the plagioclase transformation progress in the Noble Hills granite. (**a**) Progressive illitization of plagioclase initiating in the core of the mineral under optical microscope in polarized—analyzed light. (**b**) Needles shape illite replacing plagioclase in polarized—analyzed light under optical microscope. (**c**) Back-scattered electron image of needles/flakes shape illite replacing plagioclase. (**d**) Plagioclase completely replaced by illite, kaolinite and calcite and non-altered K-feldspar under optical microscope in polarized—analyzed light. (**e**) Back-scattered electron image showing a fan shape kaolinite, calcite and oxide veinlet. (**f**) Back-scattered electron image showing a magnified view of a single mineral of kaolinite under SEM. Abbreviations (except for "Ox") after [45]: Ap—Apatite, Bt—Biotite, Cal—Calcite, Ill—Illite, Kln—Kaolinite, Kfs—K-feldspar, Ox—Oxides, Pl—Plagioclase, Qtz—Quartz.

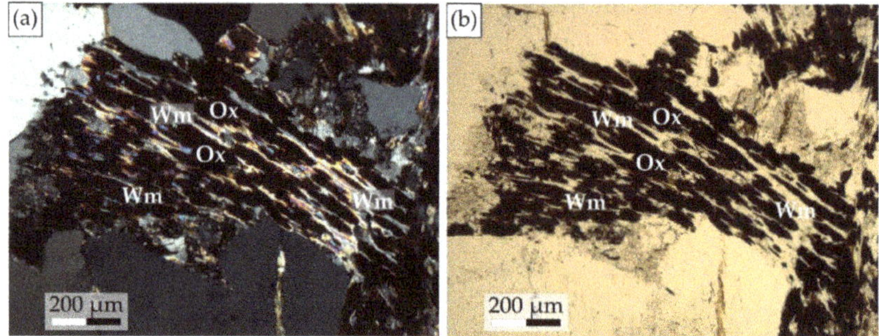

Figure 6. Photomicrographs showing a biotite completely altered and replaced by oxides and K-white mica (essentially illite) according to the cleavage planes (**a**) in polarized—analyzed light under optical microscope and (**b**) in polarized—analyzed light under optical microscope. Abbreviations (except for "Ox and Wm") after [45]: Bt—Biotite, Cal—Calcite, Kfs—K-feldspar, Wm—K-white mica, Ox—Oxides, Pl—Plagioclase, Qtz—Quartz.

All the samples are plotted in the Streckeisen ternary diagram (Figure 7) [68]. The OM and NH granites of this study are both defined as monzogranites. The OM sample analyzed by [34] (Figure 2a, blue star) shows a different composition (Figure 7, blue star). It is rather a monzonite, as defined by [34], than a monzogranite, as defined in this study.

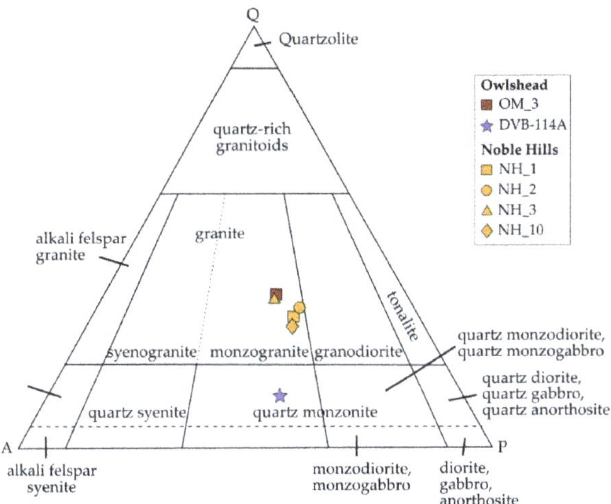

Figure 7. Normative composition of one sample from the OM and 4 samples from the NH in a QAP (Quartz-Alkali-feldspar-Plagioclase) ternary diagram [68].

4.1.3. Clay Minerals Identification and Kübler INDEX

Clay minerals from the <2 μm and 2–6 μm fractions were separated. The clay composition of both fractions of each studied sample (OM and NH) is given in Table 3, based on [51].

Owlshead Mountains

The OM sample clay analyses confirm the presence of illite observed under optical microscope. They also reveal the pattern of montmorillonite (Mnt) [51], a common smectite (Figure 8). Its very intense 001 peak allows the determination of its amount representing 15% of the clay fraction. This peak is characterized by a shift from 14.87 Å (air-dried) to 16.90 Å after glycol solvation and it collapses at 10.09 Å after heating. After glycol solvation, new peaks appear at 5.61 Å and 8.46 Å.

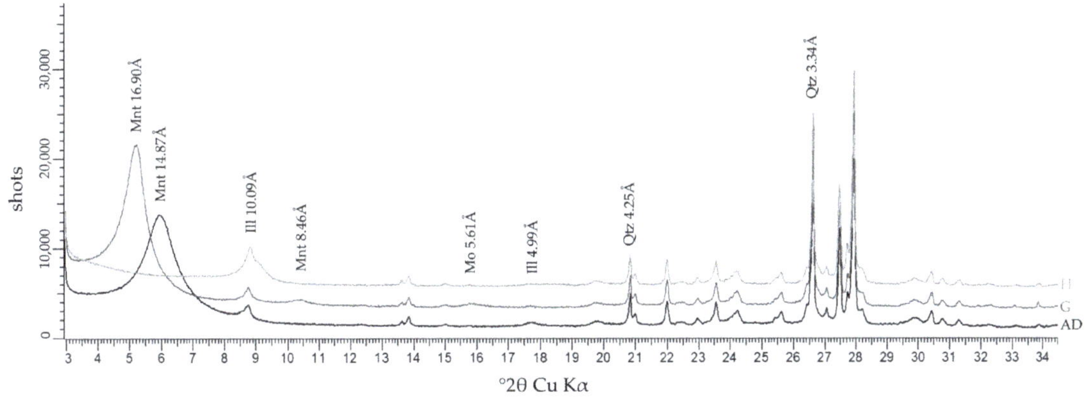

Figure 8. XRD result obtained for the clay fraction < 2 μm in Air-Dried (AD), Glycolated (G) and Heated (H) of the samples in the OM granite and showing a montmorillonite (Mnt) pattern.

Noble Hills

Two typical XRD patterns of clay minerals are identified in the NH samples for the <2 μm fraction (Figure 9):

- Corrensite, kaolinite and illite are identified in the first pattern, where corrensite is well known as the trioctahedral variety of regular 50:50 mixed-layer chlorite/smectite [69] (Figure 9a). It is characterized by (1) the peak at 13.60 Å in air-dried conditions, shifting to 15.62 Å after glycol solvation and collapses to 11.72 Å after heating, and (2) new peaks at 7.78 Å, 5.15 Å and 3.44 Å appear after glycol solvation and disappear after heating. The corrensite found in the NH granite is considered as a low charge corrensite after [51].

- Illite/smectite (I/S) mixed-layer, kaolinite and illite (Figure 9b) are identified in the second pattern, where I/S is illite-rich (R3), with more than 90% of illite and R representing the Reichweite parameter [70]. I/S is characterized by a large peak at 10.08 Å in air-dried, becoming narrower when it collapses to 9.93 Å after glycol solvation and by a peak at 5 Å swelling after glycol solvation.

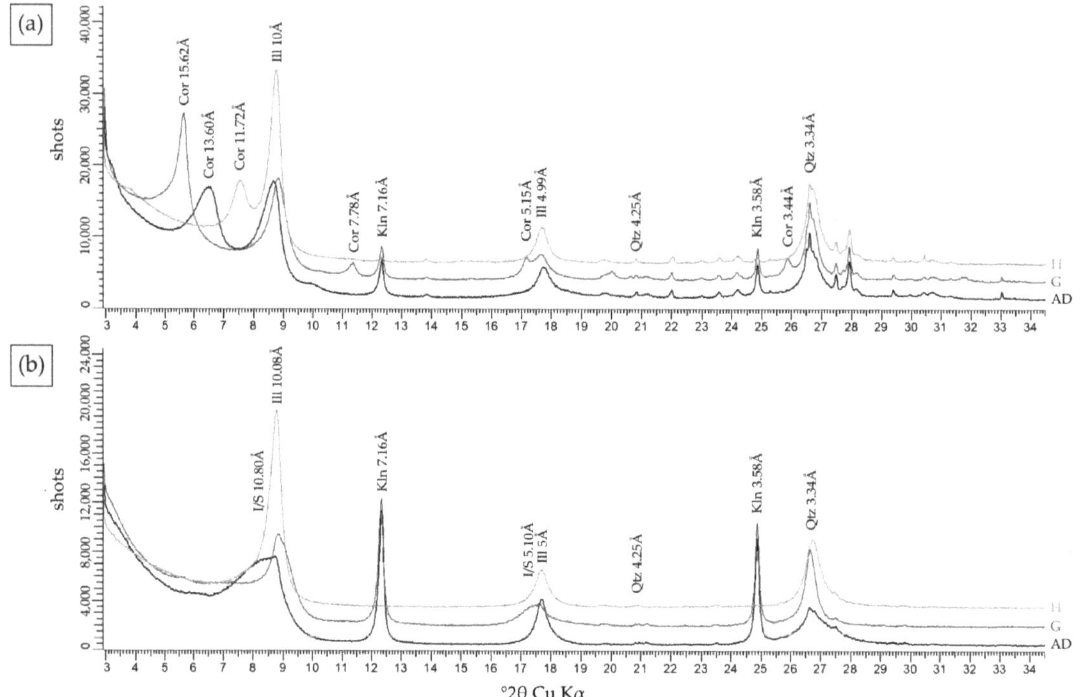

Figure 9. XRD patterns obtained for the clay fraction < 2μm in Air-Dried (AD), Glycolated (G) and Heated (H) for the NH granite and showing the clay composition of the samples in the NH granite: (**a**) corrensite (Cor), a chlorite/smectite mixed-layer, and (**b**) illite/smectite mixed-layer (I/S). Abbreviations (except for "Cor and I/S") after [45]: Bt—Biotite, Kln—Kaolinite, Qtz—Quartz, Ill–Illite.

In the NH KI values range from 2.25Δ°2θ to 0.43Δ°2θ for the <2 μm fraction and from 1.93Δ°2θ to 0.44Δ°2θ for the 2–6 μm fraction (Table 3). The spatial distribution of the samples and the KI values are shown in Figure 10. A NW to SE decrease in KI values in the main granitic body is distinguishable in the <2 μm fraction (Figure 10a) and confirmed in the 2–6 μm fraction (Figure 10b).

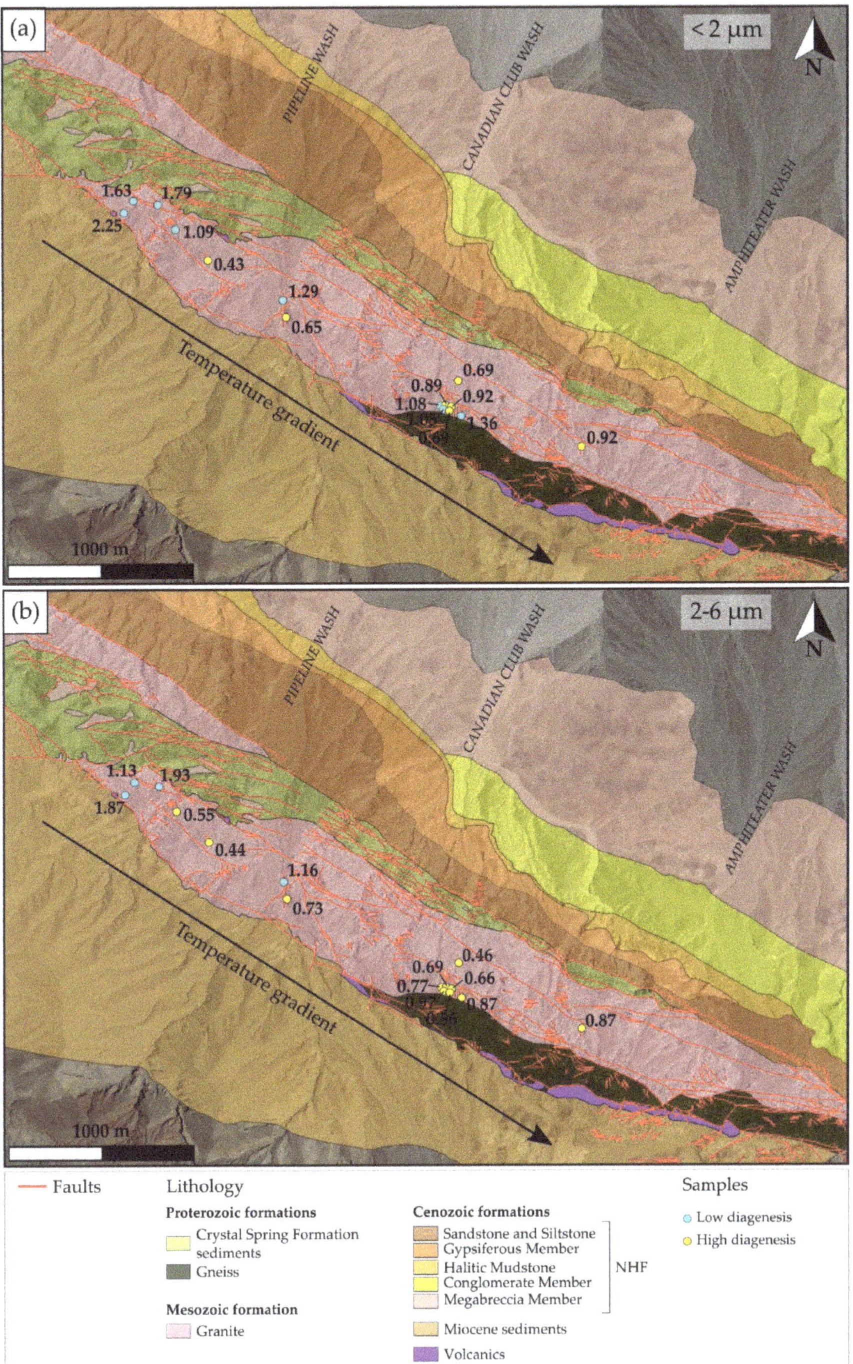

Figure 10. Geological map (see Figure 2b) representing the Kübler Index (KI) in air dried conditions of each sample and the corresponding metamorphic zone showed by the color of the dots for (**a**) the fraction < 2μm and (**b**) the fraction 2–6 μm.

4.2. Geochemical Analyses

4.2.1. Major Element Bulk Rock Chemistry

Major element geochemistry (Table 4) allows classifying the granite samples in different diagrams (Figures 11 and 12).

Table 4. Major elements of the Owlshead and Noble Hills granites.

Sample ID	OM_3	NH_10	NH_1	NH_2	NH_3
	Oxides (weight %)				
SiO_2	74.65	68.54	69.56	68.69	68.78
Al_2O_3	12.90	15.36	14.93	14.53	14.99
Fe_2O_3	1.83	3.02	2.83	4.29	2.95
MgO	0.31	0.57	0.56	1.15	0.78
CaO	1.09	2.62	2.20	2.50	1.99
Na_2O	3.51	3.31	3.42	3.19	2.78
K_2O	4.43	4.31	4.10	3.60	4.22
TiO_2	0.20	0.29	0.26	0.46	0.28
P_2O_5	0.07	0.14	0.13	0.20	0.14
MnO	0.08	0.09	0.08	0.12	0.08

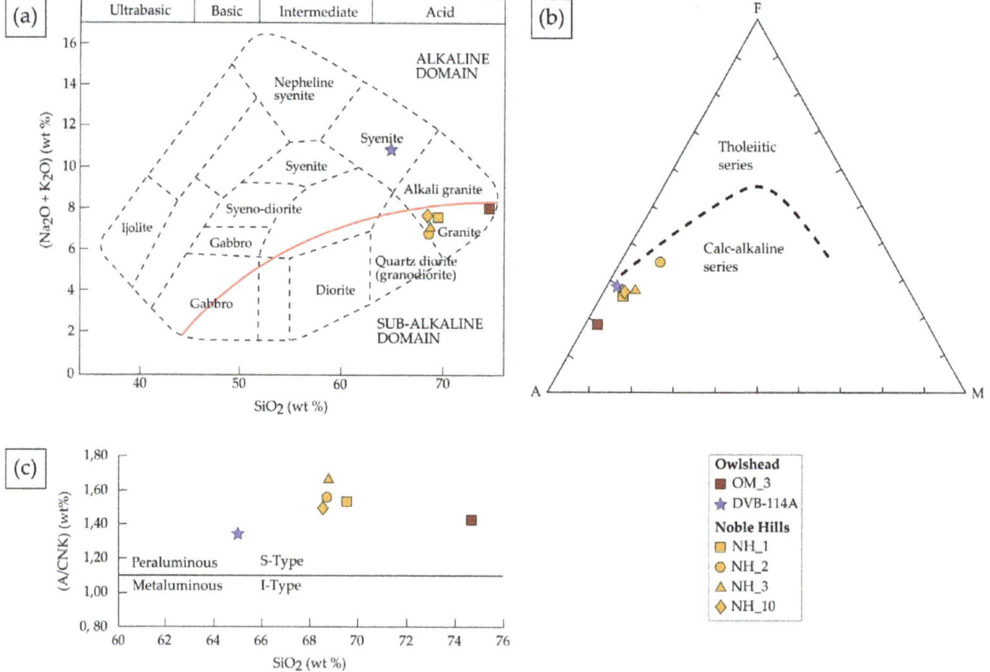

Figure 11. (**a**) Total alkali versus silica diagram ([71] adopted for plutonic rocks by [72]). (**b**) AFM (Alkali-Fe_2O_3-MgO) classification diagram established by [73]. (**c**) A/CNK (Al_2O_3/(CaO + Na_2O + K_2O)) versus silica diagram. The blue star corresponds to a sample from the OM analyzed by [34].

In the Na_2O + K_2O versus SiO_2 classification (called TAS) diagram of [71], modified after [72] for plutonic rocks, all the samples plot in the sub-alkalic domain. They all have a granite composition close to the granodiorite zone (Figure 11a). However, the sample from [34] plots in the syenite field, while it plots in the monzonite field in the Streckeisen diagram (Figure 7). The AFM triangular plot classifies all the samples as calc-alkaline

(Figure 11b). The silica content of the OM granite 74.65 wt.% is higher than that of the NH granite which ranges from 68.54 to 69.56 wt.%. According to the SiO_2 versus A/CNK diagram (Figure 11c) [74], all the samples are peraluminous rocks, and are found rather far from the boundary with metaluminous S-type granites, with a A/CNK between 1.4 and 1.6 wt.%. Harker diagrams complete the information by showing a high content of K_2O ranging from 3.6 to 4.43 wt.%, and plot the samples in the high-K calc-alkaline domain (Figure 12a) which limits where defined by [75]. However, the OM sample has a lower Al_2O_3 content (12.9 wt.%) than the NH samples (14.53–15.36 wt.%) forming a separate cluster (Figure 12b). Again, the sample from [45] (DVB-114A—blue star) shows a large difference in composition compared to OM_3.

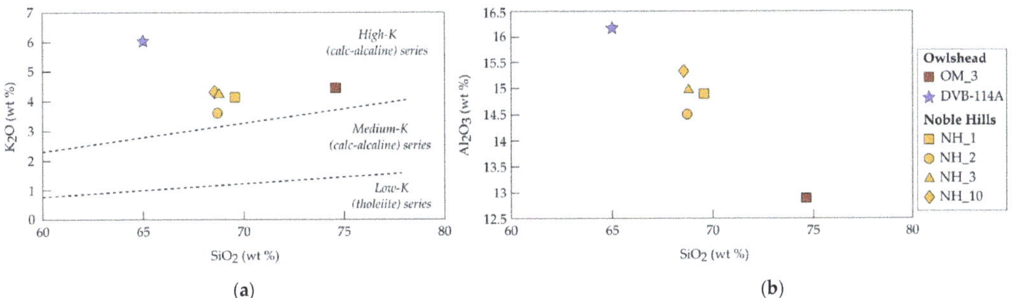

Figure 12. Harker diagrams showing the variation of (**a**) K_2O [75] and (**b**) Al_2O_3 in the OM and NH granites. A sample from the OM analyzed by [34] is represented by the blue star.

The loss on ignition (LOI) is of 0.7 wt.% for the OM granite and ranges from 1.1 to 2.8 wt.% for the NH granite (Table 5), showing that the OM granite contains less volatile elements than the NH granite, even for the freshest samples (NH_10, NH_1 and NH_2). By comparing the alteration degree of each sample estimated from optical observations of plagioclase and biotite (Table 5) and the LOI values, we note that a low LOI corresponds to a slight alteration, e.g., OM_3 or NH_2 and a high LOI corresponds to a more pronounced alteration, e.g., NH_3.

Table 5. Loss on ignition values compared to the alteration degree of plagioclase and biotite estimated under optical microscope.

Sample ID	OM_3	NH_10	NH_1	NH_2	NH_3
Loss on ignition (LOI) (wt.%)	0.7	1.5	1.7	1.1	2.8
Alteration degree (%)	2	10	9	4	18
Plagioclase alteration (%)	*5*	*20*	*20*	*10*	*40*
Biotite alteration (%)	*0*	*15*	*10*	*0*	*20*

In diagrams representing selected major elements versus the LOI (Figure 13), LOI values of NH samples show a positive correlation with K_2O and a negative correlation with Na_2O and CaO. SiO_2 and Al_2O_3 contents are approximately constant, 68.54–69.56 wt.% and 14.53–15.36 wt.% respectively. MgO content varies a lot (0.56–1.15 wt.%) showing no clear correlation with the LOI. K_2O and Na_2O contents of the OM granite are equivalent to those of the freshest NH granites (NH_2, NH_10 and NH_1). However, the OM granite sample presents a higher amount of SiO_2 and a lower amount of Al_2O_3, CaO and MgO. The chemistry of the OM granite seems different from that of the NH granite regarding Figures 12 and 13 even though, according to [11], the NH granite derived from the OM granite.

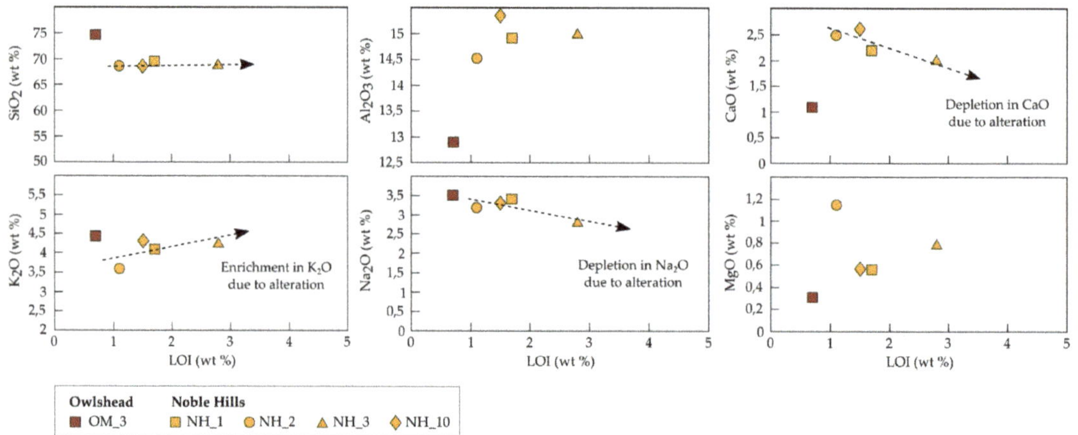

Figure 13. Plot of selected major element oxides (SiO$_2$, K$_2$O, Al$_2$O$_3$, Na$_2$O, CaO and MgO) versus LOI (loss on ignition).

4.2.2. Trace Element and REE Bulk Chemistry

Trace elements and Rare Earth Elements (REE) (Table 6) analyses were performed on the same samples as for major elements.

Table 6. Trace elements chemical composition of the Owlshead and Noble Hills granites.

Sample ID	OM_3	NH_10	NH_1	NH_2	NH_3
	Trace elements (ppm)				
Be	2	2	2	3	2
Co	1.5	4.2	2.9	6.4	3.5
Cs	1.6	2.4	1.1	3.4	1.8
Ga	15.6	16.7	16.0	17.2	14.1
Hf	4.3	3.6	4.1	4.6	4.6
Nb	22.3	13.8	11.4	18.9	10.5
Rb	154.7	144.9	103.0	139.9	115.7
Sn	2	1	<1	2	<1
Sr	116.2	275.2	237.5	268.2	182.7
Ta	1.6	0.9	0.7	1.3	0.6
Th	19.3	11.4	10.2	9.0	11.0
U	1.8	2.2	1.5	1.4	1.5
V	16	33	22	40	25
W	<0.5	<0.5	0.7	<0.5	<0.5
Zr	139.8	134.2	156.6	180.6	159.0
Y	19.7	22.5	17.5	21.4	17.0
Ba	554	1009	845	704	772
Ni	<20	<20	<20	<20	<20
Sc	4	5	4	8	4
Cr$_2$O$_3$	<0.002	<0.002	<0.002	<0.002	<0.002
Be	2	2	2	3	2

Table 6. *Cont.*

Sample ID	OM_3	NH_10	NH_1	NH_2	NH_3
	Rare Earth Elements (ppm)				
La	52.1	28.3	26.2	21.7	26.8
Ce	97.0	53.0	48.1	40.8	52.2
Pr	9.97	5.93	5.31	4.59	5.62
Nd	31.7	20.6	18.8	16.6	19.5
Sm	5.29	3.82	3.51	3.56	3.48
Eu	0.53	0.89	0.89	0.81	0.81
Gd	4.24	3.66	3.14	3.60	3.22
Tb	0.65	0.59	0.51	0.59	0.51
Dy	3.66	3.63	3.05	3.65	2.96
Ho	0.69	0.78	0.62	0.73	0.59
Er	2.05	2.25	1.87	2.18	1.81
Tm	0.30	0.33	0.28	0.33	0.27
Yb	2.08	2.32	1.77	2.16	1.86
Lu	0.33	0.37	0.28	0.36	0.28
TOT/C	0.02	0.14	0.21	0.04	0.26
TOT/S	<0.02	<0.02	<0.02	<0.02	<0.02

Chondrite-normalized REE patterns (Figure 14a) present enriched samples in light rare earth elements (LREE) relative to heavy rare earth elements (HREE) in both granites. However, the OW granite shows a higher abundance of LREE and a stronger negative Eu anomaly than the NH samples. In spite of different degrees of mineral alteration, all the NH samples follow the same trend. The primitive mantle-normalized multi-element diagram (Figure 14b) is characterized by distinct negative anomalies for Nb and Sr and high Th and U contents, typical for upper crustal composition [76]. They also show variable but high Cs, Rb and Ba contents. More generally, samples are relatively rich in large-ion lithophile elements (LILEs) such as Rb, Ba, Th and U, with Sr having the greatest depletion relative to the LILEs. High field strength elements (HFSEs) such as Ta, Nb, Zr and Hf are depleted compared to the LILEs. The OW granite shows, once again, a different trend compared to the NH samples. It has higher Th, Ta, Nb, La, Ce, Nd and Sm contents and a lower Ba content. On both diagrams of Figure 14, the sample DVB-114A [34] shows a different pattern compared to OM_3.

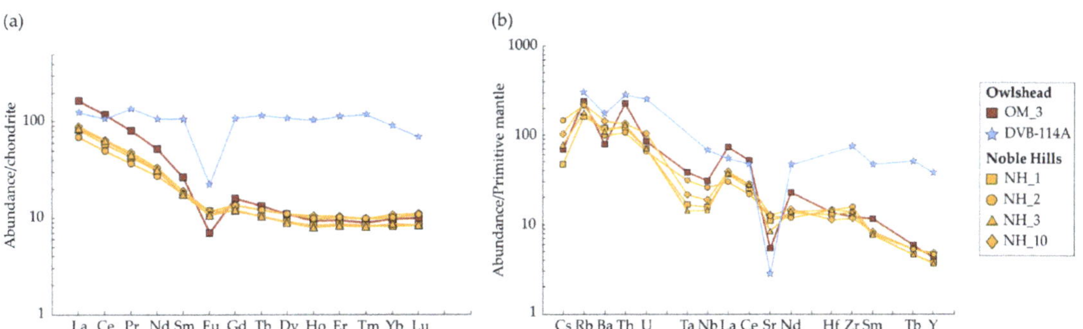

Figure 14. (**a**) Chondrite-normalized (values from [77]) rare earth element (REE) patterns of samples from the OM and the NH. (**b**) Primitive mantle-normalized (values from [78]'s slight revision of [79]) multi-element diagram showing trace element patterns of samples from the OM and the NH.

4.3. Calcimetry and Porosimetry

Calcite is present in all the samples, essentially linked to the alteration of the plagioclase. The calcite content is of 0.55% for the OM granite and ranges from 0.55 to 6.53% for the NH granite, with an average of 2.2% and a standard deviation of 1.64%. The porosity is of 2.28% for the OM granite and ranges from 2.21 to 5.17% for the NH granite. A clear positive correlation can be seen between the calcite content and the LOI (Figure 15a). Positive correlations are also visible between the porosity and the LOI and between the porosity and the calcite content (Figure 15b,c). However, the sample NH_7 in Figure 15c is different. Having a low calcite content, it presents a higher porosity compared to the others. Those diagrams show that a low LOI corresponds to a low amount of calcite and a low porosity and that a low amount of calcite fits with a low porosity except for NH_7.

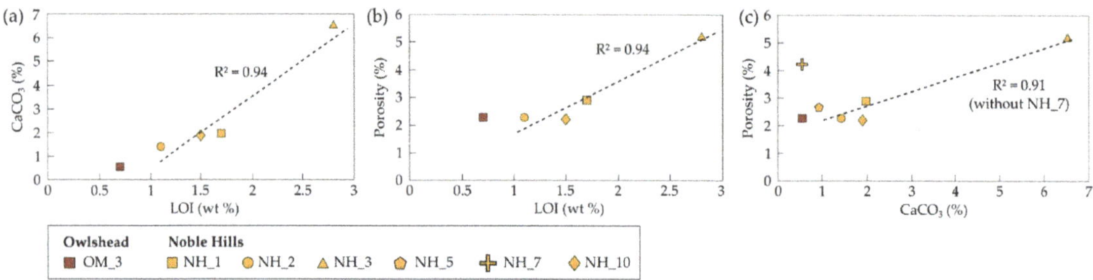

Figure 15. (**a**) Plot of the porosity versus the calcite content. (**b**) Plot of calcite content versus LOI (Loss on ignition). (**c**) Plot of porosity versus LOI.

5. Discussion

5.1. Petrogenesis of the OM and NH Granites

Bulk rock analyses of the OM and NH granites indicate for both of them a calc-alkaline, monzogranite composition, S-type in character (Figure 11). Rämö et al. (2002) [34] also investigated the OM in the same batholith, but south of our sampling area (blue star in Figure 2a). The comparison of geochemical data from their sample with the samples studied here, shows differences in terms of chemical composition (Figures 7, 11, 12 and 14). The OM seems to be a heterogenous pluton.

Chondrite-normalized REE patterns for the NH granite samples show the same trend (Figure 14a), but are different from the OM granite one. REE are relatively immobile during low-grade metamorphism and hydrothermal alteration [76]. The enrichment of the LREE and the depletion of Eu indicate the degree of magmatic differentiation the rock underwent. The Eu negative anomaly is attributed to the plagioclase fractionation involved in the setting of granites [80]. However, the breakdown of plagioclase and biotite can release some REE except Eu, which can be accommodated by illite formation [81].

The same observations were done through the primitive mantle-normalized spider diagrams, which shows variations in Ta and Nb contents of the NH samples (Figure 14b). Those variations are in good correlation with the LOI: the higher the LOI, the lower the Nb and Ta contents. Li et al. (2013) [82] observed Nb/Ta ratios in altered domains in granitic rocks due to the Nb and Ta decreasing content during the illitization stage. The LOI is directly related to the percentage of alteration, the same phenomenon is observed in our samples. The depletion of Sr in the NH granite might be due to the alteration of magmatic primary plagioclase [82]. However, Sr content in OM is relatively low.

Those diagrams show and confirm the difference in terms of chemical composition between the NH and the OM, which was suspected during thin sections observations. A comparison between our own chemical data within the OM and the data presented by [34] show substantial differences meaning that the OM granite is not homogeneous. Given the fact that the NH granite is considered as a transported part of the OM [11], it is therefore not surprising that the NH granite presents local composition variations. The OM samples are considered as representing the protolith prior to the mineral transformations identified in the NH granite.

Microcline was identified in the OM granite and only in the southern part of the NH indicating that this part of the range might have undergone a slower cooling than the northern part of the range or a warming.

5.2. Thermal Evolution of the NH Granite

Petrographic investigations have shown that plagioclase and sometimes biotite are replaced by illite. A recent study from [83] has confirmed that IC provides a useful method for characterizing regional grades of diagenesis and low-grade metamorphism. As already mentioned, in some conditions the alteration of a granite, resulting from fluid circulation, shows similarities with diagenetic reactions present in feldspathic sandstones [19], meaning that it is possible to characterize a regional grade by using the KI values obtained in granitic rocks.

As shown in Figure 2, the NH range is a structure stretched NW-SE. The KI values display a trend following this direction with decreasing values towards the south-east (Figure 10). KI values can be associated to different metamorphic zones corresponding to ranges of temperature (Table 1, [60]). The northwestern part of the range, characterized by high KI values, reveals mostly low-grade diagenesis tending progressively to high-grade diagenesis roughly toward the SE. A decrease of KI values indicates an increase in temperature [59]. Thus, this tendency reflects a temperature gradient increasing from the NW toward the SE. The elevation being higher in the south-east, with a higher temperature in this zone might indicate that the southern part of the range was more buried than the rest of the range and has been exhumed. A northeast-vergent contractional deformation is well expressed along the NH range in addition to the strike-slip deformation. It increases in intensity where the NH range converges with the Avawatz mountains ([38] and references therein), at the intersection between the SDVFZ and the GFZ [30]. Chabani et al. (2021) [12] have identified E-W structures in this area and they suppose that these structures are linked to the activity of the GFZ and the convergence of the Avawatz and the NH range. It is tempting to propose that this convergence implied the elevation of the topography at the southern part of the NH range. A work is in progress about exhumation history reconstruction by means of isotopic dating.

5.3. Alteration Parageneses

The OM and NH granites characterization reveals the presence of various secondary minerals as oxides, epidote, corrensite, K-white mica, calcite, kaolinite, illite/smectite mixed-layer and montmorillonite. Those minerals are well known as being the product of alteration processes due to the interaction of a circulating fluid with the surrounding rock. They are in equilibrium with the new environment in response to temperature, pressure and composition of the altering fluid [6,84–86]. In the OM and the NH granites, only plagioclase and biotite are affected. In the case of intense circulation, some primary minerals like plagioclase or biotite may be completely replaced. The newly formed minerals consist mainly in clay minerals [9]. According to optical observations, SEM-EDS and XRD analyses, two types of alteration processes have been identified in the OM and NH granites which are classified as (1) propylitic alteration and (2) argillic alteration. The characterization of the paragenesis and the alteration processes of the granites investigated in this study help to refine a part of the history of the OM and NH granitic basements (Figure 16).

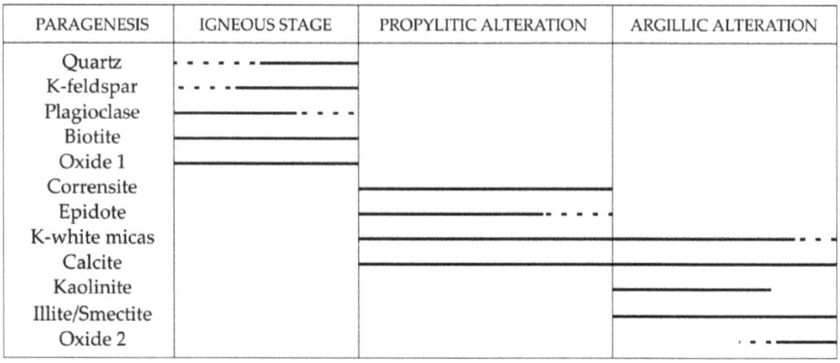

PARAGENESIS	IGNEOUS STAGE	PROPYLITIC ALTERATION	ARGILLIC ALTERATION
Quartz	·····——		
K-feldspar	····——		
Plagioclase	——····		
Biotite	——		
Oxide 1	——		
Corrensite		——	
Epidote		——····	
K-white micas		————————····	
Calcite		————————————	
Kaolinite			——
Illite/Smectite			——
Oxide 2			··——

Figure 16. Paragenesis sequence and alteration evolution in the OM and NH granite.

5.3.1. Propylitic Alteration

The newly formed minerals depend on the composition of the host mineral. Corrensite, epidote, K-white micas, calcite and iron oxides were identified in the OM and NH granites (Figure 16) by means of optical observations, SEM-EDS and XRD analyses:

1. The calcite as in the OM granite occurs as infills of the microcracks without interacting with the surrounding rock, as well as at grain boundaries.
2. Mixed-layer clay minerals are the intermediate products of reactions involving end-member clays [87]. Corrensite, a chlorite/trioctahedral smectite mixed-layer phyllosilicate is considered as a stable mineral and also as an indicator of propylitic alteration [86,88,89]. It replaces partially biotite and occurs between 160–250 °C in geothermal fields [6,84,90].
3. Epidote crystallization occurs around 220 °C [86]. It is also one major indicator of the propylitic alteration with corrensite [89].
4. The presence of K-white mica flakes allows to fix temperatures around 230 °C up to 350 °C [84,91].

All those secondary minerals occur at temperatures between approximately 160 °C and 350 °C. The presence of corrensite and epidote is the major indicator of a stage of propylitic alteration. The propylitic alteration (Figure 17b) is considered as an earlier pervasive alteration stage. It is common at the margins of alteration zones produced at low fluid/rock ratio [6] and it takes place at the end of the crystallization of the granite [92]. The propylitic alteration results in the partial recrystallization of primary minerals (biotite and plagioclase) in secondary propylitic assemblages by interstitial fluids trapped into the grain boundaries during the cooling of the pluton [88]. Its effects are discrete, but both the OM and NH massifs are affected.

5.3.2. Argillic Alteration

Other newly formed mineral assemblages were identified only in the NH granite. They consist in illite + kaolinite + mixed-layers as illite/smectite (I/S) + calcite and oxide (Figure 16):

1. Illitic minerals are well known to be indicators of fluid circulation as well as paleo-circulation systems [93]. [19,94] show that illite crystallization episodes can occur, for example, in a temperature range of 120 to 160 °C, corresponding to the argillic alteration facies. The illitization process mainly develops in plagioclase and biotite. It is a form of alteration product found extensively in granitoids, and felsic rocks, whereas K-feldspar remains relatively unaltered [8,67,85].
2. According to [84], the presence of kaolinite in alteration paragenesis indicates a fluid temperature lower than 200–150 °C. Kaolinite is stable under more acidic conditions than illite, with pH values ranging from about 4.5 to 6. It also represents a more

advanced product of hydrolysis reaction due to a high H$^+$ activity in hydrothermal fluids.

3. The illite-rich (R3) I/S mixed-layer form around 150 °C [84,87], with more than 90% of illite based on [51].
4. Plagioclase, oligoclase in composition, presents patches of calcite. Those patches are interpreted as a product of Ca release due to plagioclase alteration.
5. Oxides can be present along the cleavages of the altered biotite. They are interpreted as the result of Mg and Fe release during biotite alteration.

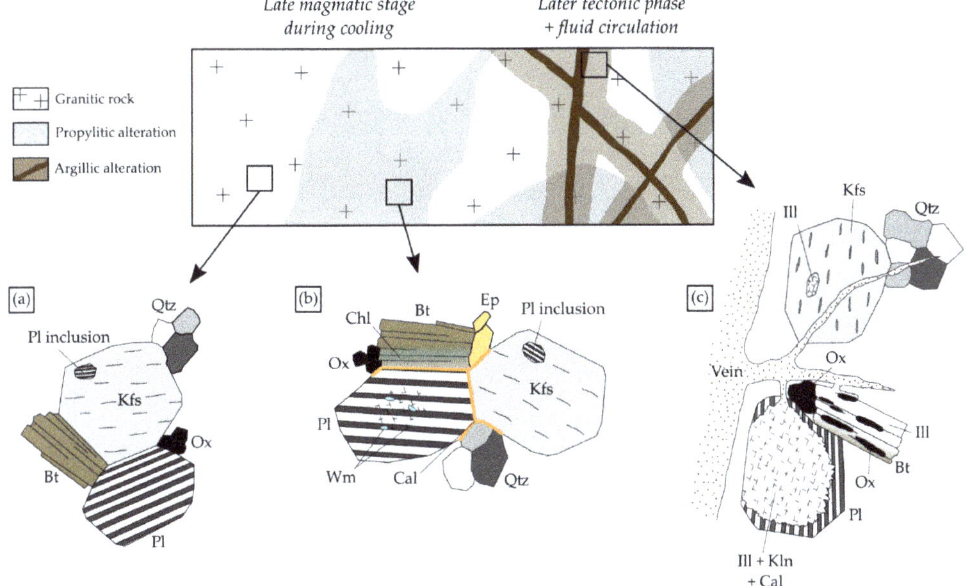

Figure 17. Schematic representation of the different alteration processes which can be observed in the OM and NH granites. (**a**) Fresh granite non-affected by alteration. (**b**) Granite affected by the propylitic alteration showing the slight chloritization of biotite and the crystallization of K-white micas in the core of plagioclase. (**c**) Granite affected by the argillic alteration and localized along fractures where fluid has circulated. Biotite are locally completely transformed into illite and plagioclase into illite, calcite and/or kaolinite. Perthites and inclusions in K-feldspars can also be altered. Abbreviations (except for "Ox and Wm") after [45]: Bt—Biotite, Cal—Calcite, Chl—Chlorite, Ep—Epidote, Ill—Illite, Kln—Kaolinite, Kfs—K-feldspar, Ox—Oxide, Pl—Plagioclase, Qtz—Quartz, Wm—K-white mica.

Plagioclase and quartz form an interconnected skeleton through the texture. The difference of physical and chemical behaviors between both minerals results in different types of porosity. Quartz shows microcracks and plagioclase shows dissolution pits (Figure 18). Ref. [95] observed the same in the Soultz-sous-Forêts granite. They also showed that the exchange surface between plagioclase and a fluid is around 20 times higher than in quartz. As a consequence, they assume that all the pores are interconnected in plagioclase. They considered this mineral as the main path for fluid flow. Thus, this can be also available in the NH granite. K-feldspar is not affected by the alteration, but perthites and mineral inclusions present in the K-feldspars can be dissolved.

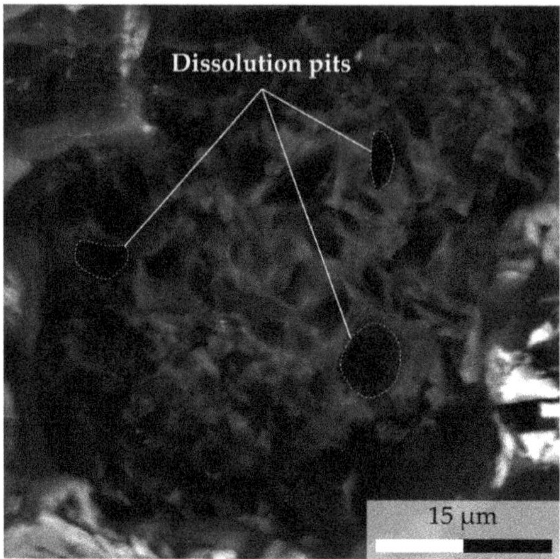

Figure 18. Back-scattered image showing a magnified view of plagioclase dissolution pits under SEM (15 kV 5.3 mm × 1.7 k BSECOMP).

The crystallization of these minerals occurs at temperature between 130 and 200 °C. This range of temperature corresponds to the stage of argillic alteration [84]. The argillic alteration (Figure 17c) consists into the chemical leaching and clay enrichment processes produced at lower temperatures [19,88,96]. This alteration is known as a vein alteration organized in the vicinity of fractures where fluids have circulated [86,88].

5.3.3. Evidences of Weathering

XRD analyses on an oriented sample of the clay fraction from the OM reveal the presence of Montmorillonite (Mnt) (Figure 8). Mnt, a common smectite pattern, results either from argillic (vein) alteration [84,89] or from weathering [97,98]. As described above, the OM granite appears as "fresh" from (1) field observations: no veins were observed at outcrop scale and macro-scale and the granite appears competent, (2) thin-section analyzes: no veins or veinlets, calcite crystallizations are only present at grain boundaries, and (3) geochemical analyzes, which indicate a very low LOI (0.7 wt.%). No vein was observed at any scale and secondary minerals are all characteristic of propylitic alteration. Those data show that the Mnt cannot result from argillic alteration. In the case of the OM granite, the Mnt is interpreted as a signature of weathering. As its amount is low (15% of the clay fraction), the global alteration of the granite is considered as being dominated by the propylitic alteration. With the OM being considered as the protolith, weathering can be considered negligible in both OM and NH granites.

5.3.4. Alteration Stage Occurrences

The OM granite presents evidence of propylitic alteration only, while the NH granite presents both propylitic and argillic facies. The argillic facies seems most of the time to overprinting the propylitic facies. However, some samples present only evidence of argillic alteration. In those cases, either the propylitic facies have either completely disappeared, or some zones were not initially affected by the propylitic alteration. Some minerals, such as illite and kaolinite, crystallize during the argillic alteration under different conditions [84]. This suggests that several episodes of alteration could have occurred in the NH granite. Afterward, weathering can occur at ambient temperature, when the granite was exhumed. These alteration overprints result also in a retrograde evolution in terms of temperature (T) withT propylitic > T argillic > T weathering.

In the NH granite, the argillic alteration seems to be associated to fracturing. A fresh granitic rock is not permeable enough to allow fluid circulation [6]. This suggests that the development of fractures is an important factor allowing fluid to circulate [6], the observation that the freshest NH granite is more altered than the one from the OM (Figure 3) can be related to the fact that the NH lie along an important shear corridor. None of NH samples considered in the present study are fracture free, even though they were collected far from the faults, and each of them shows evidences of argillic alteration. In spite of the low fracture density the existing number of microfractures is enough to allow fluid to significantly percolate through the host rocks and chemically interact with it. A second paper (PART 2, this issue) focusses on the role of the fracture system on the granite alteration processes.

5.4. Effects of Alteration on Petrogaphic and Petrophysical Behaviour

Petrographic observations and the range of LOI (0.7–4.1 wt.%) values confirm that the NH granite experienced alteration. LOI is so directly related to the degree of mineral alteration as done by [99,100]. Mineral changes being related to the propylitic and argillic alterations, the LOI can be defined in this study as a monitor for alteration processes. The effects of alteration on element transfers can be evaluated thanks to the diagrams plotting SiO_2, Al_2O_3, CaO, K_2O, Na_2O and MgO against LOI (Figure 13). In Figure 13, the K_2O, Na_2O, CaO and MgO define a broad correlation with LOI, indicating that they may have been mobile during alteration [99]. The NH granite shows that Na_2O and CaO have a negative correlation which can be related to the alteration of the plagioclase (oligoclase initial composition). Indeed, plagioclase can be depleted in Na and Ca mobile elements when it interacts with a fluid [101]. Likewise, the K_2O is observed in alteration products of plagioclase as illite and shows an enrichment with the LOI increase. Newly formed minerals, as presented above, are related to the recrystallization of the plagioclase as a result of its interaction with the fluid. As for the CaO and MgO contents, they remain high compared to the OM granite. This can be linked to an external contribution. In a whole, the NH granite was affected by an alkali alteration, which results in major compositional changes. Thus, all the elements, Si and Al excepted, have partly left the system, meaning that the system is open. Their content differences in the NH granite compared to the OM granite could be explained by the depletion of the most mobile elements during the alteration of the NH granite. This can so influence the percentage of each element. Otherwise, it can be suggested that the visible variations of SiO_2 and Al_2O_3 were controlled by protolith composition rather than alteration processes.

Link alteration degree and amount of calcite is difficult since petrographic analyses of the samples is qualitative. However, Figure 15a shows a positive correlation between LOI and calcite content. Ledésert et al. (2009) [61] show that calcite can be encountered in high amount in altered zones. Thus, in the NH, the correlation shows that the higher the calcite content, the higher the degree of alteration and vice versa. In the same way, LOI can be related to porosity (Figure 15b) with a low LOI linked to low porosity (NH_1, NH_2 and NH_10) and high LOI linked to as high porosity (NH_3). Therefore, the increase of porosity is also linked to the increase of alteration. Studies [102,103] showed that the porosity increases from unaltered to altered granite. Figure 15c shows a similar correlation between the calcite content and the porosity. Samples having a low calcite content (NH1, NH_2, NH_5 and NH_10) present a low porosity, while NH_3 has a high calcite content associated with a high porosity. One exception can be observed. The sample NH_7 presents a low calcite content for a high porosity. This can be explained by the presence of larger microfractures compared to the other samples, which increase the porosity. Thus, it is not always possible to link the porosity to the calcite content. According to [104], the average calcite content of a fresh granite is 0.252 wt.%, and does not exceed 1.8 wt.%. As a consequence, measurements over this last value can be regarded as a calcite anomaly, and so are representative of a granite affected by argillic alteration, which may be a sign of paleo fluid flow [94]. Ten samples out of fifteen from the NH have a calcite content higher than 1.8 wt.%. This suggests that the OM and some zones of the NH were not affected by the argillic alteration, but only by the propylitic one. By considering the calcite content average of 2.2% for the NH, those data also indicate that even if the samples were collected far from the major faults, the granite can be affected by argillic alteration.

5.5. The NH: A Paleo-Geothermal Reservoir?

In the present contribution we have given numerous pieces of evidences for a pervasive alteration of the NH granite. Ubiquitous argillic alteration affecting plagioclase and biotite is present. The K-feldspar being unaltered, potassium enrichment by incoming fluids is necessary to produce abundant amount of illite. The high concentration of calcite, in some samples, requires an external input of Ca, which cannot come only from the plagioclase alteration. Due to this alteration, the rock porosity was drastically enhanced by dissolution of the plagioclase. Porosity is also enhanced by the microfracturing well visible in quartz and K-feldspar, which is related to the activity of the SDVFZ. It is believed that this microfracturing drastically enhanced interaction surfaces between minerals and fluids allowing chemical elements exchanges between hydrothermal fluids and the granite. The strain is not homogeneously distributed in the NH. The NH granite is affected either by non-localized deformation (samples from this study) or by strain concentrations along fault zones that will be presented in the accompanying contribution (PART 2, this issue).

Our results have shown that an alteration of vast volumes of rocks is thus possible, even outside localized high strain zones. All these data are in favor of a hydrothermal fluid percolation, which is encountered in an exploited geothermal reservoir [6,19]. A geothermal reservoir consisting into a flow system where a high amount of hot fluid is stored and circulates through the rock, and our results encourage to consider the NH as a paleo-geothermal reservoir.

6. Conclusions

The NH were chosen in the framework of the H2020 MEET project as being an opportunity to characterize a granitic paleo-geothermal reservoir in a trans-tensional context. Arid conditions and a 3D exposure were important criteria allowing the thorough characterization of the granite. The aim of this study was to characterize the granite pervasive alteration processes, away from the fractures, in terms of changes of mineralogical, geochemical and related petrophysical properties.

Illite was identified by XRD in all the NH samples, allowing to obtain KI values which revealed a NW-SE temperature gradient through the range. This gradient might be considered as a sign of a possible exhumation of the southern part of the range due to the interaction with the Avawatz Mountains.

The partial recrystallization of plagioclase and biotite into newly formed minerals, due to fluid/rock interactions, was identified as reflecting three types of alteration:

1. A pervasive propylitic alteration. This alteration is present in the OM granite (the freshest one considered as the protolith) and in the NH granite and characterized by the presence of corrensite and/or epidote.
2. A local argillic alteration. This alteration was identified only locally in the NH granite by the occurrence of clay minerals such as kaolinite, illite/smectite mixed-layers and illite, all of which crystallize at a lower temperature than the propylitic alteration. Kaolinite and illite might reflect a different amount of leaching or different pH, meaning that several fluids have circulated.
3. Weathering identified in the OM granite by the presence of montmorillonite, thus formed at surface temperature.

The NH granite alteration was highlighted by optical observations. In addition, geochemistry also provided data to support them. Indeed, depletion of Na and Ca was observed with the increase of LOI, considered as a good indicator of the amount of alteration of plagioclase alteration. At the same time, K enrichment was observed with the increase of LOI, and linked to illite crystallization. Calcimetry performed on the NH granite samples showed a calcite content often higher than the 1.8% value, admitted as being the maximum in a fresh granite. These values tend to increase with the LOI, confirming that the NH granite underwent alteration. Porosity also shows a positive correlation with the LOI indicating that the porosity increases with the amount of alteration. However, its correlation with the calcite content is less obvious, as calcite might crystallize in the porosity and thus reduce it.

The NH granite underwent up to two stages of alteration before being exposed to surface conditions, showing a retrograde evolution. It was observed petrographically that the argillic alteration overprinted the propylitic alteration, until its signature is lost. The weathering signal is low (~15% of the clay fraction) and considered as negligible in the OM. The OM being considered as the protolith of the NH, the contribution of weathering is also considered as negligible in the NH granite, where no montmorillonite was found. The newly formed minerals are thus considered as being the product of hydrothermal alteration and not of weathering.

This study provides multiple evidence allowing the consideration of the NH granite as a paleo-geothermal reservoir.

The activity of a geothermal reservoir is a combination of pervasive circulation within nearly strain-free zones (as shown in the present contribution), as well as fractured domains where high strain is accumulated. Our study is therefore completed by a PART 2 (this issue), in which the same investigation approach is dedicated to samples taken within visibly fractured zones.

Author Contributions: Conceptualization, J.K., S.P., B.A.L., R.L.H., A.C. and G.T.; methodology, J.K., S.P., B.A.L., R.L.H., A.C. and G.T.; software, J.K.; validation, S.P., B.A.L., R.L.H., A.C., P.B. and G.T.; formal analysis, J.K.; investigation, J.K., S.P., B.A.L., R.L.H. and G.T.; resources, J.K.; data curation, J.K., S.P., B.A.L., R.L.H., A.C. and G.T.; writing—original draft preparation, J.K.; writing—review and editing, S.P., B.A.L., R.L.H., A.C., G.T. and P.B. and H2020 MEET consortium; visualization, S.P., B.A.L., R.L.H., A.C., G.T. and P.B.; supervision, J.K.; project administration, G.T.; funding acquisition, G.T. and H2020 MEET consortium. All authors have read and agreed to the published version of the manuscript.

Funding: This project has received funding from the European Union's Horizon 2020 research and innovation program under grant agreement No 792037 (H2020 MEET project).

Data Availability Statement: Not applicable.

Acknowledgments: This manuscript was prepared as a contribution to the PhD thesis (Institut Polytechnique UniLaSalle Beauvais) of Johanne Klee, which was funded by the European Union's Horizon 2020 research and innovation program under grant agreement No 792037 (H2020 MEET project). The authors are grateful to Terry Pavlis for his knowledge and helpful discussions about the regional geology of Death Valley and the Noble Hills. We also acknowledge Albert Genter for the fruitful exchanges about granite alteration processes. We thank Thi Tuyen Nguyen, Elena Pavlovskaia, Carl Tixier and Chloé Gindrat for their help for analyses. Helpful comments and the validation of the manuscript by the H2020 MEET consortium are gratefully acknowledged. We finally would like to thank Catherine Lerouge and the anonymous reviewer for their help and remarks to improve this manuscript.

Conflicts of Interest: The authors declare no conflict of interest.

References

1. Williams, C.F.; Reed, M.J.; Anderson, A.F. Updating the classification of geothermal resources. In Proceedings of the Thirty Sixth Workshop on Geothermal Reservoir Engineering Standford University, Standford, CA, USA, 31 January–2 February 2011.
2. Moeck, I.S. Catalog of geothermal play types based on geologic controls. *Renew. Sustain. Energy Rev.* **2014**, *37*, 867–882. [CrossRef]
3. Olasolo, P.; Juárez, M.C.; Morales, M.P.; D'Amico, S.; Liarte, I.A. Enhanced Geothermal Systems (EGS): A Review. *Renew. Sustain. Energy Rev.* **2016**, *56*, 133–144. [CrossRef]
4. Trullenque, G.; Genter, A.; Leiss, B.; Wagner, B.; Bouchet, R.; Leoutre, E.; Malnar, B.; Bär, K.; Rajšl, I. Upscaling of EGS in different geological conditions: A European perspective. In Proceedings of the 43rd Workshop on Geothermal Reservoir Engineering Standford University, Standford, CA, USA, 12–14 February 2018; p. 10.

5.	Faulds, J.E.; Hinz, N.H.; Dering, G.M.; Siler, D.L. The hybrid model—The most accommodating structural setting for geothermal power generation in the Great Basin, Western USA. *Geotherm. Resour. Counc. Trans.* **2013**, *37*, 3–10.
6.	Nishimoto, S.; Yoshida, H. Hydrothermal alteration of deep fractured granite: Effects of dissolution and precipitation. *Lithos* **2010**, *115*, 153–162. [CrossRef]
7.	Dezayes, C.; Lerouge, C. Reconstructing paleofluid circulation at the hercynian basement/mesozoic sedimentary cover interface in the Upper Rhine Graben. *Geofluids* **2019**, *2019*, 1–30. [CrossRef]
8.	Plumper, O.; Putnis, A. The complex hydrothermal history of granitic rocks: Multiple feldspar replacement reactions under subsolidus conditions. *J. Petrol.* **2009**, *50*, 967–987. [CrossRef]
9.	Inoue, A. Formation of clay minerals in hydrothermal environments. In *Origin and Mineralogy of Clays*; Velde, B., Ed.; Springer: Berlin/Heidelberg, Germany, 1995; pp. 268–329. ISBN 978-3-642-08195-8.
10.	Burchfiel, B.C.; Stewart, J.H. "Pull-apart" origin of the central segment of Death Valley, California. *GSA Bull.* **1966**, *77*, 439–442. [CrossRef]
11.	Pavlis, T.L.; Trullenque, G. Evidence for 40–41 km of dextral slip on the Southern Death Valley fault: Implications for the Eastern California shear zone and extensional tectonics. *Geology* **2021**, *49*, 767–772. [CrossRef]
12.	Chabani, A.; Trullenque, G.; Ledésert, B.A.; Klee, J. Multiscale Characterization of fracture patterns: A case study of the Noble Hills Range (Death Valley, CA, USA), application to geothermal reservoirs. *Geosciences* **2021**, *11*, 280. [CrossRef]
13.	Kisch, H.J. Correlation between indicators of very low-grade metamorphism. In *Low Temperature Metamorphism*; Frey, M., Ed.; Chapman & Hall: London, UK, 1987; pp. 227–300.
14.	Árkai, P.; Sassi, F.; Desmons, J. Very low- to low-grade metamorphic rocks. In *Metamorphic Rocks A Classification and Glossary Terms*; Cambridge University Press: Cambridge, UK, 2007.
15.	Frey, M. Very low-grade metamorphism of clastic sedimentary rocks. In *Low Temperature Metamorphism*; Chapman & Hall: London, UK, 1987; pp. 9–58.
16.	Árkai, P.; Maehlmann, R.; Suchy, V.; Balogh, K.; Sykorová, I.; Frey, M. Possible Effects of Tectonic Shear Strain on Phyllosilicates: A Case Study from the Kandersteg Area, Helvetic Domain, Central Alps, Switzerland. *TMPM Tschermaks Mineral. Petrogr. Mitt.* **2002**, *82*, 273–290.
17.	Mullis, J.; Mählmann, R.F.; Wolf, M. Fluid Inclusion Microthermometry to Calibrate Vitrinite Reflectance (between 50 and 270 °C), Illite Kübler-Index Data and the Diagenesis/Anchizone Boundary in the External Part of the Central Alps. *Appl. Clay Sci.* **2017**, *143*, 307–319. [CrossRef]
18.	Ferreiro Mählmann, R.; Bozkaya, Ö.; Potel, S.; le Bayon, R.; Šegvić, B.; Nieto, F. The Pioneer Work of Bernard Kübler and Martin Frey in Very Low-Grade Metamorphic Terranes: Paleo-Geothermal Potential of Variation in Kübler-Index/Organic Matter Reflectance Correlations. A Review. *Swiss. J. Geosci.* **2012**, *105*, 121–152. [CrossRef]
19.	Ledésert, B.; Berger, G.; Meunier, A.; Genter, A.; Bouchet, A. Diagenetic-Type Reactions Related to Hydrothermal Alteration in the Soultz-Sous-Forets Granite, France. *Eur. J. Mineral.* **1999**, *11*, 731–741. [CrossRef]
20.	Wernicke, B.; Axen, G.J.; Snow, J.K. Basin and Range Extensional Tectonics at the Latitude of Las Vegas, Nevada. *GSA Bull.* **1988**, *100*, 1738–1757. [CrossRef]
21.	Wright, L. Late Cenozoic Fault Patterns and Stress Fields in the Great Basin and Westward Displacement of the Sierra Nevada Block. *Geology* **1976**, *4*, 489–494. [CrossRef]
22.	Wernicke, B.; Spencer, J.E.; Burchfiel, B.C.; Guth, P.L. Magnitude of Crustal Extension in the Southern Great Basin. *Geology* **1982**, *10*, 499–502. [CrossRef]
23.	Stewart, J.H. Extensional Tectonics in the Death Valley Area, California: Transport of the Panamint Range Structural Block 80 Km Northwestward. *Geology* **1983**, *11*, 153–157. [CrossRef]
24.	Calzia, J.P.; Rämö, O.T. Late Cenozoic Crustal Extension and Magmatism, Southern Death Valley Region, California. *GSA Field Guides* **2000**, *2*, 135–164. [CrossRef]
25.	Norton, I. Two-Stage Formation of Death Valley. *Geosphere* **2011**, *7*, 171–182. [CrossRef]
26.	Luckow, H.; Pavlis, T.; Serpa, L.; Guest, B.; Wagner, D.; Snee, L.; Hensley, T.; Korjenkov, A. Late Cenozoic Sedimentation and Volcanism during Transtensional Deformation in Wingate Wash and the Owlshead Mountains, Death Valley. *Earth Sci. Rev.* **2005**, *73*, 177–219. [CrossRef]
27.	Hill, M.L.; Troxel, B.W. Tectonics of Death Valley Region, California. *GSA Bull.* **1966**, *77*, 435–438. [CrossRef]
28.	Butler, P.R.; Troxel, B.W.; Verosub, K.L. Late Cenozoic History and Styles of Deformation along the Southern Death Valley Fault Zone, California. *GSA Bull.* **1988**, *100*, 402–410. [CrossRef]
29.	Brady, R.H., III. Cenozoic Geology of the Northern Avawatz Mountains in Relation to the Intersection of the Garlock and Death Valley Fault Zones, San Bernardino County, California. Ph.D. Thesis, University of California, Berkeley, CA, USA, 1986.
30.	Brady, R.H.; Clayton, J.; Troxel, B.W.; Verosub, K.L.; Cregan, A.; Abrams, M. Thematic Mapper and Field Investigations at the Intersection of the Death Valley and Garlock Fault Zones, California. *Remote Sens. Environ.* **1989**, *28*, 207–217. [CrossRef]
31.	Lifton, Z.M.; Newman, A.V.; Frankel, K.L.; Johnson, C.W.; Dixon, T.H. Insights into Distributed Plate Rates across the Walker Lane from GPS Geodesy. *Geophys. Res. Lett.* **2013**, *40*, 4620–4624. [CrossRef]
32.	Nagorsen-Rinke, S.; Lee, J.; Calvert, A. Pliocene Sinistral Slip across the Adobe Hills, Eastern California-Western Nevada: Kinematics of Fault Slip Transfer across the Mina Deflection. *Geosphere* **2013**, *9*, 37–53. [CrossRef]

33. Miller, M.B.; Wright, L.A. *Geology of Death Valley National Park*, 3rd ed.; Kendall Hunt Publishing Company: Dubuque, IA, USA, 2015.
34. Rämö, T.O.; Calzia, J.P.; Kosunen, P.J. Geochemistry of Mesozoic Plutons, Southern Death Valley Region, California: Insights into the Origin of Cordilleran Interior Magmatism. *Contrib. Mineral. Petrol.* **2002**, *143*, 416–437. [CrossRef]
35. Troxel, B.W.; Butler, P.R. *Rate of Cenozoic Slip on Normal Faults, South-Central Death Valley, California*; Department of Geology, University of California: Berkeley, CA, USA, 1979.
36. Butler, P.R. Geology: Structural History and Fluvial Geomorphology of the Southern Death Valley Fault Zone, Inyo and San Bernardino Counties, California. Ph.D. Thesis, University of California, Davis, CA, USA, 1984.
37. Brady, R.H.; Troxel, B.W. Stratigraphy and Tectonics of the Northern Avawatz Mountains at the Intersection of the Garlock and Death Valley Fault Zones, San Bernardino County, California. In *Quaternary Tectonics of Southern Death Valley, California—Field Trip Guide: Shoshone, California, Friends of the Pleistocene, Pacific Cell*; USGS: Shoshone, CA, USA, 1986; pp. 1–12.
38. Niles, J.H. Post-Middle Pliocene Tectonic Development of the Noble Hills, Southern Death Valley, California. Ph.D. Thesis, San Francisco State University, San Francisco, CA, USA, 2016.
39. Mahon, R.C.; Dehler, C.M.; Link, P.K.; Karlstrom, K.E.; Gehrels, G.E. Detrital Zircon Provenance and Paleogeography of the Pahrump Group and Overlying Strata, Death Valley, California. *Precambrian Res.* **2014**, *251*, 102–117. [CrossRef]
40. DeCelles, P.G. Late Jurassic to Eocene Evolution of the Cordilleran Thrust Belt and Foreland Basin System, Western USA. *Am. J. Sci.* **2004**, *304*, 105–168. [CrossRef]
41. Stamm, J.F. Geology at the Intersection of the Death Valley and Garlock Fault Zones, Southern Death Valley, California. Ph.D. Thesis, Pennsylvania State University, State College, PA, USA, 1981.
42. Klee, J.; Trullenque, G.; Ledésert, B.; Potel, S.; Hébert, R.; Chabani, A.; Genter, A. Petrographic analyzes of fractured granites used as an analogue of the soultz-sous-forêts geothermal reservoir: Noble Hills, CA, USA. In Proceedings of the Extended Abstract, Reykjavik, Iceland, 26 May 2021.
43. Troxel, B.W. Right-Lateral Offset of ca. 28 Km along a Strand of the Southern Death Valley Fault Zone. *Calif. Geol. Soc. Am. Abstr. Programs* **1994**, *26*, 99.
44. Castaing, C.; Rabu, D. Apports de La Géologie à La Recherche et à l'Exploitation de Pierres de Taille (Roches Ornementales et de Construction). Available online: https://hal.archives-ouvertes.fr/hal-01860154 (accessed on 12 July 2021).
45. Kretz, R. Symbols for Rock-Forming Minerals. *Am. Mineral.* **1983**, *68*, 277–279.
46. Bisdom, K.; Gauthier, B.D.M.; Bertotti, G.; Hardebol, N.J. Calibrating Discrete Fracture-Network Models with a Carbonate Three-Dimensional Outcrop Fracture Network: Implications for Naturally Fractured Reservoir Modeling. *AAPG Bull.* **2014**, *98*, 1351–1376. [CrossRef]
47. Gillespie, P.A.; Howard, C.B.; Walsh, J.J.; Watterson, J. Measurement and Characterisation of Spatial Distributions of Fractures. *Tectonophysics* **1993**, *226*, 113–141. [CrossRef]
48. Schmidt, D.; Schmidt, S.T.; Mullis, J.; Ferreiro Mählmann, R.; Frey, M. Very Low Grade Metamorphism of the Taveyanne Formation of Western Switzerland. *Contrib. Mineral. Petrol.* **1997**, *129*, 385–403. [CrossRef]
49. Kisch, H.J. Illite Crystallinity: Recommendations on Sample Preparation, X-ray Diffraction Settings, and Interlaboratory Samples. *J. Metamorph. Geol.* **1991**, *9*, 665–670. [CrossRef]
50. Ferreiro Mählmann, R.; Frey, M. Standardisation, Calibration and Correlation of the Kübler-Index and the Vitrinite/Bituminite Reflectance: An Inter-Laboratory and Field Related Study. *Swiss J. Geosci.* **2012**, *105*, 153–170. [CrossRef]
51. Moore, D.M.; Reynolds, R.C. X-ray diffraction and the identification and analysis of clay minerals. In *X-ray Diffraction and the Identification and Analysis of Clay Minerals*; Oxford University Press: Oxford, UK, 1989.
52. Starkey, H.C.; Blackmon, P.D.; Hauff, P.L. *The Routine Mineralogical Analysis of Clay-Bearing Samples*; United States Government Publishing Office: Washington, DC, USA, 1984.
53. Warr, L.N.; Rice, A.H.N. Interlaboratory Standardization and Calibration of Day Mineral Crystallinity and Crystallite Size Data. *J. Metamorph. Geol.* **1994**, *12*, 141–152. [CrossRef]
54. Kübler, B. La Cristallinite de l'illite et Les Zones Tout a Fait Superieures Du Metamorphisme. In *Etages Tectoniques*; van Bemmelen, R.W., Chapman, C.A., Watznauer, A., Eds.; La Baconniere: Boudry, Switzerland, 1981; pp. 105–121.
55. Warr, L.N.; Mählmann, R.F. Recommendations for Kübler Index Standardization. *Clay Minerals* **2015**, *50*, 283–286. [CrossRef]
56. Winkler, H.G.F. Anatexis, Formation of Migmatites, and Origin of Granitic Magmas. In *Petrogenesis of Metamorphic Rocks*; Winkler, H.G.F., Ed.; Springer Study Edition; Springer: New York, NY, USA, 1979; pp. 283–339. ISBN 978-1-4757-4215-2.
57. Tilley, C.E. A Preliminary Survey of Metamorphic Zones in the Southern Highlands of Scotland. *Q. J. Geol. Soc.* **1925**, *81*, 100–112. [CrossRef]
58. Barrow, G. On an Intrusion of Muscovite-Biotite Gneiss in the South-Eastern Highlands of Scotland, and Its Accompanying Metamorphism. *Q. J. Geol. Soc.* **1893**, *49*, 330–358. [CrossRef]
59. Merriman, R.J.; Frey, M. Patterns of very low-grade metamorphism in metapelitic rocks. In *Low-Grade Metamorphism*; Frey, M., Robinson, D., Eds.; Blackwell Publishing Ltd.: Oxford, UK, 1998; pp. 61–107. ISBN 978-1-4443-1334-5.
60. Abad, I. Physical meaning and applications of the Illite Kübler index: Measuring reaction progress in low-grade metamorphism. In *Diagenesis and Low-Temperature Metamorphism, Theory, Methods and Regional Aspects*; Sociedad Española de Mineralogía: Jaén, Spain, 2007; pp. 53–64.

61. Ledésert, B.; Hébert, R.L.; Grall, C.; Genter, A.; Dezayes, C.; Bartier, D.; Gérard, A. Calcimetry as a Useful Tool for a Better Knowledge of Flow Pathways in the Soultz-Sous-Forêts Enhanced Geothermal System. *J. Volcanol. Geotherm. Res.* **2009**, *181*, 106–114. [CrossRef]

62. Dullien, F.A.L. *Porous Media: Fluid Transport and Pore Structure*; Academic Press: San Diego, CA, USA, 1979; ISBN 978-0-323-13933-5.

63. Navelot, V.; Géraud, Y.; Favier, A.; Diraison, M.; Corsini, M.; Lardeaux, J.-M.; Verati, C.; Mercier de Lépinay, J.; Legendre, L.; Beauchamps, G. Petrophysical Properties of Volcanic Rocks and Impacts of Hydrothermal Alteration in the Guadeloupe Archipelago (West Indies). *J. Volcanol. Geotherm. Res.* **2018**, *360*, 1–21. [CrossRef]

64. Gates, W.P.; Nefiodovas, A.; Peter, P. Permeability of an Organo-Modified Bentonite to Ethanol-Water Solutions. *Clays Clay Miner.* **2004**, *52*, 192–203. [CrossRef]

65. Bard, J.P. *Microtextures des Roches Magmatiques et Métamorphiques*; Masson: Paris, France, 1980.

66. Goldich, S.S. A Study in Rock-Weathering. *J. Geol.* **1938**, *46*, 17–58. [CrossRef]

67. Que, M.; Allen, A.R. Sericitization of Plagioclase in the Rosses Granite Complex, Co. Donegal, Ireland. *Mineral. Mag.* **1996**, *60*, 927–936. [CrossRef]

68. Streckeisen, A.; Le Maître, R.W. A Chemical Approximation to Modal QAPF Classification of the Igneous Rocks. *Neues Jahrb. Mineral. Abh.* **1979**, *136*, 169–206.

69. Beaufort, D.; Baronnet, A.; Lanson, B.; Meunier, A. Corrensite; a Single Phase or a Mixed-Layer Phyllosilicate in Saponite-to-Chlorite Conversion Series? A Case Study of Sancerre-Couy Deep Drill Hole (France). *Am. Mineral.* **1997**, *82*, 109–124. [CrossRef]

70. Jagodzinski, H. Eindimensionale Fehlordnung in Kristallen und ihr Einfluss auf die Röntgeninterferenzen. I. Berechnung des Fehlordnungsgrades aus den Röntgenintensitäten. *Acta Cryst.* **1949**, *2*, 201–207. [CrossRef]

71. Cox, K.G. *The Interpretation of Igneous Rocks*; Springer Science & Business Media: London, UK, 1979; ISBN 978-94-017-3373-1.

72. Wilson, M. Review of Igneous Petrogenesis: A Global Tectonic Approach. *Terra Nova* **1989**, *1*, 218–222. [CrossRef]

73. Jensen, L.S. A New Plot for Classifying Subalkalic Volcanic Rocks. *Ont. Div. Mines Misc. Pap.* **1976**, *66*, 1–22.

74. White, A.J.R.; Chappell, B.W. Granitoid types and their distribution in the Lachlan Fold Belt, southeastern Australia. In *Geological Society of America Memoirs*; Geological Society of America: Boulder, CO, USA, 1983; Volume 159, pp. 21–34, ISBN 978-0-8137-1159-1.

75. Le Maître, R.W.; Bateman, P.; Dudek, A.; Keller, J.; Lameyre Le Bas, M.J.; Sabine, P.A.; Schmid, R.; Sorensen, H.; Streckeisen, A.; Woolley, A.R.; et al. *A Classification of Igneous Rocks and Glossary of Terms*; Blackwell: Oxford, UK, 1989.

76. Rollinson, H.R. *Using Geochemical Data: Evaluation, Presentation, Interpretation*; Routledge: Abingdon, UK, 1993; ISBN 978-1-317-89819-1.

77. Boynton, W.V. Cosmochemistry of the rare earth elements: Meteorite studies. In *Developments in Geochemistry*; Elsevier: Amsterdam, The Netherlands, 1984; Volume 2, pp. 63–114. ISBN 978-0-444-42148-7.

78. McDonough, W.; Sun, S.-S.; Ringwood, A.; Jagoutz, E.; Hofmann, A. Potassium, Rubidium, and Cesium in the Earth and Moon and the Evolution of the Mantle of the Earth. *Geochim. Cosmochim. Acta* **1992**, *56*, 1001–1012. [CrossRef]

79. Sun, S.; McDonough, W.F. Chemical and Isotopic Systematics of Oceanic Basalts: Implications for Mantle Composition and Processes. *Geol. Soc. Lond. Spec. Publ.* **1989**, *42*, 313–345. [CrossRef]

80. Koljonen, T.; Rosenberg, R.J. Rare Earth Elements in Granitic Rocks. *Lithos* **1974**, *7*, 249–261. [CrossRef]

81. Alderton, D.H.M.; Pearce, J.A.; Potts, P.J. Rare Earth Element Mobility during Granite Alteration: Evidence from Southwest England. *Earth Planet. Sci. Lett.* **1980**, *49*, 149–165. [CrossRef]

82. Li, X.-C.; Fan, H.-R.; Santosh, M.; Hu, F.-F.; Yang, K.-F.; Lan, T.-G. Hydrothermal Alteration Associated with Mesozoic Granite-Hosted Gold Mineralization at the Sanshandao Deposit, Jiaodong Gold Province, China. *Ore Geol. Rev.* **2013**, *53*, 403–421. [CrossRef]

83. Warr, L.N.; Cox, S.C. Correlating Illite (Kübler) and Chlorite (Árkai) "Crystallinity" Indices with Metamorphic Mineral Zones of the South Island, New Zealand. *Appl. Clay Sci.* **2016**, *134*, 164–174. [CrossRef]

84. Fulignati, P. Clay Minerals in Hydrothermal Systems. *Minerals* **2020**, *10*, 919. [CrossRef]

85. Creasey, S.C. Hydrothermal alteration. In *Geology of the Porphyry Copper Deposits Southwestern North America*; Titley & Hicks: Tucson, AZ, USA, 1966; pp. 51–74.

86. Traineau, H.; Genter, A.; Cautru, J.P.; Fabriol, H.; Chevremont, P. Petrography of the Granite Massif from Drill Cutting Analysis and Well Log Interpretation in the Geothermal HDR Borehole GPK1 (Soultz, Alsace, France). *Geotherm. Sci. Technol.* **1991**, *3*, 1–29.

87. Środoń, J. Nature of Mixed-Layer Clays and Mechanisms of Their Formation and Alteration. *Annu. Rev. Earth Planet. Sci.* **1999**, *27*, 19–53. [CrossRef]

88. Genter, A. Géothermie Roches Chaudes Sèches: Le Granite de Soultz-Sous-Forêts (Bas-Rhin, France): Fracturation Naturelle, Altérations Hydrothermales et Interaction Eau-Roche. Ph.D. Thesis, Université d'Orléans, Orléans, France, 1989.

89. Burnham, C.W. Facies and Types of Hydrothermal Alteration. *Econ. Geol.* **1962**, *57*, 768–784. [CrossRef]

90. Velde, B. *Clays and Clay Minerals in Natural and Synthetic Systems*; Development in Sedimentology; Elsevier: Amsterdam, The Netherlands, 1977; ISBN 978-0-08-086933-9.

91. Steiner, A. Clay Minerals in Hydrothermally Altered Rocks at Wairakei, New Zealand. *Clays Clay Miner.* **1968**, *16*, 193–213. [CrossRef]

92. Ledésert, B.; Hebert, R.; Genter, A.; Bartier, D.; Clauer, N.; Grall, C. Fractures, Hydrothermal Alterations and Permeability in the Soultz Enhanced Geothermal System. *Comptes Rendus Geosci.* **2010**, *342*, 607–615. [CrossRef]

93. Vidal, J.; Patrier, P.; Genter, A.; Beaufort, D.; Dezayes, C.; Glaas, C.; Lerouge, C.; Sanjuan, B. Clay Minerals Related to the Circulation of Geothermal Fluids in Boreholes at Rittershoffen (Alsace, France). *J. Volcanol. Geotherm. Res.* **2018**, *349*, 192–204. [CrossRef]
94. Ledésert, B.A.; Hébert, R.L. How Can Deep Geothermal Projects Provide Information on the Temperature Distribution in the Upper Rhine Graben? The Example of the Soultz-Sous-Forêts-Enhanced Geothermal System. *Geosciences* **2020**, *10*, 459. [CrossRef]
95. Sardini, P.; Ledésert, B.; Touchard, G. Quantification of microscopic porous networks by image analysis and measurements of permeability in the Soultz-Sous-Forêts Granite (Alsace, France). In *Fluid Flow and Transport in Rocks: Mechanisms and Effects*; Jamtveit, B., Yardley, B.W.D., Eds.; Springer: Dordrecht, The Netherlands, 1997; pp. 171–189. ISBN 978-94-009-1533-6.
96. Glaas, C.; Patrier, P.; Vidal, J.; Beaufort, D.; Genter, A. Clay Mineralogy: A Signature of Granitic Geothermal Reservoirs of the Central Upper Rhine Graben. *Minerals* **2021**, *11*, 479. [CrossRef]
97. Meunier, A.; Velde, B.D.; Dudoignon, P.; Beaufort, D. Identification of Weathering and Hydrothermal Alteration in Acidic Rocks: Petrography and Mineralogy of Clay Minerals. *Sci. Géologiques Bull. Mémoires* **1983**, *72*, 93–99.
98. Tardy, Y.; Paquet, H.; Millot, G. Trois modes de genèse des montmorillonites dans les altérations et les sols. *Bull. Groupe Français Argiles* **1970**, *22*, 69–77. [CrossRef]
99. Liu, Y.; Xie, C.; Li, C.; Li, S.; Santosh, M.; Wang, M.; Fan, J. Breakup of the Northern Margin of Gondwana through Lithospheric Delamination: Evidence from the Tibetan Plateau. *GSA Bull.* **2018**, *131*. [CrossRef]
100. Chambefort, I.; Moritz, R.; von Quadt, A. Petrology, Geochemistry and U-Pb Geochronology of Magmatic Rocks from the High-Sulfidation Epithermal Au–Cu Chelopech Deposit, Srednogorie Zone, Bulgaria. *Miner. Depos.* **2007**, *42*, 665–690. [CrossRef]
101. Garrels, R.M.; MacKenzie, F.T. Origin of the Chemical Compositions of Some Springs and Lakes. In *Equilibrium Concepts in Natural Water Systems*; Advances in Chemistry; American Chemical Society: Washington, DC, USA, 1967; Volume 67, pp. 222–242, ISBN 978-0-8412-0068-5.
102. Rosener, M.; Géraud, Y. Using Physical Properties to Understand the Porosity Network Geometry Evolution in Gradually Altered Granites in Damage Zones. *Geol. Soc. Lond. Spec. Publ.* **2007**, *284*, 175–184. [CrossRef]
103. Cassiaux, M.; Proust, D.; Siitari-Kauppi, M.; Sardini, P.; Leutsch, Y. Clay Minerals Formed during Propylitic Alteration of a Granite and Their Influence on Primary Porosity: A Multi-Scale Approach. *Clays Clay Miner.* **2006**, *54*, 541–554. [CrossRef]
104. White, A.F.; Schulz, M.S.; Lowenstern, J.B.; Vivit, D.V.; Bullen, T.D. The Ubiquitous Nature of Accessory Calcite in Granitoid Rocks: Implications for Weathering, Solute Evolution, and Petrogenesis. *Geochim. Cosmochim. Acta* **2005**, *69*, 1455–1471. [CrossRef]

 *geosciences*

Article

Fluid-Rock Interactions in a Paleo-Geothermal Reservoir (Noble Hills Granite, California, USA). Part 2: The Influence of Fracturing on Granite Alteration Processes and Fluid Circulation at Low to Moderate Regional Strain

Johanne Klee [1,*], Arezki Chabani [1], Béatrice A. Ledésert [2], Sébastien Potel [1], Ronan L. Hébert [2] and Ghislain Trullenque [1]

[1] B2R, Geosciences Department, Institut Polytechnique UniLaSalle Beauvais, 19 Rue Pierre Waguet, F-60026 Beauvais, France; arezki.chabani@unilasalle.fr (A.C.); sebastien.potel@unilasalle.fr (S.P.); ghislain.trullenque@unilasalle.fr (G.T.)

[2] Geosciences and Environment Cergy, CY Cergy Paris Université, 1 Rue Descartes, F-95000 Neuville-sur-Oise, France; beatrice.ledesert@cyu.fr (B.A.L.); ronan.hebert@cyu.fr (R.L.H.)

* Correspondence: johanne.klee@unilasalle.fr; Tel.: +33-6-06-93-90-07

Citation: Klee, J.; Chabani, A.; Ledésert, B.A.; Potel, S.; Hébert, R.L.; Trullenque, G. Fluid-Rock Interactions in a Paleo-Geothermal Reservoir (Noble Hills Granite, California, USA). Part 2: The Influence of Fracturing on Granite Alteration Processes and Fluid Circulation at Low to Moderate Regional Strain. *Geosciences* **2021**, *11*, 433. https://doi.org/10.3390/geosciences11110433

Academic Editors: Michael G. Petterson and Jesus Martinez-Frias

Received: 23 September 2021
Accepted: 18 October 2021
Published: 20 October 2021

Publisher's Note: MDPI stays neutral with regard to jurisdictional claims in published maps and institutional affiliations.

Abstract: Fracture connectivity within fractured granitic basement geothermal reservoirs is an important factor controlling their permeability. This study aims to improve the understanding of fluid–rock interaction processes at low to moderate regional strain. The Noble Hills range (Death Valley, CA, USA) was chosen as a naturally exhumed paleo geothermal reservoir. A series of petrographic, petrophysical, and geochemical investigations, combined with a fracture distribution analysis, were carried out on samples collected across fracture zones. Our results indicate that several generations of fluids have percolated through the reservoir. An increase of (1) the alteration degree; (2) the porosity values; and (3) the calcite content was observed when approaching fracture zones. No correlation was identified among the alteration degree, the porosity, or the calcite content. At a local scale, samples showed that the degree of alteration does not necessarily depend on the fracture density or on the amount of the strain. It is concluded that the combined influence of strain and coeval fluid–rock interaction processes drastically influence the petrophysical properties of fracture zones, which in turn impact geothermal production potential.

Keywords: fracturing processes; fluid circulation; granite alteration; low to moderate regional strain; geothermal reservoir

1. Introduction

Long-term exploitation of geothermal resources is closely linked to reservoir rock petrophysical properties and regional geological settings [1]. Due to the low matrix porosity and permeability of granitic rocks, underground granitic units are considered as a reservoir only when fractures are present. These fractures provide the essential reservoir permeability and porosity for fluid flow [2,3] and are therefore of primary importance regarding geothermal exploitation [4–8]. These structures control the deep flow at the reservoir scale, in case of good connectivity [2,3,9–11], which is one of the most important controls on the permeability [12]. Several granitic reservoirs, as Soultz-sous-Forêts in the Upper Rhine Graben, France [6,13] or the Cooper Basin in Australia [14], give invaluable experience in terms of exploration and exploitation feedback. In addition to these datasets gained from data limited to boreholes and indirect geophysical methods, studies of surface reservoir analogues are common [15–17]. The MEET H2020 project (Multidisciplinary and multi-context demonstration of EGS exploration and Exploitation Techniques and potentials) [18] aims to develop enhanced geothermal systems throughout Europe. Within this project, the Noble Hills (NH) range, located in the southern termination of the Death

Valley (DV, CA, USA), has been chosen as an analogue of fractured granitic basements in a context of transtensional deformation [19]. Klee et al. (2021) [19] highlighted numerous evidences of hydrothermal alterations: (1) a propylitic alteration affecting pervasively a large volume of rock during the cooling of the pluton and (2) an argillic alteration, also called vein alteration [19–23], inducing changes of the bulk-rock chemical and mineralogical compositions and of physical properties [7,24].

A vast amount of literature [9,11,25] has proven that increasing amounts of strain within fault zones drastically change their petrophysical properties. The present paper focuses on the influence of fractures on the fluid circulation and alteration processes at low to moderate regional strain.

The present study aims to:

1. Characterize the relations among the varying amounts of strain, fracture densities, and alteration degrees at the NH scale, as well as the sample scales through the case studies.
2. Characterize the variations in (1) chemical elements concentrations; (2) calcite content; (3) porosity; and (4) temperature condition variations when approaching fracture zones.
3. Identify the different fluid circulation episodes through the granite body.

Macroscopic and microscopic petrographic studies, XRD mineralogical characterization of whole rock and clay minerals, bulk rock chemical analyses by inductively coupled plasma (ICP), mass spectrometry (MS), atomic emission spectrometry (AES), scanning electron microscopy coupled with energy dispersive spectrometry (SEM-EDS) for structural observation and local chemical analyses, mass balance calculations, fracture density calculations, calcimetry, and porosimetry, were performed on samples collected in the vicinity of fracture zones within areas of the NH affected by low to moderate regional transcurrent strains. The results will be discussed and compared with the protolith studied by [19].

2. Geological Setting

The Death Valley (DV, Figure 1a) is located in the core of a Cenozoic distributed system of dextral strike-slips, about 700 km long, comprising the Walker Lane Belt and the Eastern California Shear Zone (ECSZ/WLB) [26–29]. This narrow northwest-trending system, lying between the Basin and Range region to the east and the Sierra Nevada batholith to the west [28], today accommodate ~25% of the Pacific-North America relative motion [26,30]. DV is a structural depression, NNW–SSE oriented, bounded by the Black and Funeral Mountains to the west and by the Panamint Range to the east [31]. It has been formed by a right-lateral movement giving a pull-apart structure [32].

The area of interest for this study is the Noble Hills range (NH, Figure 1b). It is located in the southernmost part of the DV region and trends parallel to the NW-striking SDVFZ at its southern end. Geological markers along the SDVFZ trace [33] suggest that the NH correspond to a transported fragment of the frontal part of the Owlshead Mountains (OM), a Cretaceous (~95 Ma, [34]) granitic pluton at a 40–41 km distance to the SE. Several attempts have been made and discussed in the literature to give insight to the geological setting of the NH and structural relationships between SDVFZ and GFZ [35–37]. Particular emphasis has been given to a detailed description of sedimentary sequence deposits on each side of the NH Crystalline Bedrock Slice (CBS) [38,39]. The CBS is composed of Proterozoic sediments with upward younging direction the Crystal Spring (CS) quartzite, CS dolomite, detrital flysh, and carbonate sequences possibly part of the Pahrump Group, intruded by 1.1 Ga diabase sills, the whole intruded by Mesozoic granite [40]. However, a detailed structural analysis of the CBS itself is missing. Thus, Section 4.1 of this study will give new elements, improving the observations made by [38,41] concerning the NH structure. A precise fracture pattern characterization of the NH was performed by [42], through a wide-ranging analysis scale from the microscopic scale to the regional scale. These authors showed that the NH fracture network geometry has been controlled by the SDVFZ and the GFZ.

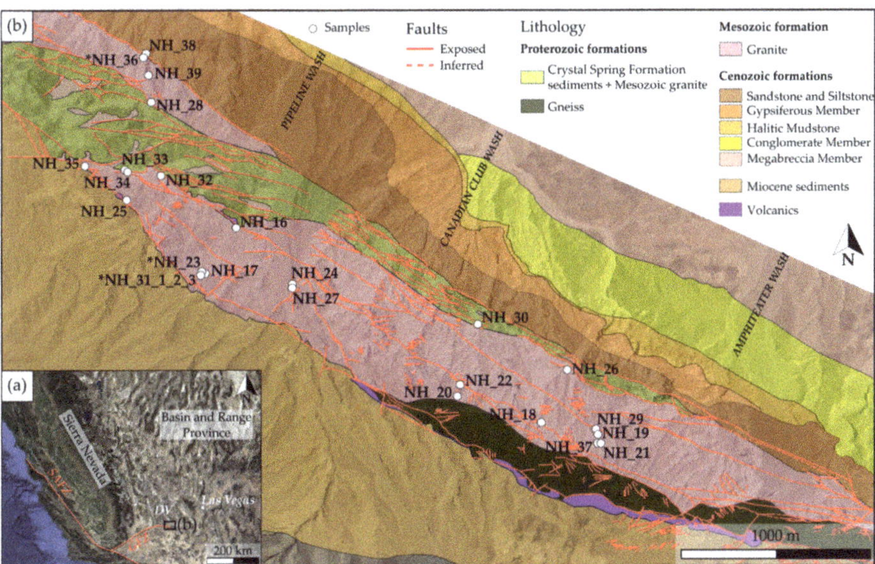

Figure 1. (**a**) Location of the Noble Hills on the western USA satellite map. SAFZ—San Andreas Fault Zone, GFZ—Garlock fault zone, and DV—Death Valley. (**b**) The geological map of the Noble Hills range, modified after [19,38,41]. The quaternary is not displayed, but available on the map provided by [38]. Sample locations are represented by the white dots. The asterisk shows the samples used as case studies in this work. NHF—Noble Hills Formation.

The NH granite was poorly studied in terms of microstructure, petrography, and geochemistry. A recent study conducted by Klee et al. (2021) [19] characterized the granite as a leucocratic equigranular monzogranite (S-type), ranging in the high-K calc-alkaline and peraluminous domains. The NH granite is composed of primary plagioclase, quartz, K-feldspar, and biotite. Klee et al. (2021) [19] have also shown that the NH granite underwent two alteration stages, forming secondary minerals that recorded the chemical and paleo-thermal conditions of the system: (1) a pervasive, propylitic alteration linked to the cooling of the pluton; and (2) a more local alteration corresponding to the argillic alteration (illitization and sometimes kaolinitization), which overprints the propylitic alteration. Among the primary minerals, optical observations and geochemical data show that only plagioclase and biotite are affected by those alteration processes.

3. Materials and Methods

3.1. Material and Sample Selection

Digital field mapping techniques have been used in the field; these include the use of portable rugged tablet laptop with internal GPS. The QGIS software has been used for sample location, geological digitization, and structural data acquisition. Georeferenced topographic maps and ortho-imagery were the initial input of the database.

A detailed petrographic, mineralogical, geochemical, and petrophysical characterization was conducted on 25 samples from the NH granite, based on fieldwork and laboratory analyses.

Samples were selected by targeting fracture zones located in CBS areas that have been affected by low to moderate strain, aimed at characterizing the influence of fracturing on fluid circulations and the associated argillic alteration, mentioned only by [19] until now. Thus, the collected samples consisted of altered granite presenting open fractures, veins, reactivated veins, and breccias. All samples were georeferenced for precise locations as well as database supply (Figure 1b). The selection took into account ranges of fracture density (F_d) defined by [19] from scanlines realized on thin sections, based on [43]:

F_d 0 < 1687 fracs/m—no to very low microfracturing;

F_d 1 = 1687 fracs/m—microfracturing of order less than the grain size;

F_d 2 = 2694 fracs/m, with a multiplicator factor of 1.6 compared to F_d1—microfracturing of grain size order with interconnections;

F_d 3 = 3549 fracs/m, with a multiplicator factor of 1.3 compared to F_d2—abundant microfracturing;

F_d 4 $\geq$ 5140 fracs/m ([42] this issue), with a multiplicator factor of 1.4 compared to F_d3—very abundant microfracturing.

Samples selected for this study had a fracture density higher than 2694 fracs/m, meaning they fell into categories from Fd2 to Fd4. Fd4 corresponds to highly strained zones in which granite becomes a breccia. In order to correlate the fracture density with the amount of alteration, three hand specimens and seven thin sections, located in the granite body, were used for fracture extraction and analysis. These fracture characterizations are based on the scanline method, described in Section 3.1. Data from Chabani et al. (this issue, under review) [44] will be used to complete the characterization.

3.2. Methods

3.2.1. Fracture Network Parameters

Fracture networks can be characterized by their spatial arrangements [45]. The fracture network geometry is used to predict fluid circulation [45] and evidence of structural growth processes [46]. Spacing measurements through the 1D-scanline method are widely used to characterize arrangements [47–50]. They consist of the digitization of fractures along those lines, in order to calculate the linear fracture density P_{10} characterized by the number of fractures per length calculated along the scanline [47,48]. The Terzaghi correction has been applied [51]. Two parallel scanlines were performed on the hand specimens, perpendicular to a major fracture. For each oriented thin section mosaic, two scanlines were realized perpendicular to the main structures. For each scanline, stick plots and cumulative frequency diagrams were realized to describe the fractures spatial distribution [50]. A coefficient of variation C_v was computed for each scanline in order to quantify the fracture distribution [48]: C_v < 1 indicates a regular fracture spacing, C_v~1 indicates a random distribution, and C_v > 1 indicates a clustered distribution. Fracture density data obtained by [42] on selected samples will be used in this study.

3.2.2. Petrographic Characterization

Twenty-four samples were selected to prepare thirty-one covered and polished thin sections, for petrographic observations. Optical microscopy was used to study the mineralogical assemblage, the alteration paragenesis, the microfabric, and the degree of microfracturing. The thin sections were observed under a Leica DM4500-P optical microscope, equipped with a Leica DFC450C camera at Institut Polytechnique UniLaSalle (ULS, Beauvais, France). Large field area imaging under polarized-analyzed and polarized-nonanalyzed light was conducted using Leica automatized stage facility and Leica Application Suite (LAS) v4.11.0 software [19].

Based on optical microscopical observations, microsites were selected on 9 thin sections for energy dispersive X-ray spectrometry (EDS) analyses [52] performed at ULS. These were conducted with a Hitachi S-3400N scanning electron microscope (SEM) equipped with a Thermo Ultradry EDS [53] and associated with NSS thermal scientific software. These analyses aimed to obtain qualitative and semi-quantitative chemical compositions and to characterize the microstructure of different selected phases. The analytical conditions consisted of a 50 µA beam current, an accelerating voltage of 20 kV, and an acquisition time of 30 s. A NORAN-type correction procedure was used.

In addition, cathodoluminescence (CL) analyses [54] was performed on 8 thin sections from 4 samples. CL imaging was performed (1) at the University of Göttingen using a "hot-cathode" cathodoluminescence microscope HC3-LM apparatus and (2) at CY Cergy Paris University in the Geosciences and Environment Cergy (GEC) laboratory, using a cathodoluminescence Olympus BX50. CL of minerals is predominantly a "defect luminescence". This allowed the detection of distribution patterns of certain trace elements, such as iron (Fe), which is the most efficient quencher element and manganese (Mn), which is the most important activator element. Many minerals show visible CL colors like quartz (blue-purple), K-feldspar (blue when fresh and brownish when weathered), plagioclase (green and blue–purple when highly affected by hydrothermal alteration), calcite (yellow or yellow–orange when resulting from hydrothermal alteration), dolomite (orange–red), magnesite (red), apatite (yellow), kaolinite (dark blue) [54,55]. Illite shows no luminescence [54,55].

Most of the mineral abbreviations used in this paper refer to [56] mineral symbols and only a few others were defined by the authors.

3.2.3. X-ray Diffraction (XRD)

XRD analyses were carried out to identify and determine the mineral phases (semi-quantitative, around 3–5%). The analyses were performed at (1) ULS using a D8-Advance Bruker-AXS (Siemens) diffractometer with a Ni-filtered CuKα radiation at 40 kV and 40 mA, a primary Soller slit of 2.5°, divergence slit of 0.6 mm, and a secondary Soller slit of 2.5°, with a detector slit of 0.1 mm and an anti-scattering slit of 0.6 mm; and (2) the University of Göttingen, using a PHILIPS PW1800 diffractometer with a Cu-anode and an automatic divergence slit. Whole rock powders and oriented clay fractions (2–6 µm and <2 µm) analyses were performed on 12 samples. Quantitative phase analysis based on reference intensity ratio values were performed on randomly oriented whole-rock powders with a step length of 0.5° and a scan speed of 0.014°/s over the range 3°–70°2θ for whole rocks composition. The uncertainty is estimated to be ±5%. Clay mineral separation was conducted using a technique described by [19], based on [57,58], according to the standard techniques suggested by [59]. For the XRD analyses of both fractions, oriented specimens were measured at air-dried (AD), solvated with ethylene glycol (G), and after heating at 550 °C (H) conditions. These three analytical steps were routinely used to better determine clay minerals content and swelling properties [7]. The clay minerals identification, which is based on d-values and the relative intensity of their 00*l* reflections, was undertaken, referencing [60,61]. These measurements were performed with a step length of 0.5° and a speed of 0.01°/s per step over the range 3°–35°2θ. The interpretations of the data obtained at ULS were performed using the DIFFRAC EVA v4.2 (©Bruker AXS) software. Two fractions were collected in order to (1) separate the last produced or neoformed clay minerals (fraction <2 µm), which could be assimilated to the result of the last fluid circulation event, from the fraction 2–6 µm assimilated to old grain recrystallization or to a possible mix between detrital and neoformed clay mineral; (2) obtain the respective clay composition; and (3) obtain the temperature conditions.

Kübler Index and Kaolinite Crystallinity Index determination

The Kübler Index (KI) was used to define the limits of metamorphic zones (diagenetic zone, anchizone, and epizone) [19,62], following the recommendations for Kübler Index calibration of [59] and the CIS-KI transformation formalism of [63], as well as the temperature of illite formation [64,65]. The KI was calculated from the illite crystallinity (IC), which is defined as the full width at half maximum (FWHM) of the (001) 10Å peak of illite, on the AD oriented clay fractions. A recent study [66] has confirmed that IC provides a useful method for characterizing regional grades of diagenesis and low-grade metamorphism. Temperature is thought to be the main factor controlling IC, but other parameters, such as lithology, also have important effects [67–69]. Working at constant lithology allows this effect to be avoided. In some conditions, an altered granite shows similarities with diagenetic reactions present in feldspathic sandstones [20]. Thus, it is possible to characterize temperature ranges by using the KI values obtained for granitic rocks. The values obtained by IC were thus standardized using the crystallinity index-standard (CIS) samples provided by [70] in order to calculate the KI. The KI values of raw data expressed in Δ°2θ, were measured into three slots, corresponding to different campaigns, which induced three standardizations given by [19]. The Kaolinite Crystallinity Index (KCI) was calculated and put in parallel with the KI in order to identify any correlation between both, and to determine temperature evolution as an indicator for samples free of illite. The KCI is defined as the FWHM of the (001) 7Å peak of kaolinite on the AD oriented clay fraction.

3.2.4. Manocalcimetry

Manocalcimetry is used as an indicator of the total calcite content in a rock sample and to assess its possible influence on permeability and consequently better understand the hydrothermal sealing of a reservoir [71]. Calcite content was determined using an OFITE 152–95 manocalcimeter composed of a glass flask and a high precision manometer. The analyses were performed on 15 samples at CY-GEC. High quality results were obtained with an accuracy of ± 0.5 wt.%. Prior to measurement, sample preparation was conducted according to [71]. Two replicates were performed for each sample, following the protocol described by [19], in order to check the reproducibility of the results. This procedure was considered to be good when the difference between the two results was lower than 0.5 wt.%, corresponding to the precision mentioned above. The $CaCO_3$ percentage was calculated according to [19]. The average calcite content of a fresh granite is 0.252 wt.% and does not exceed 1.8 wt.% [72]. As a consequence, measurements above this last value can be regarded as a calcite anomaly [71], due to hydrothermal alteration.

3.2.5. Ethanol Saturation Porosimetry

Porosity measurement quantify the available volume for fluid storage [73]. The connected porosity was measured on three samples by the triple weighing method [19,74] defined by the RILEM standard (test n°I.1, 1978). The measurements were carried out at CY Cergy Paris University in the Mechanics and Material for Civil Engineering laboratory (L2MGC). They consist in the saturation of the samples after vacuum degassing. The analyses and calculations were conducted using the technique described by [19]. Ethanol has been chosen instead of water in classical methods, in order to avoid possible clay swelling [75], which could lead to the destruction of the sample and bias to porosity values. Even though ethanol (0.469 nm) is a larger molecule than water (0.343 nm), the pore volume is not estimated to be under-evaluated, because the pore size is likely much larger than that of ethanol molecules.

3.2.6. Bulk Rock Geochemical Analyzes

Chemical analyzes of the major element oxides, rare earth elements, and trace elements were conducted on five samples, selected as case studies in this paper. Analyzes were performed at the Bureau Veritas Minerals (Vancouver, BC, Canada) using inductively coupled plasma emission spectrometry or mass spectrometry (ICP-ES and ICP-MS, respectively). Samples were crushed and mixed with $LiBO_2/LiB_4O_7$ flux. Crucibles were fused in a furnace at 980 °C. The obtained cooled bead was dissolved in ACS grade nitric acid and analyzed. Loss on ignition (LOI) was determined by igniting the samples split then measuring the weight loss. These chemical analyses were performed in order to determine the element transfers and the degree of alteration.

3.2.7. Mass-Balance Using Gresens' Method

Bulk-rock mass-balance were used to establish element transfers during hydrothermal alteration by applying Gresens' mass-balance procedure [76], consisting of the quantification of gains and losses of major elements by comparing unaltered and altered samples. NH_2 is considered as the reference because it is the freshest sample collected in the area by [19] and 3 altered samples were selected from this study (NH_20, NH_28, NH_32_3, and NH_37). The following equation defined by [76] relates the composition and volume of altered rocks to unaltered/fresh rocks:

$$X_n = F_v \times (d_A/d_F) \times C_n^A - C_n^B \tag{1}$$

With X_n corresponding to the gain or loss of a considered element n (absolute mobility (%)), F_v the volume factor, d_A and d_F the density (in g/cm^3) of the altered (A) and the fresh (F) rocks, respectively, and $C_n{}^A$ and $C_n{}^B$, the oxide percentage of the element n for the altered (A) and the fresh (F) rocks, respectively, which were given by the bulk geochemical analyses, recalculated without the LOI. The densities were calculated from the sample masses obtained during the porosity measurements. The F_v was calculated for each altered sample by considering $X_{Al2O3} = 0$ following the analysis of the values. Xn was calculated for each element. Then, the obtained Fv values being very close to 1, the Xn values were recalculated by considering $F_v = 1$ (constant volume). Hence, X_n values > 0 (positive values) represent the gains and X_n values < 0 (negative values) represent the losses for the considered elements. This calculation was applied for the major elements in each sample.

4. Results

4.1. Noble Hills Structural Overview

We investigated the geometry of outcropping Proterozoic sequences between Pipeline and Cave Spring washes (Figure 2a) and confirm at first the plutonic nature of the contact at the base of the Proterozoic sequence with the underlying Mesozoic granite by metamorphic halos [40]. The whole architecture of the NH presents signs of deformation affecting both Proterozoic and granitic units. Proterozoic units present much evidence of deformation with (1) several tectonically induced duplications of the stratigraphic sequence (Figure 2b) and (2) a lateral bending, stretching, and boudinage of this Proterozoic cover nappe stack [41]. Boudinage was identified in the field and in map view where progressive dismantling of stratigraphic markers is observed. (Figure 2b,c). Development of synthetic, oblique dextral shears (Figure 2c), offsetting along the CBS strike and accommodating the deformation, have also been identified. The age and tectonic significance of the nappe stack described in (1) is still unclear at present and possibly predates Mesozoic magmatic intrusion, as it does not appear intensively deformed along the basal contacts of the nappes. Lateral bending and stretching mentioned in (2) and depicted in Figure 2b,c are related to SDVFZ activity, since all kinematic indicators are consistent with progressive southeast oriented dextral shearing. This feature is ubiquitous within the NH as described by [33]. Given these new findings dealing with internal CBS structural organization, it is expected that areas along the rear southern limit of the CBS have recorded less transcurrent shear compared to areas situated along its northern front. Thus, the CBS gives an opportunity to study the effects of progressive transcurrent shearing within a granitic geothermal analogue. The above considerations have important implications regarding strain distribution within the NH and, in turn, concerning the sampling strategy. Given the above-mentioned findings, we consider the CBS as an exceptional example of large-scale cataclastic flow in which rock masses of the northern domain are dragged as a continuous body into a transcurrent deformation corridor (Figure 2a). No signs of a discrete, large-scale deformation structure appear on the map presented in Figure 2, where yellow dashed lines show instead a continuous flow accommodated by brittle deformation processes.

4.2. Petrographic Description

As described by [19], the primary assemblage of the NH granite is made of plagioclase (mainly oligoclase in composition), quartz, K-feldspar (perthitic orthoclase and sometimes microcline), and biotite. The granitic samples collected for this study show strong evidence of intense alteration, especially on plagioclase and biotite. Plagioclase transformed into illite, kaolinite and/or calcite and biotite into illite and oxides. In rare samples, K-feldspar perthite could be partially altered into illite. However, the magmatic texture is almost preserved, except in the case of breccias, which are not necessarily highly altered. For all samples, the alteration product is always the same, but the relative proportions of the different secondary phases might vary. The mineralogical composition and the degree of fracturing of each sample is given in Table 1.

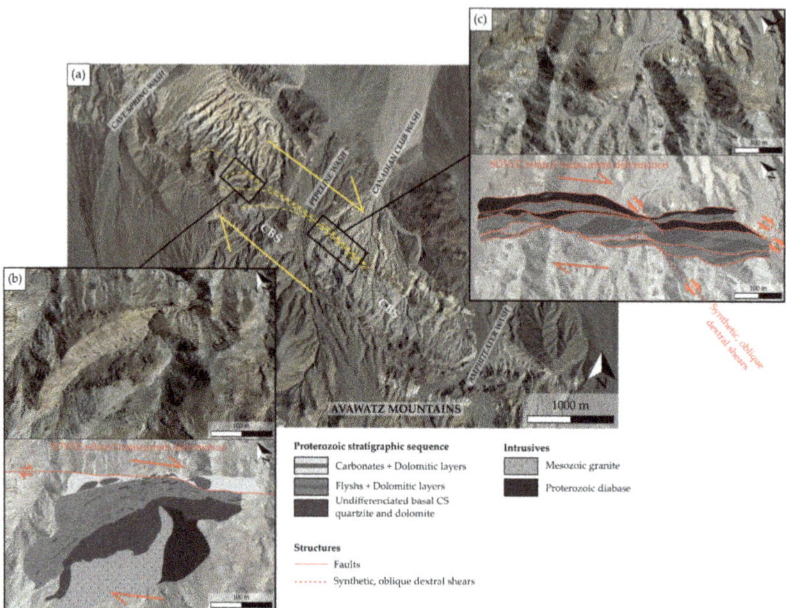

Figure 2. (**a**) Satellite view of the NH range showing a deformation gradient increasing southeastward, represented by the yellow dashed lines. (**b,c**) Map view zooms on the Proterozoic units along the NH strike showing the strain increase from (**b**) to (**c**), where basal CS sedimentary sequence appears increasingly stretched and boudinated.

The clay mineral composition of some samples has been precisely determined for the <2 μm and 2–6 μm (Table 1). It reveals the presence of illite/smectite (I/S) mixed-layers in three samples (NH_17, NH_24, and NH31_1_2) in both fractions when it was measurable. The I/S identified is always illite-rich (R3), with more than 90% of illite [60], where R represents the Reichweite parameter [60]. A small amount of corrensite (trioctahedral variety of regular 50:50 chlorite/smectite mixed-layer (C/S) [77]) was also identified in the samples NH_16, NH_27, and NH_32. Both mixed-layers (I/S and C/S) were already identified by [19] in the sampled granite away from the fracture zones. Two new patterns were identified in this study (Figure 3):

- The first pattern (Figure 3a) shows illite characterized by peaks at 10.1 Å and 4.99 Å, which support the previous results [19]. Illite is present in almost all the samples except NH_28, NH_36, NH_38, and NH_39. Kaolinite is also present and could be associated to illite or not.
- The second pattern (Figure 3b) shows the presence of dioctahedral vermiculite. It is characterized by a peak at 14.32Å in air-dried condition, which slightly swells up to 14.59Å after glycol solvation and slightly collapses to 12.26Å after heating. It was identified only in samples NH_24, NH_26, and NH_33, where it is associated to illite and kaolinite.

Table 1. List of samples collected in the field with their mineralogical composition determined after microscopic observations (optical-microscope and SEM-EDS) and XRD analysis (primary minerals and secondary minerals), their degree of microfracturing after [43], and their clay mineral composition for the fractions <2 µm and 2–6 µm. Abbreviations after [56], except for the following: Olg—oligoclase, Ox—oxide, I/S—illite/smectite, C/S—corrensite, n.m.—not measured, n.a.—not analyzed.

Sample Name	Primary Minerals	Secondary Minerals	Clay Fraction < 2 µm						Clay Fraction 2–6 µm						Microfracturing
			Ill/Ms	Kln	C/S	I/S	Vrm	Bt	Ill/Ms	Kln	C/S	I/S	Vrm	Bt	
NH_16	Qtz, Or, Bt, Olg	Cal, Clays	+	-	–				+	+	–				Fd2
NH_17	Qtz, Or, Ab, Ap	Gp, Clays			n.m.				+	-		–			Fd2
NH_18	Qtz, Or, Olg, Bt	Clays	++	–					++	–					Fd2
NH_19	Mc, Or, Olg, Bt, Qtz, Ms	Dol, Cal, Clays	+						+						Fd2–3
NH_20	Qtz, Or, Olg, Bt	Cal, Ox, Clays			n.a.						n.a.				Fd2–3
NH_21	Qtz, Or, Olg	Cal, Dol, Sd, Ox, Clays			n.a.						n.a.				Fd3
NH_22	Qtz, Or, Bt, Olg	Cal, Hem, (Dol), Clays	+	-				-	+	-				-	Fd3
NH_23	Qtz, Or, Olg, Bt	Cal, Ox, Clays			n.a.						n.a.				Fd3
NH_24	Qtz, Or, Olg, Bt	Cal, Clays	+	+		-	-		-	+		-	-		Fd3
NH_25	Qtz, Or, Bt, Ep	Cal, Ox, Clays	++						++						Fd3
NH_26	Qtz, Or, Bt, Olg, Mc	Clays	-	++			+	+	-	++			+	+	Fd3
NH_27	Qtz, Or, Bt	Ank, Cal, Clays	+	++	-			-	+	++	-			-	Fd3
NH_28	Qtz, Or, Mc, Olg, Bt, Ms	Cal, Ox, (Dol), Clays		++						++					Fd3
NH_29	Qtz, Or, Olg, Ab, Bt	Cal, Dol, Clays			n.a.						n.a.				Fd3
NH_30	Qtz, Or, Mc, Olg, Ab, Bt, Ms	Cal, (Ox, Dol), Clays			n.a.						n.a.				Fd3
NH_31_1_2	Qtz, Or, Olg, Bt	Clays	+	+		-			+	++		-			Fd3
NH_31_3	Qtz, Or, Ab, Bt	Gp, Cal, Clays			n.m.				++	–					Fd3
NH_32	Qtz, Or, Bt	Cal, Ox, (Dol), Clays	+	++	-				+	++	-				Fd4
NH_33	Qtz, Or, Bt, Ab	Cal, Clays	–	++		–	-		–	++			–	-	Fd4
NH_34	Qtz, Or, Mc, Ab, Ap, Bt	Gp, Cal, Hem, Clays			n.m.				–	++					Fd4
NH_35	Qtz, Or, Mc, Bt	Cal, Clays			n.a.						n.a.				Fd4
NH_36	Qtz, Or, Bt, Olg	Cal, Gp, Hl, Clays		+				+		++			-		Fd4
NH_37	Qtz, Or	Cal, Dol, Ank, Ox, Clays			n.a.						n.a.				Fd4
NH_38	Qtz, Or, Mc, Olg, Bt	Cal, Clays		++						++					Fd4
NH_39	Qtz, Or, Mc, Olg, Bt	Cal, Clays		++						++					Fd4

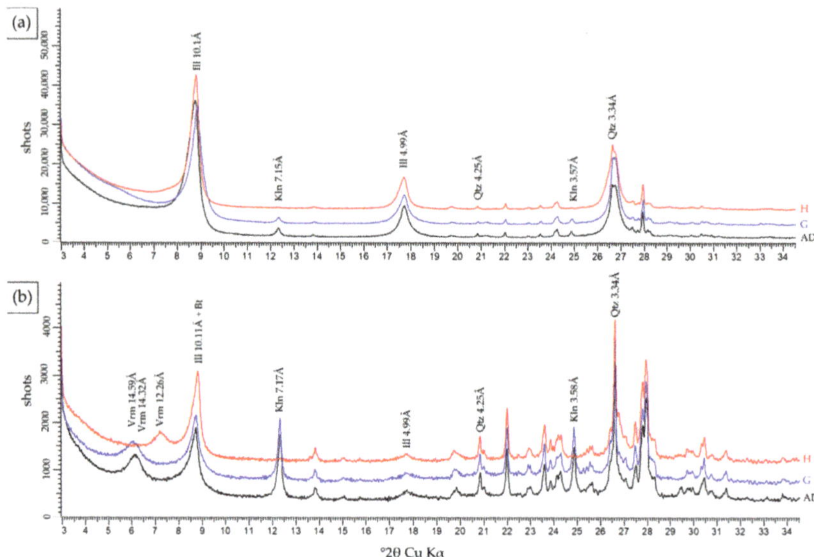

Figure 3. Two examples of XRD patterns obtained for the clay fraction <2 μm in air-dried (AD–dark), after glycol solvation (G-blue) and heated (H-red) conditions for the NH granite, which were not identified in the fresh granite described by [19], and showing different clay compositions. (**a**) Illite, kaolinite, and quartz. (**b**) Vermiculite, illite, kaolinite, and quartz.

The high proportions of illite and kaolinite are characteristic of the argillic alteration [19]. Argillic alteration is prevalent in the vicinity of fracture zones (Table 1). Vermiculite can be interpreted as either hydrothermal alteration or weathering. This will be discussed in Section 5.1.

Microscopic observations and SEM-EDS analyses have shown that, in almost all the samples, carbonates (dolomite, siderite, ankerite, and calcite) and oxides were present as veins and or in replacement of plagioclase. Moreover, XRD analyses revealed the presence of gypsum and/or halite in minor amounts in few samples (Table 1). Halite only occurred in sample NH_36 and gypsum was found in samples NH_17, NH31_3, NH_34, and NH_36.

4.3. Fracturing and Fluid Circulation

The NH granite presents ubiquitous signs of both fracturing and alteration, due to its location along the SDVFZ major fault corridor. The area lies along a deformation corridor governed by transcurrent deformation along the SDVFZ. In addition to the present contribution, Klee et al. [41] argue in favor of a deformation gradient affecting the CBS. At low to moderate strains, the system presents fracture zones composed of a fault core in which fractures are branching and where most of the displacement is concentrated (Figure 4a). These fracture zones affect the surrounding rock, creating an important fracture network (the damage zone, DMZ). Open fractures and veins are observed in this DMZ (Figure 4b).

Fault zones may act as a channel when connected and open fractures are present or as a barrier when fracture are sealed by mineralization. The distinction is made between (1) zones of opening related structures (i.e., where no signs of displacement are recognized between the two borders of the vein, Figure 4b); and (2) zones of lateral displacement (i.e., indicative of a shear displacement is observable) through three samples used as case studies (NH_31 (Figure 5a), NH_36, and NH_23 (Figure 5b)). Note that these samples are spaced at 2 m and are almost perpendicular to each other.

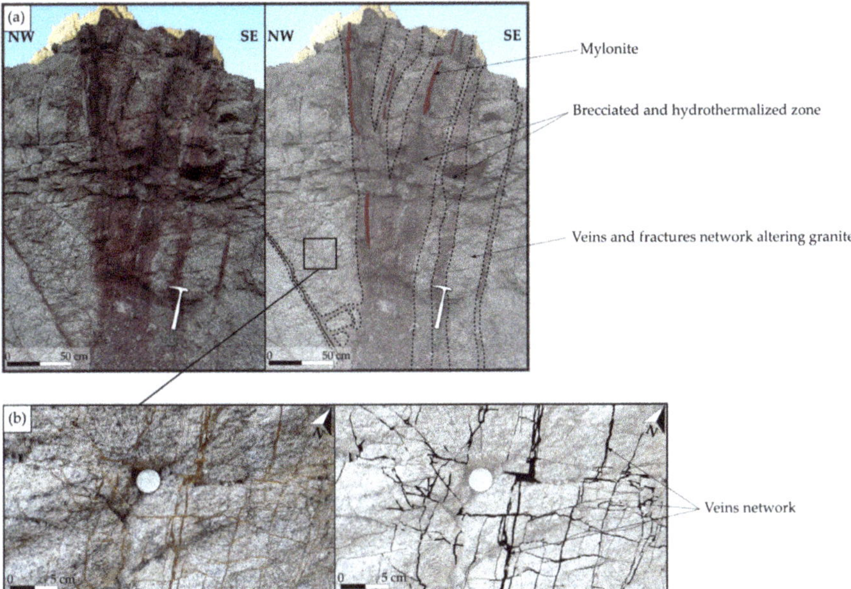

Figure 4. (**a**) Photograph and digitization showing representative hydrothermalized fracture zones observed on the field. (**b**) Photograph and digitization of granite crosscutting by multiple carbonate veins (veins network) altering pervasively the rock.

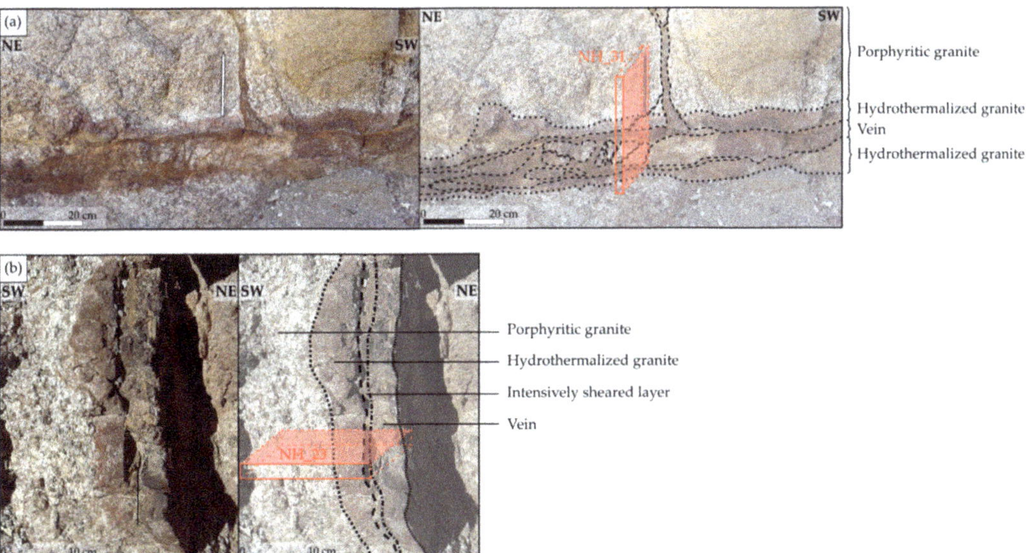

Figure 5. (**a**) Photograph and digitization of a brecciated and recrystallized vein also developing a gradient of alteration. Location of the sample NH_31 used as a case study in this section. (**b**) Photograph and digitization of a mylonitic vein developing a gradient of alteration. Location of the sample NH_23 used as a case study in this section.

4.3.1. Opening Related Structures with Minimum Shear Displacement

A significant number of veins, veinlets, and microfractures were identified in the different samples and outcrops investigated. Veinlets of various mineralization natures ($\leq$100 µm wide) were also identified in the samples. This variety of fracture infills shows different fluid circulation episodes. Rare veinlets of quartz were observed and intersected by illite veinlets (Figure 6a), which can be contemporaneous to kaolinite veinlets. Illite development is dependent on host mineral properties in terms of mechanical resistance and chemical stability. In that sense, quartz and K-feldspar, which remains unaltered, presents sharp open fractures in which illite precipitates (Figure 6b). Plagioclase, which is altered, presents blurred vein borders and start to spread out pervasively, due to the secondary porosity created by the alteration process (Figure 6b). These veinlets are generally intersected by carbonate veins (Figure 6c). These veins, the sizes of which range from 500-µm widths up to centimetric scales, are composed of dolomite, siderite, ankerite, and calcite. Most of the time, dolomite and ankerite appear contemporaneous. When siderite is present, it alters dolomite borders and crystallizes in it or crosscut it (Figure 6d). Then, calcite veins intersect the dolomite/ankerite and siderite veins (Figure 6d). Calcite can crystallize around angular clasts arranged in a fan shape characteristic of hydraulic fracturing (Figure 6e). Open microfractures present in the samples can show altered walls with kaolinite (Figure 6f) and/or illite (Figure 6g) indicating an influence of microfractures on mineral alteration.

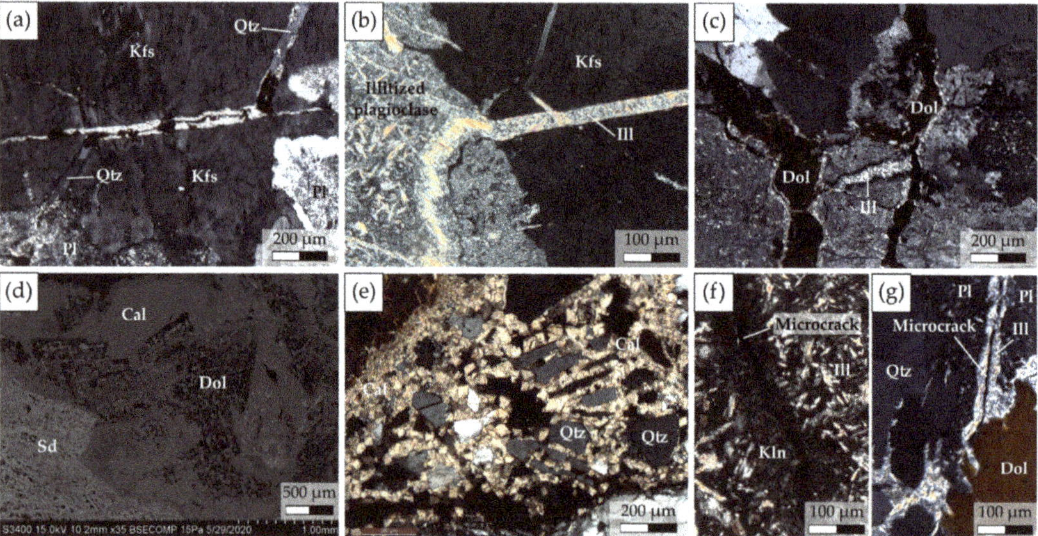

Figure 6. (**a**) Photomicrograph of a quartz veinlets intersected by an illite veinlets under optical microscope in polarized–analyzed light. (**b**) Photomicrograph showing an illite vein crosscutting K-feldspar and illitized plagioclase under optical microscope in polarized–analyzed light. (**c**) Back-scattered image showing the different phases presented in (**b**). (**c**) Photomicrograph showing dolomite intersecting illite under optical microscope in polarized–analyzed light. (**d**) Back-scattered image of a dolomite, siderite, and calcite showing their chronology. (**e**) Hydraulic fracturing with calcite precipitating around clasts of quartz placed like a fan under optical microscope in polarized–analyzed light. (**f**) Photomicrograph under optical microscope in polarized–analyzed light of a microfracture intersecting an altered plagioclase which borders are made of pure kaolinite, whereas the plagioclase was recrystallized into kaolinite and illite. (**g**) Photomicrograph under an optical microscope in polarized–analyzed light, showing illite developing on the walls of an open microfracture and dolomite filling this open space.

4.3.2. Infill of Fractures Developed with a Confirmed Lateral Shear Displacement

The previous section shows that all of the samples in the area present many fractures where fluid has circulated, precipitating or not various secondary minerals. However, the area is constrained by a shear component. Three samples were selected in order to describe the relationship between the amount of the strain and fluid circulation.

Sample NH_31

The outcrop of the sample NH_31 (Figure 5a) presents a vein around 10 cm thick crosscutting the granite and with a pinkish halo developed on each border. This halo represents the hydrothermalized zone of the granite. The sample NH_31 (Figure 7a) is composed of (1) a brecciated vein presenting brecciated quartz remnants, which were overprinted by a carbonate matrix and granitic clasts transported by the carbonate matrix; (2) a hydrothermalized zone, which corresponds to the pinkish halo; and (3) the porphyritic granite crosscut by veinlets filled by carbonates coming from the brecciated vein. Quartz veins are scarce in the NH granite. Fractures are dominantly rich in carbonates. It can be observed through this sample that carbonates crosscut the quartz.

- Spatial fracture distribution: two scanlines were realized in the granitic part of the sample (Figure 7a), from the brecciated vein towards the host rock, in order to evaluate the evolution of the fracture density along those two profiles. The spatial analysis is summarized in Table 2.

The cumulative frequency plotted against the fracture-projected position is presented in Figure 7b. The fracture frequency regularly and slowly increases over the first half part of the diagram curve, then increases regularly but more significantly. The Cv of 0.67 and 0.60 for each scanline shows a regular to random fracture arrangement, which is confirmed by the stick plots in Figure 7c. Both scanlines present an equivalent fracture density of 829 and 882 fracs/m.

Moreover, three thin sections were gained from the sample and one scanline was realized for each thin section (Figure 8a,b). As for the sample scanlines, the thin section scanlines were realized perpendicular to the vein and towards the host rock (Figure 8c).

SL_NH_31_1 shows a fracture frequency that increases following a random arrangement (Cv = 1.03) (Figure 8c,d).

The SL_NH_31_2, parallel to the previous scanline, also presents an irregular distribution. The frequency increases slowly at the beginning, followed by a high fracture density zone around 0.7–1.05 (fracture cluster), then a more significant increase comprising a new fracture cluster around 2–2.2 (Figure 8c). The slightly higher Cv compared to the previous scanline (Cv = 1.07) and the fracture distribution (Figure 8c) indicate a random to clustered arrangement of the fractures.

The last scanline SL_NH_31_3, which is the furthest from the vein, shows a greater overall increase of the fracture frequency compared to the two previous scanlines (Figure 8c). Only one fracture cluster was identified around 1.7–1.9 cm even if the Cv is higher (Cv = 1.11). The fracture arrangement is again considered as random to clustered.

Cv values, as well as fracture density values, increase from the vein towards the host rock, from 1.03 to 1.11 and from 2997 to 5084 fracs/m, respectively (Figure 8d).

By comparing fracture analyses between sample scale and thin section scale, a clear difference is observed among P_{10}, the mean spacing, and Cv values (Table 2). At thin section scale, the fracture density is significantly higher ($\times$4.7) than at sample scale and the mean spacing is six times lower. At sample scale, the Cv is lower than 1 indicating an almost regular spacing between the fractures, even if there is a change of the slope (Figure 7c). Whereas, at the thin section scale, the Cv is around 1 or slightly higher, indicating a global random distribution of the fractures along the scanlines with the appearance of a few clusters (Figure 7c). For both, differences in the fracture distribution can be linked to a subjective bias during the data collection [78], but also to the image resolution, which prevents from seeing all the fractures and so induced a bias. Thin section fracture analyses are, thus, more precise.

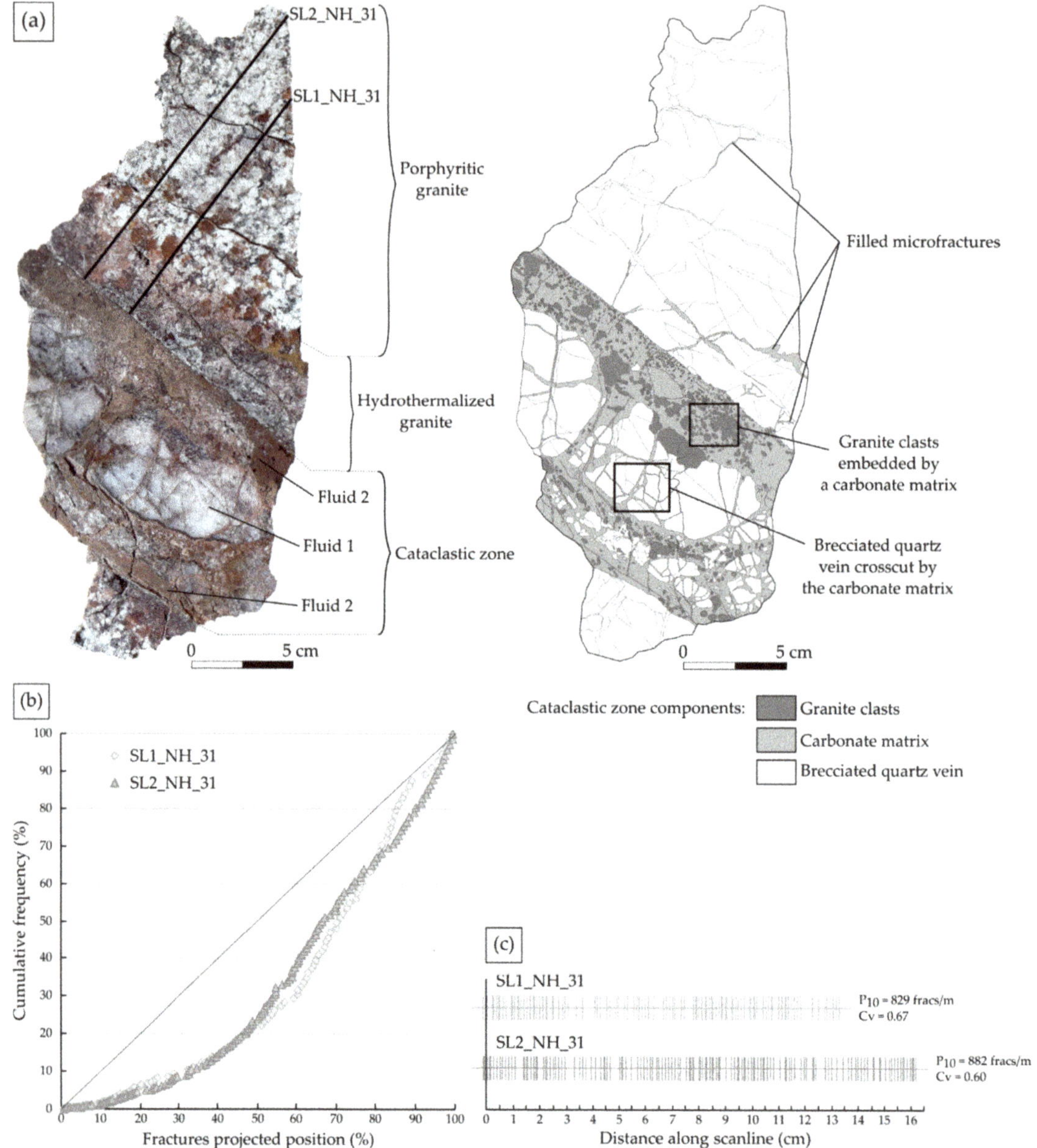

Figure 7. (**a**) Photo and the respective digitalization of the sample NH_31, gathering NH_31_1_2 and NH_31_3, showing the different compartments that compose it, as well as the two scanlines realized in the host rock of the vein. (**b**) Plots of the cumulative frequency percentage against distance percentage for both scanlines. The diagonal represents a uniform distribution. (**c**) Stick plots showing the fracture position along the scanlines and for which the fracture density (P_{10}) and the coefficient of variation (Cv) are given.

Table 2. Spatial fractures analysis. Scanline length: total length of each scanline; fracture number, mean spacing, P_{10}, and Cv value by scanline and fracture distribution tendency by scanline.

Sample Name	Scale	Scanline Name	Scanline Length (cm)	Fracture Number	Mean Spacing (cm)	P_{10} (fracs/m)	Cv	Fracture Distribution
NH_31	Sample	SL1_NH_32	13.51	111	0.12	829	0.67	Regular—Random
		SL2_NH_32	16.44	144	0.11	882	0.60	Regular—Random
	Thin sections	SL_NH_32_1	1.60	47	0.03	2997	1.03	Random
		SL_NH_32_2	2.21	87	0.02	3988	1.07	Random—Clustered
		SL_NH_32_3	2.34	118	0.02	5084	1.11	Random—Clustered
NH_36	Samples	SL_NH_36	15.8	53	0.3	342	1.03	Random
	Thin sections	SLH_NH_36_1	2.87	153	0.02	4636	1.49	Clustered
		SLV_NH_36_1	2.06	133	0.02	6312	1.01	Random
		SLH_NH_36_2	3.35	184	0.02	5231	1.11	Random—Clustered
		SLV_NH_36_2	1.88	74	0.03	3930	0.92	Regular—Random
NH_23	Sample	SL1_NH_24	14.33	197	0.07	1382	0.68	Regular—Random
		SL2_NH_24	12.93	170	0.07	1322	0.66	Regular—Random
	Thin sections	SL_NH_24_1	2.48	117	0.02	4798	0.77	Regular—Random
		SL_NH_24_2	2.37	91	0.02	3879	1.19	Random—Clustered

- Petrographic and petrophysical characterization: the brecciated vein presents a large variety of mineralogical phases. As shown in Figure 7a, this zone shows a carbonate matrix containing clasts of granite, which overprints a brecciated quartz vein remnant showing an undulatory extinction but no evidence of dynamic recrystallization. The carbonate matrix is complex, composed of several phases (Figure 9a–g).

Focusing on the veins crosscutting the quartz porphyroclasts, three phases are identified (Figure 9a–e):

1. Phase 1 is composed of calcite veins crosscutting the quartz. These veins, appearing as a single phase under SEM and optical microscope (Figure 9a,c), present two phases under cathodoluminescence (CL): a dark phase (Cal A) in the center of the vein and an orange bright phase (Cal B) (Figure 9b,d). Cal B phase seems to dissolve or corrode the Cal A phase (Figure 9b,d). A zonation corresponding to calcite growth halos are visible in the phase A, which are used as weakness zones in which the phase B can penetrate by dissolving the phase A.
2. Phase 2 is composed of a matrix rich in carbonates (calcite, dolomite, ankerite) transported clasts of quartz, K-feldspars and few biotite, but also presents barite precipitation patches (Figure 9e). This phase crosscuts the quartz and the calcite veins.
3. Phase 3 consists again of calcite veins. However, they crosscut the whole rock, and appear as a dark single phase (Cal C) under CL (Figure 9b).

Evidence of hydraulic fracturing, as presented in Figure 6e, have also been observed in this part of the sample.

By focusing in the carbonate matrix, at the contact with the granite, a new phase is observed, composed of dolomite embedding small clasts (mainly quartz). This phase is intersected by the phase 2, which is intersected by the phase 3 (Figure 9e,f). Examining the hydrothermalized granite along the brecciated vein (NH_31_1 and NH_31_2), the granite shows a preserved magmatic texture with plagioclase fully altered into illite, ± kaolinite, and calcite, completely altered biotite into illite and oxides, and unaltered perthitic K-feldspar (Figure 9h,i). Hematite and numerous dolomite/ankerite and calcite veinlets are present. Moving away from the brecciated vein (NH_31_3), the granite appears less affected by the alteration. Plagioclase and biotite are only partially altered and K-feldspar and quartz are unaltered (Figure 9j,k). Unlike the hydrothermalized zone, the granite presents only a few veinlets of calcite.

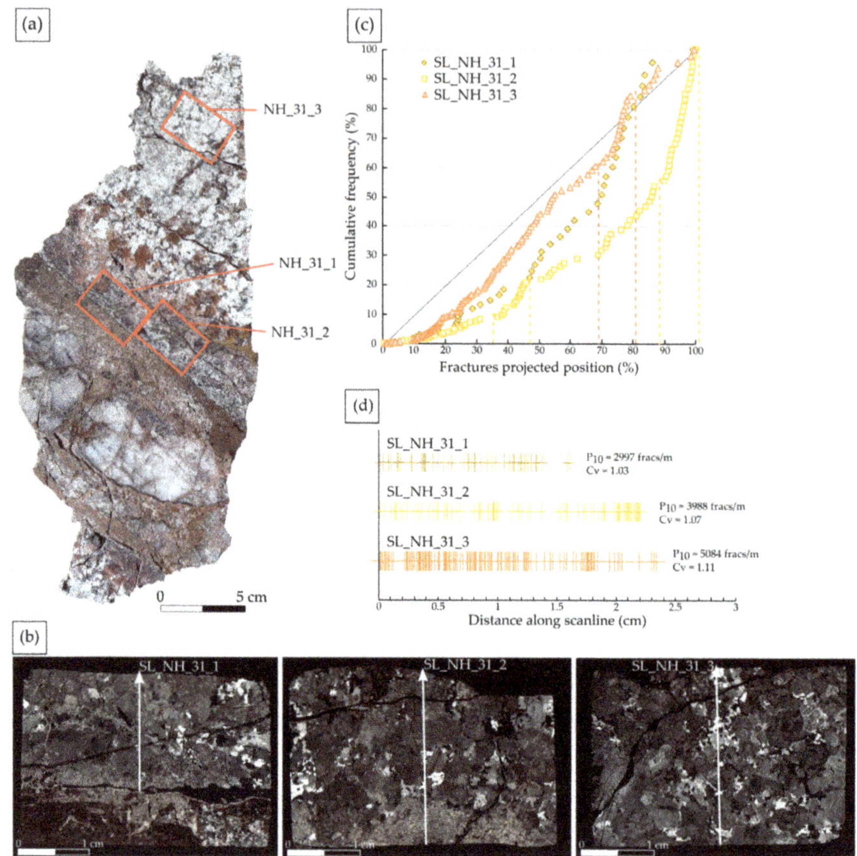

Figure 8. (**a**) Thin sections location on the sample NH_31. (**b**) Thin section mosaics showing the position and orientation of the scanlines. (**c**) Plots of the cumulative frequency percentage against distance percentage for each scanline. The diagonal represents a uniform distribution. Dashed lines indicate a zone where a rapid increase of the number of fractures is observed (slope threshold >2). (**d**) Stick plots showing the position of the fracture along the scanlines and for which the fracture density (P_{10}) and the coefficient of variation (Cv) are given.

Bulk chemical analyses were performed on the granitic part of the sample and show a low LOI of 2.5%, a porosity of 10.1%, and a calcite content of 1.86%, close to the brecciated vein (NH_31_1_2), and a calcite content of 1.37% away from it (NH_31_3).

Sample NH_36

This sample (Figures 1b and 10a) shows a fractured granite with a preserved magmatic texture intersected by brecciated zones. As for sample NH_31, the fracture distribution analyses, as well as a petrographic description, were realized.

- Spatial fracture distribution: the spatial fracture distribution analysis was realized based on data from [42]. One scanline was realized through the sample in order to evaluate the evolution of the fracture distribution and density (Table 2, Figure 10a). The cumulative frequency against the fractures projected position presented in Figure 10b shows a fracture frequency slowly increasing, as well as a fracture cluster. The Cv of 1.03 indicates a random arrangement of the fractures along the scanline (Figure 10c). A fracture density of 342 fracs/m was compiled (Figure 10c).

Two thin sections were made from the sample (Figure 10a,d). Two perpendicular scanlines were realized on each thin section (Figure 10d).

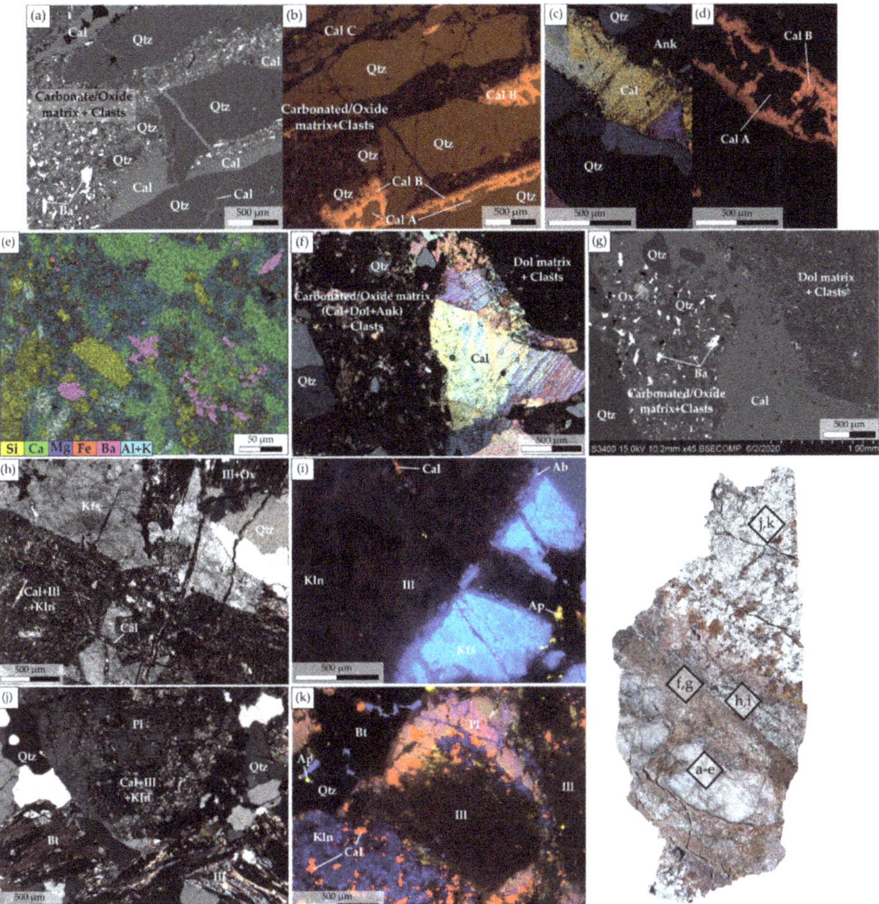

Figure 9. Photomicrographs of the different compartments of the sample NH_31 and their locations in the sample. (**a,b**) Respectively back-scattered image and CL images of calcite veins composed of two phases (Cal A and Cal B) cross-cutting quartz, crosscut by a phase made of a carbonate/oxide matrix (calcite, dolomite, ankerite) with barium patches and transported clasts of quartz, K-feldspars, and few biotites, crosscut by later calcite veins (Cal C). (**c,d**) Zoom on the first generation of calcite vein and crosscutting quartz under optical microscope in polarized–analyzed light and showing two phases under CL. (**e**) A chemical quantification map realized under SEM of the carbonate/oxide matrix with barium and clasts. Different colors show a relative abundance of specific chemical elements (counts). (**f,g**) A photomicrograph under optical microscope in polarized–analyzed and under SEM, respectively, of the contact with the granite showing a dolomite matrix with transported clasts intersected by a calcite vein, the whole intersected by a carbonate/oxide matrix with transported clasts and barium precipitation patches. (**h**) The altered granite close to the brecciated vein showing a completely altered plagioclase replaced by calcite, kaolinite, and illite, a completely altered biotite replaced by illite and oxides and non–altered K-feldspar and quartz under an optical microscope in polarized–analyzed light. (**i**) The altered granite close to the brecciated vein presented in (**h**) under CL showing illite with no luminescence, in light blue an unaltered K-feldspar, in red some calcite and apatite in yellow. (**j**) The altered granite, away from the brecciated vein, showing a plagioclase partially replaced by calcite, kaolinite, and illite, biotites partially replaced by illite, and oxides, and unaltered quartz under optical microscope in polarized–analyzed light. (**k**) The altered granite away from the brecciated vein presented in (**j**) under CL showing kaolinite in dark blue, calcite in red, apatite in yellow and illite, quartz and biotite with no luminescence.

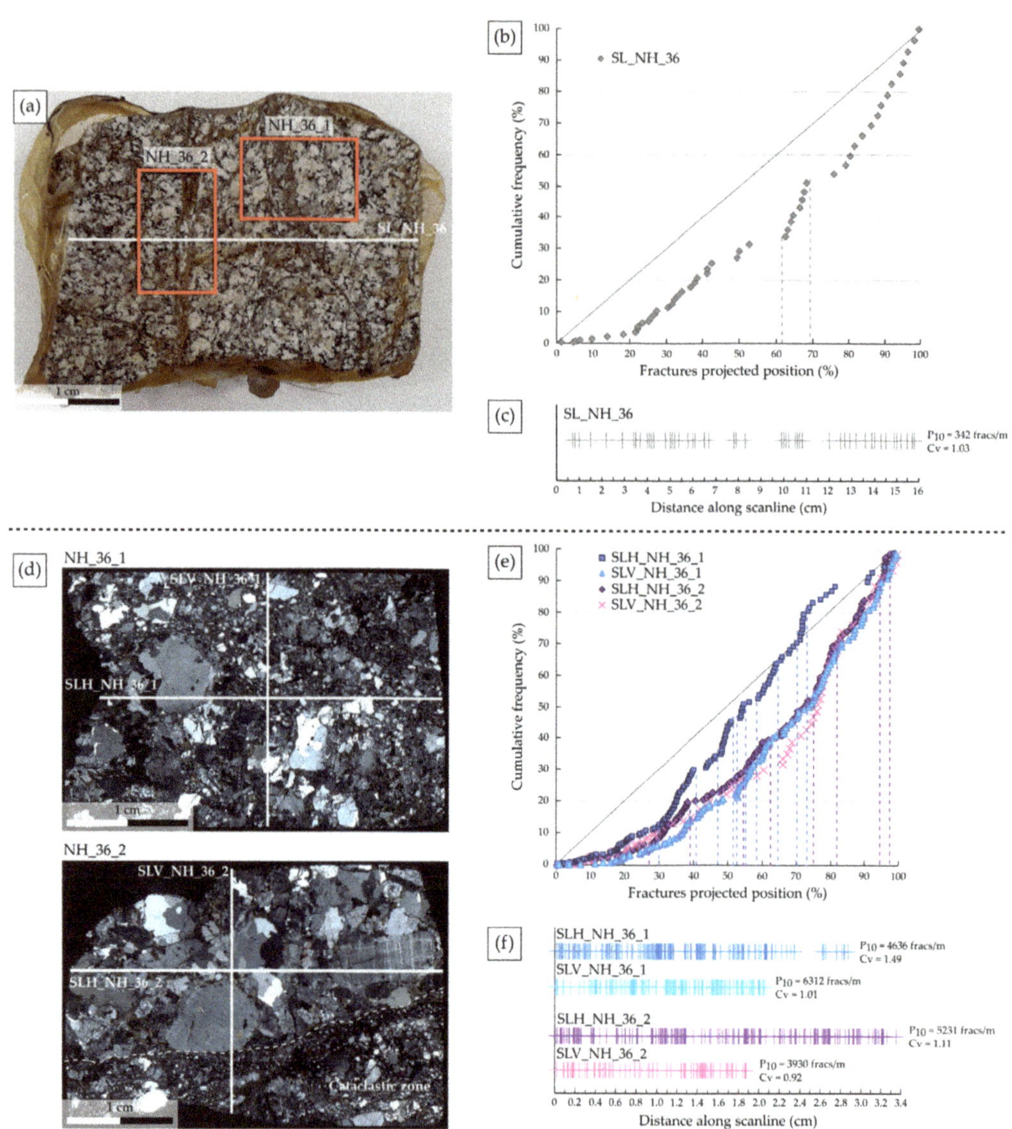

Figure 10. (**a**) Photograph of the sample NH_36 showing the scanline and the emplacement of the two thin sections made in the sample. (**b**) Plot of the cumulative frequency percentage against distance percentage for the scanline. The diagonal line represents a uniform distribution. Dashed lines indicate a cluster, meaning a zone where a rapid increase of the number of fractures is observed (slope threshold >2). (**c**) Stick plot showing the fracture position along the scanline and for which the fracture density (P_{10}) and the coefficient of variation (Cv) are given. (**d**) Thin section mosaics showing the position and the orientation of the scanlines. (**e**) Plots of the cumulative frequency percentage against distance percentage for each scanline. The diagonal line represents a uniform distribution and the dashed lines delimit a fracture cluster. (**f**) Stick plot showing the fracture position along the scanline and for which the fracture density (P_{10}) and the coefficient of variation (Cv) are given. Data were provided by [42].

Regarding the thin section NH_36_1, the scanline SLH_NH_36_1, perpendicular to the main fractures, shows a fracture frequency increases slowly the first 30%, then increases more strongly by presenting five fracture clusters (Figure 10e). This fracture distribution is highlighted by a Cv = 1.49, confirming a clustered arrangement of the fractures (Figure 10f). The fracture clusters are respectively comprised between, 0.9–1.2 cm, 1.4–1.5 cm, 1.55–1.6 cm, 1.7–1.85 cm, and 2.05–2.1 cm intervals (Figure 10e,f). The scanline SLV_NH_36_1, parallel to the main fractures, shows a different trend. The fracture frequency increases slowly and irregularly (Figure 10e). This distribution is highlighted by a Cv = 1.01, indicating a random arrangement of the fractures (Figure 10f). Considering both scanlines, the average fracture density is of 5474 fracs/m.

Regarding thin section NH_36_2, the scanline SLH_NH_36_2, parallel to the main fractures shows a fracture frequency trending similarly to SLV_NH_36_1. Four fracture clusters were identified (Figure 10e), also visible in the stick plot between 0.95–1.3 cm, 1.85–2.05 cm, 2.5–2.7 cm, and 3.1–3.2 cm (Figure 10f). The Cv = 1.11 confirmed a random to clustered arrangement of the fractures. The scanline SLV_NH_36_2, perpendicular to the main fractures, crosscut the magmatic preserved texture of the granite and a breccia. A change of the curve inclination is visible around 65% of the scanline, where the fracture frequency strongly increases (Figure 10e). It corresponds to the limit between both zones (Figure 10d). The Cv = 0.92 indicates a global regular to random fracture arrangement along this scanline (Figure 10f). Considering both scanlines, the average fracture density is of 4581 fracs/m.

Because fractures are difficult to recognize in the breccia of the sample NH_36_2, the fracture density calculated for both thin sections may be underestimated.

As for sample NH_31, an important difference between the P_{10} values of the sample and the thin sections is observed. However, a global random arrangement of the fractures at both scales is observed. Only SLH_NH_36_1 presents a clustered distribution.

- Petrographic and Petrophysical Characterization

The two thin sections (Figure 11a–c) allow petrographic characterization of the sample. The first thin section, NH_36_1 (Figure 11a,b), shows a breccia of the granite with significant variations of the grain size. No crystallographic preferential orientation (CPO) is observed. Some zones show brecciated minerals with small precipitation of carbonates as calcite and dolomite in the microfractures (Figure 11d). Primary minerals are only slightly altered but intensively deformed. Quartz shows a pronounced undulatory extinction and starts to dynamically recrystallize by means of sub-grain rotation processes (Figure 11d). Other zones show preserved minerals, but also a significant amount of carbonate, mainly calcite, between the clasts (Figure 11e). The second thin section, NH_36_2 (Figure 11c), shows a granite with a preserved magmatic texture, but fractured, and a breccia composed of a carbonate matrix and transported clasts coming from the granite with no CPO. The preserved granite part shows a low alteration degree. Quartz, K-feldspars (orthoclase and microcline), and biotite are fractured and unaltered. Plagioclase is fractured, slightly altered, and replaced by kaolinite (Figure 11f). The breccia shows a matrix composed of dolomite and calcite flowing through transported clasts of unaltered quartz and feldspar and lenses of kaolinite (Figure 11g).

Bulk analyses show a low LOI (2.6%), a calcite content of 5.95%, and a porosity of 11.5%. XRD bulk and clay fraction analyses also reveal the presence of gypsum and halite.

Sample NH_23

At outcrop (Figure 5b) and sample (Figure 12a) scales (Figure 1b), three compartments were identified: (1) a vein composed of a microcrystalline quartz-feldspathic unit, showing signs of foliation with clear shear sense indicators, bordered by a calcium rich mylonite and containing clasts of it, (2) a hydrothermalized zone in the granite along the vein, represented by a pinkish halo and (3) the porphyritic granite (Figure 12a).

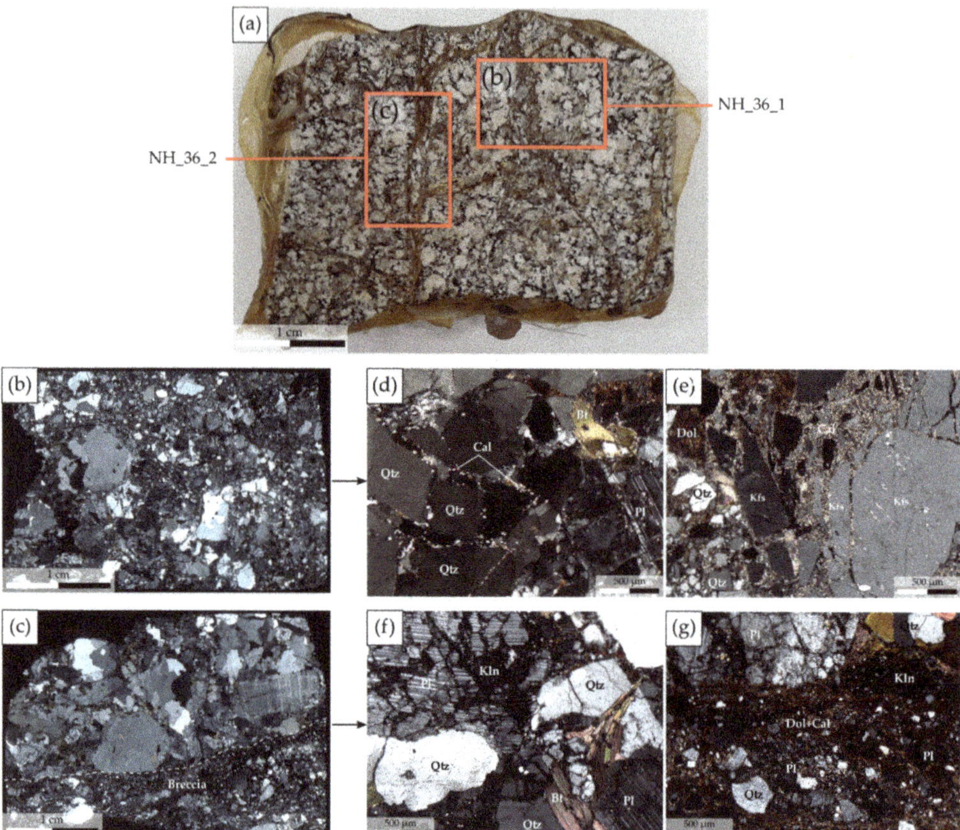

Figure 11. (**a**) Sample NH_36 collected close to a fault zone showing the location of the two thin sections. (**b**) A mosaic of the thin section NH_36_1 showing a breccia. (**c**) A mosaic of the thin section NH_36_2 showing the limit between a breccia and the almost preserved granite. (**d**) A microphotograph of the NH_36_1 thin section showing a brecciated and slightly altered plagioclase, a brecciated quartz with a significant undulatory extinction, starting to recrystallize into subgrains, a brecciated biotite, and calcite crystallizing in the microfractures. (**e**) A micrograph of thin section NH_36_1 showing brecciated and non-altered K-feldspar and quartz with dolomite and calcite deposits between the clasts. (**f**) A microphotograph zooming in the preserved granitic zone visible of the thin section NH_36_2 and showing a brecciated and partially replaced plagioclase into kaolinite and brecciated, but unaltered, quartz and biotite. (**g**) A microphotograph zooming in the brecciated part of the thin section NH_36_2 and showing a carbonate matrix composed of dolomite and calcite transported clasts from the granite.

- Spatial fracture distribution: two scanlines were realized in the granitic part of the sample NH_23 to evaluate the fracture arrangement from the vein towards the host rock (Figure 12a). The spatial analysis is summarized in Table 2. Fracture distributions presented in Figure 12b,c for each scanline show an evolution in two steps of the fracture cumulative frequency (Figure 12b).

  Along the SL1_NH_23 scanline, the frequency slowly increases and shows a regular fracture distribution. At 40% of the scanline, the frequency slightly increases, with a regular arrangement of the fractures. The stick plot Figure 12c highlights this distribution. The SL2_NH_23 scanline presents a similar evolution as SL1_NH_23 scanline, with an increase in the frequency at ¾ of the scanline. This slope change is expressed at the end by a fracture cluster around 8.7–10 cm, clearly visible on the stick plot (Figure 12c). The Cv of 0.68 and 0.67 for each scanline shows a regular to random arrangement of the fractures along both scanlines (Figure 12c, Table 2). Both scanlines fracture densities are equivalent, with respectively 1382 and 1322 fracs/m (Figure 12c, Table 2).

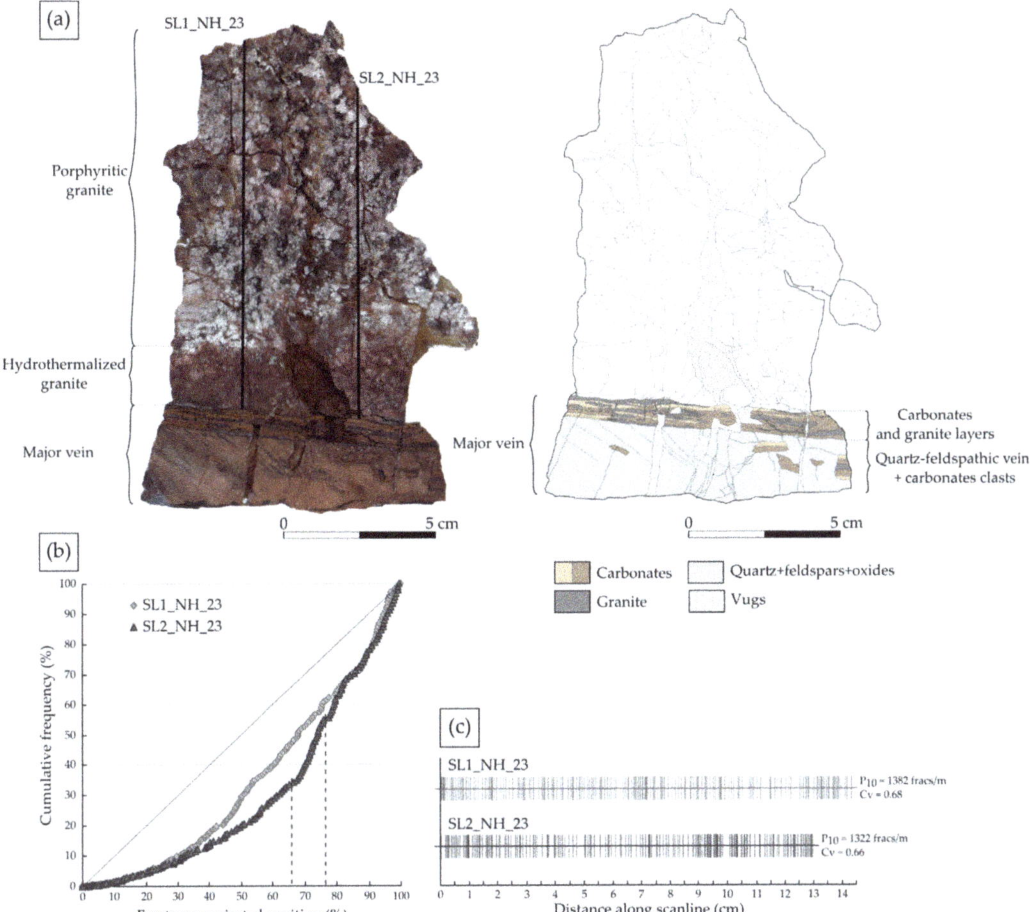

Figure 12. (**a**) A photo and the corresponding digitalization of the sample NH_23 showing the different compartments that compose it, as well as the two scanlines realized in the host rock of the vein. (**b**) Plots of the cumulative frequency percentage against distance percentage for both scanlines. The diagonal represents a uniform distribution. Dashed lines indicate a zone where a rapid increase of the number of fractures is observed (slope threshold >2). (**c**) Stick plots showing the fracture position along the scanlines and for which the fracture density (P_{10}) and the coefficient of variation (Cv) are given.

In order to complete the dataset, the method was repeated on two thin sections from the granite. One is at the border of the vein (NH_23_1) and the second away from it (NH_23_2) (Figure 13a). A scanline was realized in both thin sections (Figure 13b), perpendicular to the major vein. The spatial analysis is presented in Table 2.

Along the SL_NH_23_1 scanline (Figure 13c), a low fracture density zone is present at the beginning of the scanline, then there is an increase of frequency between 0.9–1.4 cm (fracture cluster, Figure 13d), followed by an irregular fracture distribution until the 0.4 last cm where a fracture cluster was identified. Three fracture clusters are present in the SL_NH_23_2 scanline around 1.5 cm, between 1.7–2.1 cm, and around 2.3 cm. From the first fracture cluster, the frequency increases significantly and irregularly. The scanline SL-NH_23_1, with the Cv = 0.77, shows a regular to random arrangement of the fractures, while the Cv = 1.19 of the scanline SL_NH_23_2 indicate a random to clustered arrangement (Figure 13c). The fracture distribution varies with the position of the scanline and shows a P_{10} slightly higher close to the vein (4798 fracs/m) than away from it (3879 fracs/m) (Figure 13d).

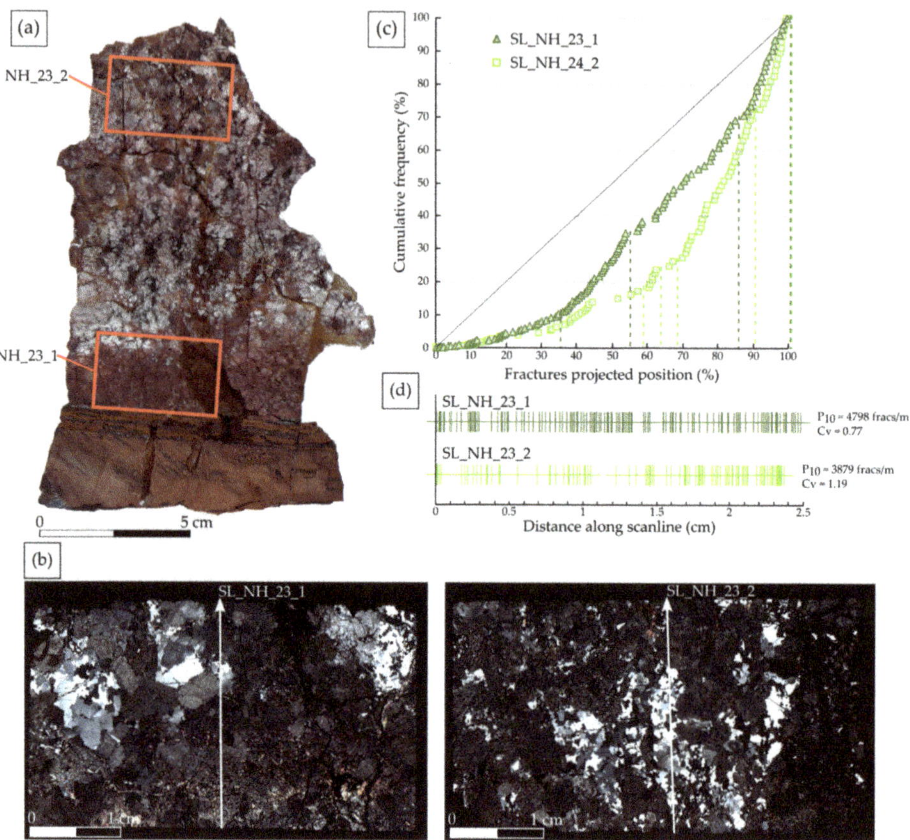

Figure 13. (**a**) Thin section location on the sample NH_23. (**b**) Thin section mosaics showing the position and orientation of the scanlines. (**c**) Plots of the cumulative frequency percentage against distance percentage for each scanline. The diagonal represents a uniform distribution. Dashed lines indicate a zone where a rapid increase of the number of fractures is observed (slope threshold >2). (**d**) Stick plots showing the position of the fractures along the scanlines and for which the fracture density (P_{10}) and the coefficient of variation (Cv) are given.

- Petrographic and petrophysical characterization: the vein of the sample NH_23 consists of two parts. The major part is mainly composed of quartz, few feldspars, and oxide minerals having no CPO. Ankerite veinlets crosscut it, and are also intersected by calcite veinlets (Figure 14a,b). This vein includes clasts of carbonates having the same characteristics as the carbonate vein at the contact with the granite (Figure 12b). This thin carbonate vein is composed of ankerite layers and granite lenses (Figure 14c). By zooming in on the ankerite layer, some part of it appears as lenses with automorphic crystals in the swell, which are crushed and stretched through the pinches (Figure 14d). Granite is pinched between these ankerite layers. Quartz clasts present in the granite lenses show an undulatory extinction, as well as some evidences of subgrain rotation phenomena. Carbonates can deform plastically and accumulate large amount of strain at relatively low P–T conditions, from a minimum temperature of 180 °C [78–81]. Quartz crystal plastic deformation is dominant from 600 °C [82]. Hence, in the present situation, carbonates accommodate large amounts of deformation by means of crystal plasticity, while granite is being deformed by cataclastic flow, i.e., a process accommodated by brittle processes. A localized and significant amount of deformation occurred in this vein. The major vein of this sample is thus composed of (1) a highly deformed zone made of ankerite and granite lenses and (2) a siliceous vein, which contains clasts of similar composition to the carbonate vein. We suggest that the carbonate vein was the first event, which was intensively deformed, and then intersected by the siliceous vein, tearing off pieces of the carbonate vein. Calcite and ankerite veins identified in Figure 14a,b intersect the carbonate vein and the siliceous vein.

Figure 14. Photomicrographs of the different compartments of the sample and their location in the sample. (**a**) Veins of ankerite and calcite crosscutting the major quartzitic/feldspathic vein of the sample under optical microscope in polarized–analyzed light. (**b**) Back-scattered electron image of the photomicrograph (**a**). (**c**) Photomicrograph of the ankerite–granite interlayering zone under optical microscope in polarized–analyzed light. (**d**) A zoom image of the ankerite vein, under optical microscope in polarized–analyzed light, presented in (**c**) and showing a sigmoid of ankerite with a swell of automorphic crystals crushed and stretched in the pinches. (**e**) The altered granite close to the vein showing a completely altered plagioclase and biotite both replaced by illite and oxides and partially altered K-feldspar under optical microscope in polarized–analyzed light. (**f**) The altered granite away from the vein showing a completely altered plagioclase replaced by illite and kaolinite, biotite partially replaced by illite and oxides, an unaltered K-feldspar, and calcite veinlets under optical microscope in polarized–analyzed light.

As presented above, the magmatic texture of the granitic part of the sample is preserved. However, a color change of the granite is observed close to the vein. At a sample scale, the granitic part along the vein (NH_23_1) shows a pronounced pinkish color (Figure 13a). In the thin section, the granite appears almost completely altered. Plagioclase and biotite are no longer recognizable and are replaced by illite and oxides. Perthitic K-feldspar are not completely affected by the alteration, but are partially replaced by illite (Figure 14e). Away from the vein (NH_23_2), the granite appears whitish at a sample scale (Figure 13a). Compared to NH_23_1, plagioclase are completely replaced by illite and kaolinite, but biotite are only partially altered and K-feldspar are unaltered (Figure 14f). This indicates that the granite is more affected by the alteration along the vein, in the hydrothermally altered zone. Bulk analyses performed on the granitic part of the sample show a LOI of 5.2%. In all the granite, calcite veinlets are observed coming from the major vein.

4.4. Geochemical Analyses

The geochemistry of major elements, presented in Table 3, allows establishing diagrams representing selected elements versus LOI (Figure 15) in order to study the alteration degree of the selected samples.

Table 3. Major elements and loss on ignition (LOI) weight percentage (wt%) for the NH granite.

Sample ID	NH_19	NH_23	NH_27	NH_31_3	NH_36
SiO_2	67.18	60.84	56.26	69.29	66.67
Al_2O_3	15.03	17.31	14.05	15.42	15.30
Fe_2O_3	2.93	4.61	3.17	2.69	3.65
MgO	0.71	1.06	0.89	0.24	0.97
CaO	1.81	2.83	9.10	2.34	3.80
Na_2O	2.20	2.45	0.27	2.89	3.12
K_2O	5.45	4.69	4.48	4.05	3.05
TiO_2	0.23	0.39	0.29	0.24	0.34
P_2O_5	0.12	0.18	0.14	0.13	0.17
MnO	0.11	0.23	0.23	0.06	0.12
LOI	4.10	5.20	10.90	2.50	2.60
Sum	99.91	99.88	99.90	99.88	99.87

The LOI ranges from 2.50 to 10.90 wt% with an average around 5 wt%. The LOI values show a negative correlation with Na_2O and SiO_2 and a positive one with K_2O. The CaO content varies slightly between 1.81 wt% and 3.12 wt%, except for the sample NH_27, which CaO content is about 9.10 wt%. Al_2O_3, MgO, and Fe_2O_3 content vary more widely in the ranges 14.05–17.31 wt%, 0.24–1.06 wt%, and 2.69–4.61 wt%, respectively. However, a positive correlation is observed between Al_2O_3 and LOI if the sample NH_27 is rejected. These diagrams show that carbonate (represented by Ca and Mg) and oxide (represented by Fe) precipitations do not depend on the LOI.

4.5. Calcite Content

Calcite is present in the majority of the samples. Samples homogeneously distributed in the studied area were selected for calcite content analyses. The calcite content is represented on the map in Figure 16 by the white dots and figures. The values range between 1.2% and 8.7%. Fracture zones, as well as the surrounding granite, present variable calcite content. However, the calcite content appears higher in the southeastern part of the range than in the northwest. All of the samples, except NH_17, NH_26, and NH_31_3, have a calcite content higher than 1.8%, meaning that they were affected by hydrothermal alteration, according to White et al. (2005) [72].

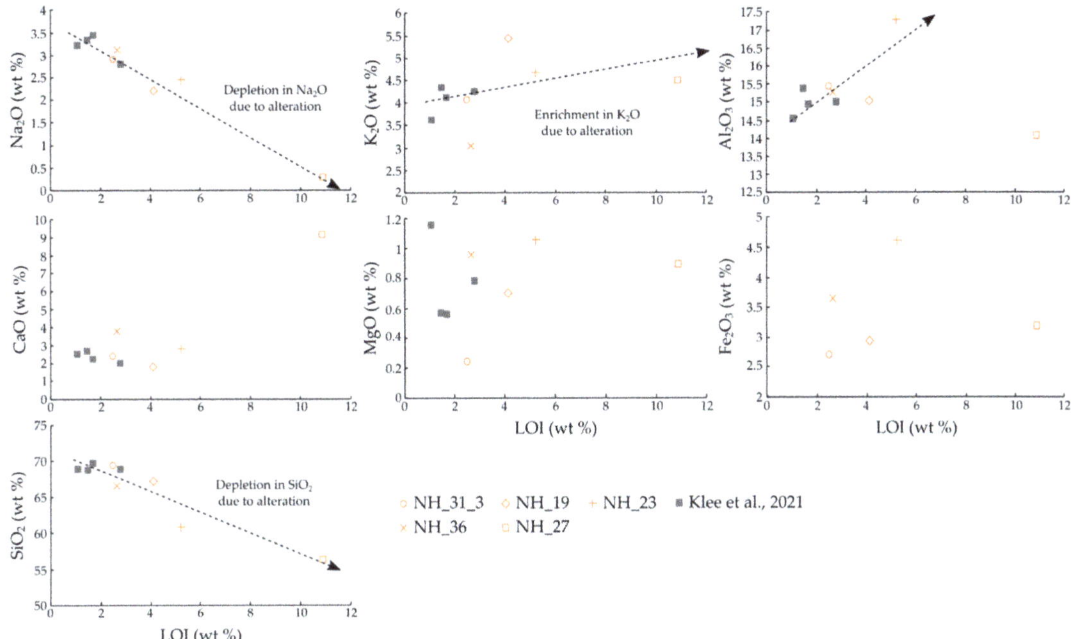

Figure 15. Plot of selected major element oxides (Na$_2$O, K$_2$O, Al$_2$O$_3$, CaO, MgO, Fe$_2$O$_3$, and SiO$_2$) versus loss on ignition (LOI) of samples from this study in orange and from [19] in grey. The dashed arrows show the different trends observed.

4.6. Porosity and Mass-Balance

Porosimetry measurement was only performed on samples NH_27, NH_31_3, and NH_36 (Table 4). NH_36 is the closest to a fracture zone and has the highest porosity (11.5%). This suggests that the porosity increases with proximity to fracture zones. Porosity measurements were used to better decipher the petrophysical properties of the rock for geothermal purpose and to realize mass-balance calculations. The chemical compositions recalculated after suppression of the LOI, the densities, and volume factors (F$_V$), assuming X$_{Al2O3}$ = 0 of the rock samples used for mass balance calculations, are given in Table 4. The F$_V$ shows values very close to 1, indicating a small change in volume between the freshest sample and the altered ones.

The results of mass-balance calculations, corresponding to the absolute mobility values of each element for each sample, are given in Table 5, by considering a constant volume, F$_V$ = 1, as exposed by [76]. Positive values represent the gains and negative values the losses of the considered elements. Absolute mobility values indicate almost immobility of Al$_2$O$_3$, TiO$_2$, and MnO. They also show losses of SiO$_2$ and Na$_2$O, which are three times more important in NH_27 than in NH_31-3, and twice that of NH_36. The losses of Fe$_2$O$_3$ and MgO are more important in NH_31_3 and less important in NH_36. CaO losses are observed in NH_31_3, whereas CaO gains are observed in NH_36 and a high gain is observed in NH_27. Finally, the results show a gain of K$_2$O in samples NH_27 and NH_31_3 and a loss in NH_37.

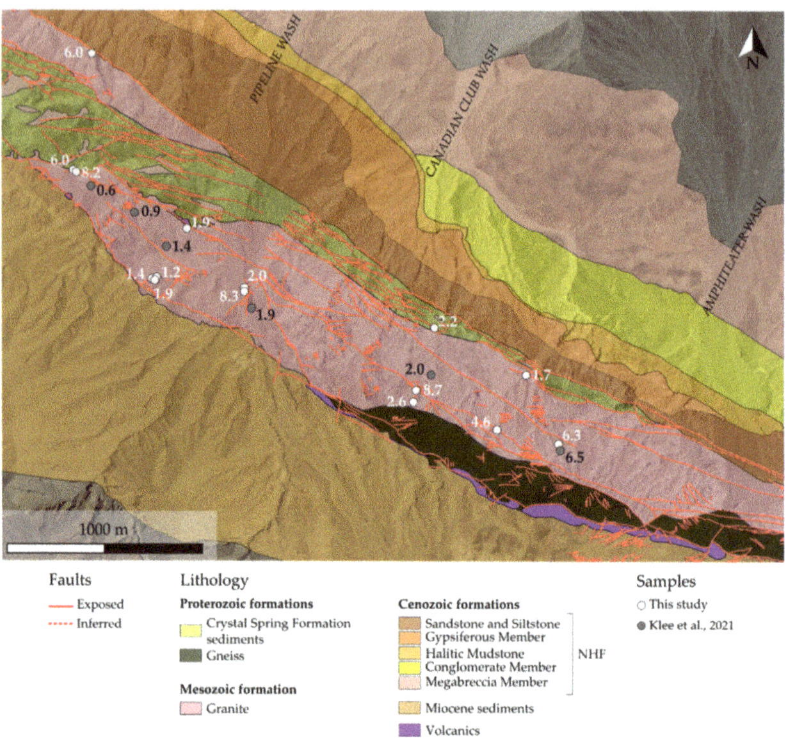

Figure 16. NH geological map showing the location and the associated calcite content of samples from this study represented by the white dots and of samples from [19] by the grey dots.

Table 4. Porosity values, recalculated bulk-rock major element weight percentage, altered rock density/fresh rock density ratio, volume factor (F_v) assuming the immobility of Al_2O_3 ($X_{Al2O3} = 0$).

Sample ID	NH_2 [1]	NH_27	NH_31_3	NH_36
Porosity (%)	2.29	10.1	10.1	11.5
Oxides (weight %)				
SiO_2	69.57	63.3	71.18	68.6
Al_2O_3	14.72	15.81	15.84	15.74
Fe_2O_3	4.35	3.57	2.76	3.76
MgO	1.16	1	0.25	1
CaO	2.53	10.24	2.4	3.91
Na_2O	3.23	0.3	2.97	3.21
K_2O	3.65	5.04	4.16	3.14
TiO_2	0.47	0.33	0.25	0.35
MnO	0.12	0.26	0.06	0.12
Density ratio (d_A/d_F)		0.93	0.92	0.92
F_v ($X_{Al2O3} = 0$)	1.00	1.00	1.01	1.02

[1] Geochemical data recalculated from [19].

Table 5. Results of mass-balance calculations assuming Fv = 1.

Sample ID	NH_27	NH_31_3	NH_36
Absolute mobility (%)			
SiO_2	−10.39	−3.54	−5.32
Al_2O_3	0.06	−0.18	−0.29
Fe_2O_3	−1.01	−1.77	−0.82
MgO	−0.23	−0.92	−0.22
CaO	7.04	−0.30	1.10
Na_2O	−2.95	−0.47	−0.22
K_2O	1.06	0.20	−0.70
TiO_2	−0.16	−0.24	−0.14
MnO	0.12	−0.06	−0.01

4.7. Temperature Conditions

The clay composition of the fractions <2 μm and 2–6 μm was given in Table 1. For some of the samples containing enough illite, Kübler Index (KI) was measured for both fractions (Table 6). The fraction <2 μm likely is supposed to represent neoformed clay minerals, thus corresponding to the youngest event. By contrast, the fraction 2–6 μm might contain either more developed illite crystals by inherited grains recrystallization, or a mix of detrital and neoformed minerals. The KI gives information concerning the degree of metamorphism for each fraction. In some fraction, it was not possible to measure the KI due to the too low amount of illite.

Table 6. List of samples with their respective FWHM and Kübler Index (KI) values in air-dried (AD) conditions for the fractions <2 μm and 2–6 μm. n.m.—not measured.

Sample ID	Illite Peak (10) <2 μm		Illite Peak (10) 2–6 μm	
	FWHM AD	KI AD	FWHM AD	KI AD
NH_16	0.62	0.97	0.43	0.65
NH_17		n.m.	0.80	1.27
NH_18	0.41	0.61	0.34	0.49
NH_19	0.52	0.75	0.51	0.74
NH_22	0.53	0.78	0.58	0.84
NH_24	0.83	1.32	0.70	1.11
NH_25	0.59	0.59	0.63	0.63
NH_27	0.65	0.96	0.55	0.80
NH_31_1_2	0.72	1.14	0.64	1.01
NH_31_3		n.m.	0.34	0.50
NH_32	0.29	0.29	0.21	0.21
NH_34		n.m.	0.28	0.39

KI values range from 1.32Δ°2θ to 0.29Δ°2θ for the <2 μm fraction and from 1.27Δ°2θ to 0.21Δ°2θ for the 2–6 μm fraction (Table 6). The spatial distribution of the samples and their associated KI values are shown in Figure 17 by the dots. In both fractions, most of the samples present KI values corresponding to the diagenetic zone, except for two samples (NH_32 and NH_34). These samples are located in the northwestern part of the main granitic body (Figure 7a,b), at the contact with the Proterozoic basement. They present lower KI values of low anchizone and epizone. No significant variations in KI or trends are observed between both fractions.

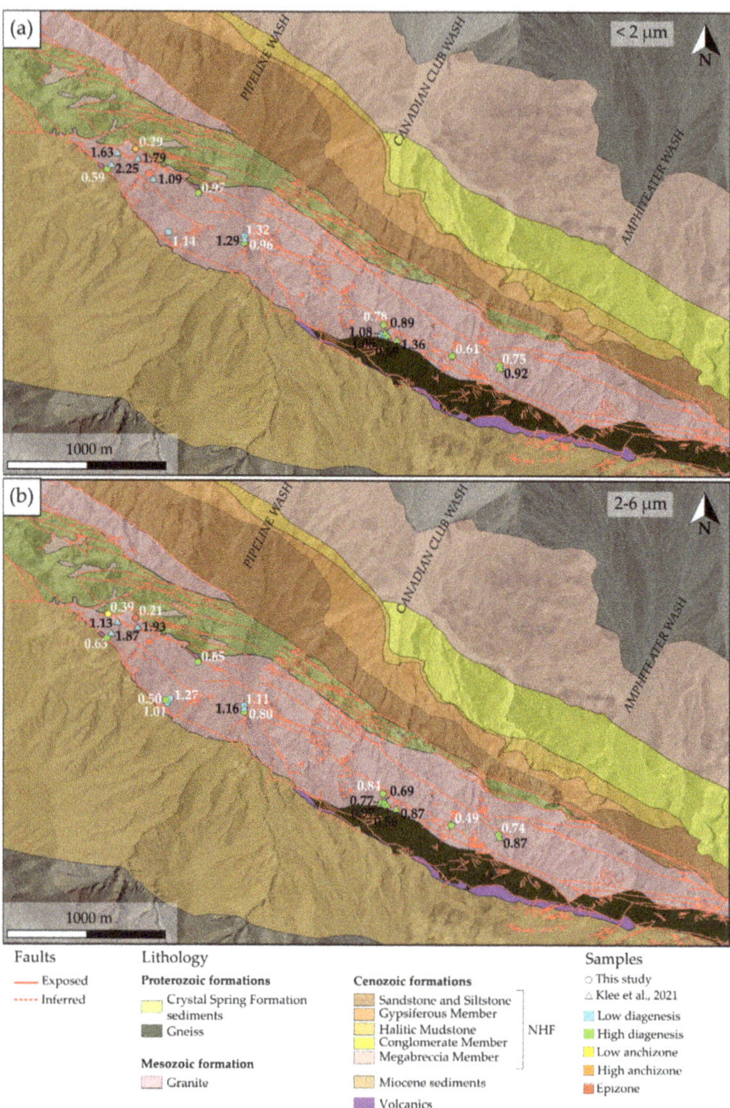

Figure 17. Geological map representing the Kübler Index (KI) in air-dried conditions of each sample from this study (dots) and from [19] (triangles) and the corresponding metamorphic zone represented by the different colors indicated in the legend for (**a**) the fraction <2 μm and (**b**) the fraction 2–6 μm.

Some samples contain only kaolinite and no illite. Positive trends have been identified between KI and KCI for both clay fractions <2 μm and 2–6 μm (Figure 18). Samples NH_6, NH_8, and NH_31_1_2, with out of range values, due to the presence of I/S and/or corrensite interfering with the 7Å peak, were excluded from the correlation. The trends observed in the two different fractions are similar and give more confidence into the concomitant evolution of KI and KCI. The correlation provides temperature indications for illite-free samples. Thus, samples NH_26, NH_33, and NH_36 show values equivalent to diagenetic zone and anchizone conditions.

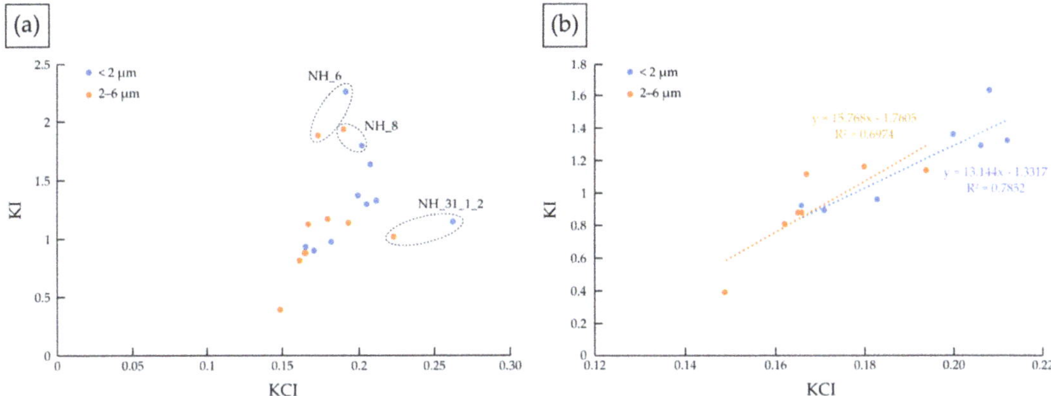

Figure 18. (**a**) Diagrams showing Kübler Index (KI) values versus the Kaolinite Crystallinity Index (KCI) for the clay fractions <2 μm and 2–6 μm. (**b**) Diagram showing positive correlations between KI and KCI for the clay fractions <2 μm and 2–6 μm, excluding samples NH_6, NH_8, and NH_31_1_2.

5. Discussion

5.1. Argillic Alteration Dominance

The characterization of the NH granite alteration processes in the vicinity of fracture zones reveals the presence of various secondary minerals, such as clay minerals (kaolinite, illite, I/S mixed layer, corrensite, and vermiculite), carbonates (calcite, dolomite, ankerite, siderite), and oxides. Thus, two successive types of hydrothermal alteration events, characterized by [19] in the protolith, were observed:

1. The propylitic alteration, which is an early stage of alteration affecting pervasively the granite during the cooling of the pluton [83]. It involves mainly the formation of corrensite and epidote considered as the major indicators of the propylitic alteration [19,22,84]. Only plagioclase and biotite are partially affected.

2. The argillic alteration, associated with fluid circulation through a fracture network. Thus, the argillic alteration is also called "vein" alteration [13,21,85]. It is characterized by (1) a high water/rock ratio in the fractures/veins walls, due to fluid circulating within the fracture network [20,21,24,86–89]; and (2) illite + kaolinite + illite/smectite mixed layers + carbonates + oxides replacing plagioclase, biotite and, more rarely, partially K-feldspar [19]. Fractures enhance the fluid circulation and, thus, the fluid–rock interaction. Alteration gradients are visible, increasing toward the fracture (Figure 7a, Figure 9h–k, Figures 12a and 14e,f). An alteration zoning around microfractures is also presented in Figure 6f,g, which could correspond to a time dependent process controlled by a sequence of interrelated mineral reactions [15].

However, compared to [19], the samples are more altered and corrensite is not as evident as in the samples from this study. Kaolinite and illite are the prevailing clay minerals. This confirms a significant fluid–rock interaction near fractures and a predominant argillic alteration, which has overprinted almost completely the propylitic alteration. Kaolinite, being the dominant clay mineral, indicates either a more important leaching of the rock or the circulation of a more acidic fluid [90].

The samples NH_24, NH_26, and NH_33 contain a small amount of vermiculite, which can be associated to hydrothermal alteration at low temperature or to weathering. It is commonly thought that most of the vermiculite is formed under supergene conditions [91,92]. In these samples, the amount of vermiculite is low and the weathering contribution already discarded by [19].

5.2. Thermal Evolution toward Fracture Zones

Studied samples present KI values characteristic of the diagenetic zone, meaning temperatures lower than 200 °C [93]. However, KI values of samples NH_32 and NH_34 have lower KI values with anchizonal to epizonal conditions, meaning temperatures around 300 °C [93]. These two samples present the highest fracture density (Fd4) suggesting that shear heating could contribute to a local increase of temperature [94]. Moreover, KI values are similar between both clay fractions indicating the predominance of neoformed illite crystallization close to fracture zones. It is highlighted by the presence of illite veins observed microscopically. By comparing KI values from this study and from [19], the KI tends to decrease approaching fracture zones. Thus, the greater the fracturing, the higher was the temperature, except for NH_31_1_2 and NH_24. These two samples present higher KI values, meaning lower temperatures. It can be explained by the presence of I/S and vermiculite, forming at lower temperature than illite and kaolinite [91,92,95]. The samples NH_26, NH_33, and NH_36 present only kaolinite. KCI values correlating with KI values, these samples show temperatures estimated to be around 200 °C or less, which correspond to the temperature obtained for the other samples. Likewise, the presence of kaolinite in hydrothermal alteration paragenesis indicates temperature lower than 200 °C [90].

5.3. A Multi-Stage Paleo-Fluid Circulation

Hydrothermal alteration in crystalline basement rocks induces the precipitation of secondary minerals that can seal the fractures [2]. The argillic facies, described above, is also characterized by fractures filled by various kinds of secondary minerals, as a result of different fluid generations. Veins are formed from fluids that had reacted with granite [96] and transported various chemical elements. Through the petrographic analyses of the whole samples, a relative chronology between the different veins can be determined. Six vein generations, following each other, were identified (Figures 6 and 9):

1. Quartz veins resulting in the precipitation of secondary quartz due to primary silicate partial dissolution.
2. Illite veinlets, which have different behaviors, according to the mineral crossed. Indeed, as presented by [19,97], plagioclase is the main pathway for fluid flow due to their abundance in the rock and to dissolution pit porosity, allowing the interconnection between the pores. In quartz and K-feldspar; however, only microfractures create the porosity allowing the fluid to circulate. This explains why veinlets look like straight lines in quartz and K-feldspar and are twisted and blurred lines in altered plagioclase. Illite veinlets can be contemporary to kaolinite veinlets.
3. A dolomitic brecciated vein embedding essentially quartz clasts, such as in sample NH_31.
4. Carbonate veins with different compositions: contemporary ankerite/dolomite veins, intersected by siderite veins and the whole intersected by later calcite veins. They are preferentially oriented NW–SE, according to the direction of the NH. It is suggested that the fluids, having precipitated these carbonate veins, have circulated through the fracture network formed by the activity of the SDVFZ.
5. A carbonate brecciated vein that is likely to have transported clasts of quartz, K-feldspar, and biotite, and is presenting precipitations of barite. This phase, but also barite itself, is only present in sample NH_31. This phase was probably due to a later event that reactivated the main fracture composing the sample and let a new fluid circulate.
6. Calcite veins, as shown in the sample NH_31, which are of a different compositions from generation 4.

Two types of calcite veins and the carbonate vein of phase 5 were, thus, identified through the sample NH_31, allowing the definition of two generations of calcite. However, they do not allow a decision on if the calcite veins identified in the other samples correspond to the generation 4 or 6. CL analyses are ongoing to identify the type of calcite generation. Sample NH_31 and the multiple vein generations show how complex the fluid circulation history was in this area.

5.4. Fluid Circulation and Argillic Alteration Effects on Petrographic and Petrophysical Behavior

The effects of the alteration on the element transfers are described by the diagrams plotting major element oxides against LOI (Figure 15) and by mass-balance calculations (Table 5) using the Gresens method [76] to quantify the losses and gains of elements during hydrothermal alteration. To this end, data from this study and from [19] were used. LOI has been defined as an indicator of the alteration degree [19,98,99] supporting optical observations. In this study, LOI values range from 2.5% to 10.9% (Table 3), whereas, Klee et al. (2021) [19] obtained values from 1.1% to 2.8% for the protolith. These values confirm a more pronounced alteration of the granite in the vicinity of fracture zones.

Concerning element transfers, only minor variations of Al_2O_3 were observed compared to the LOI, also confirmed through mass-balance calculations. Al is thus considered as immobile. This study has shown a significant SiO_2 negative correlation with LOI. This loss is confirmed by mass-balance calculations. It indicates the partial silicate alteration, explaining the presence of quartz veins and veinlets through the granite. A negative correlation with LOI was observed for Na_2O, confirmed by the loss of Na_2O obtained by mass-balance calculation. Indeed, altered plagioclase are depleted in Na_2O. No Na-bearing newly formed minerals (except a little I/S and C/S) was encountered. It is likely that Na was exported out of the alteration zone. When the alteration of plagioclase is less pronounced, as in samples NH_31_3 and NH_36 (Table 3), the Na_2O depletion is lower. However, a global enrichment of K_2O is observed linked to illite crystallization. A positive correlation is observed between the amount of alteration and illite formation. K-feldspar being rarely altered, chemical modelling would be necessary to determine whether the amount of K release by biotite alteration would be sufficient to allow the rather important formation of illite. MgO and Fe_2O_3 present no clear correlation with LOI. However, low depletion of Fe_2O_3 and MgO compared to the protolith were identified by mass-balance calculations. It is suggested that these depletions are linked to the alteration of biotite but they are compensated by their precipitation as oxides or in carbonates. Thus, it is likely that no Fe or Mg is exported out of the alteration zones. CaO and calcite content present no correlation with the LOI. Samples NH_19, NH_36 (this study), and NH_3 (in [19]) present a similar calcite content, respectively, of 6.3%, 6.0%, and 6.5%, whereas they present different LOI values (4.1%, 2.6%, and 2.8%, respectively). The case of a high calcite content associated to a low LOI can be explained by (1) physicochemical fluid composition bringing Ca, and not allowing a complete alteration of plagioclase and biotite into clay minerals [100]; (2) the residence time of the fluid not being sufficient [101]; or (3) a non-sufficient amount of fluid to allow the transformation of plagioclase and biotite into clay minerals. CaO losses are attributed to the alteration of plagioclase. However, the amount of Ca released by the alteration of plagioclase is certainly not sufficient to explain the crystallization of large amounts of carbonate into the granite veins. We infer that Ca is related to an external source as it is gained by the system (open system). As carbonate veins are related to the SDVFZ activity, we suggest that the source of the Ca is the Proterozoic series, which contain dolomite and carbonate sequences.

Fracture zones created porosity through microfractures and spaces between secondary minerals (Figure 6f) precipitated during primary mineral alteration. When fracturing increases, alteration increases too, as well as porosity [10]. However, the subsequent mineralization and chemical alteration can either decrease or increase the porosity of the rock [2]. For a similar LOI, porosity values vary significantly (e.g., NH_3 [19], NH_23, and NH_31_3). No correlation is visible between porosity and alteration or calcite content.

5.5. Relation among Fluid Circulation, Alteration, and Fracture Density

It is expected that areas along the rear southern limit of the CBS have recorded less transcurrent shear than areas situated along its northern front (Section 4.1). Recent work by Chabani et al. (2021) [42] has shown a structural compartmentalization occurring at the NH scale with a varying intensity that may influence fluid flow through the fracture network. Likewise, they identified a complex network of joints at outcrop scale, also encountered in this study at the sample and thin section scales, playing a key role in the fracture connectivity. The authors identified several fracture sets, among which, the NW/SE and the E/W oriented sets are predominant. The NW/SE trend, following the SDVFZ direction, controls the geometry in the whole range, whereas the E/W set, characterized by several short fractures, is mostly present in the central and southeastern part of the NH. The geometrical analysis of both fracture sets showed that the connectivity is ruled by fractures of different sizes. Regarding the fracture distribution, the highest fracture density is recorded within the rear southern part with values five times higher than in the northern part. This can be explained by the complex tectonic setting and the gradient of deformation observed in the entire NH range, with evidence of extreme shearing in the southern part as already presented. Moreover, through scanline analyses along drone photogrammetric outcrop profiles and ground outcrop profiles, Chabani et al. (this issue, under review) [44] show a clear correlation between fracture density and distance from a fault zone. The closer the fault zone, the greater the fracture density. There is no straightforward correlation between amounts of deformation and fracture density. Fracture density depends on a series of parameters notably rock competence, anisotropy of units and possible presence of detachment zones concentrating deformation in narrow areas.

This study, together with that of Klee et al. (2020) [19], confirm the increase of fracture density towards fault zones through several samples collected at different distances from the faults. Moreover, at the NH scale, the LOI increases as the fracture zones are approached (Section 5.4). The more intense the fracturing, the higher the alteration. Likewise, porosity values obtained for samples from Klee et al. (2021) [19] range from 2.2% to 5.2%, whereas, porosity values for this study samples range from 10.1% to 14.0%. Close to fracture zones, samples have more than twice the porosity than the samples away from them. Calcite content (Figure 16) also show higher values near to fracture zones, highlighting the importance of the presence of fractures allowing fluid circulations and thus carbonate precipitation, which mainly filled and sealed the fractures [71,102]. Carbonate veins are present in the whole range, preferentially orientated following the NH strike, probably related to the SDVFZ activity. However, calcite content is generally higher in the southeastern part of the range. This part is characterized by a high fracture density due to complex tectonic setting, related to the GFZ activity [42]. The connectivity within this part is ruled by the small and large fractures, and an additional complex joint network leading the fluid supply toward the fault zone.

The comparison of the data from Klee et al. (2021) [19] with those in this study at the NH scale shows that the increase in fracture density is correlated with the increases in alteration degree, porosity, and calcite content, respectively. However, no correlation is observed between alteration, porosity, and calcite content all together. The sample NH_36 (Figure 11) has shown that calcite content is not related to the intensity of alteration (Section 5.4). This highly fractured sample has been characterized by a specific fracture arrangement, composed mainly of fracture clusters, following the SDVFZ direction. It allows the fluid to circulate following these clusters, and then creating a flow anisotropy. However, only a low alteration degree is observed, which can either be due to (1) specific physicochemical fluid composition [100]; (2) an equilibrium between the fluid and the surrounding rock; (3) an insufficient residence time of the fluid [101]; or (4) a non-sufficient amount of fluid. Fracture density increase does not necessarily induce the increase of alteration (Figure 19a,b), whereas at the NH scale, a general correlation between fracture density and alteration was shown. Sample NH_36 shows that fluid can circulate through a highly fractured granite without necessarily producing alteration (Figure 19b).

Numerous fractures and veins, showing a high fluid–rock ratio, crosscut the granite altering it consequently (Figure 20a,b). Faulkner et al. (2010) [11] have shown that within crystalline rocks, the flow can be ruled by a small number of fractures within the rock surrounding a fracture zone. The fracture interconnections constitute the main parameter, which can enhance the fluid circulation. Samples NH_31 and NH_23 showed the relation between fracturing and alteration at local scale (Section 4.3.2). Thin section fracture analyses showed that the fracture density average is similar in both samples. NH_31 consists of a brecciated vein developing an alteration gradient but no fracturing gradient (Figure 20c), while NH_23, perpendicular to NH_31, consists of a mylonite developing an alteration and a fracturing gradient (Figure 20d). Regarding the stress axis orientation, it is suggested that the main stress axis σ3 is at a very high angle to the NH_31 initial quartz vein. The close to normal angle between the initial quartz vein and the stress axis σ3 was favorable to reactivation, inducing a brecciated vein without developing a significant additional fracture gradient in the surrounding rock (Figure 20c). It has been shown that remnants of a quartz vein composed the NH_31 brecciated vein. This quartz vein, developed during a past event of unknown age and origin, possibly magmatic (last crystallization fluids). It is suggested that this vein was reactivated, letting a new fluid rich in Ca circulates. Then carbonates crystallized giving a brecciated vein as a final product. The alteration gradient observed in the wall rock is especially pronounced at the borders of the brecciated vein. The proximity with the main fluid pathway induced an important leaching by the fluid. Within NH_23 (Figure 20d), it is supposed that the quartz-feldspathic unit was initially a dyke crosscutting the granitic pluton. Such dykes have been observed in the surroundings. During SDVFZ activity, reactivation of this zone of weakness permitted percolation of a Ca-rich fluid. An intensive shearing induced dynamic recrystallization developing a mylonite in the borders and its associated fracturing gradient in the wall rock. In that case, the main stress axis σ1 would be characterized by an intermediate angle to the shear plane. A fracturing gradient is observed in the surrounding rock. The fracture density induced in the surrounding rock does not further increase. Asymmetric clasts and shear bands are observed within this mylonitic layer.

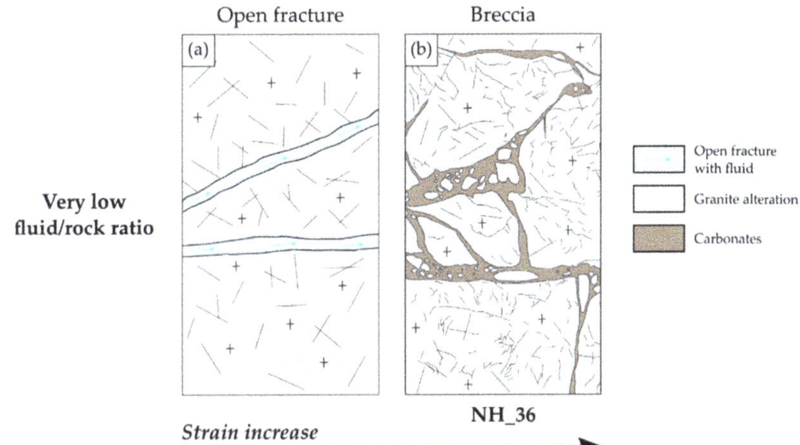

Figure 19. Synthetic scheme showing, from low to moderate strain, the interaction between the fluid and the granite in the cases of a very low fluid/rock ratio. (**a**) Open fractures with an incipient alteration of the surrounding rock. (**b**) Brecciated zones with a carbonated matrix and no change of alteration degree. No scale is given for this scheme, because it can represent fracture zones of several order of magnitude.

Fluid circulation and induced fracturing have promoted fluid–rock interactions and the granite alteration, which is more pronounced in NH_23 than in NH_31 (Figures 9h–k and 14e,f). Those two samples showed that at sample and outcrop scales, fracturing and alteration are heterogeneous, but also that the alteration degree is not always related to the fracture density (Figure 20c,d).

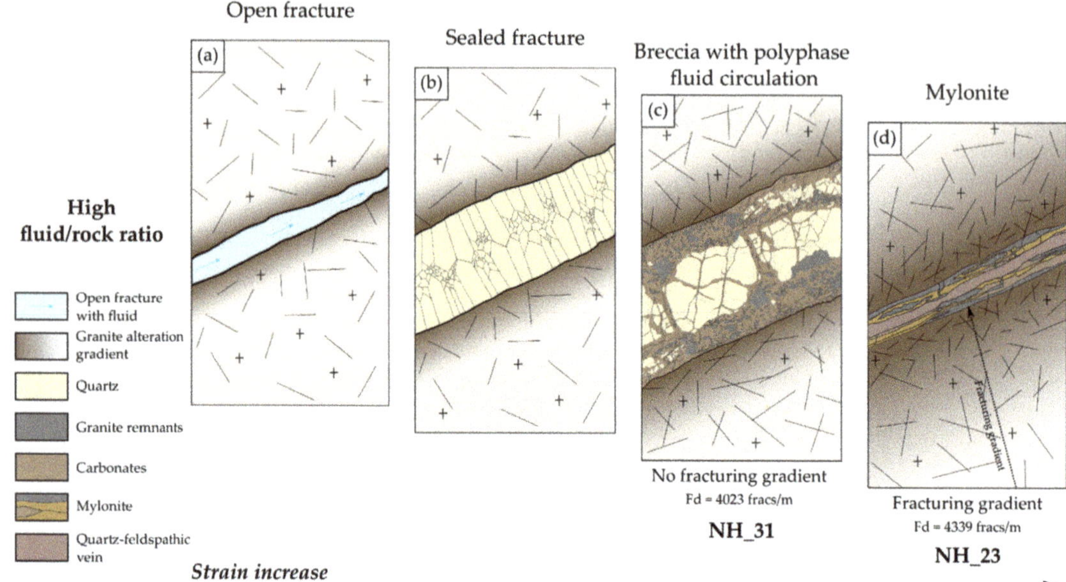

Figure 20. Synthetic scheme showing, from low to moderate strain, the interaction between the fluid and the granite in the cases of a high fluid/rock ratio observed in the field. (**a**) An open fracture with an incipient alteration gradient at the border. (**b**) A sealed fracture with a slight alteration gradient. (**c**) A brecciated vein induced from fracture reactivation and showing an alteration gradient, but no fracturing gradient. (**d**) A mylonite made of carbonates and granitic lenses showing an alteration gradient and a fracturing gradient. NH_31 and NH_23 show a similar fracture density (Fd). No scale is given for this scheme, because it can represent fracture zones of several order of magnitude.

These sample analyses allowed us to study a fault zone development depending on strain, fluid–rock ratio, and material rheology. It is shown that at low to moderate regional strain, incipient local high strain concentration is present (Figure 20). It has been observed that:

1. The fluid interacts with the surrounding rock, altering it pervasively, and fractures are sealed by secondary minerals (Figure 20a,b). Those veins can be reactivated by shearing [103]. Thus, it creates a brecciated vein (e.g., NH_31) or even a mylonite (e.g., NH_23), developing an alteration gradient. Sample NH_23 shows that mylonitic deformation appears at low to moderate regional strain when carbonates are involved. Fault reactivation tends to cause a mineral fill breakage and reopens the fracture [2,104].
2. Open fractures let fluid circulate without interaction with the surrounding rock. Shearing creates a breccia without further alteration within the granite, such as in NH_36 (Figure 19).

At any scales, fracture density promotes fluid flow through the fracture network provided a fluid is present [5,11]. However, the alteration degree is not correlated to the fracture density.

5.6. Implication of Alteration in Terms of Geothermal Reservoir Properties

Fresh granite has a very low initial matrix porosity (<1%), which "does not allow" the fluid circulation, hence inducing a low permeability [9]. A granitic geothermal reservoir is considered as exploitable when it presents a connected fracture network increasing the permeability and in which a sufficient amount of hot fluid circulates [2,3,9,11]. By flowing through fractures, the fluid interacts with the rock, increasing the matrix porosity and permeability promoted by the dissolution of the primary minerals and crystallization of newly-formed phases [13]. However, newly-formed minerals can also crystallize in the fractures and seal them, reducing the porosity and the permeability [105,106]. Thus, sealed fractures are transformed from conduits into barriers to fluid flow [83,107,108] and impact the geothermal production by decreasing the connectivity between the fractures and hence the permeability [2,109].

The NH altered granite shows an important fracture network and a high matrix porosity, which allowed fluids to circulate and to interact with the surrounding rock. Indeed, numerous veins with various infills crosscut the granite indicating several successive fluid circulation events. Carbonates occupy a prominent part of the fracture system, which can be easily dissolved thanks to acid injections in order to connect the boreholes to major conductive fractures for geothermal exploitation. It is suggested that, when NH were an active reservoir at depth, fractures were only partially filled, and the fluid could flow through the fracture network. It has been seen that the stress field has a major impact on the fluid circulation. Indeed, drilling into fractures at a very high angle from the main stress axis σ1, such as exposed by sample NH_31, is favorable to allow fluid to circulate. In the event of geothermal exploitation, only minor chemical stimulation, if any, would have been necessary [13,71,110]. Moreover, the alteration amount is not necessarily related to the degree of fracturing and, therefore, the fracture network influence on fluid circulation is hardly predictable. Thus, due to the conduit-barrier role, the deformation gradient, the degree of fracturing, the alteration processes, and the relationship between them, have a large impact on geothermal production [109].

6. Conclusions

Noble Hills (NH) is a newly studied area, in terms of an exhumed granitic geothermal reservoir. It provides an excellent opportunity to give fundamental scientific input in 3D, allowing for better understanding of granitic reservoir behavior in a trans-tensional context. This analogue shows how complex a granitic reservoir can be, in terms of structures, fluid circulation, and fluid–rock interactions. This study proposes a geometric, petrographic, petrophysical, and geochemical description in order to characterize the influence of fracturing on fluid circulation and alteration processes.

Approaching fracture zones at large scale, the NH granite shows signs of several generations of fluid circulations resulting in successive veins of various mineralization. Fluid circulation being more important in the vicinity of fracture zones, a stronger fluid–rock interaction is observed. Thus, argillic alteration prevails compared to in the protolith. It is highlighted by the increase of the LOI correlated to a Na depletion due to plagioclase alteration and a K enrichment associated to illite precipitation. Likewise, the porosity, the calcite content, and the temperature increase nearer to fracture zones. However, no correlation exists among LOI, porosity, and calcite content altogether. Moreover, a high fracture density does not necessarily imply a strong alteration (e.g., sample NH_36).

This relation among fluid circulation, alteration, and fracturing is also visible at a sample scale. The higher the fracture density, the more pronounced the alteration. However, samples NH_31 and NH_23 showed how complex this relation could be:

- NH_31 shows a reactivated vein giving a brecciated vein (quartz and granitic clasts in a carbonate matrix), which induced no fracture gradient in the surrounding porphyritic granite, but developed an alteration gradient.
- NH_23 shows a carbonate mylonite creeping around deformed granite lenses, which induced a fracture gradient, as well as an alteration gradient in the surrounding porphyritic granite.

These two samples, showing different deformation features as a result of a different orientation within the stress field, present a similar fracture density. Strain was accumulated within a carbonate mylonite within sample NH_23. The alteration gradient is more pronounced in this sample, but the fracture density does not change. Sample NH_36, consisting of a granite composed of a cohesive breccia, whose matrix is made of dolomite, shows a low alteration. Even locally, alteration does not always depend on the deformation gradient or on the fracture density.

This study shows that the deformation gradient observed at a large scale is also visible locally in the context of a low to moderate strain. It also highlights the importance of mineral crystallization in a geothermal reservoir. Fluid flow depends on a connected network of open and permeable fractures. In reservoirs at depth, fractures can be sealed and act as a barrier. Stimulation techniques are needed in order to reopen them so that a drain allowing a new fluid circulation is created. The influence of fracture zones on fluid flow and alteration is difficult to predict and yet impacts the production from the reservoir.

In order to complete this work, a future study will focus on the influence of a fault zone on fluid circulation and alteration processes, in the NH, at high strain conditions. Moreover, laboratory investigations could be performed to better understand the controlling parameters of the ongoing processes and fracturing of rocks.

Author Contributions: Conceptualization, J.K., A.C., B.A.L., S.P., R.L.H. and G.T.; methodology, J.K., A.C., B.A.L., S.P., R.L.H. and G.T.; software, J.K.; validation, A.C., B.A.L., S.P., R.L.H. and G.T.; formal analysis, J.K.; investigation, J.K., A.C., B.A.L., S.P., R.L.H. and G.T.; resources, J.K.; data curation, J.K., A.C., B.A.L., S.P., R.L.H. and G.T.; writing—original draft preparation, J.K.; writing—review and editing, A.C., B.A.L., S.P., R.L.H. and G.T.; and H2020 MEET consortium; visualization, A.C., B.A.L., S.P., R.L.H. and G.T.; supervision, J.K.; project administration, G.T.; funding acquisition, G.T. and H2020 MEET consortium. All authors have read and agreed to the published version of the manuscript.

Funding: This project has received funding from the European Union's Horizon 2020 research and innovation program under grant agreement no. 792037 (H2020 MEET project).

Data Availability Statement: Not applicable.

Acknowledgments: This manuscript was prepared as a contribution to the PhD thesis (Institut Polytechnique UniLaSalle Beauvais) of Johanne Klee, which was funded by the European Union's Horizon 2020 research and innovation program under grant agreement no. 792037 (H2020 MEET project). The authors greatly thank Terry Pavlis for his support and helpful exchanges about the regional geology of Death Valley and the Noble Hills. Albert Genter is thanked for the helpful discussions and knowledge about fractured granitic reservoirs. We are grateful to Klaus Wemmer for his help on XRD analyses and the fruitful discussions. We also thank Graciela Sosa and Alphonse M. Van den Kerkhof for their help on cathodoluminescence observations and acquisitions. We thank Mahdi Chettabi, Thi Tuyen Nguyen, Elena Pavlovskaia, Carl Tixier, and Chloé Gindrat for their help with the analyses. We acknowledge the H2020 MEET consortium for their helpful comments and the validation of the manuscript. We would like also to thank the Hauts-de-France region for its help in setting up the MEET project, in particular the FRAPPE (Fonds Regional d'Aide aux Porteurs de Projets Euopéens). We finally would like to thank the two anonymous reviewers and the editors for their help and remarks in improving this manuscript.

Conflicts of Interest: The authors declare no conflict of interest.

References

1. Williams, C.F.; Reed, M.J.; Anderson, A.F. Updating the Classification of Geothermal Resources. In Proceedings of the Thirty Sixth Workshop on Geothermal Reservoir Engineering Stanford University, Stanford, CA, USA, 31 January–2 February 2011.
2. Gillespie, P.A.; Holdsworth, R.; Long, D.; Williams, A.; Gutmanis, J. Introduction: Geology of fractured reservoirs. *J. Geol. Soc.* **2021**, *178*, 2020–2197. [CrossRef]
3. Gentier, S.; Hopkins, D.; Riss, J. Role of fracture geometry in the evolution of flow paths under stress. In *Geophysical Monograph Series*; Faybishenko, B., Witherspoon, P.A., Benson, S.M., Eds.; American Geophysical Union: Washington, DC, USA, 2000; Volume 122, pp. 169–184, ISBN 978-0-87590-980-6.
4. Curewitz, D.; Karson, J.A. Structural settings of hydrothermal outflow: Fracture permeability maintained by fault propagation and interaction. *J. Volcanol. Geotherm. Res.* **1997**, *79*, 149–168. [CrossRef]
5. Dezayes, C.; Lerouge, C. Reconstructing Paleofluid Circulation at the Hercynian Basement/Mesozoic Sedimentary Cover Interface in the Upper Rhine Graben. Available online: https://www.hindawi.com/journals/geofluids/2019/4849860/abs/ (accessed on 21 June 2019).

6.   Vidal, J.; Genter, A. Overview of naturally permeable fractured reservoirs in the central and southern Upper Rhine Graben: Insights from geothermal wells. *Geothermics* **2018**, *74*, 57–73. [CrossRef]

7.   Callahan, O.A.; Eichhubl, P.; Olson, J.E.; Davatzes, N.C. Fracture Mechanical Properties of Damaged and Hydrothermally Altered Rocks, Dixie Valley-Stillwater Fault Zone, Nevada, USA. *J. Geophys. Res. Solid Earth* **2019**, *124*, 4069–4090. [CrossRef]

8.   Bauer, J.F.; Meier, S.; Philipp, S. Architecture, fracture system, mechanical properties and permeability structure of a fault zone in Lower Triassic sandstone, Upper Rhine Graben. *Tectonophysics* **2015**, *647–648*, 132–145. [CrossRef]

9.   Géraud, Y.; Rosener, M.; Surma, F.; Place, J.; Le Garzic, É.; Diraison, M. Physical properties of fault zones within a granite body: Example of the Soultz-sous-Forêts geothermal site. *Comptes Rendus Geosci.* **2010**, *342*, 566–574. [CrossRef]

10.   Géraud, Y.; Surma, F.; Rosener, M. *Porosity Network of Soultz-Sous-Forets Granite: The Importance of the Damaged Zone around Faults and Fractures*; Ecole et Observatoire des Sciences de la Terre: Strasbourg, France, 2005.

11.   Faulkner, D.; Jackson, C.; Lunn, R.; Schlische, R.; Shipton, Z.; Wibberley, C.; Withjack, M. A review of recent developments concerning the structure, mechanics and fluid flow properties of fault zones. *J. Struct. Geol.* **2010**, *32*, 1557–1575. [CrossRef]

12.   Long, J.C.S.; Witherspoon, P.A. The relationship of the degree of interconnection to permeability in fracture networks. *J. Geophys. Res. Space Phys.* **1985**, *90*, 3087–3098. [CrossRef]

13.   Ledésert, B.; Hebert, R.; Genter, A.; Bartier, D.; Clauer, N.; Grall, C. Fractures, hydrothermal alterations and permeability in the Soultz Enhanced Geothermal System. *Comptes Rendus Geosci.* **2010**, *342*, 607–615. [CrossRef]

14.   Kuncoro, G.B. Fluid-Rock Interaction Studies on an Enhanced Geothermal System in the Cooper Basin, South Australia. Ph.D. Thesis, University of Adelaide, Adelaide, SA, Australia, 2015.

15.   Turpault, M.-P.; Berger, G.; Meunier, A. Dissolution-precipitation processes induced by hot water in a fractured granite Part 1: Wall-rock alteration and vein deposition processes. *Eur. J. Miner.* **1992**, *4*, 1457–1476. [CrossRef]

16.   Dezayes, C.; Lerouge, C.; Innocent, C.; Lach, P. Structural control on fluid circulation in a graben system: Constraints from the Saint Pierre Bois quarry (Vosges, France). *J. Struct. Geol.* **2021**, *146*, 104323. [CrossRef]

17.   Dalmais, E.; Genter, A.; Trullenque, G.; Leoutre, E.; Leiss, B.; Wagner, B.; Mintsa, A.C.; Bär, K.; Rajsl, I. MEET Project: Toward the Spreading of EGS across Europe. In Proceedings of the European Geothermal Congress, Den Haag, The Netherlands, 11–14 June 2019.

18.   Trullenque, G.; Genter, A.; Leiss, B.; Wagner, B.; Bouchet, R.; Leoutre, E.; Malnar, B.; Bär, K.; Rajšl, I. Upscaling of EGS in Different Geological Conditions: A European Perspective. In Proceedings of the 43rd Workshop on Geothermal Reservoir Engineering Stanford University, Stanford, CA, USA, 12–14 February 2018.

19.   Klee, J.; Potel, S.; Ledésert, B.; Hébert, R.; Chabani, A.; Barrier, P.; Trullenque, G. Fluid-Rock Interactions in a Paleo-Geothermal Reservoir (Noble Hills Granite, California, USA). Part 1: Granite Pervasive Alteration Processes away from Fracture Zones. *Geosciences* **2021**, *11*, 325. [CrossRef]

20.   Ledésert, B.; Berger, G.; Meunier, A.; Genter, A.; Bouchet, A. Diagenetic-type reactions related to hydrothermal alteration in the Soultz-sous-Forêts granite, France. *Eur. J. Miner.* **1999**, *11*, 731–742. [CrossRef]

21.   Genter, A. Géothermie Roches Chaudes Sèches: Le Granite de Soultz-Sous-Forêts (Bas-Rhin, France): Fracturation Naturelle, Altérations Hydrothermales et Interaction Eau-Roche. Ph.D. Thesis, Université d'Orléans, Orléans, France, 1989.

22.   Traineau, H.; Genter, A.; Cautru, J.P.; Fabriol, H.; Chevremont, P. Petrography of the Granite Massif from Drill Cutting Analysis and Well Log Interpretation in the Geothermal HDR Borehole GPK1 (Soultz, Alsace, France). *Geotherm. Sci. Technol.* **1991**, *3*, 1–29.

23.   Meunier, A. Hydrothermal Alteration by Veins. In *Origin and Mineralogy of Clays: Clays and the Environment*; Velde, B., Ed.; Springer: Berlin/Heidelberg, Germany, 1995; pp. 247–267. ISBN 978-3-662-12648-6.

24.   Bonorino, F.G. Hydrothermal alteration in the front range mineral belt, colorado. *GSA Bull.* **1959**, *70*, 53–90. [CrossRef]

25.   Caine, J.S.; Bruhn, R.L.; Forster, C.B. Internal structure, fault rocks, and inferences regarding deformation, fluid flow, and mineralization in the seismogenic Stillwater normal fault, Dixie Valley, Nevada. *J. Struct. Geol.* **2010**, *32*, 1576–1589. [CrossRef]

26.   Norton, I. Two-stage formation of Death Valley. *Geosphere* **2011**, *7*, 171–182. [CrossRef]

27.   Dokka, R.K.; Travis, C.J. Role of the Eastern California Shear Zone in accommodating Pacific-North American Plate motion. *Geophys. Res. Lett.* **1990**, *17*, 1323–1326. [CrossRef]

28.   Stewart, J.H.; Ernst, W.G. Tectonics of the Walker Lane Belt, Western Great Basin: Mesozoic and Cenozoic Deformation in a Zone of Shear. *Metamorph. Crustal Evol. West. United States* **1988**, *7*, 683–713.

29.   Lifton, Z.M.; Newman, A.V.; Frankel, K.L.; Johnson, C.W.; Dixon, T.H. Insights into distributed plate rates across the Walker Lane from GPS geodesy. *Geophys. Res. Lett.* **2013**, *40*, 4620–4624. [CrossRef]

30.   Miller, M.M.; Johnson, D.J.; Dixon, T.H.; Dokka, R.K. Refined kinematics of the eastern California shear zone from GPS observations, 1993–1998. *J. Geophys. Res. Space Phys.* **2001**, *106*, 2245–2263. [CrossRef]

31.   Hill, M.L.; Troxel, B.W. Tectonics of Death Valley region, California. *GSA Bull.* **1966**, *77*, 435–438. [CrossRef]

32.   Burchfiel, B.C.; Stewart, J.H. "Pull-Apart" Origin of the Central Segment of Death Valley, California. *GSA Bull.* **1966**, *77*, 439–442. [CrossRef]

33.   Pavlis, T.L.; Trullenque, G. Evidence for 40–41 km of dextral slip on the southern Death Valley fault: Implications for the Eastern California shear zone and extensional tectonics. *Geology* **2021**, *49*, 767–772. [CrossRef]

34.   Rämö, T.O.; Calzia, J.P.; Kosunen, P.J. Geochemistry of Mesozoic plutons, southern Death Valley region, California: Insights into the origin of Cordilleran interior magmatism. *Contrib. Miner. Pet.* **2002**, *143*, 416–437. [CrossRef]

35. Reinert, E. Low-Temperature Thermochronometry of the Avawatz Mountains, California: Implications for the Inception of the Eastern California Shear Zone. Ph.D. Thesis, University of Washington, Seattle, WA, USA, 2004.
36. Chinn, L.D. *Low-Temperature Thermochronometry of the Avawatz Mountains*; Implications for the Eastern Terminus and Inception of the Garlock Fault Zone; University of Washington: Seattle, WA, USA, 2013; 48p.
37. Spencer, J.E. Chapter 15: Late Cenozoic extensional and compressional tectonism in the southern and western Avawatz Mountains, southeastern California. In *Geological Society of America Memoirs*; Geological Society of America: Boulder, CO, USA, 1990; Volume 176, pp. 317–334.
38. Niles, J.H. Post-Middle Pliocene Tectonic Development of the Noble Hills, Southern Death Valley, California. Ph.D. Thesis, San Francisco State University, San Francisco, CA, USA, 2016.
39. Brady, R.H. Neogene stratigraphy of the Avawatz Mountains between the Garlock and Death Valley fault zones, southern Death Valley, California: Implications as to late Cenozoic tectonism. *Sediment. Geol.* **1984**, *38*, 127–157. [CrossRef]
40. Troxel, B.W.; Butler, P.R. *Rate of Cenozoic Slip on Normal Faults, South-Central Death Valley, California*; Department of Geology, University of California: Berkeley, CA, USA, 1979.
41. Klee, J.; Trullenque, G.; Ledésert, B.; Potel, S.; Hébert, R.; Chabani, A.; Genter, A. Petrographic Analyzes of Fractured Granites Used as An Analogue of the Soultz-Sous-Forêts Geothermal Reservoir: Noble Hills, CA, USA. In Proceedings of the World Geothermal Congress 2020+1, Reykjavik, Iceland, 24–27 October 2020.
42. Chabani, A.; Trullenque, G.; Ledésert, B.; Klee, J. Multiscale Characterization of Fracture Patterns: A Case Study of the Noble Hills Range (Death Valley, CA, USA), Application to Geothermal Reservoirs. *Geoscience* **2021**, *11*, 280. [CrossRef]
43. Castaing, C.; Rabu, D. Apports de La Géologie à La Recherche et à l'exploitation de Pierres de Taille (Roches Ornementales et de Construction). *BULL BRGM* **1981**, *3*, 1.
44. Chabani, A.; Trullenque, G.; Klee, J.; Ledésert, B.A. Fracture Spacing Variability and the Distribution of Fracture Patterns in Granitic Geothermal Reservoir: A Case Study in the Noble Hills Range (Death Valley, CA, USA). *Geosciences* **2021**, *11*, 280. [CrossRef]
45. Laubach, S.; Lamarche, J.; Gauthier, B.; Dunne, W.; Sanderson, D. Spatial arrangement of faults and opening-mode fractures. *J. Struct. Geol.* **2018**, *108*, 2–15. [CrossRef]
46. Olson, J.E. Joint pattern development: Effects of subcritical crack growth and mechanical crack interaction. *J. Geophys. Res. Space Phys.* **1993**, *98*, 12251–12265. [CrossRef]
47. Bisdom, K.; Gauthier, B.; Bertotti, G.; Hardebol, N. Calibrating discrete fracture-network models with a carbonate three-dimensional outcrop fracture network: Implications for naturally fractured reservoir modeling. *AAPG Bull.* **2014**, *98*, 1351–1376. [CrossRef]
48. Gillespie, P.; Howard, C.; Walsh, J.; Watterson, J. Measurement and characterisation of spatial distributions of fractures. *Tectonophysics* **1993**, *226*, 113–141. [CrossRef]
49. Priest, S.; Hudson, J. Discontinuity spacings in rock. *Int. J. Rock Mech. Min. Sci. Géoméch. Abstr.* **1976**, *13*, 135–148. [CrossRef]
50. Sanderson, D.J.; Peacock, D.C. Line sampling of fracture swarms and corridors. *J. Struct. Geol.* **2019**, *122*, 27–37. [CrossRef]
51. Terzaghi, R.D. Sources of Error in Joint Surveys. *Géotechnique* **1965**, *15*, 287–304. [CrossRef]
52. Zhou, W.; Wang, Z.L. *Scanning Microscopy for Nanotechnology: Techniques and Applications*; Springer Science & Business Media: Berlin/Heidelberg, Germany, 2007; ISBN 978-0-387-39620-0.
53. Kanda, K. Energy Dispersive X-Ray Spectrometer. U.S. Patent 5,065,020, 1991.
54. Götze, J. Cathodoluminescence in Applied Geosciences. In *Cathodoluminescence in Geosciences*; Pagel, M., Barbin, V., Blanc, P., Ohnenstetter, D., Eds.; Springer: Berlin/Heidelberg, Germany, 2000; ISBN 978-3-662-04086-7.
55. Götze, J. Application of Cathodoluminescence Microscopy and Spectroscopy in Geosciences. *Microsc. Microanal.* **2012**, *18*, 1270–1284. [CrossRef] [PubMed]
56. Kretz, R. Symbols for Rock-Forming Minerals. *Am. Mineral.* **1983**, *68*, 277–279.
57. Schmidt, D.; Schmidt, S.T.; Mullis, J.; Mählmann, R.F.; Frey, M. Very low grade metamorphism of the Taveyanne formation of western Switzerland. *Contrib. Miner. Pet.* **1997**, *129*, 385–403. [CrossRef]
58. Kisch, H.J. Illite crystallinity: Recommendations on sample preparation, X-ray diffraction settings, and interlaboratory samples. *J. Metamorph. Geol.* **1991**, *9*, 665–670. [CrossRef]
59. Mählmann, R.F.; Frey, M. Standardisation, calibration and correlation of the Kübler-index and the vitrinite/bituminite reflectance: An inter-laboratory and field related study. *Swiss J. Geosci.* **2012**, *105*, 153–170. [CrossRef]
60. Moore, D.M.; Reynolds, R.C. *X-ray Diffraction and the Identification and Analysis of Clay Minerals*; Oxford University Press: New York, NY, USA, 1989.
61. Starkey, H.C.; Blackmon, P.D.; Hauff, P.L. *The Routine Mineralogical Analysis of Clay-Bearing Samples*; U.S. G.P.O.: Washington, DC, USA, 1984.
62. Kübler, B. La Cristallinite de l'illite et Les Zones Tout a Fait Superieures Du Metamorphisme. In *Etages Tectoniques*; La Baconniere: Boudry, Switzerland, 1967; pp. 105–121.
63. Warr, L.; Mählmann, R.F. Recommendations for Kübler Index standardization. *Clay Miner.* **2015**, *50*, 283–286. [CrossRef]
64. Kisch, H.J. Correlation between indicators of very low-grade metamorphism. In *Low-Temperature Metamorphism*; Frey, M., Ed.; Blackie and Son Ltd.: Glasgow, UK, 1987; pp. 227–300.

65. Árkai, P.; Sassi, F.; Desmons, J. Very Low- to Low-Grade Metamorphic Rocks. In *Metamorphic Rocks A Classification and Glossary Terms*; Cambridge University Press: Cambridge, UK, 2003.

66. Warr, L.N.; Cox, S.C. Correlating illite (Kübler) and chlorite (Árkai) "crystallinity" indices with metamorphic mineral zones of the South Island, New Zealand. *Appl. Clay Sci.* **2016**, *134*, 164–174. [CrossRef]

67. Frey, M. Very low-grade metamorphism of clastic sedimentary rocks. In *Low-Temperature Metamorphism*; Frey, M., Ed.; Blackie: Glasgow, UK, 1987; pp. 9–58.

68. Árkai, P.; Maehlmann, R.; Suchy, V.; Balogh, K.; Sykorová, I.; Frey, M. Possible Effects of Tectonic Shear Strain on Phyllosilicates: A Case Study from the Kandersteg Area, Helvetic Domain, Central Alps, Switzerland. *TMPM Tschermaks Mineral. Petrogr. Mitt.* **2002**, *82*, 273–290.

69. Mullis, J.; Mählmann, R.F.; Wolf, M. Fluid inclusion microthermometry to calibrate vitrinite reflectance (between 50 and 270 °C), illite Kübler-Index data and the diagenesis/anchizone boundary in the external part of the Central Alps. *Appl. Clay Sci.* **2017**, *143*, 307–319. [CrossRef]

70. Warr, L.N.; Rice, A.H.N. Interlaboratory standardization and calibration of day mineral crystallinity and crystallite size data. *J. Metamorph. Geol.* **1994**, *12*, 141–152. [CrossRef]

71. Ledésert, B.; Hébert, R.L.; Grall, C.; Genter, A.; Dezayes, C.; Bartier, D.; Gérard, A. Calcimetry as a useful tool for a better knowledge of flow pathways in the Soultz-sous-Forêts Enhanced Geothermal System. *J. Volcanol. Geotherm. Res.* **2009**, *181*, 106–114. [CrossRef]

72. White, A.F.; Schulz, M.S.; Lowenstern, J.B.; Vivit, D.V.; Bullen, T.D. The ubiquitous nature of accessory calcite in granitoid rocks: Implications for weathering, solute evolution, and petrogenesis. *Geochim. Cosmochim. Acta* **2005**, *69*, 1455–1471. [CrossRef]

73. Dullien, F.A.L. *Porous Media: Fluid Transport and Pore Structure*; Academic Press: San Diego, CA, USA, 1979; ISBN 978-0-323-13933-5.

74. Navelot, V.; Géraud, Y.; Favier, A.; Diraison, M.; Corsini, M.; Lardeaux, J.-M.; Verati, C.; de Lépinay, J.M.; Legendre, L.; Beauchamps, G. Petrophysical properties of volcanic rocks and impacts of hydrothermal alteration in the Guadeloupe Archipelago (West Indies). *J. Volcanol. Geotherm. Res.* **2018**, *360*, 1–21. [CrossRef]

75. Gates, W.P.; Nefiodovas, A.; Peter, P. Permeability of an Organo-Modified Bentonite to Ethanol-Water Solutions. *Clays Clay Miner.* **2004**, *52*, 192–203. [CrossRef]

76. Gresens, R.L. Composition-volume relationships of metasomatism. *Chem. Geol.* **1967**, *2*, 47–65. [CrossRef]

77. Beaufort, D.; Baronnet, A.; Lanson, B.; Meunier, A. Corrensite; a single phase or a mixed-layer phyllosilicate in saponite-to-chlorite conversion series? A case study of Sancerre-Couy deep drill hole (France). *Am. Miner.* **1997**, *82*, 109–124. [CrossRef]

78. Andrews, B.J.; Roberts, J.J.; Shipton, Z.K.; Bigi, S.; Tartarello, M.C.; Johnson, G. How do we see fractures? Quantifying subjective bias in fracture data collection. *Solid Earth* **2019**, *10*, 487–516. [CrossRef]

79. Bestmann, M.; Kunze, K.; Matthews, A. Evolution of a calcite marble shear zone complex on Thassos Island, Greece: Microstructural and textural fabrics and their kinematic significance. *J. Struct. Geol.* **2000**, *22*, 1789–1807. [CrossRef]

80. Burkhard, M. Ductile deformation mechanisms in micritic limestones naturally deformed at low temperatures (150–350 °C.). *Geol. Soc. London Spéc. Publ.* **1990**, *54*, 241–257. [CrossRef]

81. Schmid, S.; Panozzo, R.; Bauer, S. Simple shear experiments on calcite rocks: Rheology and microfabric. *J. Struct. Geol.* **1987**, *9*, 747–778. [CrossRef]

82. Stipp, M.; Kunze, K. Dynamic recrystallization near the brittle-plastic transition in naturally and experimentally deformed quartz aggregates. *Tectonophysics* **2008**, *448*, 77–97. [CrossRef]

83. Glaas, C.; Patrier, P.; Vidal, J.; Beaufort, D.; Genter, A. Clay Mineralogy: A Signature of Granitic Geothermal Reservoirs of the Central Upper Rhine Graben. *Minerals* **2021**, *11*, 479. [CrossRef]

84. Burnham, C.W. Facies and types of hydrothermal alteration. *Econ. Geol.* **1962**, *57*, 768–784. [CrossRef]

85. Dubois, M.; Ledesert, B.; Potdevin, J.-L.; Vançon, S. Détermination des conditions de précipitation des carbonates dans une zone d'altération du granite de Soultz (soubassement du fossé Rhénan, France): L'enregistrement des inclusions fluides. *Comptes Rendus l'Académie Sci. Ser. II A Earth Planet. Sci.* **2000**, *331*, 303–309. [CrossRef]

86. Marqués, J.M.; Matias, M.J.; Basto, M.J.; Carreira, P.M.; Aires-Barros, L.A.; Goff, F.E. Hydrothermal alteration of Hercynian granites, its significance to the evolution of geothermal systems in granitic rocks. *Geothermics* **2010**, *39*, 152–160. [CrossRef]

87. Pauwels, H.; Fouillac, C.; Fouillac, A.-M. Chemistry and isotopes of deep geothermal saline fluids in the Upper Rhine Graben: Origin of compounds and water-rock interactions. *Geochim. Cosmochim. Acta* **1993**, *57*, 2737–2749. [CrossRef]

88. Parneix, J.; Petit, J. Hydrothermal alteration of an old geothermal system in the Auriat granite (Massif Central, France): Petrological study and modelling. *Chem. Geol.* **1991**, *89*, 329–351. [CrossRef]

89. Berger, G.; Velde, B. Chemical parameters controlling the propylitic and argillic alteration process. *Eur. J. Miner.* **1992**, *4*, 1439–1456. [CrossRef]

90. Fulignati, P. Clay Minerals in Hydrothermal Systems. *Minerals* **2020**, *10*, 919. [CrossRef]

91. Fordham, A.W. Weathering of Biotite into dioctahedral Clay Minerals. *Clay Miner.* **1989**, *25*, 51–63. [CrossRef]

92. Kajdas, B.; Michalik, M.; Migoń, P. Mechanisms of granite alteration into grus, Karkonosze granite, SW Poland. *Catena* **2017**, *150*, 230–245. [CrossRef]

93. Abad, I. Physical Meaning and Applications of the Illite Kübler Index: Measuring Reaction Progress in Low-Grade Metamorphism. In *Diagenesis and Low-Temperature Metamorphism, Theory, Methods and Regional Aspects, Seminarios, Sociedad Espanola: Sociedad Espanola Mineralogia*; University of Jaen: Jaen, Spain, 2007; pp. 53–64.

94. Camacho, A.; McDougall, I.; Armstrong, R.; Braun, J. Evidence for shear heating, Musgrave Block, central Australia. *J. Struct. Geol.* **2001**, *23*, 1007–1013. [CrossRef]

95. Środoń, J. Nature of mixed-layer clays and mechanisms of their formation and alteration. *Annu. Rev. Earth Planet. Sci.* **1999**, *27*, 19–53. [CrossRef]

96. Bruhn, R.L.; Parry, W.T.; Yonkee, W.A.; Thompson, T. Fracturing and hydrothermal alteration in normal fault zones. *Pure Appl. Geophys.* **1994**, *142*, 609–644. [CrossRef]

97. Sardini, P.; Ledesert, B.; Touchard, G. Quantification of Microscopic Porous Networks By Image Analysis and Measurements of Permeability in the Soultz-Sous-Forêts Granite (Alsace, France). In *Fluid Flow and Transport in Rocks*; Springer: Berlin/Heidelberg, Germany, 1997; pp. 171–189, Chapter 10.

98. Liu, Y.; Xie, C.; Li, C.; Li, S.; Santosh, M.; Wang, M.; Fan, J. Breakup of the northern margin of Gondwana through lithospheric delamination: Evidence from the Tibetan Plateau. *GSA Bull.* **2019**, *131*, 675–697. [CrossRef]

99. Chambefort, I.; Moritz, R.; von Quadt, A. Petrology, geochemistry and U–Pb geochronology of magmatic rocks from the high-sulfidation epithermal Au–Cu Chelopech deposit, Srednogorie zone, Bulgaria. *Miner. Deposita* **2007**, *42*, 665–690. [CrossRef]

100. Glassley, W.; Crossey, L.; Montañez, I. Fluid–Rock Interaction. In *Encyclopedia of Geochemistry*; Springer Iinternational Publishing: Cham, Switzerland, 2016.

101. Kadko, D.; Butterfield, D.A. The relationship of hydrothermal fluid composition and crustal residence time to maturity of vent fields on the Juan de Fuca Ridge. *Geochim. Cosmochim. Acta* **1998**, *62*, 1521–1533. [CrossRef]

102. Hebert, R.L.; Ledesert, B.; Bartier, D.; Dezayes, C.; Genter, A.; Grall, C. The Enhanced Geothermal System of Soultz-sous-Forêts: A study of the relationships between fracture zones and calcite content. *J. Volcanol. Geotherm. Res.* **2010**, *196*, 126–133. [CrossRef]

103. Dhansay, T.; Navabpour, P.; de Wit, M.; Ustaszewski, K. Assessing the reactivation potential of pre-existing fractures in the southern Karoo, South Africa: Evaluating the potential for sustainable exploration across its Critical Zone. *J. Afr. Earth Sci.* **2017**, *134*, 504–515. [CrossRef]

104. Moir, H.; Lunn, R.; Shipton, Z.; Kirkpatrick, J. Simulating brittle fault evolution from networks of pre-existing joints within crystalline rock. *J. Struct. Geol.* **2010**, *32*, 1742–1753. [CrossRef]

105. Laubach, S.E. Practical approaches to identifying sealed and open fractures. *AAPG Bull.* **2003**, *87*, 561–579. [CrossRef]

106. Woodcock, N.H.; Dickson, J.A.D.; Tarasewicz, J.P.T. Transient permeability and reseal hardening in fault zones: Evidence from dilation breccia textures. *Geol. Soc. London, Spéc. Publ.* **2007**, *270*, 43–53. [CrossRef]

107. Vidal, J.; Patrier, P.; Genter, A.; Beaufort, D.; Dezayes, C.; Glaas, C.; Lerouge, C.; Sanjuan, B. Clay minerals related to the circulation of geothermal fluids in boreholes at Rittershoffen (Alsace, France). *J. Volcanol. Geotherm. Res.* **2018**, *349*, 192–204. [CrossRef]

108. Liotta, D.; Brogi, A.; Ruggieri, G.; Rimondi, V.; Zucchi, M.; Helgadóttir, H.M.; Montegrossi, G.; Friðleifsson, G. Ómar Fracture analysis, hydrothermal mineralization and fluid pathways in the Neogene Geitafell central volcano: Insights for the Krafla active geothermal system, Iceland. *J. Volcanol. Geotherm. Res.* **2020**, *391*, 106502. [CrossRef]

109. Griffiths, L.; Heap, M.; Wang, F.; Daval, D.; Gilg, H.; Baud, P.; Schmittbuhl, J.; Genter, A. Geothermal implications for fracture-filling hydrothermal precipitation. *Geothermics* **2016**, *64*, 235–245. [CrossRef]

110. Ito, H. Inferred role of natural fractures, veins, and breccias in development of the artificial geothermal reservoir at the Ogachi Hot Dry Rock site, Japan. *J. Geophys. Res. Space Phys.* **2003**, *108*. [CrossRef]

 *geoscienes*

Article

Fracture Spacing Variability and the Distribution of Fracture Patterns in Granitic Geothermal Reservoir: A Case Study in the Noble Hills Range (Death Valley, CA, USA)

Arezki Chabani [1,*], Ghislain Trullenque [1], Johanne Klee [1] and Béatrice A. Ledésert [2]

[1] UniLaSalle, UPJV, B2R UMR 2018.C100, U2R 7511, 19 Rue Pierre Waguet, F-60026 Beauvais, France; ghislain.trullenque@unilasalle.fr (G.T.); johanne.klee@unilasalle.fr (J.K.)

[2] Geosciences and Environnent Cergy (GEC), CY Cergy Paris Université, 1 Rue Descartes, F-95000 Neuville-sur-Oise, France; beatrice.ledesert@cyu.fr

* Correspondence: arezki.chabani@unilasalle.fr; Tel.: +33-6-58-07-14-14

Abstract: Scanlines constitute a robust method to better understand in 3D the fracture network variability in naturally fractured geothermal reservoirs. This study aims to characterize the spacing variability and the distribution of fracture patterns in a fracture granitic reservoir, and the impact of the major faults on fracture distribution and fluid circulation. The analogue target named the Noble Hills (NH) range is located in Death Valley (DV, USA). It is considered as an analogue of the geothermal reservoir presently exploited in the Upper Rhine Graben (Soultz-sous-Forêts, eastern of France). The methodology undertaken is based on the analyze of 10 scanlines located in the central part of the NH from fieldwork and virtual (photogrammetric models) data. Our main results reveal: (1) NE/SW, E/W, and NW/SE fracture sets are the most recorded orientations along the virtual scanlines; (2) spacing distribution within NH shows that the clustering depends on fracture orientation; and (3) a strong clustering of the fracture system was highlighted in the highly deformed zones and close to the Southern Death Valley fault zone (SDVFZ) and thrust faults. Furthermore, the fracture patterns were controlled by the structural heritage. Two major components should be considered in reservoir modeling: the deformation gradient and the proximity to the regional major faults.

Keywords: fracture network variability; Death Valley; granite; spacing distribution; fracture intensity P_{10}; geothermal reservoir characterization

Citation: Chabani, A.; Trullenque, G.; Klee, J.; Ledésert, B.A. Fracture Spacing Variability and the Distribution of Fracture Patterns in Granitic Geothermal Reservoir: A Case Study in the Noble Hills Range (Death Valley, CA, USA). *Geosciences* **2021**, *11*, 520. https://doi.org/10.3390/geosciences11120520

Academic Editors: Giovanni Barreca and Jesus Martinez-Frias

Received: 21 September 2021
Accepted: 12 December 2021
Published: 17 December 2021

Publisher's Note: MDPI stays neutral with regard to jurisdictional claims in published maps and institutional affiliations.

1. Introduction

In deep geothermal systems, many studies have been undertaken to better understand the importance of the natural fractures in various contexts [1,2]. In granitic basement rocks, the permeability is mostly increased by the fracture network and faults [3–7], while the porosity is increased by both the alteration (e.g., dissolution of primary minerals) and the proximity to fracture zones ([8], this issue). The low rock matrix permeability and porosity allow the fluid flow within fracture networks [9–11]. The understanding of the spatial arrangement of the fracture network constitutes the main issue in fractured reservoirs [1,3,7,12,13].

A fracture network is characterized by geometrical parameters such as lengths, spacings, widths, orientations, fracture distributions, and the relationships between them significantly affect the connectivity within the reservoir [9,14–17]. Among these parameters, spacing between fractures is a well-considered parameter, because it controls the probability of intersecting fractures during drilling [18]. Statistic parameters that describe fracture spacing include: (1) The mean, which characterizes the global expected frequency of fracture intersection, and (2) the standard deviation, which describes the distribution of the fractures around the mean.

Regarding the spacing distribution, fractured zones can be classified into: (1) Fracture corridor, a term usually used for a dominant set of fractures displaying an important variation of fracture intensity. It can represent the main drains for fluid circulation in various reservoir contexts such as geothermal fields [19]; (2) fracture arrays, a term generally used to define a dominant set of fractures which forms an angle to the swarm (area in which any kind of fracture appears) [20]; (3) shear zone, which is defined as a continuous deformed zone with a high strain [21], accommodated by a cataclastic process in granitic rocks and crystalline plasticity in (e.g., carbonate rocks); and (4) fault zone characterized by a fault and its associated damage zone which can act as a barrier or a drain for flow, depending on its intrinsic properties [5,20,22]. In crystalline basement rocks, the fracture distribution within the damage zone is influenced by the distance from the fault core and its displacement along the fault plane [23]. Indeed, the strong variation in fracture distribution (fracture densities) is commonly observed near active faults. Ostermeijer et al. [23] add that the pattern is mainly ruled by the distribution of macro-damage induced on shear-accommodating subsidiary fractures.

Spatial organization of fracture systems became an important studied topic in the recent decades because of the necessity to better understand the architecture of fractured reservoirs [7,24–26]. The spatial arrangement can be quantified using statistical laws (e.g., power law, log normal, and exponential law) [24,27,28] or statistical parameters, such as the coefficient of variability (Cv) along 1D [20] and the normalized correlation count method [27]. The main goal is to enhance the fracture distribution understanding (clustered, random, or uniform distribution) and its effects on connectivity [29]. In that case, many studies are focused on fracture networks characterization in various settings and at different scales [3,12,16,20,30,31]. They commonly used field analogues at surface to resolve the challenge of lack of sub-surface information in reservoirs [32]. The characterization of heterogeneity of the fracture spacing and the fracture abundance at any scale may be performed using line sampling method along one dimension (1D) named scanline (e.g., [26]). The present study combines the spacing data of joints (opening-mode fractures), veins (partially or fully filled), and faults to highlight the spatial arrangement of the fracture patterns in granitic rocks and the influence of the regional major faults. In the present study, the measured fractures, whatever their filling have been classified according to their orientation.

The present work is part of the European MEET project (multidisciplinary and multi-context demonstration of EGS exploration and exploitation techniques and potentials, [33]). This study proposes to (1) describe fracture system distribution at outcrop scale, based on fracture network parameters; (2) shows the role of the regional major faults proximity on the fracture patterns evolution in the basement rocks; and (3) highlights the impact of the deformation at outcrop scale. The present study was performed in the desert environment of Noble Hills (NH) fractured granitic basement, located in the southern termination of Death Valley (Death Valley, CA, USA), and is considered as a paleo geothermal analogue of the Upper Rhine Graben (URG, Alsace area situated in the eastern of France) ([25,34,35], this issue) geothermal systems producing electricity, because of the similarities in the basement rock nature (granite), hydrothermal alterations and the trans-tensional tectonic setting [25,34–36]. However, the geological history of the NH range is rather different from that of to those in the URG, but numerous pieces of evidence of analogy have been highlighted by recent work of Klee et al., [37], which addressed a list of similarities between the URG reservoir targets (exploited geothermal present-day reservoir) and the NH ranges:

- Pervasive alteration of the NH granite;
- Ubiquitous argillic alteration affecting plagioclase and biotite is present;
- Unaltered K-feldspar;
- Porosity is enhanced by the alteration and microfracturing;
- Evidence of the hydrothermal fluid percolation, as identified in an exploited geothermal reservoirs;
- Fluid circulation in open system such as in EGS systems (input of potassium and carbonates).

Based on scanline methodology (e.g., [26]), this work has been performed using a spacing measurement, compiled from four different canyons located in the central part of NH. This central part has been characterized as having a distinct spatial arrangement of fractures at different scales in comparison with the northeast and southeast parts (for additional explanations, see [25]). The scanlines are located in the granitic part of the NH range, the so called crystalline basement slice (CBS) according to Brady et al. [38]. Some of the measures were acquired directly from the field and others from virtual scanlines (method detailed in Section 3.2). Apertures of fractures have also been measured directly in the field. Numerous fractures were filled by newly formed minerals such as carbonates, oxides, and sometimes barite.

In this study, scanline methodology and statistical tools are used to better understand the fracture spacing variability and the distribution of fracture patterns at depth. This allows better characterizing reservoirs in response to the developing geothermal exploration and exploitation by EGS in basement rock context. This study was conducted:

1. Through a description of the fracture system using orientation, density, spacing and aperture parameters;
2. By highlighting the role of the proximity to the regional major faults on the fracture patterns;
3. By highlighting the role of the deformation gradient and structural heritage at outcrop scale.

2. Geological Setting

The NH structurally belongs to the DV region (Figure 1a), which is characterized by a complex tectonic history (e.g., [39]), starting with late Cenozoic extensional and trans-tensional structures which overprint the Mesozoic to Early Cenozoic contractional structures [39–41]. The extensional regime of the DV has begun around 16 Ma [42,43], and is shifted to a trans-tensional regime around 5 Ma [39,44–46]. Recent work by Pavlis and Trullenque, [36] reconsider the age of the transcurrent deformation in DV around 12 Ma.

The NH ranges forms the principal physiographic feature aligned with segments of the right-lateral Southern Death Valley Fault Zone (SDVFZ) [25,47] (Figure 1b). The SDVFZ net dextral strike-slip displacement has been estimated around 40–41 km [36]. A whole compressional region was created by the interaction between the SDVFZ and the Garlock Fault (GF) system (see Figure 1b for location), which leads to shortening within the Avawatz Mountains (Figure 1b) [48–50].

The exhumation history of NH range is poorly described in the literature. Based on KI/temperature of illite crystals, recent work by Klee et al. [37] highlighted that the southeastern of NH is much elevated, with a higher temperature which could indicate that the south-eastern part of NH was buried deeper than its north-western part and has been exhumed.

Recent work by Chabani et al., [25] highlights the structural organization of the NH range according to the orders of fault magnitude classification by analyzing 2D maps at different scales. These orders consist in (1) second order scale, referring to the faults comprised between 20 and 30 km length; (2) third order scale, referring to faults around 10 km length; and (3) fourth order, referring to the faults under 1 km length. The first order referring to the crustal faults (higher than 100 km length) is not observed within the NH range.

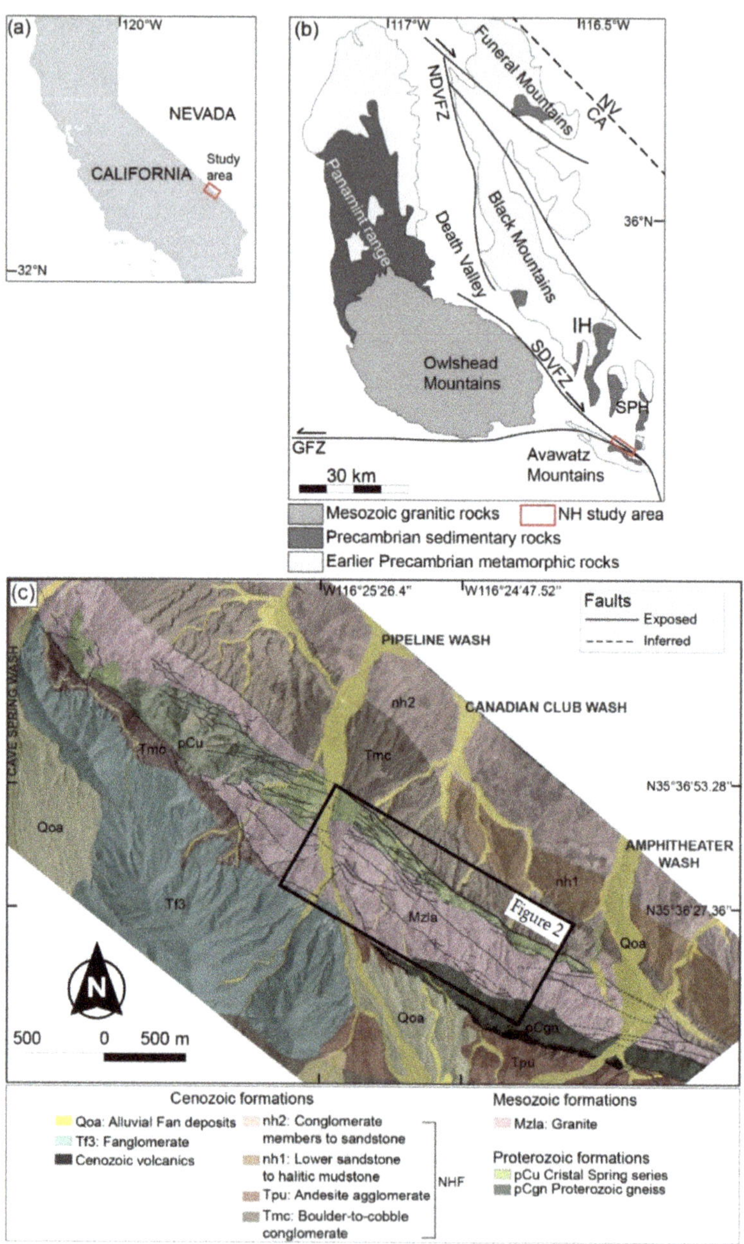

Figure 1. (**a**,**b**) Location and geological setting of DV and NH studied area modified after [50], NDVFZ: Northern Death Valley Fault Zone, SDVFZ: Southern Death Valley Fault Zone, GFZ: Garlock fault zone, CA: California, NV: Nevada, IH: Ibex Hills, SPH: Saddle Peak Hills. (**c**) Structural scheme of the NH range performed thanks to high-resolution digital mapping techniques (see below) modified after [34,47]. Additional digitized fractures were obtained using orthophotos. NHFs: Noble Hills formations.

The SDVFZ trending NW/SE and the GFZ trending E/W controlled NH range geometry. Indeed, the SDVFZ controls the NH geometry at large scale within the second order scale, while the GFZ trending E/W controls the NH geometry within the third order scale. Chabani et al., [25] add that the NH are divided into three internal structural domains: (1) Domain A, to the north, is characterized by the dominance of the NW/SE direction at the fourth order scale; (2) domain B, central, is marked by the dominance of the E/W and the NW/SE directions at respectively the fourth and third order scales; and (3) domain C, to the south, is also characterized by the E/W and NW/SE directions dominance but at the third and fourth order scales, respectively.

Numerous episodes of deformations have been highlighted during the fieldwork campaigns. Indeed, the SDVFZ fault segments act with a dextral movement (black lines in Figure 2b), highlighting an intense deformation with local evidences of extreme shearing. These structures are contemporaneous with the syn-kinematic dextral strike faults (orange lines in Figure 2b), highlighted in the recent work done by Klee et al. [8]. In addition, compressional structures like thrust faults crosscut outcrops 6 and 8 (red lines in Figure 2b). A clear overprinting has been recognized between SDVFZ (which also crosscut the OT2, OT6, and OT7) and the compressive structures are due to the GFZ, which acts with a sinistral movement. Furthermore, the thrusting highlighted in the present study postdates the activity of the SDVFZ. According to Chabani et al. [25], it is tempting to relate the thrust structures to the activity along the frontal termination of GFZ. Furthermore, the compressive structures are related to the interaction of the NH ranges with the Avawatz mountains during the GFZ movement.

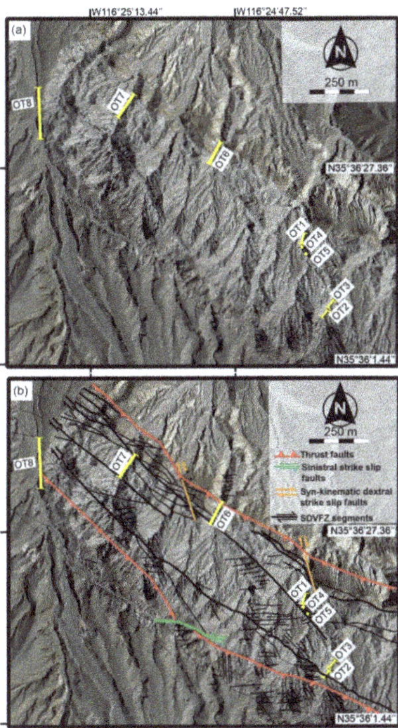

Figure 2. (**a**) Map highlighting the structural position of the central part of NH, including the outcrops (OT) location and (**b**) SDVFZ segments position in black, syn-kinematic dextral strike slip faults in orange, sinistral strike-slip faults in green lines, and thrust faults in red. Dextral strike slip faults are syn-kinematic with SDVFZ episode, those systems are followed by thrust faults, which are contemporaneous with the GFZ orientated globally E/W.

A gradient of deformation has been highlighted in the central part of NH, with evidence of extreme shearing, close to OT6 and OT7. Boudinage structures and brittle shearing are highlighted within the Crystal Spring series (pCu, Figure 1c). In that case, the new geological map (Figure 1c) built using the high-resolution mapping techniques on the ground, using a tablet and QGIS software by Klee et al. [34] revealed a stacking of the Crystal Spring series, intruded by the Mesozoic granite (Mzla, Figure 1c). Laterally, the thickness of Crystal Spring series was reduced, as they were dragged and stretched against the granite due to the SDVFZ activity, especially in OT6 to OT8 areas.

Cenozoic volcanic series have also been highlighted in the southeastern end of the NH [47]. Cenozoic formations have been characterized by Niles, [47] outside the center part of the NH. They are mainly composed of fanglomerate, alluvial fan deposits, lacustrine deposits, sabkha, evaporitic rocks, carbonate units, and megabreccia.

3. Methodology

Scanlines are commonly used to describe the reservoir properties and the fracture systems from analogues of hydrocarbon and groundwater reservoirs [20,30,51–54], and of geothermal reservoirs [55]. The scanline methodology, widely described in the literature [2,26,54,56], helps the understanding of the fractured reservoir geometry.

3.1. Scanline Data Acquisition

In the present work, the geometrical parameters of fractures, such as orientation, spacing, and aperture, were acquired directly from scanlines in the field. A decameter was installed horizontally along the outcrop (Figure 2a). Note that data about every fracture (e.g., joint, vein or fault) intersected by the scanline were collected, whatever its orientation class or filling. The cross-cutting relationships between the studied fractures are difficult to observe in the field, as intersections rarely occur along the scanline. Then, the fracture parameters were acquired by reporting the successive position of each fracture along the scanline. The projected positions were then collected and reported in Excel software v.2019. The spacing between two consecutive fractures is given by [20]:

$$S_A = P_n - P_{n-1} \tag{1}$$

S_A is the apparent spacing of fractures calculated from the fracture positions measured from field, P_n refers to the position of fracture n, and P_{n-1} refers to the position of fracture n–1, both expressed in meters, the location of the beginning of the measurement line being the reference. During the data analysis step, the fractures were filtered by orientation classes in order to discuss the effect of the regional directions on the local fracturing heterogeneity.

One to two scanlines were acquired from each outcrop. Five scanlines were performed along the outcrops OT 1, OT2, and OT3.

Fracture spacings were also calculated from virtual scanlines based on photogrammetric models and on fracture maps. The photogrammetric models were performed using two drones: 3DR Solo drone and DJI Phantom. These drones were loaned by University of Texas at El Paso (UTEP), TX, USA. The videos were recorded between the late morning and the early afternoon during seven consecutive days using a manual mode camera setting to reduce the effects of lighting condition. Then, pictures were extracted from the recorded videos. To provide a sufficient overlapping, pictures were extracted every second using ffmpeg software v.4.5 (Grenoble, France). The alignment of the pictures was done in Agisoft Metashape software 2020, v.1.6.5. (Saint Petersburg, Russia). Regarding the picture resolution, we ensured that every picture had a resolution of 300 dpi (300 pixels per 300 pixels). That permitted us to digitize the maximum number of fractures of decimeter length. The size of the pixel is 10 cm per pixel.

Several processing steps were needed to build the 3D models, starting by sky removal to reduce the noises, and the creation of different picture chunks (Figure 3). The 3D models were georeferenced and then imported in open access QGIS® software 2018, v.2.18.17 (Beaverton, OR, USA) to start the fracture extraction process. To improve the accuracy, ground control points (GCP) put in the field, using differential global positioning system (DGPS) and global positioning system (GPS) integrated directly in the drones, have been used. The georeferencing of each outcrop was realized independently using DGPS. The extraction of fractures was done manually by tracing every plane from the 3D outcrop. The orientation of every extracted trace plane was done automatically, and then compiled. For further explanation, the methodology of the modeling and the fracture extraction is detailed in Chabani et al. [35]. The digitized fractures were projected on a 2D map in order to keep data consistent among the whole datasets.

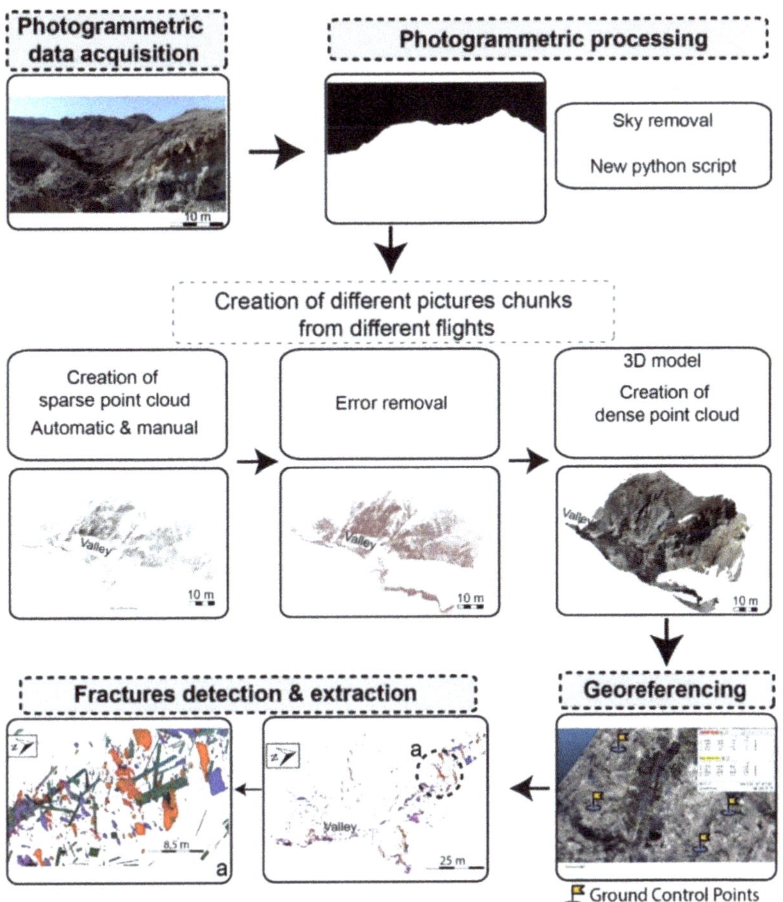

Figure 3. Workflow illustrating the different steps to build the 3D photogrammetric models, georeferencing, and fractures detection and extraction. Modified after Chabani et al. [35].

Two-dimensional fracture maps OT4 and OT5 (Figure 2) were performed from the field using a DSLR high resolution camera, with fixed focal (50 mm) in order to reduce the distortion. Furthermore, to avoid light effects, pictures were taken in the absence of direct sunlight (e.g., [16]). Several pictures were taken vertically, with the same distance, and with a sufficient overlapping. These pictures were then aligned using Agisoft Metashape software 2020 v.1.6.5 with the procedure detailed in Chabani et al. [35]. Outcrops OT6 to OT8, also located in the central part of NH, were analyzed by photogrammetric technology, and are also located in the central part of NH. In total, five scanlines were performed (Figure 4e–i).

Figure 4. Visualization of the eight outcrops studied in this work, including the scanlines (SL) location. (**a**) Outcrop 1 trending SW/NE shows heterogeneity of fracture orientation distribution. (**b**,**c**) Outcrop 2 trending E/W shows large NE/SW and NW/SE fracture planes crosscut by SL2a and SL2b. (**d**) Outcrop 3 trending NW/SE shows large E/W fracture planes crosscutting SL3a and SL3b. (**e**) Outcrop 4 consists in a fracture map; 2 scanlines were traced perpendicular to the main structures, recording then large E/W and N/S fracture planes for SL4 and SL5 respectively. (**f**) Outcrop 5 also consists in a fracture map; 2 scanlines were traced perpendicular to the main structures, crosscutting mainly fracture planes orientated NW/SE for SL6 and E/W for SL7. (**g**) Outcrop 6 consists in a canyon perpendicular to the SDVFZ segments, which crosscuts a large E/W fracture plane. This outcrop shows the transition between granitic basement and Crystal Spring sedimentary rocks. (**h**) Outcrop 7 trending NE/SW records a several fracture plane orientations. (**i**) Outcrop 8 trending NW/SE highlights mainly NE/SW and E/W fracture planes. SDVFZ: Southern Death Valley Fault Zone. For outcrops location, see Figure 2. Note that the OT6, 7 and 8 cross some major talus slopes. These outcrops have been modeled in 3D, making the fracture digitation possible to do in CloudCompare software. Then, the planes have been projected in 2D to keep data homogeneous from all fracture sets.

The fracture maps and the photogrammetric models were georeferenced and then used to extract the fractures in QGIS® software 2018, v.2.18.17. In this study, the fracture digitization was performed manually by tracing every detected fracture because of insufficient contrast between the fractures and the surrounding rock. Each digitized fracture became a georeferenced lineament in QGIS®. In the case of fracture maps, note that the fractures extending out of the sampling area were considered as one continuous feature [57–59] (Figure 2b). In QGIS®, the procedure of digitization consisted in an extraction of the end point coordinates of each fracture. Each fracture contained X and Y coordinates of each of the two end points, which helped to compute the spacings along a virtual scanline according to the following procedure:

1. Digitized fractures are loaded in shapefile format (e.g., shp format);
2. Virtual lines are traced along the georeferenced outcrop, and the intersection between the digitized fractures and the virtual line are collected. Note that, the intersection point ID must be the same as that of digitized fractures;
3. X and Y coordinates are added to the intersection points file, computed directly in QGIS®;
4. Values are classified according to X coordinate in Excel software, to ensure the right position of each intersected fracture;
5. Spacings are computed following [17]:

$$||AB|| = \sqrt{(x_b - x_a)^2 + (y_b - y_a)^2} \tag{2}$$

The calculated spacing in Equation (2) is not adjusted by Terzaghi correction. According to the scanline orientation, fracture orientations, and the position of scanline intersections for each fracture set, the fractures spacing were adjusted by applying the Terzaghi correction following [60]:

$$S = S_A \times \cos \theta \tag{3}$$

S is the true mean spacing of fractures in a set, S_A is the apparent mean spacing of fractures in a set and θ is the acute angle between the direction normal to fractures and the scanline.

Orientation bias can be minimized by drawing a scanline parallel to the normal to a fracture set, such that θ is close to 0°. All fractures intersected by scanlines were acquired. Then, all measured fractures were filtered by orientations during the analysis in order to describe the patterns of spacing according to each fracture set.

3.2. Fracture Orientation Analysis

In this study, the orientation is the first parameter analyzed to classify the fractures into fracture sets. As described before, from virtual scanlines, the fracture dip was not obtained from virtual scanlines, while it was measured for each fracture on field scanlines. The fracture dip has been measured for each fracture. Here, the classification into fracture sets is based only on strike orientation without considering the variation in dip to preserve data homogeneity. Note that, from field scanlines, the dip direction for each fracture is however provided in the Schmidt canvas.

Several software packages such as Win-tensor [61], Stereonet [62], and Digifrac [63] have been developed in order to project the structural data. Fisher distribution [64], Fisher-Bingham distribution [65], and von Mises distribution [66] are commonly used to describe symmetrical distributions of orientations in 2D, and sometimes in 3D in case of Fisher distribution. However, these distributions do not describe complex asymmetrical data. Then, the classification was performed using the mixture of von Mises distribution (MvM) [67], which is adapted to describe complex circular data and then seem relevant to model larger complex fracture networks. The methodology consists in a semi-automated approach based on appraisal tests to avoid any subjectivity in fracture set analysis. This distribution is based on three parameters: (1) Mean orientation ($\mu°$), around which the distribution is centered; (2) kappa (κ), which controls the concentration of the orientation values around the mean; and (3) weight (ω), corresponding to the relative contribution of each fracture set to the model. In addition, the best number of fracture sets is approved using the goodness of fit parameters (e.g., likelihood). The degree of precision of each mean fracture orientation is computed using the standard deviation (SD), which is of $+/-10°$. For further explanations, see Chabani et al., [68], which described and adapted the MvM methodology for structural data. To plot the orientation data in the current study, we used a rose diagram for describing data which only contain dip direction measurements, and Schmidt canvas for describing data which contain strike and dip measurements.

3.3. Analysis of Spacing

Numerous statistical tools have been developed in the recent decades specially to analyze the facture parameters such as spacing, width, length, orientation [26,51–53], and its spatial distribution (e.g., clustered, random, or uniform distribution) [26]. For spacing parameter, a coefficient of variability (Cv) has been widely described in the literature, that provides an indication of the fracture distribution [20,26,54,69]. It is given as:

$$Cv = \sigma_s/S \tag{4}$$

σ_s and S represent, respectively, the standard deviation and the mean spacing. When $Cv \approx 1$, the fractures intersected by the scanline are distributed randomly. When $Cv < 1$, fractures are more regularly spaced, while $Cv = 0$ represents uniformly spaced fractures, and $Cv > 1$ indicates fractures that are more irregularly spaced. Each Cv value can provide information about the degree of fracture clustering [26].

The heterogeneity of fracture distribution based on cumulative distribution has also been quantified using the V' statistic of [70], applied to structural geology by [20,30,31]. Indeed, the heterogeneity of distribution of the fractures and associated parameters (aperture, spacing, thickness, etc.) may be characterized from the cumulative frequency using the method described by [70]. Then, V' is defined as the measure of the heterogeneity within the scanline, which is given as:

$$V' = |D_{max}| + |D_{min}|/A \tag{5}$$

D_{max} and D_{min} are the cumulative frequency at that point if the fracture parameter was uniformly distributed [30]. D_{max} and D_{min} are positive and negative respectively. A is the total cumulative frequency of the analyzed parameter. In the present study, the V' will be used on aperture parameter in order to evaluate the strain heterogeneity. This strain heterogeneity depends on the amount of displacement (aperture or heave) and the spatial distribution of the fractures [31]. Analogical tests have been illustrated by Putz-Perrier and Sanderson [31] for two examples of the same population of extensional fractures, with different spatial arrangement, but with same strain. They obtained a fracture network uniformly distributed for the first example, and strongly clustered for the second one. Then, aperture parameter helps us to better characterize the degree of heterogeneity of every analyzed area. A perfect regular fracture distribution produces a $V' \approx 0$ as fracture sizes decrease, while the maximum heterogeneity of fractures distribution would produce a $V' = 1$ value. For further explanations, see Putz-Perrier and Sanderson [31].

3.4. Fracture Density P_{10}

The position and spacing of a set of fractures are considered whatever of their type (e.g., normal, reverse, etc.) [20]. A scanline normal to a set of fractures would intersect N fractures (number of fractures) over a length. The fracture density (P_{10}) is defined as the number of fracture intersections (N) per unit length (L) [58], following:

$$P_{10} = N/L \tag{6}$$

3.5. Cumulative Frequency Diagrams

The spacing distribution and arrangement were analyzed by using cumulative frequency. As recommended by [20,26], stick plots were used, in which the location of each fracture intersected by the scanline is mentioned. This allows to better visualize the fracture distribution. Furthermore, a plot of cumulative frequency versus distance along the scanline (from the beginning to the end) was used, in which P_{10} is proportional to the slope of the cumulative curve. The cumulative plot is normed by the maximum value, expressed in percentage (%) and it always starts at the origin (0, 0) and ends at $(d_N - d_1)$, $(N - 1)$. The parameters d_1 and d_N represents respectively the first and the last fracture. The cumulative frequency (%) against distance along the scanline provides a rapid visual comparison between datasets for each scanline and between those scanlines whatever their lengths [71].

4. Results

4.1. Description of Fracture Systems Acquired from NH Range

The studied outcrops reported in the Figure 2 were distributed homogenously within the entire central part of the NH (CBS). The structural position of each outcrop is described below.

4.1.1. Fieldwork Scanlines

The fracture network parameters were compiled from the field (Figure 2). The measurements were performed within OT1 using a scanline 1 (SL1 orientated N010) of 13.45 m length (Figure 4a), intersecting a total of 324 fractures, with a mean space of 0.04 m (Table 1). The orientation, spacing and aperture data were acquired along two scanlines with different orientations in OT2 (Figure 4b,c). The SL2a (orientated N160) of 7.5 m length intersected 109 fractures, with a mean space of 0.07 m, while the SL2b (orientated N070) of 1.64 m length intersected 37 fractures, with a mean space of 0.04 m. Within OT3, the fracture spacing and aperture were also acquired along two different orientation scanlines: SL3a and SL3b orientated N055 and N160 respectively (Figure 4d). Both scanlines intersected respectively 31 and 47 fractures, with a mean space of 0.03 and 0.06 m.

Table 1. Characteristics of the fractures acquired from each studied outcrop, along the scanlines. Outcrop 1 to 3 (without asterisk) show the characteristics of the fractures acquired directly from field scanlines. Outcrop 4 to 8 (with asterisk) show the data extracted from aligned photographs using Metashape Software v.1.6.5, with a virtual scanline. For each outcrop, number of scanlines are indicated with: Number of fractures intersected, orientation and length of scanlines, mean fractures space, density (frac/m), coefficient of variability (Cv), and V′ statistic fom [70]. The proximity to the major faults is mentioned. Mean spacing and Cv are computed with Terzaghi correction. Sgmt: segments.

Outcrop	SL	Number of Fractures	Proximity to Major Fault Segments	SL Orientation	Length (m)	Mean Spacing (m)	Density (frac/m)	Cv	V′ (95%)
1	SL1	261	10 m to SDVFZ sgmt	N010	13.45	0.04	19.4	1.2	0.29
2	SL2a	80	Crosscut by SDVFZ sgmt	N160	7.50	0.07	10.6	14.4	0.42
	SL2b	32	Crosscut by SDVFZ sgmt	N070	1.64	0.04	19.5	1.1	0.50
3	SL3a	27	40 m to SDVFZ sgmt	N055	0.85	0.03	31.8	0.7	0.32
	SL3b	38	42 m to SDVFZ sgmt	N160	4.50	0.06	8.4	1.7	0.57
4 *	SL4	28	6 m to SDVFZ sgmt	N163	0.80	0.02	42.1	0.82	/
	SL5	46	6 m to SDVFZ sgmt	N073	1.72	0.04	26.7	0.84	/
5 *	SL6	26	4 m to SDVFZ sgmt	N074	1.50	0.05	17.2	0.93	/
	SL7	31	4 m to SDVFZ sgmt	N157	0.70	0.02	44	0.89	/
6 *	SL8	171	Crosscut by SDVFZ sgmt and thrust fault	N020	109	0.64	1.55	3.22	/
7 *	SL9	188	Crosscut by SDVFZ sgmt	N132	82.22	0.5	2	3.26	/
8 *	SL10	258	Crosscut by SDVFZ sgmt and thrust fault	N154	97.50	0.4	2.66	1.67	/

4.1.2. Virtual Scanlines

The fracture variability analysis was further conducted by creating, fracture maps with a resolution of 2×10^{-4} m, OT4 and OT5. OT4 is sized 1.7 m per 1.6 m and is located in the granitic facies close to the SDVFZ segments (Figure 4e). Two scanlines: SL4 and SL5, orientated, respectively, N163 and N073, were traced perpendicular to each other in order to intersect the maximum number of fractures and avoid angular bias (e.g., [26]). In total, 28 fractures were intersected along SL4 of 0.80 m length, with a mean space of 0.02 m. Regarding the SL5, 46 fractures were intersected, with a mean space of 0.04 m (Table 1). OT5 is sized 1.8 per 1.7 m and is also located in the granitic facies, close to the SDVFZ segments (Figure 4f). Here again, two scanlines: SL6 and SL7 were traced, orientated respectively N074 and N157. They intersected, respectively, 26 and 31 fractures. The mean spacing is of 0.05 and 0.02 m for SL6 and SL7, respectively.

To perform the fracture variability study in 1D, three photogrammetric models localized only in the granitic facies were added to the present work. They are located close to the major fault segments. The fractures extracted from these models ranged from 10^{-2} to 20 m in length.

The drone photogrammetric model displayed in OT6 is sized approximately 110 per 45 m (Figure 4g); 171 fractures were traced and are intersected by the SL8. Note that OT6 presents various lithologies including granitic rocks, gneiss, gabbro, and sedimentary rocks. This may influence the spatial variability of the fractures in this area, as it will be discussed below. The drone photogrammetric model presented in OT7 is sized approximately 82 per 45 m. In this model, 188 fractures were traced and intersected by the SL9 (Figure 4h). Finally, drone photogrammetric model displayed in OT8 is sized approximatively 100 per 25 m. In total, 258 fractures were traced and intersected by SL10 (Figure 4i).

4.2. Fracture Orientation Distributions

The data acquired from scanlines in the central part of the NH show a heterogenous fracture distribution. SL1 is located near (around 10 m) a SDVFZ major segment (bold black line in Figure 2b), which acts following dextral strike-slip movement. Several lineaments are identified from the high-resolution field mapping, striking E/W (GFZ signature) to NW/SE (SDVFZ signature). From the SL1, the mean fracture orientations (μ) are striking N026, N062, N092, N130, and N171 (Figure 5a, Table 2). Fracture abundances for each fracture set are characterized by a slight dominance of the N062 and N092 fracture set with, respectively, 21% and 37% (Table 2). N026, N130, and N171 fracture sets represent, respectively, 17%, 15%, and 10% of the whole fracture set. Then, NE/SW trend appears at outcrop scale and is equivalent to E/W fractures in term of density. However, SL2a crosscut by SDVFZ major segment and close to thrust fault (Figure 2b), and a significant difference was observed in terms of fracture abundance with respectively 80 fractures in comparison with SL1 (261 fractures). The most recorded fracture sets are striking N014, N025, and N102 with, respectively, 18%, 46%, and 27%. The other fracture sets do not exceed 10% (Figure 5b, Table 2). N152 is the most dominant fracture orientation highlighted along the SL2b, representing 38%. N003 and N112 both represent 31% (Figure 5c, Table 2). Both SL3 scanlines are located far from the influence of the SDVFZ segments, thrust faults and sinistral strike-slip faults (Figure 2b). Then, SL3a, much smaller in length, intersected three fracture sets: N077, N098, and N135 with, respectively, 49%, 25%, and 26% abundances (Figure 5d, Table 2). Three fracture sets were also highlighted from the SL3b striking N040 (24%), N081 (65%), and N150 (11%). The orientations of fractures detected in SL3 scanlines are less heterogenous than in SL2 and SL1. The structural position of SL3 far from major faults (around 40 m in distance) very likely impacts the fracturing at outcrop scale.

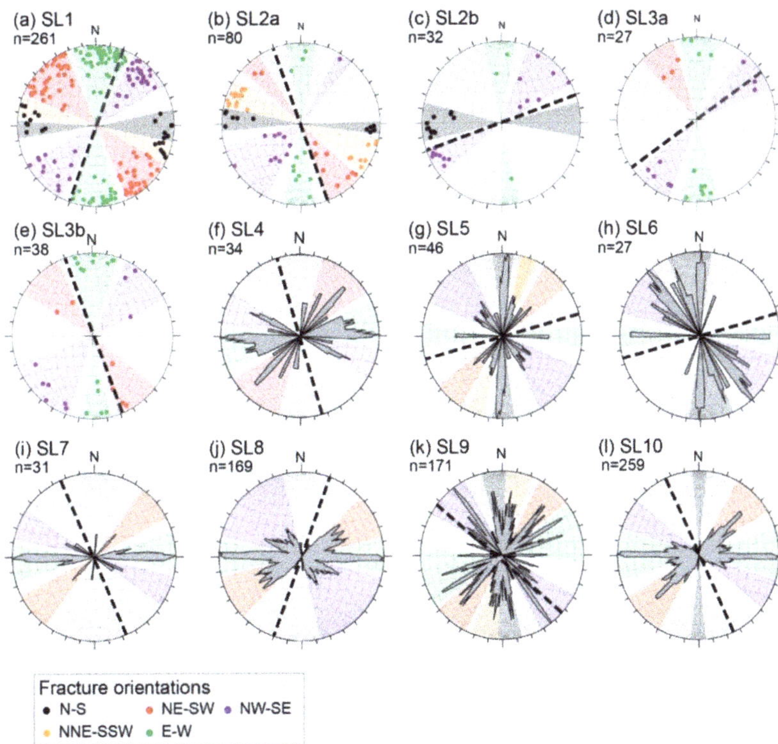

Figure 5. Fracture orientation distributions of the NH studied outcrops (**a–l**). (**a–e**) The fracture orientation distributions are represented into Schmidt canvas, lower hemisphere because they contain strike and dip measurements. (**f–l**) Fracture orientation distributions are represented by rose diagrams because they contain only dip direction measurements. Each direction rose diagram of direction is expressed with classes of 5°. The dashed lines indicate the direction of the scanline. n: Number of data. See legend in the figure for colors.

Regarding the virtual scanlines, the structural position of SL4 (OT4 orientated N163) and SL5 (orientated N073) is the same as SL1. Indeed, both scanlines are located near SDVFZ major segment (around 6 and 4 m in distance, respectively), highlighted by dominance of NW/SE and E/W structures. SL4 intersected much less fractures in comparison with the perpendicular SL5 (Table 1). SL4 displayed three fracture sets striking N034 (34%) and N086 (53%), and N123 (13%), while SL5 highlighted five fracture sets striking: N001 (39%), N018 (12%), N040 (14%), N113 (9%), and N144 (26%) (Figure 5f,g, Table 2). SL6 and SL7 orientated, respectively, N074 and N157, acquired from OT5, showed a heterogeneous fracture set, and are located near SDVFZ segments (4 m distance). Indeed, SL6 highlighted three fracture sets striking N004, N100, N161 with, respectively, 35%, 49%, and 16%, while SL7 highlighted N053, N087, and N116 fracture sets with, respectively, 13%, 81%, and 6% (Figure 5h,i, Table 2).

Table 2. Output parameters obtained from the MvM distribution fitting to fracture orientation data. Each scanline dataset was analyzed separately. Each simulation provides the number of fracture sets with their corresponding mean orientation μ (°), kappa (κ) which corresponds to the orientation variance around the mean, and weight (ω) corresponding to the proportion of each fracture set. Cv: Coefficient of variability.

Number of Fractures in Each Set	Parameters				Number of Fractures in Each Set	Parameters			
	μ (°)	k	ω (%)	Cv		μ (°)	k	ω (%)	Cv
	SL1					SL5			
44	N026	5.07	17	1.03	18	N001	44.83	39	0.79
55	N062	11.30	21	0.53	6	N018	57.7	13	0.79
97	N092	5.73	37	1.44	6	N040	31.58	13	0.86
39	N130	6.77	15	1.43	4	N113	4.95	9	0.74
26	N171	5.11	10	1.37	12	N144	15.13	26	0.95
	SL2a					SL6			
14	N014	27.98	18	1.97					
35	N025	1.19	46	1.67	10	N004	23	35	0.54
8	N050	29.67	8	1.26	13	N100	6.91	49	1.35
18	N102	4.21	27	1.12	4	N161	29.66	16	0.83
5	N143	8.91	1	1.35					
	SL2b					SL7			
10	N003	11.4	31	1.05	4	N053	3.54	13	0.88
10	N112	0.09	31	1	22	N087	46.2	81	0.83
12	N152	18	38	0.9	5	N116	33.5	6	0.84
	SL3a					SL8			
13	N077	12.06	49	1.07	54	N058	10.09	32	1.3
6	N098	14.22	25	0.6	61	N090	21.05	36	3.45
8	N135	14.1	26	0.8	54	N132	3.1	32	2.82
	SL3b					SL9			
9					34	N029	7.17	20	2.97
25	N040	4.57	24	1.23	46	N060	10.45	27	1.93
4	N081	5.05	65	1.23	15	N096	6.25	9	3.83
	N150	1.47	11	1.04	38	N132	9.85	22	2.31
					38	N176	5.38	22	2.62
	SL4					SL10			
11	N034	2.18	34	0.82	8	N004	75	3	0.81
18	N086	11.4	53	0.85	18	N040	75	7	1.68
5	N123	11	13	0.7	189	N091	43.23	73	1.7
					44	N118	35.3	17	1.35

Crosscutting the NW/SE thrust faults (in red lines, Figure 2b) and SDVFZ segment faults (in black bold lines, Figure 2b), the OT6 to OT8 displayed heterogeneous fracture orientation distribution from scanline measurements. SL8 scanline acquired within OT6 is characterized by a special geological setting. Indeed, compressional structures like thrust faults crosscut this outcrop (Figure 2b), close to the Canadian Club Wash (Figure 1c). SL7 is however crosscut by 2 SDVFZ major segments. Regarding the orientation distributions, SL8 of 109 m length, orientated N020, recorded three fracture sets, striking N058, N090, and N132 with an equivalent fracture abundance (between 32% and 36%) (Figure 5j, Table 2). However, five fracture sets were highlighted from SL9, striking N029 (20%), N060 (27%), N096 (9%), N132 (22%), and N176 (22%) (Figure 5k,l). SL10 recorded N004 (3%), N040 (7%),

N091 (73%), and N118 (17%) fracture sets. This scanline crosscut a thrust fault and several secondary SDVFZ segments (Figure 2b).

4.3. Spatial Distribution of Fractures

The spatial organization of fractures was highlighted from scanlines, whatever the fracture orientation, and is summarized in Table 1. In addition, the spatial organization was also studied following the fracture orientations in order to determine the impact of some regional directions on the NH structural organization.

From SL1, fracture density (P_{10}) and Cv are around, respectively, 19.4 frac/m and 1.2, indicating a random fracture distribution. Equivalent P_{10} and Cv were measured from SL2b with, respectively, 19.5 frac/m and 1.1, also indicating a random distribution. While SL2a recorded a P_{10} of 10.6 frac/m and Cv of 14.4, indicating a less abundant and highly clustered fracture system. Values of P_{10} of 31.8 and of 8.4 frac/m, Cv of 0.7 and 1.7 were highlighted from SL3a and SL3b, respectively, (Table 1), indicating very abundant fractures with a random distribution.

Three different spacing organizations were identified from fracture sets in SL1 (Table 2):

- Fractures distributed randomly in fracture set N026 with Cv = 1.03;
- Fractures uniformly spaced in fracture set N062 with Cv = 0.53;
- Fractures more irregularly spaced or clustered in fracture sets N092, N0130, and N171 with, respectively, Cv = 1.44, 1.43, and 1.37.

The fracture distribution presented in Figure 6 shows the layout of fractures and their location along each scanline. For SL1, the fracture distribution is showed using a cumulative frequency plot against the position of fractures along the scanline for each fracture set (Figure 6a). Two major trends are distinguished: (1) Regular fracture distribution from the beginning to the end of the scanline especially for fracture sets N026 and N171 (respectively in black and orange colors with diamond and square symbols in Figure 6a); and (2) a regular fracture distribution at the beginning, then clustered (increased frequency), followed by another regular distribution at the end for fracture sets N062, N092, and N130 (respectively in red, green and purple colors in Figure 6a). Three fracture clusters having a slope higher than 2 are highlighted in N062 fracture set, around 5.5, 8, and 10 m (Figure 6a), highlighting a clustered spacing. Note that, this slope corresponds to the shape of the cumulative frequency compared to the uniform distribution highlighted in the plots below (Figure 6). The cumulative frequency plot highlighted one fracture cluster around 12–13 m in the N092 fracture set. Within the N130 fracture set, two fracture clusters are characterized around 3–4 and 10 m.

SL1 is in moderate deformation zone, affected mainly by SDVFZ major segment and some E/W fault segments. The position of SL1 close to SDVFZ major segment, trending NW/SE creates these irregularities in fracture distribution. Indeed, the NW/SE (N130) fracture set is one of the most clustered fracture sets. The intermediate (in term of length) E/W segments shown in the Figure 2b have a strong impact on N092 fracture set distribution, making it the more clustered distribution with Cv = 1.44.

A high fracture density is observed in the central part of the SL2a profile (Figure 6b). Indeed, five fracture clusters were identified in which the fracture density is increased around 3–4 m in N143 fracture set (purple color in Figure 6b), 4.5 m in N014 and N025 fracture sets (black and orange colors, respectively, for N014 and N025 in Figure 6b), and around 6 m in N014 and N050 fracture sets, displayed, respectively, in black and red colors for N014 and N050 in Figure 6b. The SDVFZ crosscuts the SL2a profile and introduces some irregularities in fracture distribution since the beginning, mainly in N143 and N014 fracture sets, highlighting then an anisotropy following some directions.

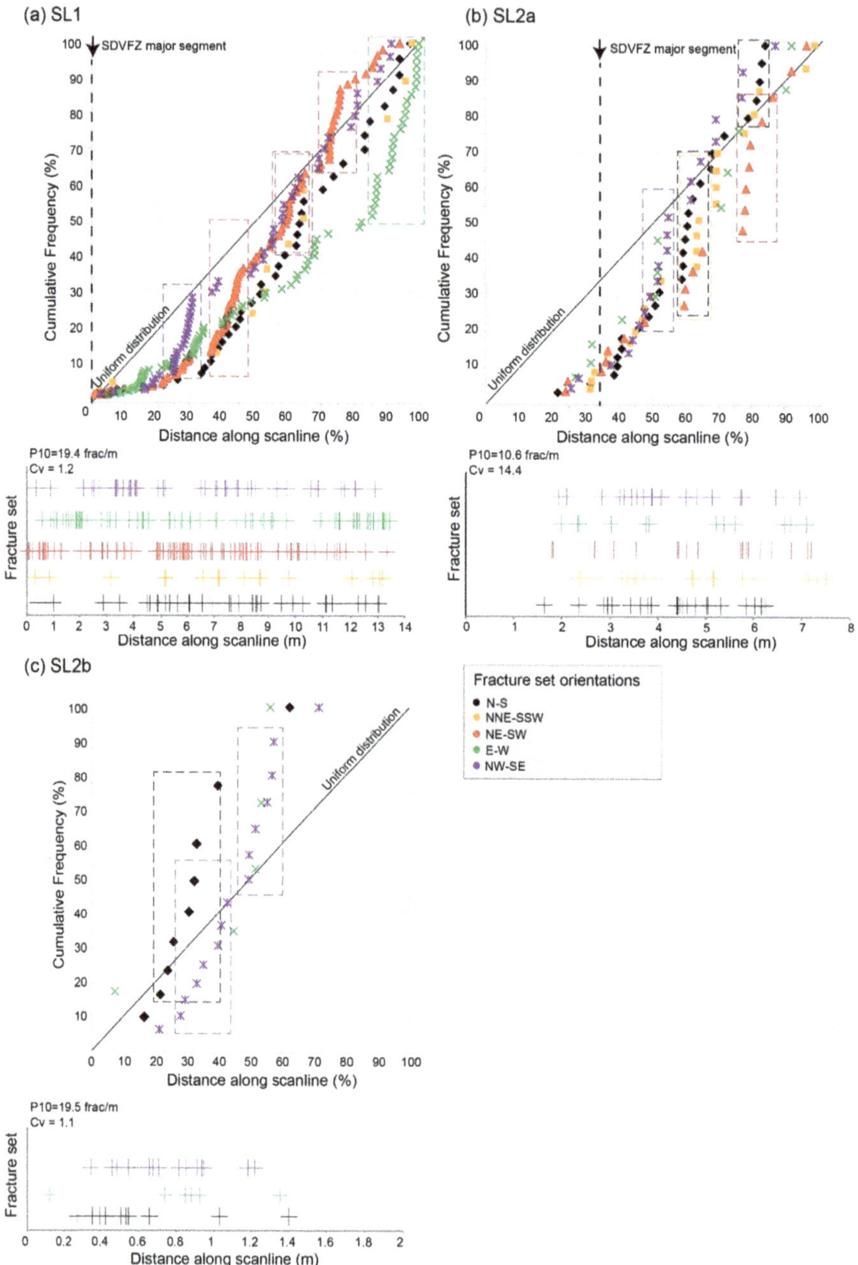

Figure 6. Cumulative frequency diagram along each scanline and following each fracture set (in colors). Plot of cumulative frequency expressed in % versus distance along scanline for: (**a**) SL1, (**b**) SL2a, and (**c**) SL2b. Dashed lines indicate a potential cluster for each fracture distribution, indicating a rapid increase in the number of fractures with slope threshold > 2. The diagonal line in each plot of cumulative frequency defines a uniform or regular distribution. Visual location of fractures expressed in stick plots for each fracture set distribution, highlighting the fracture position along the scanline, including the fracture density (P_{10}) and the coefficient of variation (Cv) values. The structural position is done in each diagram. See the color legend for the orientation of fracture sets.

Fractures intersected by the SL2b are distributed randomly whatever the direction with Cv close to 1 (Table 2). N098 and N0135 fracture sets intersected by SL3a displayed a Cv of 0.6 and 0.8, indicating a uniform spacing organization, while Cv of 1.07 measured from the N077 fracture set indicates a clustered organization. However, the beginning and the end of the SL2a profile showed a regular fracture distribution. Along SL2b, a random fracture distribution was observed whatever the fracture orientation with Cv around 1 (Figure 6c). Three fracture clusters are, however, detected within N003, N112, and N152 fracture sets (displayed respectively in black, green, and purple colors), respectively, around 0.4–0.6, 0.8–1, and 1.2 m (Figure 6c).

A regular fracture distribution was highlighted within fracture orientations N098 and N135 along SL3a profile, with Cv of 0.6 and 0.8, respectively (Figure 7a). For N077 fracture set (red color with triangle symbol), a clustered distribution was shown with two fracture clusters around 0.3–0.4 and 0.7–0.8 m. However, along SL3b scanline profile, one fracture cluster which corresponds to an increase in fracture density was shown around 2 and 2.25 m, respectively, in N040 and N150 fracture sets (respectively in red and purple colors in Figure 7b). Only three fracture clusters were identified around 1.5, 2.5–3, and 4 m in N081 fracture set (green color). A regular fracture distribution was characterized outside these fracture clusters. Located away from major fault segments, around 40 m distance, the SL3a and SL3b profiles displayed a very poor fracture system organization and fracture density. In addition, these profiles are in a moderate deformation zone, which can impact drastically the fracture distribution and density. Boudinage structures and brittle shearing are not observed within this area.

OT4 and OT5 are in a moderate deformation zone, affected mainly by SDVFZ major segment and some E/W fault segments. From virtual scanlines, three fracture sets were recorded in SL4. The N034 fracture set highlighted one fracture cluster around 0.5–0.6 m (red rectangle in Figure 8a). However, four fracture clusters were highlighted at the beginning and at the end of the SL4 profile within N086 fracture set (green color with cross mark symbol in Figure 8a), around 0.01, 0.25, 0.65, and 0.7–0.8 m. The rest of the intervals are characterized by a regular fracture distribution. The SL5 scanline recorded five fracture sets and are characterized by a regular fracture distribution whatever the fracture set (Figure 8b). An increased fracture density is identified within the N001 fracture set (black color) around 0.7 and 1.2 m.

Fractures detected within SL6 are mainly distributed regularly for N004 and N161 fracture sets (displayed respectively in black and purple colors in Figure 8c), with Cv of 0.88 and 0.84, respectively. Cv of 1.35 was, however, computed from the N090 fracture orientation (green color with cross mark symbol, Figure 8c), indicating highly clustered fractures. Then, one fracture cluster was identified around 0.5–0.6 m (Figure 8c). The three fracture sets striking N053, N087, and N132 are recorded mainly with fractures regularly distributed along the SL7, with Cv ranging from 0.83 to 0.84 (Figure 8d). Two fracture clusters are then identified within the N087 fracture set (green color with cross mark symbol in Figure 8d) around 0.5 and 0.6–0.7 m.

The influence of the regional directions was observed on E/W fractures orientation, with a higher density and clustering within this fracture set (Table 2). However, the NW/SE direction is less expressed in comparison with SL1 and SL2. The small length of OT4 and OT5 scanlines may influence the result and then introduce some bias to the analysis.

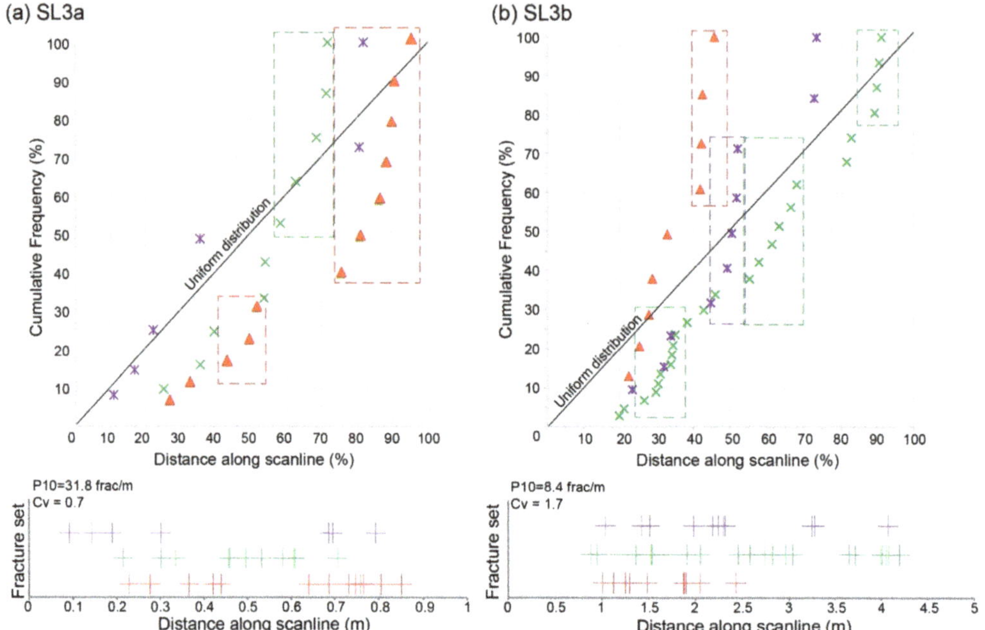

Figure 7. Cumulative frequency diagram along each scanline and following each fracture set (in colors). Plot of cumulative frequency expressed in % versus distance along scanline for: (**a**) SL3, and (**b**) SL3a. Dashed lines indicate a potential cluster for each fracture distribution, indicating a rapid increase in the number of fractures with slope threshold > 2. The diagonal line in each plot of cumulative frequency defines a uniform or regular distribution. Visual location of fractures expressed in stick plots for each fracture set distribution, highlighting the fracture position along the scanline, including the fracture density (P_{10}) and the coefficient of variation (Cv) values. See the color legend in Figure 6.

Near the Canadian wash (see Figure 1c for location), OT6 is located close to the highly deformed zone (for location, see Figure 2). Two major faults crosscut this outcrop: SDVFZ major fault segment and thrust fault. SL8 intersected three fracture sets with a high fracture density interval (Figure 9a). Three fracture clusters were identified in N058 fracture sets, respectively, around 10–20, 40, and 75 m (red color with triangle symbol legend, Figure 9a). The fracture clusters in N090 fracture set (green color with cross mark symbol, Figure 9a) are around 40, 50–60, 75, and 80 m. Finally, within N132 fracture set (purple color with strikethrough cross, Figure 9a), two fracture clusters comprised between 10–15 and 50–70 m were identified. We are fully aware that the canyon presents a complex structuration due to various lithologies and the overprinting faults.

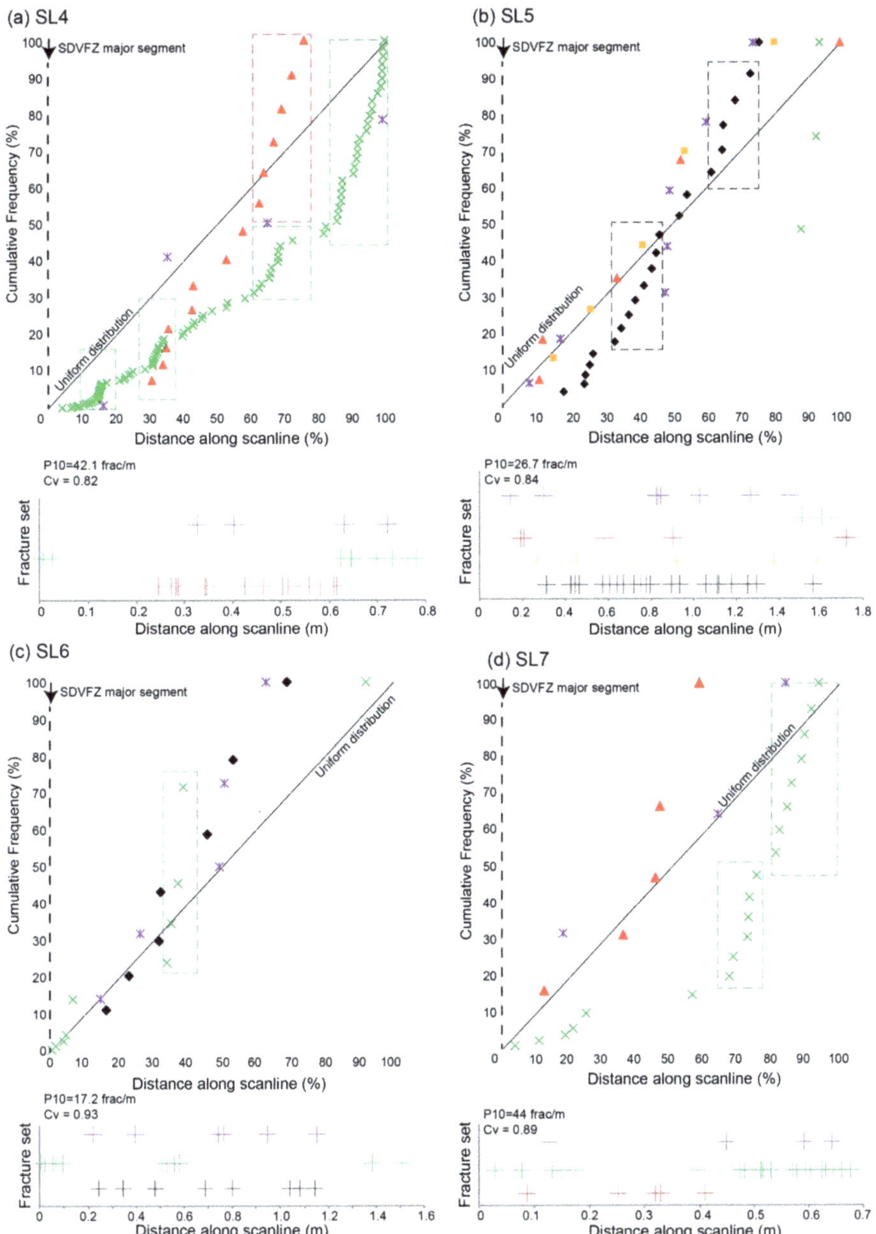

Figure 8. Cumulative frequency diagram along each scanline and following each fracture set (in colors). Plot of cumulative frequency expressed in % versus distance along scanline for: (**a**) SL4, (**b**) SL5, (**c**) SL6, and (**d**) SL6. Dashed lines indicate a potential cluster for each fracture distribution, indicating a rapid increase in the number of fractures with slope threshold > 2. The diagonal line in each plot of cumulative frequency defines a uniform or regular distribution. Visual location of fractures expressed in stick plots for each fracture set distribution, indicating the fracture position along the scanline, including the fracture density (P_{10}) and the coefficient of variation (Cv) values. The structural position is done in each diagram. See the color legend in Figure 6.

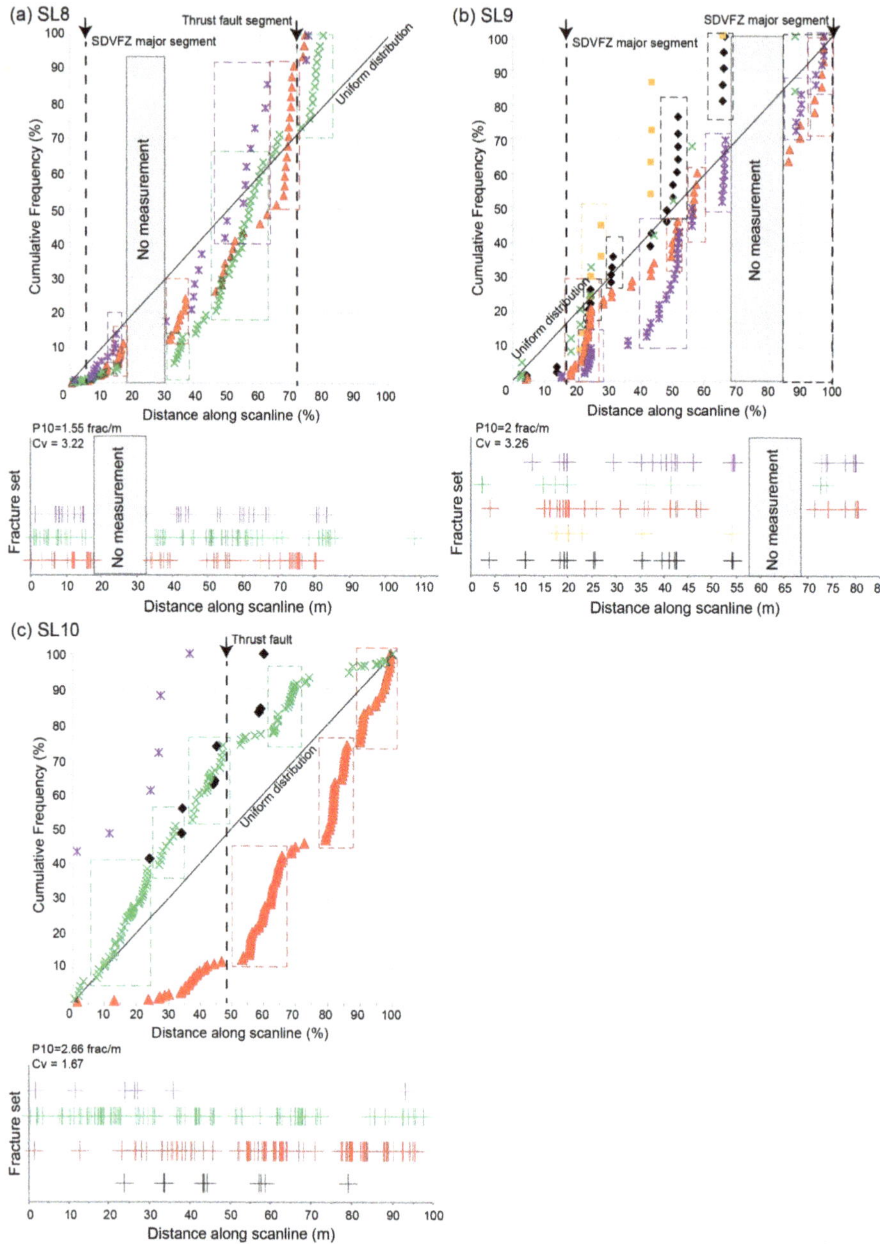

Figure 9. Cumulative frequency diagram along each scanline and following each fracture set (in colors). Plot of cumulative frequency expressed in % versus distance along scanline for: (**a**) SL8, (**b**) SL9, and (**c**) SL10. Dashed lines indicate a potential cluster for each fracture distribution, indicating a rapid increase in the number of fractures with slope threshold > 2. The diagonal line in each plot of cumulative frequency defines a uniform or regular distribution. Visual location of fractures expressed in stick plots for each fracture set distribution, showing the fracture position along the scanline, including the fracture density (P_{10}) and the coefficient of variation (Cv) values. See the color legend in Figure 6.

Additionally located in the highly deformed zone, SL9 is crosscut by two SDVFZ major segments. Several fracture clusters are observed as following (Figure 9b):

- N029 fracture set with one fracture cluster identified around 20 m;
- N060 fracture set with four fracture clusters characterized around 20, 42, 47, and 80 m;
- N096 fracture set with one fracture cluster identified around 15–20 m;
- N132 fracture set with five fracture clusters identified around 15–20, 35, 40–45, 55, and 80 m;
- N176 fracture set with four fracture clusters identified around 20, 25, 35–40, and 55 m.

SL10 scanline crosscuts several secondary SDVFZ segments and a thrust fault, in moderate deformation domain. A regular distribution of fractures was encountered in N004 fracture set (black color with diamond symbol legend in Figure 9c). Three fracture clusters were detected around 30–40, 55–65, and 80–90 m in N040 fracture set (data in red with triangle symbol legend, Figure 9c), while in N091 fracture set, the fracture clusters are around 10–20, 28, 40, 60–70, and 90 m (green color with cross mark symbol, Figure 9c). The E/W and NE/SW fractures orientations are more expressed in terms of fracture density which can be enhanced by the position of the major faults in this area.

4.4. Fracture Aperture Distribution

The heterogeneity of the fractures distribution and associated parameters (e.g., aperture, spacing, etc.) can be determined using the Kuiper's method [20,30,31,70]. In this study, the V' is used to better precise the degree of heterogeneity of aperture distribution along scanlines. V' values are summarized in Table 1 for only data acquired from the fieldwork. Confidence interval of 95% was used to compile V'. Then, a cumulative aperture was plotted against fracture location along the scanline (Figure 10). Application of the Kuiper test shows $V' = 0.29$ for SL1, meaning that the apertures are distributed randomly (Figure 10a). For SL2a and SL2b, V' is, respectively, 0.42 and 0.50 despite the difference in term of scanline length and orientation (Table 1, Figure 10b,c). These values indicate that the apertures are distributed randomly along each scanline. The same result of aperture distribution is shown along the SL3a and SL3b (Figure 10d,e), with V' of 0.32 and 0.57, respectively. The fracture aperture distribution is more heterogeneous in SL3b and SL2b, than in SL1, SL2a, and SL3a.

The possibility of any correlation between the fracture orientation and the aperture was tested using a plot of aperture against fracture set for each scanline (Figure 11). The mean fracture aperture in SL1 is about 2.5 mm and, therefore, lower than in fractures of SL2a and SL2b at, respectively, 2 and 1.7 mm. In SL1, the widest fracture apertures are around 10 mm, and occurred in fractures orientated N092 (Figure 11a). The most common observed fractures in the field were affected by mineralization of carbonates mostly (Figure 12). The widest fracture apertures detected in fracture orientation N092 are classified as fully sealed by carbonates (Figure 12a). In SL2a, the maximum fracture aperture is also 10 mm and belongs to the N050, N150, and N170 striking sets (Figure 11b). The widest fracture apertures are classified as fully sealed by carbonates and oxides (Figure 12a,b). In SL2b, the maximum fracture aperture does not exceed 8 mm, and occurred in fractures striking N009 and N090 (Figure 11c) and classified as fully sealed by carbonates and oxides.

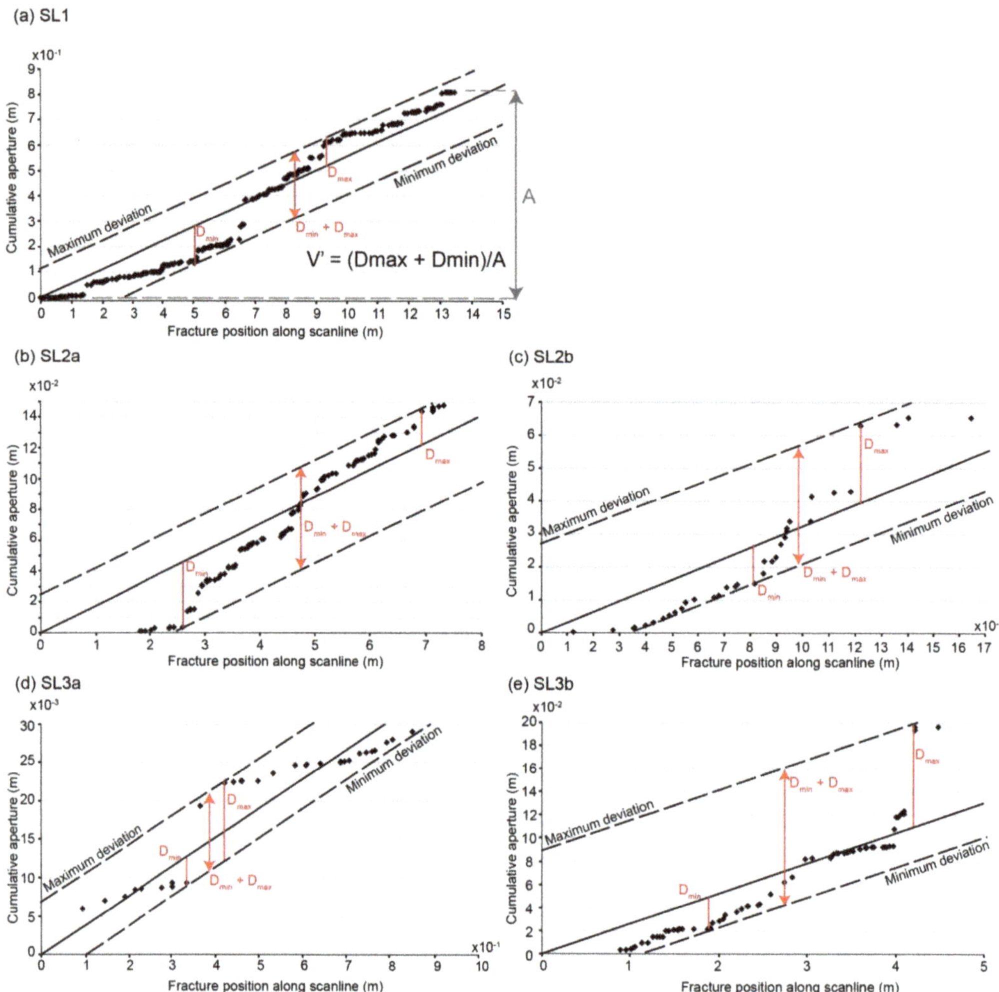

Figure 10. Plot of the sum of all apertures (cumulative fracture aperture) intersected along a scanline to show V′ for (**a**) SL1, (**b**) SL2a, (**c**) SL2b, (**d**) SL3a, and (**e**) SL3b. The diagonal of each plot (black line) represents the uniform distribution linking the origin to the cumulative aperture to the end of the scanline (for further explanations, see [20,30]). D_{max} and D_{min} correspond to the maximum and minimum difference between the cumulative aperture and the uniform strain line, respectively. V′ quantifies the heterogeneity of the strain distribution, varying from 0 (uniform distribution) to 1 (maximum possible hetrogeneity).

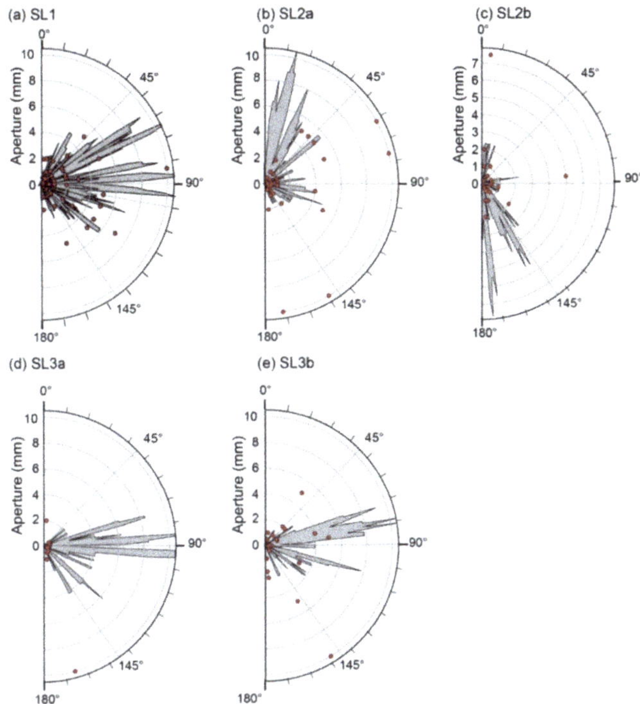

Figure 11. Half circular diagram showing the relation between fracture strike and apertures. (**a**) SL1, (**b**) SL2a, (**c**) SL2b, (**d**) SL3a, and (**e**) SL3b. Apertures from SL4 to SL10 are not available. Rose diagram scaled for mean aperture with classes of 5°.

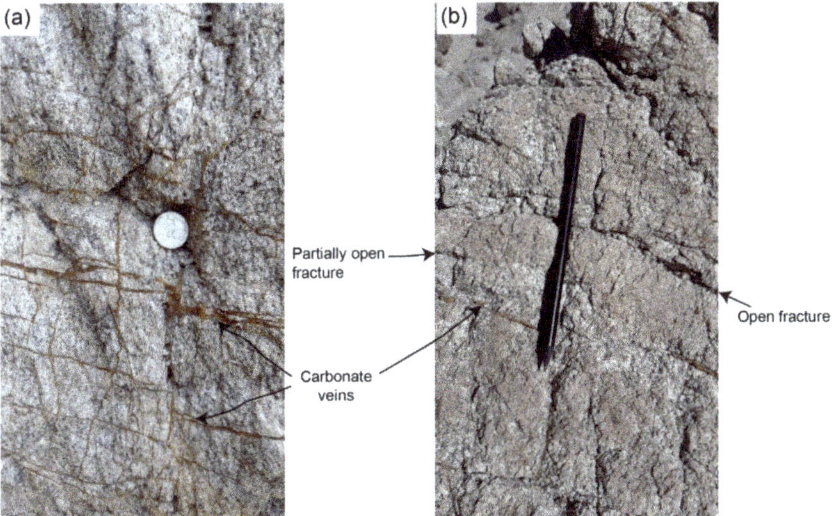

Figure 12. Illustration of the observed fractures observed in the field, mostly affected by mineralization. (**a**) Picture taken near OT6 illustrating veins fully sealed by carbonates. (**b**) Picture taken near the Canadian Wash illustrating open and partially open fractures. The carbonates are ubiquitous following SDVFZ trend (i.e., NW/SE). The fractures striking E/W have been observed with a wider aperture and were filled by barite and oxides.

The mean fracture aperture in SL3a and SL3b is about 0.9 and 2.8 mm, respectively. The maximum fracture apertures are around 10 mm. They occur in fractures striking N162 and N150 for SL3a and SL3b, respectively (Figure 11d,e). The maximum fracture apertures identified in SL3a occurred in fractures classified as partially open (Figure 12b), while in SL3b, the maximum fracture apertures occurred in fractures classified as fully sealed by carbonates and oxides. The fracture aperture distribution are more heterogenous in SL3b than in SL3a, in which apertures are generally lower than 2 mm.

As a summary, the distribution of the apertures in the central part of NH is ruled by a random behavior. Regarding the relationship between apertures and the orientations, an anisotropy was observed following E/W trend, which presents fractures with a wider aperture.

5. Discussion

5.1. Representativness of the Fieldwork and Virtual Scanlines

The present study covers a sampled fracture network from fieldwork and virtual scanlines obtained from photogrammetry. The moderate linear density compiled from the virtual scanlines, especially in SL8, SL9, and SL10, does not exceed 3 frac/m. The highest linear density is detected in SL4 and SL7 with, respectively, 42.1 and 44 frac/m. From fieldwork scanlines, the linear densities ranged between 8.4 frac/m in SL3b and 31.8 frac/m in SL3a.

Regarding the sampling resolution, in case of virtual scanlines, the shorter fracture measurement is around the decimeter to meter and depends on the photogrammetric image resolution, while in the case of fieldwork scanlines, the measurement resolution is around the millimeter. The difference in resolution strongly impacts the fracture density and its representativeness. The low density of the fractures detected from virtual scanlines might not be fully considered as representative of the whole granitic rock. However, it is very useful in zones with difficult access.

In the case of sampling in 1D, a scanline perpendicular to the fractures will detect a maximum of sub-set of all fractures. Then, the probability that a fracture is detected within a scanline is proportional to the fracture surface area [72]. In basement rocks, fractures that span the sampling study area limits whatever its lengths and heights will reflect 3D sampling. The same case is observed in sedimentary rocks when fractures span the layers and the limits of the studied area [72].

5.2. Consistency of the Recorded Fracture Sets

The main dominant orientations recorded from the different scanlines displayed in the central part of the NH highlight the fact that NH geometry is controlled by NE/SW, E/W, and NW/SE trends (Figure 13). From fieldwork scanlines, the E/W and NW/SE fracture sets are the most consistent orientations whatever the scanline. The N/S, NNE/SSW, and NE/SW fracture sets are only recorded in the SL1 and SL2a fieldwork scanlines (Figure 13). SL2b, SL3a, and SL3b detected mostly E/W and NW/SE directions with a heterogeneous recording (Figure 13). The E/W is the most dominant trend within SL3b, while the NW/SE trend dominates within SL2b, and constitutes the second and third dominant fracture sets within SL3a and SL3b, respectively. The variability in fracture orientation may be related to the influence of the major faults. Indeed, as showed in Figure 2, the position of SL1 and SL2a close to the SDVFZ segment interfering with the thrust fault induces an additional complexity in the structural signature. As mentioned before, a clear overprinting has been recognized between primary SDVFZ and syn-kinematic dextral strike slip faults related to the transcurrent movements followed in time by compressive structures. Thrusting is postdating the activity of the SDVFZ, and it is tempting to relate the thrust structures to activity along the frontal termination of GFZ which acts with a sinistral strike slip movement.

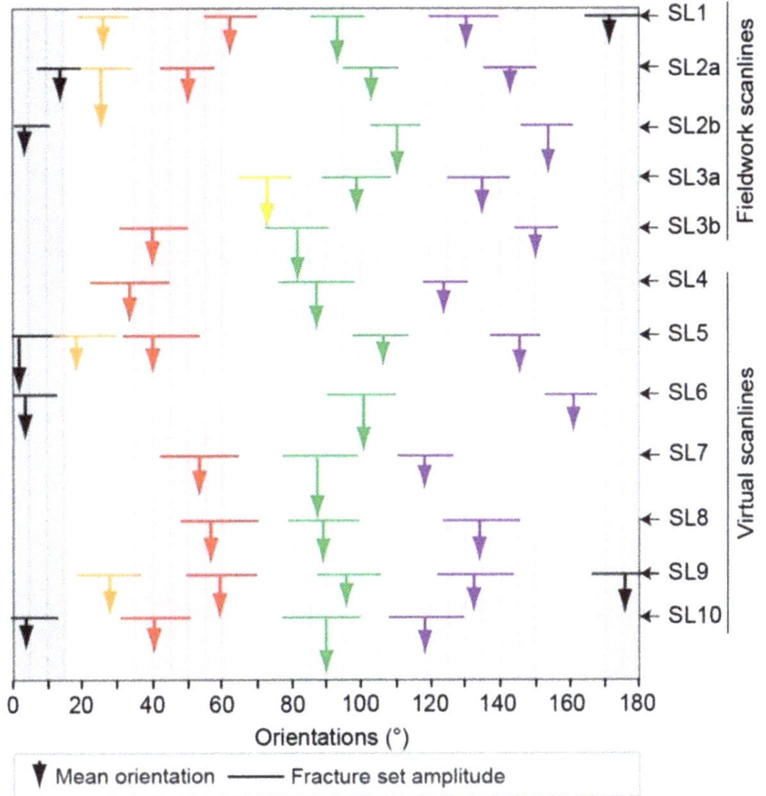

Figure 13. Plot displaying the mean orientation with arrows of each collected fracture set from NH scanlines. Each arrow was characterized by its own length, which corresponds to the fracture set abundance. The horizontal line above each arrow corresponds to the standard deviation of each fracture set, using an interval of confidence of 75% (see [68] for more explanations). Each color arrow corresponds to a fracture set. Orange arrow: NNE/SSW fracture set, red arrow: NE/SW fracture set, green arrow: E/W fracture set, yellow arrow: NNW/SSE fracture set, purple arrow: NW/SE fracture set, and black arrow: N/S fracture set.

In the southern end of the NH central area, E/W trending structures showing evidences of compression again possibly related to GFZ activity are even clearer.

Regarding the fracture orientations highlighted from the virtual scanlines, a reproducible and consistent NE/SW, E/W, and NW/SE fracture sets are encountered (Figure 13). The GFZ is confirmed as a major fault which influences the NH geometry and induces the E/W fractures which are more expressed in SL4, SL6, SL7, SL8, and SL10. According to Chabani et al. [25], who describe the spatial organization of a NH fracture network at different scales from 2D maps and previous works [47–49,73], the compression episode at the southeastern end of the NH and at the front of the Avawatz mountains plays a key role in the increase of the fracture intensity in the central domain of NH, and can explain the consistency of the E/W direction in the whole internal NH domain.

The NW/SE direction is the second most reproducible direction recorded along the internal part of NH (Figure 13), related to SDVFZ activity, which acts with a dextral movement. This direction is more expressed especially near the Canadian wash, where the deformation is important, following a second order scale as described by Chabani et al. [25].

The N/S and NE/SW fracture sets are also highlighted from the virtual scanlines but are less consistent than the E/W and NW/SE directions. These directions are heterogeneous with more than 25° (dispersion of the mean orientation) of variability in case of the NE/SW fracture set (Figure 13). These trends are highly dependent on the scale of observation and have no influence on the NH geometry [25].

As mentioned before, a recent study by Chabani et al., [25] revealed the same influence of the E/W and NW/SE fracture sets in the central part of NH. Indeed, E/W and NW/SE trends control the NH geometry within the third and fourth order scale. In addition, the major faults played a key role in the fracture density increase following different areas inside the central part of NH. The gradient of deformation shown in Klee et al. [34] and Chabani et al. [25] induced fracturing notably near OT6 and OT7, in which the strain gradient is the highest.

5.3. Clustering of Fractures in Central Part of NH

Figure 14a highlighted the Cv for the relative spaces between adjacent fractures for each studied scanline, by plotting the values against the overall fracture intensity. This correlation helps to better compare the Cv values and assess the spatial organization of the fractures. A large range of behaviors were observed. A regular spacing distribution (Cv < 1) was highlighted especially from SL3a, SL4, SL5, SL6, and SL7 (Figure 14a, Table 1). However, a more clustered distributions (Cv > 1) was observed in the SL1, SL2a, SL2b, SL8, SL9, and SL10. This large variation is probably related to the prominent influence of the thrust faults and the SDVFZ major segments close to SL8, SL9, and SL10, and only the influence of the SDVFZ major segments in the case of SL1 and SL2 profiles. However, the location in moderate deformation zone and away from major faults of SL3 scanlines highlight the absence of any organization in the fracture systems. They confirm the key role of the structural heritage (regional directions) in fracture patterns at outcrop scale. Recent study by Franklin et al. [74] highlighted and confirmed that the clustered spacing distribution is pronounced at the outcrop scale and is related to the influence of the major faulting.

From all the analyzed datasets, note that the outcrops with the greatest numbers of fractures are most usually characterized by random patterns. This observed relationship could be interpreted as the result of the increasing spatial heterogeneity of the fracture pattern with increasing strain. In that case, OT6 to OT8 highlighted that the greatest number of fractures, the fracture density, and the strong clustering within fracture sets can be related to the intensity of deformation in this area and the proximity to the SDVFZ major segments and thrust faults. Indeed, the new geological map displayed in Figure 2b highlighted a stacking of the Crystal Spring series, intruded by the Mesozoic granite. In addition, the thickness of Crystal Spring series was reduced, as they were dragged, and stretched laterally against the granite due to the SDVFZ activity, especially in OT6 to OT8 areas. This intense deformation affected the entire NH range with local evidences of extreme shearing. However, we are aware that the presence of several different lithologies with varying competences within OT6 may bias the fracture distribution and its interpretation. In addition, the structural architecture of this area is the most complex within NH range as it includes overprinting deformation phases. A future publication dedicated to these overprinting issues is in preparation.

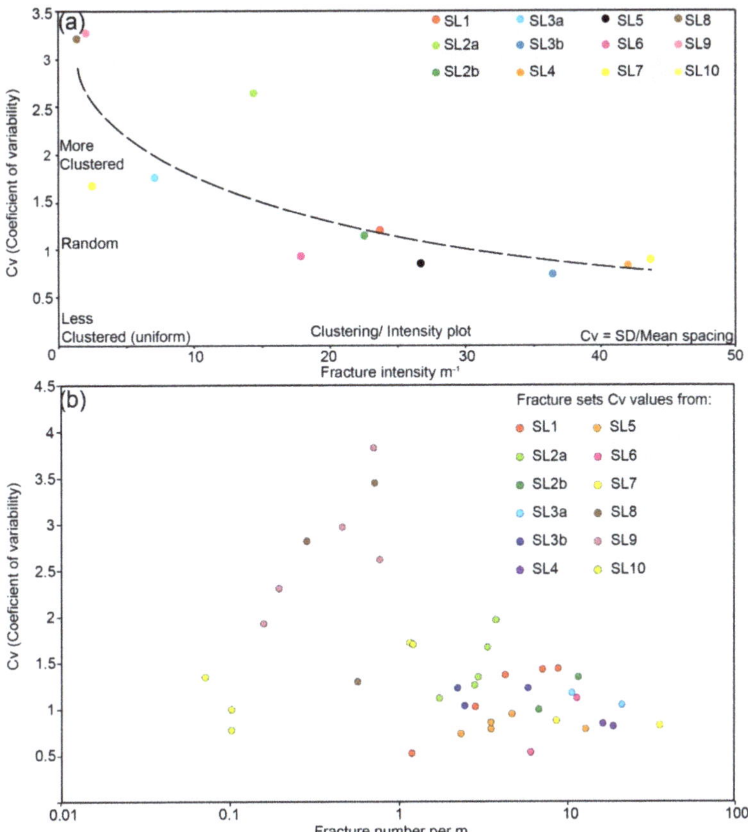

Figure 14. (**a**) Coefficient of variability (Cv) plotted against fracture intensity (P_{10}) of the NH studied outcrops. SD: Standard of deviation of spacing measured from each scanline; mean: Mean spacing measured from each scanline. (**b**) Coefficient of variability (Cv) plotted against fracture number normalized by the scanline length.

Another approach approved by Hooker et al. [30] consists of cross-examining the low number N of datasets. Cv values are commonly indistinguishable from random for low-N sets. Then, Cv increases with increasing fracture abundance (Figure 14b). This observation reflects an increase in fracture clustering with fracture abundance only in case of fracture sets detected from virtual scanlines (SL8, SL9, and SL10, Figure 14b). In case of fractures detected from fieldwork scanlines, the trend is more unclear since the Cv evolution is still comprised between 0.5 and 1.5, whatever the fracture abundance (Figure 14b). Once again, the proximity to major faults in case of SL8, SL9, and SL10 may play a key role in the fracture patterns growth. For SL1 to SL7, the fracture organization may be perturbed by the abundance of the E/W facture orientation, which can increase drastically the fracture abundance following E/W fracture set trend. This may be related to the scanline orientation which cannot intersect the rest of the fracture sets. From SL1 to SL7, fracture set trending E/W is more expressed than the rest of the sets except for SL2a, SL2b, and SL5 (Table 2). Then, the complexity in fracture orientation recording has a strong impact on the spacing variability and fracture patterns evolution, evolving a random state. In case of SL8, SL9, and SL10, the fractures are distributed homogenously following the orientations outside the fracture clusters (see Table 2 for proportion).

Regarding the clustering, relative spacing analysis shows that fracture clusters have been detected in the whole profiles, except for SL3 profiles, in which very few clusters were detected (Figure 8). Clustering depends on the fracture orientation. Two major factors are responsible for this clustering:

- Crosscutting or location close to major faults segments: This observation is supported essentially by OT1, OT6, OT7 and OT8 fracture distribution, in which SDVFZ and a thrust fault proximity enhance the considerably clustering and fracture density of each fracture set.
- The deformation gradient: Based on fieldwork observations, the most deformed zone is located close to OT6 and OT7 and strongly impacts the fracture distribution. Indeed, the fracture density is much higher in these areas, and the fracture patterns is arranged into clusters following E/W and NW/SE directions.

The regular spacing distribution exist whatever the analyzed direction. This observation is more highlighted within SL4, SL5, and SL7 profiles (Figure 8a,b,d). In these cases, clustering depends on the orientation. Few fracture clusters were detected within the E/W direction (data in green color with cross mark symbol, Figure 8).

However, the lower fracture density compiled from SL8, SL9, and SL10, due to the sampling bias relative to the virtual scanlines, nuances the real influence of this clustering on the anisotropy of the possible flow.

To summarize, clustering is mostly observed for scanlines orthogonal to the observed fractures close to the major faults, such as SDVFZ and thrust faults, except for SL3 profiles which can be biased due to their small length and their location away from major faults (Figure 14a). In contrast, scanlines sub-parallel to the main trend of the major faults have a rather random distribution pattern (Figure 14a).

5.4. Conceptual Model of NH Paleo Geothermal Analogue

In fractured reservoirs, the main challenge is to understand the fracture variability from only 1D data obtained from wells. The 3D fracture networks commonly show a strong spatial heterogeneity, related in several cases to geological features, such as faults, folds, stress fields, or lithological trends, which impact large scale fracture networks [75]. The central part NH studied here is an appropriate case study to test these assumptions, such as size distribution (e.g., length, spacing, aperture etc.), fracture trends record, and the relationship between them. In the present study, the 1D measurements which can be assimilated to synthetic well data revealed the same density results as those obtained by 2D measurements in the central part of NH. Note that, the 2D study has been published in previous work [25] for the entire NH range. This previous study concluded that the center part of NH is ruled by fracture networks dominated by E/W (GFZ trend) and NW/SE (SDVFZ trend) directions. This result is also observed in this present study based on 1D measurements. Indeed, the main dominant directions (e.g., E/W and NW/SE) are also recorded at outcrop scale, meaning that independently of well measurement, whatever any sampling bias and statistic uncertainties resulting from the data measurements, the well (in case of a reservoir) or the scanline (in case of an analogue) is representative of the heterogeneity at outcrop scale and reservoir scale.

The main directions that follow the regional trend and then control the NH geometry are commonly characterized by clustered fractures. Close to the major faults, the fracture network arrangement is more clustered, especially in the high strain zone near OT6, OT7, and OT8. Then, the fracture system is marked by a strong clustering within its organization following most of the recorded fracture sets, except for NW/SE and N/S fracture sets of OT8. This clustering impacts the reservoir behavior in that fluid circulation is influenced by the role of the major faults. Outside these major faults and their associated fracture clusters, a secondary fracture network was characterized by a random distribution, which plays a key role in the fracture connectivity leading to the fluid supply toward the fractured zone.

Evidences of fluid circulations have been identified by Klee et al. (2021a) [37] and Klee et al., 2021b [8] through (1) the alteration processes that occurred in the whole granite body (propylitic and argillic), and (2) the fracture infills. Among the different natures of fractures infills that have been identified in the granite, carbonates are omnipresent [8]. They mainly filled the fracture following the SDVFZ direction (i.e., NW/SE), which is the main fracture orientation in the area. However, as presented above, E/W-oriented fractures have also been observed with a wider aperture. These fractures, related to the activity of the GFZ [25], are often filled with barite and oxides, especially at the southern rear part of the range, close to the Avawatz Mountains. This difference of infills according to two different fracture sets shows two different episodes of fluid circulation through the granite body.

Furthermore, fluid–rock interactions investigations through the NH area, by collecting samples according to profiles approaching major fault segments, have been conducted in recent previous works [8,37]. These major segments consist of the SDVFZ segments, the same ones considered in this present study, since we considered the same faults interpretation in both works based on fieldwork mapping. Porosity measurements were realized on five samples away from fracture zones and on three samples in the vicinity of fracture zones. It aimed at evaluating a possible correlation with the amount of alteration. The authors showed that the porosity increases with proximity to fracture zones. It is known that alteration can create some porosity by the dissolution of primary minerals like plagioclase and biotite. However, these studies have shown that porosity does not always correlates with the amount of alteration. Indeed, microfracturing also creates porosity.

The argillic alteration related to hydrothermal fluid circulations through the fracture network affects dominantly the NH granite [8,37]. During this fluid–rock interaction process, plagioclase and biotite are replaced by secondary minerals, such as illite and kaolinite, which crystallize between 120 and 200 °C [8]. Such temperatures are also confirmed by the Kübler Index measured from the illite crystallinity. To know more precisely the fluid temperature that has circulated, fluid inclusions measurements are required, which will be published in the future work.

As introduced before, NH is considered as an exhumed analogue for Soultz-sous-Forêts and Rittershoffen geothermal reservoirs. Both sites display a similar alteration and hydrothermal activity [8,76]. For example, the reservoir volume in Soultz-sous-Forêts is around 12 km^3 [77], and is crosscut by regional faults following the second and third orders scale according to Morrelato et al. [78] classification. The order of faults magnitude highlighted in the central part of NH is the same as that detected in the subsurface reservoirs at Soultz-sous-Forêts and Rittershoffen. Indeed, the second and third order scales control the geometry of the internal part of NH [25]. Then, the reservoir volume in case of NH paleo geothermal analogue can be equivalent to that at those of Soultz-sous-Forêts reservoir. Furthermore, during the Rittershoffen reservoir stimulation, a repercussion has been recorded in the Soultz-sous-Forêts reservoir and confirmed the kilometer faults orders (personal communication).

To further characterize a geothermal reservoir, it is important to determine the geothermal gradient and temperature at the targeted exploitation depth. However, the geothermal gradient in the NH context is difficult to highlight. Some additional data likely provided by fluid inclusion micro-thermometry and stable isotope analyses are needed to determine this gradient. In addition, it would be important to take in consideration the present-day stress field according to their orientation, the fractures might be open or closed and thus favorable or not to fluid circulation.

6. Conclusions

The present work is part of MEET project and focused on the NH range, considered as a paleo geothermal reservoir analogue, which offers a general overview of the structural organization of the fracture networks at outcrop and wider scales. This study aims at understanding the fracture spacing variability and the distribution of fracture patterns in granitic rocks, and the impact of the proximity of major faults on fracture distribution and fluid circulation. In the NH case study, the fracture patterns are controlled by the structural heritage. This heritage results from the activity of the SDFVZ and GFZ faults, which strongly control the geometry of the entire ranges. The evidence of fluid circulations has been highlighted in previous studies and consist in the granite hydrothermal alteration (propylitic and argillic), and the fracture infills. The carbonate infill is associated to the fractures striking NW/SE (SDVFZ direction), while the barite infill is related to the fractures striking E/W (GFZ direction). Hence, at least, two episodes of fluid circulation have occurred within NH ranges.

From fieldwork scanlines, the main dominant trends within the central part of NH are the NW/SE and the E/W directions, well expressed whatever the scanline. The variability between fracture orientations was probably related to the influence of the major faults. From virtual scanlines, NE/SW, E/W, and NW/SE fracture sets were the most consistent orientations. However, some sampling bias previously discussed, such as the absence of dip in case of virtual scanlines and image resolution, which can impact the fracture detection should be reconsidered in the methodological development of the future work.

The spatial organization of the NH fracture network reveals that fracture clustering increases with fracture abundance only in the case of fracture sets detected from virtual scanlines (SL8, SL9, and SL10) due to the proximity of major faults. However, from fieldwork scanlines, the correlation seems not so clear. Furthermore, clustering depends on the strike of the fractures:

- Three configurations (uniform, random and clustered distribution) were shown for SL1 spacing fractures: Regular distribution within N062, uniform distribution within N026, and clustered distribution within N092, N130, and N171 fracture sets;
- For SL3a and SL6, two configurations were identified: Uniform distribution within respectively N077 and N100 trends, clustered distribution within N098 and N135 for SL3a, and N004 and N161 fracture sets for SL6;
- SL2b, SL4, SL5, and SL7 profiles showed only one configuration, which consists in uniform distribution, not dependent on the direction;
- SL2a and SL3b, SL8, SL9, and SL10 showed a high Cv whatever the direction, which indicates a stronger clustering in the fracture system.

The most reproducible trends in the NH range are characterized by a clustered spacing distribution. The most clustered systems were identified close to or crosscutting the major faults which control the NH geometry. These faults are SDVFZ major segments and thrust faults. In addition, the deformation gradient impacts strongly the fracture patterns. The present paper highlighted that in a high deformation context, the spatial distribution of fracture become clustered, while in moderate deformation context, the fracture patterns are distributed randomly or regularly. These distributions considerably impact the fluid flow within the reservoir. The forthcoming reservoir modeling should consider the deformation gradient and the evolution of fracture patterns toward to major fault zones, as well as present-day stress field.

Author Contributions: Conceptualization, A.C., G.T., J.K. and B.A.L.; methodology, A.C.; software, A.C.; validation, G.T., B.A.L. and J.K.; formal analysis, A.C.; investigation, A.C., G.T., J.K., B.A.L.; resources, A.C.; data curation, A.C.; writing—original draft preparation, A.C.; writing—review and editing, G.T., B.A.L. and H2020 MEET consortium; visualization, G.T., B.A.L. and J.K.; supervision, G.T.; project administration, G.T.; funding acquisition, G.T. and H2020 MEET consortium. All authors have read and agreed to the published version of the manuscript.

Funding: This project has received funding from the European Union's Horizon 2020 research and innovation program under grant agreement No 792037 (MEET project).

Institutional Review Board Statement: Not applicable.

Informed Consent Statement: Not applicable.

Data Availability Statement: Not applicable.

Acknowledgments: This work is part of postdoctoral contribution, prepared at Institut Polytechnique UniLaSalle Beauvais, which was funded by European Union's Horizon 2020 research (H2020 MEET project, grant agreement No 792037). We are grateful to Terry Pavlis for fruitful discussions. We also wish to thank the H2020 MEET consortium for helpful comments and manuscript validation.

Conflicts of Interest: The authors declare no conflict of interest.

References

1. Torabi, A.; Ellingsen, T.S.S.; Johannessen, M.U.; Alaei, B.; Rotevatn, A.; Chiarella, D. Fault Zone Architecture and Its Scaling Laws: Where Does the Damage Zone Start and Stop? *Geol. Soc. Lond. Spec. Publ.* **2020**, *496*, 99–124. [CrossRef]
2. Sanderson, D.J.; Nixon, C.W. The Use of Topology in Fracture Network Characterization. *J. Struct. Geol.* **2015**, *72*, 55–66. [CrossRef]
3. Dezayes, C.; Lerouge, C.; Innocent, C.; Lach, P. Structural Control on Fluid Circulation in a Graben System: Constraints from the Saint Pierre Bois Quarry (Vosges, France). *J. Struct. Geol.* **2021**, *146*, 104323. [CrossRef]
4. Barton, C.A.; Zoback, M.D. Self-Similar Distribution and Properties of Macroscopic Fractures at Depth in Crystalline Rock in the Cajon Pass Scientific Drill Hole. *J. Geophys. Res. Solid Earth* **1992**, *97*, 5181–5200. [CrossRef]
5. Caine, J.S.; Evans, J.P.; Forster, C.B. Fault Zone Architecture and Permeability Structure. *Geology* **1996**, *24*, 1025–1028. [CrossRef]
6. Callahan, O.A.; Eichhubl, P.; Olson, J.E.; Davatzes, N.C. Fracture Mechanical Properties of Damaged and Hydrothermally Altered Rocks, Dixie Valley-Stillwater Fault Zone, Nevada, USA. *J. Geophys. Res. Solid Earth* **2019**, *124*, 4069–4090. [CrossRef]
7. Laubach, S.E.; Lamarche, J.; Gauthier, B.D.M.; Dunne, W.M.; Sanderson, D.J. Spatial Arrangement of Faults and Opening-Mode Fractures. *J. Struct. Geol.* **2018**, *108*, 2–15. [CrossRef]
8. Klee, J.; Chabani, A.; Ledésert, B.A.; Potel, S.; Hébert, R.L.; Trullenque, G. Fluid-Rock Interactions in a Paleo-Geothermal Reservoir (Noble Hills Granite, California, USA). Part 2: The Influence of Fracturing on Granite Alteration Processes and Fluid Circulation at Low to Moderate Regional Strain. *Geosciences* **2021**, *11*, 433. [CrossRef]
9. Bour, O.; Davy, P. Connectivity of Random Fault Networks Following a Power Law Fault Length Distribution. *Water Resour. Res.* **1997**, *33*, 1567–1583. [CrossRef]
10. Genter, A.; Traineau, H.; Ledésert, B.; Bourgine, B.; Gentier, S. *Over 10 Years of Geological Investigations within the HDR Soultz Project, France*; International Geothermal Association: Reykjavik, Iceland, 2000; pp. 3707–3712.
11. Renshaw, C.E. Influence of Subcritical Fracture Growth on the Connectivity of Fracture Networks. *Water Resour. Res.* **1996**, *32*, 1519–1530. [CrossRef]
12. Laubach, S.E.; Lander, R.H.; Criscenti, L.J.; Anovitz, L.M.; Urai, J.L.; Pollyea, R.M.; Hooker, J.N.; Narr, W.; Evans, M.A.; Kerisit, S.N.; et al. The Role of Chemistry in Fracture Pattern Development and Opportunities to Advance Interpretations of Geological Materials. *Rev. Geophys.* **2019**, *57*, 1065–1111. [CrossRef]
13. Lamarche, J.; Chabani, A.; Gauthier, B.D. Dimensional Threshold for Fracture Linkage and Hooking. *J. Struct. Geol.* **2018**, *108*, 171–179. [CrossRef]
14. Bour, O.; Davy, P. On the Connectivity of Three-dimensional Fault Networks. *Water Resour. Res.* **1998**, *34*, 2611–2622. [CrossRef]
15. Darcel, C.; Bour, O.; Davy, P.; de Dreuzy, J.R. Connectivity Properties of Two-Dimensional Fracture Networks with Stochastic Fractal Correlation: Connectivity of 2D fractal fracture networks. *Water Resour. Res.* **2003**, *39*. [CrossRef]
16. Odling, N.E. Scaling and Connectivity of Joint Systems in Sandstones from Western Norway. *J. Struct. Geol.* **1997**, *19*, 1257–1271. [CrossRef]
17. Chabani, A. Analyse Méthodologique et Caractérisation Multi-Échelle Des Systèmes de Fractures à l'interface Socle/Couverture Sédimentaire–Application à La Géothermie (Bassin de Valence, SE France). These de doctorat, Paris Sciences et Lettres (ComUE). 2019. Available online: http://www.theses.fr/2019PSLEM046 (accessed on 21 September 2021).
18. Narr, W. Estimating Average Fracture Spacing in Subsurface Rock. *AAPG Bull.* **1996**, *80*, 1565–1585. [CrossRef]
19. Haffen, S.; Géraud, Y.; Diraison, M.; Dezayes, C. Determination of Fluid-Flow Zones in a Geothermal Sandstone Reservoir Using Thermal Conductivity and Temperature Logs. *Geothermics* **2013**, *46*, 32–41. [CrossRef]
20. Sanderson, D.J.; Peacock, D.C. Line Sampling of Fracture Swarms and Corridors. *J. Struct. Geol.* **2019**, *122*, 27–37. [CrossRef]
21. Fossen, H.; Cavalcante, G.C.G. Shear Zones–A Review. *Earth Sci. Rev.* **2017**, *171*, 434–455. [CrossRef]
22. Kim, Y.-S.; Peacock, D.C.P.; Sanderson, D.J. Mesoscale Strike-Slip Faults and Damage Zones at Marsalforn, Gozo Island, Malta. *J. Struct. Geol.* **2003**, *25*, 793–812. [CrossRef]
23. Ostermeijer, G.A.; Mitchell, T.M.; Aben, F.M.; Dorsey, M.T.; Browning, J.; Rockwell, T.K.; Fletcher, J.M.; Ostermeijer, F. Damage Zone Heterogeneity on Seismogenic Faults in Crystalline Rock; a Field Study of the Borrego Fault, Baja California. *J. Struct. Geol.* **2020**, *137*, 104016. [CrossRef]
24. Marrett, R.; Gale, J.F.W.; Gómez, L.A.; Laubach, S.E. Correlation Analysis of Fracture Arrangement in Space. *J. Struct. Geol.* **2018**, *108*, 16–33. [CrossRef]
25. Chabani, A.; Trullenque, G.; Ledésert, B.A.; Klee, J. Multiscale Characterization of Fracture Patterns: A Case Study of the Noble Hills Range (Death Valley, CA, USA), Application to Geothermal Reservoirs. *Geosciences* **2021**, *11*, 280. [CrossRef]
26. Gillespie, P.A.; Howard, C.B.; Walsh, J.J.; Watterson, J. Measurement and Characterisation of Spatial Distributions of Fractures. *Tectonophysics* **1993**, *226*, 113–141. [CrossRef]
27. Wang, Q.; Laubach, S.E.; Gale, J.F.W.; Ramos, M.J. Quantified Fracture (Joint) Clustering in Archean Basement, Wyoming: Application of the Normalized Correlation Count Method. *Pet. Geosci.* **2019**, *25*, 415–428. [CrossRef]

28. Pavičić, I.; Dragičević, I.; Vlahović, T.; Grgasović, T. Fractal Analysis of Fracture Systems in Upper Triassic Dolomites in Žumberak Mountain, Croatia. *Rud. Geološko-Naft. Zb.* **2017**, *32*, 1–13. Available online: https://hrcak.srce.hr/ojs/index.php/rgn/article/view/4894/pdf (accessed on 21 September 2021). [CrossRef]

29. Shiri, Y.; Hassani, H. Two-Component Fluid Front Tracking in Fault Zone and Discontinuity with Permeability Heterogeneity. *Rud. Geol. Naft. Zb.* **2021**, *36*, 19–30. [CrossRef]

30. Hooker, J.N.; Laubach, S.E.; Marrett, R. Microfracture Spacing Distributions and the Evolution of Fracture Patterns in Sandstones. *J. Struct. Geol.* **2018**, *108*, 66–79. [CrossRef]

31. Putz-Perrier, M.W.; Sanderson, D.J. Spatial Distribution of Brittle Strain in Layered Sequences. *J. Struct. Geol.* **2008**, *30*, 50–64. [CrossRef]

32. Bourbiaux, B.; Basquet, R.; Daniel, J.M.; Hu, L.Y.; Jenni, S.; Lange, G.; Rasolofosaon, P. Fractured Reservoirs Modelling: A Review of the Challenges and Some Recent Solutions. *First Break* **2005**, *23*, 33–40. [CrossRef]

33. Trullenque, G.; Genter, A.; Leiss, B.; Wagner, B.; Bouchet, R.; Léoutre, E.; Malnar, B.; Bär, K.; Rajšl, I. Upscaling of EGS in Different Geological Conditions: A European Perspective. In Proceedings of the 43rd Workshop on Geothermal Reservoir Engineering, Stanford, CA, USA, 12–14 February 2018; p. SGP-TR-213.

34. Klee, J.; Trullenque, G.; Ledésert, B.; Potel, S.; Hébert, R.; Chabani, A.; Genter, A. Petrographic Analyzes of Fractured Granites Used as An Analogue of the Soultz-Sous-Forêts Geothermal Reservoir: Noble Hills, CA, USA. In Proceedings of the Extended Abstract, Reykjavik, Island, 26 May 2021.

35. Chabani, A.; Trullenque, G.; Rishi, P.; Pomart, A.; Attali, R.; Sass, I. Modelling of Fractured Granitic Geothermal Reservoirs: Use of Deterministic and Stochastic Methods in Discrete Fracture Networks and a Coupled Processes Modeling Framework. In Proceedings of the Extended Abstract, Reykjavik, Island, 2 May 2021.

36. Pavlis, T.L.; Trullenque, G. Evidence for 40–41 Km of Dextral Slip on the Southern Death Valley Fault: Implications for the Eastern California Shear Zone and Extensional Tectonics. *Geology* **2021**, *49*, 767–772. [CrossRef]

37. Klee, J.; Potel, S.; Ledésert, B.A.; Hébert, R.L.; Chabani, A.; Barrier, P.; Trullenque, G. Fluid-Rock Interactions in a Paleo-Geothermal Reservoir (Noble Hills Granite, California, USA). Part 1: Granite Pervasive Alteration Processes Away from Fracture Zones. *Geosciences* **2021**, *11*, 325. [CrossRef]

38. Brady III, R.H. Neogene Stratigraphy of the Avawatz Mountains between the Garlock and Death Valley Fault Zones, Southern Death Valley, California: Implications as to Late Cenozoic Tectonism. *Sediment. Geol.* **1984**, *38*, 127–157. [CrossRef]

39. Pavlis, T.L.; Rutkofske, J.; Guerrero, F.; Serpa, L.F. Structural Overprinting of Mesozoic Thrust Systems in Eastern California and Its Importance to Reconstruction of Neogene Extension in the Southern Basin and Range. *Geosphere* **2014**, *10*, 732–756. [CrossRef]

40. Miller, M.; Pavlis, T. The Black Mountains Turtlebacks: Rosetta Stones of Death Valley Tectonics. *Earth-Sci. Rev.* **2005**, *73*, 115–138. [CrossRef]

41. Snow, J.K. Cenozoic Tectonism in the Central Basin and Range; Magnitude, Rate, and Distribution of Upper Crustal Strain. *Am. J. Sci.* **2000**, *300*, 659–719. [CrossRef]

42. Calzia, J.; Rämö, O. Late Cenozoic Crustal Extension and Magmatism, Southern Death Valley Region, California. *GSA Field Guides* **2000**, *2*, 135–164. [CrossRef]

43. Calzia, J.; Ramo, O. Miocene Rapakivi Granites in the Southern Death Valley Region, California, USA. *Earth-Sci. Rev.* **2005**, *73*, 221–243. [CrossRef]

44. Brady, R. Cenozoic Geology of the Northern Avawatz Mountains in Relation to the Intersection of the Garlock and Death Valley Fault Zones, San Bernardino County, California. Ph.D. Thesis, University of California, Davis, CA, USA, 1987.

45. Norton, I. Two-Stage Formation of Death Valley. *Geosphere* **2011**, *7*, 171–182. [CrossRef]

46. Burchfiel, B.C.; Stewart, J.H. "Pull-Apart" Origin of the Central Segment of Death Valley, California. *Geol. Soc. Am. Bull.* **1966**, *77*, 439–442. [CrossRef]

47. Niles, J.H. Post-Middle Pliocene Tectonic Development of the Noble Hills, Southern Death Valley, California. Ph.D. Thesis, San Francisco State University, San Francisco, CA, USA, 2016.

48. Reinert, E. Low Temperature Thermochronometry of the Avawatz Mountians, California: Implications for the Inception of the Eastern California Shear Zone. Ph.D. Thesis, University of Washington, Seattle, WA, USA, 2004. Available online: https://www.ess.washington.edu/content/people/student_publications_files/reinert--erik/Reinert_2004.pdf (accessed on 21 September 2021).

49. Spencer, J.E. Late Cenozoic Extensional and Compressional Tectonism in the Southern and Western Avawatz Mountains, Southeastern California. In *Basin and Range Extensional Tectonics Near the Latitude of Las Vegas, Nevada: Geological Society of America Memoir*; Geological Society of America: Boulder, CO, USA, 1990; Volume 176, pp. 317–333. [CrossRef]

50. Mahon, R.C.; Dehler, C.M.; Link, P.K.; Karlstrom, K.E.; Gehrels, G.E. Geochronologic and Stratigraphic Constraints on the Mesoproterozoic and Neoproterozoic Pahrump Group, Death Valley, California: A Record of the Assembly, Stability, and Breakup of Rodinia. *Bulletin* **2014**, *126*, 652–664. [CrossRef]

51. Pickering, G.; Bull, J.M.; Sanderson, D.J. Sampling Power-Law Distributions. *Tectonophysics* **1995**, *248*, 1–20. [CrossRef]

52. Odling, N.E.; Gillespie, P.; Bourgine, B.; Castaing, C.; Chiles, J.P.; Christensen, N.P.; Fillion, E.; Genter, A.; Olsen, C.; Thrane, L.; et al. Variations in Fracture System Geometry and Their Implications for Fluid Flow in Fractured Hydrocarbon Reservoirs. *Pet. Geosci.* **1999**, *5*, 373–384. [CrossRef]

53. Bonnet, E.; Bour, O.; Odling, N.E.; Davy, P.; Main, I.; Cowie, P.; Berkowitz, B. Scaling of Fracture Systems in Geological Media. *Rev. Geophys.* **2001**, *39*, 347–383. [CrossRef]

54. Priest, S.D. *Discontinuity Analysis for Rock Engineering*; Springer Science & Business Media: Berlin/Heidelberg, Germany, 1993.

55. Ledésert, B.A.; Hébert, R.L. How Can Deep Geothermal Projects Provide Information on the Temperature Distribution in the Upper Rhine Graben? The Example of the Soultz-Sous-Forêts-Enhanced Geothermal System. *Geosciences* **2020**, *10*, 459. [CrossRef]

56. Priest, S.D.; Hudson, J.A. *Estimation of Discontinuity Spacing and Trace Length Using Scanline Surveys*; Elsevier: Amsterdam, The Netherlands, 1981; Volume 18, pp. 183–197. [CrossRef]

57. Hardebol, N.J.; Maier, C.; Nick, H.; Geiger, S.; Bertotti, G.; Boro, H. Multiscale Fracture Network Characterization and Impact on Flow: A Case Study on the Latemar Carbonate Platform. *J. Geophys. Res. Solid Earth* **2015**, *120*, 8197–8222. [CrossRef]

58. Bisdom, K.; Gauthier, B.D.M.; Bertotti, G.; Hardebol, N.J. Calibrating Discrete Fracture-Network Models with a Carbonate Threedimensional Outcrop Fracture Network: Mplications for Naturally Fractured Reservoir Modeling. *AAPG Bull.* **2014**, *98*, 13511376. [CrossRef]

59. Bertrand, L.; Géraud, Y.; Le Garzic, E.; Place, J.; Diraison, M.; Walter, B.; Haffen, S. A Multiscale Analysis of a Fracture Pattern in Granite: A Case Study of the Tamariu Granite, Catalunya, Spain. *J. Struct. Geol.* **2015**, *78*, 52–66. [CrossRef]

60. Terzaghi, R.D. Sources of Error in Joint Surveys. In *Geotechnique*; Road Research Laboratory: Crowthorne, UK, 1965; Volume 15, pp. 287–304. Available online: http://worldcat.org/issn/00168505 (accessed on 21 September 2021).

61. Delvaux, D.; Sperner, B. Stress Tensor Inversion from Fault Kinematic Indicators and Focal Mechanism Data: The TENSOR Program. *New Insights Struct. Interpret. Model.* **2003**, *212*, 75–100. [CrossRef]

62. Cardozo, N.; Allmendinger, R.W. Spherical projections with OSXStereonet. *Comput. Geosci.* **2013**, *51*, 193–205. [CrossRef]

63. Hardebol, N.J.; Bertotti, G. DigiFract: A Software and Data Model Implementation for Flexible Acquisition and Processing of Fracture Data from Outcrops. *Comput. Geosci.* **2013**, *54*, 326–336. [CrossRef]

64. Fisher, R. Dispersion on a Sphere. *Proc. R. Soc. Lond. Ser. A Math. Phys. Sci.* **1953**, *217*, 295–305. Available online: http://palaeo.spb.ru/pmlibrary/pmpapers/fisher_1953.pdf (accessed on 21 September 2021). [CrossRef]

65. Kent, J.T. The Fisher-Bingham Distribution on the Sphere. *J. R. Stat. Soc. Ser. B* **1982**, *44*, 71–80. Available online: http://www.jstor.org/stable/2984712 (accessed on 21 September 2021). [CrossRef]

66. Von Mises, R. Uber Die "Ganzzahligkeit" Der Atomgewicht Und Verwandte Fragen. *Physikal. Z* **1918**, *19*, 490–500.

67. Lark, R.M.; Clifford, D.; Waters, C.N. Modelling Complex Geological Circular Data with the Projected Normal Distribution and Mixtures of von Mises Distributions. *Solid Earth* **2014**, *5*, 631–639. [CrossRef]

68. Chabani, A.; Mehl, C.; Cojan, I.; Alais, R.; Bruel, D. Semi-Automated Component Identification of a Complex Fracture Network Using a Mixture of von Mises Distributions: Application to the Ardeche Margin (South-East France). *Comput. Geosci.* **2020**, *137*, 104435. [CrossRef]

69. Gillespie, D.T. Approximate Accelerated Stochastic Simulation of Chemically Reacting Systems. *J. Chem. Phys.* **2001**, *115*, 1716–1733. [CrossRef]

70. Kuiper, N.H. Tests Concerning Random Points on a Circle. In *Proceedings of the Koninklijke*; Nederlandse Akademie van Wetenschappen: Amsterdam, The Netherlands, 1960; Volume 63, pp. 38–47. [CrossRef]

71. Sanderson, D.J.; Roberts, S.; Gumiel, P.; Greenfield, C. Quantitative Analysis of Tin-and Tungsten-Bearing Sheeted Vein Systems. *Econ. Geol.* **2008**, *103*, 1043–1056. [CrossRef]

72. Ortega, O.J.; Marrett, R.A.; Laubach, S.E. A Scale-Independent Approach to Fracture Intensity and Average Spacing Measurement. *AAPG Bull.* **2006**, *90*, 193–208. [CrossRef]

73. Chinn, L.D. Low-Temperature Thermochronometry of the Avawatz Mountains; Implications for the Eastern Terminus and Inception of the Garlock Fault Zone. 2013. Available online: https://www.ess.washington.edu/content/people/student_publications_files/chinn--logan/chinn--logan_ms_2013.pdf (accessed on 21 September 2021).

74. Franklin, B. Characterising Fracture Systems within Upfaulted Basement Highs in the Hebridean Islands: An Onshore Analogue for the Clair Field. Ph.D. Thesis, Durham University, Durham, UK, 2013. Available online: http://etheses.dur.ac.uk/7765/ (accessed on 21 September 2021).

75. Bourbiaux, B.; Basquet, R.; Cacas, M.-C.; Daniel, J.-M.; Sarda, S. An Integrated Workflow to Account for Multi-Scale Fractures in Reservoir Simulation Models: Implementation and Benefits. In Proceedings of the Abu Dhabi International Petroleum Exhibition and Conference, Abu Dhabi, United Arab Emirates, 13–16 October 2002.

76. Ledésert, B.; Hebert, R.; Genter, A.; Bartier, D.; Clauer, N.; Grall, C. Fractures, Hydrothermal Alterations and Permeability in the Soultz Enhanced Geothermal System. *Comptes Rendus Geosci.* **2010**, *342*, 607–615. [CrossRef]

77. Dorbath, L.; Cuenot, N.; Genter, A.; Frogneux, M. Seismic Response of the Fractured and Faulted Granite of Soultz-Sous-Forêts (France) to 5 Km Deep Massive Water Injections. *Geophys. J. Int.* **2009**, *177*, 653–675. [CrossRef]

78. Morellato, C.; Redini, F.; Doglioni, C. On the Number and Spacing of Faults. *Terra Nova* **2003**, *15*, 315–321. [CrossRef]

Article

Multiscale Characterization of Fracture Patterns: A Case Study of the Noble Hills Range (Death Valley, CA, USA), Application to Geothermal Reservoirs

Arezki Chabani [1],*, Ghislain Trullenque [1], Béatrice A. Ledésert [2] and Johanne Klee [1]

1   UniLaSalle, UPJV, B2R UMR 2018.C100, U2R 7511, 19 rue Pierre Waguet, F-60026 Beauvais, France; ghislain.trullenque@unilasalle.fr (G.T.); johanne.klee@unilasalle.fr (J.K.)

2   Geosciences and Environnent Cergy, CY Cergy Paris Université, 1 rue Descartes, F-95000 Neuville sur Oise, France; beatrice.ledesert@cyu.fr

*   Correspondence: arezki.chabani@unilasalle.fr; Tel.: +33-6-58-07-14-14

Abstract: In the basement fractured reservoirs, geometric parameters of fractures constitute the main properties for modeling and prediction of reservoir behavior and then fluid flow. This study aims to propose geometric description and quantify the multiscale network organization and its effect on connectivity using a wide-ranging scale analysis and orders scale classification. This work takes place in the Noble Hills (NH) range, located in the Death Valley (DV, USA). The statistical analyses were performed from regional maps to thin sections. The combination of the length datasets has led to compute a power law exponent around -2, meaning that the connectivity is ruled by the small and the large fractures. Three domains have been highlighted in the NH: (1) domain A is characterized by a dominance of the NW/SE direction at the fourth order scale; (2) domain B is characterized by a dominance of the E/W and the NW/SE directions at respectively the fourth and third order scales; (3) domain C is also marked by the E/W direction dominance followed by the NW/SE direction respectively at the fourth and third order scale. The numerical simulations should consider that the orientation depends on scale observation, while the length is independent of scale observation.

Keywords: fracture network; Death Valley; Noble Hills; power law distribution; multiscale analysis; geothermal reservoir characterization

Citation: Chabani, A.; Trullenque, G.; Ledésert, B.A.; Klee, J. Multiscale Characterization of Fracture Patterns: A Case Study of the Noble Hills Range (Death Valley, CA, USA), Application to Geothermal Reservoirs. *Geosciences* **2021**, *11*, 280. https://doi.org/10.3390/geosciences11070280

Academic Editors: Jesus Martinez-Frias and Gianluca Groppelli

Received: 10 June 2021
Accepted: 30 June 2021
Published: 3 July 2021

Publisher's Note: MDPI stays neutral with regard to jurisdictional claims in published maps and institutional affiliations.

1. Introduction

Fluid flow in fractured rocks of very low matrix permeability is localized mainly in few fractures [1]. The complex geometry of fracture and fault patterns is the main cause of the complexity of fluid flow. In that case, numerous studies have been undertaken worldwide to show the control of the fracture network on the fluid circulations especially in hydrocarbon and aquifers reservoirs [2–5], in heat transfer [6], and thus also in geothermal reservoirs [7–9].

A fracture system is characterized by geometrical parameters as fracture lengths, spacings, orientations, and relations between them [1,10–12]. In order to access hierarchical and mechanical relationships between fracture systems, many authors undertook a multi-scale approach, in sedimentary [2,12–15], and in Crystalline [16–20] rocks. This approach allows to model and predict hydraulic reservoir properties, by studying several geometric attributes, such as the distribution of orientations, lengths, widths, spacings and densities classically considered in spatial arrangement analysis [21–24].

The geometric parameters are commonly collected for (1) explicitly constructing deterministic models (Discrete Fracture Network: DFNs, [2]) or (2) ensuring inputs of stochastic simulations by determining fracture distribution functions from sampled fracture networks [23,25–27]. The main goal is to better understand the fracture network connectivity and then the fluid flow patterns [3,22,28]. Subsequently, the main question is whether data correspond to scale-limited lognormal or exponential distributions, or scale-invariant power laws, corresponding to fractal patterns? Several orders of magnitude based on length and spacing characteristics are thus mandatory to establish scaling laws from statistical distributions [29]. These orders of magnitude have been widely described in the literature in extensional [17,20,30] and trans-tensional [17] contexts. It consists of: (1) first order scale related to the crustal faults larger than 100 km length, (2) second order scale refers to the faults comprised between 20 and 30 km length, (3) third order scale refers to faults around 10 km length, (4) and the fourth order refers to faults under 1 km length [20]. In the extensional regime, [30] having defined a spacing characteristic for the two first order scales, with a 10 to 15 km spacing for first order and 3 to 8 km for the second order scale; while [20] have defined 0.8 to 1.5 km spacing for the third order scale. However, the fourth order spacing characteristic is not defined in the literature.

Fracture networks impact the fluid flow in reservoirs [23]. The 2D/3D seismic lines and 1D borehole data cannot detect respectively the fracture geometries and the spatial arrangement at the reservoir scale due to the lack of information [31]. Then, the spatial arrangement of fracture networks are widely studied from field analogues [16,23,29,32], as they give access to 2D and 3D distributions. In geothermal basement setting, the analogues are chosen according to the lithology and geological context to get closer to the reservoir conditions. Sometimes, the analogue is chosen in desert conditions, without vegetation, perfectly suited for realistic multiscale fracture network reconstructions in 2D, and in 3D in case of modeling canyons with photogrammetric [33] or lidar [34] approaches.

The present work is part of the MEET project (Multidisciplinary and multi-context demonstration of EGS exploration and Exploitation Techniques and potentials, [35]), which aims to develop geothermal exploitation at European scale by applying Enhanced Geothermal System (EGS) technology to different geological contexts. This study aims to propose geometric description and quantify the multiscale network organization and its effect on connectivity. A wide-ranging scale analysis from the microscopic scale to the regional scale was conducted in the desert environment of Noble Hills (NH) fractured granitic basement. It is located in the southern termination of Death Valley (DV, California, USA) and assimilated to a paleo geothermal analogue. The NH range is considered as analogue to the Soultz-sous-Forêts (SsF) geothermal electricity producing system, due to the granitic nature, the alteration and the trans-tensional tectonic setting of the DV region [36,37].

Measurements have been performed at various scales using the DV regional map [38] at 1:250,000, NH regional map based on previous studies undertaken by [37,39] and orthophoto images taken for the present day. At outcrop scale, fractures are digitized thanks to the photogrammetric models. Several 2D fracture maps are also used, as well as additional scales on samples and in thin sections. This allows to integrate fracture lengths ranging from micrometer to kilometer scales.

In this study, the multiscale approach is used to better understand the spatial arrangement of the fracture networks which aims at producing the necessary data for DFN modeling. This helps to better characterize reservoirs in response to the developing geothermal exploration and exploitation by EGS. This study is conducted through:

1. a 2D characterization of the NH fracture network;
2. a multiscale evolution of length distributions;
3. evaluate the fracture system in the complex tectonic and geometrical setting;
4. present a conceptual scheme of the NH fracture network organization, and the role of each fracture order of magnitude on the connectivity.

2. Geological Setting

The present study takes place in the NH fractured granitic zone, located at the southern termination of the Death Valley (DV, California, USA) (Figure 1a). The NH range is assimilated to a paleo geothermal analogue [37]. Indeed, Reference [37] confirmed the analogy between the DV region and Soultz-sous-Forêts reservoir (Rhine graben, East of France), with many similarities, especially: (1) the trans-tensional tectonic setting of the DV, and (2) the granitic nature and hydrothermal alteration of the central part of the NH range. Furthermore, the desert conditions of the NH range make this analogue perfectly suited for multiscale characterization of fracture networks dedicated to the global understanding of the spatial arrangement in granitic reservoir affected by trans-tensional tectonics [35].

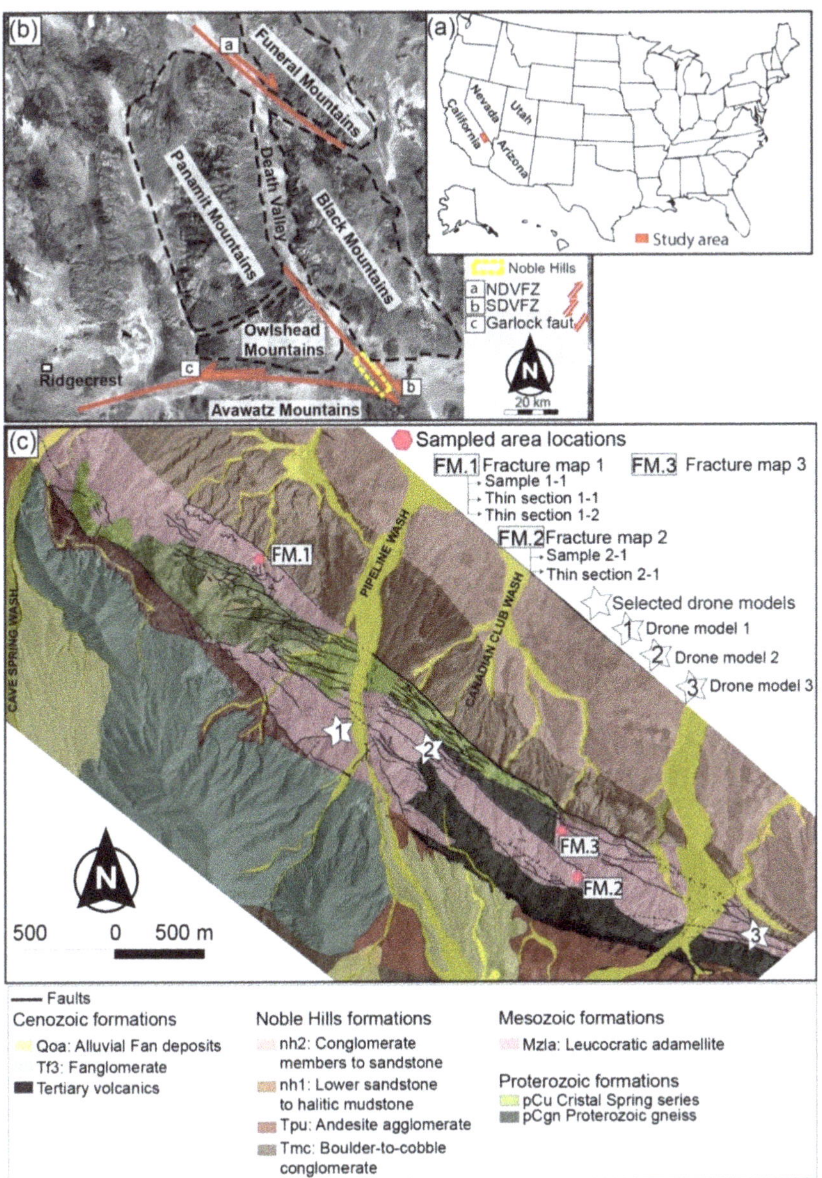

Figure 1. (**a**,**b**) Location and geological setting of DV, NDVFZ: Northern Death Valley Fault Zone, SDVFZ: Southern Death Valley Fault Zone; (**c**) Structural scheme of the NH range built using high-resolution digital mapping techniques modified after [37,39]. Additional digitized fractures were performed using orthophotos. The aerial picture in (**b**) was extracted from Google Earth®. The samples location was reported in the structural scheme. FM: fracture map.

The DV region has been characterized by a complex structural and tectonic history which includes the overprinting of Mesozoic to early Cenozoic contractional structures by late Cenozoic extensional and trans-tensional features [40–42]. The northwest trending contractional structures of the DV are at the origin of the Cordilleran orogenic belt of North America, which extends more than 6000 km from southern Mexico to the Canadian Arctic and Alaska [43]. Their age is estimated around 100 Ma and coincides with the Sevier orogenic belt [43,44]. The northeast trend is related to a thrust faults system [41].

The Mesozoic period was marked by the beginning of pluton emplacement in DV along the development of contractional structures. Then, during the Cenozoic period, the DV extension episode occurred [45,46], but there is no general agreement about its timing, e.g. [45,47]. The Miocene period was characterized by a trans-tensional regime. The opening of the DV region as a pull-apart basin began around 5 Ma [41,48–50]. A recent study done by [51] challenges this, mentioning that the opening of DV region into pull-apart basin started around 12 Ma ago.

Located in the southernmost part of the DV, the NH range trends parallel to, and then forms, the main topographic features aligned with components of the right-lateral Southern Death Valley Fault Zone (SDVFZ) [37] (Figure 1b). This SDVFZ is part of the Death Valley fault system (DVFS) [52], and its net dextral strike-slip displacement has been estimated around 40–41 km [51]. The northeast-vergent contractional deformation has been characterized along the length of the NH and constitutes the dominant structural style [48,53,54]. In the NH foreland basin, compressional structures have been characterized by [39] near the Canadian Club Wash (Figure 1c). An entire compressional region has also been created by the interaction between the SDVFZ and the GF zone, which is responsible for shortening within the Avawatz Mountains [55,56]. Furthermore, the reverse displacement of the GF was identified in the front of the Avawatz [57]. The E/W fractures observed mainly in the southeast end of the NH show an identical orientation to GF. Given this similarity in orientation and the close vicinity to the GF, it is tempting to relate the E/W trending fracturing to GF zone activity.

Recent work by [37] using high-resolution digital mapping techniques reveals the dominance of basement rocks in the central part of the NH (Figure 1c). Indeed, the granitic rocks are mostly represented and related to the Mesozoic granitic intrusion (Mzla, Figure 1c) [58]. Proterozoic formations characterized by gneiss in the bottom have been defined in the center part and the southeast end of the NH (pCgn, Figure 1c). The Proterozoic Crystal Spring sedimentary series defined in the northwestern part of the NH are mainly composed of carbonates and quartzites facies (pCu, Figure 1c). Some tertiary volcanic series have also been mapped in the southeastern end of the NH.

Outside the center part, Reference [39] has widely studied the Cenozoic NH formations mainly composed of Fanglomerate and alluvial fan deposits.

3. Materials and Methods

3.1. Materials

3.1.1. Large Scale Characterization

The analysis of large scale fracture attributes was performed on fractures digitized from DV regional map at 1:250,000 scale (geological map of California, Trona Sheet [38]) (Figure 2a). In the central part of the NH range, the main structures were mapped during 2018 and 2019 field campaigns published by Klee et al. [37]. In addition, in this present study, orthophoto images of 1 m resolution were used to digitalize and then complete the fractures sampling (Figure 2b).

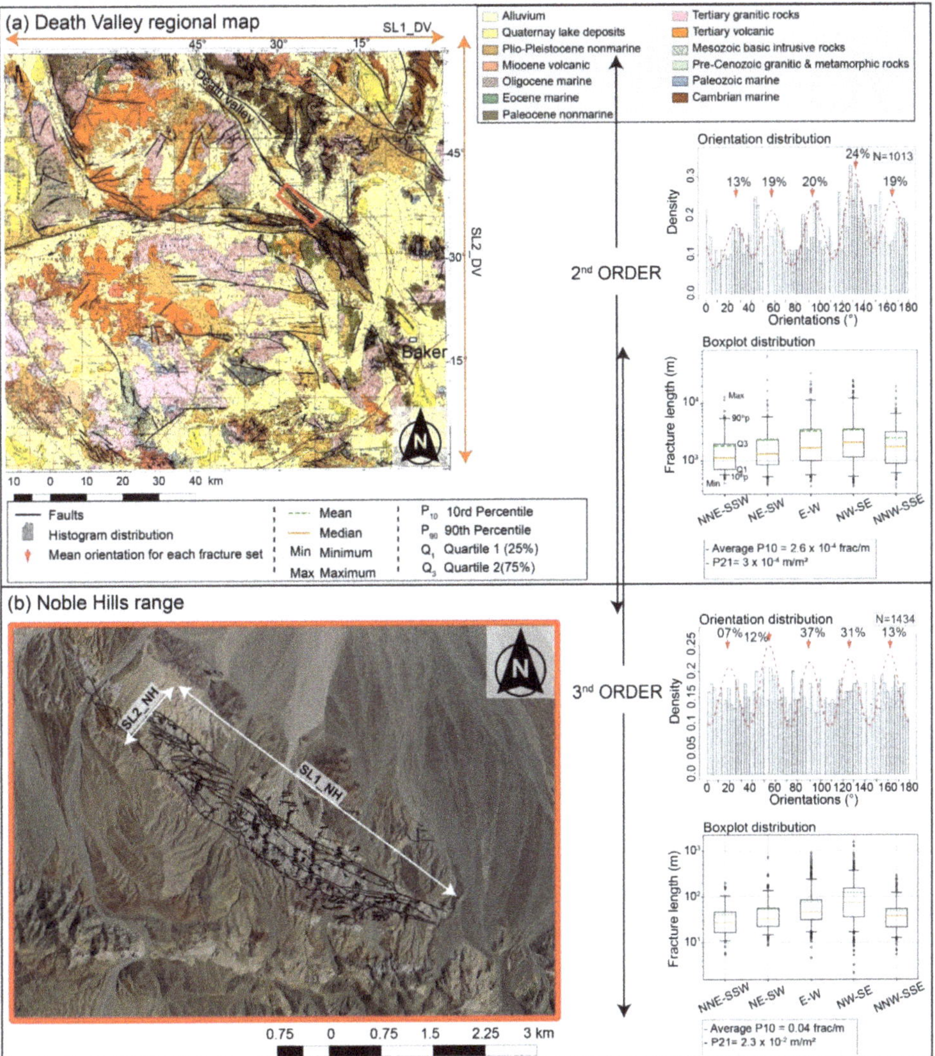

Figure 2. Fracture traces, orientation, and boxplot distributions from (**a**) DV geological map and (**b**) NH range map (in red rectangle for large scale location). The order of fractures magnitude were referred to the classification detailed in [30]. The first order referred to the faults larger 100 km length is not detected in this study. Each fracture map dataset is expressed in histogram for orientation distribution, and into boxplot for length distribution, according to the fracture orientations. Two scanlines (SL) were taken for every map.

With the aim of pinpointing the likely fracture variability, the NH range map has been divided into three domains (Figure 3). The sampling strategy along domains is detailed below. In total, 1013 fractures are extracted from the DV geological map, 1434 fractures from the NH map with 306 from domain A, 522 from domain B, and 606 from domain C.

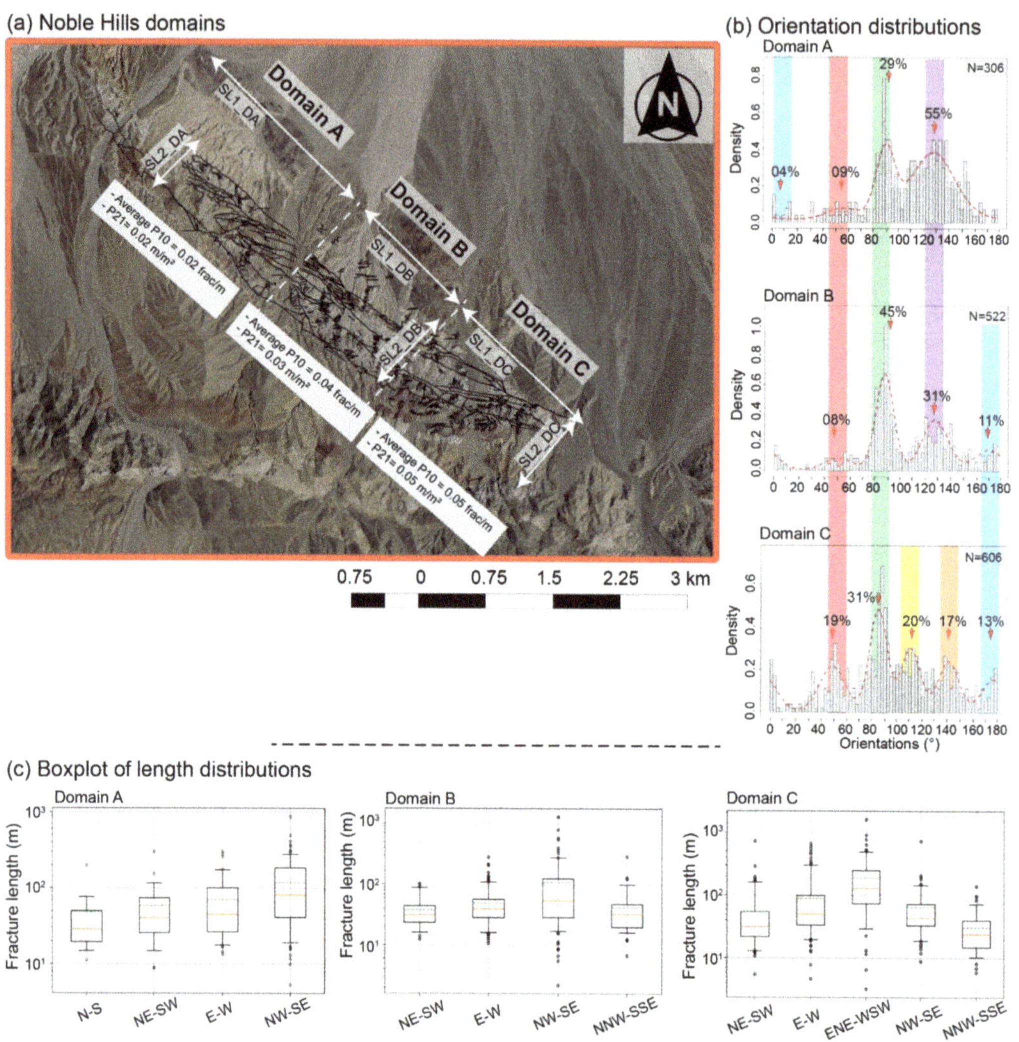

Figure 3. Fracture traces from the NH range divided into 3 domains. (**a**) NH domains: A, B, and C, (**b**) orientation distributions, and (**c**) boxplot of length distributions according to the fracture orientations. Two scanlines (SL) were taken for every domain. For boxplot and orientation distribution legend, see Figure 2.

Orientation and length distributions were collected and 1D/2D densities were calculated from them. All of these data are summarized in Table 1.

Table 1. Spatial statistics of 2D samples acquired at large scale (DV and NH regional maps) and outcrop scale (Drone models). Two Scanlines (SL) at least were generated for calculating the fracture density (P10). The length average is provided.

	Nb of Fractures	Mean Orientation	Average SL Length (m)/Area (m^2)	Fracture Density (P10) (frac/m)	Fracture Density (P21) (m/m^2)
DV regional map	133 188 199 248 197	N025 N058 N096 N132 N167	$10^5/1.3 \times 10^{10}$	2×10^{-4}	3×10^{-4}
NH regional map	97 168 535 441 192	N020 N054 N090 N126 N163	$4.7 \times 10^3/4.4 \times 10^6$	2.6×10^{-4}	3×10^{-4}
NH domain A	13 26 89 168	N010 N055 N091 N132	$1.5 \times 10^3/1.9 \times 10^6$	2×10^{-2}	2×10^{-2}
NH domain B	43 233 164 59	N050 N088 N128 N171	$1.7 \times 10^3/1.2 \times 10^6$	4×10^{-2}	3×10^{-2}
NH domain C	113 191 119 103 77	N049 N085 N111 N141 N177	$2 \times 10^3/1.2 \times 10^6$	5×10^{-2}	5×10^{-2}
Drone model 1	235 208 183 112 281 214	N025 N055 N085 N115 N142 N174	$3.5 \times 10^2/1.4 \times 10^4$	6.5×10^{-1}	2×10^{-1}
Drone model 2	681 767 642 616 722 637	N022 N043 N085 N108 N142 N171	$1.7 \times 10^2/5.8 \times 10^3$	1.28	7×10^{-1}
Drone model 3	2968 2531 1993 2127 2232 3084	N018 N038 N079 N109 N140 N169	$2.8 \times 10^2/1.7 \times 10^4$	9×10^{-1}	1

3.1.2. Photogrammetric Study/Analysis

In order to combine the widest fracture network analysis, three photogrammetric models, localized only in the granitic facies and alongside the major faults that structure the NH range (see Figure 1 for the location), are considered in this study to extract the fractures ranging approximatively from 10^{-2} m to 20 m. A photogrammetric campaign was carried out in October 2018 with a high-resolution 3D point cloud and large collection of photographs.

Drone model 1 is sized approximately 350 m per 46 m and is located in the granitic facies (Figure 1); 1271 fractures are traced. Drone model 2 is located in the granitic facies within a high deformation zone, and is sized approximately 172 m per 44 m (Figure 1). In total, 4196 fractures are traced. Finally, drone model 3 is located in the southeast termination of NH, and is sized approximately 280 m per 70 m. 15,594 fractures are traced.

The main statistical characteristics are summarized in Table 1.

3.1.3. Fracture Map Characterization

To complete the multiscale analysis with the small fractures, fracture maps were created with a resolution of 2×10^{-4} m. The fracture map 1 (see Figure 1c for location) is sized at 3.4 m per 3.1 m and is located between the granitic facies and the tertiary sediments, near the influence of the SDVFZ (see Figure 1b for the SDVFZ location). It is composed of quartzite boudin facies (part of Crystal Spring series) and granitic rocks, which respectively represent approximately 20% and 80% of occurrences.

In order to include the very small fractures in this study, additional fracture sets were collected from this fracture map with sample 1-1 sized 17 cm per 12 cm (resolution of 10^{-4} m) and thin sections 1-1 and 1-2 (resolution of 10^{-5} m), both sized 3.5 per 2.5 cm (see Figure 1c for location). Both sections are composed mainly of Plagioclase, Quartz, K-Feldspar, and Biotite representing respectively 35%, 30%, 25%, and 10% of occurrences.

Fracture map 2 (Figure 1c) is sized 1.7 m per 1.6 m and is located in the granitic facies, near the SDVFZ deformation corridor. This map includes sample 2-1 and thin section 2-1 which size respectively 15 cm per 5.2 cm and 3.5 cm per 2.5 cm. Thin section 2-1 represents the same mineral composition as thin sections 1-1 and 1-2 with approximately the same occurrences. Fracture map 3 (Figure 1c) is sized 1.8 m per 1.7 m and is located also in the granitic facies.

Table 2 summarizes the main statistical characteristics.

Table 2. Spatial statistics of 2D samples acquired from outcrop scale (fracture maps) to thin section scale. Two Scanlines (SL) at least were generated for calculating the fracture density (P10); the length average is provided.

	Nb of Fractures	Mean Orientation	Average SL Length (m)/Area (m^2)	Fracture Density (P10) (frac/m)	Fracture Density (P21) (m/m^2)
Fracture map 1	1133 631	N005 N083	3.1/7.94	12	12
Sample 1-1	104 115 117 51	N018 N053 N090 N126	$15 \times 10^{-2}/2.2 \times 10^{-2}$	184	118
Thin section 1-1	723 754 751 730	N018 N053 N090 N126	$3 \times 10^{-2}/8 \times 10^{-4}$	5140	715
Thin section 1-2	612 659 444 512 623	N020 N051 N088 N128 N162	$3 \times 10^{-2}/8 \times 10^{-4}$	4525	702
Fracture map 2	36 34 98 32 83	N005 N038 N070 N110 N145	1.7/1.91	18	13

Table 2. *Cont.*

	Nb of Fractures	Mean Orientation	Average SL Length (m)/Area (m^2)	Fracture Density (P10) (frac/m)	Fracture Density (P21) (m/m^2)
Sample 2-1	75	N017	$10^{-1}/10^{-2}$	177	240
	145	N057			
	101	N092			
	116	N130			
	135	N157			
Thin section 2-1	201	N014	$3 \times 10^{-2}/8 \times 10^{-4}$	517	1064
	213	N059			
	170	N097			
	221	N140			
Fracture map 3	149	N009	1.7/3	20	12
	88	N033			
	119	N071			
	70	N104			
	146	N149			

3.2. Methods

3.2.1. Fractures Acquisition

In this study, fracture digitization was performed manually by tracing every fracture segment from entire 2D maps in QGIS® software v.2.18.17. Only fractures reported as full solid lines on the DV geological map were considered for statistical analysis (Figure 2a).

Regarding the procedure of fracture digitization, the fractures extending out of the sampling domain were considered as one continuous feature [16,23,59] (Figure 2b). The digitization of fractures consists of an extraction of the end point coordinates of each fracture using the QGIS® software v.2.18.17. Each extracted fracture is characterized by the X and Y coordinates of each of the two end points, which help to compute the orientations and the length of segments. In case of curved fractures, a straight segment line was automatically traced between the end of point coordinates, and then the mean directions and lengths were computed.

As described above, the NH fracture data were acquired during the field mapping campaign and then completed by orthophoto image analysis. The data were analyzed in the entire area (Figure 2b). However, the complexity in loading history within NH widely discussed in the literature [48,54,60] makes the analysis of the fracture geometries difficult. Furthermore, the field observations and the new geological map done by [37] reveal additional complexities inside this area, and show a stacking of different Crystal Spring series (pCu, Figure 1c), intruded by the Mesozoic granite (Mzla, Figure 1c). This observation was highlighted in the northwest part of the NH. In center part, Crystal Spring series thicknesses were reduced, dragged, and stretched against the granite due to the SDVFZ activity (pCu, Figure 1c). The southeast of the NH is characterized by the absence of the Crystal Spring series. To better characterize the fracture variability, it is prime of interest to consider the facies variability and the deformation intensity within NH, because they influence the fracture evolution [23,59]. In this sense, we decided to divide the NH range into three domains according to the facies variability and change in range trend orientation variation (Figure 3): (1) Domain A is located northwestward, highlighted by thick Crystal Spring sediments and granitic rocks; (2) Domain B is in the central part of NH, highlighted by reduced thickness of Crystal Spring sediments, granitic and gneissic rocks, and (3) Domain C is located in the southeastern part of the NH and is characterized by granite, gneiss, and Tertiary volcanism. Note that Domain B and C are composed approximately of the same facies. However, range orientation in domain B is NW/SE, while in domain C it is toward E/W. This general E/W trend is puzzling as it is similar to the GF (see Figure 1b for location). Therefore, this difference in orientation constitutes an additional argument to better delimitate domain B and C. Figure 3 shows the fracture distributions within each domain. Some more work based on field analyses could more precisely delimitate the domains.

The photogrammetric models undertaken in this work were performed using a 3DR Solo drone with 4k cameras and DJI Phantom drone provided by University of Texas at El Paso (UTEP), Texas, USA. The videos were recorded between the late morning and the early afternoon during seven days with a manual mode camera setting to reduce the effects of lighting condition. Pictures were then extracted every second from the recorded drone videos using ffmpeg software v.4.5. The extracted pictures with a sufficient overlapping were aligned in Agisoft Metashape software (2020, Version 1.6.5). The obtained 3D models are georeferenced and then imported in CloudCompare software (version 2.9.1; [GPL software] (2017)) to begin the fracture extraction process. A workflow explaining the procedure of the models construction and the fractures extraction is detailed in [36]. In this study, the digitized fractures are projected on a 2D map to make the work only in 2D in order to keep data homogenous at different scales.

The 2D fracture maps at outcrop scale were performed in the field using a DSLR high resolution camera provided by UniLaSalle (Beauvais, France), with a 50 mm focal lens to minimize distortion. Pictures were taken vertically, with the same distance, and in the absence of sunlight to avoid the light effects in the fracture digitization [12]. From each fracture map, several pictures were taken with a sufficient overlapping. Then, pictures were aligned using the Agisoft Metashape software with the same procedure detailed in [36]. The maps are georeferenced and then used to extract the fractures in QGIS® with the same technique as for large scale maps.

Additional fracture datasets are used in this study by adding samples. These samples were taken directly from fracture maps. Then, a high-resolution picture was taken for each sample using a DSLR camera. With the same procedure as large-scale fractures acquisition, the sample was added in QGIS® to extract the fractures.

Thin section mosaics are also added to this study, made from samples as described earlier to keep a consistent sampling strategy. These mosaics were taken under an optical microscope Leica DM4500-P provided also by UniLaSalle (Beauvais, France), using a ×5 magnification and polarized non-analyzed light mode. A Leica DFC450C high resolution camera provided also by UniLaSalle (Beauvais, France) and Leica application Suite v.4.11.0 were used to take pictures. Every mosaic was relatively oriented using the corresponding sample. Then, with the same fracture digitization procedure described earlier, fracture parameters were extracted under QGIS® v.2.18.17.

3.2.2. Fractures Analysis

Different geometrical parameters were then collected from each fracture database at different scales. Detailed orientation distributions and classification into fracture sets were performed using the mixture of von Mises distribution (MvM) [61]. This approach consists of a semi-automated procedure based on appraisal tests in order to avoid any subjectivity in fracture sets analysis. For each fracture set, three output parameters were considered: (1) mean orientation (μ) around which the distribution is centered, (2) kappa (κ) which controls the concentration of the orientation's values around the mean, and (3) weight (ω) corresponding to the relative contribution of each fracture set to the model. Then, we are able to check the best number of fracture sets from the distributions using the goodness of fit parameters (e.g., Likelihood). For more details, see [62] who describe and adapt the methodology for structural data. The standard of deviation of $+/-10°$ was calculated for each given mean orientation value in this study.

Fracture length is the most used geometrical parameter to characterize the spatial organization in the natural fracture networks [4,25,63]. This characterization is performed using a statistical distribution as the Power law distributions widely used in structural data from field analogue [4,25,26,63,64]. The power law is often used to describe the distribution of fracture parameters such as length and aperture [25]. It is recognized that power law and fractal geometry provide widely applicable descriptive tools for fracture system characterization. This is due to the absence of characteristic length scales in the fracture growth process. The power law exponent α provides a real significance on the fracture connectivity. Indeed, for an exponent comprised between 2 and 3, fracture connectivity is ruled by both small and large fractures [65]. In addition, fractal analyses using Cantor's dust method allow the quantification of fracture distribution in clusters or, in the opposite, with a homogeneous distribution [66] and prediction of the fracture occurrence [7]. The exponential law is also used to describe the size of discontinuities in rocks [67–69], and to incorporate a characteristic scale that reflects a physical length in the system, such as thickness of a sedimentary layer [70].

However, fitting the geometric attributes to the statistical distributions suffers from a lot of biases, such as truncation and censoring biases [25]. Truncation effects are caused by resolution limitations of a field observation such as from satellite images, the human eye, or microscopes [63]. Censoring effects are associated with the probability that a long fracture intersecting the boundary of the sampling area is not sampled, and to the subjective choice of the sampling area which tends to exclude very long fractures [63,71]. Then, censoring effects cause an overestimation of fracture density. In this study, the truncation and censoring effects were automatically excluded from the distributions. Length distribution in Figure 4 shows the data excluded from the distribution (grey color) due to the impact of truncation and censoring effects.

The fracture length distributions are analyzed using the cumulative distribution. Statistical distributions used in this study were adjusted after corrections from the truncated and censoring effects. Research for truncation and censoring thresholds was performed using the « poweRlaw: Analysis of Heavy Tailed Distributions » package from Rstudio® v.1.3.1056 [72]. We also used the coefficient of determination (R^2) to estimate the goodness of fit of the power law to the length distribution. This coefficient is comprised between 0 and 1. A high R^2 for a given number of degrees of freedom means that the regression is a statistically meaningful description of the data.

Fracture densities are computed in this study. They consist of fracture density (P_{10}), defined as the number of fracture intersections along a 1D virtual scanline traced on every analyzed 2D map [59,73]. The scanline methodology is widely used to describe the fracture variability in 1D [73]. In this study, the linear fracture density P_{10} was obtained from two virtual scanlines, oriented perpendicular to the main structures.

Finally, the surface fracture density (P_{21}) is defined as the total sum of fracture lengths within the area [59,73].

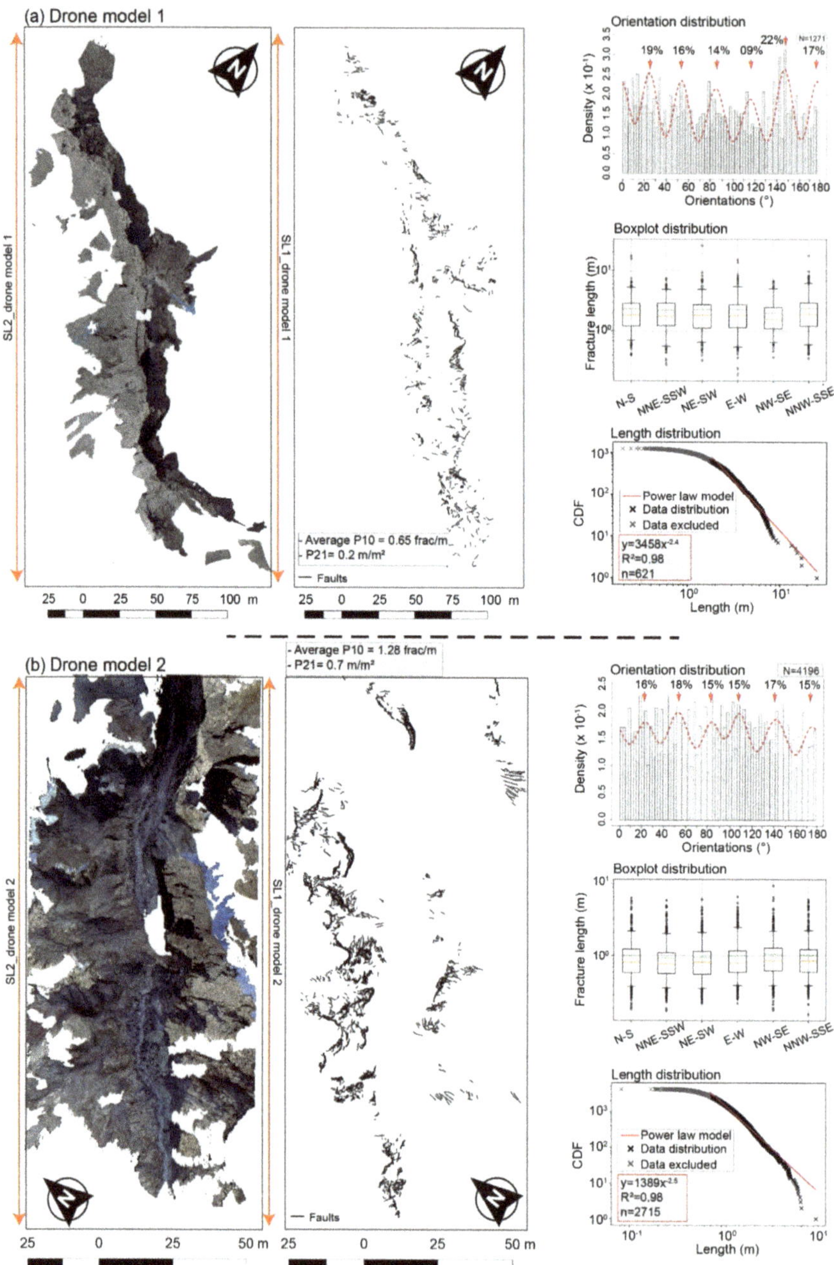

Figure 4. Fracture traces, orientation, boxplot length distributions from (**a**) drone model 1 and (**b**) drone model 2. Each fracture map data is expressed into histogram for orientation distribution, boxplot of length distribution according to the fracture orientations, and cumulative distribution function (CDF) of lengths. Two scanlines (SL) were taken along the left and the right side of the canyon. For the boxplots and orientation distributions legend, see Figure 2.

4. Results

4.1. Large Scale Domains

The mean orientation (μ) N095° (E/W trend) and N132° (NW/SE trend) trending fractures occur at the largest length scale, with a 20 and 24% of whole fractures, and about 3300 to 4400 m mean length (Figure 2a). Both directions are well expressed following the DV and Garlock strike slip trend faults (See Figure 1b for location) [39,48]. The largest fracture length is recorded within the NE/SW fracture set with 65 km length. The other recorded fracture sets are relatively equivalent (around 20%), except for the N025° fracture set which is less dominant, with 13% and 1900 m mean length.

At NH scale, the same dominance of N090° and N126° fracture sets with respectively 37% and 31% are observed (Figure 2b). The other fracture sets are poorly expressed (<15%). The boxplot distribution in Figure 2b showed also the dominance of the E/W and NW/SE fracture sets with respectively 80 m and 125 m mean length. The other fracture sets did not exceed the 50 m mean length.

The occurrence of two prominent fracture sets is shown in the NH domains A and B (Figure 3a,b). Domain A is characterized by two main fracture sets striking N091° and N132° representing respectively 29% and 55% of occurrences (Figure 3b). The N091° fracture set lengths range from 30 m to 300 m, while the N132° fracture set length ranges from 5 m to 800 m, meaning that the NW/SE is the most dominant fracture set in domain A (Figure 3b).

Orientation and length data of domain B shows an equivalent distribution as in domain A (Figure 3b). Indeed, two main fracture sets striking also N088° and N128° are mostly representative with respectively 45% and 31%. The fractures are much longer within NW/SE trend, ranging from 2 m to 1300 m (mean is of 108 m). The other fracture sets show an intermediate length with a 50 m mean value.

Fracture geometry in Domain C shows immediately a different distribution with four main fracture sets striking N050° (19%), N085° (31%), N111° (20%), and N141° (17%) (Figure 3b). The NW/SE fracture set mainly sampled and representative in the domain A and B is split into N111° and N141° fracture sets in domain C, with respective lengths ranging from 3 to 1600 m (mean is 190 m) and from 8 m to 730 m (mean is 65 m).

The fracture densities are increasing considerably from domain A to C with $2 \times 10^{-2}/3 \times 10^{-2}$ m/m^2 in the domain A and B, and 5×10^{-2} m/m^2 in the domain C.

4.2. Photogrammetric Models

NH canyons modeled with photogrammetric method provide fractures from centimetric to decametric scale. Fracture distributions are displayed by drone model 1 for both sides of the canyon walls (Figure 4a). Six fracture sets are recorded and strike N025°, N055°, N085°, N115°, N142°, and N174°. The N142° is also the most dominated fracture set with 22%. The E/W fracture set is the less expressed (14%) at outcrop scale. An equivalent mean length of fractures is shown by the boxplot distribution graphic with 2.3 m, whatever the fracture set (Figure 4a).

The exponent of power law (red line in the length distribution, Figure 4) is of $\alpha = -2.4$ for length distribution of the drone model 1, outside lengths affected by truncation and censoring bias (data mentioned in the length distribution graphic with grey color, Figure 4). The range of length values performed by the power law is 1.85 m to 25 m.

Finally, fracture densities P_{21} and P_{10} are of respectively 2×10^{-1} m/m^2 and 6.5×10^{-1} frac/m.

Fractures from drone model 2 are mainly distributed along the left side of the canyon (Figure 4b). The same fracture sets as drone model 1 are recorded with an equivalent percentage (between 15 and 18%) (Figure 4b). Fracture lengths are highlighted by the boxplot distribution and are also equivalent with 1 m mean length, whatever the fracture set. The largest fractures are recorded within the NW/SE fracture set (maximum of 8.5 m), and then confirm the consistency of the NW/SE direction at outcrop scale.

The fracture length range modeled by the power law is of 0.69 m to 8.5 m, for exponent $\alpha = -2.5$ and $R^2 = 0.97$ (Figure 4b). Fracture lengths under 0.69 m are excluded automatically from the distribution (truncation effects).

Fracture densities are quite high in comparison with Drone model 1, with 7×10^{-1} m/m^2 and 1.28 frac/m for respectively P_{21} and P_{10}.

Drone model 3 is the largest studied model with 15594 traced fractures (Figure 5). Six fracture sets are highlighted also in this canyon. Their proportions range from 13% (N079° and N109°), 14–16% (N140° and N038°), to 19–20% (N018 and N169°). The fracture lengths are equivalent, once again, with a 1.2 m mean length.

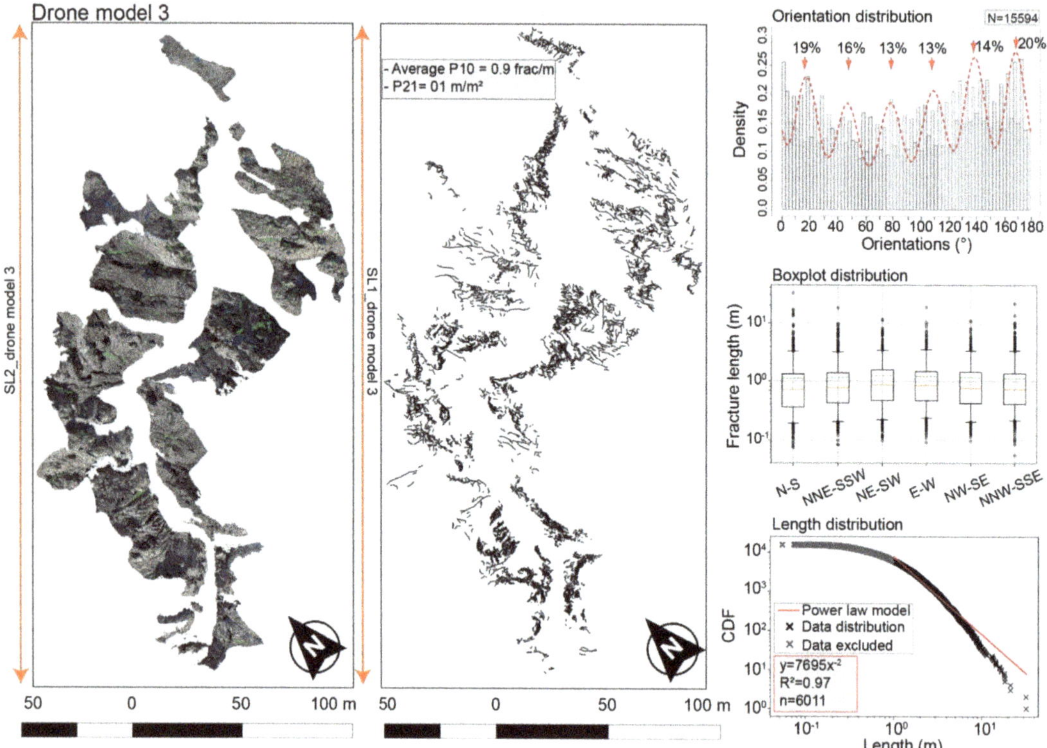

Figure 5. Fracture traces, orientation, boxplot length distributions from drone model 3. Different Datasets are expressed into histogram for orientation distribution, boxplot of length distribution according to the fracture orientations, and cumulative distribution function (CDF) of lengths. Two scanlines (SL) were taken along the left and the right side of the canyon. For Boxplot and orientation distribution legend, see Figure 2.

The power law exponent for drone model 3 length distribution is of $\alpha = -2$, outside lengths affected by truncation and censoring effects (Figure 5). R^2 value is of 0.97, indicating a good fit of the power law to the data. Fracture density P_{21} is of 1 m/m^2, increasing by a factor 5 compared to drone model 1 fracture density.

4.3. Fracture Maps

Fracture map 1 is characterized by intense fracture areas, especially along the center part (Highly fractured area, Figure 6). The high fracture density is located mainly in the quartzite boudin, while the less fractured zone is related to the granitic rocks.

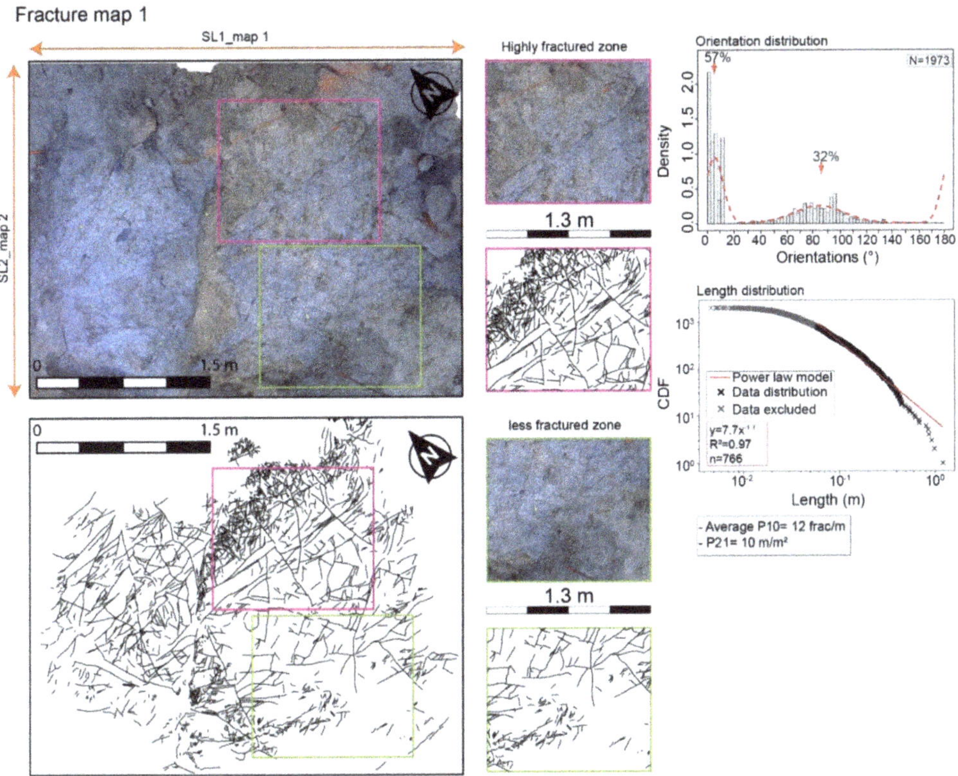

Figure 6. Fracture traces, orientation and length distributions from fracture map 1. Different datasets are expressed into histogram for orientations distribution, and into cumulative distribution function (CDF) of lengths. Two scanlines (SL) were taken for every domain. For orientation distribution legend, see Figure 2.

For the rest of the fracture map, average P_{10} of 12 frac/m is computed from two scanlines (SL1_map1 and SL2_map 2, Figure 6), while P_{21} is of 10 m/m^2. Two main fracture sets are highlighted: N005° and N083°, with respectively 57% and 32% (Figure 6). Outside fracture lengths affected by truncation and censoring bias, the power law exponent is of $\alpha = -1.7$ for values ranging from 6×10^{-2} m to 1.2 m.

Sample 1-1 is characterized by homogeneous fracture distribution with four main fracture sets: N018°, N053°, N090°, and N126° representing respectively 27%, 30%, 30%, and 13% (Figure 7a). The largest veins visible in the sample map are characterized by carbonate mineralizations in two directions: N053° and N126°. P_{10} and P_{21} fracture densities are of 184 frac/m and 118 m/m^2 respectively. Regarding the length distribution, the power law exponent is of $\alpha = -1.43$, with thresholds of 3×10^{-3} m and 7×10^{-2} m.

Four main fracture sets are highlighted in thin section 1-1: N018°, N053°, N090°, and N126° with a comparable fracture proportion (around 25%). Additional fracture set is detected in thin section 1-2: N162°. Regarding the densities, P_{21} is of 715 m/m^2 and 702 m/m^2, while P_{10} is of 5140 frac/m and 4525 frac/m respectively for thin section 1-1 and 1-2 (Figure 7b,c). The power law exponents are quite higher with $\alpha = -1.95$ (modeled range from 2.2×10^{-4} m to 4.63×10^{-3} m) and $\alpha = -1.7$ (modeled range from 2.2×10^{-4} m to 7×10^{-3} m) respectively for thin section 1-1 and 1-2.

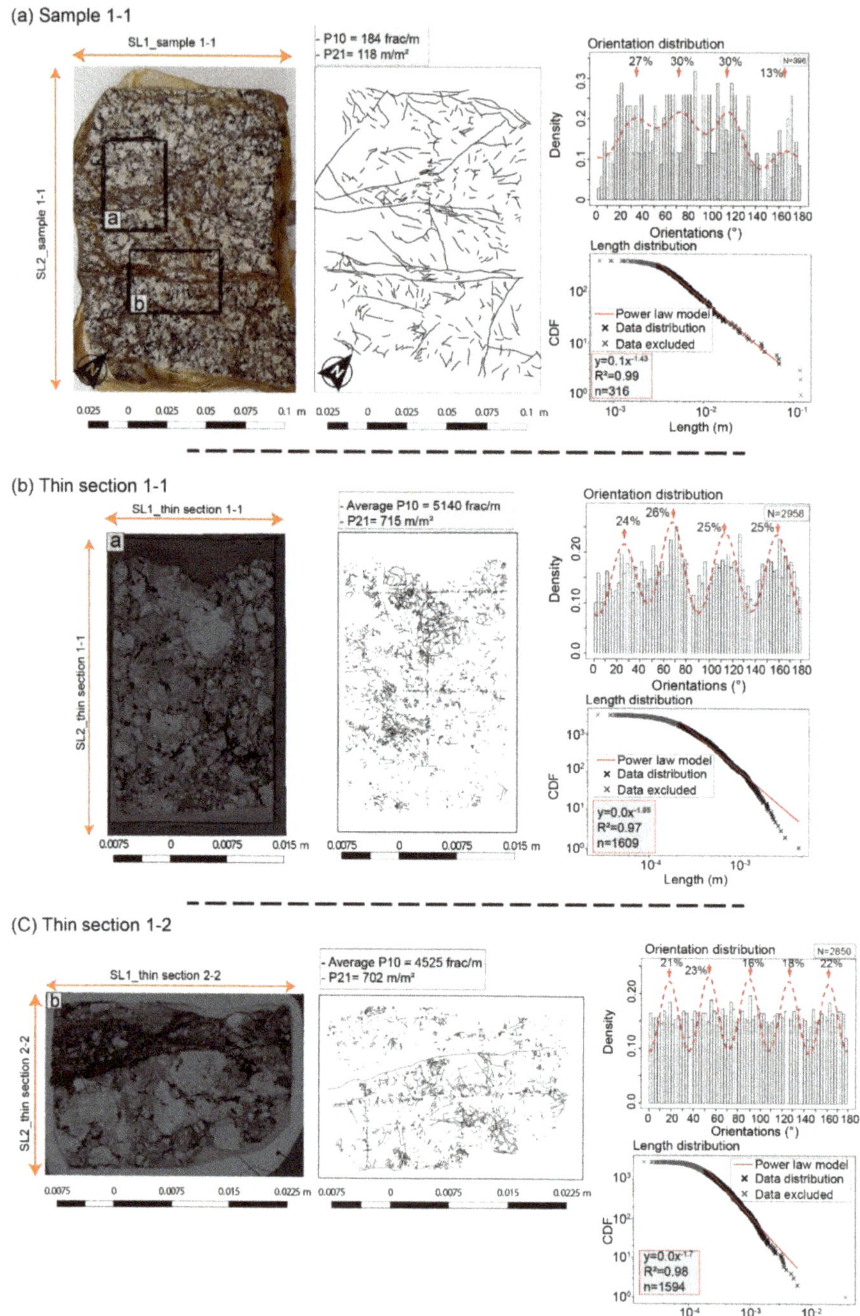

Figure 7. Fracture traces, orientation and length distributions from (**a**) sample 1-1, (**b**) thin section 1-1, and (**c**) thin section 1-2 sampled from fracture map 1. Different datasets are expressed into histograms for orientations distribution, and into cumulative distribution function (CDF) of lengths. Two scanlines (SL) were taken for each map. For orientation distribution legend, see Figure 2.

Fracture densities of fracture map 2 are comparable with fracture map 1 with average P_{10} = 18 frac/m and P_{21} = 13 m/m^2 (Figure 8a). Five fracture sets are highlighted: N005°, N038°, N070°, N110°, and N145°. The N070° and N145° are the most recorded fracture sets with respectively 32% and 27%.

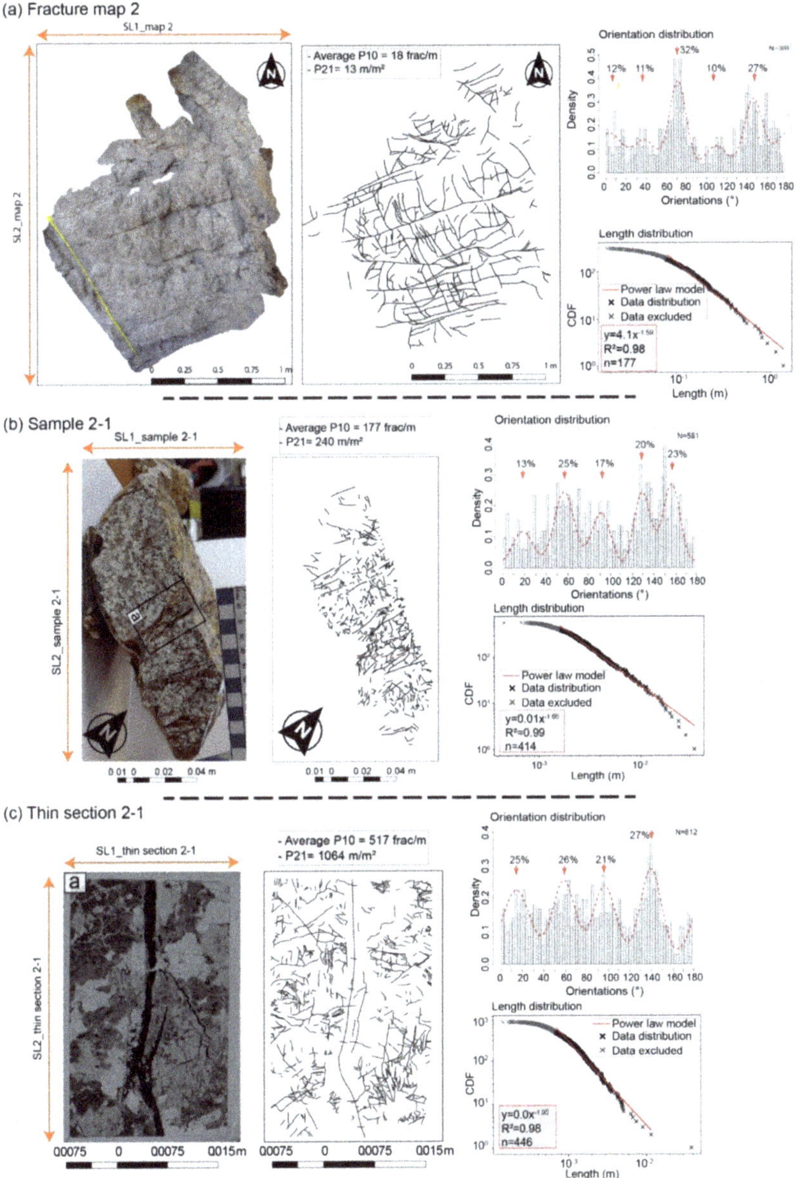

Figure 8. Fracture traces, orientation, and length distributions from (**a**) fracture map 2, (**b**) sample 2-1, and (**c**) thin section 2-1, located in sample 2-1. Different datasets are expressed into histogram for orientations distribution, and into cumulative distribution function (CDF) of lengths. Two scanlines (SL) were taken for each map. For orientation distribution legend, see Figure 2.

The power law exponent is also comparable to that of fracture map 1 ($\alpha = -1.59$), with the thresholds of 8×10^{-2} m and 1.43 m (Figure 8a).

Sample 2-1 is characterized by a fracture density of $P_{10} = 177$ frac/m and $P_{21} = 240$ m/m^2 (Figure 8b). In total, five fracture sets are highlighted: N017°, N057°, N092°, N130°, and N157°. Three fracture sets are the most recorded: N057°, N130° and N157°, with 20% to 25%. The power law exponent is of $\alpha = -1.66$, thresholds from 1.6×10^{-3} m to 3.33×10^{-2} m.

Thin section 2-1 trans-cutting a large carbonate vein in the central part (Figure 8c), is characterized by a fracture density $P_{21} = 1064$ m/m^2 higher than in thin section 1-1 and 1-2. Only four fracture sets are highlighted: N014°, N059°, N097° and N140°, with an equivalent fracture proportion. Outside the lengths affected by sampling bias, the exponent is of $\alpha = -2$, thresholds of 7×10^{-4} m to 3.5×10^{-2} m.

Finally, fracture map 3 is characterized by a comparable fracture density as in fracture map 2 with $P_{10} = 20$ frac/m and $P_{21} = 12$ m/m^2 (Figure 9). Five fracture sets are also detected with a high proportion of 26% for N009° and N149°. Fracture sets N033°, N077°, N104° showed an abundance of 15%, 21%, and 12% respectively. The power law exponent is of $\alpha = -1.53$ for lengths data ranging from 4×10^{-2} m to 1.21 m.

Fracture map 3

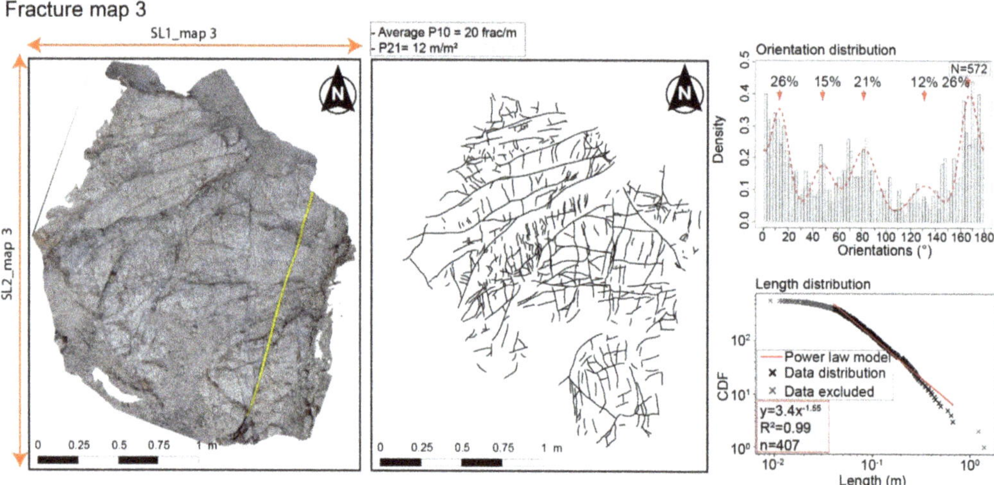

Figure 9. Fracture traces, orientation and length distributions from fracture map 3. The computed dataset is expressed into histogram for orientations distribution, and into cumulative distribution function (CDF) of lengths. Two scanlines (SL) were taken. For orientation distribution legend, see Figure 2.

5. Discussion

The combination of data at various scales in this study allowed to propose a new interpretation of the fracture network in the NH range, based on multiscale evolution of fractures length and orientation. A conceptual scheme of the NH structuration is created and discussed in terms of fracture network connectivity and its influence of fluid flow.

5.1. Multiscale Length Characterization

The cumulative length distributions of each fracture map from the DV regional map scale to thin section scale are plotted in Figure 10. All length datasets have been fitted to the power law distribution, outside lengths affected by truncation and censoring bias.

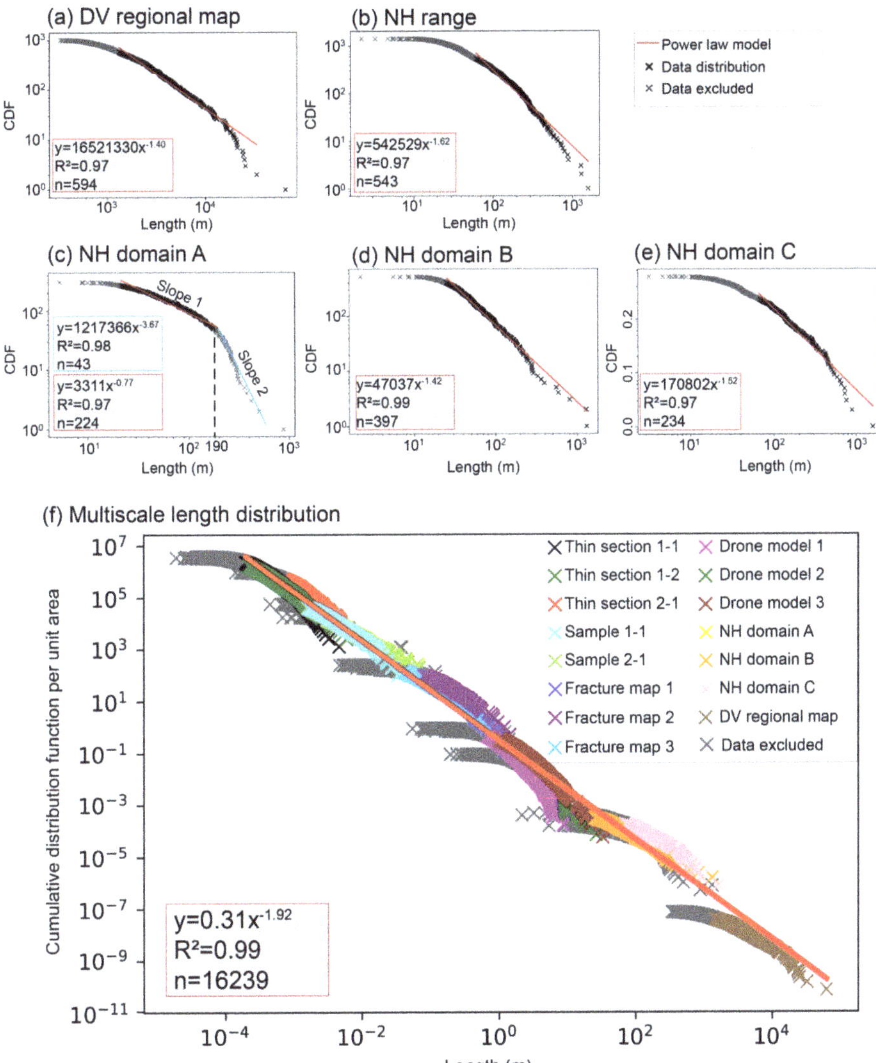

Figure 10. Fracture length distribution from (**a**) DV geological map, (**b**) NH range, (**c**) NH map domain A, (**d**) NH map domain B, and (**e**) NH map domain C, and (**f**) Multiscale length distribution obtained from the combination of the whole data.

DV regional map exponent is quite low, around $\alpha = -1.40$, and gradually increase to around $\alpha = -1.62$ in NH map (Figure 10a,b). The division of the NH into 3 domains according to the facies variability and the change in range trend orientation showed a disparity in power law exponent values. Indeed, the exponent $\alpha = -0.83$ in the domain A (values ranging from 21 m to 230 m), while $\alpha = -1.42$ (values ranging from 25 m to 1283 m) and $\alpha = -1.52$ (values ranging from 70 m to 1500 m) respectively in domain B and C (Figure 10c–e). The lowest α value in domain A can be explained by the dependence of fracture lengths on scale of observation for fractures ranging from 5 m to 870 m. In addition, the domain A is composed of granitic rocks and Crystal Spring sedimentary series with carbonates and quartzites (Figure 1), meaning that the sedimentary rocks can have a different behavior from that of basement rocks during the fracture growth and propagation process.

Cumulative distribution of fracture lengths in the domain A showed also two different trend slopes (Figure 10c), with a failure slope quantified around 230 m. Distribution into two slopes has been largely discussed in the literature [25,67,70], and can be explained by the fracture growth process which has been divided into two trends. However, this result and interpretation cannot allow to determine if these two trends correspond to a single or to two different regional directions. Cumulative distribution of fracture lengths according to their orientations and the nature of rocks could help to resolve the episodes of deformation challenge, together with some more work in the field (e.g., looking for evidences of displacement).

The cumulative fracture length from all maps has been plotted in a single graphic, normalized by surface area of each map [63] (Figure 10f). The power law distribution included fracture lengths from 2.2×10^{-4} m to 65 km scale. Power law distribution was performed over 6 orders of magnitude. Two length ranges were not represented in this study. The first one corresponds to the fracture lengths over 100 km and related to the first order fracture defined by [30]. This absence can be explained by: (1) a different evolution stage and opening mechanism of faults [30,74], (2) a structural heritage which controls the reactivation during the DV trans-tensional tectonic setting. The second one corresponds to the length ranging from 1 km to 5 km length (Figure 10f). The Basin and Range regional map and the geological map at 1:25,000 scale could help to resolve both gaps.

The power law distribution gives an exponent $\alpha = -2$ for the whole 2D fracture lengths analyzed in this study (Figure 10f), which suggests the 2D representation is self-similar [4,16,23,26]. The probability to detect fracture of the size of the sampling window is the same at all scales. Then, the fracture connectivity is ruled by the small and large fractures [65], meaning that the large fractures detected and studied at large scale in the NH play the same role as any fracture at millimeter scale in fracture network connectivity.

5.2. Spatial Organization of NH Fracture Network

The fracture densities and the main dominant orientations on the regional maps show the control of NH geometry by NW/SE and E/W trends. Both trends are associated to the second order scale faults over 20 km length, well expressed in terms of abundance (DV regional map, Figure 11). The third order scale detected along NE/SW trend is characterized by faults around 10 km length. Spacing values were computed and correspond to 5 km and 1 km respectively for the second and the third order scale (Figure 12a). The same geometrical characteristics (detailed in the introduction) were approved in the extensional regime [20,30], meaning that the second and third order scale spacing classification can be generalized to the trans-tensional regime studied in the present work.

At NH scale, the fracture densities and the main dominant orientations approve the separation of the NH into three domains:

1. The first domain referred to the NH domain A is characterized by a specific spatial arrangement with a NW/SE direction dominance (Domain A, Figure 11). This direction is marked by the SDVFZ bordering fault of the strike-slip corridor, and is mostly dominated the NH fracture network in terms of length and abundance. It approved the importance of the NW/SE deformation episode at the fourth order scale. These fractures are the longest ones and also control the internal structuration inside the domain (Figure 12b). The E/W direction is less represented in comparison with domains B and C (Figure 11), but still the second most dominated fracture set, with less control on the NH geometry. Regarding the spacing characteristics, the NW/SE fractures set are regularly spaced from 0.1 to 0.2 km, while the E/W fractures set did not exceed 0.1 km spacing (Figure 12b).
2. The central domain (domain B, Figure 3) is characterized by several short segments of fractures. Indeed, the E/W direction is also dominated by short fractures with 60 m mean length, while the NW/SE direction is the second dominated system with long

fracture of 100 m mean length (Figure 11). Then, NW/SE and E/W directions control the NH geometry respectively following the third and fourth order scale length. In this case, the spacing related to the third order scale is around 1 km between the SDVFZ bordering faults and regularly spaced at 0.2 km (Figure 12b). A spacing of 0.05 km is defined along the E/W fracture set (Figure 12b).

3. The southeastern end area corresponding to the domain C (Figure 3) highlights a specific spatial arrangement with an additional fracture set (yellow color, Figure 11), the longest one with 200 m mean length. Indeed, the NW/SE direction is split into N111° and N141° fracture sets thus highlighting the influence of both NW/SE and E/W deformation episodes in this area. Once again, the E/W direction is more expressed, with a 100 m mean length (Figure 11). The ENE/WSW, E/W and also NW/SE directions control the NH geometry and structuration respectively following the third and fourth order scale. The spacing of 0.3 km defined for the third order scale in domain C between the SDVFZ bordering faults is under the third order spacing characteristic. Indeed, the NW/SE direction has been deviated and then the relative spacing is reduced. Regarding the internal organization, the NW/SE and E/W fractures are also regularly spaced with 0.1–0.2 km and 0.05 km, respectively (Figure 12b).

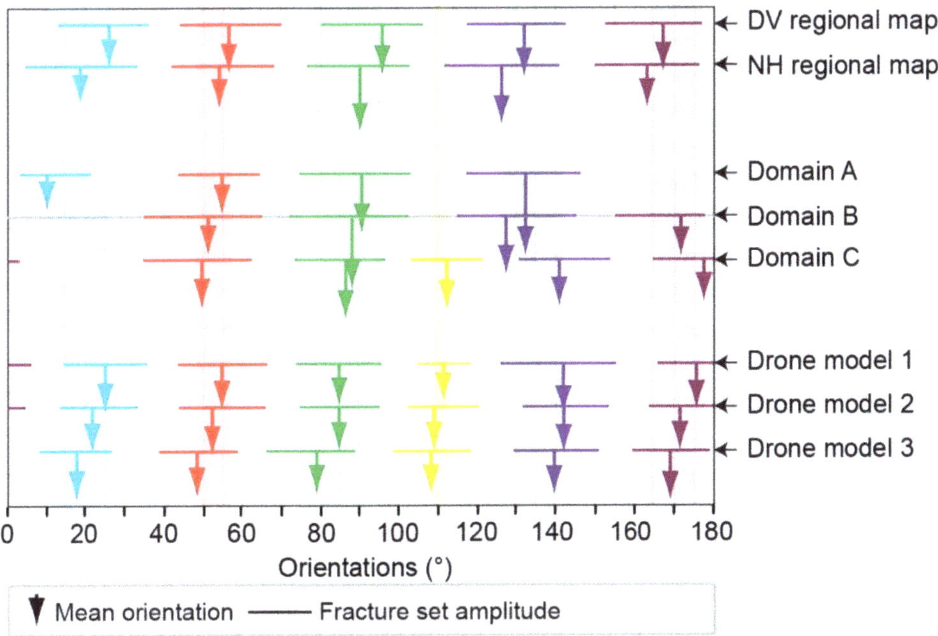

Figure 11. Orientation plot representing mean orientation of each fracture set from DV regional map, NH regional map, NH domains and drone models at outcrop scale. Each arrow corresponds to the mean orientation. The length of each arrow corresponds to the fracture set abundance. The horizontal line above each arrow corresponds to the fracture set standard deviation within an interval of confidence of 75% (see [62] for more explanations). Each color arrow corresponds to fracture set: Blue arrow corresponds to NNE/SSW fracture set, red arrow corresponds to the NE/SW fracture set, green arrow corresponds to the E/W fracture set, yellow arrow corresponds to the NNW/SSE fracture set, purple arrow corresponds to the NW/SE fracture set, and marron arrow corresponds to the N/S fracture set.

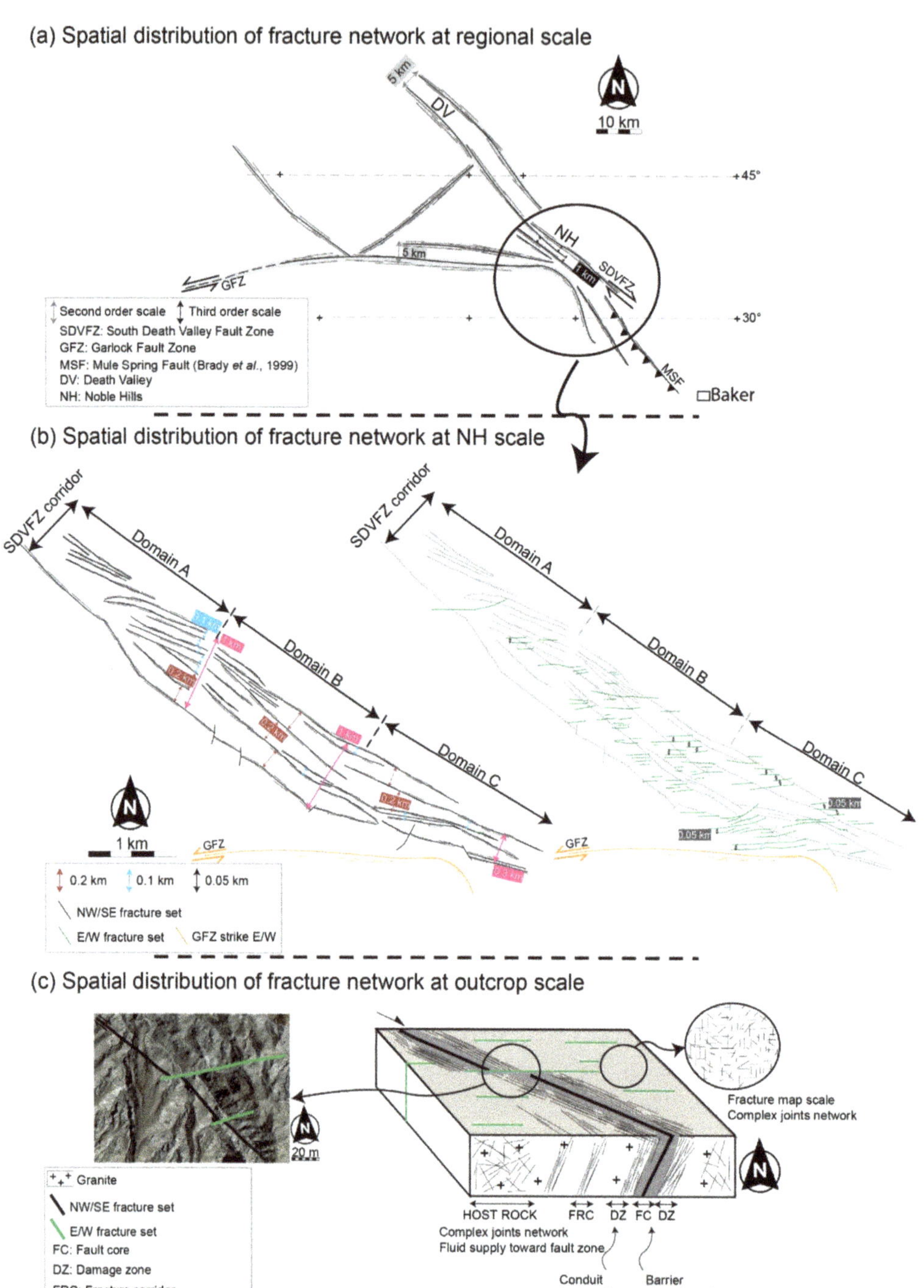

Figure 12. Conceptual scheme presenting the spatial arrangement of the NH fracture network at (**a**) regional scale, (**b**) NH scale, and (**c**) outcrop scale. The organization is based on scale orders referring to the [16,17,20,30] classification, and their associated spacing characteristics referring to [20,30]. The picture presented in (**c**) is provided by GoogleEarth®.

In the present work, the spacing characteristics of 0.05 km to 0.1–0.2 km computed in the three internal domains are attributed to the fourth order scale (Figure 12b). These values vary according to their directions and also lengths. Two spacing characteristics can be highlighted from this study: (1) spacing of 0.05 km corresponding to the fractures comprised between 0.1 and 0.35 km (percentile 90 values), (2) spacing of 0.1–0.2 m corresponding to the fractures comprised between 0.35 and 0.5 km. Further work on the fourth order scale spacing characteristics will obviously be needed in order to generalize them to the trans-tensional regime.

At outcrop scale, the drone models highlighted the reproducibility and the consistency of the whole fracture sets defined at large scale (Figure 11). However, N111° fracture set, detected only in domain C, is less expressed in domain A (drone model 1, 9%) (Figure 11). The E/W direction trend is more expressed in the drone models 2 and 3 respectively in the domain B and C. This observation is confirmed by the fracture maps 2 and 3, sample 2-1 and thin section 2-1 (Figures 8 and 9), highlighting a consistent fracturing episode. In the previous works, [39,55–57] mentioned the presence of the compression signature at the southeastern end of the NH and the front of the Avawatz mountains. According to these studies and our current observations, a hypothesis of fracturing intensity which affected up to the internal domain can explain the consistency of the E/W direction.

A special fracture arrangement was observed in fracture map 1 (Figure 6). The length distribution is characterized by a high density of short fractures in the quartzite boudin rocks, while the granitic part is characterized by a lower density of long fractures. This leads to a change in the power law and orientation distributions and then makes a bias. Indeed, only N/S and E/W directions are recorded, while the sample 1-1, thin sections 1-1 and 1-2 have recorded the whole fracture sets.

Fractures are organized along SDVFZ direction as fault zone segments with a highly mineralized (clay minerals) fault core (FC) which can act as a barrier and damage zone (DZ), acting as a conduit (Figure 12c). A complex joint network that can play a key role in the fracture connectivity is highlighted in the host rock. A petrographic study focusing on these fault zones is required to better characterize the fluid circulation potential.

The N/S to NNE/SSW and the NE/SW fracture sets are less expressed and are attributed to the fourth order scale with 40 m mean length on NH map, while they are still very important and consistent at outcrop, sample, and thin section scales, regardless of domain (Figures 2–5). These trends are highly dependent on the scale of observation and have no influence on the NH geometry.

To summarize, a new spatial organization of the NH range based on second, third, and fourth order scales has been proposed with length, spacing, density, and orientation distributions. Each NH domain shows its own internal organization. Variability in between recorded fracture sets from different areas is a marker of a complex tectonic and geometrical setting. One main deformation episode played a key role in the NH structuration: SDVFZ trending NW/SE affected whole NH and then controlled its geometry. The second main deformation episode is in the GF system trending E/W, likely responsible for E/W fracturing episode, and controlled the NH geometry mostly in its internal part and the southeastern end.

The identical orientation of GF and the E/W structures could suggest that the GF was responsible for the E/W fracturing in the internal and southeastern end domains (Figure 12b). Furthermore, the E/W structures are characterized by a sinistral strike-slip displacement, highlighted by the movement of the tertiary volcanic blocks in the back of the NH. This sinistral strike-slip movement has been characterized along the GF zone, widely described in the literature [55,57]. The similarities on the strike-slip nature can help to consider that the E/W fractures are related to GF activity which plays a key role on the NH geometry. In addition, the lower dominance of the NW/SE direction in the central and the southeastern end of the NH can be explained by the overprinting of the E/W fracturing, highlighting that these structures are posterior to the SDVFZ system (Figure 12c).

Regarding the fracture distributions, the fracture density is higher in the domain C with a factor 2.5 and 1.5, in comparison respectively with domain A and B (Figure 3). At outcrop scale, fracture densities raise from domains A and B to domain C with a factor 05 and 1.4, approving the localization of the fracture intensity in this area due to the complex tectonic setting [39]. Furthermore, a gradient of deformation has been observed in the field along the entire NH length with evidence of extreme shearing, notably in the internal domain. Boudinage structures and brittle shearing are prevalent within Crystal Spring series. A future publication will be planned to better characterize this deformation.

5.3. Fracture Network Impact on Flow

The deformation occurred with a different intensity at the whole NH scale. This can influence the fluid circulation through the fracture network following the domains. Even if the connectivity in the reservoir analogue is ruled by the large and small fractures (Figure 10f), some domains, such as domain A, had a different behavior insofar as the connectivity was ruled by only the large fractures. A complex joint network is highlighted at outcrop scale in the host rock and plays a key role in the fracture connectivity leading the fluid supply toward the fault zone.

The fluid flow modeling in this reservoir analogue should consider that the orientation parameter depends on scale observation and then should be modeled differently at each scale, while the length parameter is independent of the observation scale in the case of NH geothermal reservoir analogue.

Flow simulations will be planned in the future publications, based on photogrammetric models acquired from different NH domains. This study can help to approve the sensitivity of the fluid circulations to the gradient of strike-slip deformation and the spatial arrangement of NH fracture network. The sensitivity of the deterministic models on resulting permeability will be tested according to the directional dependence and their corresponding length distribution.

6. Conclusions

This study is part of the MEET project (Multidisciplinary and multi-context demonstration of EGS exploration and Exploitation Techniques and potentials), aiming at developing geothermal exploitation at European scale by applying Enhanced Geothermal System (EGS) technology to different geological settings. NH range, assimilated to a paleo geothermal reservoir analogue, gives an opportunity to study the basement rocks in trans-tensional context. This study proposes geometric description and quantifies the multiscale network organization and its effect on connectivity using a wide-ranging analysis scale from the microscopic scale to the regional scale. We used 2D fracture maps at different scales. We have shown a power law distribution for six orders of scales. Then, a power law exponent of $\alpha = -2$ was computed by combining the whole datasets, meaning that the connectivity is ruled by the small and the large fractures.

A new spatial arrangement of the fracture network at different scales has been proposed for the NH range, based on densities, spacings, orientations, and length distributions. The SDVFZ controls the NH geometry at large scale within the second order scale, while the E/W directions—whose origin remains to be determined—controls the NH geometry within the third order scale. The spacing characteristics are of 5 and 1 km for respectively the second and the third order scale, and correspond to the spacings highlighted in the extensional regime.

The division into three domains according to the facies variability and the change in range trend orientation has been approved by statistical analysis. Indeed, domain A is characterized by a dominance of the NW/SE direction in terms of length and abundance at the fourth order scale with a regular fracture spacing of 0.1–0.2 km. Then, NW/SE direction controls the internal structuration of this domain. Domain B is structured by the second and third orders of fractures scales following the E/W and NW/SE direction with regular spacing of 0.2 km and 0.005 km, respectively. Domain C was also characterized by the structuration of a specific spatial arrangement with the dominance of short fractures following E/W direction, which highlight the persistence of the E/W fracturing episode. The ENE/WSW, E/W, and also NW/SE directions control the NH geometry and structuration in this area. A regular spacing of 0.2 km and 0.005 km is computed inside the domain at fourth order scale. Therefore, two spacing characteristics have been highlighted in this study at fourth order scale: 0.1–0.2 km and 0.05 km spacing for fractures length of 0.1–0.35 km and 0.005 km, respectively.

Each internal domain described in this study proposes its own spatial arrangement of fracture network. Indeed, two main deformation episodes referring to the SDVFZ striking NW/SE and fractures system striking E/W play a key role in the structuration of the NH range. However, the SDVFZ deformation affected the whole NH, whereas the E/W fracturing affected only the central and the southeastern area.

Fluid flow modeling will be planned and will take into consideration that orientation parameters should be modelled differently at each scale, while length parameters modelled by the power law should be considered as homogenous at different scales.

Author Contributions: Conceptualization, A.C., G.T. and B.A.L.; methodology, A.C.; software, A.C.; validation, G.T., B.A.L. and J.K.; formal analysis, A.C.; investigation, A.C., G.T., B.A.L. and J.K.; resources, A.C.; data curation, A.C.; writing—original draft preparation, A.C.; writing—review and editing, G.T., B.A.L. and H2020 MEET consortium; visualization, G.T., B.A.L. and J.K.; supervision, G.T.; project administration, G.T.; funding acquisition, G.T. and H2020 MEET consortium. All authors have read and agreed to the published version of the manuscript.

Funding: This project has received funding from the European Union's Horizon 2020 research and innovation program under grant agreement No 792037 (MEET project).

Data Availability Statement: Not applicable.

Acknowledgments: This work is part of postdoctoral contribution, prepared at Institut Polytechnique UniLaSalle Beauvais, which was funded by European Union's Horizon 2020 research (H2020 MEET project). We are grateful to Armand Pomart and Ruben Attali for their help for analyses. We also wish to thank the H2020 MEET consortium for their helpful comments and manuscript validation. We also thank the Assistant Editor Sorin Hadrian Petrescu and the anonymous reviewers for the relevant remarks that significantly approved the article.

Conflicts of Interest: The authors declare no conflict of interest.

References

1. Bour, O.; Davy, P. Connectivity of Random Fault Networks Following a Power Law Fault Length Distribution. *Water Resour. Res.* **1997**, *33*, 1567–1583. [CrossRef]
2. Gillespie, D.T. Approximate Accelerated Stochastic Simulation of Chemically Reacting Systems. *J. Chem. Phys.* **2001**, *115*, 1716–1733. [CrossRef]
3. Berkowitz, B. Characterizing Flow and Transport in Fractured Geological Media: A Review. *Adv. Water Resour.* **2002**, *25*, 861–884. [CrossRef]
4. Odling, N.E.; Gillespie, P.; Bourgine, B.; Castaing, C.; Chiles, J.P.; Christensen, N.P.; Fillion, E.; Genter, A.; Olsen, C.; Thrane, L. Variations in Fracture System Geometry and Their Implications for Fluid Flow in Fractured Hydrocarbon Reservoirs. *Pet. Geosci.* **1999**, *5*, 373–384. [CrossRef]
5. Johnston, J.D.; McCaffrey, K.J.W. Fractal Geometries of Vein Systems and the Variation of Scalingrelationships with Mechanism. *J. Struct. Geol.* **1996**, *18*, 349–358. [CrossRef]
6. Geiger, S.; Emmanuel, S. Non-Fourier Thermal Transport in Fractured Geological Media. *Water Resour. Res.* **2010**, *46*, W07504. [CrossRef]
7. Ledésert, B.; Dubois, J.; Genter, A.; Meunier, A. Fractal Analysis of Fractures Applied to Soultz-Sous-Forets Hot Dry Rock Geothermal Program. *J. Volcanol. Geotherm. Res.* **1993**, *57*, 1–17. [CrossRef]
8. Dezayes, C.; Genter, A.; Valley, B. Structure of the Low Permeable Naturally Fractured Geothermal Reservoir at Soultz. *Comptes Rendus Géosciences* **2010**, *343*, 517–530. [CrossRef]
9. Valley, B.; Dezayes, C.; Genter, A.; Maqua, E.; Syren, G. Main Fracture Zones in GPK3 and GPK4 and Cross-Hole Correlations. In Proceedings of the EHDRA Scientific Conference, Soultz-sous-Forêts, France, 17–18 March 2005.
10. Bour, O.; Davy, P. On the Connectivity of Three-dimensional Fault Networks. *Water Resour. Res.* **1998**, *34*, 2611–2622. [CrossRef]

11. Darcel, C.; Bour, O.; Davy, P.; de Dreuzy, J.R. Connectivity Properties of Two-Dimensional Fracture Networks with Stochastic Fractal Correlation: Connectivity of 2D Fractal Fracture Networks. *Water Resour. Res.* **2003**, *39*, 1272. [CrossRef]

12. Odling, N.E. Scaling and Connectivity of Joint Systems in Sandstones from Western Norway. *J. Struct. Geol.* **1997**, *19*, 1257–1271. [CrossRef]

13. Castaing, C.; Halawani, M.A.; Gervais, F.; Chilès, J.P.; Genter, A.; Bourgine, B.; Ouillon, G.; Brosse, J.M.; Martin, P.; Genna, A. Scaling Relationships in Intraplate Fracture Systems Related to Red Sea Rifting. *Tectonophysics* **1996**, *261*, 291–314. [CrossRef]

14. Marrett, R.; Ortega, O.J.; Kelsey, C.M. Extent of Power-Law Scaling for Natural Fractures in Rock. *Geology* **1999**, *27*, 799–802. [CrossRef]

15. Soliva, R.; Schultz, R.A.; Benedicto, A. Three-dimensional Displacement-length Scaling and Maximum Dimension of Normal Faults in Layered Rocks. *Geophys. Res. Lett.* **2005**, *32*, L16302. [CrossRef]

16. Bertrand, L.; Géraud, Y.; Le Garzic, E.; Place, J.; Diraison, M.; Walter, B.; Haffen, S. A Multiscale Analysis of a Fracture Pattern in Granite: A Case Study of the Tamariu Granite, Catalunya, Spain. *J. Struct. Geol.* **2015**, *78*, 52–66. [CrossRef]

17. Bertrand, L.; Jusseaume, J.; Géraud, Y.; Diraison, M.; Damy, P.-C.; Navelot, V.; Haffen, S. Structural Heritage, Reactivation and Distribution of Fault and Fracture Network in a Rifting Context: Case Study of the Western Shoulder of the Upper Rhine Graben. *J. Struct. Geol.* **2018**, *108*, 243–255. [CrossRef]

18. Ehlen, J. Fractal Analysis of Joint Patterns in Granite. *Int. J. Rock Mech. Min. Sci.* **2000**, *37*, 909–922. [CrossRef]

19. Genter, A.; Castaing, C. Effets d'échelle Dans La Fracturation Des Granites. *Comptes Rendus de l'Académie des Sci. Ser. IIA Earth Planet. Sci.* **1997**, *325*, 439–445. [CrossRef]

20. Le Garzic, E.; de L'Hamaide, T.; Diraison, M.; Géraud, Y.; Sausse, J.; De Urreiztieta, M.; Hauville, B.; Champanhet, J.-M. Scaling and Geometric Properties of Extensional Fracture Systems in the Proterozoic Basement of Yemen. Tectonic Interpretation and Fluid Flow Implications. *J. Struct. Geol.* **2011**, *33*, 519–536. [CrossRef]

21. De Dreuzy, J.-R.; Davy, P.; Bour, O. Hydraulic Properties of Two-dimensional Random Fracture Networks Following Power Law Distributions of Length and Aperture. *Water Resour. Res.* **2002**, *38*, 12-1–12-9. [CrossRef]

22. Davy, P.; Le Goc, R.; Darcel, C.; Bour, O.; de Dreuzy, J.R.; Munier, R. A Likely Universal Model of Fracture Scaling and Its Consequence for Crustal Hydromechanics. *J. Geophys. Res.* **2010**, *115*, B10411. [CrossRef]

23. Hardebol, N.J.; Maier, C.; Nick, H.; Geiger, S.; Bertotti, G.; Boro, H. Multiscale Fracture Network Characterization and Impact on Flow: A Case Study on the Latemar Carbonate Platform. *J. Geophys. Res. Solid Earth* **2015**, *120*, 8197–8222. [CrossRef]

24. McCaffrey, K.J.W.; Sleight, J.M.; Pugliese, S.; Holdsworth, R.E. Fracture Formation and Evolution in Crystalline Rocks: Insights from Attribute Analysis. *Geol. Soc. Lond. Spec. Publ.* **2003**, *214*, 109–124. [CrossRef]

25. Bonnet, E.; Bour, O.; Odling, N.E.; Davy, P.; Main, I.; Cowie, P.; Berkowitz, B. Scaling of Fracture Systems in Geological Media. *Rev. Geophys.* **2001**, *39*, 347–383. [CrossRef]

26. Bour, O.; Davy, P.; Darcel, C.; Odling, N. A Statistical Scaling Model for Fracture Network Geometry, with Validation on a Multiscale Mapping of a Joint Network (Hornelen Basin, Norway). *J. Geophys. Res. Solid Earth* **2002**, *107*, ETG 4-1–ETG 4-12. [CrossRef]

27. Bonneau, F.; Caumon, G.; Renard, P.; Sausse, J. Stochastic Sequential Simulation of Genetic-like Discrete Fracture Networks. *Proceeding 32nd Gocad Meeting, Nancy, France.* 2012. Available online: https://www.ring-team.org/component/liad/?view=pub&id=2214 (accessed on 3 July 2021).

28. de Dreuzy, J.-R.; Méheust, Y.; Pichot, G. Influence of Fracture Scale Heterogeneity on the Flow Properties of Three-Dimensional Discrete Fracture Networks (DFN): 3D Fracture Network Permeability. *J. Geophys. Res.* **2012**, *117*, B11207. [CrossRef]

29. Chabani, A. Analyse Méthodologique et Caractérisation Multi-Échelle Des Systèmes de Fractures à l'interface Socle/Couverture Sédimentaire Application à La Géothermie (Bassin de Valence, SE France). These de Doctorat, Paris Sciences et Lettres (ComUE), Paris, France, 2019. Available online: http://www.theses.fr/2019PSLEM046 (accessed on 3 July 2021).

30. Morellato, C.; Redini, F.; Doglioni, C. On the Number and Spacing of Faults. *Terra Nova* **2003**, *15*, 315–321. [CrossRef]

31. Bourbiaux, B.; Basquet, R.; Daniel, J.M.; Hu, L.Y.; Jenni, S.; Lange, G.; Rasolofosaon, P. Fractured Reservoirs Modelling: A Review of the Challenges and Some Recent Solutions. *First Break* **2005**, *23*. [CrossRef]

32. Gillespie, P.; Monsen, E.; Maerten, L.; Hunt, D.; Thurmond, J.; Tuck, D.; Martinsen, O.J.; Pulham, A.J.; Haughton, P.D.W.; Sullivan, M.D. Fractures in Carbonates: From Digital Outcrops to Mechanical Models. *Outcrops Revital.—Tools Tech. Appl. Tulsa Okla. SEPM Concepts Sedimentol. Paleontol.* **2011**, *10*, 137–147.

33. Vollgger, S.A.; Cruden, A.R. Mapping Folds and Fractures in Basement and Cover Rocks Using UAV Photogrammetry, Cape Liptrap and Cape Paterson, Victoria, Australia. *J. Struct. Geol.* **2016**, *85*, 168–187. [CrossRef]

34. Biber, K.; Khan, S.D.; Seers, T.D.; Sarmiento, S.; Lakshmikantha, M.R. Quantitative Characterization of a Naturally Fractured Reservoir Analog Using a Hybrid Lidar-Gigapixel Imaging Approach. *Geosphere* **2018**, *14*, 710–730. [CrossRef]

35. Trullenque, G.; Genter, A.; Leiss, B.; Wagner, B.; Bouchet, R.; Léoutre, E.; Malnar, B.; Bär, K.; Rajšl, I. Upscaling of EGS in Different Geological Conditions: A European Perspective. In Proceedings of the Proceedings 43rd Workshop on Geothermal Reservoir Engineering, Stanford, CA, USA, 12–14 February 2018.

36. Chabani, A.; Trullenque, G.; Rishi, P.; Pomart, A.; Attali, R.; Sass, I. Modelling of Fractured Granitic Geothermal Reservoirs: Use of Deterministic and Stochastic Methods in Discrete Fracture Networks and a Coupled Processes Modeling Framework. In Proceedings of the Extended Abstract, Reykjavik, Iceland, 2 May 2021.

37. Klee, J.; Trullenque, G.; Ledésert, B.; Potel, S.; Hébert, R.; Chabani, A.; Genter, A. Petrographic Analyzes of Fractured Granites Used as An Analogue of the Soultz-Sous-Forêts Geothermal Reservoir: Noble Hills, CA, USA. In Proceedings of the Extended Abstract, Reykjavik, Iceland, 26 May 2021.
38. Jennings, C. Geologic Map of California, Scale 1: 250,000. Olaf P. Jenkins Edition, Long Beach Sheet. *Calif. Div. Mines Geol, Serie No 33, List No 6347.033*; CA, USA, 1962. Available online: https://www.davidrumsey.com/luna/servlet/detail/RUMSEY~8~1~324920~90093990:Geologic-Map-of-California,-Long-Be?sort=Pub_List_No_InitialSort%2CPub_Date%2CPub_List_No%2CSeries_No (accessed on 3 July 2021).
39. Niles, J.H. Post-Middle Pliocene Tectonic Development of the Noble Hills, Southern Death Valley, California. Doctoral Dissertation, San Francisco State University, San Francisco, CA, USA, 2016.
40. Miller, M.; Pavlis, T. The Black Mountains Turtlebacks: Rosetta Stones of Death Valley Tectonics. *Earth-Sci. Rev.* **2005**, *73*, 115–138. [CrossRef]
41. Pavlis, T.L.; Rutkofske, J.; Guerrero, F.; Serpa, L.F. Structural Overprinting of Mesozoic Thrust Systems in Eastern California and Its Importance to Reconstruction of Neogene Extension in the Southern Basin and Range. *Geosphere* **2014**, *10*, 732–756. [CrossRef]
42. Snow, J.K.; Wernicke, B. Uniqueness of Geological Correlations: An Example from the Death Valley Extended Terrain. *Geol. Soc. Am. Bull.* **1989**, *101*, 1351–1362. [CrossRef]
43. DeCelles, P.G. Late Jurassic to Eocene Evolution of the Cordilleran Thrust Belt and Foreland Basin System, Western U.S.A. *Am. J. Sci.* **2004**, *304*, 105–168. [CrossRef]
44. Walker, J.D.; Burchfiel, B.; Davis, G.A. New Age Controls on Initiation and Timing of Foreland Belt Thrusting in the Clark Mountains, Southern California. *Geol. Soc. Am. Bull.* **1995**, *107*, 742–750. [CrossRef]
45. Stewart, J.H. Possible Large Right-Lateral Displacement along Fault and Shear Zones in the Death Valley-Las Vegas Area, California and Nevada. *Geol. Soc. Am. Bull.* **1967**, *78*, 131–142. [CrossRef]
46. Stewart, J.H. Extensional Tectonics in the Death Valley Area, California: Transport of the Panamint Range Structural Block 80 Km Northwestward. *Geology* **1983**, *11*, 153–157. [CrossRef]
47. Snow, J.K. Cenozoic Tectonism in the Central Basin and Range; Magnitude, Rate, and Distribution of Upper Crustal Strain. *Am. J. Sci.* **2000**, *300*, 659–719. [CrossRef]
48. Brady, R. Cenozoic Geology of the Northern Avawatz Mountains in Relation to the Intersection of the Garlock and Death Valley Fault Zones, San Bernardino County, California. Doctoral Dissertation, University of California, Davis, CA, USA, 1987.
49. Norton, I. Two-Stage Formation of Death Valley. *Geosphere* **2011**, *7*, 171–182. [CrossRef]
50. Burchfiel, B.C.; Stewart, J.H. "Pull-Apart" Origin of the Central Segment of Death Valley, California. *Geol. Soc. Am. Bull.* **1966**, *77*, 439–442. [CrossRef]
51. Pavlis, T.L.; Trullenque, G. Evidence for 40–41 Km of Dextral Slip on the Southern Death Valley Fault: Implications for the Eastern California Shear Zone and Extensional Tectonics. *Geology* **2021**, *49*, 767–772. [CrossRef]
52. Machette, M.N.; Klinger, R.E.; Knott, J.R.; Wills, C.J.; Bryant, W.A.; Reheis, M.C. *A Proposed Nomenclature for the Death Valley Fault System. Quaternary and Late Pliocene Geology of the Death Valley Region: Recent Observations on Tectonics, Stratigraphy, and Lake Cycles (Guidebook for the 2001 Pacific Cell—Friends of the Pleistocene Fieldtrip)*; US Department of the Interior, US Geological Survey: Reston, VA, USA, 2001; p. 173.
53. Troxel, B.W.; Butler, P.R. Rate of Cenozoic Slip on Normal Faults, South-Central Death Valley, California. Doctoral Dissertation, Departement of Geology, University of California, Office of Scholarly Communication (OSC), Irvine, CA, USA, 1979.
54. Brady III, R.H. Neogene Stratigraphy of the Avawatz Mountains between the Garlock and Death Valley Fault Zones, Southern Death Valley, California: Implications as to Late Cenozoic Tectonism. *Sediment. Geol.* **1984**, *38*, 127–157. [CrossRef]
55. Reinert, E. Low Temperature Thermochronometry of the Avawatz Mountians, California: Implications for the Inception of the Eastern California Shear Zone. Doctoral Dissertation, University of Washington, Washington, DC, USA, 2004. Available online: https://www.ess.washington.edu/content/people/student_publications_files/reinert--erik/Reinert_2004.pdf (accessed on 3 July 2021).
56. Spencer, J.E. *Late Cenozoic Extensional and Compressional Tectonism in the Southern and Western Avawatz Mountains, Southeastern California. Basin and Range Extensional Tectonics near the Latitude of Las Vegas, Nevada: Geological Society of America Memoir*; Geological Society of America: Las Vegas, NV, USA, 1990; Volume 176, pp. 317–333. [CrossRef]
57. Chinn, L.D. *Low-Temperature Thermochronometry of the Avawatz Mountains; Implications for the Eastern Terminus and Inception of the Garlock Fault Zone*; University of Washington: Washington, DC, USA; Available online: https://www.ess.washington.edu/content/people/student_publications_files/chinn--logan/chinn--logan_ms_2013.pdf (accessed on 3 July 2021).
58. Rämö, T.O.; Calzia, J.P.; Kosunen, P.J. Geochemistry of Mesozoic Plutons, Southern Death Valley Region, California: Insights into the Origin of Cordilleran Interior Magmatism. *Contrib. Mineral. Petrol.* **2002**, *143*, 416–437. [CrossRef]
59. Bisdom, K.; Gauthier, B.D.M.; Bertotti, G.; Hardebol, N.J. Calibrating Discrete Fracture-Network Models with a Carbonate Threedimensional Outcrop Fracture Network: Mplications for Naturally Fractured Reservoir Modeling. *AAPG Bull.* **2014**, *98*, 13511376. [CrossRef]
60. Brady, R.H.; Troxel, B.W.; Wright, L.A. The Miocene Military Canyon Formation: Depocenter Evolution and Constraints on Lateral Faulting, Southern Death Valley, California. *Spec. Pap. Geol. Soc. Am.* **1999**, 277–288. [CrossRef]
61. von Mises, R. Uber Die "Ganzzahligkeit" Der Atomgewicht Und Verwandte Fragen. *Physikal. Z.* **1918**, *19*, 490–500.

62. Chabani, A.; Mehl, C.; Cojan, I.; Alais, R.; Bruel, D. Semi-Automated Component Identification of a Complex Fracture Network Using a Mixture of von Mises Distributions: Application to the Ardeche Margin (South-East France). *Comput. Geosci.* **2020**, *137*, 104435. [CrossRef]
63. Pickering, G.; Bull, J.M.; Sanderson, D.J. Sampling Power-Law Distributions. *Tectonophysics* **1995**, *248*, 1–20. [CrossRef]
64. Pickering, G.; Peacock, D.C.; Sanderson, D.J.; Bull, J.M. Modeling Tip Zones to Predict the Throw and Length Characteristics of Faults. *AAPG Bull.* **1997**, *81*, 82–99.
65. Davy, P.; Bour, O.; De Dreuzy, J.-R.; Darcel, C. Flow in Multiscale Fractal Fracture Networks. *Geol. Soc. Lond. Spec. Publ.* **2006**, *261*, 31–45. [CrossRef]
66. Ledésert, B.; Dubois, J.; Velde, B.; Meunier, A.; Genter, A.; Badri, A. Geometrical and Fractal Analysis of a Three-Dimensional Hydrothermal Vein Network in a Fractured Granite. *J. Volcanol. Geotherm. Res.* **1993**, *56*, 267–280. [CrossRef]
67. Cruden, D.M. *Describing the Size of Discontinuities*; Elsevier: Amsterdam, The Netherlands, 1977; Volume 14, pp. 133–137.
68. Priest, S.D.; Hudson, J.A. *Estimation of Discontinuity Spacing and Trace Length Using Scanline Surveys*; Elsevier: Amsterdam, The Netherlands, 1981; Volume 18, pp. 183–197.
69. Priest, S.D.; Hudson, J.A. Discontinuity Spacings in Rock. *Int. J. Rock Mech. Min. Sci. Geomech. Abstr.* **1976**, *13*, 135–148. [CrossRef]
70. Cowie, P.A.; Sornette, D.; Vanneste, C. Multifractal Scaling Properties of a Growing Fault Population. *Geophys. J. Int.* **1995**, *122*, 457–469. [CrossRef]
71. Priest, S.D. *Discontinuity Analysis for Rock Engineering*; Springer Science & Business Media: Berlin/Heidelberg, Germany, 1993.
72. Gillespie, C.S. The PoweRlaw Package: A General Overview. Newcastle, UK. Available online: https://cran.r-project.org/web/packages/poweRlaw/vignettes/a_introduction.pdf (accessed on 3 July 2021).
73. Gillespie, P.A.; Howard, C.B.; Walsh, J.J.; Watterson, J. Measurement and Characterisation of Spatial Distributions of Fractures. *Tectonophysics* **1993**, *226*, 113–141. [CrossRef]
74. Bertrand, L. Etude Des Réservoirs Géothermiques Développés Dans Le Socle et à l'interface Avec Les Formations Sédimentaires. Doctoral Dissertation, Université de Lorraine, Lorraine, France, April 2017; p. 492.

geosciences

MDPI

Article

Lab-Scale Permeability Enhancement by Chemical Treatment in Fractured Granite (Cornubian Batholith) for the United Downs Deep Geothermal Power Project, Cornwall (UK)

Katja E. Schulz *, Kristian Bär * and Ingo Sass

Department of Geothermal Science and Technology, Technische Universität Darmstadt, Schnittspahnstraße 9, 64287 Darmstadt, Germany; sass@geo.tu-darmstadt.de
* Correspondence: kontakt@schulzks.de (K.E.S.); baer@geo.tu-darmstadt.de (K.B.)

Abstract: A hydrothermal doublet system was drilled in a fault-related granitic reservoir in Cornwall. It targets the Porthtowan Fault Zone (PTF), which transects the Carnmenellis granite, one of the onshore plutons of the Cornubian Batholith in SW England. At 5058 m depth (TVD, 5275 m MD) up to 190 °C were reached in the dedicated production well. The injection well is aligned vertically above the production well and reaches a depth of 2393 m MD. As part of the design process for potential chemical stimulation of the open-hole sections of the hydrothermal doublet, lab-scale acidification experiments were performed on outcrop analogue samples from the Cornubian Batholith, which include mineralised veins. The experimental setup comprised autoclave experiments on sample powder and plugs, and core flooding tests on sample plugs to investigate to what degree the permeability of natural and artificial (saw-cut) fractures can be enhanced. All samples were petrologically and petrophysically analysed before and after the acidification experiments to track all changes resulting from the acidification. Based on the comparison of the mineralogical composition of the OAS samples with the drill cuttings from the production well, the results can be transferred to the hydrothermally altered zones around the faults and fractures of the PTF. Core Flooding Tests and Autoclave Experiments result in permeability enhancement factors of 4 to >20 and 0.1 to 40, respectively. Mineral reprecipitation can be avoided in the stimulated samples by sufficient post-flushing.

Keywords: enhanced geothermal systems (EGS); fractured granite; core flooding experiments; autoclave experiments; Cornubian Batholith

Citation: Schulz, K.E.; Bär, K.; Sass, I. Lab-Scale Permeability Enhancement by Chemical Treatment in Fractured Granite (Cornubian Batholith) for the United Downs Deep Geothermal Power Project, Cornwall (UK). *Geosciences* **2022**, *12*, 35. https://doi.org/10.3390/geosciences12010035

Academic Editors: Béatrice A. Ledésert, Ronan L. Hébert, Ghislain Trullenque, Albert Genter, Eléonore Dalmais, Jean Hérisson and Jesús Martínez-Frías

Received: 29 September 2021
Accepted: 10 January 2022
Published: 12 January 2022

Publisher's Note: MDPI stays neutral with regard to jurisdictional claims in published maps and institutional affiliations.

1. Introduction

The geothermal doublet at United Downs in Redruth, Cornwall, consists of the production well UD-1 and the injection well UD-2. UD-1 reaches a depth of 5058 m TVD (true vertical depth)/5275 m MD (measured depth), with an approximately 1.2 km long open-hole section, while the injection well UD-2 reaches 2214 m TVD/2393 m MD, with an approximately 550 m long open-hole section. Geothermal Engineering Limited (GEL) plans the establishment of a geothermal power plant with 1 to 3 MW$_{el}$ net energy production [1]. The reservoir rock is the fractured Carnmenellis granite, which is one of the onshore plutons comprising the Cornubian Batholith. It is characterised by a strong geothermal anomaly caused by radioactive decay of U, Th and K in the granite [2]. The geological target structure is the Porthtowan fault zone (PTF), which vertically links the two wells [1]. Since a sufficient reservoir temperature of ~190 °C was reached in the production well, the limiting factor for the project is the hydraulic productivity of the reservoir. For this reason, the analysis of permeability enhancement by chemical treatment in the fractured Cornish granite is the prerequisite for an assessment of the potential effectivity of such stimulation in the pilot project United Downs Deep Geothermal Power (UDDGP).

Around fault zones, a complex permeability pattern develops, which is determined by the characteristics of the fracture network. Permeability can vary by up to five orders of magnitude between unaffected granite and fractured and altered granite [3]. Under reservoir conditions, open fractures will be filled with secondary minerals over time if no tectonic reactivation occurs. These (partly) mineralised fractures can be a relevant target for chemical treatment.

Within the scope of the EU-Horizon2020-project Multidisciplinary and multi-context demonstration of EGS exploration and Exploitation Techniques and potentials (MEET), the elaborated results will be applied at transfer sites in similar geological settings within the Variscan basement of Europe.

Geological Setting

The geology of Cornwall is dominated by the Cornubian Batholith, which intruded into metasediments during the late phase of the Variscan orogeny in the Early Permian period. The metasediments consist in Devonian and Carboniferous passive continental margin successions, locally termed 'Killas', which were affected by regional low-grade metamorphism during Variscan shortening in early Carboniferous (prehnite–pumpellyite facies [4]). During the succeeding post-collisional extension 292–270 Ma ago (Early Permian period) the Cornubian Batholith was emplaced in at least two major phases. It is a peraluminous S-type granite. The source rock of the melt is a meta-greywacke [5]. The emplacement occurred mainly along preexisting fault- and weakness zones, possibly including fault zones related to the extensional reactivation of the Rhenohercynian suture [6]. Whereas the rock exposed at the surface was mapped in detail during the intense Cornish mining activity of the last few centuries [7], the composition and fracturation of the granite in the subsurface are the subject of current investigations. Detailed information can be found, e.g., in the work of Beamish and Busby, Shail and Leveridge, Willis-Richards, Edmunds et al., Yeomans et al., and Ghosh [2,6,8–12].

Simons et al. [5] have classified the intrusion stages of the Cornubian Batholith and provide detailed information about the general mineralogical composition, possible accessory minerals and grain size distribution of the respective granite types (Figure 1). The rock samples analysed in the present study were therefore classified according to Simons et al. [5], although the samples are hydrothermally influenced. Information on the mineralogical composition of the relevant samples is provided in Table 1 and Figure 1.

Table 1. Origin of the six analysed samples.

Sample No. (MEETCW)	1 (001004)	2 (010003)	3 (010006)	4 (011003)	5 (017001)	6 (023001)
Granite Complex	St. Austell	Land's End	Land's End	Land's End	Tregonning Godolphin	Cligga
Granite Type, after Simons et al. [5]	G3/G4/G5	G3/G2	G3/G2	G3 (G2)	G5 (G1)	G2
Location Name	IMERYS China clay mining area, Karslake	Castle an Dinas Quarry	Castle an Dinas Quarry	Rosewall Hill	Rinsey Cove	Cligga Head
Location (Brit. Nat. Grid)	SW 98080 57596	SW 48662 34678	SW 48885 34735	SW 46973 38510	SW 59343 26881	SW 73799 53720
Sample Quality, Alteration, Weathering	Fresh quarry sample	Fresh quarry sample	Fresh quarry sample, slightly kaolinised, slightly weathered	Boulders on hill, close to outcrop, partly iron strained	Loose rocks	Loose rocks around abandoned quarry, weathered, greisenisation present in outcrop

Cornwall is intersected by several fault systems, such as the Porthtowan Fault Zone (PTF), which are mainly strike slip faults, roughly trending NW–SE (NNW–SSE) and locally referred to as 'cross courses' [9], connecting ENE–WSW-striking extensional faults [13] or dykes, called 'Elvans'.

The PTF is a NNW–SSE striking strike–slip fault zone (Figure 1). It consists of several fault strands and thus is 200 to 500 m wide at the surface and assumed to be more than 15 km long [1]. The northwestern part was mapped geologically [14] and remotely [11], while the southeastern part is not exposed at the surface and its extent is only presumed from indirect hints (e.g., morphology).

Along the PTF, hydrothermal alteration has affected the mineralogical composition of the reservoir rocks. Two main phases of alteration can be distinguished: younger alteration related to circulation of (partly) meteoric water in structural discontinuities and older alteration related to hydrothermal convection cells around the cooling granitic bodies, with: (a) fluids possibly linked with the first marine incursions in late Triassic and (b) fluids with magmatic origin [4].

Products of hydrothermal alteration are present, such as Kaolinite (likely Dickite at reservoir depth [15–18], greisenisation (paired with tourmalinisation and cassiterite veins [19], Tourmaline veins (related to the circulation of boron-enriched fluids in hydrothermal convection cells [20], Chlorite [20], hydrothermal W/Sn/Cu/As/Zn ore minerals (mainly in the metasediments [4,13] and Quartz, Fluorite or Barite veins [15,21].

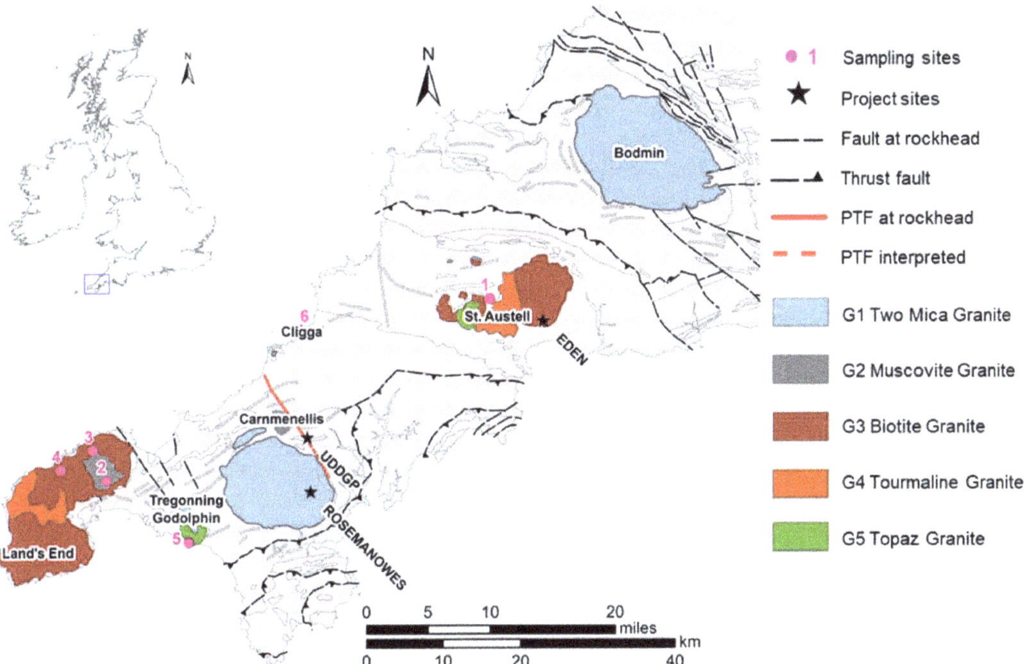

Figure 1. Simplified geological map of Cornwall, showing granite types after Simons et al. [5], sampling locations and relevant project locations; PTF simplified after Reinecker et al. [22]; meta sediments, dykes and non-magmatic geological units in grey.

2. Materials and Methods

The analysed rock samples are outcrop analogue samples (OAS). They were collected during fieldwork in June 2019 from six different locations of the Cornubian Batholith (selected from a total of 47 samples from 35 different locations—sea cliffs, major and minor natural outcrops, rock dumps and active and abandoned quarries) [23]. They have a relatively high geochemical conformity with the Carnmenellis Granite (based on data from Simons et al. [5]) and include mineralised veins, which allow analogies with hydrothermally altered zones around the faults and fractures of the PTF encountered in UD-1. As a result of a lack of sampling material from the granites of the Carnmenellis, Carn Marth or Carn Brea plutons, which contain veins, the representativity for the Carnmenellis granite is limited to far-field analogues (Land's End, Tregonning Godolphin, Cligga and St. Austell plutons). Nevertheless, the present samples display an analogy to the fault and fracture zone itself. For analysis, the samples were cored into plugs 40 mm in diameter and ranging in length from 20 to 80 mm. Each core targets a vein, with the core axis drilled parallel to the vein (Figure 2). Cutoffs were ground to powder finer than 63 μm for chemical and mineralogical analysis.

For comparison, drill cuttings from production well UD-1 from the open-hole section between 4 and 5 km MD are analysed. The cuttings, which were sampled by GEL/ Geosciences Limited, represent 44 depth intervals of 10 m in length, between 4050 m TVD and 4930 m MD.

Figure 2. Top view on representative cores (40 mm diameter) from the six samples; the mineralogical composition of the sample and the veins is given in Table 1.

Besides an accompanying petrophysical analysis, the effect of chemical treatment on the permeability of the rock matrix and fracture systems due to acidification is investigated under approximation of reservoir conditions. This implies the acidification of samples using two methods (Core Flooding Tests, CFT; Autoclave Experiments, AE), a petrophysical before/after comparison and a geochemical analysis of the fluid–rock interaction.

A set of different analytical methods, such as XRF, XRD, AAS, IC, ICP-MS, thermoscanning, helium pycnometry, helium permeametry, water permeametry and sonic velocity, is used to characterise the samples petrologically and petrophysically before and after chemical treatment in AEs and CFTs at reservoir temperatures and pressures, as well as the spent acids from the AEs (Table 2, Figure 3).

In the chemical treatment experiments (AEs), the samples are placed in autoclaves together with acids at 150 °C for 24 h. The utilised acids are: (a) 15% HCl and (b) 'white acid', which consists of 12% HCl and 3% HF and resembles Regular Mud Acid (RMA), which is commonly employed in chemical well stimulation.

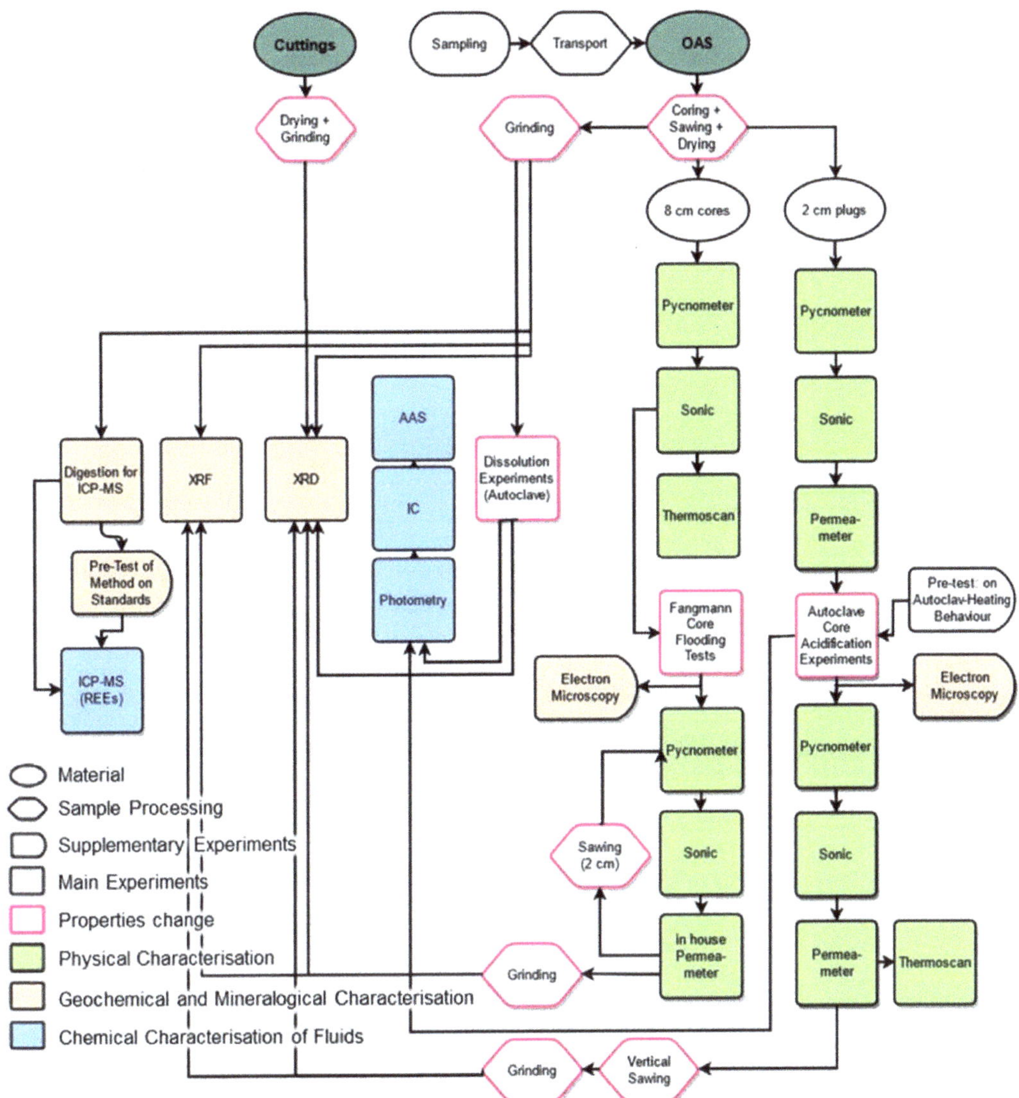

Figure 3. Flow chart of the lab work.

In the CFTs, which were conducted at the labs of Fangmann Energy Services (FES), the chemical blends SSB-007 and SFB-007 were circulated through the cores at a temperature of 150 °C and a confining pressure of 172 bar (2500 psi). During the flow rate-controlled tests, the differential pressure is logged to calculate the permeability development over time [24]. SSB-007 and SFB-007 have relatively similar characteristics to the fluids used in the autoclave experiments.

Table 2. Applied methods and analysed sample sets. Abbreviations: CFT = core flooding tests; AE = autoclave experiments; PEF = permeability enhancement factor; OAS = outcrop analogue samples.

Method	Sample Condition	Derived Parameters	Analysed Sample Sets
Gas pycnometry	Dry	Grain volume, grain density, porosity	All cuttings, all 8 cm cores, all 2 cm plugs, before and after acidification
Sonic	Dry	Seismic velocities, Young's modulus, Poisson ratio	All 8 cm cores, all 2 cm plugs and CFT cores, before and after acidification
Thermoscanning	Dry	Thermal conductivity, thermal diffusivity, specific heat capacity	All 8 cm cores, acidified 2 cm plugs
Gas Permeametry	Dry	Intrinsic permeability, permeability enhancement factor (PEF)	18 plugs before acidification, 12 plugs after AE, 5 plugs after CFT
Water Permeametry	Water-saturated	Permeability	5 plugs, sawn from the 3 acidised CFT cores
CFT	NH_4Cl-saturated	Macroscopic changes, PEF	3 cores
AE: HCl, HCl + HF (powder)	HCl + HF saturated	Weight loss of sample	1 OAS powder sample with HCl, 6 OAS powder samples with HCl + HF, 1 OAS power sample with HCl
AE: HCl (plugs)	Partly HCl-saturated	Macroscopic changes	6 OAS plugs + 1 Blank
AE: HCl + HF (plugs)	Partly HCl + HF saturated	Macroscopic changes	6 OAS plugs + 1 Blank
REE-Analysis: Digestion, ICP-MS	Fluid	REE concentration	9 digested OAS powder samples + 2 Blanks + 13 Standards
AAS	Fluid	Element concentration	22 fluid samples (AE set 1, 2, 3)
IC	Fluid	Ion concentration	22 fluid samples (AE set 1, 2, 3)
Photometry	Fluid	Silicate concentration	22 fluid samples (AE set 1, 2, 3)
XRD	Powder tablets	Normalised mineralogical composition	9 representative OAS samples, 36 cuttings, 22 powdered samples after acidification (AE set 1, 2, 3 and CFT)
XRF	Pressed powder tablets	Normalised weight % of element or oxide concentration	9 OAS samples before acidification, 15 powdered plugs after acidification (AE set 2, 3 and CFT)

2.1. Sample-Preparation

All samples were dried to mass constancy (cuttings at 105° for 24 h, OAS at 65 °C for >48 h to avoid low-temperature alteration processes of clay minerals) and stored in closed containers at ~20 °C with silica gel. OAS were cored, sawn and burnished at the Institute of Applied Geosciences in Darmstadt. All cores have a diameter of 40 mm and include veins of different mineralogical composition. The initial length of the 'cores' was set to 80 mm (optimum: 2:1 length to width ratio), which had to be reduced to 20 mm 'plugs' for further experiments. The core axis was oriented parallel to the strike of the vein; the cores had plan-parallel surfaces. Grinding was performed in a tungsten carbide disc mill (two minutes at a speed of 1000 rpm) following Ferreiro Mählmann and Frey [25]. Grinding was performed on cuttings and core-cutoffs from sawing, which allow best comparability with OAS cores. The cutting samples lack the sub-63 μm-fraction, which includes the clay minerals (due to washing at the project site: rinsing 2–3 times and decantation of excess water through a 63 μm sieve without containment).

2.2. Petrophysical Methods

Gas pycnometry (determination of the effective porosity, which is an approximation of the total porosity, due to He diffusion) was conducted on all samples according to the recommendations of Micromeritics [26] as described in Weydt et al. [27] with the two-chamber systems AccuPyc 1330, AccuPyc II 1340 and CorePyc 1360 (Micromeritics, Germany) using He. The bulk volume was calculated from the geometry of the cylindrical cores.

Sonic measurements were conducted as described in Weinert et al. [28] with an ultrasonic pulse generator (USG40) (Geotron-Elektronik, Germany), a PicoScope Detector and a UPG-S/UPE-S emitter/receiver set (pushed against the sample with a pressure of 1 bar, coupled by Magnaflux 54-T04 shear gel) on 80 mm cores and 20 mm plugs to calculate the dynamic Poisson ratio and dynamic Young's Modulus.

Thermal conductivity and thermal diffusivity were measured as described by Mielke et al. [29] on the 80 mm cores before treatment and on 20 mm plugs after treatment with a Thermal Conductivity Scanner (Lippmann and Rauen, Germany), using gabbro standards (TC) and Quartz/Titan and Titan/Steel standards (TD) [30]).

The permeability was analysed with a column gas permeameter (stationary air permeameter with a Hassler type cell [31]). A verification of the results was performed with a water permeameter.

The gas permeameter measurements were conducted according to Filomena et al. [31], using the different pressure levels 1050–1250–1500–2000–3000–5000 mbar and a mantle pressure of 10 bar for each experiment. The pressure difference was set between up to 1000 mbar (samples with permeabilities in the range of 10^{-17} m^2) and only 50 mbar (permeabilities below 10^{-14} m^2). Further information about the method is provided by Weydt et al. [27]. Since the sample axis was parallel to the vein, the measured permeability was a combined vein and matrix permeability and the permeability of the whole sample was highly anisotropic.

2.3. Acidification

CFTs were conducted at the labs of Fangmann Energy Services, using a Manual Reservoir Permeability Tester (OFI Testing Equipment) to induce and analyse changes to the sample properties caused by acidification (Figure 4). The relative permeability changes over time were analysed by the measurement of the pressure difference at the in- and outlet of the sample (constant flow rate; measurement similar to a permeameter). The CFT procedure can be summarised in the following steps:

1. 18 h pre-saturation of the core with NH$_4$Cl (Ammonium-chloride solution with 50 kg NH$_4$Cl per m^3 water) under vacuum conditions; installation of the sample in the permeability tester.
2. Pre-flush: Core flooding with NH$_4$Cl and initial permeability measurement.
3. Acidification 1: Core flooding with SSB-007.
4. Acidification 2: Core flooding with SFB-007.
5. Post-flush: Core flooding with NH$_4$Cl (flushing the spent acid and particles out of the core) and final permeability measurement.
6. Deinstallation.

The NH$_4$Cl-brine was used because it has relatively similar flow properties compared to the applied acids and does not react with the acids, cores or the device. Besides, since NH$_4^+$ is not included in the geochemical analysis (and would be falsified anyway by the use of ammonium bifluoride as a constituent for the preparation of HF), the falsification of the analytical results was limited to Cl$^-$.

The composition of Fangmann's acid blends SSB-007 and SFB-007 is confidential. According to general information from Fangmann, SSB-007 is based on HCl or on a Strong Organic Acid (SOA) with comparable reaction kinetics. SFB-007 contains additional HF and is relatively similar to Regular Mud Acid (RMA) [24]. RMA is a common acid used for sandstone matrix acidising [32]. Recalde Lummer and Rauf [24] state the applicability of Fangmann's product SFB-007 as well for granite.

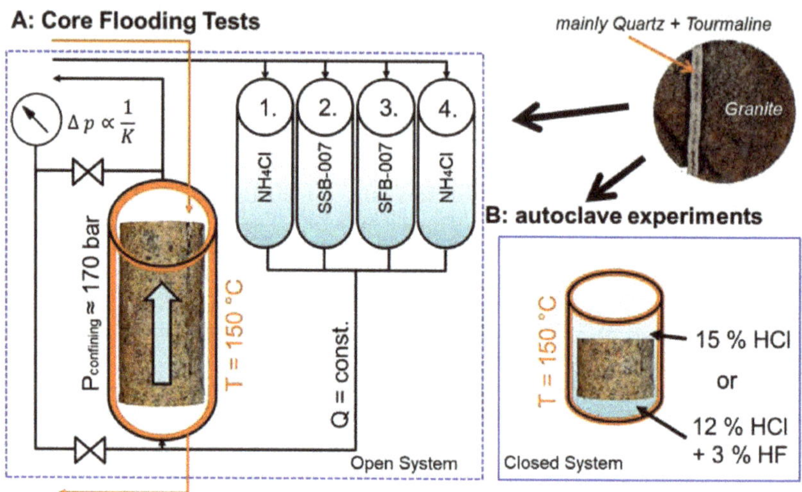

Figure 4. Simplified sketch of the setup and conditions of core flooding tests (**A**) and autoclave experiments (**B**).

The applied test conditions were temperatures of 150 °C, confining pressure (mantle pressure, prevents lateral flow around the sample) of 172.4 bar and back pressure (system pressure, prevents boiling of the fluids) of 34.5 bar. A drive pressure pushed the fluid through the core and depended on the flow rate (sample 010006004: 0.9 mL/min through artificial fissure; 017001005: 1.36 mL/min through artificial fissure; 23001002: 0.665 mL/min through natural vein) and the permeability of the sample [24]. As the initial permeability of two samples was too low for the circulation of the brine, those samples were sawn along the original vein. This joint was compressed by the confining pressure. Further information on the permeability measurement before and after the CFTs is given in Appendix A.

Due to cost- and time-intensity and unknown composition of the acids, only three cores were analysed in CFTs and AEs were used to resemble the CFTs.

During the AEs, the acid and sample were encased in high-pressure autoclaves (Parr Acid Digestion Bombs, 302AC T304, Bomb No. 4748, stainless-steel autoclave system with a cylindrical Teflon (PTFE) container) and heated at 150 °C for 24 h (closed system). In contrast to the CFTs, no circulation of fluids though the sample was realised, implying that no pre- or post-flush was conducted to cleanse the samples of any reaction products after reaction (Figure 4).

In the autoclave experiments, two different acids were used to resemble Fangmann's acid blends SSB-007 and SFB-007:

- 15% HCl.
- 12% HCl with 3% HF (White acid, common basis for RMA, produced by mixing 1000 mL HCl (15%) and 50 g ammonium bifluoride (ammonium hydrogen fluoride, F_2H_5N, purchased as granulate) [33].

HF in a concentration range between 1 and 7% is dangerous to life in case of skin contact, among other hazards, and is class 2 in the German Water Hazard Classification System [34]. It is not biodegradable.

- For powder acidification experiments (set 1: 12% HCl + 3% HF) the autoclaves were loaded with 6 g of powder (from ground OAS) and 60 mL of acid (ratio of powder mass to acid volume: 1:10, assuming that the entire powder participates in chemical reactions with the acid) and rotated and shaken carefully.
- For core acidification experiments (set 2: 15% HCl and set 3: 12% HCl + 3% HF) the autoclaves were loaded with 40 mm diametral, 16–27 mm-high OAS plugs and

a volume of acid corresponding to the surface area of the plug in a ratio 1:1. This approach assumed that mainly the sample surface, plus permeable parts of the sample, mainly along the vein–rock matrix interface, would contribute to chemical reactions.

- To saturate the samples with the acids, the loaded Teflon cylinders were evacuated in a desiccator with a water-jet vacuum pump for 30 min.

After the heating phase, the autoclaves were cooled with wet tissue until they were cold enough to be touched and opened (approximately one hour, but as short a time as possible, to reduce precipitation). The liquid (spent acid) was then separated from the solid sample remnants for AAS and IC analysis via pipetting. The plugs were then rinsed with deionised water to allow safe handling and to reduce precipitation. For the powder samples, the non-dissolved solid fraction of the sample was repeatedly centrifugated with deionised water to increase the pH from pH < 1 to pH > 5, to allow safe handling of the sample and enable further analysis.

The following bullet points summarise the most important conditions of the autoclave experiments:

- Firstly, 15% HCl or 12% HCl + 3% HF.
- Plugs (surface to acid ratio 1:1) or powder samples (mass to acid ratio 1:10).
- 150 °C for 24 h.
- Ambient pressure (pressure increase only due to fluid expansion and reaction processes).
- No flushing of the samples during the experiment.

As the applied acids, especially at high temperatures, are very corrosive, the materials of the autoclaves and the Manual Reservoir Permeability Tester, as well as any used lab equipment in contact with the acids, had to be chosen accordingly and all devices and tools were visually controlled before and after use. With respect to the CFTs, the used acids contained corrosion inhibitors.

2.4. Fluid Chemistry

Atomic Absorption Spectroscopy (AAS) was conducted with a ContrAA 300 (Analytic Jena, Germany, Xenon continuum source lamp plus monochromator, PTFE impact bead; acetylene/air-flame (2200 °C) or N_2O/acetylene flame (2750 °C), measurement duration (3 to 4 s), flame height and the characteristic lines used for analysis depend on the element and are selected according to Analytic Jena recommendations, based on the software database; software Aspect CS, version 1.5.6.0). Further details on the method can be found in Broekaert, Skoog and Leary as well as Welz and Sperling [35–37]. The spent acids from the autoclave experiments were diluted by a factor of 50 or 100 as a reasonable compromise between the fitting with the calibration range and a reduction in the dilution error. This affects the measurement quality for Si, Al, Fe and partly Ca, K and Na. A six (or seven)-point calibration was conducted, with calibration standards from single-element standards (Carl Roth) and a LaCl-CsCl-solution (Merck KGaA, Germany) in a 2% HCl (supra quality) matrix.

Ion exchange chromatography was conducted with a Compact IC (Metrohm, Germany; Compact IC autosampler plus, two separation columns for an- and cations with organic polymer resin (both Compact IC plus; software MagIC Net™; cation-eluents: Pyridine-2,6-dicarboxylic acid ($C_7H_5NO_4$) and nitric acid (HNO_3), anion eluent: Sodium carbonate (Na_2CO_3)/Sodium hydrogen carbonate ($NaHCO_3$). A 16-point calibration of the device is performed every three to four months, a control measurement is conducted within the analysis.

The silica content of the spent acid was determined with a Photometer Specord 200+ (Analytik Jena, Germany), and the reagents 'Merck 14,794 (Silicate Test)' using the software WinAspect+, according to DIN 38405-21 [38]. A nine-point calibration (0, 0.5, 1, 2, 3, 4, 5, 6 and 7 mg/L) was performed. Although the spent acid had a yellowish colour, the optical measurement was possible for set 1 and 3, as these samples were diluted by a factor of 1000;

set 2 was not measured because the required dilution factor was too high, as the silicate content is very low.

2.5. Mineralogy and Geochemistry

X-ray diffractometry (XRD) was conducted on powder tablets (low-texture samples) by Dr. R. Petschick (Goethe University of Frankfurt) with a Panalytical X'Pert Pro Powder X-ray-diffractometer (Kassel, Germany; Cu-ray tube with Cu radiation: 40 kV, 30 mA; Ni-filter with $\lambda(CuK\alpha)$ = 1.542 Å (no separate α-peaks), automatic divergence slits, X'Cellerator line counter) with the following goniometer settings: two hours per sample with a detection angle from 2.5–70° 2θ, emitter and detector circuit step length of 0.0083° 2θ and 100 s measurement time. The software X'Pert Data Collector (Panalytical, German<), X'Pert Highscore Pro (Panalytical, Germany) and MacDiff [39] were used to evaluate the measurements; measured intensities were normalised with an external standard (Corundum), approximating the natural concentration. Fluorescence effects were reduced by a baseline correction and a peak-position correction was conducted based on the Quartz peaks. Further details on the method are described by Petschick [40].

For geochemical analysis the internal method 'Quant Express, Best Detection' (wavelength-dispersive, no external standards) of the X-ray-fluorescence-spectrometer S8 Tiger 4 kW (Bruker, Karlsruhe, Germany, scintillation counter for heavy elements, proportional counter for light elements (Bruker [41]) was used. For the evaluation, the total sum of measured elements was normalised to 100%.

For the powder tablets, a ground <63 μm sample was combined with Hoechst wax C micro powder ($C_{38}H_{76}N_2O_2$) (Merck, Germany) in a ratio 1:4, homogenised in a rotator mixer for 30 min and pressed to tablets in a stainless-steel cylinder under pressure of 160 bar.

3. Results

Both acidification methods, CFTs and AEs, were used to analyse the permeability enhancement by chemical treatment (Appendix B, Figure A1) in samples from the Cornubian Batholith. Additionally, mineralogical, geochemical, petro- and thermophysical rock properties were determined, allowing before–after comparison and the quantification of the acidification effects. The OAS from the Cornubian Batholith were selected with focus on their comparability with the Carnmenellis granite and the presence of veins. The veins allowed comparison with hydrothermally altered zones in the well. Additionally, cuttings from the geothermal well UD-1 were used for a preliminary transfer of results from lab scale to the reservoir.

3.1. Sample Composition

All six sampled outcrops or quarries were granitic. However, as veins (mainly filled with Quartz) were targeted for sampling, all samples had elevated Quartz content compared to the cuttings from UD-1 and the regional chemistry, as described by Simons et al., (2016). In three cases the sample composition was classified as quartzolitic, as the volumetric overrepresentation of the veins caused elevated Quartz (and Tourmaline) content (Table 3). In relation to Plagioclase, the K-Feldspar content was also slightly elevated. As the resilience of Plagioclase is usually lower, compared to K-Feldspar, this may be an indicator for weathering effects on the samples. Macroscopically, Muscovite and Biotite were detected in all samples (Ms $\geq$ Bt). As accessory minerals, Tourmaline, Chlorite and Cassiterite were identified in the OAS and Hornblende and Chlorite in the cuttings. All veins consisted of Quartz, plus Tourmaline in most cases and rarely ore minerals. The general mineralogical evaluation of the samples with XRD confirms the macroscopically detected minerals, except for Biotite, which is underrepresented in the XRD results. The conducted REE analysis [42] generally correlates with the trends described by Simons et al. [5]. Detected deviations in REE may also be explained by weathering processes in the samples, because an inverse

Europium anomaly could be related to weathering of plagioclase. The REE evaluation might thus be useable as a tool to detect weathering.

Table 3. Composition of the six analysed samples. Mineral abbreviations according to Kretz [43].

Sample No. (MEETCW)	1 (001004)	2 (010003)	3 (010006)	4 (011003)	5 (017001)	6 (023001)
Minerals: matrix(macroscopic)	Qtz + Kfs + Pl + Ms + Bt	Qtz + Kfs + Pl + Ms + Bt	Qtz + Kfs + Pl + Ms + Bt + clay minerals	Qtz + Kfs + Pl + Bt + Ms	Qtz + Kfs + Pl + Ms + Bt + Toz?	Qtz + Kfs + Pl + Ms + Bt
Matrix grainsize	fine to medium	Fine	fine	fine	fine	fine to medium
Micas	Ms + Bt	Ms > Bt	Ms > Bt	Bt > Ms	Ms + Bt	Ms + Bt
Minerals: Vein(macroscopic)	Qtz/Tur/(Hem, reddish colour)	Qtz + Tur	Qtz	Qtz + Tur	Qtz (+ Tur)	Qtz, ore mineral (Cst)
Minerals: entire sample (XRD)	Qtz > Pl > Kfs > Ms > Tur	Qtz > Kfs > Ms > Pl > Tur	Qtz > Kfs > Ms > Pl > Tur	Qtz > Ms > Kfs > Tur	Qtz > Ms > Tur > Pl > Bt	Qtz > Ms > Bt > Tur > Chl (> Cst)

3.2. Acidification

The CFTs, conducted and evaluated by Fangmann Energy Services [44], resulted in a significant increase in permeability. To quantify the effectivity of the acidification, we defined the permeability enhancement factor (PEF), which is the ratio between the permeability before (K_{pre}) and after (K_{post}) treatment:

$$PEF = K_{post}/K_{pre} \tag{1}$$

The CFTs resulted in a PEF between 4 and 50. On the treated samples, a white precipitate was visible after the acidification.

In contrast to the CFTs, the trend in permeability variation was more ambiguous in the AEs. The AEs induced a permeability variation by a PEF between 0.1 and 40 (for 15% HCl: 0.8–39.44; for 12% HCl + 3% HF: 0.13–40.48). A permeability decrease (PEF < 1) was, in most cases, accompanied by (and likely caused by) the precipitation of a yellowish and a white mineral, which were macroscopically visible on the plugs (Figure 5). After treatment with HCl + HF, the bottom of the samples showed more intense changes: a relief was created, which was especially visible in samples 2, 3, 5 and 6 (Figure 5). In this relief, the Quartz veins acted as the resistant parts and remained almost unaffected. In sample 3, this effect was visible for the Quartz grains in the matrix. On the top of the sample, these effects were less intense. In addition, samples 2 to 6 showed yellow or white precipitation, also mainly on the bottom of the samples. After treatment with HCl, a relief was only created in sample 2, 5 and 6, mainly close to the veins. No white precipitates were visible, but yellow precipitates occurred on sample 2 to 6.

The yellow mineral does not resemble any original crystal structures and can partially be dissolved or washed away, but reprecipitates during the drying process (65 °C). This is most likely caused by the evaporation of the remaining pore fluid and indicates that the acid penetrated the sample surface. This was confirmed by a reaction front, which was visible in most samples in the cross section of the plugs after sawing them vertically. This reaction front was commonly located around the vein (sample 1, 4, 5 and 6) or in the entire sample (sample 2 and 3), but with a gradual increase towards the vein. The white mineral mainly occurred on or along pre-existing minerals (Feldspars and Quartz), especially on the bottom of the plugs. In difference to the yellow reaction front, it did not seem to be related to the veins and occurred only on the outside of the samples. The precipitate occurred in spent 15% HCl as well as 12% HCl + 3% HF.

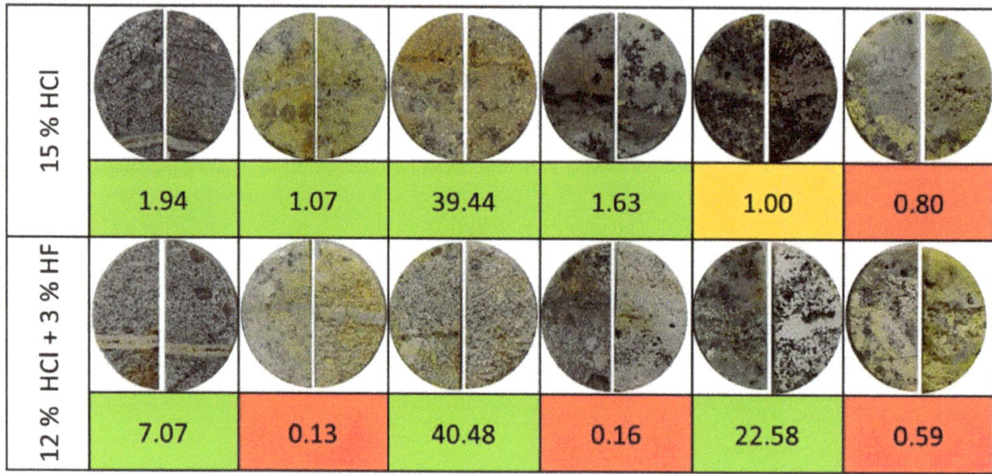

Figure 5. Photomontage of the plugs after acidification, combined with the permeability enhancement factor resulting from treatment with the respective acids 15% HCl or 12% HCl + 3% HF in autoclave experiments: top and bottom view of the sawn plugs after the autoclave experiments illustrating the stronger degree of precipitation on the samples bottom surfaces.

In the fluid samples, no precipitation processes were observed, even after several weeks of storage. During the AEs, the acid colour changed from clear transparent to an intense yellow. Based on the hydrochemical analysis of the spent acid, showing elevated iron concentrations, most likely these minerals were partly iron hydroxides (FeO(OH), Goethite or Limonite). Macroscopically this was indicated by the yellowish colour of the fluids, which could be interpreted as an indicator for hydrated trivalent iron, as the iron(III)complex $[FeCl(H_2O)_5]^{2+}$(aq) (ligand Cl^-) was yellow [45]. The precipitate may also include silica gel. According to Portier and Vuataz [46] this is formed by a secondary substitutionary reaction of hexafluorosilicic acid (high solubility) (which is formed from HF-acidification of Quartz) where Si ions are exchanged by Al^{3+} and precipitate as SiO_2, while the fluoaluminates remain in the solution. This reaction is triggered when HF is nearly consumed. As clays or Micas are the main source for the Al^{3+}, they increase the probability of silica gel precipitation.

Permeability tracking during the CFTs, as displayed in Figure 6, shows the processes during the acidification. Regarding the evaluation of the CFTs, the tracked pore volume can be considered proportional to time, since the flow rate was constant. As visible in Equation (A1) in Appendix A, the pressure difference was inversely proportional to the permeability and the maximum measurement range of the device is approx. 19 bar. The sudden drop in measured pressure difference at the transitions between the different fluids, which was visible in all three CFTs, was caused by an inlet pressure drop when the fluid valves were changed. Regarding sample 6 (Figure 6), the effect of SSB-007 was insignificant in the sample (no calcite or well-soluble minerals), as the differential pressure did not vary before and after SSB-007. During the application of SSB-007, differential pressure decreased, implying a high efficiency of HF in this sample. The shape of the curve resembled an exponential decrease, implying a fast efficiency decrease in HF. In the sample this effect could be due to an initial dissolution phase, which mainly affects pore throats and macropore contaminants. Combined with the flushing of the macropores and smaller pores, now with extended pore throats, the removal of particles increases the permeability with high effectivity in this initial phase. Another explanation might be that the exponential decrease was an effect of grain integrity and grain size distribution. Because this samples showed signs of weathering, the trend could express a significant reaction with fine- to very fine pre-damaged grains. After this, the grains with higher integrity showed a lower reactivity with SFB-007. In the post-flush phase, the pressure difference increased after the circulation of 22 pore volumes. This behaviour might indicate the collapse of a fracture under the applied confining pressure during the flushing process. This implies the risk of a permeability decrease by the reduction in rock stability during chemical treatment. In further CFTs, permeability increased during the post-flush phase displayed the relevance of the post-flush, which seemed to mobilise particles that cause pore clogging.

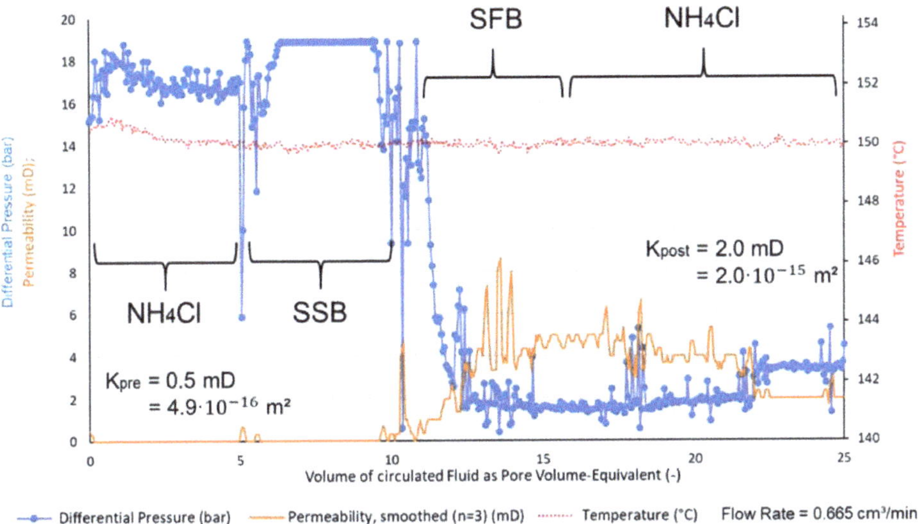

Figure 6. Pore volume—differential pressure diagram showing the development of the CFT on sample 6, applied flow rate: 0.665 mL/min; modified after [44].

In difference to sample 6, the samples 5 (Figure 7) and 3 (Figure 8) were sawn vertically along the natural vein to increase the permeability, because the execution of the CFTs was not possible as the initial permeability was too low for circulation of ammonium chloride.

The documented pressure differences in the CFT on sample 5 resemble sample 6. A major difference is that during the long post-flush period, the pressure difference decreased significantly after a total of approximately 275 PV of circulated fluids and the permeability increased, respectively (doubling of the permeability from 1 mD (9.87×10^{-16} m²) to 4 mD (3.95×10^{-15} m²)). This implies a high relevance of the post-flush, which seems to mobilise particles, that cause pore clogging.

The evaluation of sample 3 shows major difficulties during the CFT. The permeability of the sample, even after sawing an artificial vein, was too low to allow the analysis of the pressure differences, because the device's measurement range is exceeded. The given diagram cannot be analysed with regard to the effectivity of the acid blends, but a general statement about the acidification is still possible: the permeability increases significantly, since a low-rate circulation of ammonium chloride becomes possible after the acidification.

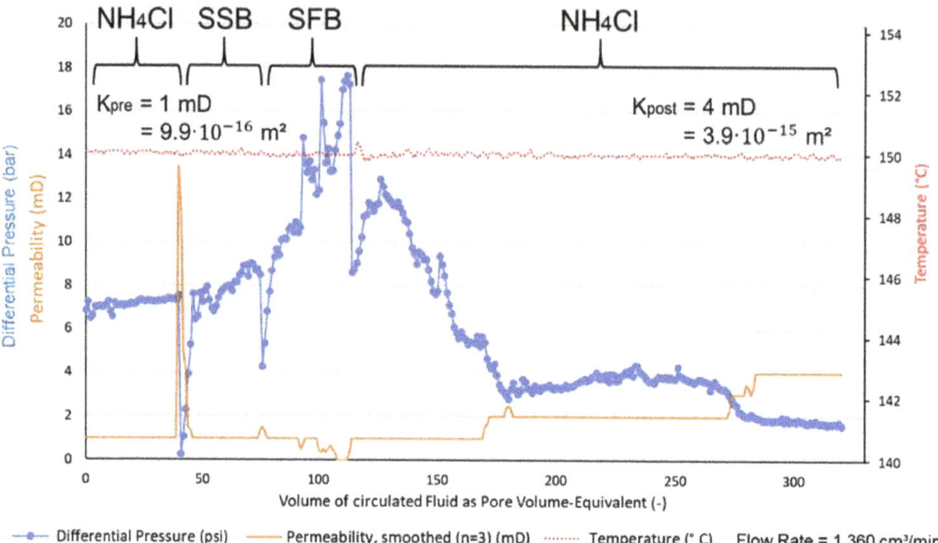

Figure 7. Pore volume—differential pressure diagram showing the development of the CFT on sample 5, applied flow rate: 1.360 mL/min; modified after [44].

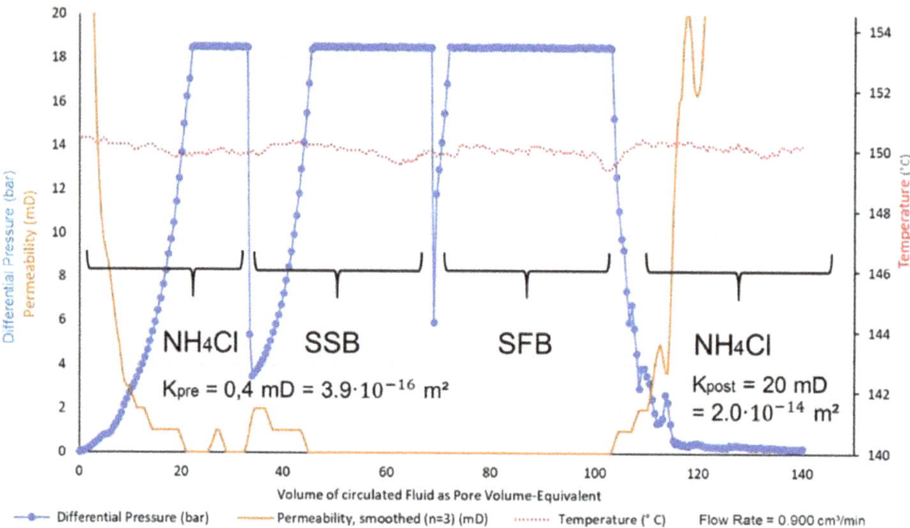

Figure 8. Pore volume—differential pressure diagram showing the development of the CFT on sample 3, applied flow rate: 0.900 mL/min; modified after [44].

This implies a significant permeability increase in all three cases, which is almost exclusively caused by SFB-007. The flushing of the pores was identified as an important factor to increase the permeability after the acidification but may result in a pore collapse due to the confining pressure.

The three CFT diagrams show that the permeability increase could account for the HF-bearing acid blend SFB-007, while the effect of SSB-007 was insignificant in the samples.

The relevance of SSB-007 was mainly as a pre-treatment to avoid unwanted chemical reactions with HF, such as the precipitation of CaF_2. In the AEs, the PEF was higher if HF was involved, but the number of cases with permeability decreased as well. Hydrochemical analyses, using AAS, IC and Si-Photometry, show that the total ion load of the spent HF-based acid was far higher compared to spent 15% HCl. The precipitation of the iron hydroxide was most intense in samples with the highest ion load in the fluid.

3.3. Petrophysical Methods

Permeametry, Sonic, Thermoscanning and Pycnometry were used to assess the petro- and thermophysical rock properties before and after treatment. The initial gas-permeabilities of the untreated samples ranged between 1.3×10^{-18} and 3.1×10^{-14} m^2. After treatment, the gas permeabilities ranged between 1.2×10^{-17} m^2 and 1.8×10^{-14} m^2. The permeability correlated with the porosities, which ranged between 0 and 12.02% before treatment and increased to 0.92 to 14.52% after treatment. The largest porosity increase was achieved during the CFTs (max. plus 5.03%). In the AEs the treatment 12% HCl + 3% HF was more effective than 15% HCl. Thermal conductivity (2.5–4.5 W·m^{-1}·K^{-1})) and thermal diffusivity (1.2–3.9×10^{-6} m^2/s) resulted in specific heat capacities between 572 and 827 J·kg^{-1}·K^{-1}) and were not significantly affected by acidification. Sonic velocities (p-wave velocities between 3095 and 6360 m/s, s-wave velocities between 1909 and 4447 m/s, mean value per sample before treatment) resulted in high Young's moduli and low Poisson ratios. These deviated from the literature, possibly because Quartz veins provide preferential wave propagation paths. While the Poisson ratio decreased after acidification, Young's modulus increased, indicating that precipitation might have increased the mineral interconnectivity.

4. Discussion

As a major difference between the CFTs and the AEs, in the CFTs the applied acids SSB-007 (based on a strong organic acid), followed by SFB-007 (based on hydrochloric acid plus hydrofluoric acid) were circulated through up to 80 mm long cores for up to 2.5 h under approximation of reservoir conditions (150 °C, 172.4 bar confining pressure). In contrast, in the AEs, the acids 15% HCl or 12% HCl + 3% HF, together with 20 mm long rock plugs or powder from ground samples, were heated to 150 °C for 24 h without circulation. This implies that no pre- or post-flush was conducted.

The analysis of the effectivity of the acids regarding different minerals, quantified by using XRD and XRF as well as AAS, IC and Si-Photometry on the spent acids, displayed similar trends, which generally correlate with earlier studies [46,47] as well as with results from Economides and Nolte [48].

The most relevant chemical equations for reactions with HCl + HF are listed below to give a preliminary understanding of the results discussed in the following section. Chemical reactions with only HCl are not displayed separately since, chemically, Cl⁻ has comparable characteristics to F⁻ and can therefore be involved in similar chemical reactions, but with reaction rates that are several magnitudes lower. The following chemical equations were documented by Economides and Nolte [48]:

- Quartz $4HF + SiO_2 \Leftrightarrow 2H_2O + SiF_4$ (silicon tetrafluoride)
- $SiF_4 + 2HF \Leftrightarrow H_2SiF_6$ (hexafluorosilicic acid)
- Na-feldspar $NaAlSi_3O_8 + 14HF + 2H^+ \Leftrightarrow Na^+ + AlF_2^+ + 3SiF_4 + 8H_2O$
- K-feldspar $KAlSi_3O_8 + 14HF + 2H^+ \Leftrightarrow K^+ + AlF_2^+ + 3SiF_4 + 8H_2O$
- Calcite $CaCO_3 + 2HF \Leftrightarrow CaF_2 + H_2O + CO_2$

Depending on the respective stoichiometry, clay minerals reacted with HF + H+ under the creation of AlF_2^+, SiF_4 and H_2O in different ratios. As the composition of Tourmaline is extremely variable, no chemical reaction was provided.

While the reactions between HF and silicates generally resulted in the creation of liquid hexafluorosilicic acid, which, depending on the pH of the resulting fluid, could precipitate as amorphous silica gel, a reaction between HF and Calcite would immediately result in the precipitation of CaF_2 and cause pore clogging.

The XRD results show that on ground samples 12% HCl + 3% HF, the Quartz content in the treated samples increased to up to 100%. This implies that every other mineral reacted entirely. An exception is Tourmaline (Schorl), which was partly persistent in samples with an initial concentration above 5.5%. As a product of the HF acidification, Hieratite ($K_2[SiF_6]$) crystallised in several samples. In accordance with results from Sclar and Fahey [49], it was most likely a product of the acidification of K-Feldspars, since it contains K and F and was found in the samples with initially highest K-Feldspar concentrations. In the powder samples, Micas and Plagioclase were dissolved completely, but no correlation reaction products could be identified. The crystallisation of further minerals is indicated with additional spectral peaks in the XRD-results, but those minerals could not be identified, as well as amorphous phases. In the acidified rock samples (plugs) trends were less distinctive. For 12% HCl + 3% HF as well as 15% HCl, Quartz and Tourmaline were least affected, while Micas, especially Fe-Chlorite, were statistically most affected. The Feldspar contents did not display clear trends, but in relative numbers Plagioclase seemed to be less affected than K-Feldspar. If 15% HCl was applied, the effect on Quartz was insignificant. In all samples the effectivity of the acidification regarding the relative mineral concentrations increased if HF was involved. Comparing the SSB-007 plus SFB-007 acidification in CFTs with 12% HCl + 3% HF in AEs, the trends were similar, except for Muscovite, which seemed to be less affected. This may be related to additives with the purpose of inhibiting the participation of clay minerals.

The XRD results were generally confirmed by the XRF results, which displayed low to no effectivity regarding Quartz, depending on the acid, (relative SiO_2 increase) and higher effectivity regarding Micas and Feldspars or other Al-bearing silicates (relative Al_2O_3, K_2O or Na_2O decrease). The effectivity regarding Schorl was low to intermediate (Fe_2O_3 concentration decreased or did not change in samples without Fe-Chlorite and Biotite). Again, the trends accounted for 15% HCl as well as for 12% HCl + 3% HF, but the magnitude was larger for 12% HCl + 3% HF. Further trends in other main elements (TiO_2, MnO, MgO, CaO, Na_2O, K_2O, P_2O_5) or trace elements cannot be generalised.

In sample 6 (023001, Cligga granite), as an exception, indicators for Cassiterite, Galena, Pyrite or other ore minerals were detected. This confirms the XRD results and macroscopic observations for the specific sample.

The chemical analysis of the spent acids, using AAS, IC and Si-Photometry also confirmed the trends from XRD and XRF. In most cases, relative concentrations can be generalised as trends for the respective acidification set:

- 12% HCl + 3% HF, 150 °C, powder: Al > Si > Na > Ca > K > Ti > Mg > Fe > Mn
- 15% HCl, 150 °C, plugs: Mn > Al > Fe > Ti > Ca > K > Mg > Si
- 12% HCl + 3% HF, 150 °C, plugs: Al > Na > Fe > Si > Ca > K > Mg > Ti > Mn

Regarding 12% HCl + 3% HF as well as 15% HCl acidification on plugs, the relative participation rates of Feldspars are low (relatively low K^+, Na^+ and Ca^{2+} concentrations), while Micas, especially Muscovite, are strongly affected (relatively high Al^{3+}-concentrations). Tourmaline and other Fe-bearing minerals have intermediate relative participation rates, if present (relatively high Mn^{2+} and $Fe^{2+/3+}$ concentrations). If 15% HCl is used, Si is absent in the fluid, implying zero participation of silicates. This is a contradiction to the abovementioned results unless assuming precipitation of silica gel, as described above. In contrast, applying 12% HCl + 3% HF (comparable trends on plugs as on powder), the participation of Feldspars and Micas, as well as mafic minerals, is magnitudes higher compared to 15% HCl (extremely high Al^{3+}, high Si and high $Fe^{2+/3+}$, Mg^{2+}, Ca^{2+}, Na^+ and K^+ concentrations). Total ion loads reach up to approx. 11,000 mg/L in spent 12% HCl + 3% HF, but only approx. 550 mg/L in spent 15% HCl.

Comparing the acidification on plugs and powder (with 12% HCl + 3% HF in both cases), the data show similar trends. In relative numbers, the ratio of the dissolved elements fits quite well between acidised OAS powder and plugs. The main difference is an extremely low $Fe^{2+/3+}$ concentration in the fluid from powder acidification experiments (maximum $Fe^{2+/3+}$ concentration in powder experiment: 16 mg/L; in plug experiment:

1355 mg/L). This might indicate the precipitation of an iron mineral on the large powder surface and separation during centrifugation of the powder sample. In XRD, the respective mineral would likely not be detected as the precipitate would most likely precipitate as an amorphous mineral. No XRF measurements of the powder samples were performed, because the remaining material mass was too small for the method.

Summarising the effects on specific minerals, the effectivity of the acidification depends primarily on the composition of the spent acid. Theoretically, the effect of the temperature is relevant, but since all experiments were conducted at the same temperature, this cannot be validated by the data. The effect of the grainsize is minor, as powder and plug acidification had comparable effectivities regarding different minerals, although the grainsize was homogenised by pulverisation. It also depends on geochemical variations in the sample, such as varying concentrations of ore minerals, Chlorite, Muscovite and Biotite or Tourmaline. Feldspars and Mica are mainly affected by the chemical reactions. The effect on Quartz is far lower, even if HF is involved. Since the AEs last 24 h and near-equilibrium conditions are established, reaction kinetics are not as important for the resulting effects on the samples as they would be in a real stimulation or in CFTs, in which the reaction time is much lower (0.5 to 6 h). However, since all relevant reactions imply an activation energy threshold, reaction kinetics still influence the results. Equilibrium constants in a multi-phase—multi-component system can be derived from reaction rate constants. Portier and Vuataz [46] use respective parameters for a granite system and state that, using 12% HCl + 3% HF, the reaction rates for Feldspars and Micas are two magnitudes higher than for Quartz. Their results also confirm a higher reactivity of K-Feldspar compared to Plagioclase. Reaction rates for HCl and HCl + HF in varying concentrations are also provided by Economides and Nolte [48] and confirm the present results. If the effectiveness of the acids is summarised in terms of the estimated solubility of the minerals, which are predominantly involved in the analysed samples, the following order can be derived:

- Micas (especially Fe-Chlorite) > K-Feldspar > Plagioclase > Tourmaline > Quartz

Further quantifications of the reaction kinetics in the autoclave experiments would require a larger dataset and more detailed analysis. Regarding SSB-007 and SFB-007, reaction kinetics are influenced by the addition of retardants and corrosion inhibitors. As the exact composition of the acid blends has not been provided, no further interpretation of the reaction processes during the CFTs is possible.

Regarding the comparability of the acidification methods (CFTs versus AEs; 15% HCl versus SSB-007; SFB-007 versus 12% HCl + 3% HF), it is very important to consider the different reaction times, as the SSB-007 flush lasted less than one hour in the CFTs, while the AEs were conducted during 24 h. As a summarising comparison, 15% HCl has a weak, but evident effect on the AE samples and in the fluid samples, while SSB-007 did not cause any effect on permeability that could be identified during the CFTs. All effects related to geochemical variations induced by the CFTs could solely account for SFB-007. Comparing SFB-007 and 12% HCl + 3% HF, the effects seem to be quite comparable, except for the slightly lower effectivity of SFB-007 on Micas, detected in the XRF data. From several papers published by Fangmann Energy Services, only Recalde Lummer and Rauf [24] unambiguously describe the application of SFB-007, stating that K-Feldspars of the analysed samples are not affected by conducted experiments. This is not confirmed by the present results. The absolute effectivity of both methods is not comparable due to the differences in duration, pressure and the acid circulation in the CFTs.

The most important limitations of methods are related to the acidification experiments: in the CFTs, no absolute permeability is measured since the fluid properties (viscosity of the acid, as well as the exact composition of the acid blends) are unknown and likely change during the experiment. Thus, unfortunately, the permeability determination is only reliable during the ammonium chloride flush. The dynamic viscosity of ammonium chloride under test conditions would need to be determined, as well as the fluid properties of SSB-007 and SFB-007, which are required for hydraulic modelling. The flow behaviour, and thus also the retardation and reaction behaviour, is highly dependent on viscosity and density. In the

AEs, the quantification of the mineral dissolution is corrupted by precipitates, which do not occur in the CFTs to a comparable extent and would not occur in the near-well regions of the reservoir (metres to tens of metres), if an adequate post-flush is conducted to displace the reaction products deeper into the formation. This displays the main limitation of the AEs.

Indicators for the relevance of ore minerals in acidification are found in sample 6 (023001), where precipitates are most abundant. The thickness and the extent of the yellow to white precipitate is highest. Uniquely, sample 6 contains macroscopically visible fine-grained ore minerals. Cassiterite was identified by XRD; further ore minerals (possibly Galena and Pyrite) are indicated in XRF data. The AEs on sample 6 result in a PEF of 0.8 for 15% HCl and 0.59 for 12% HCl + 3% HF. It is the only sample with a permeability decrease after acidification with 15% HCl. Fe was identified to cause the yellow precipitate, which occurs not only in the presence of ore minerals but also in samples that contain mafic or Fe-rich minerals, such as Schorl, Hornblende or Fe-Chlorite. Assuming the presence of Pyrite, the abundance of the precipitate can be explained. In the XRF data, the relative Fe_2O_3 content increases, indicating that Fe-ions reprecipitate. To assess the presence of ore minerals for a chemical stimulation, XRF is recommended instead of XRD, as XRD is not a well-suited method to identify ore minerals.

Comparing OASs and cuttings, only limited transferability is given between the two sample types. This highlights the significant effect of surface weathering or alteration processes on the samples, which can be excluded by the drilling of Side Wall Cores for a high-quality assessment of the reservoir. The reduced normalised Plagioclase content in the OAS, in comparison to the cuttings, could be another indicator—besides the REE pattern—for weathering, because Plagioclase has higher weathering rates than K-Feldspars. Nevertheless, the mineralogical composition of the OAS resembles hydrothermally altered fracture zones and is therefore a relevant approach for deep geothermal systems in fractured granite.

5. Conclusions

Regarding the results of the present study, the most efficient acid for United Downs is clearly HF-based. In other wells, when calcite is present in veins, HCl or an Organic Acid is usually sufficient for stimulation, but this is not the case in UDDGP. In case chemical treatment is not an option, e.g., due to due to regulatory requirements regarding the applied acids, pulse fracturing treatment—as described by Tariq et al. [50]—might be an option. Thermochemical acid fracturing, as described by Tariq et al. [51], which is based on the pressure increase during specific chemical reactions and can be based on less hazardous chemicals, may also display an alternative. HF is not biodegradable and not environmentally friendly, although Recalde Lummer and Rauf state that the treatment system SSB-007 + SFB-007 is biodegradable [24], which is confirmed by the safety sheets of the single components. To reduce safety and environmental risks, the HF-based acid is prepared on site from ammonium bifluoride and strong organic acids, which are less hazardous. This allows safer handling and a demand-controlled preparation of the required acid volume. Technical restrictions regarding the project site, such as swelling of clays and compatibility with borehole and reservoir fluids, as well as casing steel and cements were analysed and show no signs of incompatibility. Technical constraints from the environmental side include the presence of a hydraulic barrier, such as a customised barrier (e.g., inert, impermeable textiles or foils) on top of a concrete pad of sufficient extent around the well, the integrity of the well cementation, the exclusion of any leaks in or around the well or project site and the exclusion of any other hydraulic links between the reservoir and non-target areas.

The CFTs were conducted to approximate a chemical reservoir stimulation on lab scale. A major difference between the lab-scale experiments and a reservoir stimulation is that in a reservoir stimulation, chemical stimulation is accompanied by hydraulic and thermal stimulation effects.

As relatively cold fluids, compared to the reservoir temperature, are injected, thermoelastic effects may cause fracture opening or fracturing as well, as significant contraction effects may occur due to the large reservoir volume and will strongly be influenced by injection volume and the duration of injection. In the CFTs, the sample is heated to 150 °C, while the fluid is at room temperature before the contact with the sample. For the small core volume, thermal contraction is not relevant. Effects of thermal contraction would most likely occur during the ammonium chloride pre-flush but have not been detected.

The hydraulic pressure is at least the hydrostatic pressure of the fluid column in the well, but may be artificially increased, depending on the stimulation approach.

Regarding the pressure conditions, the applied confining pressure in the CFTs was set to 172.4 bar and a back pressure of 34.5 bar was applied to prevent boiling of the fluids. The pore pressure was in a dynamic relation with the measured differential pressure and depended on the pump pressure, which varied to maintain a constant flow rate (OFITE 2019). In comparison, the stress conditions in the production well were strike–slip stress conditions (max. horizontal pressure > vertical pressure > min. horizontal pressure), critically stressed for shearing on pre-existing fractures with matching orientation [22]. Indicators for a change of stress regime in the open-hole sections were found and are described by Reinecker et al. [22].

While the CFTs affect a specific fissure or vein, a reservoir stimulation is conducted with packers or coiled tubing, affecting entire well sections. The fluid follows preferential flow paths, which implies the risk of creating wormholes along the most permeable fractures, while closed fractures might only be affected by the acid to a very low degree. This issue should be investigated in further studies, e.g., by digital rock imaging, as well as quantification of the changes in roughness.

In the CFTs, during the flushing process under the applied confining pressure, a fracture collapse was detected, as described in Section 3.2. This risk is transferable to the reservoir, as it implies the risk of a permeability decrease by the reduction in vein or fissure stability by particle washout or during chemical treatment. In hydraulic stimulations, proppants are commonly used to avoid fracture collapse. In the case of the granitic reservoir rocks, the Quartz grains, however, might act as natural proppants and keep the fractures open, while Micas and ore minerals in the fractures are preferentially dissolved.

The dissolution characteristics for a HF based acid are good, but imply the precipitation of pore-clogging minerals. A post-flush needs to be executed, which ensures that precipitation occurs only in the reservoir at greater distance of the well, minimising negative hydraulic effects to the well productivity. As an attempt to quantify the required post-flush volume for chemical treatment, it should correspond to at least one times the pore volume of the sample. With respect to the possible fracture collapse and pressure limitations due to technical and safety constraints, the maximum post-flush volume is limited as well. In view of these limitations, a post-flush of three to five times the pore volume is recommended.

Chemical stimulation with HF is common in petroleum projects and has also been applied in geothermal projects [52–54], but open-access datasets are scarce. HF-treatment involves severe risks for human health and the environment. The company and operator have to ensure that during chemical treatment, all risks to health and the environment are considered. Procedures to handle the fluids have to be designed accordingly and countermeasures need to be in place to be applied immediately if needed. The acid blend SFB-007 has already been applied in geothermal projects with granitic reservoir rocks as Soultz sous Forêts, Rittershoffen and Vendenheim [24]. An intense pre-stimulation assessment of the reservoir is required to search for the optimum acid blends and to define required volumes or technical alternatives. The presented research contributes to the optimisation of stimulation pre-assessment, which is transferable to other projects in fractured crystalline rock. The present work contributes significantly to the stimulation pre-assessment for the United Downs Deep Geothermal Power project and therefore improves the planning process of the reservoir treatment. It complies with the MEET project objective to optimise chemical treatment and reservoir assessment. SSB-007 + SFB-007 are innovative, state of the art acid blends, which are applied internationally. The approximation of these acid blends with 15% HCl and 12% HCl + 3% HF has proven to be a suitable approach. Elaborated results from the selected samples can be transferred to further project sites in a fractured granitic basement. The applied methods in the compiled workflow are an innovative, well-suited tool set for a generalised approach to the lab scale assessment of rock acidification, particularly for application in a crystalline basement.

Using a set of laboratory tests that, to our knowledge, have never before been performed with similar acid combinations on granitic rocks for geothermal purposes, the present study provides new insights into the efficiency as well as possible difficulties of chemical stimulation in such rock types.

Supplementary Materials: The following are available online at https://doi.org/10.48328/tudatalib-629 and https://www.mdpi.com/article/10.3390/geosciences12010035/s1, Table S1: Pn2MEET_KSchulz: petrophysical and petrological characterisation of outcrop analogue samples before and after acidification; Table S2: Fluid Chemistry-AAS_IC_Photometry: chemical characterisation of spent acids after autoclave experiments, Table S3: Core Flooding Tests: permeability tracking during Core Floodings Tests.

Author Contributions: Conceptualisation, K.B. and K.E.S.; methodology, K.E.S.; validation, K.B. and K.E.S.; writing—original draft preparation, K.E.S.; writing—draft review, K.B., I.S.; visualisation, K.E.S.; supervision, K.B., I.S.; project administration, K.B.; funding acquisition, K.B., I.S. All authors have read and agreed to the published version of the manuscript.

Funding: The work is conducted as part of the MEET project that has received funding from the European Union's Horizon 2020 research and innovation programme under grant agreement No 792037.

Institutional Review Board Statement: Not applicable.

Informed Consent Statement: Not applicable.

Data Availability Statement: Supplementary Materials are published by Schulz et al. [42] and can be found at https://doi.org/10.48328/tudatalib-629 (access date: 25 November 2021).

Acknowledgments: The present work was elaborated at the Institute of Applied Geosciences, Technical University of Darmstadt in the framework of the projects MEET (grant agreement No. 792037) and UDDGP. We would like to thank the supporting staff from the Department of Geothermal Science and Technology, namely G. Schubert, S. Schmidt, R. Seehaus, D. Scheuvens and R. Petschik from Goethe University of Frankfurt for the XRD measurements, as well as our project partners related to MEET/UDDGP, to mention A. Genter, A. Turan, J. Reinecker, J. Gutmanis, R. Shail, B. Simons, C. Dalby. We would also like to acknowledge the editorial support of Mayes Nie and Aleksandar Pavlovic, as well as two anonymous reviewers, who allowed us to substantially improve our manuscript.

Conflicts of Interest: The authors declare no conflict of interest.

Appendix A

Calculation of permeability from the measured pressure difference before and after the CFTs: the differential pressure and the runtime, which—under constant flow rate—can be expressed as multiples of the pore volume, are measured automatically during the CFTs. The evaluation of the measured pressure difference (given in psi) is based on Darcy's law. An additional multiplication factor, which depends on the device and on the conversion from bar (or Pa) to psi (1 bar = 14,504 psi), is used to calculate the permeability (OFITE, 2019). With k: permeability (mD); μ: viscosity of fluid (cP); Q: flow rate (cm^3/s); l: core length (cm); A: cross sectional area of the core (cm^2); Δp: differential pressure (psi), the resulting equation is:

$$k = 14{,}700 \cdot \mu \cdot Q \cdot l / (A \cdot \Delta p) \qquad (A1)$$

As described in Section 3.2, the permeability can only be measured during the pre- and post-flush. Since the viscosity of the brine (NH$_4$Cl) is temperature dependent, it had to be determined in additional experiments, conducted by Fangmann Energy Services. Nevertheless, to our knowledge, in the data provided by Fangmann Energy Services, which are included in the Supplementary Materials, the viscosity of the fluid was set to 1 cP.

To avoid falsification of the results, our evaluation of the CFTs is based on the PEF, which is introduced in Section 3.2 and allows a relative quantification of the permeability increase. The permeability changes were verified by additional gas permeability measurements, as described in Section 3.3.

The viscosity of the acids, considering the chemical reactions of the acids and under the applied temperatures, is unknown. Therefore, the tracking of the differential pressure—which is inversely proportional to the permeability—during the CFTs, as discussed in Section 3.2, is only a qualitative approach to the permeability changes during the CFTs.

Appendix B

	Powder HCl+HF						Powder HCl	Plug HCl						Plug HCl+HF						CFT		
	1	2	3	4	5	6	3	1	2	3	4	5	6	1	2	3	4	5	6	3	5	6
Quartz	↑	↑	↑	↑	↑	↑	↑	↑	↑	↑	↑	↑	↑	↓	↑	o	o	↓	↓	↓	o	o
K-Feldspar	↓	↓	↓	↓	o	↓	↑	↓	↓	↑	↓	o	o	↓	↓	↑	↓	o	o	↓	↓	↓
Plagioclase	↓	↓	↓	o	↓	↓	o	↑	↑	↑	↑	o	o	↑	↓	o	↑	o	↑	↑	o	↑
Muscovite	↓	↓	↓	↓	↓	↓	↓	↓	↓	↓	↓	o	o	↓	↓	↓	↓	o	↑	o	o	↑
Biotite	o	o	o	o	↓	↓	o	o	o	o	o	↓	↓	o	o	o	o	↓	↓	o	↓	↓
Chlorite	o	o	o	o	o	↓	↓	o	o	o	o	o	↓	o	o	o	o	o	↓	o	↓	↓
Schörl	↓	↓	↓	↓	↓	↓	↓	o	↑	↑	↑	o	o	↑	o	o	↑	o	↑	↑	o	↑
Cassiterite	o	o	o	o	o	↑	o	o	o	o	o	o	o	o	o	o	o	o	o	o	o	o
Hieratite	o	↑	↑	↑	o	o	o	o	o	o	o	o	o	o	o	o	o	o	o	o	o	o

relative increase (↑), decrease (↓) or no significant changes (variation < 0,3 %) (o) after acidification

Figure A1. Comparison of the relative mineralogical composition before and after chemical treatment in AEs and CFTs.

References

1. Law, R.; Cotton, L.; Ledingham, P. The United Downs Deep Geothermal Power Project. In Proceedings of the European Geothermal Congress, Den Haag, The Netherlands, 11–14 June 2019.
2. Beamish, D.; Busby, J. The Cornubian geothermal province: Heat production and flow in SW England: Estimates from boreholes and airborne gamma-ray measurements. *Geotherm. Energy* **2016**, *4*, 1–25. [CrossRef]
3. Staněk, M.; Géraud, Y. Granite microporosity changes due to fracturing and alteration: Secondary mineral phases as proxies for porosity and permeability estimation. *Solid Earth* **2019**, *10*, 251–274. [CrossRef]
4. Gleeson, S.A.; Wilkinson, J.J.; Stuart, F.M.; Banks, D.A. The Origin and Evolution of Base Metal Mineralising Brines and Hydrothermal Fluids, South Cornwall, UK. *Geochim. Cosmochim. Acta* **2001**, *65*, 2067–2079. [CrossRef]
5. Simons, B.; Shail, R.K.; Andersen, J.C.Ø. The petrogenesis of the Early Permian Variscan granites of the Cornubian Batholith: Lower plate post-collisional peraluminous magmatism in the Rhenohercynian Zone of SW England. *Lithos* **2016**, *260*, 76–94. [CrossRef]
6. Shail, R.K.; Leveridge, B.E. The Rhenohercynian passive margin of SW England: Development, inversion and extensional reactivation. *Comptes Rendus Geosci.* **2009**, *341*, 140–155. [CrossRef]
7. Bromley, A. (Ed.) *Water-Rock Interaction in Southwest England: The Evolution of the Cornubian Orefield*; International Association of Geochemistry and Cosmochemistry: Lawrence, KS, USA, 1989; 58p.
8. Willis-Richards, J. Three-dimensional modelling from gravity of the Carnmenellis granite pluton, West Cornwall. Paper presented at the Tenth UK Geophysical Assembly. *Geophys. J. Int.* **1986**, *92*, 323–334.
9. Edmunds, W.M.; Kay, R.L.F.; McCartney, R.A. Origin of Saline Groundwaters in the Carnmenellis Granite (Cornwall, England): Natural Processes and Reaction during Hot Dry Rock Reservoir Circulation. *Chem. Geol.* **1985**, *49*, 287–301. [CrossRef]

10. Yeomans, C.; Middleton, M.; Shail, R.; Grebby, S.; Lusty, P. Regional Lineament Detection Using Bottom-Up Object-Based Image Analysis Methods. In Proceedings of the 80th EAGE Conference and Exhibition 2018, Copenhagen, Denmark, 11–14 June 2018; EAGE Publications BV: Houten, The Netherlands, 2018.
11. Yeomans, C.M.; Middleton, M.; Shail, R.K.; Grebby, S.; Lusty, P.A.J. Integrated Object-Based Image Analysis for semi-automated geological lineament detection in southwest England. *Comput. Geosci.* **2019**, *123*, 137–148. [CrossRef]
12. Ghosh, P.K. The Carnmenellis Granite: Its Petrology, Metamorphism and Tectonics. *Q. J. Geol. Soc.* **1934**, *90*, 240–276. [CrossRef]
13. Shail, R.K.; Alexander, A.C. Late Carboniferous to Triassic reactivation of Variscan basement in the western English Channel: Evidence from onshore exposures in south Cornwall. *J. Geol. Soc.* **1997**, *154*, 163–168. [CrossRef]
14. British Geological Survey. BGS Geology 625k I DiGMapGB-625: Digital Geology Data. BGS. 2007. Available online: https://www.bgs.ac.uk/datasets/bgs-geology-625k-digmapgb/ (accessed on 25 November 2021).
15. Manning, D.A.C.; Hill, P.I.; Howe, J.H. Primary Lithological Variation in the Kaolinized St Austell Granite, Cornwall, England. *J. Geol. Soc.* **1996**, *153*, 827–838. [CrossRef]
16. Psyrillos, A.; Manning, D.; Burley, S. Geochemical constraints on kaolinization in the St Austell Granite, Cornwall, England. *J. Geol. Soc.* **1998**, *155*, 829–840. [CrossRef]
17. Tierney, R.L.; Glass, H.J. Modelling the structural controls of primary kaolinite formation. *Geomorphology* **2016**, *268*, 48–53. [CrossRef]
18. Zotov, A.; Mukhamet-Galeev, A.; Schott, J. An experimental study of kaolinite and dickite relative stability at 150–300 °C and the thermodynamic properties of dickite. *Am. Mineral.* **1998**, *83*, 516–524. [CrossRef]
19. Sanderson, D.J.; Roberts, S.; Gumiel, P.; Greenfield, C. Quantitative Analysis of Tin- and Tungsten-Bearing Sheeted Vein Systems. *Econ. Geol.* **2008**, *103*, 1043–1056. [CrossRef]
20. Geothermal Energy Limited. United Downs Deep Geothermal Power Project—Confidential Geology Summary Report. In Proceedings of the Redruth, 44th Workshop on Geothermal Reservoir Engineering Stanford University, Stanford, CA, USA, 11–13 February 2019.
21. Scrivener, C.; Darbyshire, D.P.; Shepherd, T. Timing and significance of crosscourse mineralization in SW England. *J. Geol. Soc.* **1994**, *151*, 587–590. [CrossRef]
22. Reinecker, J.; Gutmanis, J.; Cotton, L.; Foxford, A.; Dalby, C.; Law, R. Geothermal Exploration and Reservoir Modelling of the United Downs Deep Geothermal Project, Cornwall (UK). *Geothermics* **2021**, *97*, 102226. [CrossRef]
23. Bär, K.; Arbarim, R.; Turan, A.; Schulz, K.; Mahmoodpour, S.; Leiss, B.; Wagner, B.; Sosa, G.; Ford, K.; Trullenque, G.; et al. *Database of Petrophysical and Fluid Properties and Recommendations for Model Parametrization of the Four Variscan Reservoir Types*; MEET Report, Deliverable D5.5; Technical University of Darmstadt: Darmstadt, Germany, 2020; 110p.
24. Recalde Lummer, N.; Rauf, O. Premium Treatment System for Granite and Sandstone Formations: Fluid Development and Field Trial in a Geothermal Well. In Proceedings of the European Geothermal Congress, Den Haag, The Netherlands, 11–14 June 2019.
25. Ferreiro Mählmann, R.; Frey, M. Standardisation, Calibration and Correlation of the Kübler-Index and the Vitrinite/Bituminite Reflectance: An Inter-Laboratory and Field Related Study. *Swiss J. Geosci.* **2012**, *105*, 153–170. [CrossRef]
26. Micromeritics. Product Showcase AccuPyc II 1340: Specification Sheet. Available online: https://www.micromeritics.com/Product-Showcase/AccuPyc-II-1340.aspx (accessed on 12 November 2019).
27. Weydt, L.M.; Heldmann, C.-D.J.; Machel, H.G.; Sass, I. From oil field to geothermal reservoir: Assessment for geothermal utilization of two regionally extensive Devonian carbonate aquifers in Alberta, Canada. *Solid Earth* **2018**, *9*, 953–983. [CrossRef]
28. Weinert, S.; Bär, K.; Sass, I. Thermophysical rock properties of the crystalline Gonghe Basin Complex (Northeastern Qinghai–Tibet-Plateau, China) basement rocks. *Environ. Earth Sci.* **2020**, *79*, 77. [CrossRef]
29. Mielke, P.; Weinert, S.; Bignall, G.; Sass, I. Thermo-physical rock properties of greywacke basement rock and intrusive lavas from the Taupo Volcanic Zone, New Zealand. *J. Volcanol. Geotherm. Res.* **2016**, *324*, 179–189. [CrossRef]
30. Popov, Y.; Lippmann, E.; Rauen, A. TCS–Manual: TCS Thermal Conductivity (TC) and Thermal Diffusivity (TD) Scanner. Version 9.3.2020. Available online: www.geophysik-dr-rauen.de/tcscan/technical_data.html (accessed on 9 March 2020).
31. Filomena, C.M.; Hornung, J.; Stollhofen, H. Assessing Accuracy of Gas-Driven Permeability Measurements: A Comparative Study of Diverse Hassler-Cell and Probe Permeameter Devices. *Solid Earth* **2014**, *5*, 1–11. [CrossRef]
32. Van Leong, H.; Ben Mahmud, H. A preliminary screening and characterization of suitable acids for sandstone matrix acidizing technique: A comprehensive review. *J. Pet. Explor. Prod. Technol.* **2019**, *9*, 753–778. [CrossRef]
33. Bauer, M.; Freeden, W.; Jacobi, H.; Neu, T. (Eds.) *Handbuch Tiefe Geothermie: Prospektion, Exploration, Realisierung, Nutzung*; Springer: Berlin/Heidelberg, Germany, 2014; 867p.
34. BGRCI, BGHM. GisChem, BG Gefahrstoffinformationssystem Chemikalien: Datenblatt Fluorwasserstoffsäure ab 1% bis unter 7%. 2020. Available online: https://www.gischem.de/download/01_0-007664-39-3-000400_5_1_2510.PDF (accessed on 13 August 2021).
35. Broekaert, J.A. *Analytical Atomic Spectrometry with Flames and Plasmas*; John Wiley & Sons: Hoboken, NJ, USA, 2006.
36. Skoog, D.A.; Leary, J.J. *Instrumentelle Analytik: Grundlagen, Geräte, Anwendungen*, 1st ed.; Springer: Berlin/Heidelberg, Germany, 1996.
37. Welz, B.; Sperling, M. *Atomabsorptionsspetrometrie*; Wiley-VCH: Weinheim, Germany, 1997.
38. Deutsches Institut für Normung e.V. *Deutsche Einheitsverfahren zur Wasser-, Abwasser- und Schlammuntersuchung; Anionen (Gruppe D); Photometrische Bestimmung von Gelöster Kieselsäure (D 21)*; Beuth Verlag GmbH: Berlin, Germany, 1990.

39. Petschick, R. MacDiff. 2017. Available online: http://www.geol-pal.uni-frankfurt.de/Staff/Homepages/Petschick/classicsoftware.html (accessed on 13 August 2021).
40. Petschick, R. Röntgendiffraktometrie in der Sedimentologie. Schriftenreihe der DGG. *Sediment* **2002**, *18*, 99–118.
41. Bruker AXS GMBH. *S8 Tiger—Introductory User Manual*; Bruker AXS GmbH: Karlsruhe, Germany, 2010.
42. Schulz, K.; Bär, K.; Sass, I. Petrophysical and Hydrochemical Dataset of Lab-Scale Permeability Enhancement Tests by Chemical Treatment in Fractured Granite (Cornubian Batholith, Cornwall, UK). 2021. Available online: https://tudatalib.ulb.tu-darmstadt.de/handle/tudatalib/2925 (accessed on 25 November 2021). [CrossRef]
43. Kretz, R. Symbols for rock-forming minerals. *Am. Mineral.* **1983**, *68*, 277–279.
44. Fangmann Energy Services. *Permeability Tests with Granite Cores under Borehole Conditions. Lab Report 19-033 for Institute of Applied Geosciences*; Technical University of Darmstadt: Darmstadt, Germany, 2019.
45. Lever, A.B. *Inorganic Electronic Spectroscopy*; Wiley: Amsterdam, The Netherlands, 1983.
46. Portier, S.; Vuataz, F.D. Developing the ability to model acid-rock interactions and mineral dissolution during the RMA stimulation test performed at the Soultz-sous-Forêts EGS site, France. *Comptes Rendus Geosci.* **2010**, *342*, 668–675. [CrossRef]
47. Portier, S.; André, L.; Vuataz, F.D. *Review on Chemical Stimulation Techniques in Oil Industry and Applications to Geothermal Systems: Deep Heat Mining Association, Engine, Work Package 4*; Technical Report; DHM: Neuchâtel, Switzerland, 2007; 34p.
48. Economides, M.J.; Nolte, K.G. *Reservoir Stimulation*; Prentice Hall: Englewood Cliffs, NJ, USA, 1989; Volume 2.
49. Sclar, C.B.; Fahey, J.J. The staining mechanism of potassium feldspar and the origin of hieratite. *Am. Mineral.* **1972**, *57*, 287–291.
50. Tariq, Z.; Mahmoud, M.; Abdulraheem, A.; Al-Nakhli, A.; Bataweel, M. A review of pulse fracturing treatment: An emerging stimulation technique for unconventional reservoirs. In Proceedings of the SPE Middle East Oil and Gas Show and Conference, Manama, Bahrain, 15 March 2019. [CrossRef]
51. Tariq, Z.; Aljawad, M.S.; Mahmoud, M.; Abdulraheem, A.; Al-Nakhli, A.R. Thermochemical acid fracturing of tight and unconventional rocks: Experimental and modeling investigations. *J. Nat. Gas Sci. Eng.* **2020**, *83*, 103606. [CrossRef]
52. Baujard, C.; Genter, A.; Dalmais, E.; Maurer, V.; Hehn, R.; Rosillette, R.; Vidal, J.; Schmittbuhl, J. Hydrothermal characterization of wells GRT-1 and GRT-2 in Rittershoffen, France: Implications on the understanding of natural flow systems in the rhine graben. *Geothermics* **2017**, *65*, 255–268. [CrossRef]
53. Nami, P.; Schellschmidt, R.; Schindler, M.; Tischner, T. Chemical Stimulation Operations for Reservoir Development of the Deep Crystalline HDR/EGS System at Soultz-Sous-Forêts (France). In Proceedings of the 32nd Workshop on Geothermal Reservoir Engineering, Stanford, CA, USA, 22–24 January 2008; pp. 28–30.
54. Schill, E.; Genter, A.; Cuenot, N.; Kohl, T. Hydraulic performance histpory at the Soultz EGS reservoirs from stimulation and long-term circulation tests. *Geothermics* **2017**, *70*, 110–124. [CrossRef]

Article

Hydrothermal Numerical Simulation of Injection Operations at United Downs, Cornwall, UK

Saeed Mahmoodpour [1,*], Mrityunjay Singh [2,*], Christian Obaje [3], Sri Kalyan Tangirala [4], John Reinecker [5], Kristian Bär [5] and Ingo Sass [2,6]

1 Group of Geothermal Technologies, Technical University Munich, 80333 Munich, Germany
2 Group of Geothermal Science and Technology, Institute of Applied Geosciences, Technische Universität Darmstadt, 64287 Darmstadt, Germany
3 Fachbereich Geowissenschaften, Universität Bremen, 28359 Bremen, Germany
4 Department of Applied Geophysics, Indian Institute of Technology, Indian School of Mines, Dhanbad 826004, India
5 GeoThermal Engineering GmbH, 76133 Karlsruhe, Germany
6 Helmholtz Centre Potsdam, GFZ German Research Centre for Geosciences, 14473 Potsdam, Germany
* Correspondence: saeed.mahmoodpour@tu-darmstadt.de (S.M.); mrityunjay.singh@tu-darmstadt.de (M.S.)

Citation: Mahmoodpour, S.; Singh, M.; Obaje, C.; Tangirala, S.K.; Reinecker, J.; Bär, K.; Sass, I. Hydrothermal Numerical Simulation of Injection Operations at United Downs, Cornwall, UK. *Geosciences* 2022, 12, 296. https://doi.org/10.3390/geosciences12080296

Academic Editors: Béatrice A. Ledésert, Ronan L. Hébert, Ghislain Trullenque, Albert Genter, Eléonore Dalmais, Jean Hérisson and Jesus Martinez-Frias

Received: 17 May 2022
Accepted: 27 July 2022
Published: 29 July 2022

Publisher's Note: MDPI stays neutral with regard to jurisdictional claims in published maps and institutional affiliations.

Abstract: The United Downs Deep Geothermal Project (UDDGP) is designed to utilize a presumably permeable steep dipping fault damage zone (constituting the hydrothermal reservoir in a very low permeability granitic host rock) for fluid circulation and heat extraction between an injection well at 2.2 km depth (UD−2) and a production well at 5 km depth (UD−1). Soft hydraulic stimulation was performed to increase the permeability of the reservoir. Numerical simulations are performed to analyze the hydraulic stimulation results and evaluate the increase in permeability of the reservoir. Experimental and field data are used to characterize the initial reservoir static model. The reservoir is highly fractured, and two distinct fracture networks constitute the equivalent porous matrix and fault zone, respectively. Based on experimental and field data, stochastic discrete fracture networks (DFN) are developed to mimic the reservoir permeability behavior. Due to the large number of fractures involved in the stochastic model, equivalent permeability fields are calculated to create a model which is computationally feasible. Hydraulic test and stimulation data from UD−1 are used to modify the equivalent permeability field based on the observed difference between the real fractured reservoir and the stochastic DFN model. Additional hydraulic test and stimulation data from UD−2 are used to validate this modified permeability. Results reveal that the equivalent permeability field model derived from observations made in UD−1 is a good representation of the actual overall reservoir permeability, and it is useful for future studies. The numerical simulation results show the amount of permeability changes due to the soft hydraulic stimulation operation. Based on the validated permeability field, different flow rate scenarios of the petrothermal doublet and their respective pressure evolution are examined. Higher flow rates have a strong impact on the pressure evolution. Simulations are performed in the acidized enhanced permeability region to make a connection between the ongoing laboratory works on the acid injection and field response to the possible acidizing stimulation.

Keywords: United Downs; EGS; hydraulic stimulation; equivalent permeability field

1. Introduction

Petrothermal systems are reservoirs with preferably high heat flow, geothermal gradient, and temperature, but lack sufficient natural fluid flow (in most cases) and permeability to accommodate flow rates required for economic geothermal systems. In such geothermal play types, enhanced geothermal systems (EGS) can be implemented by creating new fractures or stimulating naturally existing ones to achieve the permeabilities needed for economical fluid flow. This is essential to create a connected network of pathways between the injection well and production well [1]. Achieving economic flow rates at wellhead is the goal of any stimulation operation [2]. There are different techniques of stimulating fractures: (i) hydraulic stimulation, where water is injected at high pressure to enhance the permeability; (ii) chemical stimulation, where acids are injected to dissolve fracture mineralization and thus increase the permeability; and (iii) thermal stimulation, where cold water injection creates thermal stress in the reservoir rock which helps in increasing the permeability of the fracture network [3]. The latter is usually best applicable in very high-temperature reservoirs. In this paper, the impact of hydraulic stimulation on the permeability enhancement of naturally occurring fractures in the fault zone through Thermo-Hydraulic (TH) modeling of the United Down Deep Geothermal Project (UDDGP) at Cornwall, UK, is discussed.

UDDGP is geographically located near Redruth in Cornwall, southwest England. This project is the first in the UK where both injection and production wells are drilled into the highly fractured and steeply dipping Porthtowan Fault Zone (PTF), cross-cutting the Carnmenellis Granite of the Cornubian Batholith. For selecting the geological targets of UDDGP, previous experiences from Rosemanowes HDR wells are considered where the hydraulically active fractures are mainly oriented parallel or at an oblique angle to the maximum horizontal stress direction (S_{Hmax}) oriented in the NW-SE direction. To use these favorable flow paths, the geological target selection is focused on faults oriented more or less parallel to S_{Hmax}. Furthermore, hot springs in mines confirm this fracture permeability in the NW-SE direction [4]. Evidence from mining activities in the area of interest covered only the top 400 m, but it provides insights regarding fault dip, direction, damage zone thickness, displacement, and hydraulic activity. Combining these data with surface geological maps, the fault model is generated [5]. For the well trajectories, two criteria are considered: first, longer-length and larger displacement faults, because they are most likely to penetrate to the deeper zones, and second, the selected area should be underlain by granite with high heat production. Given the lack of available significant geophysical data to detect the fault structures at depth, the combination of the surface geology, mining data, few seismic profiles, and analog structures are used to make a drilling prognosis [6–10]. The injection well UD−2 was drilled to a depth of 2393 m MD (2214 m TVD), and the production well UD−1 was drilled to 5275 m MD (5058 m TVD) with a horizontal spacing of 8 m between the two wellheads and a vertical spacing of around 2000 m between the ends of the wells (to prevent thermal short-circuiting). The production well UD−1 is the deepest onshore well drilled in the UK. Both wells were directionally drilled towards the WSW of the well heads to intersect the NW-SE striking PTF (Figure 1). The bottom hole temperature at a depth of 5 km was found to be > 180 °C. The open fractures in the PTF are considered to be the main pathways for fluid flow. However, only a small group of fractures were proven to be hydraulically active by indicators, e.g., mud losses during drilling, geothermal gradient anomalies, induced seismic events, and borehole image analysis [11].

The geology of Cornwall is characterized by the Cornubian batholith, which can be distinguished into several plutons. The Carnmenellis granite is one such pluton that formed around 293 Ma ago [12]. The region also comprises metamorphosed Devonian sediments, which are locally known as "killas". Granites in this region have high contents of uranium (U), thorium (Th), and potassium (K), which result in high heat flow values and the high geothermal gradient. There are a few thermal springs in the vicinity of the site. The granitic outcrops, mines with U deposits, and thermal springs, along with faults, are shown in Figure 2. The Carnmenellis granite was estimated to be about 10 km deep, and the temperature at that depth was 650 °C [11,13].

The PTF is a set of strike-slip structures with the strike in the NW-SE direction and a width of around 300 m which can be traced as various fault segments on the surface (Figure 1). The PTF accommodates periods of extensional (Devonian, Permo-Trias) and compressional (Variscan, Alpine) tectonics of both dextral and sinistral movements. More information on regional geology and tectonics is given in Reinecker et al. [11].

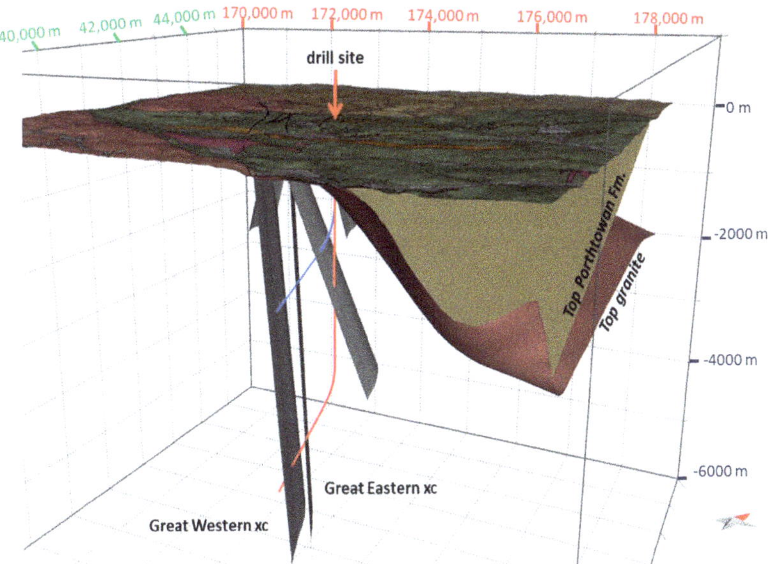

Figure 1. Structural model of UDDGP with injection (UD−2, blue) and production (UD−1, red) wells and their intersection with the Porthtowan Fault Zone (PTF) constituting the two branches of Great Western cross-courses (xc) and Great Eastern cross-courses (xc).

Ultrasonic imaging of the wellbore has given a very reliable understanding of the natural fracture distribution in UD−1 from 906 m to 5206 m MD (see Figure 3). The fractures can be divided into two sets: fractures trending in the NW-NNW direction and the ENE striking fractures. For more geological details, readers are suggested to read Reinecker et al. [11]. Both these sets consist of mineralized and open fractures (a few of them are hydraulically active). The probable number of fractures in the former set is higher than that of the latter set (based on possible geometrical sampling bias from the log data). The intensity of both the fracture sets generally decreases with depth, a behavior which has also been demonstrated for comparably deep wells in the Upper Rhine Graben by Afshari et al. (2022) [14]. Fractures favorably oriented with the in-situ stress field are critically stressed, and their permeability can be enhanced by slip initiated through fluid-driven stimulation. The fracture sets were further divided on the basis of orientation, intensity, and depth into four domains [11]. Additionally, two large-scale fractures were detected based on a combination of fluid losses during drilling, geothermal anomalies in the temperature logs, by the borehole image logs, and induced seismic events during drilling and subsequent hydraulic testing. Both fractures are oriented more or less in parallel to another, but slightly oblique to the PTF, and are critically stressed in the in-situ stress field [11].

In this study, the thermo-hydraulic process is numerically simulated based on the reported fracture and hydraulic testing/stimulation data. The fractured reservoir for this study is represented by a stochastic DFN-based equivalent continuum model with both homogeneous and anisotropic units representing the host rock and the fault zone, respectively. This study was conducted within the framework of the EU-funded Horizon 2020 project MEET (Multidisciplinary and multi-contact demonstration of EGS exploration and Exploitation Techniques and potentials), which is described in general by Trullenque et al. [15]. The fluid flow and heat transfer during the hydraulic testing and stimulation operations were numerically simulated to estimate the permeability field. Furthermore, the pressure dependency on different injection flow rates is presented and analyzed. Finally, a preliminary model based on the conceptual impact of chemical stimulation on the reservoir permeability is developed, and consequently, its impact on reservoir pressure development is estimated.

Due to confidentiality clauses from the industrial partner, all data are reported based on a function of x.

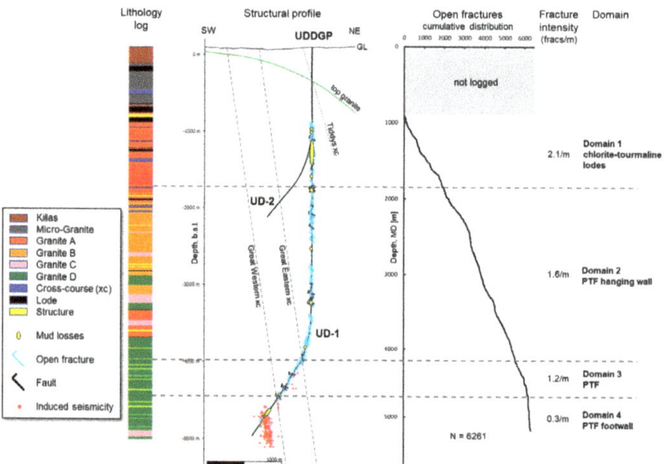

Figure 2. Geological map of the region. Isolines of top granite are not displayed on the western and northern sides of the Carnmenellis granite. The red square indicates the UDDGP model outline. Ptn: Porthtowan Formation, MrSl: Mylor Slate Formation, PBr: Porthleven Breccia Member, Pto: Portscatho Formation, all Devonian in age. Towns: T: Truro, F: Falmouth, R: Redruth, C: Camborne, H: Helston [11].

Figure 3. Fracture distribution interpreted from ultrasonic logs in UD−1. Note that the depth scale of the structural profile is the true vertical depth below sea level. Cumulative fracture distribution and lithology log originally in measured depth below ground level are shifted and stretched to fit the true vertical depth scale [11]. Two large-scale fractures were detected by induced seismic events during drilling and subsequent injection testing.

2. Methodology

Well data, seismological data, surface mapping, data from nearby mines, outcrop analysis, and data from the Rosemanowes project, along with gravity analysis, are all used to define the target faults, well trajectories, and fractures. For setting up a static structural reservoir model, the Move 2018 software [16] developed by Midland Valley Explorations Ltd. was used [11]. Fracture network data obtained from fieldwork, borehole logging, and literature are summarized in Tables 1 and 2 for the unfaulted host rock zone and the fault zone, respectively. Discrete fracture networks (DFNs) are generated for the host rock zone and the fault zone using the FracMan 7 [17] software. Due to unavailability of mean fracture half-length and mean hydraulic fracture aperture data for the fracture set 2, corresponding data from fracture set 1 is assumed in Tables 1 and 2. Furthermore, this geometrical model (see Figure 4) is imported into COMSOL Multiphysics [18] for the hydraulic stimulation simulation.

To enhance the hydraulic connection of the open hole sections of both wells to the fault-bounded reservoir, hydraulic stimulation operations have been performed on UD−1 and UD−2. The injectivity of UD−2 is within the targets set for the power plant operation. To increase the hydraulic connectivity of UD−1 to the reservoir and the near wellbore reservoir hydraulic conductivity, three stages of hydraulic stimulation were performed [11]: (a) Phase 1—step rate injection testing during August 2020, (b) Phase 2—extended injection testing between September and October 2020 and (c) Phase 3—low pressure extended injection testing between October 2020 and February 2021. All phases can be characterized as soft hydraulic stimulation operations due to the low injection volumes and flow rates but sufficient pressures to induce micro-seismic events. Figure 5 shows the operational details of phase 3 hydraulic stimulation for UD−1. This data is used to validate and adjust the anisotropic permeability field of the stochastic DFN models based on the fracture network characteristics for the UDDGP geothermal field. In this data, some changes need to be highlighted: (i) in the first part, from steps 0 to 15, the injection volumes were increased while flow rates and pressures were kept at a low level, followed by (ii) a set of injections with significantly higher injection pressures and considerably higher flow rates from step 16 until step 21 accompanied by increasing induced micro-seismicity (not shown due to confidentiality) leading into (iii) a third part where flow rates increase further with significantly lower injection pressures. These changes indicate that the hydraulic injections are increasing the reservoir injectivity and prove that they can be defined as soft hydraulic stimulation.

Table 1. DFN characteristics of host rock zone.

DFN Parameter	Fracture Set 1	Fracture Set 2	Comment	Reference
Mean fracture orientation, strike [deg]	130–310	50–230	Mean orientations from Rosemanowes wells RH12 and RH15; set 1 from approx. 80% of the total fracture surface area	[19]
Mean fracture orientation, dip [deg]	80–90	70–90		[20]
Mean fracture half-length [m]	5.5		Log-normal distribution with $\mu = 1.7$ and $\sigma = 0.45$ [ln(m)]	[20]
Mean hydraulic fracture aperture [μm]	59		Between 31 and 65 μm	[19]
Fracture density [m^{-1}]	5	0.8	Density is decreasing with depth	[21]
Producing fracture spacing [m]	10		Spacing between producing fractures is in the order of 10 m	[21]
Fracture area density [m^{-1}]	0.9	0.2	Only 10–15% of the fractures carry appreciable flow	

Furthermore, the hydraulic testing/stimulation operational data for UD−2, conducted from the 17th until the 20th of November 2020, is used to validate the calculated permeability field (Figure 6). Based on this validated permeability field, different operational scenarios of flow conditions were simulated to evaluate reservoir performance.

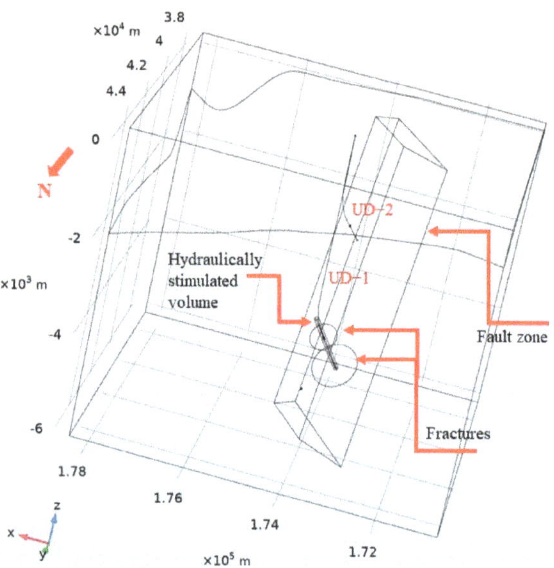

Figure 4. Schematic geometry used to simulate a cubic space of the UDDGP geothermal system. The hydraulically stimulated region is depicted by a cylinder around UD−1. Two critically stressed large-scale fractures, subparallel to the PTF, are implemented based on the seismic events as documented by, e.g., [11] and indicated with a disc shape.

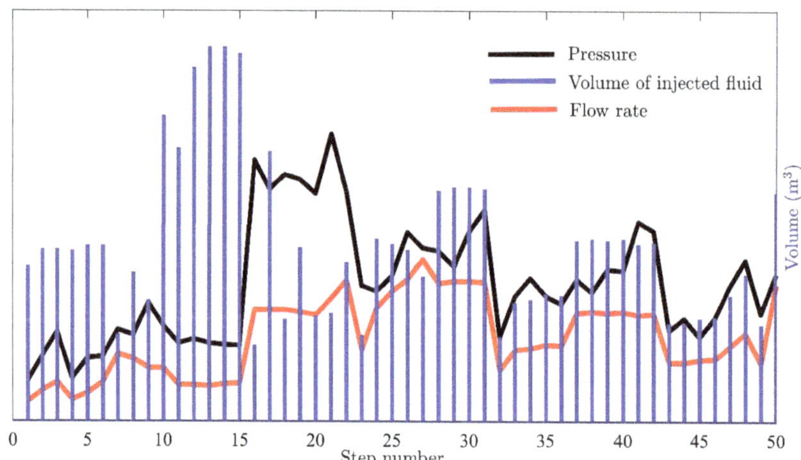

Figure 5. Daily flow rate, injection pressure, and injected volume of the hydraulic testing/stimulation operation on UD−1 from September 2020 until January 2021. Here flow rate, pressure, and injection volume are hidden for commercial issues. The black and blue colors indicate the pressure and volume of the injected fluid.

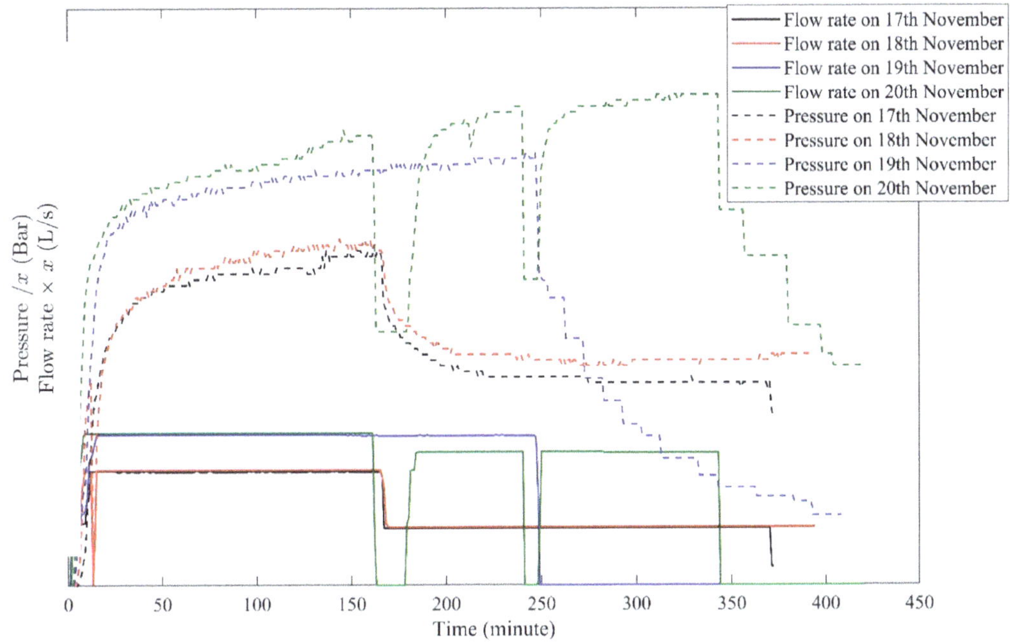

Figure 6. Pressure and flow rates of the four days of hydraulic testing/stimulation operation performed in UD−2 from 17th to 20th November 2020. Due to confidentiality, exact flow rate and pressure values are hidden and are shown only in terms of multiplication of x.

Table 2. DFN characteristics within the fault zone.

DFN Parameter	Fracture Set 1	Fracture Set 2	Comment	Reference
Mean fracture orientation, strike [deg]	157–337	50–230	From UD−1 borehole logging, Sub-parallel to the main fault	
Mean fracture orientation, dip [deg]	80–90	70–90	As above.	[22]
Mean fracture half-length [m]	10	10 (assumed)	Meters to tens of meters; may be scaled to the length of the fault and offset	[22]
Mean hydraulic fracture aperture [μm]	95	95 (assumed)	assumed	[23]
Fracture density [m^{-1}]	6	0.8	assumed	[23]
Fracture area density [m^{-1}]	0.2 (assumed)	0.2		

Using the following equation [18], heat and mass transfer in a porous media is coupled to simulate the fluid flow during the hydraulic stimulation operation:

$$\rho_1(\phi_m S_1 + (1-\phi_m)S_m)\frac{\partial p}{\partial t} - \rho_1(\alpha_m(\phi_m \beta_1 + (1-\phi_m)\beta_m))\frac{\partial T}{\partial t} = \nabla.\left(\frac{\rho_1 k_m}{\mu}\nabla p\right) \quad (1)$$

In this equation, p, T, ϕ_m, k_m, S_1, S_m, α_m, β_1 and β_m are pressure, temperature, porous media porosity, permeability, storage coefficient of rock, storage coefficient of fluid, Biot's coefficient of porous media, the thermal expansion coefficient of fluid, and thermal expansion coefficient of the porous medium, respectively. The fluid dynamic viscosity and density are denoted by ρ_1 and μ, respectively.

Heat exchange between the rock matrix and water is modeled based on the local thermal non-equilibrium approach, and the governing equation for the rock can be written as:

$$(1 - \phi_m)\rho_m C_{p,m}\frac{\partial T_m}{\partial t} = \nabla.((1 - \phi_m)\lambda_m\nabla T_m) + q_{ml}(T_l - T_m) \tag{2}$$

Here, T_m, T_l, ρ_m, $C_{p,m}$, λ_m and q_{ml} are rock temperature, fluid temperature, rock density, rock-specific heat capacity, rock thermal conductivity, and the rock-fluid heat transfer coefficient, respectively. If $C_{p,l}$ and λ_l are the heat capacity and thermal conductivity of fluid, then the governing equation of heat transfer for the fluid is:

$$\phi_m\rho_l C_{p,l}\frac{\partial T_l}{\partial t} + \phi_m\rho_l C_{p,l}\left(-\frac{k_m\nabla p}{\mu}\right).\nabla T_l = \nabla.(\phi_m\lambda_l\nabla T_l) + q_{ml}(T_m - T_l) \tag{3}$$

These equations are fully coupled and solved in COMSOL Multiphysics [18], where the necessary thermophysical properties of water (dynamic viscosity (μ), specific heat capacity (C_p), density (ρ), and thermal diffusivity (κ)) are given below:

$$\begin{aligned}
\mu = 1.38 &- 2.12 \times 10^{-2} \times (T + 273.15)^1 + 1.36 \times 10^{-4} \times (T + 273.15)^2 - 4.65 \times 10^{-7} \times (T + 273.15)^3 \\
&+ 8.90 \times 10^{-10} \times (T + 273.15)^4 - 9.08 \times 10^{-13} \times (T + 273.15)^5 + 3.85 \\
&\times 10^{-16} \times (T + 273.15)^6 \ (0 - 140\,^\circ\text{C})
\end{aligned} \tag{4}$$

$$\begin{aligned}
\mu = 4.01 \times 10^{-3} &- 2.11 \times 10^{-5} \times (T + 273.15)^1 + 3.86 \times 10^{-8} \times (T + 273.15)^2 \\
&- 2.40 \times 10^{-11} \times (T + 273.15)^3 \ (140 - 280\,^\circ\text{C})
\end{aligned} \tag{5}$$

$$\begin{aligned}
C_p = 1.20 \times 10^4 &- 8.04 \times 10^1 \times (T + 273.15)^1 + 3.10 \times 10^{-1} \times (T + 273.15)^2 \\
&- 5.38 \times 10^{-4} \times (T + 273.15)^3 + 3.63 \times 10^{-7} \times (T + 273.15)^4
\end{aligned} \tag{6}$$

$$\rho = 1.03 \times 10^{-5} \times (T + 273.15)^3 - 1.34 \times 10^{-2} \times (T + 273.15)^2 + 4.97 \times (T + 273.15) + 4.32 \times 10^2 \tag{7}$$

$$\begin{aligned}
\kappa = -8.69 \times 10^{-1} &+ 8.95 \times 10^{-3} \times (T + 273.15)^1 - 1.58 \times 10^{-5} \times (T + 273.15)^2 \\
&+ 7.98 \times 10^{-9} \times (T + 273.15)^3
\end{aligned} \tag{8}$$

A finite element discretization approach is adopted for this simulation. The complete mesh consists of 135,748 domain elements, 14,320 boundary elements, and 670 edge elements. Backward Differential Formula (BDF) with variable time step is used. The hydrothermal model used here is validated against Bai [24] in previous works by the authors [25,26].

Table 3 shows the rock properties and operational conditions for the numerical simulations. These parameters are obtained from the experimental tests and field works.

Table 3. Rock and fluid properties for the United Downs geothermal system, as defined for the numerical simulations.

Parameter	Value	Parameter	Value
Injection rate	Case dependent	Matrix porosity	0.005
Injection temperature	70 °C	Matrix permeability	Case dependent
Well diameter	0.2159 m	Fault porosity	0.02
Fluid properties	Dynamic (T)	Fault permeability	Case dependent
Side boundaries	Open mass flux, open heat flux	Top and bottom boundaries	No heat, No flow
Thermal Gradient	33.3 °C/km	Pressure gradient	9.79 MPa/km
Rock density	2620 kg/m^3	Rock thermal conductivity	3 W/(m × °C)
Rock specific heat capacity	960 J/kg/°C		

3. Results and Discussions

The present section is organized in the following approach: initially, the equivalent permeability field is calculated from the discrete fracture network, followed by updating the permeability field with the operational data from UD$-$1. The reason behind this step is to compensate for the permeability loss due to closed fractures or uneven morphology of the fracture surfaces, or resulting error from the stochastic approach (from FracMan). Next, the obtained permeability field is validated with the operation data from UD$-$2, assuming a constant fracture density in the fault zone. Finally, simulations for the cold-water injection for a longer time period are shown to capture the pressure evolution and temperature reduction for different flow rates.

Within the fault zone, 10 million fractures were produced with FracMan based on the fracture network characteristics presented in Tables 1 and 2. This output is difficult to present in a DFN framework included in numerical simulation software. Hence the upscaling of the permeability of the single fractures of the corresponding networks into an equivalent permeability field becomes a necessity. To calculate the equivalent permeability field, the Oda approach [27], as implemented in FracMan, was used. Oda [27] employed the geometry of fractures to calculate effective permeability for a given grid with a given pressure gradient. Inside the FracMan, Poiseuille law is used to calculate the fracture permeability from the fracture aperture. The roughness of the fracture is considered one. Using the orientation of individual fractures present in a grid cell and assigning each fracture as a unit normal vector n, a tensor depicting the mass moment of inertia of fractures normally distributed over a unit sphere was obtained through integration of the fractures over all of the unit normal [17]. Figure 7 shows the equivalent permeability field estimation using the Oda method, and it is obvious that the heterogeneity of the final permeability field is a function of the grid size. A smaller grid size results in a more accurate representation of the heterogeneous permeability distribution of the fractured rock mass than a large grid size, where single permeable fractures are less well represented. However, reducing the grid size increases the computational cost. While the permeability range of the permeability field created with the fine grid size is larger than for the coarse grid size, the mean permeability value is approximately the same. Therefore, the coarse grid size, which still provides a sufficiently accurate representation of the fractured reservoir, is used in this study to minimize the computational cost. The permeability field resulting from the Oda methodology for the fault zone is a heterogeneous field that is implemented through the TH calculations in the COMSOL. Consider that meshing for the equivalent permeability from FracMan (structured rectangular mesh) is different than the simulation meshes in the COMSOL (unstructured tetrahedral meshes). Therefore, permeability values mapped from the structural rectangular to the unstructured tetrahedral inside the COMSOL. The average value of the permeability field in anisotropic matrix is reported as below with the highest value in the z direction and the second highest value in the y-direction, both subparallel to the fault zone strike, which shows the effect of the fracture set orientation (see Table 2):

$$\begin{bmatrix} 5.5 \times 10^{-15} & 0 & 0 \\ 0 & 1.7 \times 10^{-14} & 0 \\ 0 & 0 & 2 \times 10^{-14} \end{bmatrix}$$

The reported data from the field work, logging and literature (see Table 2) shows that fractures are mainly vertically oriented, which aligns with higher permeability in the z-direction. Strike values with the fracture orientation in the x-y space indicate that their tendency towards the y-axis is greater, resulting in a higher permeability in the y-direction. While the fracture density slightly decreases with depth, a single averaged fracture density is used to simplify reservoir permeability for the entire fault zone. A similar approach is used to obtain the host rock permeability, and the resulting value is two orders of magnitude smaller than those of the fault zone. This limits the host rock's contribution to the convective heat and mass flux while its effects are still considered in the numerical simulation. The possible discrepancy between actual field and simulated permeability fields is solved with a correction factor to the permeability fields of the Oda methodology based on the hydraulic testing data. Technically, the permeability obtained from the hydraulic test data could have been directly matched with the numerical model to obtain a realistic value, but in this case, the resultant reservoir permeability from the Oda method is heterogeneous. The DFN generation and Oda upscaling were chosen specifically to better reflect the anisotropic permeability field for the faulted reservoir. From Figure 8, it is clear that the pressure increases at a higher rate by increasing the peak flow rate until step 21. The obtained permeability field till step 21 is:

$$\begin{bmatrix} \frac{5.5\times10^{-15}}{38} & 0 & 0 \\ 0 & \frac{1.7\times10^{-14}}{38} & 0 \\ 0 & 0 & \frac{2\times10^{-14}}{38} \end{bmatrix}$$

The above permeability field is thus 38 times smaller compared to the permeability obtained by the Oda approach. Possible differences between the actual fractures and the stochastic fractures can be attributed to differences in the fracture aperture caused by chemical reactions, physical precipitations, rough fracture surfaces, and stress reorientation or the general share of open vs. closed fractures.

After reaching the maximum injection pressure at step 21, Figure 5 indicates that the flow rates beyond step 21 do not increase the pressure in a similar magnitude with respect to steps prior to 21. This suggests that some pre-existing fractures are propagated to reduce the incremental pressure rate with increasing the peak flow rate. The initiation of new fractures can be ruled out due to the injection pressure being too small to exceed the tensile strength of the granite and due to the high fracture density of the formation. Therefore, step 22 is seen as proof of a significant hydraulic stimulation of the reservoir. Data matching is performed as shown in Figure 8 for two periods (before and after step 21), and the permeability tensor obtained by averaging the permeability field is 1.7 times higher than the initial rock permeability:

$$1.7 \times \begin{bmatrix} \frac{5.5\times10^{-15}}{38} & 0 & 0 \\ 0 & \frac{1.7\times10^{-14}}{38} & 0 \\ 0 & 0 & \frac{2\times10^{-14}}{38} \end{bmatrix}$$

The obtained permeability field through the UD$-$1 well data is validated with the hydraulic testing and stimulation data of well UD$-$2. Figure 9 demonstrates a good match between the operational data and numerical simulation of UD$-$2 for the first three stages of injection. However, there is a discrepancy between the measured and modeled data, as shown in Figure 6 for 20 November. Based on this figure, the flow rate decreases to zero in the middle of the test and increases afterward. The operational data shows that the injection pressure increases following the initial trend without experiencing an impact due to the intermittent flow rate breaks. The numerical simulation for UD$-$2, on the contrary, shows that if the flow rate decreases only for a short time interval, pressure decreases for the rest of the entire injection step and does not reach the previous pressure level. The observed low pressure after a short gap in the injection is not due to the permeability change. Rather the pressure diffuses inside the system leading to pressure drop. It, therefore, seems likely that the flow rate stops and pressure drops are rather a measurement error while reporting the pressure and flow rate data or that the wellbore effect [26] between the wellhead and the reservoir section mitigates the flow rate and pressure drops.

It should be noted that these short-term hydraulic stimulation results are restricted to the near wellbore area, and do not influence the far-field of the reservoir. Therefore, uncertainty remains regarding the volume of the reservoir, which is affected by this operation. To examine the near wellbore effect, a cylindrical zone around UD$-$1 is considered in which the permeability is changed by hydraulic stimulation. Two radii of 25 and 50 m are considered for this zone, and the behavior of the system is modeled at three fluid rates of x,

$2x$, and $3x$ L/s. Figure 10 shows the resulting pressure data on the bottom of the UD$-$1 and UD$-$2 based on different flow rates and hydraulically affected area radius. There is no data to show which radius around the wellbore is affected by the hydraulic stimulation. By comparing 25 and 50 m, it is shown that the region size is not important for the studied time periods, but the amount of permeability changes is important. Another model without considering this cylindrical region is run, and the results are different than the operational observations (see Figure 8). This permeability variation importance around the production wellbore is not exactly for this studied short time period, but for the long-term operation, authors showed in another study that permeability around the production wellbore is an important factor [26].

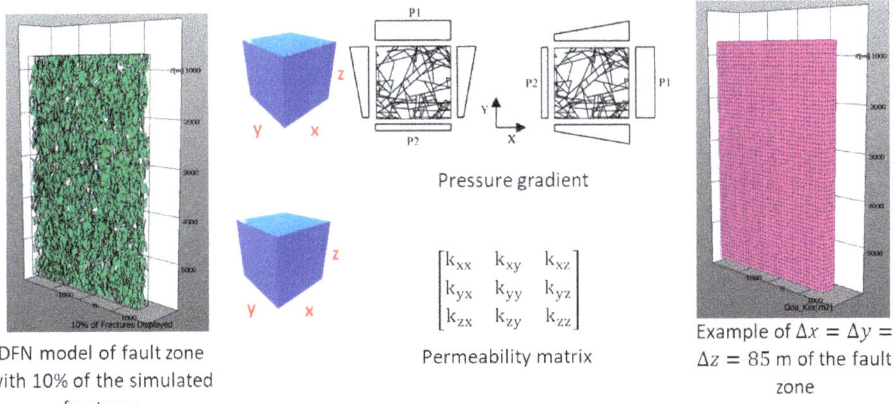

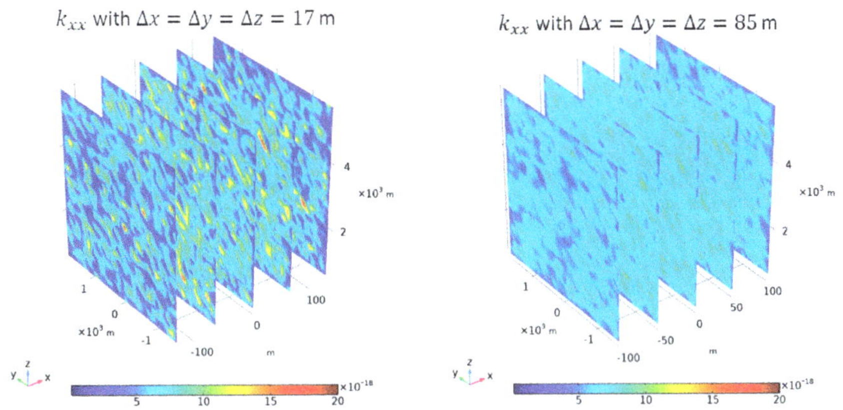

Figure 7. Oda methodology [27] to calculate the equivalent permeability field and its sensitivity to the two different mesh sizes (17 and 85 m, respectively). Upper left: first, a stochastic DFN of the fault zone and host rock is created based on the fracture network characteristics; Upper middle: then, using the Oda approach, a permeability field is calculated for each cell of a given grid (upper right) separately. The bottom panels show the heterogeneous permeability field for the matrix zone for two different mesh sizes for the equivalent permeability estimation (lower right and lower left). The unit of permeability is m^2.

Initially, the simulated pressures of the UD$-$1 are higher than UD$-$2 due to the hydrostatic pressure. By injecting fluid at UD$-$2 and producing from UD$-$1, they experience pressure buildup and drawdown, respectively. Obviously, by increasing the flow rate, the pressure separation between the two wells increases at later times. Pressure changes are greater at early times, and as time goes on, pressure show semi-steady behavior, and the operator can decide regarding the optimal flow rate. For example, for the cases with 25 m of hydraulically stimulated area, the pressure after ten days reaches $4.78x$, $7.39x$, and $10.05x$ MPa with flow rates of x, $2x$, and $3x$ L/s, respectively, whereas after 150 days it becomes $5.8x$, $9x$ and $12.1x$ MPa respectively. However, there is not much change between 150 and 300 days, where in the latter case, the final pressure values are $6.1x$, $9.34x$, and $12.6x$ MPa for the flow rates of x, $2x$, and $3x$ L/s, respectively. Correspondingly for UD-1, a decreasing trend is observed at a gentle gradient. For the flow rates of x, $2x$, and $3x$ L/s, observed pressure values after ten days are $4.05x$, $3.72x$ and $3.38x$ MPa, respectively, which becomes $3.38x$, $2.86x$ and $2.35x$ MPa, respectively, after 150 days which finally reaches to $3.08x$, $2.51x$ and $1.93x$ MPa, respectively, after 300 days. This analysis shows that pressure differences between the two wells are $3.02x$, $6.83x$, and $10.67x$ MPa after 300 days for three flow rates, x, $2x$, and $3x$ L/s, respectively. It is interesting to note that from an economical point of view, the higher flow rate is attractive, but it may eventuate the seismic events. During the examined period, the radius of the hydraulically stimulated area near the UD-1 has a negligible effect on the pressure profiles in comparison to the flow rate. The pressure differences between the wellbore for a 50 m radius of hydraulically stimulated zone reach $2.97x$, $6.71x$, and $10.56x$ MPa after 300 days for x, $2x$, and $3x$ L/s injection rates, respectively. Therefore, in the remaining parts, a small radius of 25 m is examined. Similar behavior is recognized by Mahmoodpour et al. [26] that the permeability field in the vicinity of the production well is an important factor.

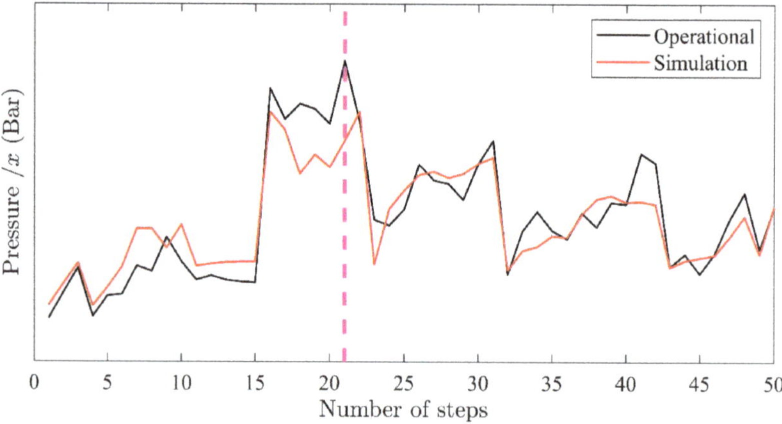

Figure 8. Matching operational and simulation pressure data to obtain the permeability field before and after the hydraulic stimulation process. The pink dotted line marks the occurrence of a micro-seismic event, which significantly changed reservoir injectivity.

Figure 11 shows the pressure plume development at iso-surface of $3.5x$ MPa after 100, 200, and 300 days of the injection. At the early time, the pressure around the UD-2 is lower than $3.5x$ MPa, and around the UD-1, pressure is higher than $3.5x$ MPa. Therefore, small pressure iso-surfaces around the wells being observed. As time goes on, the pressure near UD-2 increases and near the UD-1 decrease, and the trend of these changes in Figure 11 is clearly obvious. Obviously, with increasing the injection and production flow rates, the plume size increases. Furthermore, due to the higher permeability of the fault zone, pressure changes mainly happen inside the fault zone and two large-scale fractures, which are detected through seismic events. These sharp pressure changes (and consequently the stress changes not discussed in this paper) inside the fault zone requires special attention to examine the possibility of the seismic events.

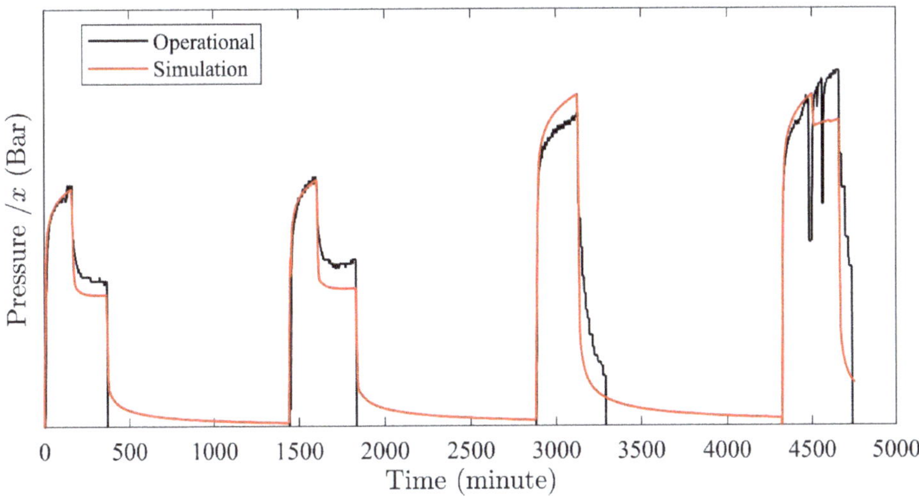

Figure 9. Matching of the operational and simulation pressure data of the UD−2 hydraulic injection/Stimulation tests to validate the simulated permeability field obtained from the UD−1 operational data.

During the 300 days, the temperature changes are limited near the injection well. Therefore, to examine the effects of different flow rates on the temperature, temperature changes alongside the open hole section of the injection well and inside the fault zone are being considered, which is very close to the wellbore in Figure 12 for different flow rates. On the horizontal axis, 0 m shows the top point of the open hole section of the injection well, and 300 m shows the bottom point of this section. Initially (0 days in the graphs), the temperature distribution follows up the temperature gradient of the system. As time goes on, the cold front propagates to the deeper zones of the reservoir. An increase in the flow rate increase the speed of the temperature front propagation in the farther regions, and it shows that the heat transfer process is mainly controlled by the convective mechanism. The possibility of higher fluid injection should be examined through the techno-economic aspects. The increasing flow rate will increase the heat extraction rate at early times. On the other hand, it increases the possibility of seismic events and early breakthrough time [25].

The acid injection may increase the permeability to a higher extent than the hydraulic stimulation. Preliminary studies in the core scale samples from the rocks of this reservoir show a good outcome of the acid injection and the possibility of permeability increase between 4 and 50 times in core-flooding tests, but the autoclave experiments show this increment between 0.1 (this shows that permeability decreases up to the 10 % of the initial value and it explains the possible precipitation which is visible in the microscopic scale) and 40 times [28]. Therefore, we assumed imaginary cases of permeability alteration by ten times, and 100 times in comparison to the initial state due to the possible acid injection process at the reservoir scale using a cylindrically affected region surrounding the wellbore with a radius of 25 m. Figure 13 (left column) shows the results for ten times of the permeability enhancement where the pressure of UD−2 after ten days reaches $2.92x$, $3.22x$ and $3.51x$ MPa with flow rates of x, $2x$ and $3x$ L/s, respectively, whereas after 150 days it becomes $3.6x$, $3.93x$ and $4.26x$ MPa, respectively. While, there is not much change between 150 and 300 days, where in the latter case, the final pressure values are $3.06x$, $4x$, and $43.3x$ MPa for the flow rates of x, $2x$, and $3x$ L/s, respectively. Correspondingly for UD−1, these values are $4.02x$, $3.96x$, and $3.9x$ MPa for the flow rates of x, $2x$, and $3x$ L/s. After 150 days, pressure values are $3.32x$, $3.26x$, and $3.2x$ MPa, respectively, which becomes $3.34x$, $3.28x$, $3.22x$ and $3.22x$ MPa, respectively, after 300 days. Therefore, the pressure difference between the two wells is $0.32x$, $0.72x$, and $1.11x$ MPa after 300 days for three flow rates,

x, $2x$, and $3x$ L/s. From this pressure behavior, it is clear that the acidizing process for ten times permeability increment has a great potential to increase the flow rate with a huge decrease in the pressure gradient between the two wells, compared to only hydraulic stimulation. Experimental studies show that this permeability enhancement is attainable at the core level [26]. For the case of 100 times in the permeability enhancement, all flow rates show the same pressure behavior indicating the presence of a highly conductive region between the injection and production wells, which has a significant impact on the pressure gradient. At later times for both permeability enhancement scenarios, the pressure of UD−1 increases, demonstrating the production of injected fluid rather than the production of fluid initially residing in the fault zone.

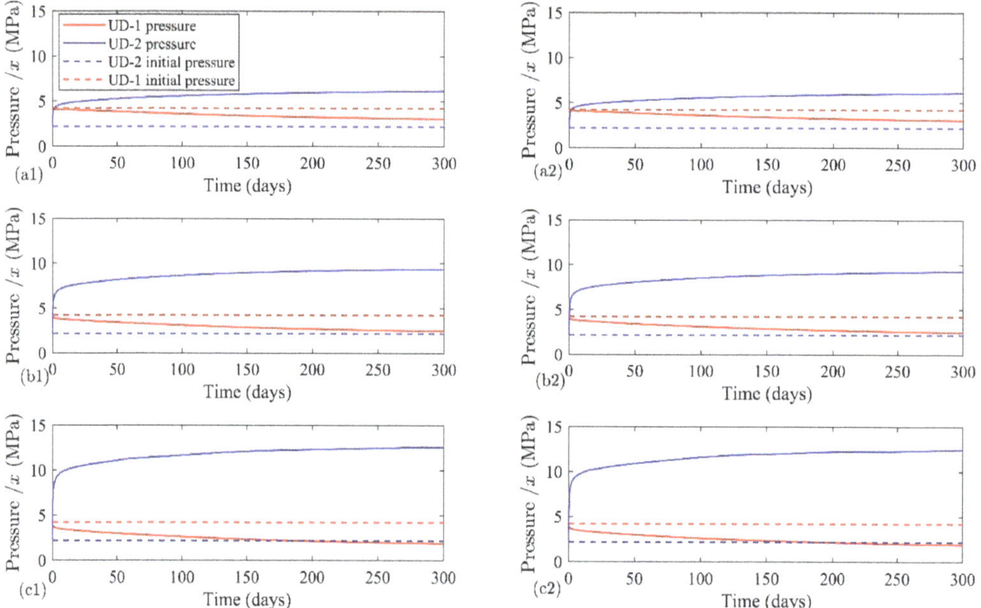

Figure 10. Simulated pressures at the injection and production wells for injection flow rate of (**a1,a2**) x L/s, (**b1,b2**) $2x$ L/s and (**c1,c2**) $3x$ L/s. The left panels are plotted for a hydraulically stimulated region that spans over 25 m, whereas it is 50 m for the right panel.

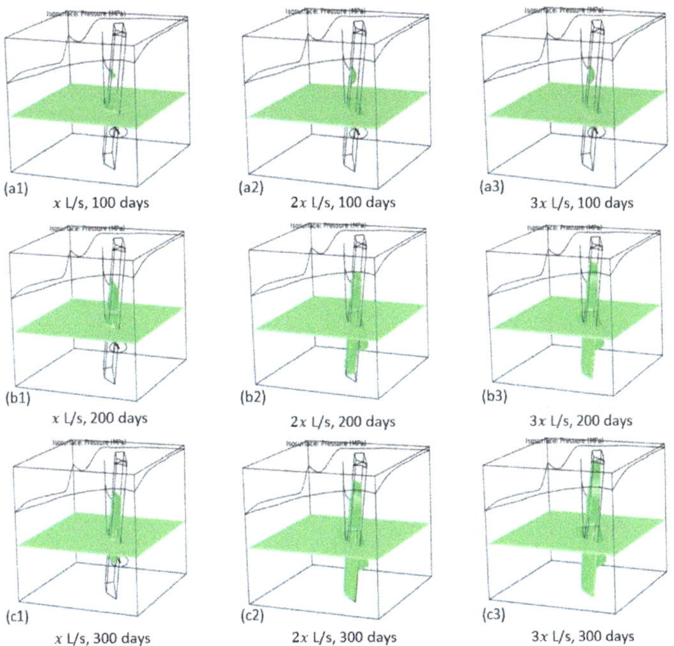

Figure 11. Pressure plume development for three different injection rates at three different times. The iso-surfaces are plotted for $3.5x$ MPa.

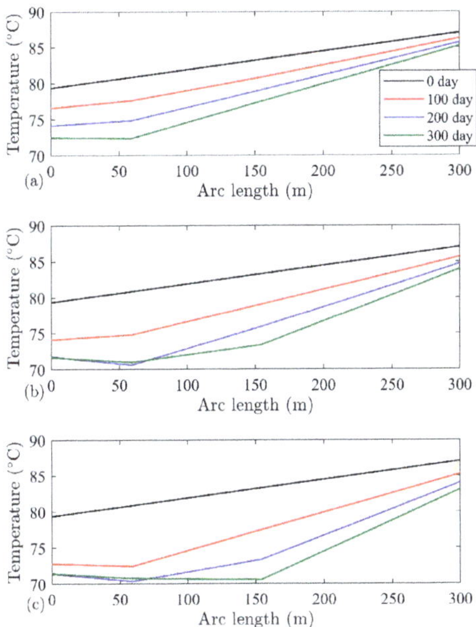

Figure 12. Simulated temperature alongside the open hole section of the injection well UD-2 for the injection flow rates (**a**) x L/s, (**b**) $2x$ L/s, and (**c**) $3x$ L/s. The arc length is measured from the top of the open hole section of the well that falls inside the Porthtowan Fault zone.

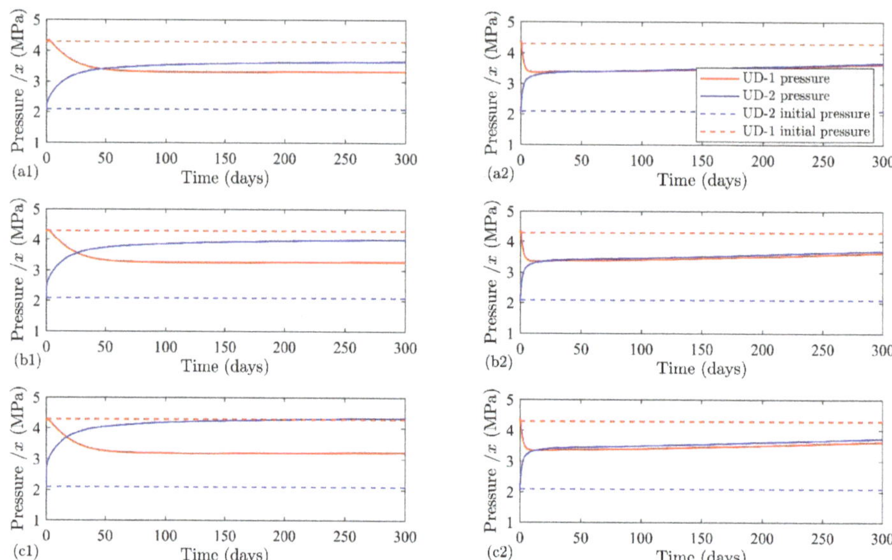

Figure 13. Simulated pressures at the injection and production wells for the injection flow rate of (**a1,a2**) x L/s, (**b1,b2**) $2x$ L/s, and (**c1,c2**) $3x$ L/s after the permeability enhancement through the considered chemical stimulation. The left column shows a permeability enhancement by a factor of 10, and the right column shows a permeability enhancement by a factor of 100 with respect to the initial values.

4. Conclusions

In this study, hydraulic injection testing/stimulation operations for both the wells of UDDGP are numerically simulated. Results are used to characterize the permeability field of the system and examine pressure development based on different injection flow rates. Results show that pressure changes are occurring primarily in the early time, with later time showing a semi-steady pressure increase. With the current permeability field after the hydraulic stimulation operation, huge pressure differences develop by increasing the flow rates. The possible effect of chemical stimulation operation on the permeability is considered through the data obtained from the lab-scale models, and reservoir response after the permeability modification is numerically simulated. Results for the permeability enhancement of the 10-time compared to the initial value due to the chemical stimulation (it is attainable as shown by experiments) enable us to triple the flow rate with one-third of the initial pressure difference without chemical stimulation. The reported values and the reservoir characterization performed during this study will build a basis for future numerical simulations for the United Down geothermal site.

Author Contributions: Conceptualization, S.M., M.S. and K.B.; methodology, S.M. and M.S.; software, S.M. and M.S.; validation, S.M. and M.S. writing—original draft preparation, S.M., M.S., C.O. and S.K.T.; writing—review and editing, J.R. and K.B.; visualization, S.M., M.S. and C.O.; supervision, K.B. and I.S.; project administration, K.B. and I.S.; funding acquisition, K.B. and I.S. All authors have read and agreed to the published version of the manuscript.

Funding: The work is conducted as part of the MEET project that has received funding from the European Union's Horizon 2020 research and innovation program under grant agreement No 792037.

Data Availability Statement: Due to the confidential nature of the field data, the exact units of the data are hidden.

Acknowledgments: Group of Geothermal Science and Technology, Institute of Applied Geosciences, Technische Universität Darmstadt has provided institutional support to the authors. The authors would like to acknowledge Geothermal Engineering Limited (Ryan Law and Hazel Farndale) for providing the well-testing data for the hydraulic stimulation operations conducted from September 2020 until February 2021. Additionally, we would like to thank two anonymous reviewers for their helpful and constructive remarks, which greatly helped to improve the manuscript. Additionally, we would like to acknowledge Golder Associate Inc for providing academic licenses of FracMan v7.

Conflicts of Interest: The authors declare no conflict of interest.

References

1. Tester, J.W.; Brown, D.W.; Potter, R.M. *Hot Dry Rock Geothermal Energy—A New Energy Agenda for the 21st Century*; Report LA-11514-MS; Los Alamos National Laboratory: Santa Fe, NM, USA, 1989.
2. Willis-Richards, J. Assessment of HDR reservoir stimulation and performance using simple stochastic models. *Geothermics* **1995**, *24*, 385–402. [CrossRef]
3. Zhao, Y.S.; Feng, Z.J.; Zhao, Y.; Wan, Z.J. Experimental investigation on thermal cracking, permeability under HTHP and application for geothermal mining of HDR. *Energy* **2017**, *132*, 305–314. [CrossRef]
4. Pine, R.J.; Batchelor, A.S. Downward migration of shearing in jointed rock during hydraulic injections. *Int. J. Rock Mech. Min. Sci. Geomech. Abstr.* **1984**, *21*, 249–263. [CrossRef]
5. LeBoutillier, N.G. The Tectonics of Variscan Magmatism and Mineralisation in Southwest England. Ph.D. Thesis, University of Exeter, Exeter, UK, 2002; p. 712.
6. Kim, Y.S.; Andrews, J.R.; Sanderson, D.J. Extension fractures, secondary faults, and segment linkage in strike-slip fault systems at Rame Head, Southern Cornwall. *Geosci. SW Engl. Proc. Ussher Soc.*. **2001**, *10*, 123–133.
7. Holloway, S.; Chadwick, R.A. The Sticklepath-Lustleigh fault zone: Tertiary sinistral reactivation of a Variscan dextral strike-slip fault. *J. Geol. Soc.* **1986**, *143*, 447–452. [CrossRef]
8. Faulkner, D.R.; Mitchell, T.M.; Rutter, E.H.; Cembrano, J. (Eds.) On the structure and mechanical properties of large strike-slip faults, in Wibberley. In *The Internal Structure of Fault Zones: Implications for Mechanical and Fluid Flow Properties*; Geological Society London Special Publication: London, UK, 2008; Volume 299, pp. 139–150.
9. Martel, S.J. Formation of compound strike-slip fault zones, Mount Abbot quadrangle, California. *J. Struct. Geol.* **1990**, *12*, 869–882. [CrossRef]
10. Massart, B.; Paillet, M.; Henrion, V.; Sausse, J.; Dezayes, C.; Genter, A.; Bisset, A. Fracture characterization and stochastic modeling of the granitic basement in the HDR Soultz project, France. In Proceedings of the World Geothermal Congress, Bali, Indonesia, 25–29 April 2010.
11. Reinecker, J.; Gutmanis, J.; Foxford, A.; Cotton, L.; Dalby, C.; Law, R. Geothermal Exploration and reservoir modelling of the United Downs deep geothermal project, Cornwall (UK). *Geothermics* **2021**, *97*, 102226. [CrossRef]
12. Chen, Y.; Clark, A.; Farrar, E.; Wasteneys, H.; Hodgson, M.; Bromley, A. Diachronous and independent histories of plutonism and mineralization in the Cornubian Batholith, southwest England. *J. Geol. Soc.* **1993**, *150*, 1183–1191. [CrossRef]
13. Charoy, B. The Genesis of the Cornubian Batholith (South-West England): The example of the Carnmenellis Pluton. *J. Petrol.* **1986**, *27*, 571–604. [CrossRef]
14. Afhsari Moein, M.J.; Evans, K.F.; Valley, B.; Bär, K.; Genter, A. Fractal characteristics of fractures in crystalline basement rocks: Insights from depth-dependent correlation analysis to 5 km depth. *Int. J. Rock Mech. Min. Sci.* **2022**, *155*, 105138. [CrossRef]
15. Trullenque, G.; Genter, A.; Leiss, B.; Wagner, B.; Bouchet, R.; Léoutre, E.; Malnar, B.; Bär, K.; Rajšl, I. Upscaling of EGS in Different Geological Conditions: A European Perspective. In Proceedings of the 43rd Workshop on Geothermal Reservoir Engineering Stanford University, Stanford, CA, USA, 12–14 February 2018. SGP-TR-213.
16. MOVE Structural Geology Software MOVE Core (Computer Software). 2020. Available online: https://www.petex.com/products/move-suite/ (accessed on 5 April 2021).
17. Golder Associate Inc. *FracMan7 User Documentation*; Golder Associate Inc.: Seattle, WA, USA, 2011.
18. COMSOL. *Multiphysics® v. 5.5*; COMSOL AB: Stockholm, Sweden. Available online: www.comsol.com (accessed on 20 December 2020).
19. Lanyon, G.W.; Kingdon, R.D.; Herbert, A.W. The application of a three-dimensional fracture network model to a hot-dry-rock reservoir. In *The 33rd US Symposium on Rock Mechanics (USRMS)*; OnePetro: Santa Fe, NM, USA, 1992.
20. Lanyon, G.W.; Batchelor, A.S.; Ledingham, P. Results from a discrete fracture network model of a Hot Dry Rock system. In Proceedings of the 18th Workshop on Geothermal Reservoir Engineering, Stanford University, Stanford, CA, USA, 26–28 January 1993. SGP-TR-145.
21. Parker, R. The Rosemanowes HDR Project 1983–1991. *Geothermics* **1999**, *28*, 603–615. [CrossRef]
22. Cotton, L.; Gutmanis, J.; Shail, R.; Dalby, C.; Batchelor, T.; Foxford, A.; Rollinson, G. Geological Overview of the United Downs Deep Geothermal Power Project, Cornwall, UK. In Proceedings of the World Geothermal Congress 2020+1, Reykjavik, Iceland, 24–27 October 2021.

23. GeoScience Ltd. *Identification of Potential Deep Geothermal Sites in the Vicinity of the Carnmenellis Granite, West Cornwall, Unpublished Report to Geothermal Engineering Limited*; GeoScience Ltd.: Falmouth, UK, 2009.
24. Bai, B. One-dimensional thermal consolidation characteristics of geotechnical media under non-isothermal condition. *Eng. Mech.* **2005**, *22*, 186–191.
25. Mahmoodpour, S.; Singh, M.; Turan, A.; Bär, K.; Sass, I. Simulations and global sensitivity analysis of the thermo-hydraulic-mechanical processes in a fractured geothermal reservoir. *Energy* **2022**, *247*, 123511. [CrossRef]
26. Mahmoodpour, S.; Singh, M.; Bär, K.; Sass, I. Impact of Well Placement in the Fractured Geothermal Reservoirs Based on Available Discrete Fractured System. *Geosciences* **2022**, *12*, 19. [CrossRef]
27. Oda, M. Permeability tensor for discontinuous rock masses. *Geotechnique* **1985**, *35*, 483–495. [CrossRef]
28. Schulz, K.E. Analysis of Permeability Enhancement by Chemical Treatment in Fractured Granite for the United Downs Deep Geothermal Power Project (Cornubian Batholith). Master's Thesis, Technical University Darmstadt, Darmstadt, Germany, 2020; p. 121.

Article

Use of Analogue Exposures of Fractured Rock for Enhanced Geothermal Systems

D. C. P. Peacock [1,2,*], David J. Sanderson [3] and Bernd Leiss [1,2]

1. Geoscience Centre, Department of Structural Geology and Geodynamics, Georg-August-Universität Göttingen, Goldschmidtstraße 3, 37077 Göttingen, Germany
2. Universitätsenergie Göttingen GmbH, Hospitalstraße 3, 37073 Göttingen, Germany
3. Ocean and Earth Science, University of Southampton, National Oceanography Centre, Southampton SO14 3ZH, UK
* Correspondence: peacock@uni-goettingen.de

Abstract: Field exposures are often used to provide useful information about sub-surface reservoirs. This paper discusses general lessons learnt about the use of deformed Devonian and Carboniferous meta-sedimentary rocks in the Harz Mountains, Germany, as analogues for a proposed enhanced geothermal reservoir (EGS) at Göttingen. The aims of any analogue study must be clarified, including agreeing with people from other disciplines (especially reservoir modellers) about the information that can and cannot be obtained from surface exposures. Choice of an analogue may not simply involve selection of the nearest exposures of rocks of a similar age and type, but should involve consideration of such factors as the quality and geological setting of the analogue and reservoir, and of any processes that need to be understood. Fieldwork should focus on solving particular problems relating to understanding the EGS, with care being needed to avoid becoming distracted by broader geological issues. It is suggested that appropriate questions should be asked and appropriate analyses used when planning a study of a geothermal reservoir, including studies of exposed analogues.

Keywords: exposed analogue; enhanced geothermal system; fractures

Citation: Peacock, D.C.P.; Sanderson, D.J.; Leiss, B. Use of Analogue Exposures of Fractured Rock for Enhanced Geothermal Systems. *Geosciences* **2022**, *12*, 318. https://doi.org/10.3390/geosciences12090318

Academic Editors: Ronan L. Hébert, Ghislain Trullenque, Albert Genter, Eléonore Dalmais and Jean Hérisson

Received: 30 June 2022
Accepted: 25 August 2022
Published: 26 August 2022

Publisher's Note: MDPI stays neutral with regard to jurisdictional claims in published maps and institutional affiliations.

1. Introduction

Rock exposures are commonly used as analogues to understand and make predictions about sub-surface geology. Exposed analogues have been used for a range of applications, including in the geothermal (e.g., [1]), mining (e.g., [2]), petroleum (e.g., [3]), hydrogeology (e.g., [4]), nuclear waste disposal (e.g., [5]) and carbon sequestration (e.g., [6]) industries. Many of these applications require understanding or prediction of fluid flow in the sub-surface, which may be either through pores or through fractures within the rock.

Use of exposed analogues is relatively simple in cases in which there are not significant differences between rocks at the surface and sub-surface. For example, sedimentological analogues relate to depositional processes. While compaction and diagenesis during burial may produce changes in the fabrics and thicknesses of sedimentary rocks, the arrangement of depositional facies are largely unchanged. This means that sequence stratigraphy (e.g., [7]), building on simple principles such as Steno's "Law of Superposition" (e.g., [8]) and Walther's "Law of Facies Adjacency" (e.g., [9]) still apply. Sedimentary analogues are therefore routinely used to understand sub-surface sedimentary and reservoir geology (e.g., [10]). Exposed analogues can be useful to study aspects of structural geology that are not affected by significant changes as the rocks are exhumed. For example, exposed analogues have been used to analyse fault geometries, kinematics and mechanics (e.g., [11]), and to characterise vein networks (e.g., [12]). There are, however, problems in using exposed analogues in situations where there are significant changes to key aspects of the structural geology as the rocks are exhumed (e.g., [13]). This is particularly the case with joints, which are commonly created during exhumation (e.g., [14]). This means that joints seen at the surface do not necessarily occur in the sub-surface.

This paper describes lessons we have learnt, while undertaking an analogue study in the Harz Mountains as part of the EU-Horizon project MEET (Multidisciplinary and multi-context demonstration of Enhanced Geothermal Systems exploration and Exploitation Techniques and potentials). MEET was aimed at developing cost-effective techniques for Enhanced Geothermal Systems (EGS) in a variety of geological settings across Europe. The western Harz Mountains were selected as an analogue for a proposed EGS project at Göttingen (Lower Saxony, Germany), where the reservoir rocks are expected to be folded and thrusted Devonian and Carboniferous metasedimentary rocks. EGS involves stimulation, which is the artificial increase in fluid flow through the reservoir rocks. Our fieldwork therefore focussed on collecting structural data that would help model the effects of hydraulic [15] and thermal stimulation on the Devonian and Carboniferous rocks beneath Göttingen. During the MEET project, we attempted to use analogue exposures of fractures constrain flow models and make predictions about enhanced geothermal systems.

The aim of this paper is to pass on our experiences and the knowledge gained. While scientific papers tend to highlight successes, here we pass on some of the problems we encountered, lessons we learnt and advice on how to avoid such problems.

2. Rationale for Using the Harz Mountains as an Analogue

This section explains the rationale behind the use of the Harz Mountains as an analogue for the proposed EGS at Göttingen, gives an introduction to the geology of the region, and compares the analogue with the sub-surface geology we expect at Göttingen. In the absence of well data and with limited seismic data for the sub-Zechstein rocks, the rocks beneath the Permian Zechstein evaporites at Göttingen must be predicted. Devonian quartzitic sandstones and slates, and Carboniferous greywackes and slates (predominantly Culm flysch deposits) are exposed in the western Harz Mountains ~40 km to the NE and in the Rhenish Massif ~70 km to the SW (e.g., [16–18]), as shown in Figure 1. We have selected analogue sites in the Oberharz Anticline and the Culm Fold Zone, which belong to the parautochthonous domain of the Harz Mountains (e.g., [19]) because we consider these to be areas to have the best exposures of the sub-Zechstein rocks that are most likely to occur beneath Göttingen [20].

The Devonian and Carboniferous rocks of the region are part of the Variscan fold-thrust belt (e.g., [21]). Later deformation events in the region include Late Cretaceous and Cenozoic uplift and exhumation in the Harz Mountains (e.g., [22]) and Cenozoic rifting in the Göttingen area (the Leinetal Graben, e.g., [23]). The fractures visible in the exposures of the Harz Mountains include veins and joints [24,25], with cleavage being well-developed in the slates [26]. Thrusts that appear to have displacements of up to a few metres are exposed [27], but larger thrusts are not well-exposed. The north-eastern boundary the Harz Mountains is marked by the Harznordrand Fault, which was active during the Late Cretaceous and Tertiary (e.g., [28]). NW-SE striking Mesozoic normal and oblique-slip faults occur in the Harz Mountains, some of which contained economic Pb-Zn and Ba-F deposits [29].

Table 1 shows seven key parameters that control deformation of rocks and our predictions for the Variscan rocks that occur beneath Göttingen. Table 1 also includes comments on what can and cannot be learnt from the Harz Mountains. While the exposed analogues give useful information about lithologies, the conditions that existed during the Variscan Orogeny and about Variscan structures, they tend to give less information about subsequent conditions and deformation. For example, it is difficult to determine absolute ages of joints and to determine the tectonic conditions under which they formed. Differences between the rocks exposed in the Harz Mountains and those at a depth of more than 1.5 km beneath Göttingen include:

1. The Göttingen area has undergone post-Variscan rifting (the Leinetal Graben) and still has a cover sequence of Permian and Triassic sedimentary rocks about 1.5 km

thick. The Harz Mountains, however, have undergone several kilometres of exhumation since the Late Cretaceous and have been fully exhumed (i.e., there is no cover sequence).

2. The different overburden and tectonic regimes of the Harz Mountains and the Göttingen area mean that the post-Variscan (and present-day) stress magnitudes and orientations were different, potentially leading to different types and orientations of structures.

3. For example, the graben formation at Göttingen is related to a set of N-S trending fault zones that consist of NNE-SSW striking en echelon segments, which may extend into the Variscan rocks. This set of faults is not observed in the Harz Mountains but may be comparable with a set of NW-SE striking Mesozoic normal and oblique-slip faults developed during the uplift of the Harz Mountains.

4. Exhumation means that many of the post-Variscan structures observed in the Harz Mountains, especially the joints, may not occur at reservoir depths at Göttingen. We suggest, however, that the joints observed in the Harz Mountains give us strong indications of the patterns of induced fractures that would be created by stimulation of Variscan rocks beneath Göttingen. Both natural jointing and fracture development during artificial stimulation occurs as the rock responds to changes in stresses, fluid pressures and/or temperatures. Such observations as whether or not joints cross bedding planes or follow pre-existing veins therefore give indications about likely fracture behaviour during stimulation. As such, they are useful for predicting patterns of induced fractures and for use in DFN modelling.

5. The cover sequence at Göttingen includes the Zechstein salt deposits, which may mean both that the pore fluids in the Variscan rocks may be saline and that those fluids could be overpressured.

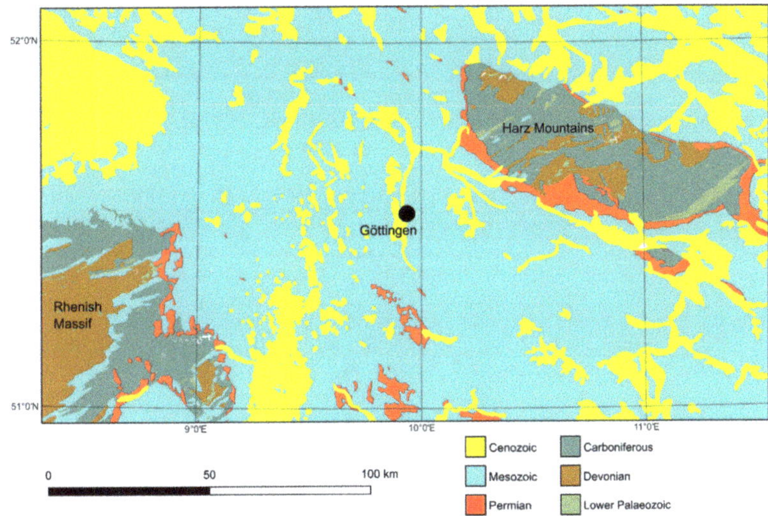

Figure 1. Geological map showing the location of the Harz Mountains, where the exposed Devonian and Carboniferous rocks are used as an analogue for a proposed EGS at Göttingen. Additionally, shown is the Rhenish Massif, which also shows exposures of Devonian and Carboniferous rocks. Göttingen along strike of the Variscan belt between the Harz Mountains and the Rhenish Massif. The dominant lithologies are as follows. Lower Palaeozoic, Devonian and Carboniferous: metasedimentary sandstones, slates and carbonates. Permian: sandstones, marls and evaporites. Mesozoic: shales, sandstones and limestones. Cenozoic: unconsolidated siliciclastic sediments. Data source: GK1000© BGR, Hannover, 2013. Downloaded on 6 June 2022, from https://gdk.gdi-de.org/geonetwork/srv/api/records/5f77d681-b7e4-4dd0-8f15-7b93744450b0.

Table 1. Key parameters that control rock deformation with predictions for the Variscan rocks beneath Göttingen and comments on what can and cannot be learnt from the exposed analogues in the Harz Mountains. For a list of parameters to be characterised in a geothermal reservoir, see [30].

Factor	Meaning	Significance	Variscan Rocks Beneath Göttingen	What Can Be Learned from Exposed Analogues
Lithologies	Rock types, their porosities and mechanical behaviour	Controls the mechanical behaviour of the rock. This parameter can change significantly through time, especially as deformation occurs	Devonian and Carboniferous greywackes and slates	Information about rock types and their mechanical properties (from deformation tests of samples)
Fluid type	The chemistry and phase of the palaeo- and present-day fluid(s)	Controls the fluid pressure gradient and mineralisation events	Present-day: water. Mineralising fluid and remobilised sediments during the Variscan Orogeny	Information about fluids before, during and after the Variscan Orogeny
Stress	Magnitudes and orientations of the applied stresses, including the vertical stress (overburden) and horizontal stresses. Horizontal stresses are related to the geostatic pressure ratio, applied tectonic stresses and to internal stresses (e.g., related to temperature changes)	Along with fluid pressure, controls the effective stresses, which control the deformation	While the vertical stress can be calculated using the mean density of the overburden, the magnitudes and orientations of the horizontal stresses are uncertain	Information about the orientations and possibly magnitudes of stresses during the Variscan Orogeny but limited information about subsequent stresses, including present-day stresses
Fluid pressure	Palaeo- and present-day fluid pressures	Along with the stresses, controls the effective stresses, which control the deformation	Presently probably hydrostatic. Veins, breccias and possible remobilised sediments indicate phases of overpressure during the Variscan Orogeny	Information about fluid pressure during the Variscan Orogeny and during later mineralisation events, but not about present-day fluid pressure. The Zechstein Salt may enable overpressure to occur in the underlying Variscan rocks
Temperature	Palaeo- and present-day temperatures	Influences the style of deformation, with present-day temperature controlling commercial viability	Depends on the geothermal gradient (~30 °C per km), but likely to be reduced because of the overlying salt. Possibly elevated by Tertiary igneous activity	Indications about temperatures during the Variscan Orogeny but limited information about later temperatures
Strain	The amount of strain and the existing structures	Influences fluid flow in the sub-surface and present-day mechanical behaviour of the rocks	Dominated by: (1) Variscan Orogeny, with folds, thrusts and veins; (2) Late Cretaceous and Tertiary rifting and/or uplift, with normal faults and joints developing	Information about strains and structures caused by the Variscan Orogeny, but limited information about later deformation. Göttingen shows post-Variscan rifting, while the Harz Mountains underwent Late Cretaceous and Cenozoic uplift and exhumation

Table 1. *Cont.*

Factor	Meaning	Significance	Variscan Rocks Beneath Göttingen		What Can Be Learned from Exposed Analogues
History	The relative and absolute timing of deformation (including mineralisation) events and structures	Controls the types of fractures (faults, veins, joints, etc.) and therefore their effects on fluid flow in the sub-surface	(1)	Sedimentation during the Devonian and Carboniferous.	Information about the pre- and syn-Variscan history, but not about subsequent deformation. While the Variscan rocks beneath Göttingen have undergone rifting and have not been fully exhumed, the Harz Mountains have undergone several kilometres of Late Cretaceous and Cenozoic exhumation
			(2)	Variscan Orogeny.	
			(3)	Permian and Mesozoic sedimentation and basin development.	
			(4)	Cretaceous and Tertiary regional uplift.	
			(5)	Tertiary rifting and volcanism	

3. Lessons Learnt about the Use of Analogues

We experienced the following problems and learnt the following lessons about the use of analogues for fractured geothermal reservoir rocks during the course of the MEET project.

3.1. Aims of Using an Exposed Analogue

A problem we had with the fieldwork in the Harz Mountains arose because it was undertaken for several different reasons. Although at the start of the project the aim was to develop a numerical model for fluid flow through the rocks [31], various other aims were introduced. These included: to create general and conceptional structural models for the Harz Mountains, to determine the histories of deformation and fluid flow, to make predictions about the sub-surface at Göttingen, and to provide data that could be used by colleagues to perform discrete fracture network (DFN) modelling of fluid flow in the sub-surface. While these different aims were not necessarily conflicting, they did lead to some confusion in deciding what types of structures were important to analyse, how best to analyse them, and what data were needed. For example, one of the methods used was to map traces of veins, joints and other fractures exposed on rock surfaces to provide inputs into DFN models. We realised, however, that these maps lacked vital information that would enable realistic DFN models for the sub-surface at Göttingen to be created (Section 3.3), although they did help us interpret the histories of deformation and fluid flow in the Harz Mountains.

A study of an exposed analogue needs to be set up carefully, including definition of the problems being addressed and the aims of the work. These aims then need to be translated into specific objectives, with careful consideration of the priorities of each, determination of the data needed, and establishment of the appropriate methods. It is also important to keep evaluating the work, to determine whether the approach is working, and to make changes as needed. General questions that may typically be asked about a potential exposed analogue include:

- Is the aim to make specific predictions about the rocks and structures in the sub-surface or is it to understand certain processes?
- What lithologies and lithological relationships occur?
- Can rock samples be collected that are suitable for determining geomechanical and petrophysical properties, and for geochemical analyses?
- What structures occur in the exposed analogues?
- What are the spatial and temporal relationships between those structures?
- What frequencies and patterns of open fractures occur in different lithologies, and what controls their development?

- Is there information about active fluid flow in the sub-surface, such as evidence of hydrothermal fluids in spring water?

An important lesson from our fieldwork in the Harz Mountains was not just that we needed to be clear at the start of the work what problems were being addressed, but that we needed to frequently review our progress in solving (and ability to solve) those problems and be prepared to adjust the approaches used.

3.2. Choice of Exposed Analogue

The Harz Mountains were an obvious choice for an analogue study because similar rocks are expected to occur in the sub-surface at Göttingen and are likely to have similar structures and deformation histories (Section 2 and Table 1). Furthermore, geographical proximity meant that long journeys and overnight accommodation were not needed for fieldwork. There are, however, two main problems with the Harz Mountains as an analogue study area. Firstly, the exposure quality is commonly poor, being restricted to steep road-cuts, abandoned quarries and natural exposures that are typically up to a few tens of metres long and a few metres high. Many of the best exposures are cliff faces formed by the dominant set of joints, which are steeply dipping and strike at a high angle to the Variscan folds, thrusts and cleavage. While these exposure surfaces are ideal for analysing the Variscan structures, they are poor for analysing the fracture networks, especially the joint systems, because it is difficult to analyse those structures in three dimensions. Secondly, while the Harz Mountains have undergone several kilometres of exhumation since the Late Cretaceous, the Göttingen area is in a Late Cretaceous and Cenozoic rift and has undergone far less exhumation (Table 1). Both areas, however, underwent post-Variscan, pre-Zechstein exhumation.

Lessons learnt about the choice of field areas from use of the Harz Mountains include the following:

- Göttingen lies between exposures of Devonian and Carboniferous metasedimentary rocks in the Harz Mountains and the Rhenish Massif (e.g., [21]). The field area and the geothermal reservoir rocks are therefore assumed to show similar lithologies (including lithological relationships, bed thicknesses, weathering, etc.), structural geometries, age relationships between those structures, tectonic histories, kinematics, mechanical properties and fluid flow histories. Although no exposed analogue is likely to show fractures (especially joints) that are a perfect match to the sub-surface, the analogue should show enough commonality to the reservoir to enable reasonable comparisons to be made about aspects of the geology.
- Exposure quality is important in controlling the data that can be obtained. The Harz Mountains are only approximately 50 km away from the proposed geothermal reservoir at Göttingen, and we aimed to correlate our results with those of partners in the MEET project who were carrying out fieldwork in the Rhenish Massif. An alternative, however, would have been to study more distance but better-exposed areas to make it easier to investigate fracture systems in three dimensions. For example, the Devonian and Carboniferous rocks in Cornwall, UK, are very well exposed in coastal cliffs and on wave-cut platforms (e.g., [32]). Even though the exposures on the coast of Cornwall are approximately 1000 km west of Göttingen, they may provide additional valuable information about the structures developed in Upper Palaeozoic sedimentary rocks in the Variscan Belt.
- There may be greater flexibility on the choice of analogue if the aim of the fieldwork is to understand processes rather than to obtain specific data about such information as the lithologies and types of structures that are likely to occur in a geothermal reservoir. For example, one question the fieldwork in the Harz Mountains addressed was whether slates can fracture, which would help us determine their reservoir properties and behaviour during stimulation. Veins and joints in the Upper Palaeozoic slates of the Harz Mountains (Figure 2) show that the slates have been prone to fracturing at different times in their history, so they have potential as reservoir rocks. The same

observations could have been made on slates in other regions. Indeed, observations from a range of slates of different compositions, of different ages and subjected to different deformation histories, are likely to have improved knowledge about the reservoir properties of slates, thereby improving predictions for the sub-surface of Göttingen.

Figure 2. Photograph of a syncline in Upper Palaeozoic slates at Schulenberg im Oberharz, Harz Mountains, Germany. Most of the exposure surface is defined by a set of steeply dipping joints that strike at a high angle to the fold.

3.3. What Exposed Analogues Can and Cannot Tell You

One of the main aims of the fieldwork in the Harz Mountains was to provide data about fracture networks and their relationships to folds to create a conceptual model that could be used in DFN modelling to predict fluid flow rates in the sub-surface at Göttingen. Such modelling requires such data as the in situ stresses as well as the apertures and connectivities of open fractures in the sub-surface (e.g., [33]). This information cannot be provided from rocks exposed at the surface. For example, joints commonly develop during exhumation (e.g., [34]), with the apertures, frequencies, sets and patterns of joints seen at the surface not necessarily being the same as those at reservoir depths. While the joints analysed in the Harz Mountains may give some indications about open fractures in the sub-surface at Göttingen, such as whether open fractures may be expected to occur in slates, they did not provide sufficient information about sub-surface fractures to enable meaningful DFN modelling to be undertaken.

Exposed analogues may provide information about such characteristics of sub-surface geology as lithologies, the geometries and kinematics of certain structures (e.g., folds, faults, veins), the histories of deformation and fluid flow, and aspects of mechanical behaviour (e.g., mechanical stratigraphy). Exposed analogues cannot, however, provide with much certainty information that is important for reservoir modelling, such as the presence and geometries of open fractures, fracture porosity, stresses, temperatures and fluid flow. Some certainty about conditions in the sub-surface will only be obtained when well data become available.

Fieldwork in the Harz Mountains, and our subsequent attempts to use the data to make predictions about the sub-surface geology at Göttingen taught us the importance of understanding what you can and cannot do with field data, and of conveying that information clearly to other people who may rely on those data. Both field geologists and reservoir modellers need to be aware of the limitations of field data, as to be aware of the uncertainty limitations of subsequent models. We found that field data are useful for developing initial models and for testing concepts (e.g., [35]), and we used data and observations from the field to make simple predictions about the effects of hydraulic stimulation in the potential geothermal reservoir rocks at Göttingen [15].

An important consideration in the comparison between an exposed analogue and a reservoir is that they have different histories, especially their burial and exhumation histories. Many of the structures that occur in the surface exposures in the Harz Mountains that are likely to be important in a geothermal reservoir at Göttingen probably post-date the Variscan Orogeny. These include the joints, many of which are likely to have been caused by exhumation (e.g., [36]). Such differences must be considered when comparing an exposed analogue with a reservoir.

3.4. Avoid Distracting Topics

Fieldwork in the Harz Mountains taught us the importance of focussing on the key aims of the study and of being careful to avoid spending too much time on by other interesting topics. Two aims of the project were: (1) discovering the ways and methods to use an exposed analogue for predicting the properties of enhanced geothermal systems, and (2) to use the exposures in the Harz Mountains to help us predict the characteristics and behaviour of the reservoir rocks beneath Göttingen. Fieldwork often leads to new, unexpected discoveries, and these should not be ignored. It is important, however, that such discoveries do not cause too much distraction from the aims of the field study. We were occasionally distracted by discussions about Variscan tectonics, which were not directly related to the aims of the fieldwork but did lead us towards new research projects. While an academic environment provides freedom to follow new ideas and research interests, it is still important to do the work expected by academic partners and by funding bodies. The need to stay focused on solving particular problems is more intense in industry, where managers and clients will expect particular outcomes from a study.

While the Harz Mountains were selected as an analogue because they show similar rocks and structures that are expected to occur beneath Göttingen, and these were compared with exposures in the Rhenish Massif, a more distant analogue site could have been selected, for example if the more distance analogue has better exposure. Key factors to consider in the choice of an analogue site include how well the lithologies and structures match the geothermal reservoir, and the quality of the exposure. As such, the ages of the rocks and of the deformation may be irrelevant.

3.5. Use of the Term "Fracture"

Our work in the Harz Mountains, and the subsequent attempts to use this work to make predictions about the sub-surface at Göttingen, showed the importance of specifying fracture types in the field. Reservoir modellers commonly use the general term "fracture", and often do not distinguish between faults, veins, joints and other types of fracture. When field geologists also use "fracture", they cannot properly understand the geometries, age relationships, tectonics, mechanics and fluid flow histories of the rocks and the fractures they contain [37,38]. Different fracture types have different origins, distributions and properties, hence different significance in predicting the engineering behaviour of a reservoir. For example, veins are fractures that are partly or completely sealed by minerals, while joints are not mineralised (Figure 3). Thus, veins will tend to have greater mechanical cohesion and are less likely to be conduits for present-day fluid flow, than joints. Veins also tend to be clustered, such as in fold hinges (e.g., [39]) and around faults (e.g., [40]), whilst joints tend to be more widely distributed in a rock mass (e.g., [41,42]). Such differences have significance for the properties of an EGS and on the ways the reservoir rocks respond to stimulation. Merging these different types of structures together as "fractures" during initial fieldwork in the Harz Mountains meant that important interpretations about the histories and significance of those structures could not be made.

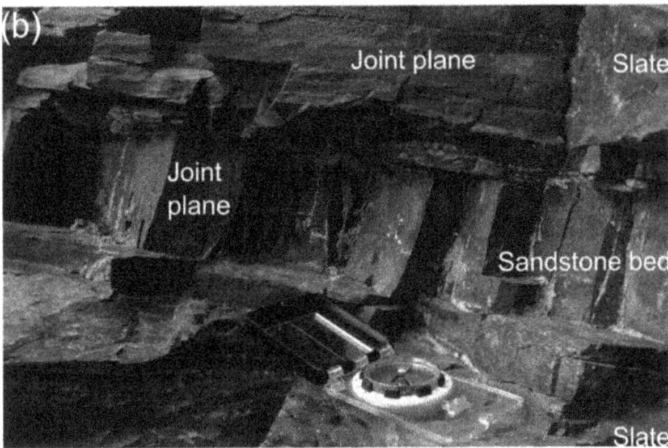

Figure 3. Photographs of different types of "fractures" in the Harz Mountains. (**a**) Partly filled quartz vein in a sandstone, which resulted from and gives information about palaeo fluid flow. The vein was either never completely cemented or has been subjected to weathering. The quartz will hinder present-day fluid flow through the vein. (**b**) Joints in sandstones and slates. Veins are (by definition) not mineralised, so are potential pathways for present-day fluid flow in the sub-surface.

Whilst geologists are generally well-aware of the significance of different types of fracture, engineers are more likely to think of a fracture simply as a surface across which the rock mass may lose cohesion or frictional resistance. To a reservoir engineer, a fracture may assume the role of a large void that facilitates fluid flow. A major problem occurs when these different groups start to communicate referring to everything as a "fracture", mainly in the interest of finding a common language. In the MEET project, geologists investigating faults, veins, joints and other types of fracture would present their field data as measurements of the geometries and topologies of "fractures". This can easily result in vein data being used to estimate fluid flow in DFNs (i.e., treated as "open" fractures, modelled by flow between parallel plates; see Section 3.7).

3.6. Palaeo Fluid Flow vs. Present-Day Fluid Flow

The fieldwork in the Harz Mountains taught us the importance of differentiating between studies of palaeo fluids and studies to make predictions about present-day fluid flow. This issue is related to use of the term "fracture" (Section 3.5), because the minerals within veins tend to give information about palaeo fluid flow (e.g., [43]), while knowledge about joints and other open fractures helps in predicting present-day fluid flow (Figure 3).

For example, while the maps of fracture traces we created (Section 3.1) turned out to be of little use for DFN modelling, they did provide a basis for understanding the palaeo fluid flow through veins and the potential for present-day fluid flow through joints.

When the aims of the fieldwork are decided, it is important to study the appropriate fractures using appropriate methods. For example, a study of an ancient hydrothermal system should focus on veins and the minerals they contain, while a study to predict present-day fluid flow would tend to focus on joints and other open fractures.

3.7. Understanding the Needs of Other Disciplines

One of the original aims of the fieldwork in the Harz Mountains (and of other analogue areas used in the MEET project) was to obtain field data that could be used to create models to predict fluid flow in the sub-surface. A particular problem, however, was that the field geologists were not able to collect key data that the modellers, who were not geologists, needed to make realistic DFN models (Section 3.3). Such data included information about the apertures and connectivities of open fractures (e.g., joints) in the sub-surface. This led to mutual frustration. It is therefore important that field geologists are clear about what can and cannot be learnt from analogues, including what data can be collected. Another problem was that the field geologists would use terminology that the reservoir modellers did not understand and vice versa. An example of such miscommunication was the use of "fracture" by the reservoir modellers to mean a discontinuity along which fluids can flow in the present-day, while the geologists would use the term for any brittle discontinuity, regardless of the potential for fluid flow (Section 3.5). For example, if a field geologist reports an abundance of veins with apertures of 1–10 mm, this does not justify a modeller using open fractures with apertures of 1–10 mm in a DFN model.

Lessons from our experience of working with reservoir modellers to make predictions about the sub-surface at Göttingen include: (1) it is important to understand the data they need to develop their models; (2) modellers must be informed about what data can and cannot be provided from field data, and this should include information about the uncertainties in the field data; (3) a common understanding of the terms used by both geologists and non-geologists must be established.

4. Integration of Exposed Analogues in the Analysis of Geothermal Reservoirs

This section discusses how exposed analogue studies may be best integrated in the analysis of geothermal reservoirs.

4.1. Improved Understanding of the Contribution of Exposed Analogue Studies

The starting point of any scientific study is generally a question or problem that needs to be solved. The initial aims of the use of exposed analogues in the MEET project are described by [31]. Table 2 shows some of the questions that should be asked when setting up a study, either of a geothermal reservoir or an exposed analogue, along with elements that need to be considered in such studies. It also shows how exposed analogues may help answer those questions. Table 2 aims to show the importance of deciding what problems are to be addressed and determining the appropriate ways to solve those problems. We suggest four steps in determining how best to use an exposed analogue:

1. Decide whether the analogue is being used to understand a palaeo-hydrothermal system or to make predictions about present-day conditions in the sub-surface (Section 3.6).
2. Determine what aspects of the analogue need to be characterised. Table 1 gives information about key parameters that control rock deformation, about which information is likely to be needed.
3. Determine what data are needed and that can be obtained.
4. Determine the appropriate analysis type or types. Analysis types commonly used to analyse geothermal reservoirs are shown in Table 2.

Table 2. Examples of questions that should be asked when planning a study of a geothermal reservoir or of an exposed analogue. Additionally, shown are typical elements that need to be considered and comments about how exposed analogues can be used to study these elements.

Questions	Elements	Use of Exposed Analogues
What type of system is being studied?	Palaeo hydrothermal Modern geothermal	Information about a palaeo hydrothermal system can be obtained from minerals, from either exposed analogues or from well data. Information about present-day fluid flow in a reservoir requires information about porosity and open fractures, ideally supported by production data. There are problems involved in using exposed analogues to predict fluid flow in the sub-surface (e.g., Section 3.3)
What aspect of the system is of interest?	Rock Fractures Pressure regime Temperature regime Fluid phase Gaseous phase	Ref. [30] present the "hexagon concept" to show the six elements of a geothermal reservoir that need to be analysed. Exposed analogues can provide important information about the "rock" and "fracture" elements (next two rows)
What aspects of the rocks are of interest?	Lithologies Ages Tectonostratigraphy Diagenesis, metamorphism Geometries and structures Mechanical behaviour Porosity and permeability	Exposed analogues can give vital information about the rocks in a reservoir, although care is needed. The reservoir rocks and exposed analogues will have different amounts of burial and exhumation, so some elements may differ
What aspect of the fractures are of interest?	Basic geology Geometry and topology Age relationships Kinematics Tectonics Mechanics Fluid flow	Ref. [38] describe seven types of analysis that can be used to study fractures in rock. The appropriate analysis type is needed to answer specific questions about fractures in a geothermal reservoir or in exposed analogues
What information is needed?	Geophysical data Well data Exposed analogues Similar reservoirs Reservoir models Conceptual models Geochemical data	Typical information that may be available, obtainable or desired to study geothermal reservoirs, including exposed analogues
What disciplines are appropriate?	Geophysics Rock mechanics and petrophysics Geochemistry Sedimentology Igneous geology Metamorphic geology Structural geology Numerical modelling	Typical types of study (disciplines) used to study the geology of geothermal reservoirs, including exposed analogues

4.2. Modelling Techniques

As discussed in Section 3.3, exposed analogues cannot provide some information that reservoir modellers may need to create realistic DFN models. To make predictions about the sub-surface at Göttingen before well data are available, we therefore had to use modelling techniques. We used Mohr diagrams with realistic ranges of input parameters (mostly using data from the exposed analogues in the Harz Mountains) to make predictions about how the Devonian and Carboniferous rocks expected to occur beneath Göttingen will respond to stimulation [15]. Modelling techniques need to be used that are suited to answering the right questions at early stages of geothermal assessment, and that are feasible based on the data available at the pre-drilling stage.

4.3. Ranges and Variabilities of Parameters

The modelling approach we used to predict the response of rocks in the sub-surface at Göttingen used information obtained from the exposed analogues in the Harz Mountains (Section 4.2; [15]). This included information about the ranges and variabilities of such factors as rock mechanical properties and the geometries of structures (e.g., folds, veins and joints). Exposed analogues are therefore useful for providing realistic ranges of input parameters for modelling. In our experience, non-geologists tend to want a single value or answer for any given parameter. Exposed analogues are therefore particularly useful for informing people from other disciplines (such as reservoir modellers and drilling engineers) about such geological variabilities.

4.4. Knowledge Transfer

Another way exposed analogues can be used in the analysis of geothermal reservoirs is in improved communication and knowledge transfer between scientists from different disciplines (e.g., between geologists, reservoir modellers and drilling engineers). Discussions about exposed analogues, especially when undertaken in the field, can help geologists and non-geologists appreciate potential and actual links between surface and sub-surface geology, and between geology and other disciplines. Exposed analogues can be used to help answer a series of key questions, such as:

1. What measurements are (and are not) needed for a specific model?
2. Can geologists make such measurements?
3. How much certainty do we have in those measurements?
4. How can measurements be scaled for modelling purposes?
5. Does the parameter-space in modelling adequately encompass the geological variability and uncertainty?

Expressing a model purely in terms of black-box equations will not usually help a geologist understand what the model does or how different parameters are used. Similarly, field data fed to modellers needs to be accompanied by information that captures potential factors that are important in its use (e.g., "open" vs. "closed" fractures, scale ranges over which data were collected).

5. Discussion of Other Uses of Analogue Exposures of Fractured Rock

Although we have used exposures in the Harz Mountains as analogues to make predictions about a proposed EGS, our experiences can give useful insights into the use of exposed analogues for other situations in which it may be important to predict fluid flow through fractures in the sub-surface. This includes use of analogues for studies on other types of geothermal systems (e.g., [44]), hydrogeology (e.g., [45]), nuclear waste storage (e.g., [46]), CO_2 sequestration (e.g., [47]), mineral extraction (e.g., [48]), and petroleum resources (e.g., [49]). Any study using analogues has to be carefully planned and focussed so that specific questions about the sub-surface can be answered. The fieldwork must be of use to, and understandable by end-users from other disciplines.

As discussed in Section 3.3, it is important to understand what can and cannot be learnt from exposed analogues, and this will change based on the intended use of the analogue. For example, if the fieldwork is to collect information about veins that formed during particular time periods or tectonic events, then this information is likely to be directly applicable to understanding those vein systems that occur in the sub-surface (e.g., [50]). Note that certain minerals can be utilised for chemical stimulation (e.g., [51]). More care is needed, however, if the exposed analogue is used to map joint systems for direct use in DFN models of the sub-surface, especially if those joints formed during exhumation. For example, while the Devonian and Carboniferous rocks beneath Göttingen may have joints related to exhumation during and immediately after the Variscan Orogeny, they are less likely than the Harz Mountains to have joints related to Late Cretaceous and Cenozoic exhumation. Joints exposed at the surface may therefore not equate in a simple way to open fractures in the sub-surface.

6. Conclusions

While exposed analogues can provide critical information that can be used to make predictions about the sub-surface, including the behaviour of enhanced geothermal systems (EGS), care is needed in planning and executing fieldwork. We highlight key learning gained during our use of the Harz Mountains (Germany) as an analogue for an EGS at Göttingen, which will be applicable to studying analogues elsewhere:

1. The aims of the analogue study need to be clearly established, with focus being placed on what questions will be addressed by the fieldwork (Table 2).
2. A field area needs to be selected that is best suited to solving the problems being addressed. For example, an analogue should not necessarily be selected just because it is the nearest exposure of the rocks that are expected to occur in the EGS. While more distant field areas may show better-quality exposure, they must provide useful insights into the geothermal reservoir.
3. It is important to understand, and to be clear with people from other disciplines, what information exposed analogues can and cannot give about the sub-surface. Exposed analogues may not provide the specific data required by, for example, dynamic modelling. Fieldwork can, however, provide critical insights into the likely behaviour of rocks in the sub-surface, such as whether natural or induced fractures will cut across bedding planes.
4. Science is generally about solving problems, and exposed analogues for EGS must answer particular questions about the behaviour of reservoir rocks. It is therefore important to avoid letting interesting new topics cause too much distraction from the main aims of the fieldwork.
5. It is unhelpful to use the term "fracture" as a field description, and the types of "fractures" must be defined during fieldwork. While veins give useful information about palaeo fluids, they are less likely to contribute to fluid flow in the sub-surface than are such open fractures as joints.
6. A key decision about the exposed analogue is whether it is being used to obtain information about a palaeo hydrothermal system or to provide information about present-day fluid flow in the sub-surface.
7. Because exposed analogues are often used to provide information to people from other disciplines, it is important to understand what those people need and to gain a common understanding between field geologists and the people who will be using the field data.

We make several suggestions for how studies of exposed analogue studies can be better integrated into the analysis of geothermal reservoirs:

1. Improved understanding of the contributions exposed analogues can and cannot make to predicting the behaviour of geothermal reservoirs will help analogue studies be used more effectively.
2. Modelling techniques need to be used that are appropriate to the data that are available.
3. The fieldwork must establish the ranges and variabilities of parameters that will be used to understand an EGS. This information can then be used to predict the ranges of structures that are likely to be encountered in the sub-surface, and to help explain well data in the reservoir evaluation stage of EGS development.
4. Exposed analogues can play a vital role in knowledge transfer between different disciplines, especially between field geologists, modellers and drilling engineers.

While we focus on the use of exposed analogues for EGS, much of this paper is applicable to other situations in which it is important to make predictions about fluid flow in the sub-surface. For example, understanding the behaviour of rocks, the structures within those rocks and their effects on fluid flow is vital to nuclear waste storage, CO_2 sequestration, hydrogeology and the petroleum industry. In each of these applications, an appropriate exposed analogue should be studied in suitable ways to answer key questions that will enable predictions to be made about the sub-surface.

Author Contributions: Conceptualization, D.C.P.P., D.J.S. and B.L.; methodology, D.C.P.P., D.J.S. and B.L.; validation, D.C.P.P., D.J.S. and B.L.; resources, B.L.; data curation, B.L.; writing—original draft preparation, D.C.P.P.; writing—review and editing, D.J.S. and B.L.; supervision, B.L.; project administration, B.L.; funding acquisition, B.L. All authors have read and agreed to the published version of the manuscript.

Funding: This work has received funding from the European Union's Horizon 2020 research and innovation programme (grant agreement No. 792037-MEET Project).

Institutional Review Board Statement: Not applicable.

Informed Consent Statement: Not applicable.

Data Availability Statement: No new data were created or analysed in this study. Data sharing is not applicable to this article.

Acknowledgments: Carl-Heinz Friedel is thanked for his helpful advice and discussions. We thank Paolo Pace and two anonymous reviewers for their helpful comments.

Conflicts of Interest: The authors declare no conflict of interest.

References

1. Weydt, L.M.; Bär, K.; Colombero, C.; Comina, C.; Deb, P.; Lepillier, B.; Mandrone, G.; Milsch, H.; Rochelle, C.A.; Vagnon, F.; et al. Outcrop analogue study to determine reservoir properties of the Los Humeros and Acoculco geothermal fields, Mexico. *Adv. Geosci.* **2018**, *45*, 281–287. [CrossRef]
2. Chappell, D.; Craw, D. Geological analogue for circumneutral pH mine tailings: Implications for long-term storage, Macraes Mine, Otago, New Zealand. *Appl. Geochem.* **2002**, *17*, 1105–1114. [CrossRef]
3. Hodgetts, D. Laser scanning and digital outcrop geology in the petroleum industry: A review. *Mar. Pet. Geol.* **2013**, *46*, 335–354. [CrossRef]
4. Klingbeil, R.; Kleineidam, S.; Asprion, U.; Aigner, T.; Teutsch, G. Relating lithofacies to hydrofacies: Outcrop-based hydrogeological characterisation of Quaternary gravel deposits. *Sediment. Geol.* **1999**, *129*, 299–310. [CrossRef]
5. Krauskopf, K.B. Geology of high-level nuclear waste disposal. *Annu. Rev. Earth Planet. Sci.* **1988**, *16*, 173–200. [CrossRef]
6. Bickle, M.; Kampman, N. Lessons in carbon storage from geological analogues. *Geology* **2013**, *41*, 525–526. [CrossRef]
7. Christie-Blick, N.; Driscoll, N.W. Sequence stratigraphy. *Annu. Rev. Earth Planet. Sci.* **1995**, *23*, 451–478. [CrossRef]
8. Guntau, M. Concepts of Natural Law and Time in the History of Geology. *Earth Sci. Hist.* **1989**, *8*, 106–110.
9. Middleton, G.V. Johannes Walther's Law of the Correlation of Facies. *Geol. Soc. Am. Bull.* **1973**, *84*, 979–988. [CrossRef]
10. Alexander, J. A discussion on the use of analogues for reservoir geology. In *Advances in Reservoir Geology*; Ashton, M., Ed.; Geological Society of London: London, UK, 1993; Volume 69, pp. 175–194.
11. Morley, C.K. Developments in the structural geology of rifts over the last decade and their impact on hydrocarbon exploration. In *Hydrocarbon Habitat in Rift Basins*; Lambiase, J.J., Ed.; Geological Society of London: London, UK, 1995; Volume 80, pp. 1–32.
12. Gillis, K.; Roberts, M. Cracking at the magma–hydrothermal transition: Evidence from the Troodos Ophiolite, Cyprus. *Earth Planet. Sci. Lett.* **1999**, *169*, 227–244. [CrossRef]
13. Sanderson, D.J. Field-based structural studies as analogues to sub-surface reservoirs. In *Value of Outcrop Studies in Reducing Subsurface Uncertainty and Risk in Hydrocarbon Exploration and Production*; Bowman, M., Smyth, H.R., Good, T.R., Passey, S.R., Hirst, J.P.P., Jordan, C.J., Eds.; Geological Society of London: London, UK, 2016; Volume 436, pp. 207–217.
14. Engelder, T.; Gross, M. Pancake joints in Utica gas shale: Mechanisms for lifting overburden during exhumation. *J. Struct. Geol.* **2018**, *117*, 241–250. [CrossRef]
15. Peacock, D.; Sanderson, D.J.; Leiss, B. Use of Mohr Diagrams to Predict Fracturing in a Potential Geothermal Reservoir. *Geosciences* **2021**, *11*, 501. [CrossRef]
16. Fielitz, W. Variscan transpressive inversion in the northwestern central rhenohercynian belt of western Germany. *J. Struct. Geol.* **1992**, *14*, 547–563. [CrossRef]
17. Franke, W. The mid-European segment of the Variscides: Tectonostratigraphic units, terrane boundaries and plate tectonic evolution. In *Orogenic Processes: Quantification and Modelling in the Variscan Belt*; Franke, W., Haak, V., Oncken, O., Tanner, D., Eds.; Geological Society of London: London, UK, 2000; Volume 179, pp. 35–61.
18. Franke, W. Rheno-Hercynian belt of central Europe: Review of recent findings and comparisons with south-west England. *Geosci. South West Engl.* **2007**, *11*, 263–272.
19. Friedel, C.-H.; Huckriede, H.; Leiss, B.; Zweig, M. Large-scale Variscan shearing at the southeastern margin of the eastern Rhenohercynian belt: A reinterpretation of chaotic rock fabrics in the Harz Mountains, Germany. *Int. J. Earth Sci.* **2019**, *108*, 2295–2323. [CrossRef]

20. Leiss, B.; Friedel, C.H.; Heinrichs, T.; Tanner, D.C.; Vollbrecht, A.; Wagner, B.; Wemmer, K. Das Rhenohercynikum als geothermischer Erkundungshorizont im Raum Göttingen. In *Harzgeologie 2016. 5. Workshop Harzgeologie—Kurzfassungen und Exkursionsführer*; Friedel, C.H., Leiss, B., Eds.; Universitätsverlag Göttingen: Göttingen, Germany, 2016; p. 31. [CrossRef]

21. Brink, H.-J. The Variscan Deformation Front (VDF) in Northwest Germany and Its Relation to a Network of Geological Features Including the Ore-Rich Harz Mountains and the European Alpine Belt. *Int. J. Geosci.* **2021**, *12*, 447–486. [CrossRef]

22. Von Eynatten, H.; Dunkl, I.; Brix, M.; Hoffmann, V.-E.; Raab, M.; Thomson, S.N.; Kohn, B. Late Cretaceous exhumation and uplift of the Harz Mountains, Germany: A multi-method thermochronological approach. *Int. J. Earth Sci.* **2019**, *108*, 2097–2111. [CrossRef]

23. Tanner, D.; Leiss, B.; Vollbrecht, A.; Kallweit, W.; Meier, S.; Oelrich, A.; Reyer, D. The Role of Strike-Slip Tectonics in the Leinetal Graben, Lower Saxony. *Geotecton. Res.* **2008**, *95*, 166–168. [CrossRef]

24. Friedel, C.H.; Hoffmann, C. Stopp 4: Schichtgebundene Deformationsstrukturen in der Kulmgrauwacke des Oberharzes (Bundesstraße 242). In *Harzgeologie 2016. 5. Workshop Harzgeologie Kurzfassungen und Exkursionsführer*; Friedel, C.H., Leiss, B., Eds.; Universitätsverlag Göttingen: Göttingen, Germany, 2016; pp. 85–90.

25. Lippold, T. Vein Formation in the Variscan Slates and Greywackes of the Upper Harz Mountains—An Attempt of a Macro- and Microscopic Characterization and Interpretation. Master's Thesis, Georg-August-Universität Göttingen, Göttingen, Germany, 2020.

26. Zeuner, M. Conceptual 3D-Structure Model of the Variscan "Culm Fold Zone" in the Vicinity of the Oktertal dam (Harz Mountains, Germany) in Terms of Geothermal Reservoir Development. Master's Thesis, Georg-August-Universität Göttingen, Göttingen, Germany, 2018.

27. Wagner, B.; Leiss, B.; Tanner, D.C. Stopp 2: Gefalteter unterkarbonischer Kieselschiefer am Bielstein nördlich Lautenthal (Innerstetal). In *Harzgeologie 2016. 5. Workshop Harzgeologie Kurzfassungen und Exkursionsführer*; Friedel, C.H., Leiss, B., Eds.; Universitätsverlag Göttingen: Göttingen, Germany, 2016; pp. 69–78.

28. Franzke, H.J.; Hauschke, N.; Hellmund, M. Spätpleistozäne bis holozäne Tektonik an der Harznordrand-Störung bei Benzingerode (Sachsen-Anhalt). In *Harzgeologie 2016. 5. Workshop Harzgeologie Kurzfassungen und Exkursionsführer*; Friedel, C.H., Leiss, B., Eds.; Universitätsverlag Göttingen: Göttingen, Germany, 2016; pp. 13–17.

29. De Graaf, S.; Lüders, V.; Banks, D.A.; Sośnicka, M.; Reijmer, J.J.G.; Kaden, H.; Vonhof, H.B. Fluid evolution and ore deposition in the Harz Mountains revisited: Isotope and crush-leach analyses of fluid inclusions. *Miner. Depos.* **2020**, *55*, 47–62. [CrossRef]

30. Wagner, B.; Günther, S.; Ford, K.; Sosa, G.M.; Leiss, B. The "Hexagon concept": A fundamental approach for the geoscientific spatial data compilation and analysis at European scale to define the geothermal potential of Variscan and pre-Variscan low- to high-grade metamorphic and intrusive rocks. In Proceedings of the World Geothermal Congress 2021, Reykjavik, Iceland, 24–27 October 2021; pp. 1–10.

31. Trullenque, G.; Bär, K.; Turan, A.; Schulz, K.; Leiss, B.; Ford, F.; Reinecker, J.; Vanbrabant, Y. Evaluation of Outcrop Analogues of the Four Variscan Reservoir Types, MEET Report, Deliverable D5.4. 2020; 48p. Available online: https://www.meet-h2020.com/project-results/deliverables/ (accessed on 29 June 2022).

32. Lloyd, G.E.; Chinnery, N. The Bude Formation, SW England—A three-dimensional, intra-formational Variscan imbricate stack? *J. Struct. Geol.* **2002**, *24*, 1259–1280. [CrossRef]

33. Lei, Q.; Latham, J.P.; Tsang, C.F. The use of discrete fracture networks for modelling coupled geomechanical and hydrological behaviour of fractured rocks. *Comput. Geotech.* **2017**, *85*, 151–176. [CrossRef]

34. Loope, D.B.; Burberry, C.M. Sheeting joints and polygonal patterns in the Navajo Sandstone, southern Utah: Controlled by rock fabric, tectonic joints, buckling, and gullying. *Geosphere* **2018**, *14*, 1818–1836. [CrossRef]

35. Romanov, D.; Leiss, B. Analysis of Enhanced Geothermal System Development Scenarios for District Heating and Cooling of the Göttingen University Campus. *Geosciences* **2021**, *11*, 349. [CrossRef]

36. Nadan, B.J.; Engelder, T. Microcracks in New England granitoids: A record of thermoelastic relaxation during exhumation of intracontinental crust. *Geol. Soc. Am. Bull.* **2009**, *121*, 80–99. [CrossRef]

37. Manda, A.K.; Horsman, E. Fracturesis Jointitis: Causes, symptoms, and treatment in groundwater communities. *Groundwater* **2015**, *53*, 836–840. [CrossRef]

38. Peacock, D.C.P.; Sanderson, D.J. Structural analyses and characterising fracture networks: Seven pillars of wisdom. *Earth Sci. Rev.* **2018**, *184*, 13–28. [CrossRef]

39. Narahara, D.K.; Wiltschko, D.V. Deformation in the hinge region of a chevron fold, Valley and Ridge Province, central Pennsylvania. *J. Struct. Geol.* **1986**, *8*, 157–168. [CrossRef]

40. Kenworthy, S.; Hagemann, S.G. Fault and vein relationships in a reverse fault system at the Centenary orebody (Darlot gold deposit), Western Australia: Implications for gold mineralisation. *J. Struct. Geol.* **2007**, *29*, 712–735. [CrossRef]

41. Gillespie, P.; Walsh, J.; Watterson, J.; Bonson, C.; Manzocchi, T. Scaling relationships of joint and vein arrays from The Burren, Co. Clare, Ireland. *J. Struct. Geol.* **2001**, *23*, 183–201. [CrossRef]

42. Peacock, D.C.P. Differences between veins and joints using the example of the Mesozoic limestones of Somerset. In *The Initiation, Propagation, and Arrest of Joints and Other Fractures*; Special Publication; Cosgrove, J.W., Engelder, T., Eds.; Geological Society of London: London, UK, 2004; Volume 231, pp. 209–221.

43. Middleton, D.; Parnell, J.; Carey, P.; Xu, G. Reconstruction of fluid migration history in Northwest Ireland using fluid inclusion studies. *J. Geochem. Explor.* **2000**, *69–70*, 673–677. [CrossRef]

44. Bauer, J.F.; Krumbholz, M.; Meier, S.; Tanner, D.C. Predictability of properties of a fractured geothermal reservoir: The opportunities and limitations of an outcrop analogue study. *Geotherm. Energy* **2017**, *5*, 24. [CrossRef]
45. Al-Ajmi, H.; Hinderer, M.; Keller, M.; Rausch, R.; Blum, P.; Bohnsack, D. The role of outcrop analogue studies for the characterization of aquifer properties. *Int. J. Water Resour. Arid. Environ.* **2011**, *1*, 48–54.
46. Follin, S.; Hartley, L.; Rhén, I.; Jackson, P.; Joyce, S.; Roberts, D.; Swift, B. A methodology to constrain the parameters of a hydrogeological discrete fracture network model for sparsely fractured crystalline rock, exemplified by data from the proposed high-level nuclear waste repository site at Forsmark, Sweden. *Hydrogeol. J.* **2013**, *22*, 313–331. [CrossRef]
47. Ogata, K.; Senger, K.; Braathen, A.; Tveranger, J.; Olaussen, S. The importance of natural fractures in a tight reservoir for potential CO_2 storage: Case study of the upper Triassic to middle Jurassic Kapp Toscana Group (Spitsbergen, Arctic Norway). In *Advances in the Study of Fractured Reservoirs*; Spence, G.H., Redfern, J., Aguilera, R., Bevan, T.G., Cosgrove, J.W., Couples, G.D., Daniel, J.M., Eds.; Geological Society of London: London, UK, 2012; Volume 374, pp. 395–415.
48. Sanderson, D.J.; Roberts, S.; Gumiel, P.; Greenfield, C. Quantitative Analysis of Tin- and Tungsten-Bearing Sheeted Vein Systems. *Econ. Geol.* **2008**, *103*, 1043–1056. [CrossRef]
49. Iñigo, J.F.; Laubach, S.E.; Hooker, J.N. Fracture abundance and patterns in the Subandean fold and thrust belt, Devonian Huamampampa Formation petroleum reservoirs and outcrops, Argentina and Bolivia. *Mar. Pet. Geol.* **2012**, *35*, 201–218. [CrossRef]
50. Garofalo, P.; Heinrich, C.; Matthäi, S.K. Three-dimensional geometry, ore distribution and time-integrated mass transfer through the quartz-tourmaline-gold vein network of the Sigma deposit (Abitibi belt, Canada). *Geofluids* **2002**, *2*, 217–232. [CrossRef]
51. Na, J.; Xu, T.; Jiang, Z.; Bao, X.; Yongdong, W.; Feng, B. A study on the interaction of mud acid with rock for chemical stimulation in an enhanced geothermal system. *Environ. Earth Sci.* **2016**, *75*, 1025. [CrossRef]

 *geosciences*

Article

Long-Term Evolution of Fracture Permeability in Slate: An Experimental Study with Implications for Enhanced Geothermal Systems (EGS)

Chaojie Cheng [1,*], Johannes Herrmann [2], Bianca Wagner [3,4], Bernd Leiss [3,4], Jessica A. Stammeier [1], Erik Rybacki [1] and Harald Milsch [1]

[1] GFZ German Research Centre for Geosciences, 14473 Potsdam, Germany; jessica.stammeier@gfz-potsdam.de (J.A.S.); uddi@gfz-potsdam.de (E.R.); milsch@gfz-potsdam.de (H.M.)
[2] Institute of Drilling Engineering and Fluid Mining, Technische Universität Bergakademie Freiberg, 09599 Freiberg, Germany; Johannes.Herrmann1@tbt.tu-freiberg.de
[3] Department of Structural Geology and Geodynamics, Geoscience Centre of the University of Göttingen, 37077 Göttingen, Germany; bwagner1@gwdg.de (B.W.); bleiss1@gwdg.de (B.L.)
[4] Universitätsenergie Göttingen GmbH, Hospitalstraße 3, 37073 Göttingen, Germany
* Correspondence: chaojie@gfz-potsdam.de

Citation: Cheng, C.; Herrmann, J.; Wagner, B.; Leiss, B.; Stammeier, J.A.; Rybacki, E.; Milsch, H. Long-Term Evolution of Fracture Permeability in Slate: An Experimental Study with Implications for Enhanced Geothermal Systems (EGS). *Geosciences* 2021, 11, 443. https://doi.org/10.3390/geosciences11110443

Academic Editors: Béatrice A. Ledésert, Ronan L. Hébert, Ghislain Trullenque, Albert Genter, Eléonore Dalmais, Jean Hérisson and Jesus Martinez-Frias

Received: 30 September 2021
Accepted: 26 October 2021
Published: 28 October 2021

Abstract: The long-term sustainability of fractures within rocks determines whether it is reasonable to utilize such formations as potential EGS reservoirs. Representative for reservoirs in Variscan metamorphic rocks, three long-term (one month each) fracture permeability experiments on saw-cut slate core samples from the Hahnenklee well (Harz Mountains, Germany) were performed. The purpose was to investigate fracture permeability evolution at temperatures up to 90 °C using both deionized water (DI) and a 0.5 M NaCl solution as the pore fluid. Flow with DI resulted in a fracture permeability decline that is more pronounced at 90 °C, but permeability slightly increased with the NaCl fluid. Microstructural observations and analyses of the effluent composition suggest that fracture permeability evolution is governed by an interplay of free-face dissolution and pressure solution. It is concluded that newly introduced fractures may be subject to a certain permeability reduction due to pressure solution that is unlikely to be mitigated. However, long-term fracture permeability may be sustainable or even increase by free-face dissolution when the injection fluid possesses a certain (NaCl) salinity.

Keywords: fracture; permeability; fluid–rock interactions; slate; temperature; time-dependent; pressure solution; dissolution

1. Introduction

Geothermal energy is ubiquitous at a certain depth of the Earth's crust and may provide great potential to meet the energy demand [1]. To utilize such hot formations for district heating or electricity generation, a sufficient amount of fluid needs to be injected and extracted into/from the reservoirs. Therefore, high hydraulic conductivity of the reservoir determines the success and economic efficiency of such utilization. However, often deep hot reservoirs are of low or no hydraulic conductivity, and thus need to be stimulated to increase their permeability by creating new fracture networks or by activating pre-existing fractures, denoted as Enhanced Geothermal Systems (EGS) [2]. Sustainable fractures, acting as the main pathways for fluid flow, are most important in this context.

Fracture permeability changes associated with, e.g., effective stress, fracture surface roughness, fracture offset/shear displacement, and the mechanical properties of the rock matrix have been widely investigated in experiments e.g., [3–11]. Such short-term experiments mainly focused on the stress-dependent permeability variation to shed light on the critical factors (e.g., shear displacement, surface roughness, mechanical properties of the reservoir rock) that determine the sustainability of fracture permeability. Most rock fractures, once created, are always conductive for fluid flow and difficult to close completely by mechanical loading [11,12], which is favorable for EGS. However, a decline in fracture permeability with time has been observed in the field and in laboratory experiments. For example, the productivity index of the geothermal reservoir in Groß Schönebeck, Germany, decreased by about one order of magnitude within two and a half years [13], which was believed to be induced by fluid–rock interactions [14]. Time-dependent fracture permeability reduction in core samples under constant pressure and temperature conditions was also documented in some previous studies e.g., [4,8,15–18], including various rock types, such as novaculite [15,17], granite [4,19–21], shale [21,22], limestone [16], dolomitic anhydrite [18], sandstone, and mudstone [23].

Typically, such permeability reduction occurs at the early stage of an experiment after pressure build-up and progressively towards a steady state with time. The mechanisms governing the fracture permeability evolution are expected to be an interplay of pressure solution, stress corrosion, and mineral dissolution/precipitation [15,17,20,24–29], which are often referred to as sealing/healing for macroscopic fracture strengthening or strength recovery [30]. Pressure solution and stress corrosion cracking are driven by the imposed (effective) stress on the fracture plane. The former is a three-step process that involves dissolution at the fracture contacts (i.e., asperities), diffusion along the contact interface, and precipitation within the unstressed fracture void space [31]. The latter mainly results in subcritical crack propagation due to fluid–rock interaction at the stressed crack tips [32–34]. Mineral dissolution/precipitation under hydrostatic pore fluid pressure is driven by chemical potential gradients and independent of the applied effective stress on the solid phases. All of the former processes imply a change in the fracture void space, i.e., the fracture aperture, available for fluid flow and thereby yield either a decrease (pressure solution, stress corrosion, and precipitation) or increase (free-face dissolution) in fracture permeability. In addition, fine particle migration-induced permeability reduction was commonly observed in porous media, where fines/clay particles attached to grain surfaces are susceptible to fluid flow [35]. Fines migration may also occur in (re-)activated rock fractures, where gouge material may lead to clogging of the main flow pathways, resulting in permeability reduction [5].

Depending on the dominating mechanisms, the overall process of fracture evolution, permeability may change differently. For example, for the dissolution/precipitation-dominated processes, open fractures can be sealed by mineral deposits within months to hours depending on temperature [29,36,37]. Supersaturated alkaline fluids result in permeability reduction in granite due to promoted precipitation of clay minerals, but under-saturated alkaline fluids can create cavities along the fractures by dissolution, generating sustainable and pressure-independent fracture permeability [38]. Reactive flow experiments with high and low flow rates show significant fracture permeability reduction and unchanged permeability over time, respectively, indicating that equilibrium of fluid–rock interaction processes is important [18]. By using an injection fluid close to chemical equilibrium with the rock matrix, precipitation-induced permeability reduction can be minimized [39]. In some cases, permeability can persist to hundreds of days at intermittent flow of deionized water (DI), where fluid–rock interactions tend to equilibrium during the stopped flow stages [23,24]. Even for substantial mineral precipitation in fractures, permeability can remain almost unchanged up to two months duration if the deposition is mainly located behind contact asperities with respect to the flow direction [30]. Interestingly, limestone permeability can even increase with time if flowing DI produces "wormholes" that develop due to dissolution and mass transfer [16].These findings suggest that mineral dissolution/precipitation effects on fracture permeability evolution are closely related to the fluid/rock compositions, the reaction kinetics, and how the reactive components reshape the flow channel patterns [40–42].

In addition to pure chemical effects, stress-induced fracture deformation is also an important process leading to time-dependent permeability reduction. A high fracture closure rate occurs under high effective stress, indicating the contribution of pressure solution [20,43]. The permeability reduction with time can be described by a power-law relation under constant differential stress conditions [21,22], which may result from decreasing normal stress acting on an increasing area of the contact asperities [17,19,44–46]. In the presence of stress, subcritical crack growth by stress corrosion may also contribute to the mechano-chemical processes that affect permeability evolution [28], but the effect on fluid chemistry is small [17].

In the aforementioned mechanisms, temperature plays an important role since it determines the reaction kinetics and the solubility of minerals [15,27,43,47]. Fracture permeability evolution, either increase or reduction, depends on the coupled process of mechano-chemical compaction and free mineral dissolution/precipitation under specific pressure and temperature conditions [48]. Therefore, the thermodynamic boundary conditions (pressure, temperature), fluid composition, and rock composition are important for predicting fracture permeability evolution with time.

This experimental study is part of the 'MEET' project (Multidisciplinary and multi-context demonstration of EGS exploration and Exploitation Techniques and potentials) within the framework of the European Union's Horizon 2020. With the aim of evaluating the potential of Variscan metamorphic rocks for EGS in the future, this study focuses on slate material, which is one of the target rocks for a planned EGS at the Göttingen University campus, Germany. We investigated the time-dependent fracture permeability evolution in slates, considering the effects of flow operations (continuous or intermittent flow), fluid composition (deionized water (DI) or brine), and temperature (up to 90 °C as the expected upper bound fluid reinjection temperature after energetic use of the geothermal fluid).

2. Materials and Methods

2.1. Rock Samples

The samples were extracted from a drill core made up of dark grey to black Middle Devonian Wissenbach slate from the Hahnenklee well, Harz Mountains, Germany. The original drill core was taken about 40 years ago, is 80 mm in diameter and 320 mm in length, and originates from a well depth between 1156 and 1156.32 m. The core has no macroscopic veins (i.e., a distinct sheet-like body of crystallized minerals within a rock) or obvious fractures (i.e., separated voids in rocks). It should be noted that the choice of well analogue rocks may not fully represent the rock encountered in the reservoir to be accessed and used later for geothermal energy provision. However, in the absence of an exploration well at a particular site, this is the closest one can get and, consequently, has become a standard in experimental reservoir assessment. The sample material used in this study was carefully selected in this regard and originates from a well that is located in geologically close vicinity to the well to be drilled later at the Göttingen University campus, Germany. Regarding possible aging or alteration of the cores during storage in the repository, based on our experiences, there is no indication that this were the case to a degree that would impede any conclusions drawn on the corresponding behavior of the rock encountered in situ.

To prepare three cylindrical samples, SM1, SM2, and SM3, the drill core was first cut with a saw perpendicular to its longitudinal axis to create a macroscopic fracture. Subsequently, smaller cores with a diameter of 25 mm were drilled with maintaining the saw-cut fracture in its center. The two ends of each core were cut and polished to obtain a cylinder with a length of 50 mm. Finally, the saw-cut fracture surfaces were ground using rolling grains of a defined diameter to obtain closely identical surface roughness spectra and thereby ensure comparability between the samples. Some leftover material was used for X-ray powder diffraction (XRD) analysis at the University of Göttingen, Germany, to determine the mineral composition. The main constituents are quartz, muscovite, chlorite, and feldspar (Table 1). This table also evidences the excellent mineralogical homogeneity of the samples. As sample SM3 was located in the original drill core at a cm scale distance to both SM1 and SM2, there is no reason to believe that its composition should differ significantly and thereby impede any comparability. The matrix permeability of the sample material is on the order of 10^{-19} m^2, determined by a gas permeameter at TU Darmstadt, Germany, using a fourth (neighboring) sample and argon gas and applying the Klinkenberg correction.

Table 1. Mineral composition of slate samples.

Mineral Content (wt %)	Quartz	Muscovite	Chlorite	Plagioclase	Chalcocite	Ankerite	Pyrite
SM1	36	33	12	8	5	5	1
SM2	36	35	11	8	5	4	1

2.2. Experimental Procedures

The two halves of the prepared specimens were assembled, jacketed with a heat-shrink tube (Figure 1a,b), and vacuum-saturated with DI in a desiccator for more than 24 h. The experiments were performed using three different flow-through apparatuses. For experiments with saline fluids, a device made of Hastelloy C-276 was used, while the other two devices made of stainless steel were used for tests with DI as the fluid medium. All apparatuses can apply pore fluid pressures up to P_p = 50 MPa at hydrostatic confining pressure up to P_c = 100 MPa and temperatures between room temperature and T = 200 °C (for details, see Milsch et al. [49]). Fracture permeability can be determined by monitoring the differential pressure (using a 0~0.6 MPa differential pressure transducer) between the sample's ends at a constant flow rate. The upstream pump maintains a constant flow rate, while the downstream pump is set to a constant pressure to receive the fluid volume (Figure 1c). A relief valve is connected in parallel to the downstream side, allowing sampling of the effluents for chemistry analysis under constant pore pressure.

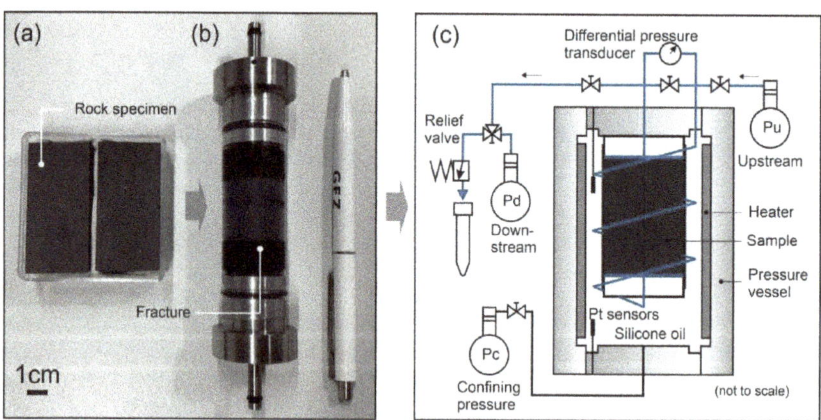

Figure 1. Workflow for the long-term flow-through experiments, (**a**) two halves of prepared slate specimens, (**b**) the assembled sample jacketed with heat-shrink tubing, and (**c**) the flow-through apparatus.

Here, all experiments were conducted at constant effective pressure (P_c = 10 MPa and P_p = 1 MPa). Each experiment included three stages: (1) initially, continuous flow-through tests after pressurization at room temperature, (2) temperature cycles between room temperature and up to 70 or 90 °C, and (3) long-term permeability measurements with the intermittent flow at 70 or 90 °C. In two flow-through experiments (SM1 and SM2), DI was used as fluid medium, and in one test (SM3), a 0.5 M NaCl solution was used to investigate the effect of fluid type on fracture permeability evolution. The salinity was chosen since fluid inclusions within the slates contain mainly water with low salinity NaCl. Note, however, that in some cases, fluid inclusions also contain CaCl$_2$. Details about applied temperatures, flow rates, duration of each stage, experimental conditions, and permeating fluid types for the three samples are listed in Table 2. The intention to divide each experiment into three different stages was the following:

(1) The purpose of stage one was to investigate the potential transient fracture permeability degradation after pressure build-up, which exerts a force on the fracture surfaces. Such initial fracture permeability decline was widely observed at the first dozens to hundreds of hours of continuous fluid flow through fractured granitic rocks [4,21], shale [22], novaculite [15,17], limestone [16], and dolomitic anhydrite [18]. However, this time-dependent fracture permeability decay did not occur in some fractured sandstones and mudstones [23,24] with the intermittent flow (flow-stop-flow with a certain time interval). To monitor the influence of pressure on fracture permeability evolution in slates, we continuously measured permeability for several to dozens of hours, followed by stopping the flow for dozens of hours and measured fracture permeability again. This initial stage was performed at room temperature immediately after pressurization.

(2) The second stage was to reveal thermal effects on fracture permeability evolution and to eliminate any irreversible fracture permeability changes upon thermal expansion of the rock matrix. The temperature was increased and decreased stepwise between room temperature and 70 °C for sample SM2 and between 25 °C and 90 °C for sample SM1 and SM3 (Table 2). Fracture permeability was measured after stabilization of temperature in each step.

(3) In the last stage, the temperature was kept at the highest value (70 or 90 °C), and permeability was measured regularly after a time interval of 6 days. In between the time intervals, the valve of the upstream pump was closed, and the downstream pump maintained constant pressure so that the pore fluid could be considered as a semi-closed system. Before each permeability measurement, the effluent was sampled through the relief valve at a constant flow rate of $Q = 0.1$ mL/min. Each time, seven to nine subsamples with a volume of $V = 1.0$ mL at constant pore fluid pressure of $P_p = 1$ MPa were collected (downstream side). Each sample was acidified by addition of 0.01 ml super-pure HNO_3 to minimize any potential precipitation or alteration of the fluid. The purpose of the chosen sampling strategy with collecting small-volume subsamples ($V = 1.0$ mL) was to better specify the fluid composition within the fracture. Otherwise, a large volume of effluent would have mixed the fluid within the fracture with the fluid in the capillaries connected to the sample.

Table 2. Experimental conditions.

Sample	Temperature (°C)	Flow Rate [a] (mL/min)	Flow Type	Duration [b] (Days)	Permeant Fluid
SM1	Room temperature $25 \rightarrow 50 \rightarrow 70 \rightarrow 90 \rightarrow 70 \rightarrow 50 \rightarrow 32 \rightarrow 90$ 90	0.3~0.5 0.3~0.5 0.05~0.3	Continuous Intermittent Intermittent	~3 3 34	DI
SM2	Room temperature $25 \rightarrow 50 \rightarrow 70 \rightarrow 50 \rightarrow 32 \rightarrow 70$ 70	0.3~0.5 0.3~0.5 0.1~0.3	Continuous Intermittent Intermittent	<1 3 34	DI
SM3	Room temperature $25 \rightarrow 50 \rightarrow 70 \rightarrow 90 \rightarrow 70 \rightarrow 50 \rightarrow 32 \rightarrow 90$ 90	0.02~0.3 0.02~0.1 0.1	Continuous Intermittent Intermittent	~3 3 34	0.5 M NaCl

[a] Flow rate used for permeability measurements, the flow rate during effluent sampling is always $Q = 0.1$ mL/min. [b] Elapsed time during the corresponding stages.

The experiments target the evaluation of a medium enthalpy EGS system with a reservoir or (fluid) production temperature of about 150 °C. After the heat exchanger, the fluid, typically, has a temperature of approximately 70 °C (lower bound experimental temperature), which increases to around 90 °C (upper bound experimental temperature) before being reinjected into the formation. Due to technical constraints, the temperature of the sample and the one of the fluid, when being injected into the sample, are identical (Figure 1c). The experiments thus replicate a reservoir scenario, where, when starting from the well-to-formation interface, rock and fluid are in thermal equilibrium. As fluid is continuously injected and transported further into the reservoir, the volume where rock and fluid are in thermal equilibrium expands. This study thus investigates processes that may alter fracture permeability in the vicinity of an injection well, where the formation is at fluid injection temperature.

During fluid flow (i.e., fracture permeability measurements and effluent sampling), the differential pressure between the sample ends was limited to $\Delta p \leq 0.5$ MPa (i.e., the maximum upstream pore fluid pressure is below 1.5 MPa) to ensure to not exceed the measurable range of the differential pressure transducer and to avoid changing effective stress beyond critical values.

Sample permeability k is calculated based on Darcy's law assuming steady-state conditions as

$$Q = \frac{k \Delta p A}{\mu L} \tag{1}$$

where Q is the flow rate, L is the sample length, $A = \pi r^2$ is the cross-sectional area of the cylindrical sample, Δp is the differential pressure over the sample length L, and μ is the dynamic fluid viscosity, which depends on the fluid type, salinity, temperature, and pore pressure [50,51] and was adjusted according to tabulated values in the cited literature when calculating permeability. By ignoring fluid flow in the rock matrix because of the low matrix permeability ($k \sim 10^{-19}$ m^2), the "cubic law", assuming laminar flow through a fracture between parallel plates, gives the expression of the separation distance, b_h, between the two smooth plates during flow-through tests as [52,53]

$$Q = \frac{b_h^3 W \Delta p}{12 \mu L} \tag{2}$$

where W is the width of the fracture (sample diameter). b_h can be considered an equivalent aperture (hydraulic aperture) in case of rough fractures. Using this approximation and assuming that all fluid flow through the fracture ($A = W b_h$), fracture permeability, k_f, can be expressed by combining Equations (1) and (2) as

$$k_f = \frac{b_h^2}{12} \tag{3}$$

2.3. Analytical Methods

2.3.1. Effluent Element Concentrations

Effluent element concentrations of Al, Ca, Fe, K, Mg, Na, Si, and Zn were determined by Inductively Coupled Plasma-Optical Emission Spectrometry (ICP-OES) analyses performed at the ElMiE Lab at the German Centre for Geosciences (GFZ, Potsdam, Germany) using a 5110 spectrometer (Agilent, Santa Clara, CA, USA). The analytical precision and reproducibility are generally better than 2%, regularly tested using certified reference material and in-house standards. Effluent samples were diluted with HNO_3.

2.3.2. Fracture Surface Topographies

The fracture surface topographies were measured before and after the experiments using white light interferometry (Keyence VR 3000). The resolution of the in-plane coordinates was 23.5 µm, and the vertical resolution was 1.0 µm. Statistical parameters, the peak height difference R_p, the mean R_m, and the root-mean-square R_{rms}, of fracture surfaces were used to compare the changes of surface roughness, expressed as

$$R_p = \max|z_i - z_a| \tag{4}$$

$$R_m = \frac{1}{n} \sum_{i=1}^{n} |z_i - z_a| \tag{5}$$

$$R_{\mathrm{rms}} = \sqrt{\frac{1}{n}\sum_{i=1}^{n}(z_i - z_a)^2} \qquad (6)$$

where z_i is the height of the ith point, and z_a is the mean height of the elevation plane, which is discretized by n points.

2.3.3. SEM-EDX

After the experiments were conducted, the fracture surfaces were coated with carbon and observed by performing scanning electron microscopy (SEM) using both the backscattered electron (BSE) and the secondary electron (SE) mode. Combined energy dispersive X-ray analysis (EDX) was performed to identify the elemental composition of representative locations.

3. Results

Fracture permeability (Equation (3)) evolution was explored upon continuous fluid flow after pressurization at ambient temperature, heating–cooling cycles, and intermittent flow over a long-time duration (Section 3.1). Effluent element concentrations were used to qualitatively analyze potential mineral reactions in rock fractures (Section 3.2). We compared fracture surface topographies before and after the experiments (Section 3.3) and explored the mechanisms governing fracture permeability evolution using microstructural observations (Section 3.4).

3.1. Variations of Fracture Permeability

3.1.1. Initial Fracture Permeability Decline with Continuous Flow (Stage 1)

All samples exhibited a progressive fracture permeability decline with time during fluid flow, particularly pronounced for SM3 with NaCl solution as permeating fluid (Figure 2). Samples SM1 and SM2 showed nearly identical fracture permeability degradation, which slowly converged over time/reaching a minimum value. After the fluid flow was stopped for about 65 h, the fracture permeability of SM1 remained unchanged (i.e., $k_f = 3.7 \times 10^{-12}$ m^2). Fracture permeability of sample SM3 was first continuously measured for 41 hours and subsequently measured again after flow was stopped for about 28 h and 16 h (see two lower right panels in Figure 2). Compared to samples SM1 and SM2, fracture permeability was more strongly reduced in the first 2 hours and progressively converged from initial 3.5×10^{-12} m^2 to 7×10^{-13} m^2 at the end of continuous flow. Fracture permeability slightly declined further to 6.7×10^{-13} m^2 and 6.6×10^{-13} m^2 after interrupting fluid flow for 28 h and 16 h, respectively.

The evolution of fracture permeability with cumulative fluid flow volume is shown in Figure 3. Samples SM1 and SM2 presented almost similar permeability decay with flow volume, independent of flow rate. In contrast, the permeability of sample SM3 decreased more drastically with the same amount of NaCl solution. Again, the flow rate was adjusted during the flow tests to ensure that the differential pressure did not reach the maximum range of the transducer, but had no effect on permeability evolution. We expect that the total flow volume determines permeability variations by interaction with the fracture surface rather than flow dynamics.

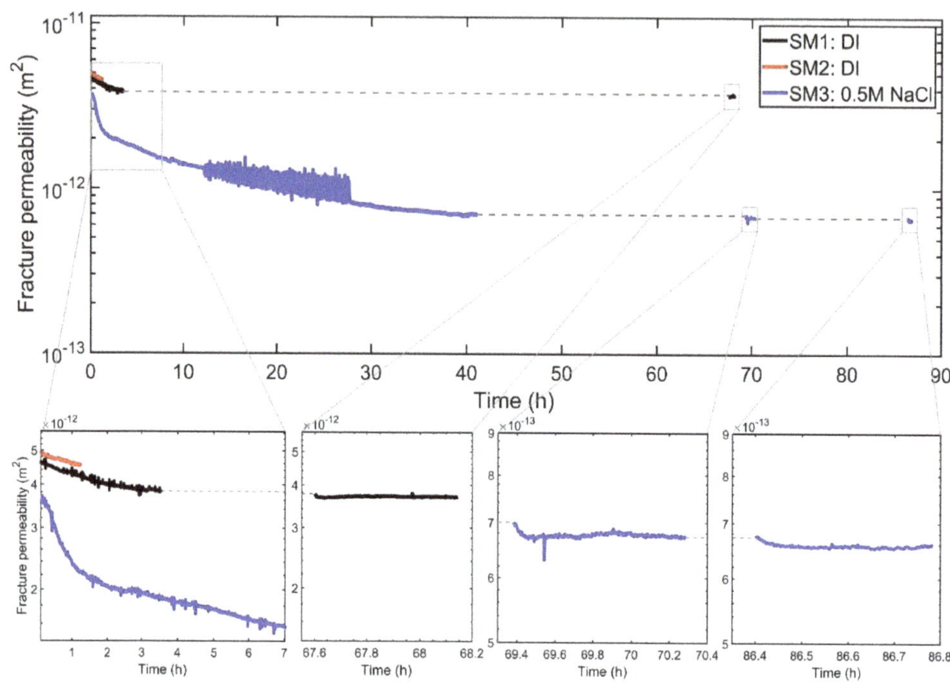

Figure 2. Fracture permeability determined by continuous flow tests after pressure build-up (P_c: 10 MPa and P_p: 1 MPa) as a function of elapsed time at room temperature. The large fluctuation of SM3 permeability (blue line) between 13 and 27 h is because the upstream pump pressure reached a set limit, and the flow rate was automatically adjusted. Panels at the bottom show measurements after fluid flow was interrupted for certain time intervals.

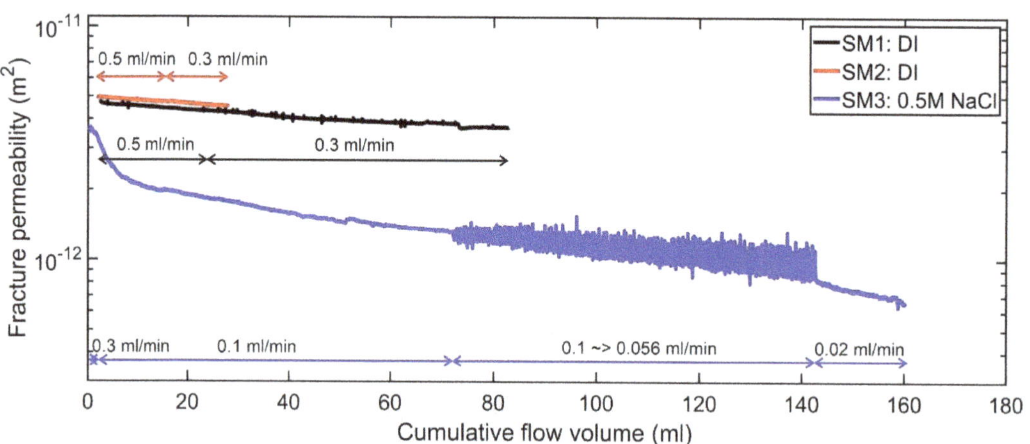

Figure 3. Fracture permeability evolution in stage 1 after pressure build-up (P_c: 10 MPa and P_p: 1 MPa) versus cumulative flow volume. Temperature is 25 °C. Flow rates are indicated.

3.1.2. Temperature Effects (Stage 2)

Using DI as permeating fluid, increasing the temperature stepwise resulted in a slight fracture permeability reduction from 3.7×10^{-12} m^2 (at 25 °C) to 3.4×10^{-12} m^2 (at 85 °C) in sample SM1, and from 4.5×10^{-12} m^2 (at 25 °C) to 3.8×10^{-12} m^2 (at 70 °C) in SM2, respectively (Figure 4). The permeability decline was not recovered after cooling to room temperature. After re-heating, the permeability of samples SM1 and SM2 was slightly reduced by about 3~6% at peak temperature (Figure 4, right-hand panels), indicating a time-dependent permeability reduction. A temperature increase from room temperature to 90 °C caused an irreversible permeability increase from 6.6×10^{-13} m^2 to 1.3×10^{-12} m^2 in sample SM3, in particular between 70 and 90°C (Figure 4). Opposite to the behaviors of the other two samples, the permeability of SM3 slightly increased after re-heating to 90 °C.

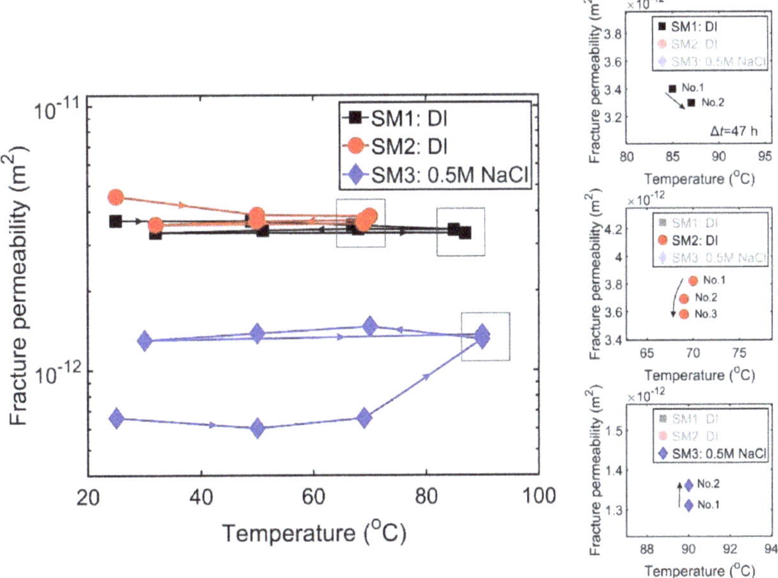

Figure 4. Fracture permeability variations during the temperature cycles between room temperature and up to 90 °C (SM1 and SM3) and 70 °C (SM2) (left panel). The arrows indicate the heating and cooling sequences. All tests started at approximately 25 °C. Details of permeability variations at the respective target temperatures (grey boxes) after first and second heating are shown in the three panels on the right. Sample SM2 was measured twice (No. 1 and No. 2) at 70 °C after first heating within a time interval of 15 h, indicating a time-dependent permeability reduction at constant temperature. After second heating of this sample, the respective permeability value is therefore labelled No. 3. Error bars are in the range of $\pm 2 \times 10^{-14}$ m^2 and hence not visible on the logarithmic y-axis.

3.1.3. Time Dependence of Permeability with Intermittent Flow (Stage 3)

For intermittent flow of DI through samples SM1 and SM2, fracture permeability progressively reduced over time, slightly more at T = 90 °C (SM1) than at T = 70 °C (SM2) (Figure 5). However, such time-dependent permeability reduction of both samples was likely to vanish or slowed down after cooling the sample to room temperature. For sample SM3 with a NaCl solution as permeating fluid, the permeability showed a 1.5-fold increase by over 34 days duration that remained constant after cooling down to 25 °C. In comparison to the initial permeability measured at room temperature in stage 1, all samples, subjected to the three test stages (i.e., the initial continuous flow, temperature cycles, and the intermittent flow), show at 25 °C fracture permeability reductions between 42 and 78% after a total test period of about 40 days (Table 3).

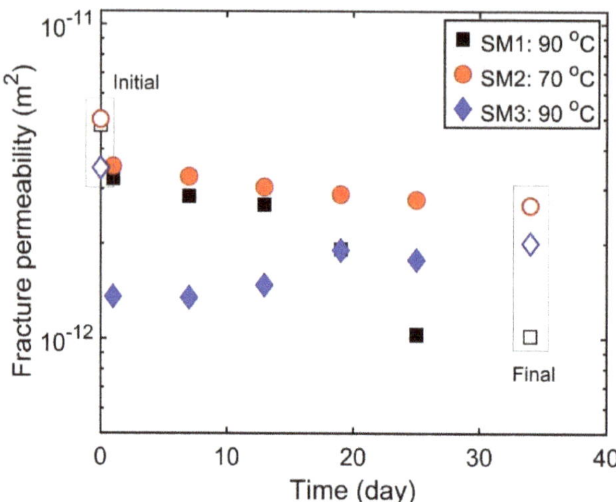

Figure 5. Time-dependent fracture permeability evolution with intermittent flow at 90 °C (SM1 and SM3) and 70 °C (SM2). Each symbol represents a steady-state permeability measurement. The open symbols indicate permeability measured at room temperature at the beginning and end of the experiments, respectively.

Table 3. Summary of experimental results.

Sample	Max T (°C)	k_f (Initial) (10^{-12} m²)	k_f (Final) (10^{-12} m²)	Duration (Days)	Fluid
SM1	90	4.76	1.02	40	DI
SM2	70	4.98	2.66	38	DI
SM3	90	3.49	2.01	40	0.5 M NaCl

The initial and final permeability indicate the first permeability after pressure build-up and the end permeability at room temperature (see open symbols in Figure 5).

3.2. Fluid Chemistry Evolution

The element concentrations of Al and Zn in all effluent samples were negligible (<0.1 mg/L). Ca, Fe, K, Mg, Na, and Si concentration changed within a set of subsamples that were collected on day 7 after the first time interval of 6 days (Figure 6). This and each subsequent sampling sequence took approximately 1.5 h. It is implied that fluid–rock reactions occurred along the fracture and that dissolved matter diffused from the fracture aperture to the capillaries connected to the sample within the semi-closed pore fluid system. We interpret the subsample with peak element concentrations as the fluid representative of the composition in the fracture aperture during the 6-day interval. Because the fracture volume (on the order of ~0.01 mL) is significantly smaller than 1.0 mL (i.e., the volume of each effluent subsample), the absolute element concentrations of the fluid within the fracture aperture may be significantly larger than the measured peak values, depending on the dissolution and diffusion rates. Therefore, the measured element concentrations indicate the degree of fluid–rock reactions, but may represent a lower bound.

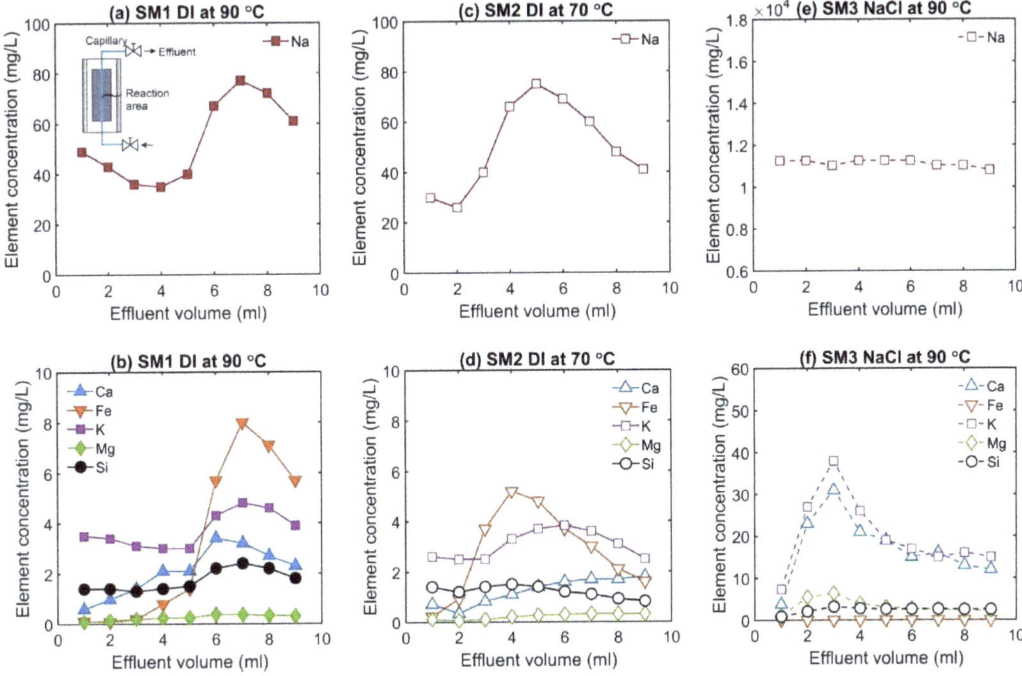

Figure 6. Effluent element concentrations of subsamples after the first time interval of 6 days in SM1 (**a,b**), SM2 (**c,d**), SM3 (**e,f**). Peak values indicate fluid composition extracted from the fracture aperture, while the other subsamples include fluid within the capillaries adjacent to the sample. Note the different scales of Na concentrations.

For samples SM1 and SM2 with DI, all element concentrations of SM1 are slightly higher than that of SM2, which is likely due to enhanced reaction rates at the higher temperature applied to sample SM1. Strikingly, the Na concentrations of both samples are about one order of magnitude higher than that of the other elements, revealing relatively strong reactions of Na-rich minerals in DI (e.g., plagioclase, c.f., Table 1). In sample SM3 with 0.5 M NaCl solution, the concentrations of Ca, K, and Mg are about one order of magnitude larger than those in SM1 and SM2 (Figure 6), and the Si concentration is ≈1.5–2 times higher than in SM1 and SM2. For sample SM3, the Fe concentration is negligible in all subsamples, possibly due to the dissolution limit of Fe-contained minerals or the lack of Fe-contained minerals (e.g., Ankerite and Chlorite) on the SM3 fracture surfaces.

By comparing all peak concentrations obtained after each time interval, we were able to evaluate the reaction rate evolution of the fluid–rock system under constant pressure and temperature conditions (Figure 7). For samples SM1 and SM2, the maximum concentrations occurred after the first time interval on day 7. Subsequently, all element concentrations progressively reduced with time, implying a reduction of fluid–rock interactions. In particular, Fe concentration reduced to nearly zero, which may manifest the reaction termination of Fe-containing minerals or the disappearance of such minerals after dissolution. In contrast, for sample SM3, the concentrations of all elements, except K, showed minor reduction after each week (Figure 7f), indicating that fluid–rock interactions in SM3 were relatively stable, although the dissolution reaction rates were faster than those in SM1 and SM2.

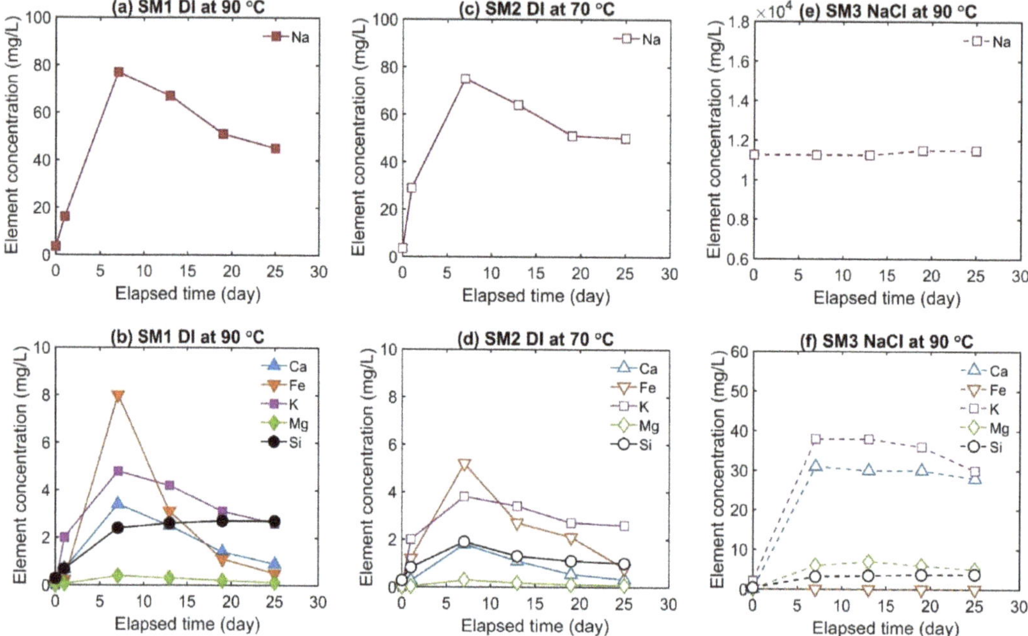

Figure 7. Peak effluent element concentrations extracted from each set of subsamples of sample SM1 (**a,b**), SM2 (**c,d**), SM3 (**e,f**), after prescribed time intervals. The data on day zero present the element concentrations of the injected fluid and data on day 1 (for SM1 and SM2) indicate the effluent sampled right after increasing the temperature to the targets. Note the different scales of Na concentrations.

3.3. Fracture Surface Topography

Fracture surface topographies before and after the experiments are shown in Figures A1–A3 (in the Appendix A) for samples SM1, SM2, and SM3, respectively. The topography of both fracture surfaces changed due to fluid flow, particularly in areas close to the fluid injection side (i.e., dark areas in maps measured after testing), and in most cases resulted in a slightly wider height distribution compared to the initial surface (Figure 8). The removal of surface heights (dark areas in Figures A1–A3), where stresses are expected to be high (stress concentrations at the tips of the surface asperities), indicates that the mass loss is induced by pressure solution of contact asperities. However, the peak height difference R_p, the mean height R_m, and the root mean square height R_{rms} of the whole surface before and after the experiments show that fracture surface roughness remains nearly unchanged or slightly increased (Table 4), which indicates that the global statistical values cannot easily reveal some localized solution/dissolution processes. It can be assumed that such fracture surface roughness changes may result from a combination of pressure solution and dissolution/precipitation reactions, which, in turn, might counteract each other to a certain degree.

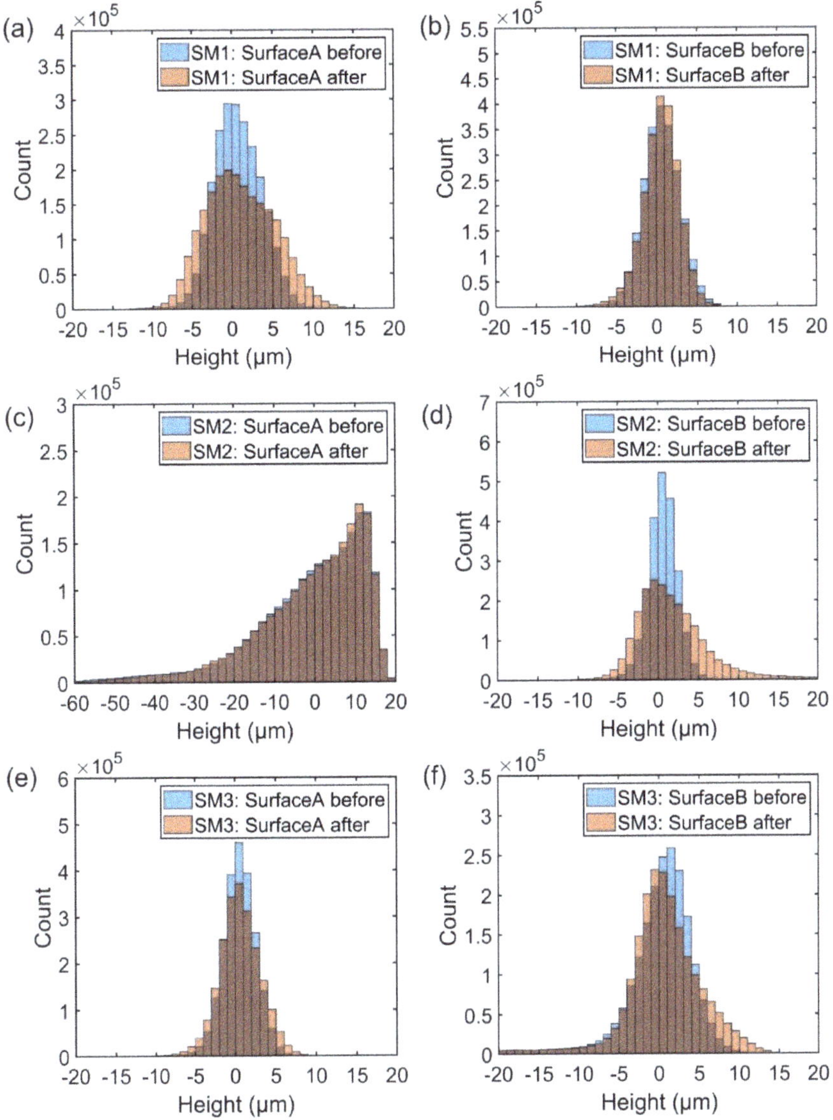

Figure 8. Comparison of fracture surface height distribution before and after the experiments, two surfaces (**a**,**b**) for sample SM1, (**c**,**d**) for sample SM2, (**e**,**f**) for sample SM3, respectively.

Table 4. Statistical parameters of fracture surface roughness before and after the flow-through experiments.

Parameters	SM1_A	SM1_B	SM2_A	SM2_B	SM3_A	SM3_B	Stage
R_p (µm)	41.18	51.08	78.13	80.97	24.00	119.26	Before
	65.45	70.98	78.57	116.80	52.96	106.55	After
R_m (µm)	2.35	1.82	10.99	1.46	1.56	10.31	Before
	3.52	1.78	10.69	3.47	1.98	9.12	After
R_rms (µm)	3.00	2.41	14.31	2.16	2.05	17.94	Before
	4.33	2.44	13.69	4.93	2.60	16.27	After

3.4. Microstructures

The post-experimental backscattered electron (BSE) micrograph and the EDX map of sample SM1 (Figure 9) taken as an example, indicate a complex mineral distribution, where pyrite and ankerite are normally located as a cluster, and other minerals are mixed without clear boundaries. Fracture surfaces are rough and attached with various particles (e.g., pyrite, plagioclase, muscovite) that are authigenic (Figure 10). The particle size is in general within 10 μm. Mineral dissolution and secondary mineral formation (precipitation) are hard to distinguish based on the micrographs. However, we observed some fiber-like texture at the edge of some crystals (Figure 10d) in sample SM3, which may imply strong reactions on the mineral surfaces.

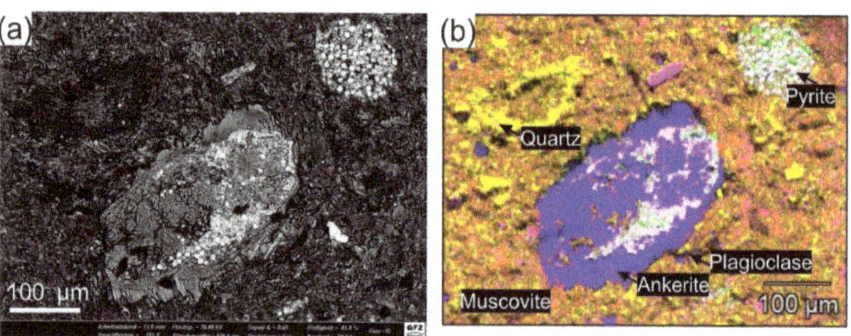

Figure 9. (**a**) Backscattered electron (BSE) micrograph and (**b**) EDX map of the fracture surface of sample SM1 after experiments.

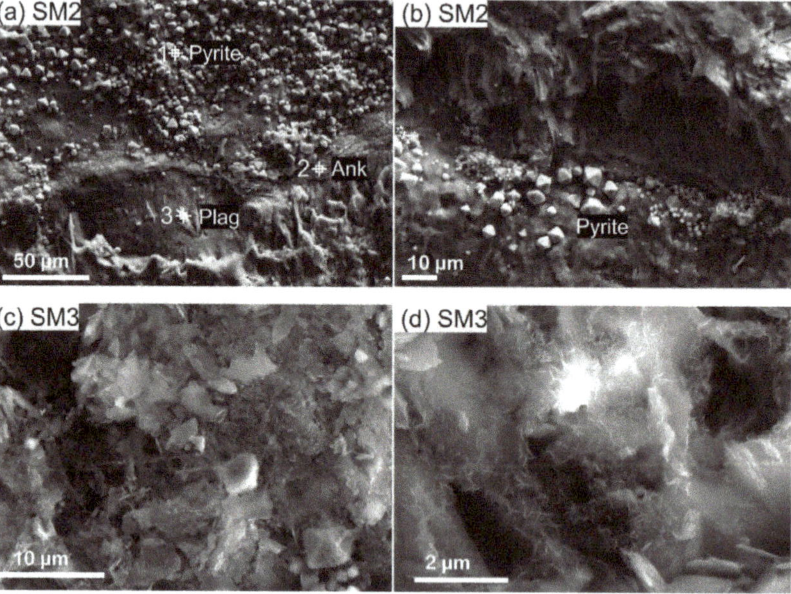

Figure 10. Scanning electron (SE) micrographs of the fracture surface of sample SM2 (**a**,**b**) and sample SM3 (**c**,**d**) after the experiments.

4. Discussion

The observed evolution of fracture permeability with time and temperature in addition to the associated change of the effluent chemistry may be explained by one or more of the following mechanisms: pressure solution, stress corrosion, free-face dissolution, and fines migration. We discuss if these mechanisms may have contributed to permeability evolution in the separated experimental stages.

4.1. Mechanism of Initial Permeability Decline under Constant Conditions

Time-dependent fracture permeability decrease was observed with DI flow under static stress conditions [15,17,21,22]. The degradation was explained by pressure solution, a continuous sealing process of the macroscopic fracture. Im et al. [21] found that fracture permeability decay during hold periods over dozens of hours in slide-hold-slide experiments can be well described by power-law compaction coupled with the "cubic law". Power-law compaction was established in indentation experiments, where the time-dependent displacement of the indenter into the crystalline mineral matrix can be described by a power-law function, induced by pressure solution [54]. Because of indentation of the contact asperities into the matrix, it is assumed that the overall fracture closure (Δb, geometrical aperture changes) also follows a power law with respect to time as

$$\Delta b = \alpha t^n \tag{7}$$

where n is the power-law exponent, t is the elapsed time, and α is the aperture change when $t = 1$. Further, by substituting Equation (1) into Equation (2), sample permeability k can be expressed with hydraulic aperture b_h as,

$$k = \frac{b_\mathrm{h}^3}{6\pi r} \tag{8}$$

Assuming equivalent hydraulic and geometrical aperture b, sample permeability evolution can be correlated to fracture closure Δb [21,52,55] as

$$k = k_0 \left(1 - \frac{\Delta b}{b_0} \right)^3 \tag{9}$$

where k_0 and b_0 are the initial permeability and aperture, respectively. Substituting Equation (7) into Equation (9) yields

$$k = k_0 \left(1 - \frac{\alpha}{b_0} t^n \right)^3 \tag{10}$$

We used the measured permeability data during continuous flow (Figure 2) to parameterize k_0, α/b_0, and n by using the least-squares method. The resulting values, provided in Table 5, fit very well to the data (Figure 11). We observed a similar permeability degradation trend but with a larger range of power exponents in comparison to similar permeability measurements in granite with saw-cut fractures under effective pressure of 3 MPa ($n = 0.3\sim0.4$) described in [21]. The power exponent n measured for sample SM3 is distinctly lower than of samples SM1 and SM2, which may imply that the permeability changes are not solely controlled by pressure solution. Because the indenter experiments indicate that pressure solution-controlled indenting displacement yields a power-law function with exponents normally larger than 0.3 [54]. Our results imply other processes (free-face dissolution) also play an important role in the overall fracture aperture variation, which will be discussed below. Evidently, when the flow was stopped for some time, measured permeability showed much less reduction than predicted for continuous flow (c.f., samples SM1 and SM3 in Figure 11). This suggests that a chemical equilibrium within the fracture aperture was attained during no flow periods leading to a high concentration of the pore fluid and less efficient pressure solution, resulting from limiting the diffusion from contacts to the pore fluids.

Table 5. Fitting parameters.

Sample	k_0	α/b_0	n	Adjusted R^2
SM1	1.95×10^{-15}	7.3×10^{-2}	0.43	0.99
SM2	2.0×10^{-15}	4.5×10^{-2}	0.61	0.99
SM3	1.13×10^{-14}	0.6	0.072	0.98

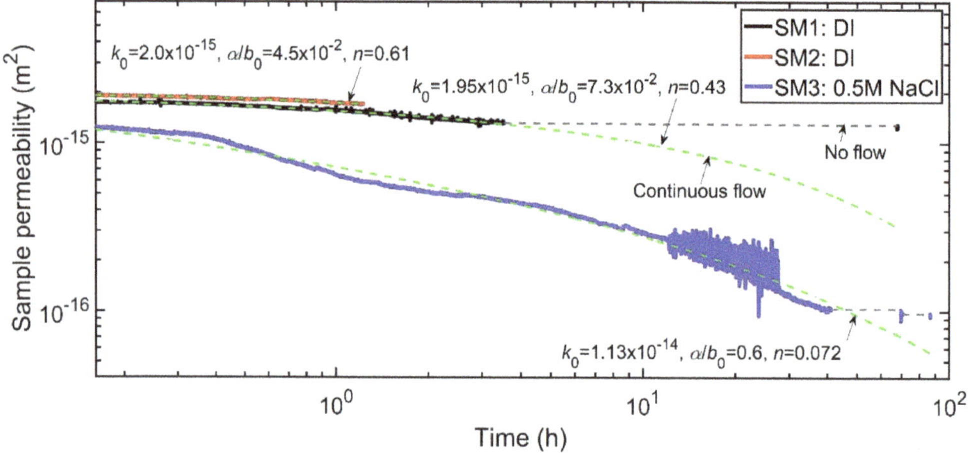

Figure 11. Sample permeability (Equation (1)) variations with time under constant pressure (P_c: 10 MPa and P_p: 1 MPa) and room temperature conditions. Fitting curves (in green) are derived based on Equation (10).

Another possibility is fines migration-induced permeability decline, commonly occurring in porous media [35]. In our samples, fine particles are attached to the fracture surfaces (Figure 10), which may have been transported upon fluid flow, causing permeability to decrease. During no flow periods, particles may not migrate and clog the flow pathways, mitigating permeability reduction. The drag force F_d, causing the particle to move, is proportional to the fluid dynamic viscosity μ, flow rate U, and particle radius r_s: $F_d \propto \mu r_s U$ [56,57]. We found that permeability reduction is relatively independent of flow rates (Figure 3). For DI and the 0.5 M NaCl solution, the dynamic viscosities at room temperature are 889.9 and 928.6 μPa*s, respectively [50,51]. For particles of similar size, this would lead to a difference of drag forces between the two fluid compositions of about 4%, which is unlikely to explain the lower permeability obtained with NaCl compared to DI. In addition, high-salinity fluid would increase the stability of mobile particles due to the increase in the electrostatic force [35]. In this case, with the assumption of fines migration, permeability should present less reduction upon fluid flow with NaCl, which is opposite to our experimental results. Therefore, fines migration is expected to be not the dominant factor in flow-dependent permeability reduction.

Assuming that the initial fracture aperture b_0 is equivalent to the initial hydraulic aperture b_{h0}, b_0 can be expressed based on Equation (8) as

$$b_0 = \sqrt[3]{6\pi r k_0} \tag{11}$$

Using fitted parameters, k_0 and α/b_0 (Figure 11 and Table 5), we can determine the aperture change as a function of time, $\Delta b = \alpha t^n$ (Figure 12a). We noticed that the initial aperture changes of sample SM3 were significantly larger than that of the other two samples.

However, the aperture closure rate $\Delta \dot{b} = n\alpha t^{n-1}$, of sample SM3 decreased much faster and was smaller than that of SM1 and SM2 after about 10 hours duration (Figure 12b). This implies that at the beginning of the experiments, the NaCl solution accelerated the pressure solution rate by increasing the dissolution kinetics of some minerals, e.g., quartz [58,59] and calcite [60]. Due to an increasing rate of free-face dissolution, which has a contrasting (i.e., enlarging) effect on fracture aperture, the total aperture closure rate may significantly reduce if pressure solution rate slows down or terminate eventually due to the continuous increase in contact areas. Therefore, we expect that the initial permeability decline observed in stage 1 of our experiments was mainly governed by pressure solution with a relatively increasing contribution of free-face dissolution with increasing time.

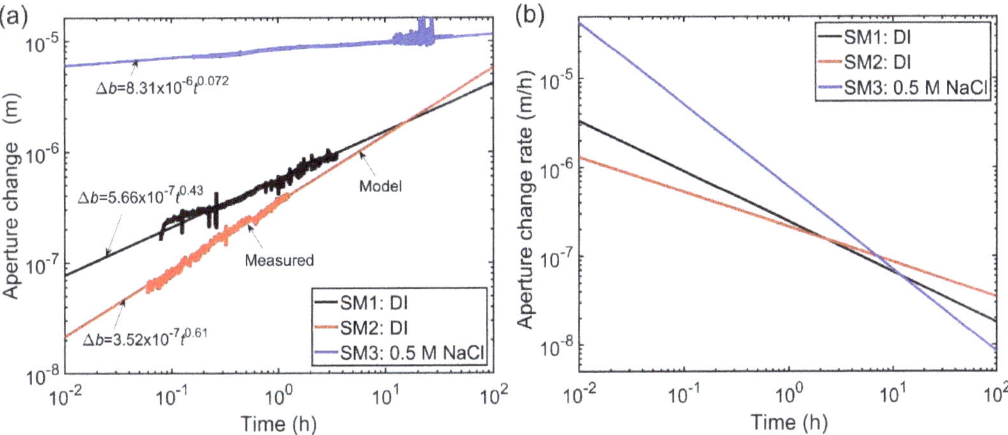

Figure 12. Measured and calculated (with Equation (7)) aperture change with time in log-log plot (**a**), where all parameters are derived based on the fitting curves in Figure 11, and (**b**) fracture aperture change rate as a function of time.

4.2. Thermal Effects on Permeability

Thermally driven hydraulic aperture reduction is believed to be induced by thermal dilation, mechanical creep, and pressure solution [19,43]. However, in our experiments, temperature cycles resulted in opposite irreversible permeability changes (i.e., decrease in sample SM1 and SM2 with DI and increase in sample SM3 with the NaCl solution) (Figure 4). Therefore, thermal dilation or thermal stress effects cannot solely control the process, otherwise similar permeability changes are expected. In addition, we observed time-dependent permeability changes at peak temperatures and after temperatures cycling (Figure 4).

In general, increasing temperature enhances fluid–rock reaction rates [17,24,61–64], leading to enhanced pore space or permeability changes. The continuous flow tests in stage 1 demonstrated an interplay of pressure solution and free-face dissolution, where the initial pressure solution rate in sample SM3 with the NaCl solution was much faster than that of the other two samples. Obviously, the rate-limiting process of pressure solution (dissolution, transport, or precipitation) was different for DI and NaCl solutions and strongly temperature-dependent between 70 and 90 °C. Although we measured the amount of species diluted in the effluent, which strongly differed if using DI or NaCl solution as permeating fluid (Figure 7), we were not able to quantify the element concentrations. We expect that temperature-dependent solubilities and reaction rates, as well as time, control the permeability evolution, but cannot specify the relative contribution based on the available data. Nevertheless, the slight permeability reduction with time of samples SM1 and SM2 after temperature cycling, and the slight increase with time for sample

SM3 may be explained by the cross-over from dominantly pressure solution controlled to free-face dissolution-controlled permeability evolution (c.f., Figure 12).

4.3. Potential Fluid–Rock Interactions on Time-Dependent Permeability Changes

Figure 13 shows potential processes that cause permeability changes with time. They may occur solely or simultaneously depending on the pressure, temperature, fluid, and rock compositions [16,17,24]. Despite fluid chemistry, pressure solution and stress corrosion require normal stress acting on the solid contacts as the driving force. Thus, they occur predominantly at the early stage when the effective stress on the contacts is relatively high. In this case, fracture permeability reduction is always accompanied by obvious fracture deformation [22]. In addition, sufficient dissolved components can be detected in the effluents due to pressure solution-enhanced solubility [17,19,24,43]. For the other mechanisms, i.e., free-face dissolution, precipitation [25,61], and fines migration [35], no obvious deformation is required in conjunction with fracture permeability changes at a relatively small scale.

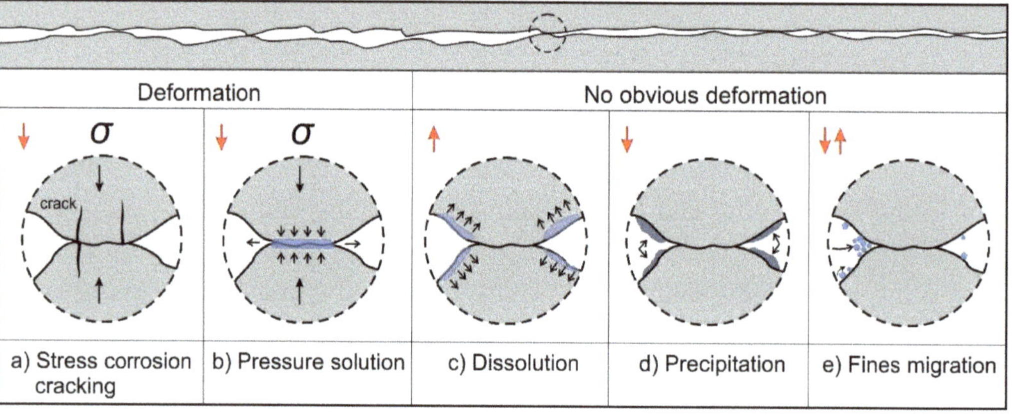

Figure 13. Potential fluid—rock interactions occurring during the long-term experiments and their effects on fracture permeability and fracture deformation. (**a**) Stress corrosion cracking, (**b**) pressure solution, (**c**) dissolution, (**d**) precipitation, and (**e**) fines migration, in which both (**b**,**c**) involve dissolution of solid matter into a fluid phase, where the former is taking place within the contact area of stressed asperities and the latter occurs on the free faces of the mineral grains.

We were not able to identify if stress corrosion cracking played a role in our experiments, but fines migration appeared to terminate towards the end of stage 1 (Figure 2) and hence to play a subordinate role during stage 2 and stage 3. Since we did not measure strain data to evaluate which mechanisms contribute to the overall process, we relied on analyses of measured permeability changes and effluent concentration variations. The element concentrations in sample SM3 were significantly larger than those of the other two samples (Figure 6), supporting the hypothesis that the NaCl solution could increase mineral reaction rates in SM3. Sodium cations would accelerate the dissolution rate of, for example, quartz [58,59] and calcite [60]. The enhanced concentrations of Ca and K in sample SM3 are possibly due to the dissolution of plagioclase ($NaAlSi_3O_8$ to $CaAl_2Si_2O_8$) and muscovite ($KAl_2(AlSi_3O_8)(OH)_2$). The large Fe concentrations in samples SM1 and SM2 may result from Ankerite ($Ca(Fe^{2+}, Mg)(CO_3)_2$) dissolution.

The element concentrations of sample SM1 at 90 °C were slightly higher than those of sample SM2 tested at 70 °C (Figure 7). This supports the hypothesis that increasing temperature enhances the fluid–rock interactions. The fact that all element concentrations of samples SM1 and SM2 showed a maximum after the first time interval and declines afterwards indicates that pressure solution was dominant at the early stage of long-term intermittent flow, but was progressively less effective in later stages when contact areas expanded. The changes in the effluent concentration (SM1 and SM2) are consistent with pressure solution evolution. Temperature-dependent kinetics is in line with our observation of stronger fracture permeability reduction with time at 90 °C compared to 70 °C (Figure 5). For sample SM3, the element concentrations (except Na) also increased after the first time interval, but remained relatively stable afterwards (Figure 7). This implies permeability changes are first mainly controlled by pressure solution and then by free-face dissolution, because the latter is directly correlated to the reactive surface areas that did not change dramatically. The increase in permeability with time for sample SM3 reveals the contribution of mass transfer forming voids.

We noticed that the stopped flow within the semi-closed pore fluid system mitigated permeability decline at room temperature (Figure 11), but the intermittent flow still led to substantial permeability changes at elevated temperatures (Figure 5). This discrepancy is likely also related to thermally enhanced fluid–rock interactions as permeability changes with the intermittent flow significantly slow down after cooling to room temperature.

In our experiments, the initial hydraulic aperture of saw-cut fractures was in the range of 6~7 μm (calculated from Equation (2)) under low-stress conditions (effective stress of 9 MPa), which is in the same order of magnitude as the fracture surface roughness (Table 4). This indicates that the fracture surface roughness is strongly correlated to the initial fracture aperture and thus represents a defining parameter with respect to fracture permeability. The surface height distributions remain unchanged or slightly rougher after the experiments (Figure 8), likely resulting in the dissolved mass on some localized areas (dark areas on the fracture surface topographies, Figures A1–A3). However, we were not able to quantify if such height reduction resulted from dissolution or pressures solution, because both mechanisms led to local mass removal from the fracture surface. In addition, we expect that the effect of mass transfer (caused by pressure solution or mineral dissolution) in such narrow fracture apertures are drastic, whereas larger fractures with high surface roughness or fractures under high-stress conditions may lead to different evolutions of fracture aperture, which need to be further investigated.

4.4. Implications for EGS

The long-term sustainability of rock fracture permeability is crucial to guarantee the lifespan of a successful EGS. Our experimental results demonstrate that newly generated fractures in slates (e.g., artificially prepared, injection created, and sheared) may be subject to large and fast permeability reduction with time under constant effective stress conditions. Such time-dependent fracture permeability reduction was also observed in granite, shale, and novaculite [4,15,17,19,21,22]. Similarly, fracture permeability reduction can slow down at the late stage during fluid flow. Moreover, the governed mechanisms in our experiments are pressure solution and free-face dissolution, where the former may slow down and terminate at some point, and the latter plays a more important role in the late stage. In the sample with NaCl solutions as the permeating fluid, permeability showed a slight increase at the late stage, where we expect that pressure solution is nearly stopped because of the enlargement of contact areas, and mineral dissolution may increase the voids. This behavior was also found in a flow-through experiment of a fractured limestone with distilled water, where free-face dissolution overtakes pressure solution, generating a "wormhole" for fluid flow [16]. Therefore, we infer that the pre-existing fractures, if hydraulically conductive, may persist for a very long term, but newly generated fractures will yield a certain reduction with time due to the initial high-stress concentration on the self-propping contacts. For assessing the performance of geothermal reservoirs after stimulation, such time-dependent permeability reduction must be borne in mind.

To reach the economic utilization of geothermal reservoirs, some indicators, such as production temperature, injection temperature, and the flow rate, have to be set within a certain range. Our laboratory investigation found that elevated temperatures lead to enhanced fluid–rock interactions, causing fracture permeability to change drastically, but such high reaction rates may be conducive to permeability enhancement if mineral dissolution dominates. Slates containing fractures, subjected to time-dependent closure, can possibly be mitigated by controlling injected fluid compositions. On the other hand, precipitation was not observed in our experiments, either due to permanent fluid under-saturation or as a result of the pore fluid exchange during the last permeability measurement before final cooling. This, however, may not be true in the field [13,14]. Dissolved minerals may increase the local permeability, where they dissolved, but may cause permeability reduction in far-field flow owing to precipitation at a large scale. Running an EGS in such metamorphic strata requires an understanding of the long-term evolution of the fractures in the host rocks, to which this study aimed to contribute. How to mitigate any fracture permeability reduction observed here, however, needs to be elucidated further.

5. Conclusions

To evaluate sustainability of fractures within slates, three long-term flow-through experiments with Wissenbach slate samples containing a macroscopic saw-cut fracture were conducted under constant pressures (i.e., P_c = 10 MPa and P_p = 1 MPa) and varying temperatures (room temperature up to 90 °C). Fracture permeability and effluent element concentrations were measured throughout the experiments. The results show that after applying effective pressure, the initial permeability reduction follows a power-law function during continuous flow, but this decrease slows down or terminates when flow is stopped. Temperature cycling causes an irreversible permeability decline when DI is used as the pore fluid, but permeability increases when the pore fluid is a 0.5 M NaCl solution. When fluid flow is intermittent, permeability shows a time-dependent reduction with DI as the permeating fluid, which is more pronounced at 90 °C compared to 70 °C. Again, permeability slightly increases when the sample is saturated with the NaCl solution. Ultimately, all samples yielded a certain and time-dependent permeability reduction. It is demonstrated that fracture permeability evolution in slates is controlled by pressure solution and free-face dissolution. Temperature cycles may affect fracture permeability by thermally enhanced fluid–rock interactions. NaCl accelerates the dissolution kinetics such that pressure solution is faster. However, pressure solution slows down drastically as the driving force (i.e., the normal stress on the contact asperities) decreases with time. The permeability degradation of fractured slates is similar to that of saw-cut granite fractures, which implies that slate reservoirs may be equally suitable for EGS as those in granites. However, for comparable effective pressures, the initial fracture aperture associated with the fracture surface roughness may determine how sensitive the aperture is to fluid–rock interactions in the long term.

Author Contributions: Conceptualization, C.C., J.H., E.R. and H.M.; methodology, C.C. and H.M.; investigation, C.C., J.H., J.A.S., E.R. and H.M.; experiments and measurements, C.C., J.A.S. and B.L.; writing-original draft preparation, C.C., B.W. and H.M.; writing-review and editing, J.H., B.L., J.A.S. and E.R.; supervision, E.R. and H.M. All authors have read and agreed to the published version of the manuscript.

Funding: This research was funded by EU—European Union's Horizon 2020 research and innovation program (MEET project, grant number: 792037).

Data Availability Statement: All data related to this study can be obtained from the corresponding author upon request.

Acknowledgments: The authors would like to thank Stefan Gehrmann for rock sample preparation, Sabine Tonn for fluid chemistry measurements with ICP-OES, Ilona Schäpan and Valerian Schuster for SEM measurements. Furthermore, Tanja Ballerstedt and Jasper Florian Zimmermann are acknowledged for their assistance with the apparatus preparation and updates. We are also grateful to David Charles Peter Peacock as well as three anonymous reviewers for their thoughtful and constructive comments that helped to improve the manuscript.

Conflicts of Interest: The authors declare no conflict of interest.

Appendix A

Fracture surface topographies before and after the entire experiments are shown below.

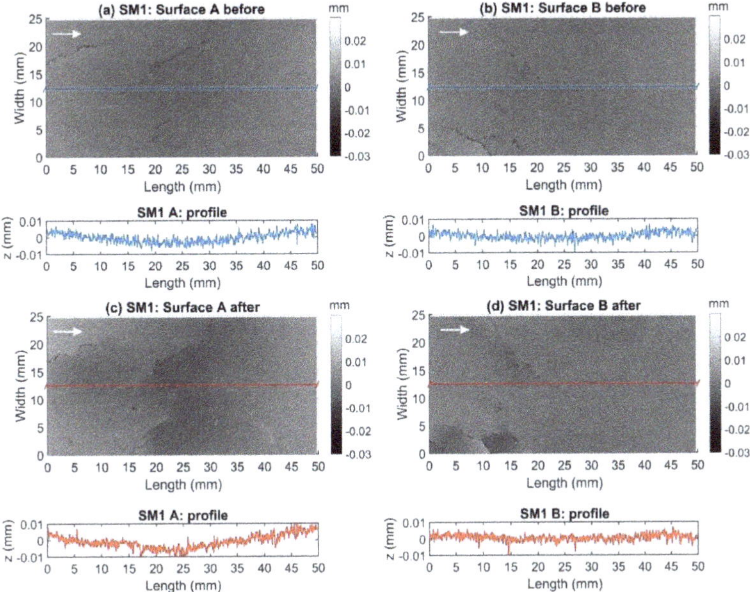

Figure A1. Fracture surface topographies and the central profiles of sample SM1 before (**a**,**b**) and after (**c**,**d**) the entire experiments, where the arrows indicate the flow direction.

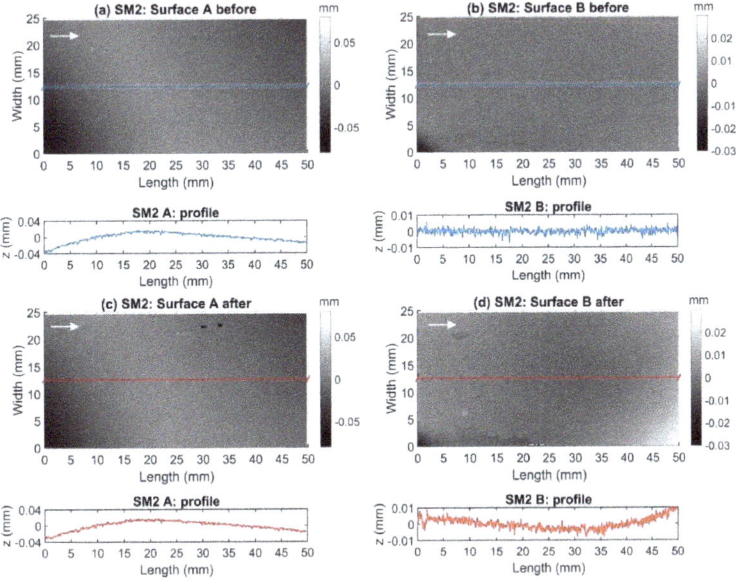

Figure A2. Fracture surface topographies and the central profiles of sample SM2 before (**a**,**b**) and after (**c**,**d**) the entire experiments, where the arrows indicate the flow direction.

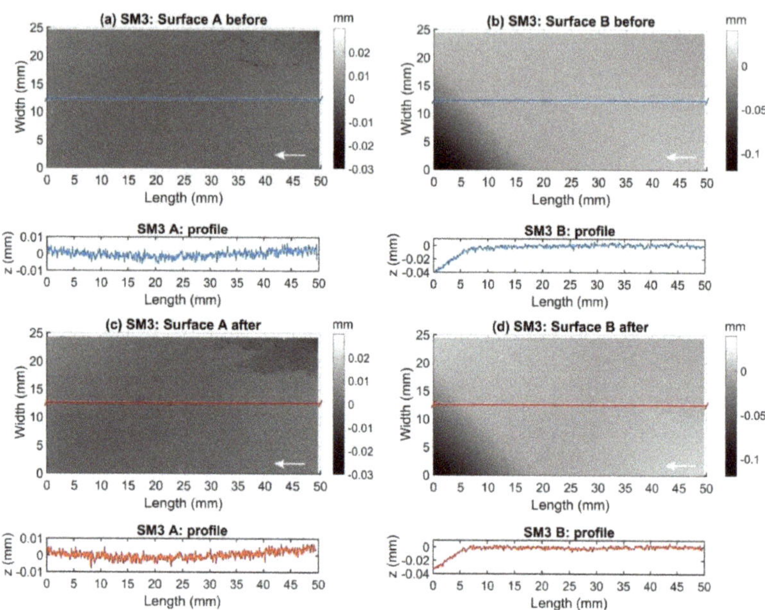

Figure A3. Fracture surface topographies and the central profiles of sample SM3 before (**a**,**b**) and after (**c**,**d**) the entire experiments, where the arrows indicate the flow direction.

References

1. Jolie, E.; Scott, S.; Faulds, J.; Chambefort, I.; Axelsson, G.; Gutiérrez-Negrín, L.C.; Regenspurg, S.; Ziegler, M.; Ayling, B.; Richter, A. Geological controls on geothermal resources for power generation. *Nat. Rev. Earth Environ.* **2021**, *2*, 5. [CrossRef]
2. Huenges, E. *Geothermal Energy Systems: Exploration, Development, and Utilization*; Wiley-VCH: Weinheim, Germany, 2010; Volume xxii, 463p.
3. Cheng, C.; Milsch, H. Hydromechanical investigations on the self-propping potential of fractures in tight sandstones. *Rock Mech. Rock Eng.* **2021**, *54*, 5407–5432. [CrossRef]
4. Hofmann, H.; Blocher, G.; Milsch, H.; Babadagli, T.; Zimmermann, G. Transmissivity of aligned and displaced tensile fractures in granitic rocks during cyclic loading. *Int. J. Rock Mech. Min. Sci.* **2016**, *87*, 69–84. [CrossRef]
5. Vogler, D.; Amann, F.; Bayer, P.; Elsworth, D. Permeability evolution in natural fractures subject to cyclic loading and gouge formation. *Rock Mech. Rock Eng.* **2016**, *49*, 3463–3479. [CrossRef]
6. Kluge, C.; Blöcher, G.; Barnhoorn, A.; Schmittbuhl, J.; Bruhn, D. Permeability evolution during shear zone initiation in low-porosity rocks. *Rock Mech. Rock Eng.* **2021**, *54*, 5221–5244. [CrossRef]
7. Milsch, H.; Hofmann, H.; Blocher, G. An experimental and numerical evaluation of continuous fracture permeability measurements during effective pressure cycles. *Int. J. Rock Mech. Min. Sci.* **2016**, *89*, 109–115. [CrossRef]
8. Dobson, P.F.; Kneafsey, T.J.; Nakagawa, S.; Sonnenthal, E.L.; Voltolini, M.; Smith, J.T.; Borglin, S.E. Fracture sustainability in Enhanced Geothermal Systems: Experimental and modeling constraints. *J. Energy Res. Technol.* **2020**, *143*, 1–37. [CrossRef]
9. Phillips, T.; Bultreys, T.; Bisdom, K.; Kampman, N.; Van Offenwert, S.; Mascini, A.; Cnudde, V.; Busch, A. A systematic investigation into the control of roughness on the flow properties of 3D-printed fractures. *Water Resour. Res.* **2021**, *57*, 4. [CrossRef]
10. Rutter, E.H.; Mecklenburgh, J. Influence of normal and shear stress on the hydraulic transmissivity of thin cracks in a tight quartz sandstone, a granite, and a shale. *J. Geophys. Res. Solid Earth* **2018**, *123*, 1262–1285. [CrossRef]
11. Gutierrez, M.; Oino, L.E.; Nygard, R. Stress-dependent permeability of a de-mineralised fracture in shale. *Mar. Petrol. Geol.* **2000**, *17*, 895–907. [CrossRef]
12. Durham, W.B.; Bonner, B.P. Self-propping and fluid flow in slightly offset joints at high effective pressures. *J. Geophys. Res.* **1994**, *99*, 9391–9399. [CrossRef]
13. Blöcher, G.; Reinsch, T.; Henninges, J.; Milsch, H.; Regenspurg, S.; Kummerow, J.; Francke, H.; Kranz, S.; Saadat, A.; Zimmermann, G.; et al. Hydraulic history and current state of the deep geothermal reservoir Groß Schönebeck. *Geothermics* **2016**, *63*, 27–43. [CrossRef]
14. Regenspurg, S.; Feldbusch, E.; Norden, B.; Tichomirowa, M. Fluid-rock interactions in a geothermal Rotliegend/Permo-Carboniferous reservoir (North German Basin). *Appl. Geochem.* **2016**, *69*, 12–27. [CrossRef]

15. Polak, A.; Elsworth, D.; Yasuhara, H.; Grader, A.S.; Halleck, P.M. Permeability reduction of a natural fracture under net dissolution by hydrothermal fluids. *Geophys. Res. Lett.* **2003**, *30*, 20. [CrossRef]
16. Polak, A.; Elsworth, D.; Liu, J.; Grader, A.S. Spontaneous switching of permeability changes in a limestone fracture with net dissolution. *Water Resour. Res.* **2004**, *40*, 40. [CrossRef]
17. Yasuhara, H.; Polak, A.; Mitani, Y.; Grader, A.; Halleck, P.; Elsworth, D. Evolution of fracture permeability through fluid–rock reaction under hydrothermal conditions. *Earth Planet. Sci. Lett.* **2006**, *244*, 186–200. [CrossRef]
18. Elkhoury, J.E.; Detwiler, R.L.; Ameli, P. Can a fractured caprock self-heal? *Earth Planet. Sci. Lett.* **2015**, *417*, 99–106. [CrossRef]
19. Yasuhara, H.; Kinoshita, N.; Ohfuji, H.; Lee, D.S.; Nakashima, S.; Kishida, K. Temporal alteration of fracture permeability in granite under hydrothermal conditions and its interpretation by coupled chemo-mechanical model. *Appl. Geochem.* **2011**, *26*, 2074–2088. [CrossRef]
20. Kamali-Asl, A.; Ghazanfari, E.; Perdrial, N.; Bredice, N. Experimental study of fracture response in granite specimens subjected to hydrothermal conditions relevant for enhanced geothermal systems. *Geothermics* **2018**, *72*, 205–224. [CrossRef]
21. Im, K.; Elsworth, D.; Fang, Y. The influence of preslip sealing on the permeability evolution of fractures and faults. *Geophys. Res. Lett.* **2018**, *45*, 166–175. [CrossRef]
22. Im, K.; Elsworth, D.; Wang, C. Cyclic permeability evolution during repose then reactivation of fractures and faults. *J. Geophys. Res. Solid Earth* **2019**, *124*, 4492–4506. [CrossRef]
23. Yasuhara, H.; Kinoshita, N.; Ohfuji, H.; Takahashi, M.; Ito, K.; Kishida, K. Long-term observation of permeability in sedimentary rocks under high-temperature and stress conditions and its interpretation mediated by microstructural investigations. *Water Resour. Res.* **2015**, *51*, 5425–5449. [CrossRef]
24. Cheng, C.; Milsch, H. Evolution of fracture aperture in quartz sandstone under hydrothermal conditions: Mechanical and chemical effects. *Minerals* **2020**, *10*, 657. [CrossRef]
25. Moore, D.E.; Lockner, D.A.; Byerlee, J.D. Reduction of permeability in granite at elevated temperatures. *Science* **1994**, *265*, 1558–1561. [CrossRef]
26. Morrow, C.A.; Moore, D.E.; Lockner, D.A. Permeability reduction in granite under hydrothermal conditions. *J. Geophys. Res.* **2001**, *106*, 30551–30560. [CrossRef]
27. Elsworth, D.; Yasuhara, H. Short-timescale chemo-mechanical effects and their influence on the transport properties of fractured rock. *Pure Appl. Geophys.* **2006**, *163*, 2051–2070. [CrossRef]
28. Yasuhara, H.; Elsworth, D. Compaction of a rock fracture moderated by competing roles of stress corrosion and pressure solution. *Pure Appl. Geophys.* **2008**, *165*, 1289–1306. [CrossRef]
29. Griffiths, L.; Heap, M.J.; Wang, F.; Daval, D.; Gilg, H.A.; Baud, P.; Schmittbuhl, J.; Genter, A. Geothermal implications for fracture-filling hydrothermal precipitation. *Geothermics* **2016**, *64*, 235–245. [CrossRef]
30. Aben, F.; Doan, M.L.; Gratier, J.P.; Renard, F. Experimental postseismic recovery of fractured rocks assisted by calcite sealing. *Geophys. Res. Lett.* **2017**, *44*, 7228–7238. [CrossRef]
31. Rutter, E. Pressure solution in nature, theory and experiment. *J. Geol. Soc.* **1983**, *140*, 725–740. [CrossRef]
32. Anderson, O.L.; Grew, P.C. Stress corrosion theory of crack propagation with applications to geophysics. *Rev. Geophys.* **1977**, *15*, 77–104. [CrossRef]
33. Atkinson, B.K. Subcritical crack growth in geological materials. *J. Geophys. Res.* **1984**, *89*, 4077–4114. [CrossRef]
34. Brantut, N.; Heap, M.; Meredith, P.; Baud, P. Time-dependent cracking and brittle creep in crustal rocks: A review. *J. Struct. Geol.* **2013**, *52*, 17–43. [CrossRef]
35. Cheng, C.; Milsch, H. Permeability variations in illite-bearing sandstone: Effects of temperature and NaCl fluid salinity. *J. Geophys. Res. Solid Earth* **2020**, *125*, 9. [CrossRef]
36. Farough, A.; Moore, D.E.; Lockner, D.A.; Lowell, R. Evolution of fracture permeability of ultramafic rocks undergoing serpentinization at hydrothermal conditions: An experimental study. *Geochem. Geophy. Geosyst.* **2016**, *17*, 44–55. [CrossRef]
37. Hilgers, C.; Tenthorey, E. Fracture sealing of quartzite under a temperature gradient: Experimental results. *Terra Nova* **2004**, *16*, 173–178. [CrossRef]
38. Sanchez-Roa, C.; Saldi, G.D.; Mitchell, T.M.; Iacoviello, F.; Bailey, J.; Shearing, P.R.; Oelkers, E.H.; Meredith, P.G.; Jones, A.P.; Striolo, A. The role of fluid chemistry on permeability evolution in granite: Applications to natural and anthropogenic systems. *Earth Planet. Sci. Lett.* **2021**, *533*, 116641. [CrossRef]
39. Kamali-Asl, A.; Ghazanfari, E.; Perdrial, N.; Cladouhos, T. Effects of injection fluid type on pressure-dependent permeability evolution of fractured rocks in geothermal reservoirs: An experimental chemo-mechanical study. *Geothermics* **2020**, *87*, 101832. [CrossRef]
40. Fazeli, H.; Vandeginste, V.; Rabbani, A.; Babaei, M.; Muljadi, B. Pore-scale modeling of fluid–rock chemical interactions in shale during hydraulic fracturing. *Energy Fuels* **2021**, *35*, 13. [CrossRef]
41. Farquharson, J.I.; Kushnir, A.R.; Wild, B.; Baud, P. Physical property evolution of granite during experimental chemical stimulation. *Geotherm. Energy* **2020**, *8*, 1–24. [CrossRef]
42. Durham, W.B.; Bourcier, W.L.; Burton, E.A. Direct observation of reactive flow in a single fracture. *Water Resour. Res.* **2001**, *37*, 1–12. [CrossRef]
43. Lima, M.G.; Vogler, D.; Querci, L.; Madonna, C.; Hattendorf, B.; Saar, M.O.; Kong, X.Z. Thermally driven fracture aperture variation in naturally fractured granites. *Geotherm. Energy* **2019**, *7*, 1–28. [CrossRef]

44. Lu, R.; Watanabe, N.; He, W.; Jang, E.; Shao, H.; Kolditz, O.; Shao, H. Calibration of water–granite interaction with pressure solution in a flow-through fracture under confining pressure. *Environ. Earth Sci.* **2017**, *76*, 12. [CrossRef]

45. Lu, R.; Nagel, T.; Shao, H.; Kolditz, O.; Shao, H. Modeling of dissolution-induced permeability evolution of a granite fracture under crustal conditions. *J. Geophys. Res.-Sol. Earth* **2018**, *123*, 5609–5627. [CrossRef]

46. Dewers, T.; Hajash, A. Rate laws for water-assisted compaction and stress-induced water-rock interaction in sandstones. *J. Geophys. Res.* **1995**, *100*, 13093–13112. [CrossRef]

47. Bijay, K.; Ghazanfari, E. Geothermal reservoir stimulation through hydro-shearing: An experimental study under conditions close to enhanced geothermal systems. *Geothermics* **2021**, *96*, 102200. [CrossRef]

48. Ishibashi, T.; McGuire, T.P.; Watanabe, N.; Tsuchiya, N.; Elsworth, D. Permeability evolution in carbonate fractures: Competing roles of confining stress and fluid pH. *Water Resour. Res.* **2013**, *49*, 2828–2842. [CrossRef]

49. Milsch, H.; Spangenberg, E.; Kulenkampff, J.; Meyhöfer, S. A new apparatus for long-term petrophysical investigations on geothermal reservoir rocks at simulated in-situ conditions. *Transp. Porous Med.* **2008**, *74*, 73–85. [CrossRef]

50. Kestin, J.; Sokolov, M.; Wakeham, W.A. Viscosity of liquid water in the range −8 °C to 150 °C. *J. Phys. Chem. Ref. Data* **1978**, *7*, 941–948. [CrossRef]

51. Kestin, J.; Khalifa, H.E.; Correia, R.J. Tables of the dynamic and kinematic viscosity of aqueous NaCl solutions in the temperature range 20–150 °C and the pressure range 0.1–35 MPa. *J. Phys. Chem. Ref. Data* **1981**, *10*, 71–88. [CrossRef]

52. Witherspoon, P.A.; Wang, J.S.; Iwai, K.; Gale, J.E. Validity of cubic law for fluid flow in a deformable rock fracture. *Water Resour. Res.* **1980**, *16*, 1016–1024. [CrossRef]

53. Zimmerman, R.W.; Bodvarsson, G.S. Hydraulic conductivity of rock fractures. *Transp. Porous Med.* **1996**, *23*, 1–30. [CrossRef]

54. Gratier, J.P.; Renard, F.; Vial, B. Postseismic pressure solution creep: Evidence and time-dependent change from dynamic indenting experiments. *J. Geophys. Res.-Sol. Earth* **2014**, *119*, 2764–2779. [CrossRef]

55. Ouyang, Z.; Elsworth, D. Evaluation of groundwater flow into mined panels. *Int. J. Rock Mech. Min. Sci.* **1993**, *30*, 71–79. [CrossRef]

56. You, Z.; Badalyan, A.; Yang, Y.; Bedrikovetsky, P.; Hand, M. Fines migration in geothermal reservoirs: Laboratory and mathematical modelling. *Geothermics* **2019**, *77*, 344–367. [CrossRef]

57. Bedrikovetsky, P.; Siqueira, F.D.; Furtado, C.A.; Souza, A.L.S. Modified particle detachment model for colloidal transport in porous media. *Transp. Porous Med.* **2011**, *86*, 353–383. [CrossRef]

58. Dove, P.M. The dissolution kinetics of quartz in aqueous mixed cation solutions. *Geochim. Cosmochim. Acta* **1999**, *63*, 3715–3727. [CrossRef]

59. Dove, P.M.; Crerar, D.A. Kinetics of quartz dissolution in electrolyte solutions using a hydrothermal mixed flow reactor. *Geochim. Cosmochim. Acta* **1990**, *54*, 955–969. [CrossRef]

60. Ruiz-Agudo, E.; Kowacz, M.; Putnis, C.; Putnis, A. The role of background electrolytes on the kinetics and mechanism of calcite dissolution. *Geochim. Cosmochim. Acta* **2010**, *74*, 1256–1267. [CrossRef]

61. Moore, D.E.; Morrow, C.; Byerlee, J. Chemical reactions accompanying fluid flow through granite held in a temperature gradient. *Geochim. Cosmochim. Acta* **1983**, *47*, 445–453. [CrossRef]

62. Morrow, C.; Lockner, D.; Moore, D.; Byerlee, J. Permeability of granite in a temperature gradient. *J. Geophys. Res.* **1981**, *86*, 445–453. [CrossRef]

63. Rimstidt, J.D.; Barnes, H. The kinetics of silica-water reactions. *Geochim. Cosmochim. Acta* **1980**, *44*, 1683–1699. [CrossRef]

64. Brantley, S.L. Kinetics of Mineral Dissolution. In *Kinetics of Water-Rock Interaction*; Springer: New York, NY, USA, 2008; pp. 151–210.

Article

Spring Water Geochemistry: A Geothermal Exploration Tool in the Rhenohercynian Fold-and-Thrust Belt in Belgium

Marina Cabidoche [1],*, Yves Vanbrabant [1],*, Serge Brouyère [2], Vinciane Stenmans [3], Bruno Meyvis [1], Thomas Goovaerts [1], Estelle Petitclerc [1] and Christian Burlet [1]

[1] Geological Survey of Belgium, Royal Belgian Institute for Natural Sciences, Rue Jenner 13, 1000 Brussels, Belgium; bmeyvis@naturalsciences.be (B.M.); tgoovaerts@naturalsciences.be (T.G.); epetitclerc@naturalsciences.be (E.P.); cburlet@naturalsciences.be (C.B.)
[2] Urban & Environmental Engineering Research Unit (UEE), ArGEnCo Department, University of Liège, Quartier Polytech 1, Bât. B52/3, Allée de la Découverte 9, 4000 Liège, Belgium; serge.brouyere@uliege.be
[3] VS.GEOFORMA, 1325 Chaumont-Gistoux, Belgium; vinciane.stenmans@gmail.com
* Correspondence: mcabidoche@naturalsciences.be (M.C.); yvanbrabant@naturalsciences.be (Y.V.); Tel.: +32-2-788-7635 (Y.V.)

Citation: Cabidoche, M.; Vanbrabant, Y.; Brouyère, S.; Stenmans, V.; Meyvis, B.; Goovaerts, T.; Petitclerc, E.; Burlet, C. Spring Water Geochemistry: A Geothermal Exploration Tool in the Rhenohercynian Fold-and-Thrust Belt in Belgium. *Geosciences* **2021**, *11*, 437. https://doi.org/10.3390/geosciences11110437

Academic Editors: Béatrice A. Ledésert, Ronan L. Hébert, Ghislain Trullenque, Albert Genter, Eléonore Dalmais, Jean Hérisson and Jesús Martínez-Frías

Received: 20 September 2021
Accepted: 15 October 2021
Published: 22 October 2021

Publisher's Note: MDPI stays neutral with regard to jurisdictional claims in published maps and institutional affiliations.

Abstract: Spring water geochemistry is applied here to evaluate the geothermal potential in Rhenohercynian fold and thrust belt around the deepest borehole in Belgium (Havelange borehole: 5648 m MD). Fifty springs and (few) wells around Havelange borehole were chosen according to a multicriteria approach including the hydrothermal source of "Chaudfontaine" (T ≈ 36 °C) taken as a reference for the area. The waters sampled, except Chaudfontaine present an in-situ T range of 3.66–14.04 °C (mean 9.83 °C) and a TDS (dry residue) salinity range of 46–498 mg/L. The processing methods applied to the results are: hierarchical clustering, Piper and Stiff diagrams, TIS, heat map, boxplots, and geothermometry. Seven clusters are found and allow us to define three main water types. The first type, locally called "pouhon", is rich in Fe and Mn. The second type contains an interesting concentration of the geothermal indicators: Li, Sr, Rb. Chaudfontaine and Moressée (≈5 km East from the borehole) belong to this group. This last locality is identified as a geothermal target for further investigations. The third group represents superficial waters with frequently high NO_3 concentration. The application of conventional geothermometers in this context indicates very different reservoir temperatures. The field of applications of these geothermometers need to be review in these geological conditions.

Keywords: blind geothermal system; compositional anomalies; hierarchical clustering; self-organizing maps; unconventional reservoirs

1. Introduction

Geothermal exploration, as any subsurface resource evaluation, requires a multidisciplinary approach (e.g., geological mapping, geophysics, geochemistry) and usually follows a downscaling approach starting from a regional scale toward a shortlist of potential sites for more detailed investigations [1,2]. These sites usually show surface indicators of the presence of a geothermal resource at depth. The nature of these indicators varies in the shape of hydrothermal springs, vapor exhalation, or sinter deposition. These indicators can even take the shape of biological manifestations as bacterial mats (e.g., [3–5], presence of bacterial community on travertine deposits [6] or even saline water tolerant plant and animal concentrations (e.g., eels) [7]). Even if such indicators are interesting clues in evaluating the geothermal fields, not all of them show obvious indicators at ground surface of their presence. An example of such blind geothermal fields is the Great Basin region in Nevada, which was recognized as a high-enthalpy important geothermal field without surface indicators [2,8]. Other blind geothermal systems were explored and/or developed in New Zealand, Hawaii, Indonesia [9,10]. Some of these blind geothermal systems were even discovered while targeting other resources (oil and gas, coal, minerals) for example in southern California [11], in Nevada [12], or in Belgium [13,14]. A common denominator of these blind geothermal systems is the presence of a cap formation, which hides many indicators of the presence of the geothermal resource.

The specificities and challenges to evaluate the geothermal potential of deep metasedimentary formations such as those encountered by the Havelange borehole (Belgium) are numerous: the Havelange borehole was drilled for gas exploration, hence, some important tasks/parameters for geothermal exploration were not conducted or recorded. The borehole site location is, therefore, fixed and not necessarily located in the most promising location for the development of a geothermal system in the region. The aforementioned downscaling approach frequently used in geothermal exploration cannot be applied. Another issue is the lack of information regarding the deep fluid flows and fluid composition since formation water inflows were thwarted by drilling mud injections as usual in drilling operation. It is important to note that two thick aquifer formations were encountered in this borehole [15]: (i) limestones and dolomites of Frasnian-Givetian ages (MD: 606–1595 m) and calcareous-sandstones of Eifelian (MD: 1610–1700 m); (ii) the fractured continuous quartzites of the Upper Praguian (MD: 4365–4554 m and 4639–4778 m). These two aquifers are separated from each other by more than 2000 m of aquiclude shales and slates of Emsian and Upper Praguian ages (from MD 1920 m to 4300 m). The aquifer character of the carbonate formations of Frasnian and Givetian is attested by significant mud losses during drilling and by few artesian events while adding drill rods. With regard to the aquifer and fractured Pragian quartzites, there are indirect indications that support this hypothesis: significant losses of drilling mud, sudden increases of mud temperature, poor Dipmeter log not showing consistent dips, Sonic log showing low speed peaks (fractures). The quarzitic Praguian, host rock for the geothermal reservoir, must be regarded as tight but permeable fracture zones that could represent potential geothermal targets. The logging data (gamma-ray, sonic, dipmeter) acquired during the 1980s provide only partial information regarding fracture networks, especially in comparison with modern logging data.

The punctual bottom hole temperature measurements indicate a mean moderate geothermal gradient of 22 °C/km but a sudden increase to 30 °C/km from 4400 m to the deepest parts of the borehole to reach a corrected value of 126 °C at 5369 m (MD-5277 VD) [15,16]. If the geothermal resource exists, it corresponds to a blind geothermal system with no obvious surface indicator of the resource.

Besides all these challenges, the existence of a deep borehole cross cutting a full (meta-)sedimentary sequence is not so common. All the cores, the cuttings, and most of the documents (e.g., logs) are still available and conserved by the Geological Survey of Belgium. The observations conducted thanks to this borehole provide important control to build a model of the deep structure of the region. Finally, the observation of significant mud losses during the drilling operations through Lower Devonian quartzite units and in the absence of borehole wall collapses, as indicated by the Caliper log, provides a promising key indication of the existence of deep fluid flows within the metasedimentary units in the region of the Havelange borehole. In this regard, the main target for the development of a deep geothermal reservoir is, therefore, these lower Devonian quarzitic units and a secondary target at shallower depth could be represented by Givetian/Frasnian limestone formations.

In this study, we apply one of the geothermal exploration tools to extend the investigation area surrounding the Havelange borehole by collecting spring and shallow borehole water samples. The geochemical composition of these samples aims to detect compositional anomalies as markers of deep fluid flows. The notion of anomaly is addressed in this paper. The signal of such flows requires, however, detecting a small anomaly, since deep-origin water is more likely mixed with superficial water during its ascending path. Our paper includes, first, the regional geological setting and key information regarding the Havelange borehole. The field sampling protocol and the analysed chemical compounds and physicochemical parameters are then listed. The data treatment and the main results are in turn presented by applying various clustering technics. Finally, we discuss the results through the application of common geothermometers to evaluate the potential reservoir temperature. This discussion also focuses on the applicability of these geothermometers in our study case; those thick meta-sedimentary formations in a fold-and-thrust belt.

2. Geological Setting

The study area and the Havelange borehole are located on the Eastern part of the Dinant Synclinorium, a sub-unit of the Ardenne Allochthon. These units compose the Belgian segment of the Rhenohercynian fold-and-thrust belt, which results from the accretion of the Rhenohercynian passive margin during the Variscan orogeny between the end of the Visean age to the Upper Carboniferous period (330–300 Ma) [17–20]. The Dinant Synclinorium (Figure 1) is bordered to the North by the trace of the Midi Thrust, which separates the Ardenne Allochthon from the Brabant Parautochthon [21].

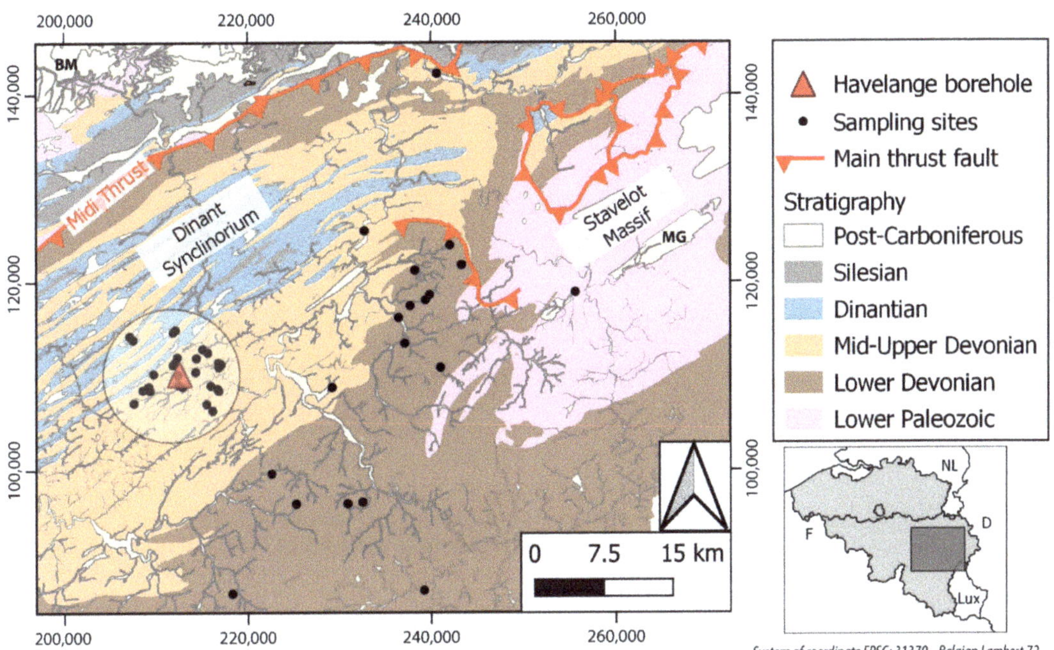

Figure 1. Simplified regional geological map around Havelange's borehole in Dinant synclinorium, Wallonia region (Eastern Belgium) and location of sampling sites for the water springs study. The Brabant Massif (BM) and Malmedy Graben (MG) are also visible in this map. The study near field (see text) is represented by a circle.

The main lithologies composing the Dinant Synclinorium are thick detrital Devonian formations combined with carbonate formations of Givetian, Frasnian, and Dinantian ages. These Devono-Carboniferous formations underwent during their evolution high diagenetic conditions and even green-schist facies conditions for the deepest formations (Lower Devonian) [22].

The Dinant Synclinorium passes eastward to the Lower Palaeozoic Stavelot Massif inlier. The contact between the metasedimentary units of the Stavelot Massif and the Lower Devonian formation of the Dinant Synclinorium is marked by an unconformity reflecting the action of a branch of the Caledonian Orogeny. The rock of the Stavelot Massif underwent, therefore, the Caledonian and Variscan orogenies.

The main faults in the region are South- to South-East dipping longitudinal thrust faults, as well as bulged faults forming the Theux and Gileppe windows to the North of the Stavelot Massif [23]. The fold structures are oriented NE-SW or even E-W depending of local conditions. Transversal or oblique faults crosscutting the folds and longitudinal thrust faults were formed either at the end of the Variscan orogeny or even during the poorly controlled age post-Variscan events. Some of these transversal and oblique fractures were filled by significant Pb-Zn deposits. Other transversal fractures are frequent targets as conduits for superficial groundwater resources, especially in carbonate formations.

Since the end of the Variscan orogeny, the tectonic activity in the study area has been quite limited. In the Stavelot Massif, the narrow Malmedy graben was formed, probably during the Permo-Triassic period. The whole region underwent a global uplift during the Quaternary. The current seismic activity in Southern Belgium is mainly located along the rim of the Brabant Parautochthon and in the Stavelot Massif [24]. Only a very limited number low magnitude events (Local Magnitude (ML) < 3.0) were recorded in the study area and the probabilistic model for a return period of 475 years predicts a maximum ground acceleration ranging between 0.04 and 0.06 g.

Following its geological characteristics, the Havelange borehole was selected in the framework of the H2020 MEET project as a demo-site to evaluate the geothermal potential of EGS development in a setting of Variscan meta-sedimentary units not affected by a younger extension period.

The Havelange borehole [16] was drilled for the Geological Survey of Belgium between January 1981 and November 1984 and reached a depth of 5648 m (MD). It aimed to investigate the potential natural gas resources under the Midi Thrust and, therefore, to drill through the Dinant Synclinorium to reach the deeply rooted Brabant Parautochthon southern extension. The site selection for the Havelange borehole results from the gas detection in another deep borehole (Focant) and the results of a seismic reflection campaign conducted in 1978. This seismic campaign indicates the presence of deep bulge-like reflectors interpreted as the presence of a local dome of the southern extension of the Midi Thrust. The methane gas inflows were too limited for any economic development program.

Briefly, the Havelange borehole encountered from top to bottom:

- Micaceous sandstone, siltstone, and shale units of the Famennian age;
- Frasnian and Givetian limestone, dolomite and shale formations;
- Sandstone, siltstone, slate-phyllite, and quartzite thick bed of the Eifelian, Emsian, Praguian, and Lochkovian stages.

The borehole allowed the detection of several thrust faults indicated by the repetition of stratigraphic units especially in the lower part of the borehole.

The main mineralogical phases encountered are: quartz > illite/muscovite > clinochlore > calcite > dolomite > pyrophyllite > feldspar > garnet > hematite > ilmenite, with obviously major variations according to the lithostratigraphic units. Another important property to mention is the high level of lithification of the studied formation. Even if Rhenohercynian fold-and-thrust resulted from the accretion of sedimentary basin, the involved lithology underwent a major lithification, reducing the porosity to a few percent and, hence, the connate water content is more likely very limited. This context differs, therefore, significantly from oil-gas sedimentary basins associated with significant brine contents.

The study area was the object of several studies to evaluate the subsurface temperature and heat flow conditions at a regional scale. The temperature values recorded in the Havelange borehole frequently serve as input data to calibrate the models [16].

Vandenberghe and Fock [25] reviewed the existing temperature measurements in wells from Belgium. They established a series of temperature distribution maps for a set of depths. They observed some trends such as the presence of a cold anomaly located in the northern part of the Stavelot Massif, while high temperature halos (~40 °C/km) are present in Western and Northern Belgium. According to their study, the Havelange borehole is located in an average geothermal gradient zone. The main limitation of such maps is that isotherms are based on widely spread data from boreholes separated to each other by a few tens of kilometres. The determined isotherms are extrapolated on a large scale without taking in consideration any thermal parameters. As a result, the isotherms are smooth and obliterate the presence of any potential thermal anomalies at a local scale (few km).

More recent approaches applied 2D numerical models to compute the subsurface temperature and the surface heat flows again at a regional scale. These studies [26,27] consider the thermal properties (i.e., thermal conductivity and radiogenic heat production) and other parameters such as the density and porosity from reference rock samples or from published data for crustal horizons located at great depth. These models also rely on a selection of published deep cross-sections to establish the structure and the composition of the upper crust. The comparison of the works of Rogiers et al. [26] and Schintgen et al. [27] is a complicated exercise since the models follow very different numerical approaches and try to answer different scientific questions. Rogiers et al.'s. [26] study aims to understand the presence of low heat flow anomaly detected in the shallow part of some boreholes in Belgium (Soumagne, Grand-Halleux, and Havelange). They consider the influence of groundwater flow at a shallow depth through various heat transfer mechanisms: conduction vs. heat advection. The influence of Quaternary paleoclimate changes is also applied through temperature variation of the model top boundary condition leading to a transient model. The lower part of the model is associated with a homogeneous and constant heat flow at depths ranging from 20 to 10 km. In the case of the Havelange site, the estimated basal heat flow would be 90 mW/m^2 at 14 km deep.

By contrast, the model by Schintgen et al. [27] aims to evaluate the heat flow with a specific focus on the situation in the Great-Duchy of Luxembourg, but the studied sections extend to the Stavelot Massif and in the neighbourhood of the Havelange borehole. Their model is a 2D steady-state conductive model. The model dimensions are also very different, since the lower boundary condition correspond to lithosphere–asthenosphere boundary (LAB) located at a depth ranging from 80 to 130 km and with a temperature of 1300 °C. The model of Schintgen et al. [27] also consider the presence of an Eifel plume in Germanym resulting in a LAB depth reduced between 40 and 60 km. The lateral extension of this plume would reach the Eastern border of Belgium [28,29]. The model results evaluate the heat flow at 1 km deep in the South-East part of the Dinant Synclinorium to a value of c. 80 mW/m^2, while in Rogiers et al. [26], the heat flow at a shallow depth of the Havelange borehole is smaller, c. 55 mW/m^2. This low value would result from a downward water flow at a shallow depth in the Havelange borehole. Besides the numerous differences between the approaches of Rogiers et al. [26] and Schintgen [30], both modelling predict significant heat flow variations at a local scale in the shallow part of the studied areas. In the approach of Schintgen et al. [27], these variations result from strong thermal conductivity contrasts observed for the rocks composing the upper crust, whereas the forced water advection in the shallow crust is the driving mechanisms of Rogiers et al.'s [26] model.

Recently, a review work on the origin of CO_2 content of Fe-rich soda springs, mainly located in the Stavelot Massif, indicates that the water is of meteoritic origin, but the CO_2-content results from the mixing of sources from hypothetic carbonate dissolution and from magmatic origins [31]. Their study focuses mainly on the isotopic signature of H, O, C, and He. The CO_2 magmatic source is regarded by Barros et al. [31] to be linked to the Eifel plume.

Most of the previous models rely on old observations acquired during the 1960s and 1980s, thanks to the exploration deep well campaigns led by the Geological Survey of Belgium. Some debate points between models result frequently because of the very limited numbers of information. An example of this issue is the use of a specific cross-section as a base for the numerical thermal models. However, these cross-sections are matter of debates and the choice of another section will fundamentally modify the results of the numerical models.

The approach followed in our paper aims to acquire new information using a cost-effective approach for the geothermal exploration thanks to the geochemical analyses of spring waters. The collected samples were acquired in two distinct regions. A first group of water samples was collected in a 5 km perimeter around the Havelange borehole head.

The subsurface composition of these springs is considered very similar to the geological formations observed in the shallow part of the borehole, that is, the mid- and upper-Devonian geological formations. This zone is referred to here as the "near field". The second group of samples were mainly acquired in the South and East borders of the Dinant Synclinorium where mid- and mainly lower-Devonian formations outcrops. The analogue formations were encountered in the deep part of the Havelange borehole. This second sampling zone is named, here, "the far field".

3. Materials and Methods

3.1. Sampling Campaign

The sampling campaign started with an initial deskwork phase for selecting potential target sites. This pre-selection includes the analysis of topographic, geological, and hydrogeological maps. A series of priorities were established based on the spring elevation, their position in a valley (e.g., lower part of the valley bank or at the valley head) or their vicinity with a fault. Springs located at an elevation lower than +250 m (Z) were considered as a priority, since this elevation corresponds to the static level of the Praguian and Lochkovian aquifer(s) in the non-cased section of the Havelange borehole. A site in the village of Moressée was also selected based on the toponymy of a spring: "La chaude Fontaine" (translated in English by "the hot fountain").

An additional site located at the north end of the study zone in the city of Chaudfontaine was chosen to collect a reference sample of hydrothermal water. The natural mineral water source (Source Astrid) captured in the Frasnian limestone at a depth of 396 m reaches a temperature of 36 °C.

The 50 samples were collected and conditioned by the team of the Geological Survey of Belgium (GSB) between the 13 November 2019 and the 5 March 2020. The ISO 5668–11 recommendations and the OFEFP guide [32] were followed during the actual water sampling. Each site was also the object a detailed description regarding its environment, its precise location, the water flow magnitude, the presence of infrastructures (e.g., metallic pipes, concrete walls, etc.). Physicochemical parameters of spring water (pH, EC, T) were also measured in each site using a multiparameter meters from Hanna Instruments.

During the field campaign, some of the pre-selected sites were not sampled as some springs were dry. The region underwent a series of drought periods especially during the three previous summers. Conversely, the field reality led us to samples other sites discovered during the campaign.

All the observations, field measurements, and analytical results were included in a relational database designed for this campaign. A total of 20 samples were collected in the far field study zone and 30 samples in the near field. A table with the data used in this article are retrievable in Supplementary Materials (Table S1).

3.2. Laboratory Analyses

The central laboratory of the "Société Wallonne des Eaux (SWDE)" performed the analyses on the water spring samples on behalf of the GSB. The following parameters were measured: pH, electrical conductivity at 20 °C (EC), colour (ISO 7887 (C Method)), turbidity (NTU) (nephelometric method), dry residue (TDS), and suspended solids (SSC) (ISO 11923).

The major and minor anions (SO_4^{2-}, Cl^-, NO_3^-, $o\text{-}PO_4^{3-}$, NO^{2-}, F^-, Br^-) were analysed by ion chromatography methods (ISO 10304-1, ISO 10304-4). The major and minor cations (Na^+, K^+, Ca^{2+}, Mg^{2+}, Ba^{2+}, Fe (total), U, Mn^+, Si^{4+}, Sr^{2+}), trace metals (Ag, Al, As, B, Be, Cd, Co, Cr, Cu, Hg, Li, Mo, Ni, Pb, Rb, Sb, Se, Sn, V, Zn) on unfiltered and filtered water were measured by Inductively Coupled Plasma Mass Spectrometry (ICP-MS) (ISO 17294-2). The concentrations of NH_4^+ and the complete alkalimetric strength (TAC) were analysed by Spectrophotometry.

Other parameters such as the Total Hardness (TH) and the TOC (Total Organic Carbon) (ISO 8245) were also acquired. The presence and the distribution between

$HCO_3^-/CO_3^{2-}/$ dissolved CO_2 are deduced from the values of the Alkalimetric strength $(TA) = (OH^-) + (CO_3^{2-})$, the $TAC = TA + (HCO^{3-})$, and pH. The details of the laboratory parameters and LOQ are presented in the additional material (Table S2).

3.3. Data Treatment Methods

The interpretation of the water spring analyses made in this paper is based on the filtered water results. The use of filtered water (filtration at 0.45 microns) allows to concentrate the interpretation on the dissolved element in water and avoid the micro particles present in non-filtered water (e.g., clay).

Before statistically processing the data, the charge balance equilibrium (CBE) of the analytical values of the 50 water samples (converted from mg/L to meq/L) were calculated, using the formula: $(CBE(\%) = ((Scat - San)/(Scat + San)) \times 100$; S, cat, and an mean total, cation and anion, respectively [33]. Applied on one hand to major ions only (those used for Piper Diagram) and on the other hand to major and some minor ions (those mentioned in annual report of SPW-Agriculture, Ressources Naturelles et Environnement: F, PO_4, NO_2–Sr, Ba, Al, Fe, Mn [34], the resulting CBE are ranging as follows: (0.02–9.89%) with an average of 3.47% for MAJOR and some MINOR ions of 49/50 water samples; only one sample with CBE > 10% sample 32 at 28.10%.

The Piper diagrams [35] and the Stiff diagrams (e.g., [36]) were also made in this unit of measurement. The total ionic salinity (TIS) with Na + K vs. Ca + Mg is expressed in meq/kg [37]. The nitrate concentration values due to potential anthropic contamination of the water composition, from superficial water samples especially, are not included in the Piper diagrams. Finally, a hierarchical clustering method was applied and specifics elements of hydrogeological or geothermal interest (Li, Rb, pH, NO_3, etc.) were selected to be represented as boxplots. In addition, several maps were made to show the locations of the clusters and the concentration in particular elements with the use of the Jenks natural breaks classification methods.

3.3.1. Hierarchical Clustering Methods

The geochemical analyses acquired during this campaign are the results of several cases of the water evolution during its subsurface transit. The latter can be short (e.g., superficial water in and out-flow during a brief period) or long (deep fluid circulation or shallow low-permeable system). The geological setting of each spring can also be influenced by numerous parameters, such as the mineralogical composition of the rocks and the specific dissolution rate of their mineral, the physicochemical parameters of water during its interactions with the minerals, amongst others. The mixing of water masses that followed different subsurface paths certainly represent a key factor influencing the spring water composition. Finally, external parameters influencing the spring water composition are the climatic conditions (rain precipitation regime) or possible natural or anthropic contaminations (e.g., mineral deposit interactions, road de-icing, or agricultural spreading). Hence, the challenge is to detect the key trends and key parameters allowing us to detect trends in a multivariate system. In this study, we have chosen to apply the hierarchical clustering method to discover a structure in the data set. The hierarchical clustering method was applied on all 50 samples and for a set 37 parameters, including the physicochemical parameters (e.g., pH, EC, TAC), the concentration of the major anions, and trace metal cations. The data set is organized in a matrix of 50 rows and 37 columns. In a first step, the Euclidean distance is computed between each row (i.e., sample) of the matrix in a m-dimension space, with m equals to 37. The output of this first operation is a distance matrix, which represents the input data for the hierarchical clustering. Concretely, the distance matrix and the hierarchical clustering were computed by applying the dist and the hclust functions from the base 'Stats' Package from the R language [38]. The hierarchical clustering outputs are commonly represented by a dendrogram where the distances between the samples is represented by a difference of height between the tree limbs.

3.3.2. Jenks Natural Breaks Classification Methods

The Jenks natural breaks classification method is used to create a map based on any kind of data with spatial attribute. This method conducts natural groupings (clusters) inherent to the chosen data (e.g., the concentration in one element). The breaks are determined to find the best group with similar values and with a marked difference between classes. This method pursued to minimize the variance within classes and maximize the variance between classes [39]. In the GIS application, the number of desired classes in a result set for the Jenks method must be added before the algorithm is applied on the dataset. In this paper, the reference separation was chosen at five, after different tests.

The Jenks method is applied in this study to create thematic maps based on the element concentration (Li, Rb and Sr) with QGIS software. It allows us to see the natural break in the concentration dataset for each element.

4. Results

4.1. Hierarchical Clustering of Water Analyses

The application of the hierarchical clustering on the spring water dataset (Figure 2) indicates that seven clusters can be reasonably distinguished. Cluster 1 contains only one sample (n°33) collected in the lower Palaeozoic Stavelot Massif in the vicinity of the Permian graben of Malmedy. This sample in the dendrogram occupies its own limb with a significant distance with all other samples. Clusters 2 to 4 include two samples each, whereas cluster 5 has only one sample (n°25). As we move down in the dendrogram, the height between the limbs is decreasing, reflecting tighter clusters or, in other words, samples with compositions more and more similar. The last two clusters (6 and 7) include 14 and 28 samples, respectively, and encompass the bulk (84%) of the collected samples.

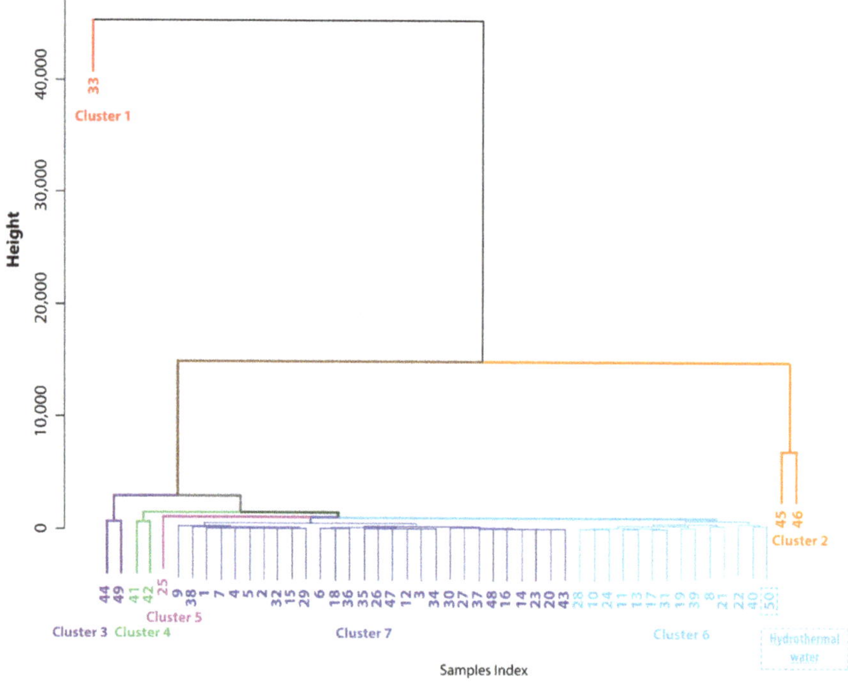

Figure 2. Dendrogram representing the result of the hierarchical clustering (R program). The colours represent the attribution of samples to their cluster. Note that the hydrothermal water from Chaudfontaine (n°50) belongs to cluster 6.

In this analysis, the hydrothermal reference sample (n°50) from Chaudfontaine is classified within cluster 6 along with 13 other samples, which show, therefore, a degree of affinity in their chemical composition with the hydrothermal reference one. A corollary of this observation is that the composition of samples from clusters 1 to 4 clearly deviates from the bulk of collected samples and from the hydrothermal reference sample. At this stage of the analysis, the results indicate that in terms of anomalies, the study area is, thus, characterized by several types of springs with "abnormal" compositions.

The samples from clusters 1 to 3 and cluster 5 (Figure 3) come from sites located in the Stavelot Massif and mainly the Eastern rim of the Dinant Synclinorium, which is the study zone "far field". In this region, the surface and subsurface formations consist primarily of Lower Palaeozoic and Lower Devonian detrital meta-sedimentary rocks (phyllite/slate, sandstone, and quartzite). By contrast, samples from cluster 4 originate from a very limited zone located 4 km to the East of the Havelange borehole site. The samples of cluster 6 were collected mainly in the near field with two other samples located on the west edge of the far field. As already stated, the reference hydrothermal water from Chaudfontaine is located at the north end of the study zone. Finally, the samples of cluster 7 cover indiscriminately the entire study zone with a common characteristic of a low salinity (TDS ranging from 46 to 276 mg/L and TIS < 9.25 meq/kg (see Figure 8 in point 4.3).

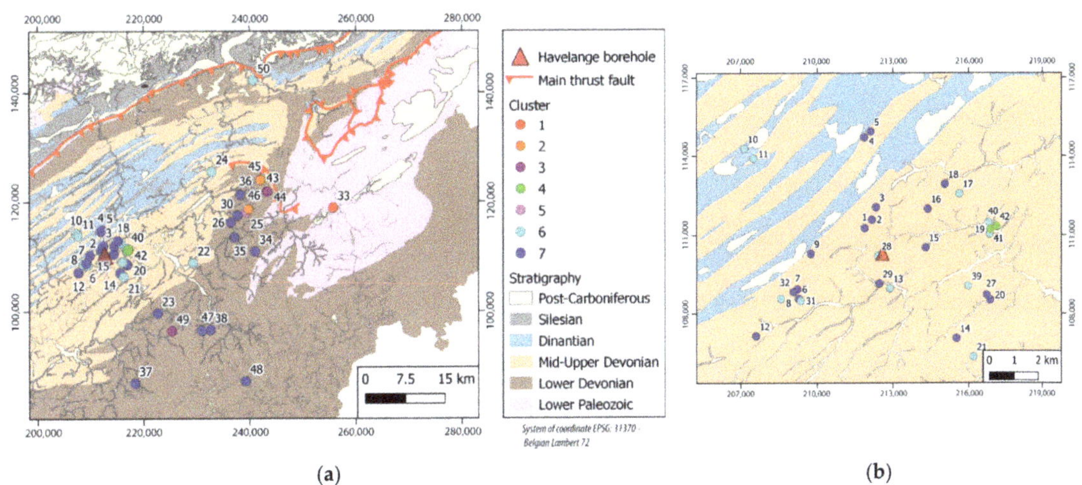

(a) (b)

Figure 3. (**a**) Geological map located the samples in the entire study zone according to their cluster attribution; (**b**) Zoom on the location of the samples with their cluster attribution in the near field around the Havelange borehole.

4.2. Heat Map and Boxplots

Even if the hierarchical cluster provides valuable information to detect compositional affinities/dissimilarities between samples, it does not directly provide the key compositional characteristics of each cluster. Regarding cluster 1, the examination of laboratory results indicates a very high concentration of iron (45,052.0 mg/L) and manganese (4321.3 mg/L), while the mean iron concentration for the dataset reaches 1462.7 mg/L and 189.5 mg/L for manganese and the median values are only 5 mg/L for Fe and 1.25 mg/L for Mn.

For the other clusters, the set of elements characterizing them is less obvious to detect. Figure 4 presents a heatmap on the same dataset as for the hierarchical clustering. Note that parameter values or concentrations for each column are normalized to avoid the predominance of very high values (e.g., iron or manganese concentration) with respect to lower values as those of trace elements. The study of the heatmap indicates three broad horizons of values. From bottom to top, one can distinguish a lower horizon gathering samples from clusters 1 to 5 that are characterised by high concentration of numerous elements, such as Fe, Mn, Ni, Be, Mg, Li. These samples have usually a low concentration of NO_3, Se, Sb. The EC and TDS are usually high, and the pH level is variable. Some sub-horizons can be distinguished with for instance the samples from cluster 4 (sample n°41 and 42), which show higher levels of Sr, Mo, and Sb. The central horizon of the heatmap (from samples 30 to 2 along the sample index axis) contrasts with, globally, a low concentration in several elements: Li, Rb, Ca, Na, K, Mo, U, HCO_3. These low levels are also reflected by low EC and TDS values. At least a part of these samples is also affected by high nitrate levels. This central horizon is occupied by samples from cluster 7. Finally, the top heatmap horizon, which corresponds to cluster 6, shows, again, higher levels in numerous elements, but not the same as in the lower horizon. The high levels of the top horizon include Bromide, Mo, U, V, Se, Sb, Cu, TOC. The EC, TDS, and pH levels are usually higher than in the central horizon.

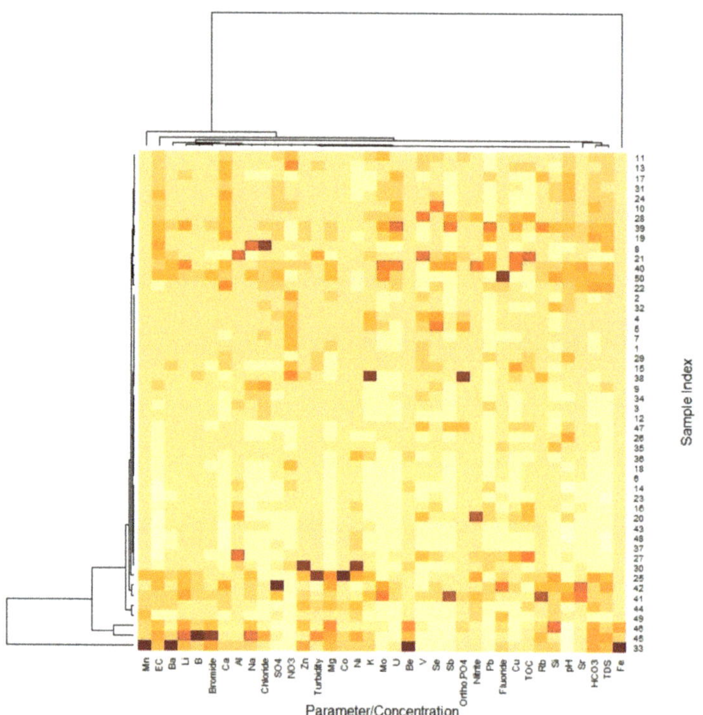

Figure 4. Cluster heatmap of the dataset of §4.2. The water samples are organized according to the hierarchical clustering along the vertical axis. Light yellow colours represent low-value parameters/concentrations, whereas the high concentrations/parameters are expressed as darker colours (orange, red, brown). Note that the parameter/concentration list is also reshuffled to highlight the similarities between the variables.

The high nitrate levels from spring water in areas with intense agricultural activities such as in the study area are commonly regarded as a signal of anthropic contamination of the aquifers. For our study, we can consider that high NO_3 levels is a good indicator of important influx of superficial water within the aquifers drained by the springs. Clusters 5 to 7 (Figure 5a) gathers, therefore, springs with water from superficial flow or at least a mix of superficial water with deeper aquifers. Interestingly, the water from Chaudfontaine well occupies the minimum quartile (Q1) of cluster 6, indicating the absence of any significant nitrate input in this hydrothermal system.

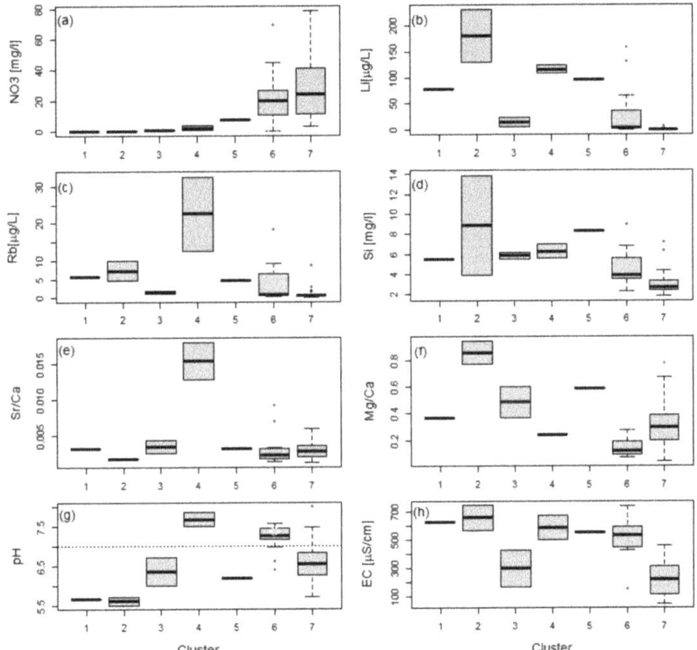

Figure 5. Boxplots presenting the concentration of (**a**) NO$_3$; (**b**) Li; (**c**) Rb; (**d**) Si; (**e**) Sr/Ca; (**f**) Mg/Ca; (**g**) pH; (**h**) EC according to the clusters. The geothermal reference water sample (n°50) is pointed by a cyan cercle in cluster 6 and a horizontal line marked the neutral pH.

Li, Rb, and Si are amongst the numerous elements that are commonly analysed to detect and to characterize the presence of geothermal water. In our dataset, the higher levels of lithium (Figure 5b) are observed in Clusters 1, 2, 4, 5, and the upper part of cluster 6. By contrast, water samples from cluster 3 and mainly 7 are lacking significant Li concentrations. A similar observation can be applied for the Rb concentration (Figure 5c). Even if the Li and Rb levels observed in cluster 3 are quite similar with those of clusters 6 and 7, the Si concentration of samples from cluster 3 is clearly larger than in the other two clusters (Figure 5d).

Element level ratios are also frequently evaluated to characterize the aquifers, such as [Sr]/[Ca] and [Mg]/[Ca] (Figure 5e,f). The results indicate that cluster 4 is characterized by high values of [Sr]/[Ca] with respect to all other clusters, while it is more difficult to detect a clear trend for the ratio [Mg]/[Ca].

The analysis of the pH levels (Figure 5g) shows that water from clusters 4 and 6 is mainly composed by water with a pH above 7, while the other clusters are primarily associated with slightly acidic water. Finally, the EC of clusters 1, 2, 4, to 6 have high values, while clusters 3 and 7 exhibit low or intermediate electrical conductivity (Figure 5h).

4.3. Water Composition

The Piper diagram of the filtered samples (Figure 6) shows that 86% of the spring water samples can be classified into three main categories, namely the "Magnesian Calcium", "Calcium and Magnesian bi-carbonate", and "Chlorinated and sulphated calcium and magnesian" types. It is observed that the samples rarely belong to a type containing high concentration of sodium and none of them are of "magnesium" type or in "sulfate" type.

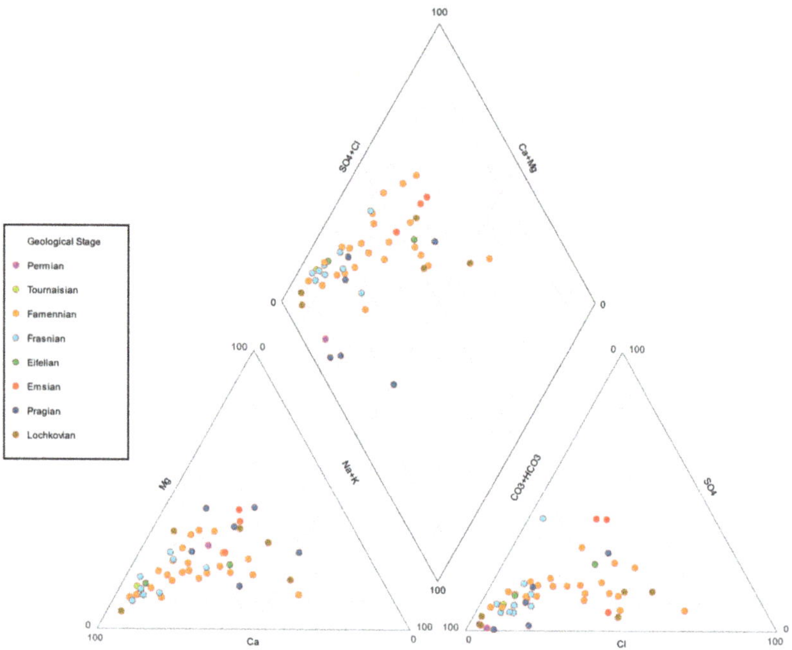

Figure 6. Piper diagram of filtered water spring analyses combined with the surface geology.

A total of 23 water samples were collected from sites where the Famennian detrital formations outcrop. Their major dissolved constituents exhibit a broad spectrum of compositions, including the three aforementioned main water types (i.e., Ca + Mg, Ca(Mg)HCO$_3$ and Ca(Mg)Cl(SO$_4$)). The dominant cation is Ca or there is no dominant cation, whereas the bicarbonate anion is dominant or no dominant anion can be identified. The nine samples from the outcropping zone of carbonate and shale formations of Frasnian age are characterized by a composition of mainly Ca + Mg type with Ca and HCO$_3$ as the dominant ions. The three samples from Emsian formation indicate a sulphated composition.

The Praguian meta-sedimentary rocks (phyllite, siltstone, sandstone, and quartzite) are of great interest in this study since they represent the main target for the geothermal development of the Havelange demo-site and they represent a cumulative apparent thickness by tectonic staking of 1565 m in the borehole. The six spring water samples from the Praguian Stage show a much broader distribution of compositions with bicarbonate as the main anion, but without any dominant cation. However, three samples from the Praguian and one from the Permian are located on the lower triangles (Na-K-HCO$_3$ type). This type can indicate a cation exchange (Na, Ca) carried out by a deep-water circulation over a long period in clay facies. It is hazardous for the other Geological Stages to detect a clear trend or to evaluate the composition coverage due to the very limited number of samples per stage.

This broad distribution of compositions and their rather weak relationships with the Geological stages reflect the difficulty to associate the composition of a spring water with its surface geology. As we will discuss in the following parts of this paper, numerous springs discharge superficial water masses with a short subsurface transit period from the recharge zone and without enough time to reach a real equilibrium with their host rocks. On the other hand, water volumes flowing from deep units are likely to be in contact with several formations of different lithologies. The attempt to link the spring water composition of the main elements with the surface geology must be regarded as an oversimplification.

The second Piper diagram represent the 50 water samples according to the defined clusters (Figure 7). For samples from clusters 1 to 5, it is not possible to define a trend due to the very limited number of samples per cluster (e.g., 1 or 2). However, samples from clusters 1, 2, and 5 are located close to perimeter of $HCO_3 + CO_3$ type water. By contrast, many samples from cluster 6 are located along the Ca + Mg water type or within the $Ca(Mg)HCO_3$ domain. Finally, the water samples for which there is an indication of superficial subsurface flow (cluster 7) spread over a broad zone, including $Ca(Mg)HCO_3$ and $Ca(Mg)Cl(SO_4)$ water types.

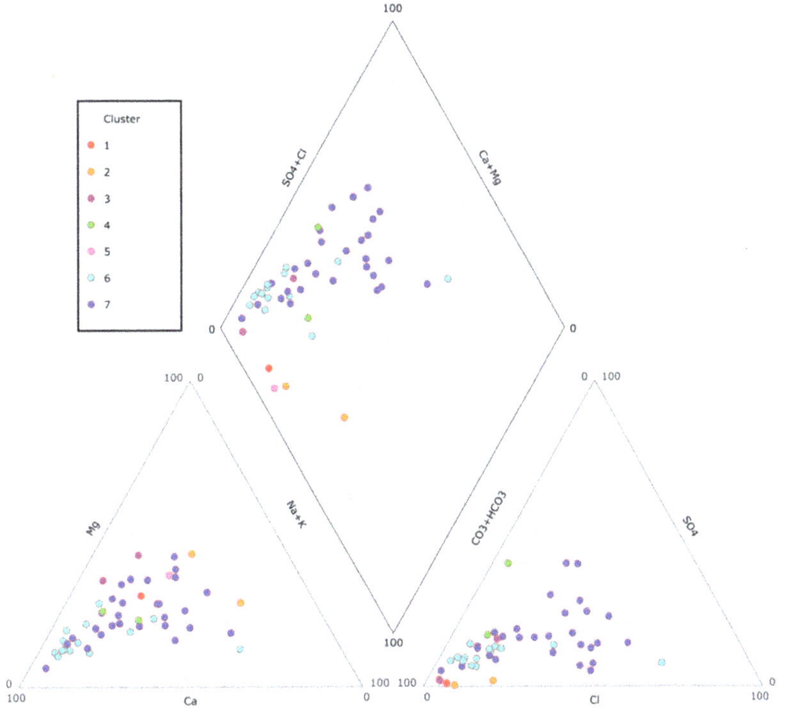

Figure 7. Piper diagram of filtered water spring analysis with hierarchical clusters (mEq) (R program).

If we compare the Piper diagram of surface geology (Figure 6) with the one of clusters (Figure 7), two relationships can be detected. First, the spring water samples collected from sites where the surface geology belongs to Famennian formations are frequently attributed to cluster 7, namely the group of superficial waters. This relationship indicates that the shale formations of Famennian age act as aquiclude leading springs discharging only very superficial water masses. A corollary of this statement is that the same shale act as a cap rock for the deeper water masses. The second relationship is indicated by a degree of correlation between water samples from cluster 6 with samples collected in the outcrop zones of the Frasnian carbonate and shale formations.

Furthermore, the spring waters show low total ionic salinity (TIS) values between 1 and 16.25 meq/kg (Figure 8). More specifically, the lowest TIS < 9.25 meq/kg are associated with cluster 7 (identified as superficial water). All the other samples, except one from cluster 6, have a TIS > 9.25 meq/kg. In addition, cluster 6 (which contains hydrothermal water) presents a higher concentration in Ca + Mg than in Na + K. On the other hand, one sample from cluster 2 and one from cluster 6 show higher concentrations in Na + K than in Ca + Mg.

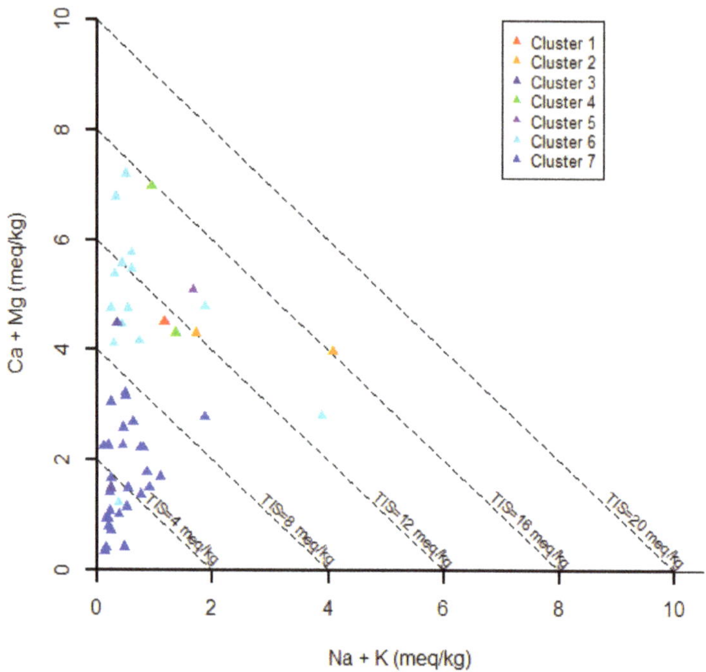

Figure 8. Correlation plot of Na + K vs. Ca + Mg (meq/kg) of filtered water spring analysis with hierarchical clusters also presenting iso-ionic-salinity (TIS) lines for reference (R program).

4.4. Stiff Diagrams and Clusters

The Stiff diagrams presented in this paper are built with the median values of the main constituents from each cluster expressed in meq/L (Figure 9).

The clusters 4 and 6 are the ones that show the most typical shapes for bicarbonate calcic groundwaters. However, cluster 4 shows enrichment in SO_4 compared to cluster 6. The clusters 1, 5 (one sample each), and the cluster 2 are characterised by an asymmetric polygon due to an elevated concentration of bicarbonate (around 7 meq/L) in comparison to cluster 4 and 6. This might reflect a deeper source of CO_2 in these waters. Moreover, cluster 2 shows a relative depletion of Ca compared to Na + K and Mg. This probably reflects the occurrence of cation exchange processes affecting groundwater, which has circulated deeper, in contact with clay materials. Cluster 3 presents intermediate mineralization with low indication of anthropic contamination (low Na, Cl and NO_3). This might be the result of groundwater circulation through geological units that contain a relatively low level of potassium. Finally, the cluster 7 displays the lowest global mineralization, reflecting surface waters or very shallow groundwaters.

The Stiff diagrams of all the samples are available in additional material (Figure S1).

4.5. Anomalies and Elementary Maps

The previous heatmap (see Figure 4) highlights three main horizons in the dataset. Each horizon reflects a group of samples with similar constituents' concentration and/or similar physicochemical parameters. At the scale of a single sample the concentration in a specific element can be regarded in some cases as an anomaly within its heatmap horizon. Sample n°5 has for instance a higher Se level with respect to all other samples in the central heatmap horizon. Such deviation from the trend/horizon, that is an anomaly, can result from a water transit through a specific subsurface environment such as the presence of mineral concentration. The evaluation of anomalies in the composition of water can be conducted in different ways. For this study, the first approach consists of the application of the Jenks natural breaks with five classes for a set of 16 elements. A sample is considered with anomalous concentration in a given element if its concentration belongs to the two upper Jenks classes. Table 1 presents the results of detected anomalies.

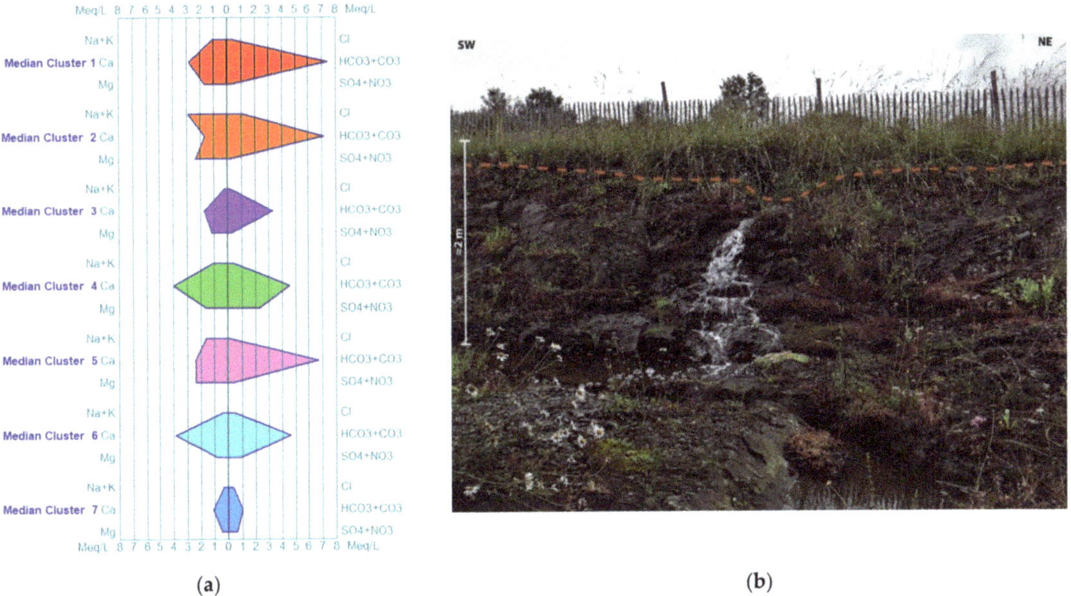

(**a**)   (**b**)

Figure 9. (**a**) Stiff diagrams of spring water sampling interpretation made with the free software "diagrams" created by Roland Simler in the hydrogeology laboratory of Avignon (France); (**b**) Picture of Jalna site (M. Cabidoche) (Sample n°27) illustrating a superficial water flowing from a spring located at the interface between the colluvium and Famennian shale and siltstone units. This interface is shown by red dashed line.

Nearly half (22) of the collected samples exhibit a least one element with a concentration anomaly. In this analysis, samples n°39, 33, 40, and 45 show the highest number of anomalies and the elements reported more frequently as anomalies are Li, Si, Mn, and Mo.

Surprisingly, some anomalies are detected even in samples related to superficial water (cluster 7). The explanation for those particular cases are not easy to address, but at least in one case (sample n°15), its copper anomaly can be linked to the presence of a former Pb-Zn-Cu mine near the village of Heure [40,41].

Another approach to evaluate the presence of abnormal compositions is to compare the measured concentrations with literature values of element concentration in aquifers. For instance, six samples in the dataset have a Co concentration > 0.33 µg/L, which is a level well above the average value generally reported for spring waters [42]. In the same way, five samples have a Ba concentration > 100 µg/L [43] and seven samples have a concentration Mn > 200 µg/L [44]. In this paper, the focus is made on Li, Rb, and Sr concentrations in the water spring samples (Figure 10).

4.5.1. Lithium

The concentration of lithium in fresh waters usually ranges between 1 µg/L and 20 µg/L [45]. In this study, 11 samples (Table 2) have a concentration of Li above 20 µg/L. Lithium deposits or concentration are usually found in salars or in pegmatites and less frequently they are found in clays and Li-rich micas. There is no observation of salars or pegmatites in the region and, hence, the Li source comes probably from clay minerals. However, mineral occurrences of Lithiophorite ((Al,Li)$MnO_2(OH)_2$) with possible traces of

Co, Ni, Cu, and Zn are reported in the mines of Rahier, Bihain, Vielsalm, and Malempré in the Stavelot massif. No similar occurrence has been reported so far in the Devonian formations. One of the hypotheses is that the high concentration of lithium in the MEET samples could come from deep fluids that crossed Devonian formation.

Table 1. Summary table of samples showing anomalies (higher than the Jenks mean concentration value for each chemical element).

Sample	Cluster	Li	B	Si	Mn	Fe	Co	Ni	Cu	Zn	Br	Rb	Sr	Mo	Ba	Pb	U	Total
10	6																X	1
11	6													X				1
15	7								X									1
17	6																X	1
19	6															X	X	2
21	6								X							X		2
22	6												X					1
25	5		X	X	X		X	X		X								6
27	7								X									1
30	7							X		X								2
33	1		X		X	X	X			X	X				X			7
35	7			X														1
36	7							X										1
39	6	X		X					X			X		X	X	X	X	8
40	6	X							X				X	X	X	X	X	7
41	4	X	X									X	X	X				5
42	4	X		X								X	X	X				5
44	3				X			X		X								3
45	2	X	X		X	X	X	X			X							7
46	2	X	X		X	X				X	X							6
49	3			X														1
50	6			X	X						X		X	X	X			6
Total		6	5	6	6	3	3	5	5	5	4	3	5	6	4	4	5	

Table 2. Table of MEET samples with the concentration Li, Sr, Rb: [Rb] > 7.6 µg/, [Li] > 20 µg/L or [Sr] > 1000 µg/L.

Sample n°	17	25	33	38	39	40	41	42	44	45	46	50
Li (µg/L)	36.6	96.4	78.5		131.6	158.9	123.8	107.6	23.3	229.9	129	65
Rb (µg/L)	9.3			8.6	18.4		32.6	12.5		9.9		7.6
Sr (µg/L)							1110	1280				

4.5.2. Rubidium

Rubidium is usually found in potassium minerals such as lepidolite, biotite, and feldspar. In natural groundwater, the Rb concentration is around 0.1 to 100 µg/L [46]. In this study, all the samples are below the highest reference values for spring water 100 µg/L. The referential sample n°50 (hydrothermal spring) shows only 7.6 µg/L of Rb. Six samples are above the referential sample n°50 and three of them (samples n°42, n°39, and n°41) are higher than 10 µg/L (Table 2). Four sampled sites are located in the near field (samples n°17, n°39, n°41, n°42) to the East of the Havelange borehole and two others are in the far field (samples n°38 and n°45). These samples are mostly located in shale, siltstone, and sandstone formations.

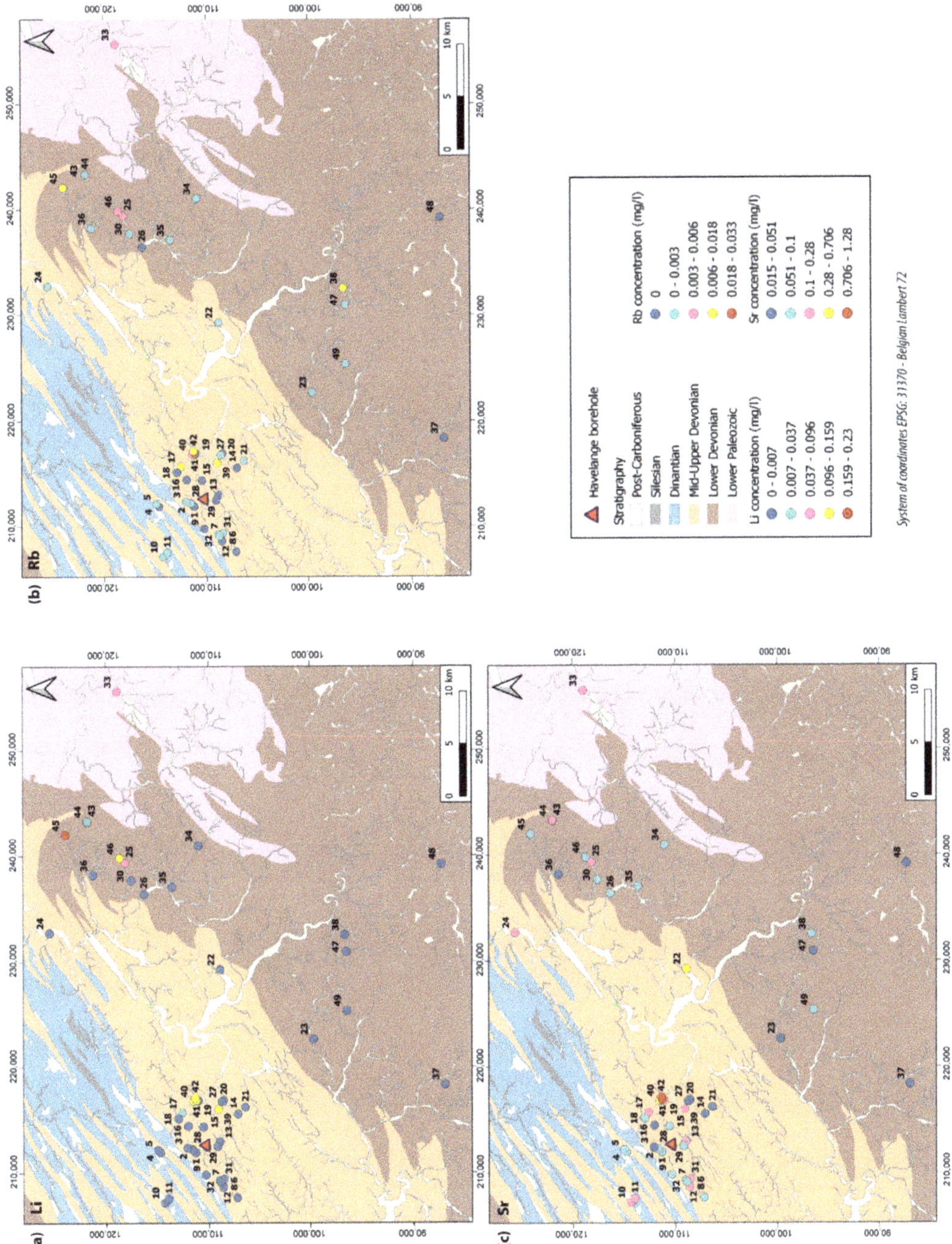

Figure 10. Maps representing the concentration in (**a**) Li, (**b**) Rb, and (**c**) Sr, following a separation in five classes with the Jenks natural breaks. The high lithium levels are observed in a few sites located in the far field, but also in a zone located 4 km to the East of the Havelange borehole (near field) near the villages of Moressée and Heure. Very similar observations can be conducted for the Rb and Sr concentrations.

4.5.3. Strontium

Strontium enters into the groundwater composition during the leaching of limestone, igneous and metamorphic rocks, especially in granites and sedimentary rocks, as hydrated Sr^{2+} [47]. The concentration of strontium in fresh water is generally present as traces below 1000 µg/L [48]. The referential sample n°50 (hydrothermal spring) shows only 455.1 µg/L of Sr.

In the spring water from this study, two samples (n°41 and n°42) are above 1000 µg/L (Table 2). The high concentration of strontium for these samples could come from the leaching of Givetian or Frasnian limestone formations. Moreover, a strontium occurrence (Celestine) has been described in the literature near the village of Verdenne, on the East of Marche-en-Famenne city.

5. Geothermometry

One of the main goals of geochemical exploration through the evaluation of spring water composition aims to quantify or at least to estimate the thermodynamic conditions encountered by water during its path from the recharge area to the potential geothermal reservoir and in turn in discharge zone. Numerous equations are published to derive the geothermal reservoir temperature, but all these equations are based on a series of assumptions leading to a specific field of applications. This field of application includes some restrictions with the main ones that are the range of temperature values within the geothermal reservoir, the assumption that the thermodynamic conditions are satisfied for a reaction equilibrium between the water and the reservoir rocks, but also the geological setting (volcanic zone, sedimentary basin, metamorphic terranes...). The available papers on the subject usually discuss the field of applications, but even in these cases it is not always straightforward to decide if a particular equation can or cannot be applied. For instance, even if we have detected geochemical anomalies of subsurface water on the East side of the study near field, we must assume that the reservoir temperature fits with the temperature range of the applied geothermometers. For our study case, the temperature constraints within the Havelange borehole are known to be limited with a maximum bottom temperature of 126 °C (after correction) measured at 5369 MD (5277 m VD) [15]. Another temperature value is related to the Chaudfontaine site where the temperature of 36 °C was recorded during the sample collection, while for the other sites the spring temperature ranges between 3.7 °C and 14 °C. As a result, the geothermometers developed for the very high temperature conditions are not necessary the most appropriate for our study.

Some geothermometers were developed for geochemical reaction of minerals as feldspars, which are present in the Havelange borehole subsurface, but in a limited proportion. The available literature abounds with geothermometer evaluations for high-enthalpy geothermal systems in magmatic and volcanic contexts, but the number of geothermometer research in fold-and-thrust belt setting is quite limited. The closest domain of such evaluation to fold-and-thrust belt corresponds to sedimentary basins, with the main source of information on the reservoir water composition and temperature coming from oil fields. In this setting, the temperature ranges are usually quite low, but the conditions are not identical to those of fold-and-thrust. Amongst the differences the rocks of the Rhenohercynian Variscan orogen are strongly lithified, the diagenetic and metamorphic reactions are completed, and the porosity/permeability is mainly controlled by fracture networks, since the primary porosity is strongly reduced. From a lithological point of view, geothermometers were published by Chiodini et al. [49] for a geological setting, including limestone and evaporite formations. Their study led to a good correlation between a theoretical model and water samples collected in the Etruscan Swell (Italy). The occurrence of Givetian-Frasnian limestones in our study area could represent a promising target for the application of the Chiodini et al.'s model [49]. However, evaporite deposits were not recognized in the subsurface of the Havelange demo-site. Hence, our water samples are strongly undersaturated in SO_4 concentration (between 0.03 and 3.41 meq/L), which is between three orders of magnitude and at best a factor 3 lower than samples from the Etruscan Swell. This SO_4

deficit induces an overestimation of the reservoir temperature (~>100 and 150 °C) even for samples of superficial waters.

Despite these unknown parameters and uncertainties on the field of application of the geothermometers, we attempt and discuss some of the published geothermometers. The most widely used geothermometer is the one based on silica concentration in water. Different equations are available according to the silica forms [50]. It is usually considered that the quartz phase is controlling the silica dissolved concentration for water temperature above 180 °C, while chalcedony is the main phase governing the water silica content for a temperature below 110 °C. Between these values, the main controlling phase is undetermined. If we applied the equations developed by Fournier [50] for the dataset of this study, several problems are encountered: firstly, the ranges of computed temperatures for amorphous quartz, α-cristobalite, β-cristobalite include only negative temperature values. If chalcedony is considered as the controlling phase of dissolved silica, negative temperatures are computed for most of the superficial waters associated with cluster 7 of the hierarchical clustering. Some negative temperatures are also evaluated for samples of cluster 6. For the other cluster, the temperature values are quite small. Secondly, another problem in this approach is that the computed temperature for the Chaudfontaine hydrothermal reservoir is only 29.9 °C, while the recorded temperature at the sampling site is 36 °C. For similar reasons, Graulich [51] applied the equation of quartz and evaluate a temperature of 50 °C for the reservoir of Chaudfontaine. If we apply the equation for quartz to the dataset of our study, we can observe that the computed temperature values vary according to the clusters (Figure 11a). Computed values for clusters 1, 3, and 4 are located near a temperature of 50 °C. The two points belonging to cluster 2 exhibit different values: 33.5 and 79.1 °C. Cluster 6 corresponds to temperatures stretching between 17.9 and 62.0 °C. The last value equals the computed temperature of the reservoir of Chaudfontaine. Finally, the superficial water samples (cluster 7) are associated with the lowest temperature except two outliners.

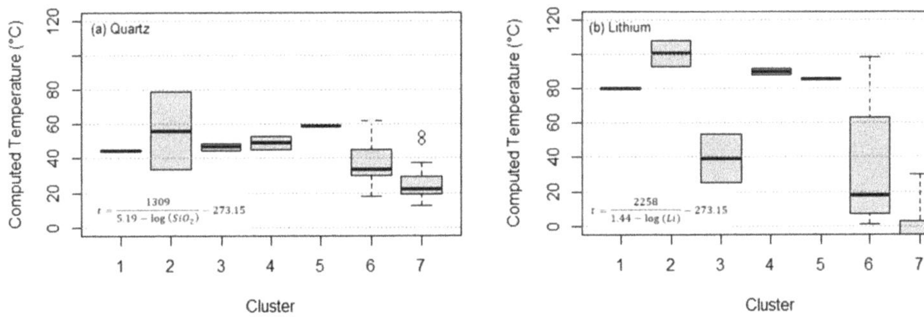

Figure 11. Computed temperature ranges for the different clusters from the geothermometers: (**a**) using the equation for quartz [50]; (**b**) lithium for a low Cl content (<0.2 M) [52]. The geothermometer equations are reported in the lower left corners. t is the temperature expressed in °C, SiO_2 concentration in mg/L and Li concentration in mol/kgw considering a low salinity (Cl < 0.2 M). The geothermal water sample (n° 50) is pointed by a green circle in cluster 6. For the sake of clarity, computed temperatures below 0 °C are not plotted.

The question of the application of the quartz phase equation for our dataset remains. It seems surprizing that the widely used silica geothermometers lead to unrealistic values if chalcedony is considered and if the quartz phase equation is applied the computed temperatures seem closer to conceivable values, but we apply an equation far below its lower temperature limit (180 °C) that is the equation is applied out of its field of application. Several remarks need to be addressed. First, Brook et al. [53] consider that in granitic massifs, the quartz is still the controlling phase for the dissolved silica down to a temperature of 90 °C. An assumption would be, therefore, to extend this lower limit even below 90 °C in the case of the rock assemblage like the one of Havelange in a setting of a fold-and-thrust belt, but the thermodynamic explanation will need to be addressed in the future. Another explanation of such paradoxical situation is that all silica geothermometers are directly related to the silica concentration. Hence, the unrealistic temperature observed with the chalcedony equation is related to abnormal low SiO_2 concentrations in the collected samples. In this case, the reasons of this SiO_2 apparent deficit are manyfold: the amount of chalcedony available for reaction with water in the Havelange subsurface is too small. Another possibility could be by the admixture of water with another water mass with a low SiO_2 concentration (e.g., superficial water). The water flow in the subsurface is too high to allow a full equilibrium between the water and the reservoir rocks. Finally, the potentially SiO_2-rich water in the near field that would flow at great depth are likely to encounter carbonate rocks during the ascending phase. The flow through the carbonate formations is likely to be associated with several reactions associated with the change in physicochemical conditions. The last two assumptions (fast flow, carbonate reactions) seem, however, unlikely since the abnormal computed temperatures are observed in both the near and far field. In this last zone, some of the geochemical anomalies are related to sites where the presence of limestone in the subsurface has never been recognized. Finally, the assumption of a fast-ascending flow is incompatible with the low temperature recorded in the springs.

The geothermometer analyses are rarely conducted with just one method, but it is common to compare the results of several methods. Another method based on a single element is associated with the Li concentration [52]. If applied on the Havelange dataset, it comes out that the computed temperature for clusters 1, 2, 4, and 5 are significantly higher than with the quartz geothermometer (Figure 11b). The computed temperature values for this group range between 80.4 and 108.3 °C. Cluster 6 includes values covering the full interval between 100 and 0 °C with the reservoir temperature in Chaudfontaine reaching a temperature of 76 °C. The lack or very low Li concentration in samples of cluster 7 (superficial waters) leads to very low or even negative value for this group.

If the observations based on lithium concentrations are crossed with the assumption of the SiO_2 concentration too low due to the admixture with superficial waters, the lithium concentrations in clusters 1 to 6 are, therefore, underestimated and as a result the computed temperatures with the Li geothermometer would be underestimated. In this approach, the computed temperatures presented in Figure 11b need to be regarded as minimum values for the reservoir.

Another widely applied geothermometer is based on the ratio Na/K and its development results from a series of empirical observations combined with experimental and thermodynamics modelling. As such, the Na/K-based geothermometers cannot be directly applied since they are valid only for high temperature conditions (>180 °C) [54–56] likely to be above the temperature conditions in the Havelange demo-site case. Furthermore, the computations are based on feldspar stability conditions, which are not the main mineralogical phases in the studied part of the Rhenohercynian fold-and-thrust belt. To extend the field of application of such geothermometer, Fournier and Truesdell [54] includes in the equation the Ca concentration. The range of temperatures covered by their geothermometer range between 4 and 340 °C. The Figure 12a shows the computed temperatures with the Fournier and Truesdell geothermometer [54]. The bulk of values is located below 60 °C with the higher temperatures for samples of clusters 1, 2, and 5. The composition of samples from cluster 3 leads to apparent temperature close to zero. The computed temperatures in cluster 6 are shifted towards low and even negative values, while the superficial water from cluster 7 is promoted to higher and more likely unrealistic values.

The Na-K-Ca geothermometer of Fournier and Truesdell [54] was criticized due to deviation for low temperature and modifications, or new geothermometers were published. Giggenbach [55] proposes the application of a K-Mg geothermometer for low temperatures. The computed temperatures based on the equation of Giggenbach on our dataset (Figure 12b) show, globally, the same trend that the Na-K-Ca geothermometer, but the temperature for samples are less spread and negative values are absent. Finally, Kharaka and Mariner [57] proposed an equation based on the ratio Mg/Li (Figure 12c). Following this model, the computed temperatures for clusters 1, 2, 4, and 5 are between 31 and 49 °C. Some of the samples from cluster 6 drops to negative values and the very low Li-content from superficial waters of cluster 7 lead to only negative values.

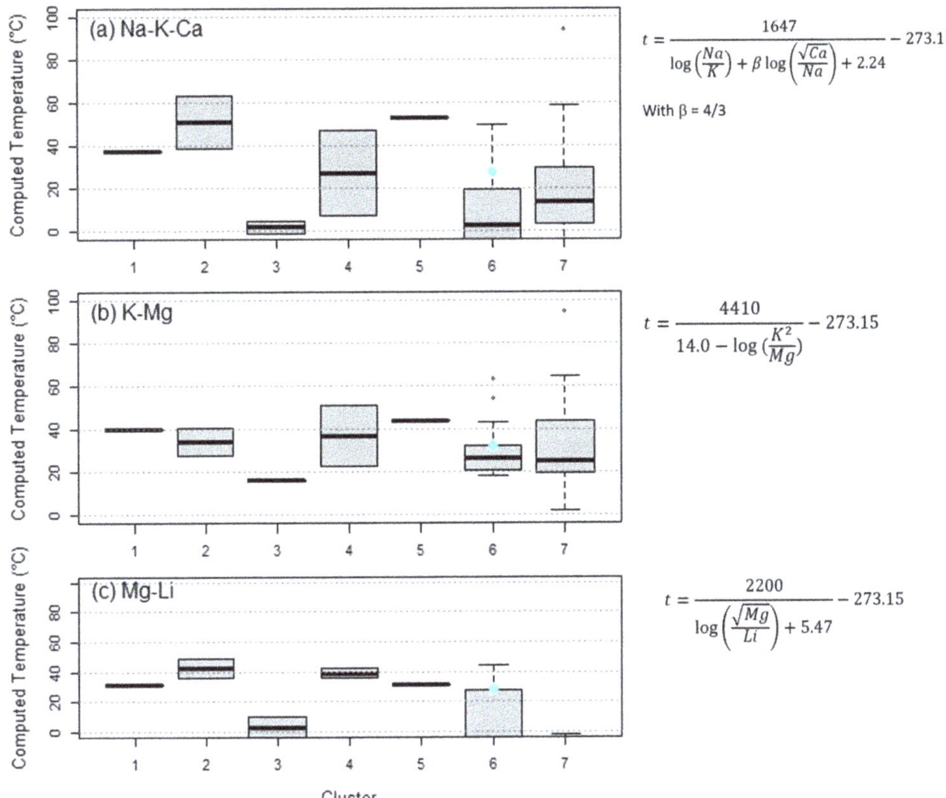

$$t = \frac{1647}{\log\left(\frac{Na}{K}\right) + \beta \log\left(\frac{\sqrt{Ca}}{Na}\right) + 2.24} - 273.15$$

With $\beta = 4/3$

$$t = \frac{4410}{14.0 - \log\left(\frac{K^2}{Mg}\right)} - 273.15$$

$$t = \frac{2200}{\log\left(\frac{\sqrt{Mg}}{Li}\right) + 5.47} - 273.15$$

Figure 12. Computed temperature ranges for the different clusters from the geothermometers: (**a**) using the equation for the ion Na-K-Ca [54]—concentrations expressed in mol/L; (**b**) K-Mg [55]—concentrations expressed in mg/L and (**c**) Mg-Li [57]—concentrations expressed in mg/L. The geothermometer equations are reported to the right. *t* is the temperature expressed in °C. The geothermal water sample (n°50) is pointed out by a cyan circle in cluster 6.

The previous three geothermometers predict a reservoir temperature for the Chaudfontaine of 27.6 °C for Na-K-Ca, 32.5 °C for K-Mg, and 28.9 °C for Mg-Li. These values are below the recorded temperature during the sampling site (36 °C), indicating that these geothermometers predicts too low values for the reservoir. A possible interpretation for such uncertainties is the impact of high-Ca and high-Mg concentrations in the studied samples. More specifically, the Ca, Mg contents in clusters 4 and 6 tends to draw downward the values for these groups. Giggenbach [55] evaluate the impact of calcium in the reactions and concluded that the problems of the Na-K-Ca geothermometers result from sensitivity on variations on CO_2-content in the geothermal waters. For the case of the Na-K-Ca geothermometer, one of the assumptions (excess in silica) for the establishment of the equation is probably not met (see the silica geothermometer).

The last geothermometers to be considered here were developed to evaluate the behaviour of, and trace elements from, major hot water reservoirs in granitic terranes [58].

This study covers several domains in Europe and showed that the concentrations in the studied elements are not significantly affected during the flow from the reservoir to the surface for alkaline water conditions (pH > 9–9.5). The water samples collected during the Havelange demo-site campaign show that alkaline conditions are encountered for samples of clusters 4 and 6, while the other clusters gather slightly acidic water. The geothermometer equation for the Rb/Na ratio corresponds to a low temperature aquifer from Bulgaria. If this geothermometer is applied to the Havelange demo-site dataset, it comes out that the computed temperatures are significantly higher than those predicted by the previous geothermometers (Figure 13a). For all clusters, except the last one, all temperatures are largely higher than 50 °C and in the case of cluster 4, temperature estimations even reach values above 200 °C. Similar observations can be conducted for some outliner values of cluster 6. For this geothermometer, the reservoir temperature in Chaudfontaine would be 110.8 °C. Interestingly, these higher temperatures are associated with the clusters of alkaline waters, which are closer to the field of application of this geothermometer. On the other hand, the evaluated temperatures for the superficial waters of cluster 7 are clearly too high.

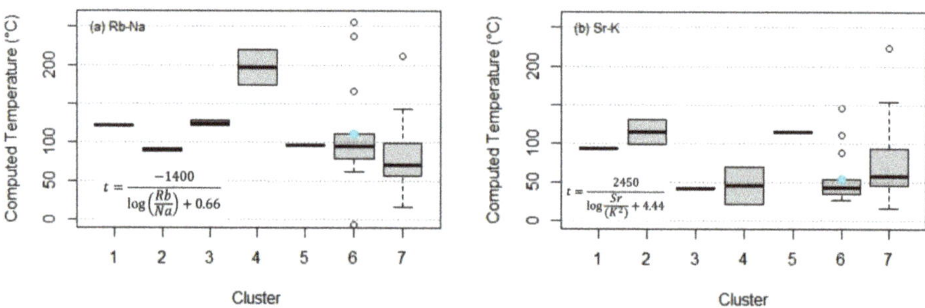

Figure 13. Computed temperature ranges for the different clusters from the geothermometers: (**a**) using the equation for the ratio Rb/Na [58]—concentrations expressed in mol/L; (**b**) the ratio Sr/K [58]—concentrations expressed in mol/L. The geothermometer equations are reported in the lower left corners. *t* is the temperature expressed in °C. The geothermal water sample (n°50) is pointed by a green circle in cluster 6.

The Sr/K geothermometer also indicates high values: c. 100 °C for samples from clusters 1, 2, 4, and some outliners of cluster 6. By contrast, temperatures for cluster 3 and 4 show a more reduced values around 50 °C (Figure 13b).

If we compare all computed temperatures with the different geothermometers, it is challenging to identify a clear trend since the temperature ranges are very broad. In a general sense, the temperatures evaluated from the concentrations of major elements (SiO_2, Ca, K, Na, Mg) indicate low temperatures for the geothermal reservoirs. Some of these values are clearly underestimated as indicated by the reservoir temperature that would be inferior to the recorded value at the collecting point (Chaudfontaine). Numerous assumptions can be considered, but a likely feature is the global low concentrations in many of these major elements, with respect to geothermal water. This deficit in major elements was already developed for the silica concentration. The values Lkm ($\log(K^2/Mg)$) and Lkc ($\log(K^2/Ca)$), as defined by Giggenbach [55], are very low (<0), indicating that the potassium concentration is larger than the square root of Mg and Ca. When reported into a triangle plot of Na, Mg, and K concentration, all points from the current study plot in the Mg corner correspond to immature waters, that is, far from the fully equilibrated waters.

The second category of tested geothermometers are based on the concentration of minor element (Li) or the ratio of a minor elements (Li, Rb, Sr) and a major element (Na, K, Mg). For those cases, the evaluated temperatures are usually higher than for the major element-only methods. The presence of a major element in low concentration associated with a minor element in the geothermometer equations has probably the reverse effect of overestimating the reservoir temperature. The unrealistic temperatures of superficial waters (cluster 7) are clearly the sign of this overestimation. Qualitatively, the real temperatures are probably between the two categories of geothermometer.

6. Discussion

The geothermal exploration must follow a multidisciplinary approach, especially in an unconventional and mainly blind geothermal system. This study case uses the geochemistry of spring water for leading an investigation around the gas exploration well of Havelange, which is here under investigation for a reconversion in a deep geothermal development project. As a result, the common procedure of first conduct exploration at the regional scale followed by a downscaling approach towards sweet spots for the evaluation and development of geothermal cannot be followed, since the borehole was already implemented and drilled in the 80's without any geothermal objective. It is, therefore, difficult to answer fundamental questions such as: are there significant deep fluid flows? Or if another borehole needs to be drilled to develop a geothermal doublet, where is the best place to intersect the deep fluid flow(s)?

On the other hand, the presence of such deep borehole crosscutting a whole sequence of Devonian metasedimentary formations is relatively unique. It allows us to investigate the geothermal potential of an unconventional reservoir, which consists of fracture tight metamorphic units in a fossil fold-and-thrust belt and in a zone where the presence of post-Variscan extensional fractures is absent or very-limited. The Havelange borehole is, therefore, a good candidate for exploring and evaluating the development of a geothermal system in such unconventional conditions. The success of such a project would unlock new possibilities for the development of the geothermal industry in regions where conventional targets are not available. These regions include many places in the Rhenish Massif in Germany, Northern Luxembourg, a large portion of southern Belgium, and the buried basement under the London–Paris basin.

In this paper, the various subsurface fluid flows around the Havelange deep borehole were analysed by studying the groundwater geochemistry composition from mainly springs and a few other infrastructures (wells, a drainage gallery). The dataset includes physicochemical parameters recorded in a laboratory and in the field of 50 sites in the near field (NF) (< 5 km from Havelange borehole) and the far field (FF). The Lower Devonian rocks from the FF represent the ground surface analogue formations to those observed in the deep part of the Havelange borehole. One of the samples was also acquired from a water catchment in the city of Chaudfontaine and is used here as a hydrothermal reference.

The analysis results treatment includes conventional data representations of aquifer composition in hydrogeology such as the Piper and Stiff diagrams along with a multivariate approach to define compositional affinities/dissimilarities between the samples. Several geothermometers such as chalcedony, quartz, lithium, Na-K-Ca, K-Mg, Mg-Li, Rb-Na, and Sr-K were also used to evaluate reservoirs temperatures.

The results show a broad variability of the compositions and various clusters, or groups can be identified. The number of clusters is, however, dependant of the applied method and of the analyst choice. For this study, seven clusters were identified with the hierarchical clustering method; the heat map representation indicates three main horizons of compositions. Even if the data-treatment methods provide different numbers of clusters, they globally tend to similar groupings with marginal variation for intermediate composition.

The largest group from the hierarchical clustering corresponds to superficial water, slightly acidic and with frequent high NO_3 levels more likely related to agricultural activities. This group is also characterised by a low electrical conductivity, a low TDS level and, hence, a narrow Stiff polygon. A strong relationship is observed between these superficial water discharges and the presence of detrital formation of Famennian age in surface geology. The known barrier or cap rock behaviour of the Famennian formations, especially the lower Famennian shale, is here confirmed.

At the other end of the composition spectrum, a few springs in the FF show acidic water and a high mineral charge. These springs are known in the region as 'pouhons' or carbogaseous springs and they are associated with detrital metasedimentary rock of either the Lower Devonian or Lower Palaeozoic age. The iron and manganese contents are very high and indicators of thermal water such as Li, Sr, and Rb are also observed in high concentration. The NO_3 levels in this group is very low.

A third important group corresponds to slightly alkaline water with a significant mineral charge and high levels of the Li, Sr, and Rb indicators. The mineral composition is, however, different from the previous carbogaseous group. Samples from this third group are acquired in the NF and especially in an area located about 4 km to the East of the Havelange borehole in the villages of Moressée and Heure. Such composition of water in the region was not yet coined and provide new valuable information. In addition, during the field campaign several springs in the NF shown the presence of gas. Besides these occurrences in the NF, the reference hydrothermal water from Chaudfontaine also belongs to this group. In other words, water samples from this group in the Havelange borehole NF have, therefore, affinities with the hydrothermal water of Chaudfontaine. The water composition of this group is interpreted as water discharges from a deep, partly confined, aquifer developed in either Givetian or Frasnian limestone formations. The discharge of this aquifer at ground surface and, thus, through the confining beds of Famennian formations is probably the result from the presence of a draining fracture zone.

Besides these three large groups of composition, other springs are characterized by intermediate composition that reflects transitions; either a local specificity (e.g., mineral deposits) or the mixing of water from different composition groups.

At this stage of the exploration, it is delicate to evaluate the significance of the fluid flow between the aquifer located in the Givetian and Frasnian limestones and those located deeper in the detrital Lower and Mid-Devonian formations observed in the deep part of the Havelange borehole. As a first approach and following the spring water composition observed in similar rocks in the FF, it is likely that the water in the deepest part of the Havelange borehole is acidic with a significant mineral charge, include significant concentration in Fe, Mn. In a hypothesis of a water flow upward from the deep subsurface, it will reach the limestone beds and their associated aquifer, and a series of reactions will take place, such as a dissolution of the limestone and the production of dissolved CO_2. The presence of gas in several springs from the NF is a probable indicator of such reactions.

The application of geothermometers in a metasedimentary context of a fold-and-thrust belt indicates various results and shows different limits. In fact, the use of one type of geothermometer rather than another is based on a supposed reservoir temperature and on the phase equilibrium. In this study, we considered that the reservoir is around 100 °C. Of the nine geothermometers used, only two seems to show convincing results (quartz and lithium). Indeed, if the Chaudfontaine site is taken as a reference (temperature of 36 °C at the surface), the results obtained with the use of chalcedony (29.9 °C), Na-K-Ca (27.6 °C), K-Mg (32.5 °C), and Mg-Li (28.9 °C) geothermometers do not seem realistic (not to mention the negative values in some clusters). The higher concentration of Ca and Mg in some clusters may explain the calculated low temperature and the presence of CO_2 and silica deficiency may interfere with the use of the Na-K-Ca geothermometer. Geothermometers based on major and trace elements indicate very high temperatures (Chaudfontaine temperature 110.8 °C (Rb-Na)). Moreover, the superficial water reservoirs would be at more than 50 °C due to the higher concentration of major elements than trace elements.

Regarding geothermometers, which seems to indicate more consistent results, quartz can be used for reservoirs above 180 °C and research in granite massifs has shown that it remains applicable below 90 °C in this environment. It would be necessary to conduct a study to verify the applicability of this method to a geological context like the Havelange.

If the quartz-based calculations are applied to Chaudfontaine, the reservoir temperature calculated in this paper is about 62 °C instead of 50 °C found by Graulich in the 1980s.

The geothermometer based on the lithium concentration also shows values that seem consistent although overestimated. However, the application of this tool should only be done when the clusters contain a significant content of Li, otherwise negative temperatures are calculated.

In view of these diverging results, the questions would be: is it possible to use these geothermometers in the geological context of the Havelange, namely meta-sediments in fold-and-thrust belt? Are these variations due to dilution effects or to the unavailability of the necessary mineral phases in the water?

7. Conclusions

In this study, the geochemical composition analysis of spring water is applied as a surface tool for deep geothermal potential exploration. It aims to detect the geochemical indications of the impact of ascending deep-water flow on the composition of spring water samples. This approach is of particular interest in blind geothermal systems where there are no clear surface manifestations of the presence of a deep geothermal system. The technique is here applied to water samples collected in the near and far field around the Havelange borehole. This site is the selected H2020 MEET project demo-site for the evaluation of the geothermal potential of the unconventional reservoirs in the Variscan metasedimentary system, unaffected later by an extensional deformation phase.

The results show that the water samples can be separated into different clusters of compositions revealing different water evolutions. The number of clusters is dependent on the statistical method applied for the data treatment. Using the hierarchical clustering, seven clusters can be evaluated and related to different combination of concentrations/values of elements or parameters such as Li, Rb, Sr, pH, and EC. The application of the heat map technique demonstrates the presence of at least three types of water. The first group consisting of water sources locally known under the name of "Pouhon", is characterized by an acidic pH, a high EC, and TDS and by the frequent presence of anomalies in Fe, Mn, Ni, Be, Mg, and Li. The second group is represented primarily by carbonate water compositions with neutral to basic pH as well as high EC and TDS. Bromide, Mo, U, V, Se, Sb, Li, Rb, Sr, and Cu anomalies are frequently observed in this category. The Chaudfontaine reference hydrothermal site and the samples located in Eastern neighbourhood (near the village of Moressée) belong to this second type. Waters from this category constitute the targets for future geothermal studies. In fact, the water samples around the village of Moressée show higher concentrations of Li, Sr, and Rb than in Chaudfontaine. The third and last group contains superficial waters with a low conductivity and TDS. They exhibit a high concentration of NO_3 and low concentrations of Li, Rb, Ca, Na, K, Mo, and HCO_3. These superficial waters are frequently associated with the outcropping zone of lower Famennian formations that represent an aquiclude and act as a cap rock for the deep-water reservoir. Future investigations will need to focus on other categories of analyses such as the isotopic and gas content.

In addition, this paper shows that the use of geothermometers in the Havelange context, and probably more broadly in metasedimentary fold-and-thrust belt domains, need to be considerate with caution. In fact, the low temperature presupposed (around 100 °C) and the lack of knowledge on the phase balance in the reservoir suggest that the method can hardly directly be applied with commonly used geothermometers (chalcedony, quartz, Li, Na-K-Ca, K-Mg, and Mg-Li). A hydrogeochemical model of the reservoir would possibly provide indication of the water composition evolution in our system and to develop geothermometers usable in this geological context.

To conclude, the indication of deep fluid flow in the region of the Havelange borehole is a significant information, but at this stage it is difficult to evaluate if such phenomenon is restricted to a small zone around the study site and if it can be extrapolated to numerous zones of occurrence of Variscan meta-sedimentary formations. If this second option could be verified, it would open new opportunities for the development of deep geothermal projects in numerous regions of North-West Europe, such as the Rheinische Schiefergebirge in Germany, the Oesling in the Great Duchy of Luxembourg, the Ardenne in Belgium to even Devon and Cornwall in SW-England.

Supplementary Materials: The following are available online at https://www.mdpi.com/article/10.3390/geosciences11110437/s1, Figure S1: Stiff diagrams of all the water springs samples classified by clusters, Table S1: Coordinates of the spring water samples and results of analyses of the physico-chemical parameters used in this article, Table S2: Detailed list of the SWDE laboratory parameters and LOQ.

Author Contributions: Conceptualization, Y.V. and M.C.; methodology Y.V., M.C., V.S., T.G., B.M., E.P. and C.B.; software, Y.V. and M.C.; validation, Y.V., M.C., S.B. and V.S.; formal analysis, Y.V., M.C., S.B. and V.S.; investigation, Y.V., M.C., S.B. and V.S.; resources, Y.V., T.G., V.S., C.B. and SWDE; data curation, Y.V. and M.C.; writing—original draft preparation, Y.V. and M.C.; writing—review and editing, Y.V., M.C., S.B., V.S. and H2020 MEET consortium; visualization, Y.V. and M.C.; supervision, Y.V.; project administration, Y.V.; funding acquisition, Y.V., E.P. and H2020 MEET consortium. All authors have read and agreed to the published version of the manuscript.

Funding: This project has received funding from the European Union's Horizon 2020 research and innovation program under grant agreement No 792037 (H2020 MEET project).

Institutional Review Board Statement: Not applicable.

Informed Consent Statement: Not applicable.

Data Availability Statement: Not applicable.

Acknowledgments: The authors greatly thank, Jean-Louis Cornet for his contribution on Chaudfontaine's hydrogeology, Jean-Benoit Charlier and Jennifer Defoux for their advice and explanations on analysis made in the SWDE laboratory. We want to thank Jean Goffaux from AIEC for is help during the sampling campaign. Thanks to Renata Barros for her help on Piper diagrams and water analyses interpretation. We are grateful to de Woot and the owners who have authorized us to sample the water on their land. Also, we wish to thank the MEET project partners and particularly Albert Genter for his help on geothermometers methods reflexion. The authors thank the reviewers who gave their time for critical reading of the manuscript and the improvement of it.

Conflicts of Interest: The authors declare no conflict of interest.

References

1. Ma, Z.Y. Unconventional Resources from Exploration to Production. In *Unconventional Oil and Gas Resources Handbook*; Elsevier: Amsterdam, The Netherlands, 2016; pp. 3–52. ISBN 978-0-12-802238-2.
2. Faulds, J.E.; Craig, J.W.; Hinz, N.H.; Coolbaugh, M.F.; Earney, T.E.; Schermerhorn, W.D.; Peacock, J.; Deoreo, S.B.; Siler, D.L. Discovery of a Blind Geothermal System in Southern Gabbs Valley, Western Nevada, through Application of the Play Fairway Analysis at Multiple Scales. *GRC Trans.* **2018**, *42*, 14.
3. Ward, D.M.; Ferris, M.J.; Nold, S.C.; Bateson, M.M. A Natural View of Microbial Biodiversity within Hot Spring Cyanobacterial Mat Communities. *Microbiol. Mol. Biol. Rev.* **1998**, *62*, 1353–1370. [CrossRef] [PubMed]
4. Lacap, D.C.; Barraquio, W.; Pointing, S.B. Thermophilic Microbial Mats in a Tropical Geothermal Location Display Pronounced Seasonal Changes but Appear Resilient to Stochastic Disturbance. *Environ. Microbiol.* **2007**, *9*, 3065–3076. [CrossRef]
5. Lezcano, M.Á.; Moreno-Paz, M.; Carrizo, D.; Prieto-Ballesteros, O.; Fernández-Martínez, M.Á.; Sánchez-García, L.; Blanco, Y.; Puente-Sánchez, F.; de Diego-Castilla, G.; García-Villadangos, M.; et al. Biomarker Profiling of Microbial Mats in the Geothermal Band of Cerro Caliente, Deception Island (Antarctica): Life at the Edge of Heat and Cold. *Astrobiology* **2019**, *19*, 1490–1504. [CrossRef]
6. Valeriani, F.; Crognale, S.; Protano, C.; Gianfranceschi, G.; Orsini, M.; Vitali, M.; Spica, V.R. Metagenomic Analysis of Bacterial Community in a Travertine Depositing Hot Spring. *New Microbiol.* **2018**, *41*, 126–135.
7. Kraml, M.; Jodocy, M.; Reinecker, J.; Leible, D.; Freundt, F.; Al, S.; Schmidt, G.; Aeschbach, W.; Isenbeck-Schroeter, M. TRACE: Detection of Permeable Deep-Reaching Fault Zone Sections in the Upper Rhine Graben, Germany, During Low-Budget Isotope-Geochemical Surface Exploration. In Proceedings of the European Geothermal Congress 2016, Strasbourg, France, 19–24 September 2016; p. 10.
8. Blackwell, D.D.; Waibel, A.F.; Richards, M. Why Basin and Range Systems Are Hard to Find: The Moral of the Story Is They Get Smaller With Depth! *GRC Trans.* **2012**, *36*, 6.

9. Hanson, M.C.; Oze, C.; Horton, T.W. Identifying Blind Geothermal Systems with Soil CO 2 Surveys. *Appl. Geochem.* **2014**, *50*, 106–114. [CrossRef]

10. Utomo, D.A.; Abiyudo, R.; Saputra, I.J.; Irfan, R.; Archady, D. Temperature Dependent Minerals as a Tool to Prove High Temperature of a Blind Geothermal System. Case Study: Well "X" at Java Island of Indonesia. *IOP Conf. Ser. Earth Environ. Sci.* **2021**, *732*, 012001. [CrossRef]

11. Garg, S.K.; Combs, J.; Pritchett, J.W. Exploring for Hidden Geothermal Systems. In Proceedings of the World Geothermal Congress 2010, Bali, Indonesia, 25–29 April 2010.

12. Kratt, C.; Coolbaugh, M.; Peppin, B.; Sladek, C. Identification of a New Blind Geothermal System with Hyperspectral Remote Sensing and Shallow Temperature Measurements at Columbus Salt Marsh, Esmeralda County, Nevada. *GRC Trans.* **2009**, *33*, 6.

13. Habets, M.A. *Revue Universelle des Mines, de la Métallurgie des Travaux Publics, des Sciences et des Arts Appliqués à l'Industrie*; 47 Eme Année—Quatr. Sér.; 1903; Tome II; p. 6.

14. Leclercq, V. Le Sondage de Douvrain. 1980, p. 64. Available online: http://biblio.naturalsciences.be/rbins-publications/professional-papers-of-the-geological-survey-of-belgium/pdfs/pp_1980_03_le-sondage-de-douvrain_leclercq.pdf (accessed on 21 October 2021).

15. Leclercq, V. *Reconversion d'un Forage de Prospection Gaz en un Forage de Reconnaissance Géothermique: Cas du Forage Profond d'HAVELANGE (Belgique)*, UNINE—CAS DEEGEOSYS. 2014; Non published confidential report (in case of interest, contact the author).

16. Graulich, J.M.; Leclercq, V.; Hancel, L. Le sondage d'Havelange—Principales données et aspects techniques. In *Memoirs of the Geological Survey of Belgium*; Geological Survey of Belgium: Brussels, Belgium, 1989; p. 65.

17. Fielitz, W. Variscan Transpressive Inversion in the Northwestern Central Rhenohercynian Belt of Western Germany. *J. Struct. Geol.* **1992**, *14*, 547–563. [CrossRef]

18. Dittmar, D.; Meyer, W.; Oncken, O.; Schievenbusch, T.; Walter, R.; von Winterfeld, C. Strain Partitioning across a Fold and Thrust Belt: The Rhenish Massif, Mid-European Variscides. *J. Struct. Geol.* **1994**, *16*, 1335–1352. [CrossRef]

19. Oncken, O.; von Winterfeld, C.; Dittmar, U. Accretion of a Rifted Passive Margin: The Late Paleozoic Rhenohercynian Fold and Thrust Belt (Middle European Variscides). *Tectonics* **1999**, *18*, 75–91. [CrossRef]

20. Vanbrabant, Y.; Braun, J.; Jongmans, D. Models of Passive Margin Inversion: Implications for the Rhenohercynian Fold-and-Thrust Belt, Belgium and Germany. *Earth Planet. Sci. Lett.* **2002**, *202*, 15–29. [CrossRef]

21. Belanger, I.; Delaby, S.; Delcambre, B.; Ghysel, P.; Hennebert, M.; Laloux, M.; Marion, J.-M.; Mottequin, B.; Pingot, J.-L. Redéfinition des unités structurales du front varisque utilisées dans le cadre de la nouvelle Carte géologique de Wallonie (Belgique). *Geol. Belg.* **2012**, *15*, 169–175.

22. Fielitz, W.; Mansy, J.-L. Pre- and Synorogenic Burial Metamorphism in the Ardenne and Neighbouring Areas (Rhenohercynian Zone, Central European Variscides). *Tectonophysics* **1999**, *309*, 227–256. [CrossRef]

23. Hance, L.; Dejonghe, L.; Ghysel, P.; Laloux, M.; Mansy, J.L. Influence of Heterogeneous Lithostructural Layering on Orogenic Deformation in the Variscan Front Zone (Eastern Belgium). *Tectonophysics* **1999**, *309*, 161–177. [CrossRef]

24. Leynaud, D.; Jongmans, D.; Teerlynck, H.; Camelbeeck, T. Seismic Hazard Assessment in Belgium. *Geol. Belg.* **2000**, *3*, 67–86. [CrossRef]

25. Vandenberghe, N.; Fock, W. Temperature Data in the Subsurface of Belgium. *Tectonophysics* **1989**, *164*, 237–250. [CrossRef]

26. Rogiers, B.; Huysmans, M.; Vandenberghe, N.; Verkeyn, M. Demonstrating Large-Scale Cooling in a Variscan Terrane by Coupled Groundwater and Heat Flow Modelling. *Geothermics* **2014**, *51*, 71–90. [CrossRef]

27. Schintgen, T.; Förster, A.; Förster, H.-J.; Norden, B. Surface Heat Flow and Lithosphere Thermal Structure of the Rhenohercynian Zone in the Greater Luxembourg Region. *Geothermics* **2015**, *56*, 93–109. [CrossRef]

28. Keyser, M.; Ritter, J.R.R.; Jordan, M. 3D Shear-Wave Velocity Structure of the Eifel Plume, Germany. *Earth Planet. Sci. Lett.* **2002**, *203*, 59–82. [CrossRef]

29. Walker, K.T.; Bokelmann, G.H.R.; Klemperer, S.L.; Bock, G. Shear-Wave Splitting around the Eifel Hotspot: Evidence for a Mantle Upwelling. *Geophys. J. Int.* **2005**, *163*, 962–980. [CrossRef]

30. Schintgen, T. Exploration for Deep Geothermal Reservoirs in Luxembourg and the Surroundings—Perspectives of Geothermal Energy Use. *Geotherm. Energy* **2015**, *3*, 9. [CrossRef]

31. Barros, R.; Defourny, A.; Collignon, A.; Jobe, P.; Dassargues, A.; Piessens, K.; Welkenhuysen, K. A Review of the Geology and Origin of CO2 in Mineral Water Springs in East Belgium. *Geol. Belg.* **2020**, *24*, 17–31. [CrossRef]

32. Thierrin, J.; Steffen, P.; Cornaz, S.; Vuataz, F.-D.; Balderer, W.; Looser, M. *Echantillonnage des Eaux Souterraines: Guide Pratique*; Office Fédéral L'environnement For. Paysage (OFEFP): Neuchâtel, Switzerland, 2003; pp. 1–83.

33. Bowell, R. C. A. J. Appelo, & D. Postma, 1993. Geochemistry, Groundwater and Pollution. xvi 536 pp. Rotterdam, Brookfield: A. A. Balkema. ISBN 90 5410 105 9; 90 5410 106 7 (pb). *Geol. Mag.* **1995**, *132*, 124–125. [CrossRef]

34. SPW. *Ressources Naturelles et Environnement Etat des Nappes d'Eau Souterraine de la Wallonie*; *Région Wallonne—SPW -Agriculture, Ressources Naturelles et Environnement*; SPW: Namur, Belgium, 2021; p. 66.

35. Piper, A.M. A Graphic Procedure in the Geochemical Interpretation of Water-Analyses. *Trans. Am. Geophys. Union* **1944**, *25*, 914. [CrossRef]

36. Briel, L.I. *Documentation of a Multiple-Technique Computer Program for Plotting Major-Ion Composition of Natural Waters*; Open-File Report; U.S Geological Survey: Richmond, VA, USA, 1993; p. 94.

37. Apollaro, C.; Vespasiano, G.; De Rosa, R.; Marini, L. Use of Mean Residence Time and Flowrate of Thermal Waters to Evaluate the Volume of Reservoir Water Contributing to the Natural Discharge and the Related Geothermal Reservoir Volume. Application to Northern Thailand Hot Springs. *Geothermics* **2015**, *58*, 62–74. [CrossRef]
38. European Environment Agency. R Core Team. 2020. Available online: https://www.eea.europa.eu/data-and-maps/indicators/oxygen-consuming-substances-in-rivers/r-development-core-team-2006 (accessed on 17 August 2021).
39. Jenks, G.F. *Optimal Data Classification for Choropleth Maps*; Occasional Paper No. 2; Department of Geography, University of Kansas: Lawrence, Kansas, 1977.
40. Dejonghe, L.; Bouckaert, J. Presence d'un exoclaste de nature ignee dans les schistes noduleux Frasniens à Nettine (Province de Namur). *Annales de la Société Géologique de Belgique* **1977**, *T.100*, 103–113.
41. De Walque, L.; Bouckaert, J.; Martin, H. *Géochimie de Surface et Minéralisations Du Paléozoïque de Belgique. III. Plomb, Zinc et Fer Au Voisinage de l'ancienne Exploitation Minière de Heure-En-Famenne*; Ministère des Affaires Économiques, Administration des Mines, Service Géologique de Belgique: Bruxelles, Belgium, 1975; p. 35.
42. Foregs—Geochemical Baseline Database: Instructions. Available online: http://weppi.gtk.fi/publ/foregsatlas/ForegsData.php (accessed on 17 August 2021).
43. Barbier, J.; Berthier, F. *Origine des Concentrations en Baryum Dissous Dans les Eaux du Captage de Crôt-Chaud, Saint-Bonnet-Troncais (Allier)*; BRGM: Orléans, France, 2001; p. 18.
44. World Health Organization. *Manganese in Drinking-Water: Background Document for Development of WHO Guidelines for Drinking-Water Quality*; World Health Organization: Geneva, Switzerland, 2004; p. 29.
45. Kavanagh, L.; Keohane, J.; Cleary, J.; Cabellos, G.G.; Lloyd, A. Lithium in the Natural Waters of the South East of Ireland. *Int. J. Environ. Res. Public Health* **2017**, *14*, 561. [CrossRef] [PubMed]
46. Harter, T. *Groundwater Quality and Groundwater Pollution*; University of California, Agriculture and Natural Resources: Davis, CA, USA, 2003; ISBN 978-1-60107-259-7.
47. Malina, G. Ecotoxicological and Environmental Problems Associated with the Former Chemical Plant in Tarnowskie Gory, Poland. *Toxicology* **2004**, *205*, 157–172. [CrossRef]
48. Skougstad, M.W.; Horr, C.A. *Occurrence of Strontium in Natural Water*; Geological Survey Circular 420; Geological Survey Circular: Washington, DC, USA, 1960.
49. Chiodini, G.; Frondini, F.; Marini, L. Theoretical Geothermometers and PCO_2 indicators for Aqueous Solutions Coming from Hydrothermal Systems of Medium-Low Temperature Hosted in Carbonate-Evaporite Rocks. Application to the Thermal Springs of the Etruscan Swell, Italy. *Appl. Geochem.* **1995**, *10*, 337–346. [CrossRef]
50. Fournier, R.O. Chemical Geothermometers and Mixing Models for Geothermal Systems. *Pergamon Press* **1977**, *5*, 41–50. [CrossRef]
51. Graulich, J.M. L'hydrogeologie thermale de Chaudfontaine. *Bulletin Société Belge Géologie* **1983**, *92*, 195–212.
52. Fouillac, C.; Michard, G. Sodium/Lithium Ratio in Water Applied to Geothermometry of Geothermal Reservoirs. *Geothermics* **1981**, *10*, 55–70. [CrossRef]
53. Brook, A.; Mariner, R.H.; Mabey, D.R.; Swanson, J.R.; Guffanti, M.; Muffler, L.J.P. Hydrothermal Convection Systems With Reservoir Temperatures > 90 °C. In *Assessment of Geothermal Resources of the United States*; Geological Survey Circular 790; Muffler, L.J.P., Ed.; Geological Survey: Arlington, VA, USA, 1978; p. 170.
54. Fournier, R.O.; Truesdell, A.H. An Empirical Na-K-Ca Geothermometer for Natural Waters. *Pergamon Press* **1973**, *37*, 1255–1275. [CrossRef]
55. Giggenbach, W.F. Geothermal Solute Equilibria. Derivation of Na-K-Mg-Ca Geoindicators. *Geochim. Cosmochim. Acta* **1988**, *52*, 2749–2765. [CrossRef]
56. Yock, A. *Short Course on Surface Exploration for Geothermal Resources: Geothermometry*; UNU-GTP: Ahuachapan, El Salvador; LaGeo: Santa Tecla, El Salvador, 17 October 2009; p. 8.
57. Kharaka, Y.K.; Mariner, R.H. Chemical Geothermometers and Their Application to Formation Waters from Sedimentary Basins. In *Thermal History of Sedimentary Basins*; Naeser, N.D., McCulloh, T.H., Eds.; Springer: New York, NY, USA, 1989; pp. 99–117. ISBN 978-1-4612-8124-5.
58. Michard, G. Behaviour of Major Elements and Some Trace Elements (Li, Rb, Cs, Sr, Fe, Mn, W, F) in Deep Hot Waters from Granitic Areas. *Chem. Geol.* **1990**, *89*, 117–134. [CrossRef]

geosciences

Article

Use of Mohr Diagrams to Predict Fracturing in a Potential Geothermal Reservoir

D.C.P. Peacock [1,2,*], David J. Sanderson [3] and Bernd Leiss [1,2]

1   Geoscience Centre, Department of Structural Geology and Geodynamics, Georg-August-Universität Göttingen, Goldschmidtstraße 3, 37077 Göttingen, Germany; bleiss1@gwdg.de
2   Universitätsenergie Göttingen GmbH, Hospitalstraße 3, 37073 Göttingen, Germany
3   Ocean and Earth Science, University of Southampton, National Oceanography Centre, Southampton SO14 3ZH, UK; d.j.sanderson@soton.ac.uk
*   Correspondence: peacock@uni-goettingen.de

Abstract: Inferences have to be made about likely structures and their effects on fluid flow in a geothermal reservoir at the pre-drilling stage. Simple mechanical modelling, using reasonable ranges of values for rock properties, stresses and fluid pressures, is used here to predict the range of possible structures that are likely to exist in the sub-surface and that may be generated during stimulation of a potential geothermal reservoir. In particular, Mohr diagrams are used to show under what fluid pressures and stresses different types and orientations of fractures are likely to be reactivated or generated. The approach enables the effects of parameters to be modelled individually, and for the types and orientations of fractures to be considered. This modelling is useful for helping geoscientists consider, model, and predict the ranges of mechanical properties of rock, stresses, fluid pressures, and the resultant fractures that are likely to occur in the sub-surface. Here, the modelling is applied to folded and thrusted greywackes and slates, which are planned to be developed as an Enhanced Geothermal System beneath Göttingen.

Keywords: stress; fluid pressure; Mohr diagrams; fracturing; geothermal; greywackes; slates

Citation: Peacock, D.C.P.; Sanderson, D.J.; Leiss, B. Use of Mohr Diagrams to Predict Fracturing in a Potential Geothermal Reservoir. *Geosciences* **2021**, *11*, 501. https://doi.org/10.3390/geosciences11120501

Academic Editors: Béatrice A. Ledésert, Ronan L. Hébert, Ghislain Trullenque, Albert Genter, Eléonore Dalmais, Jean Hérisson and Jesus Martinez-Frias

Received: 24 September 2021
Accepted: 2 December 2021
Published: 9 December 2021

Publisher's Note: MDPI stays neutral with regard to jurisdictional claims in published maps and institutional affiliations.

1. Introduction

A study is being undertaken to predict the structures and fluid flow behaviour in the sub-surface as part of a proposed geothermal project at Göttingen in Germany [1–3]. The reservoir rocks are the Devonian and Carboniferous metasedimentary sequence thought to occur at a depth of ~1.5 km, beneath a cover of Permian and Mesozoic sedimentary rocks [4]. Göttingen is one of the sites within the European funded project Multidisciplinary and multi-context demonstration of EGS exploration and Exploitation Techniques and potentials (MEET) to develop enhanced geothermal systems (EGS) across Europe (e.g., [5–7]). The Triassic Bunter Sandstone is also being considered as a potential medium-deep geothermal reservoir (i.e., at depths between about 200 m and 1000 m; e.g., [8]), including heat storage options [2], and this is also discussed here.

Only two wells have yet penetrated the metasedimentary rocks in the Göttingen area, these not extending far below the base Zechstein, and so limited well data are available. Additionally, the limited seismic data available show poor resolution beneath the Zechstein salt (e.g., [9,10]). Two seismic lines were shot in 2015, these being 10 km and 11 km long and to about 5 km deep. These seismic lines allow interpretation of the post-Carboniferous sedimentary rocks in the north-south striking Leinetal Graben [3], but do not enable reliable interpretation of the sub-Zechstein rocks. The metasedimentary rocks exposed in the Harz Mountains, ~40 km NE of Göttingen, are being used to predict the rocks and structures that may occur in the sub-surface at Göttingen, and to gain data input for discrete fracture network (DFN) modelling. Exposures of the Bunter Sandstone occur in the area around Göttingen (e.g., [11–13]).

There are significant uncertainties involved in evaluating the characteristics and mechanical properties of the rocks beneath Göttingen. Although field analogues indicate many different types of natural fractures occur (faults, veins, joints, etc.), their intensity, geometry, topology and spatial variation in the sub-surface are unknown. Many of the fractures, especially the joints, may not exist at reservoir depths (e.g., [14,15]). The proposed EGS project would involve stimulation of intensely-deformed Upper Palaeozoic rocks, and there are significant uncertainties involved in predicting the hydraulic fracturing of such anisotropic and pre-fractured rocks (e.g., [16]).

Because there is limited knowledge of the reservoir rocks in the sub-surface, including such parameters as fracture types, orientations, apertures, and connectivity, there is need for an early-stage (pre-drilling) assessment that will enable reasonable predictions to be made about natural and induced fractures in the sub-surface. Here, we use simple modelling and reasonable values of stress states, fluid pressures and rock failure criteria to make predictions about the conditions under which rocks will fracture.

The aim of this paper is to develop a simple workflow in which we assess the likely stress states, fluid pressures, and failure conditions in the sub-surface. We then use Mohr diagrams (Figure 1; e.g., [17,18]) to make predictions of the behaviour of rocks and fluids within potential reservoir rocks. The approach uses information about rock types and structures obtained from exposed analogues and rock deformation tests to predict the effects of changes in effective stresses on natural and induced fractures. In the absence of information on the geometry and topology of the fracture networks and on the conductivity of the fractures, this represents an early-stage approach and a necessary prerequisite to more detailed analysis of the contribution of fractures to fluid flow, such as through the use of DFN modelling (e.g., [19]).

Although the approach is simple, it provides understanding of the relationships between key factors controlling rock behaviour. It helps us identify uncertainties and their possible effects. It can also be considered as a "dynamic" analysis for fracturing, because it can be used to predict the reactivation and generation of fractures as fluid pressures and stresses change. Here, *reactivation* is used to mean renewed shear and/or opening displacement on a pre-existing fracture.

The methods presented here have general applicability to making predictions about fractures in the sub-surface, but here we focus on several key questions about the proposed geothermal reservoir at Göttingen:

1. Which lithologies are most likely to fracture?
2. What stresses and fluid pressures are needed for the reactivation of pre-existing fractures or the development of new fractures?
3. Which orientations and types of fractures are most likely to be reactivated?
4. Will reactivated or new fractures show shear or extension?
5. What effects do heterogeneities (veins, joints, cleavage, bedding planes) have, and what are the different mechanical significances of veins vs. joints?
6. What are the effects of Late Cretaceous and Tertiary exhumation and what amount of exhumation is needed to create joints?

2. Predictions about the Pre-Permian Geology beneath Göttingen

In the absence of well data or high-quality seismic data for the sub-Zechstein rocks, the rocks beneath the Permian Zechstein evaporites at Göttingen are predicted to be Devonian quartzitic sandstones and slates, and Carboniferous greywackes and slates (mainly Culm flysch deposits). Such rocks are exposed in the western Harz Mountains ~40 km to the NE, in the Oberharz Anticline and the Culm Fold Zone, which belong to the par-autochthonous domain of the Harz Mountains (e.g., [3,20]) and in the Rhenish Massif ~70 km to the SW (e.g., [21–23]). The fractures visible in these exposures include veins and joints [24], with cleavage being well-developed in the slates [25]. Small thrusts that appear to have displacements of up to a few metres are exposed [26], but larger thrusts are not well-exposed. The north-eastern boundary the Harz Mountains is marked by the Harznordrand

Fault, which was active during the Late Cretaceous and Tertiary [27]. NW-SE striking Mesozoic normal and oblique-slip faults occur in the Harz Mountains, some of which contained economic Pb-Zn and Ba-F deposits [28].

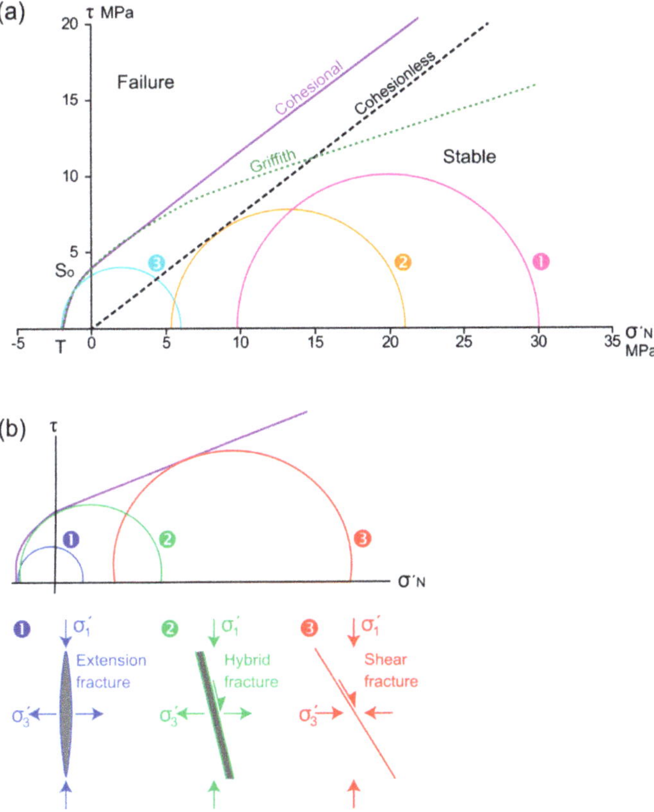

Figure 1. (**a**) Generalised Mohr diagram of shear stress (τ) against normal effective stress (σ'_N) showing failure envelopes. Continuous line ("Cohesional") = Mohr failure envelope for intact rock, with tensile strength (T = 2 MPa), cohesion (S_0 = 4 MPa) and coefficient of internal friction (μ = 0.75). Dotted (parabolic) line ("Griffith") = Griffith criterion. Dashed line ("Cohesionless") = cohesionless failure (μ = 0.75) typical of slip on faults. Three stress states are shown, representing stable (**❶**), shear failure on cohesionless fractures (**❷**) and extension fracturing (**❸**). (**b**) Mohr diagram showing the fields in which extension (**❶**), hybrid (**❷**) and shear (**❸**) fractures occur [29].

Table 1 shows features of these "basement" rocks that are likely to be important for geothermal energy. These predictions are expressed in terms of seven parameters that we consider important in describing a geothermal reservoir: lithology and inferred rheological behaviour, fluid type, stresses, fluid pressure, temperature, strains and existing structures, and geological history (cf. [30]).

More information is available about the Bunter Sandstone, which is exposed in the Göttingen area (e.g., [11,12]). Fractures exposed include normal faults [31], strike-slip faults [32], veins [33], sedimentary dykes [34], and joints [35]. There is limited well data, which suggests that the Bunter Sandstone has been affected by halokinesis of the Zechstein evaporites [1,36,37].

Table 1. Predictions about the geology of the potential geothermal reservoirs at Göttingen. See, for example, [38] for full information about deformation of the Bunter Sandstone in the Leinetal Graben. See [20,39] for further information about the geology of the Devonian and Carboniferous rocks of the Harz Mountains.

Factor	Meaning	Significance	Bunter Sandstone	Variscan rocks
Lithology	Rock types, their porosities and mechanical behaviour	Controls the thermo-mechanical behaviour of the rock. Mechanical behaviour can change significantly through time, especially as deformation occurs	Triassic sandstone (see Table 2 for mechanical properties)	Devonian and Carboniferous greywackes and slates (see Table 2 for mechanical properties)
Fluid type	The chemistry and phase (liquid or gas) of the palaeo- and present-day fluid(s)	Controls the fluid pressure gradient and mineralisation events	Present-day: water (possibly brine)	Present-day: water, probably saline because of the overlying Zechstein. No information on gas content. During the Variscan: mineralising fluids and fluidised sediments
Stress	Magnitudes and orientations of the applied stresses, including the vertical stress (overburden) and horizontal stresses. Horizontal stresses are related to the geostatic stress ratio, applied tectonic stresses and to internal stresses (e.g., related to temperature changes)	Along with fluid pressure, controls the effective stresses, which control the deformation	The vertical stress can be calculated using the mean density of the overburden, but the magnitudes and orientations of the horizontal stresses are uncertain	The vertical stress can be calculated using the mean density of the overburden, but the magnitudes and orientations of the horizontal stresses are uncertain
Fluid pressure	Palaeo- and present-day fluid pressures	Along with the stresses, controls the effective stresses, which control the deformation	Presently probably hydrostatic	The Zechstein evaporites may allow present-day overpressure. Veins, breccias and possible remobilised sediments indicate phases of overpressure during the Variscan Orogeny
Temperature	Palaeo- and present-day temperatures	Influences the style of deformation, with present-day temperature controlling commercial viability	Depends on the geothermal gradient	Depends on the geothermal gradient, but likely to be reduced because of the overlying salt. Possibly elevated by Tertiary igneous activity
Strain	The amount of strain and the existing structures	Influences fluid flow in the sub-surface and present-day mechanical behaviour of the rocks	Controlled by Tertiary rifting. Likely to be influenced by salt tectonics and possibly by Tertiary igneous activity. Steeply-dipping joints and some normal faults are likely to occur	Dominated by: (1) Variscan Orogeny, with folds, thrusts and veins; (2) Late Cretaceous and Tertiary rifting and/or uplift, with normal faults and joints developing
History	The relative and absolute timing of deformation (including mineralisation) events and structures	Controls the types of fractures (faults, veins, joints, etc.) and therefore their effects on fluid flow in the sub-surface	(1) Triassic sedimentation during Mesozoic basin development. (2) Cretaceous and Tertiary regional uplift. (3) Tertiary rifting (Leinetal Graben) and volcanism	(1) Sedimentation during the Devonian and Carboniferous. (2) Variscan Orogeny. (3) Permian and Mesozoic sedimentation and basin development. (4) Cretaceous and Tertiary regional uplift. (5) Tertiary rifting and volcanism

Table 2. Parameters used in the modelling. * Geostatic stress ratios are calculated from the Poisson ratios. ** Representative values of the coefficient of internal friction are used here, using the approximate median value of the range of internal angles of friction.

	Definition and significance	Bunter sandstone	Greywacke	Slate	Unit
Density	The mass per unit volume of the rock and/or the fluids in the rock. Mean density controls the vertical (overburden) stress	2.68 [40]	2.42 to 2.74 [41]	2.7 to 2.9 [42]	g/cm^3
Tensile strength	The stress needed to cause extension fracturing. Controls where the failure envelope intersects the zero shear stress axis of the Mohr diagram, and the magnitudes of the effective tensile stresses needed to create extension fractures	6 [43]	20.3 to 35.7 [41]	4.4 normal to cleavage,14.4 along cleavage [42]	MPa
Uniaxial compressive strength	The strength of a rock derived from a uniaxial compression test (e.g., [44])	70 to 134 [45]	Average $\approx$ 200 (range 41–209) [46]	2.33 to 151.6 [47]	MPa
Cohesion	The shear strength of a material when the stress normal to a shear surface is zero (e.g., [48,49]). Controls where the failure envelope intersects the zero normal stress axis of the Mohr diagram, and the magnitudes of the effective differential stresses needed to create shear fractures	12 [45]	49 to 51 [41]	64 normal to cleavage, 11 when σ_1 30° to cleavage [42]	MPa
Poisson ratio	The relationship between the tendency to shorten in one direction and the tendency to expand in another direction (e.g., [50])	0.16 to 0.35 [51]	0.11 to 0.29 [41]	0.22 to 0.29 [42]	
Geostatic stress ratio *	The ratio of the horizontal effective stress (σ'_H) to the vertical effective stress (σ'_V) (e.g., [52]). It gives the effect the overburden has on horizontal stresses. Influences the diameter of the Mohr circle. Values calculated using Equation (2)	0.19 to 0.54	0.125 to 0.41 [41]	0.28 to 0.41	
Angle of internal friction (φ)	The angle of the fracture to σ_1 is $\pm (45° - \varphi/2)$ (e.g., [48])	27.6 to 37.9 [51]	43 to 44 [41]	30 to 50 [53]	Degrees
Coefficient of internal friction (μ) **	Controls the slope of the failure envelope in the compressional field of the Mohr diagram (e.g., [48]). $\mu = \tan \varphi$	0.52	0.97	0.84	
Young's modulus	The stiffness of a solid material (e.g., [54])	22 to 37 [51]	2.3 to 7 [41]	12 to 56 [42]	GPa
Coefficient of thermal expansion	The extent to which a material expands when heated or contracts when cooled. Can influence the development of tensile stresses during exhumation and cooling of rocks	11.25 [55]	17 [56]	8 [57]	10^{-6} K^{-1}

3. Model Set-Up

A Microsoft Excel spreadsheet has been created to make the necessary calculations and to plot Mohr diagrams (e.g., Figure 2). For simplicity, we keep the Mohr diagram analysis two-dimensional, considering just the vertical stress and the horizontal stress in one direction. This approach is taken both because it simplifies the analysis and because the orientations and magnitudes of the horizontal stresses are presently unconstrained. We start with a reference state where the horizontal strains are zero, which is the uniaxial strain condition (e.g., [58]) and the horizontal effective stress (σ'_H) is given by:

$$\sigma'_H = k_0\, \sigma'_V \tag{1}$$

where k_0 is the geostatic stress ratio [59] or coefficient of lateral earth pressure [60]. The geostatic stress ratio is the ratio of the horizontal effective stress to the vertical effective stress (i.e., $k_0 = \sigma'_H / \sigma'_V$). For an isotropic elastic material:

$$k_0 = \nu/(1 - \nu) \tag{2}$$

where ν is Poisson's ratio, which is generally in the range 0 to 0.5 (where 0.5 is incompressible). For most water-saturated rocks, is in the range $0.15 < \nu < 0.4$ (e.g., [61]).

We consider the vertical stress to result from the weight of the overlying material and the horizontal stresses to result from the combined effects of the geostatic stress ratio and the applied tectonic stresses. We start the analysis with a base-case, in which fluid pressure is hydrostatic and there are no applied tectonic stresses. We then consider how changes in fluid pressure and/or apply tensile or compressional horizontal ("tectonic") stresses might lead to fracturing.

3.1. Mohr Diagrams, Stresses and Failure Envelopes

Mohr diagrams are commonly used to show the relationships between stresses, fluid pressure, and fractures (Figure 1a; e.g., [62]). They provide a convenient graphic representation of the effective stress states and of failure (e.g., [17]). Normal stress (σ_N) is plotted on the *x*-axis and shear stress (τ) on the *y*-axis, with the principal axes of stress (σ_1 = maximum compressive stress, σ_2 = intermediate compressive stress, σ_3 = minimum compressive stress) or principal axes of effective stress (σ'_1 = maximum effective compressive stress, etc., where $\sigma' = \sigma - B.P_F$, where P_F = fluid pressure and B is Biot's constant, which we will assume to be approximately 1). Since the magnitudes and orientations of the horizontal stresses are currently unknown, we consider stress in 2D, with σ_2 being ignored (e.g., [63]). We assume an Andersonian stress regime with one principal stress vertical (the overburden stress, σ_V) and the other horizontal stress (σ_h) [64]. In the base case models used here, the starting point is that there are no applied tectonic stresses, so $\sigma_V = \sigma_1$, and $\sigma_h = \sigma_3$ (Figure 2).

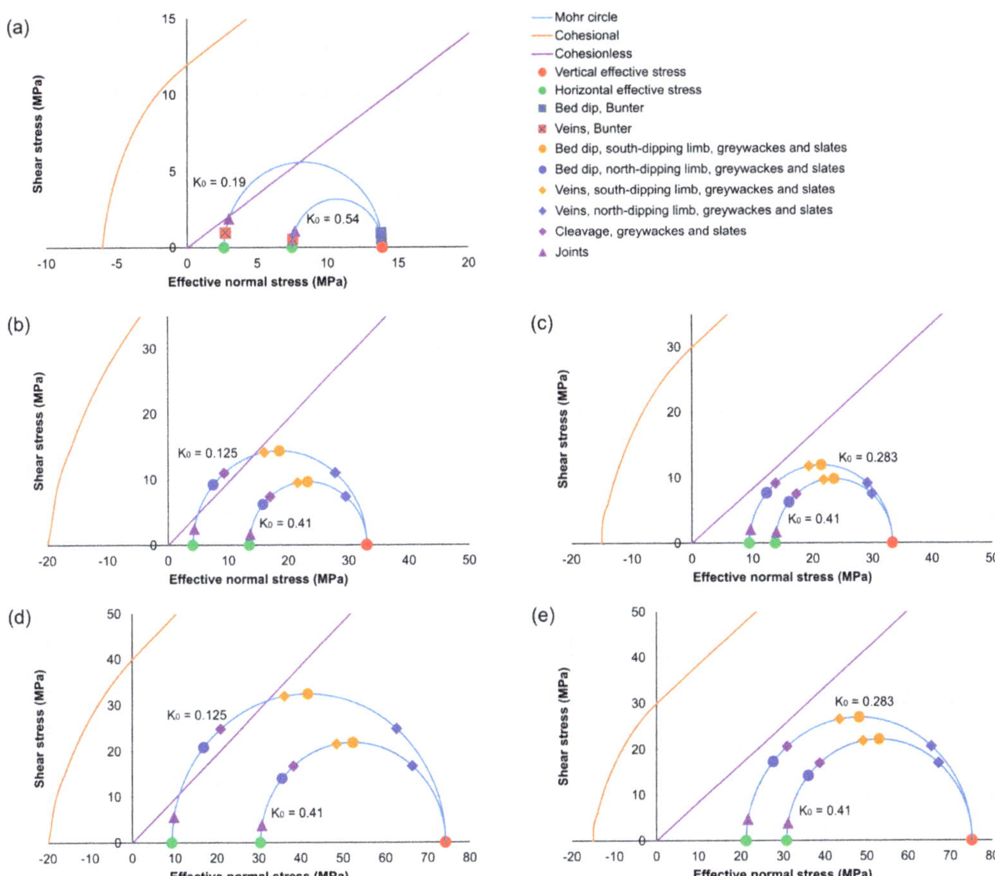

Figure 2. Base case Mohr diagrams, with Mohr circles for end-member geostatic stress ratios (k_0) shown (see Table 3 for parameters used). Representative dips of structures are shown to illustrate which of these structures might be reactivated under different stress conditions. (**a**) The Bunter Sandstone. (**b**) Devonian and Carboniferous greywackes at a depth of 2 km. (**c**) Devonian and Carboniferous slates at a depth of 2 km. (**d**) Devonian and Carboniferous greywackes at a depth of 4.5 km. (**e**) Devonian and Carboniferous slates at a depth of 4.5 km.

Table 3. Parameters used to model the base case scenarios for the Bunter Sandstone and the Devonian and Carboniferous greywackes and slates, with stresses and fluid pressures rounded to one decimal place. Values that depend on different geostatic stress ratios (k_0) are shown in different colours (red for lower k_0 values in range, blue for upper k_0 values). Values specific to a modelled depth of 2 km are shown in bold, and values specific to a modelled depth of 4.5 km are shown in italics. Representative structures are shown for the dips of structures in both the SE- and NW-dipping limbs of folds in the greywackes and slates. Effective stresses are calculated assuming a Biot coefficient of 1.

		Bunter	Greywacke	Slate	Unit
Input parameters	Depth	*1000*	**2000** *4500*	**2000** *4500*	m
	Average rock density	2.41	2.68	2.68	g/cm^3
	Fluid density	1	1	1	g/cm^3
	Overpressure	0	0	0	MPa
	Poisson's ratio (ν)	0.16	0.11, 0.29	0.22, 0.29	0.16
	Applied tectonic stress	0	0	0	MPa
	Cohesion	12	40	30	MPa
	Coefficient of internal friction (μ)	0.7	0.97	0.84	
	Friction angle	33	44	40	Degrees
	Tensile strength (T)	6	20	15	MPa
	Bed dip	5	SE limb = 45° NW limb = 70°	SE limb = 45° NW limb = 70°	Degrees
	Vein dip	85	SE limb = 45° NW limb = 25°	S limb = 45° NW limb = 25°	Degrees
	Cleavage	N/A	65	65	Degrees
	Joint dip	90	85	85	Degrees
Derived parameters	Fluid pressure	9.81	**19.6** *44.1*	**19.6** *44.1*	MPa
	Geostatic pressure ratio (k_0)	0.19	0.125, 0.41	0.283, 0.41	
	Vertical effective stress (σ'_V)	13.85	**33** *74.2*	**33.4** *75.1*	MPa
	Horizontal effective stress ($\sigma'_H = \sigma'_V k_0$)	2.63	**4.1, 13.5** *9.3, 30.4*	**9.4, 13.7** *21.2, 30.8*	MPa

Any 2D state of effective stress can be represented by a circle that intersects the *x*-axis at σ'_1 and σ'_3, with the centre of the circle representing the mean stress ($\sigma'_{Mean} = (\sigma'_1 + \sigma'_3)/2$). Mohr diagrams are useful for illustrating the conditions under which fracturing may occur in a particular rock under specific conditions (i.e., the failure envelope), but only if the effective stresses are plotted.

Fracturing is classically attributed to conditions where the effective stress components exceed some critical value, usually termed the strength, which is thought to be a property of the material (e.g., [65]). Tensile failure occurs if:

$$\sigma'_3 \leq T \tag{3}$$

where T is the tensile strength. Compressive stresses are positive. Shear failure occurs if:

$$\tau \geq S_0 + \mu \, \sigma_N \tag{4}$$

where μ is the coefficient of (internal) friction and S_0 is the cohesion. These conditions are usually linked to produce a failure envelope (e.g., [66,67]), as in Figure 1a.

The composite failure criteria used in this work (e.g., Figures 1 and 2) combine a linear, Mohr–Coulomb envelope for shear fracture under compressive effective stresses, with a parabolic, plane Griffith envelope for tensile/hybrid fractures under tensile stress conditions. The two envelopes join at the τ-axis ($\sigma' = 0$), where there is a discontinuity in the slopes of the failure envelopes. Continuity in the slope could be achieved by using the methods outlined in [68]. This type of composite failure envelope is consistent with a modification of the Griffith theory to account for closure and frictional behaviour under compression [69,70].

Following [71], the plane Griffith failure envelope in Mohr space (τ, σ') is given by:

$$\tau^2 = 4\,T\,(\sigma' + T) \tag{5}$$

and the Mohr–Coulomb envelope by:

$$\tau = S_0 + \mu\,\sigma' \tag{6}$$

where T = Tensile strength, S_0 = cohesion, and μ = coefficient of internal friction. Putting $\sigma' = 0$ in both Equations (5) and (6) gives:

$$\tau = S_0 = 2T \tag{7}$$

Hence, the cohesion would be simply twice the tensile strength. The unconfined compressive strength (C_0), is a widely measured parameter representing the stress required for failure when $\sigma_{3'} = 0$ which for Mohr–Coulomb failure is given by:

$$C_0 = 2S_0\,[(\mu^2 + 1)^{1/2} + \mu] \tag{8}$$

Substituting Equation (7) in Equation (8) gives:

$$C_0 = 4T\,[(\mu^2 + 1)^{1/2} + \mu] \tag{9}$$

Using a value of $\mu \approx 1$, gives the following approximations for rocks: (1) $S_0 \approx 0.2\,C_0$; (2) $T \approx 0.5\,S_0$; so (3) $T \approx 0.1\,C_0$ [71]. Sedimentary rocks typically show an exponential increase in C_0 with decreasing porosity [72].

We assume a lower bound to likely failure envelopes is given by the failure envelope shown by the "Cohesionless" line in Figure 1a, which is typical of slip on pre-existing cohesionless fractures (e.g., joints) (e.g., [73]). Such failure envelopes typically have coefficients of friction of 0.6 to 0.9 (average 0.75; e.g., [74]).

In Figure 1a, the effective stress state shown by Mohr circle ❶ is typical of a rock at depth of ~2 km, with the maximum compressive stress being vertical (σ_V) and caused by the weight of the overlying material. The rock is under a hydrostatic fluid pressure (where P_F/σ_V = ~0.4, based on the assumption that $P_F \sim 20$ MPa and $\sigma_V \sim 50$ MPa at 2 km) and a tectonic stress (-2 MPa, as may occur during regional extension). Note that this is in the stable region, but within ~5 MPa of the failure envelope for rocks with cohesionless fractures. Figure 1a shows stress states that are stable (❶), that would cause slip on a suitably orientated pre-existing cohesionless fractures (❷), and that would create tensile failure of the intact rock (❸).

In this paper, we use Mohr diagrams to illustrate the conditions under which a rock can go from a stable stress system (i.e., no active fracturing occurs) to an unstable stress system (i.e., fracturing occurs). The change from stable to unstable condition depends upon:

1. The rock properties used to define the failure envelope are the tensile strength (T), uniaxial compressive strength (UCS), cohesion (S_0), and coefficient of internal friction (μ) [75];
2. The stress state (σ), which is defined in terms of principal stresses, mean stress and differential stress [17]. Stresses are in turn controlled by factors, such as depth of burial (overburden), tectonic (horizontal) stresses, and other changes in the physical state of the material, such as expansion or contraction caused by temperature and volume change (e.g., [76]). Changes in stresses that lead to fracturing can either be by increasing [77] or reducing the [78] the applied compressive stresses;
3. In the upper crust, fluid pressure in the pores and cracks combines with the applied stresses to produce an effective stress, where $\sigma' = \sigma - P_F$ (e.g., [79–81]). In the absence of specific information, we use a Biot coefficient (B) of 1, where $\sigma' = \sigma - B.P_F$ [82]. Changes in fluid pressure that can lead to fracturing can either be an increase in fluid pressure (e.g., [83]) or a reduction in fluid pressure, which can cause pore collapse (e.g., [84,85]). Pore collapse is not considered further in this paper.

The type and orientation of fractures developed can be predicted from the relationship between the Mohr circle for effective stress and the failure envelope (Figure 1b). Extension fractures typically develop perpendicular to the direction of σ'_3 if σ'_3 touches the failure envelope on the $\tau = 0$ axis (Figure 1b❶), which generally requires a relatively low differential effective stress ($\sigma'_{Diff} = \sigma'_1 - \sigma'_3$). Shear fractures are predicted to develop if the Mohr circle touches the failure envelope in the compressive field (Figure 1b❸), this typically requiring a relatively high σ'_{Diff}. Hybrid fractures develop by synchronous extension and shear (e.g., [29]), and occur if the Mohr circle touches curved part of the failure envelope within the tensile field of the Mohr diagram (Figure 1b❷).

3.2. Input Data, Assumptions, and Uncertainties

Table 2 shows values from the literature for various rock mechanical parameters for the Bunter Sandstone, greywackes, and slates derived from triaxial tests. Table 2 also gives definitions and the significance of each parameter. Triaxial tests typically use intact hand specimen sized rock samples with no pre-existing fractures visible. As such, these values almost certainly over-estimate the strengths of the larger rock masses in the sub-surface, which generally have pre-existing fractures (e.g., [86]), and should be considered as upper-limit values of rock strength. We, therefore, use lower values for rock strength as inputs into the models (Table 3). The failure envelope (e.g., Figure 2) for each rock type is controlled by the tensile strength (T), cohesion (S_0) and coefficient of internal friction ($\mu = \tan \varphi$, where φ = angle of internal friction) (Equations (3) and (4)). The most widely available rock strength parameter is the uniaxial compressive strength (UCS or C_0), determined directly from triaxial tests or estimated from geophysical log data [72,73].

Note that the values presented in Table 3 are generally within the ranges of values for Devonian and Carboniferous greywackes and slates in the Harz Mountains and Belgium presented by [87]. Exceptions are the porosity (greywackes, mean value = 2.42%, range 0.04 to 7.35%, $n = 88$; slates, mean value = 2.21%, range 0.17 to 5.51%, $n = 20$), and the cohesion of the greywackes (mean value = 26.57 MPa, range 10.5 to 45.9 MPa, $n = 5$).

The Mohr circle for effective stress is controlled by the applied stresses and the fluid pressure. The vertical stress can be calculated from the densities of the overlying rocks (e.g., [73,88]). The magnitudes and orientations of these horizontal stresses are much harder to calculate (e.g., from well data; [89]) or predict (e.g., from tectonic stress tensors; [90]). Similarly, pore fluid pressure in rocks in the sub-surface is difficult to predict without well data. For example, [91] use sediment consolidation experiments and numerical models to predict fluid pressures.

In the absence of appropriate sub-surface data, the magnitudes of the horizontal stresses are considered in terms of the geostatic stress ratio of the rock (e.g., [92]). The pore pressures are discussed in terms of the hydrostatic pressure and any assumed overpressure.

The modelling presented here makes several assumptions. These assumptions are made to act as a starting point, to simplify the analysis and because of the various uncertainties:

1. An Andersonian stress system is assumed, i.e., with one of the principal axes of stress being vertical and the other two being horizontal [64];
2. The analysis is carried out in two-dimensions, considering just vertical stress and horizontal stress. This simplifies the analysis and is, we argue, justified at the pre-drilling stage of analysis because of the magnitudes and orientations of the horizontal stresses are currently unknown. Hydrofracture data from three wells in the region suggest a thrust regime with a maximum horizontal stress orientated ~WNW-ESE [93];
3. The vertical stress is produced by the weight of overburden;
4. The fluids are hydrostatically pressured;
5. The failure parameters used in the modelling (Table 3) are assumed to be representative of the rock properties in the sub-surface.

3.3. Base Case Models

We start with base case models (Figure 2) to derive threshold conditions for the reactivation of fractures or the generation of new fractures. These base cases use the uniaxial strain condition (e.g., [58]), with no applied tectonic stresses. The fluids in pore spaces and open fractures are hydrostatically pressured, with the water table being at the ground surface. In such circumstances, the vertical effective stress creates a horizontal effective stress that is given by the geostatic stress ratio (Figure 2), with the input parameters used shown in Table 3. The base case models are used as a starting point for experiments in which fluid pressure is increased (as would occur either by natural overpressure or by hydraulic stimulation), and the horizontal (tectonic) stress either decreased or increased, until the model predicts either the reactivation of existing cohesionless fractures (hydraulic stimulation) or the generation of new fractures.

For most rocks, $T \ll 100$ MPa, with poorly-consolidated sediments having $T \rightarrow 0$, whereas for shear failure $0.5 < \mu < 1.5$ (generally) and $0 < S_0 < 50$ MPa. In this paper, we use ranges of T, S_0, and μ that are appropriate for the rocks being considered.

An important point illustrated by Figure 2 is that, at typical K_0 values (0.3–0.5), the stress state is stable and no fracturing would be expected. This means that low K_0 values are needed to allow failure in the base case model, which simulates simple burial, with hydrostatic fluid pressure and no applied tectonic stresses. Any failure would largely be by shearing of pre-existing fractures, which may occur in the Devonian and Carboniferous rocks that are considered to occur beneath Göttingen.

3.4. A Range of Stress States for Fracturing

We consider a range of states of effective stress, seven of which are shown in Figure 3, together with their effect on fracturing. Note that the discussion is of changes in effective stresses, which can be caused by changes in applied stresses and/or fluid pressure. These states of stress have been modelled for the Bunter Sandstone and the Devonian and Carboniferous greywackes and slates by increasing the fluid pressure and by applying "tectonic" stresses to the base case models.

4. Effects of Key Parameters

A definition and the significance of each of the input parameters, along with representative values, are given in Table 2. For a given overburden stress, higher geostatic stress ratios produce larger horizontal stress and, hence, lower differential stress (Figure 2). The effects of other key parameters are illustrated in Figure 4. The position and shape of the failure envelope is controlled by the tensile strength, cohesion, and coefficient of internal friction of the rock (Figure 4a). Vertical, and therefore horizontal, stress increases with depth as the overburden increases (Figure 4b). High values of compressional tectonic stress can cause the horizontal stress to exceed the vertical stress, tending to lead to shear failure, while tensile tectonic stress will increase the differential stress and may lead to the development of extension fractures (Figure 4c). Increases in fluid pressure move the Mohr circle for effective stress to the left, towards more tensile parts of the Mohr diagram (Figure 4d). Note, however, that a change in fluid pressure will cause a change in differential stress that is proportional to the geostatic stress ratio, and it is a common mistake to show a constant differential stress with changing fluid pressure (e.g., [94,95]). The base cases are close to cohesionless failure if lower K_0 values are used, which implies that fracturing will generally require some tectonic stress and/or overpressure, especially to initiate the development of new fractures.

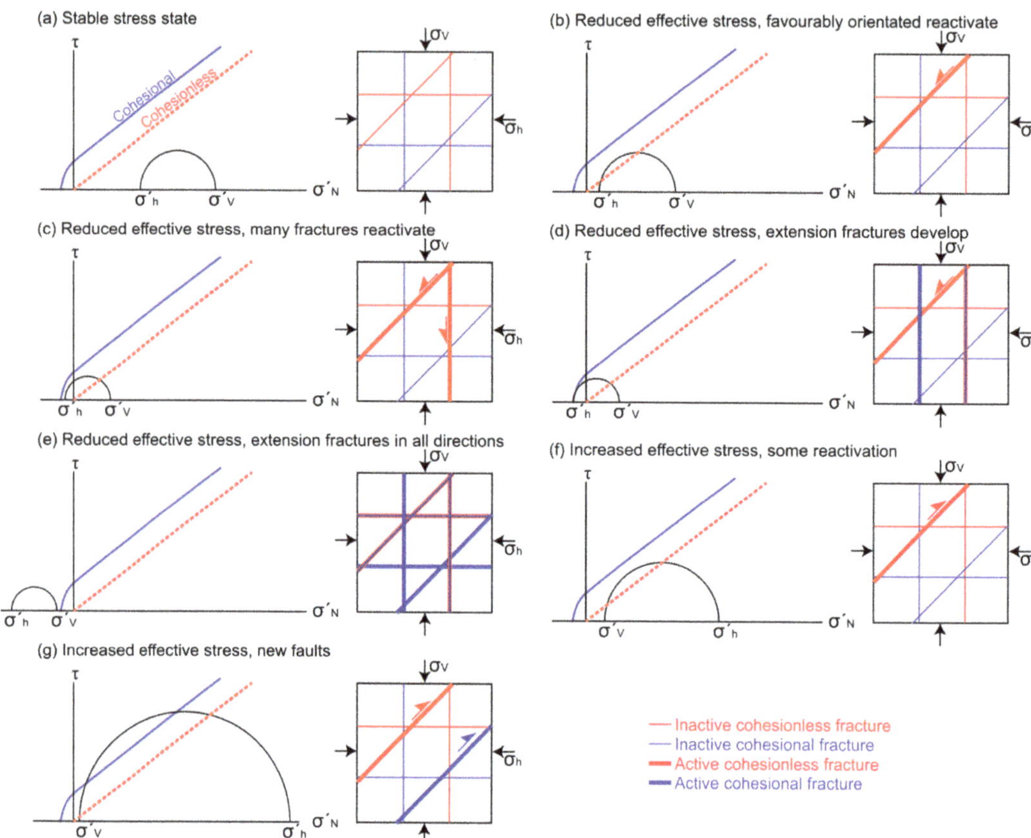

Figure 3. A range of effective stress states for fracturing illustrated using Mohr diagrams and sketch cross-sections with cohesive fractures (e.g., veins) and non-cohesive fractures (e.g., joints) dipping at 90°, 45°, and 0°. σ_V = vertical stress, σ_h = horizontal stress. (**a**) Stable stress state, in which the effective stresses are insufficient to reactivate existing fractures or create new fractures. (**b**) Reduced effective stresses, such that favourably-orientated pre-existing cohesionless fractures can be reactivated as shear fractures. (**c**) Reduced effective stresses, such that favourably-orientated pre-existing cohesionless fractures can be reactivated as shear fractures. (**d**) Reduced effective stresses, such that new extension fractures are created perpendicular to the least effective stress. (**e**) Reduced effective stresses (e.g., as fluid pressure increases), such that new extension fractures with any orientation can be created. (**f**) Increased effective stresses, such that favourably-orientated pre-existing cohesionless fractures are reactivated as shear fractures. (**g**) Increased effective stresses, such that new shear fractures are generated.

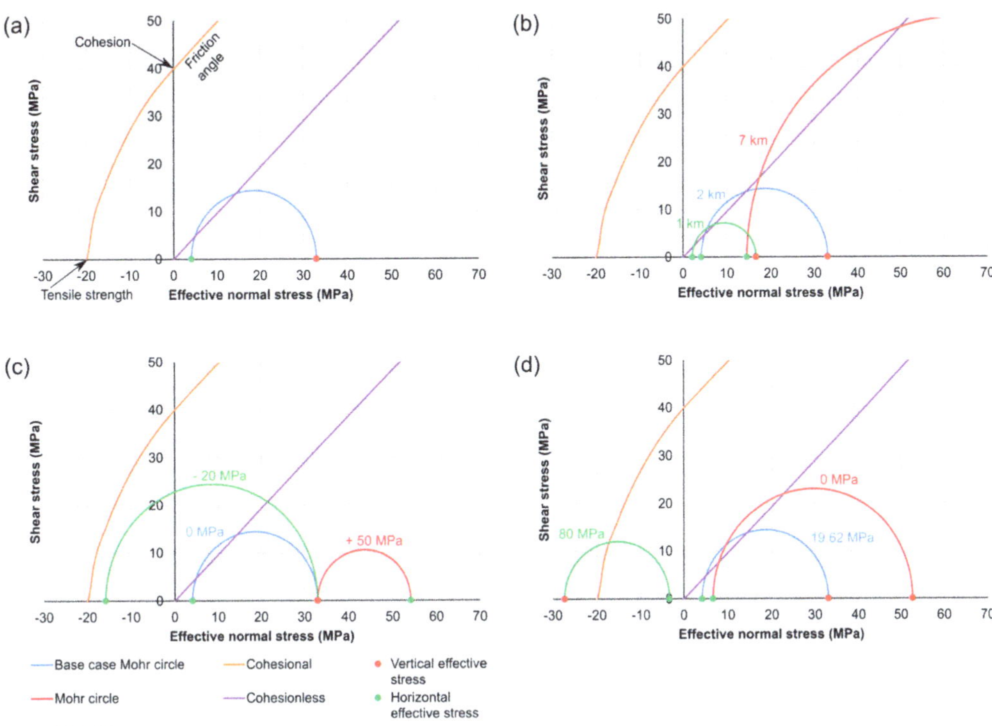

Figure 4. Mohr diagrams showing the effects of key parameters on the failure envelope and state of effective stress, all shown using the model for the Devonian and Carboniferous greywackes with a geostatic stress ratio of 0.125. The parameters used are shown in Table 3. (**a**) The base case Mohr circle (depth of 2 km, hydrostatic fluid pressure and no applied tectonic stress), with a representative cohesional failure envelope shown (Table 3, "Greywacke" and "Slate" columns). "Friction angle" refers to the slope the failure envelope. (**b**) Mohr circles for effective stress at different depths (hydrostatic fluid pressure and no applied tectonic stress). The magnitudes of the effective stresses increase with depth. (**c**) Mohr circles for a depth of 2 km, hydrostatic fluid pressure and different applied tectonic stresses. (**d**) Mohr circles for a depth of 2 km, different fluid pressures and zero applied tectonic stresses.

The Mohr diagrams presented in Figures 2–4 show two types of failure envelope. The "Cohesionless" lines on the Mohr diagrams are typical of slip on pre-existing unfilled fractures (e.g., joints; Figure 5a), while the "Cohesional" lines represent the behaviour of intact rock or fractures that are filled by cohesive material (e.g., veins or faults with cohesive gouge; Figure 5b). These different failure envelopes illustrate the different mechanical behaviours of such cohesionless fractures as joints and such cohesional fractures as fully-filled veins and faults with clay gouge, highlighting the importance of distinguishing between different fracture types when using a field analogue for a geothermal reservoir.

5. Potential for Reactivating and Generating Fractures

Results of the models can be expressed in terms of the critical values of fluid pressure or horizontal (tectonic) stresses needed to reactivate existing cohesionless fractures, or to generate new fractures (Table 4). Note that these are simply the results of the model, which is based on poorly constrained parameters and should only be taken as suggestions of what may occur under the conditions stated.

5.1. Bunter Sandstone

Critical values predicted by the modelling for the Bunter Sandstone at a depth of 1 km are presented in Table 4, from which the following predictions can be made. (1) Favourably-orientated cohesionless fractures in the Bunter Sandstone may be critically stressed without fluid overpressure or applied tectonic stresses (i.e., the base-case model) if the geostatic stress ratio is low (e.g., 0.2). (2) Shear may therefore occur on favourably-orientated cohesionless fractures (e.g., joints) if the fluid is hydrostatically pressured (i.e., fluid pressure ~10 MPa; e.g., Figure 3b) for a geostatic stress ratio ~0.2, but a fluid pressure of about 24 MPa would be needed to reactivate cohesionless fractures if the geostatic stress ratio is ~0.54. (3) Gently-dipping extension fractures would start to develop at fluid pressures of about 30 MPa. (4) Steeply-dipping extension fractures will develop if there is a tensile tectonic stress of about 8.2 to 13.5 MPa (e.g., Figure 3d). (5) Extension fractures may develop in all directions if the fluid pressure is about 34 to 54 MPa (e.g., Figure 3e). (6) Favourably-orientated cohesionless fractures may be reactivated in shear (normal faulting) in the greywackes if there is a tensile tectonic stress of 0 to 3.7 MPa (e.g., Figure 3b). (7) Favourably-orientated cohesionless fractures may be reactivated as thrusts if there is a compressive tectonic stress of 43 to 48 MPa (e.g., Figure 3f). (8) New thrusts will develop if there is a compressive tectonic stress of about 90 to 95 MPa (e.g., Figure 3g).

Figure 5. Examples of fractures in Carboniferous greywackes, cherts and slates in the Harz Mountains. (**a**) Joints (cohesionless) in thinly-bedded greywackes and slates at Okerstausee, Niedersachsen. (**b**) Example of cohesional fractures at Lautenthal, Niedersachsen. Clay has been injected along faults with centimetre-scale displacements, related to block rotation in a chert bed accommodated by thickness variations in a slate layer.

Table 4. Results of the modelling for the Bunter Sandstone and the Lower Palaeozoic greywackes and slates, showing the calculated critical values. Values specific to a modelled depth of 2 km are shown in bold, and values specific to a modelled depth of 4.5 km are shown in italics. * Shear or hybrid fractures, rather than extension fractures, may develop in the slates under these conditions.

	Bunter Sandstone $K_0 = 0.19$	Bunter Sandstone $K_0 = 0.54$	Greywacke $K_0 = 0.125$	Greywacke $K_0 = $ ratio $= 0.41$	Slate $K_0 = 0.283$	Slate $K_0 = 0.41$	Units	Notes
Fluid pressure, cohesionless	9.81	24	**19.62** *44.15*	**50** *115*	**51** *115*	**51** *115*	MPa	Shear on favourably-orientated cohesionless fractures
Fluid pressure, gently-dipping cohesional	29.5	30	**72** *138*	**72** *138*	**68** *134*	**68** *134*	MPa	Gently-dipping extension fracture develop
Fluid pressure, all orientations of extension fractures	54	34.4	**210** *278*	**100** *168*	**105** *172*	**90** *155*	MPa	Steep extension fractures develop
a. Base case	Unstable	Stable	**Unstable** *Unstable*	**Stable**	**Stable**	**Stable**	MPa	Stable stress state for higher K_0
b. Decreased tectonic stress, reactivation starts	0	−3.7	**0** *0*	**−7.5** *−17*	**−2** *−5*	**−6.5** *−14*	MPa	Shear on favourably-orientated cohension-less fractures
c. Decreased tectonic stress, many fractures reactivated	−5	−8	**−10** *−20*	**−20** *−30*	**−10** *−22*	**−14** *−31*	MPa	Shear on a cohesionless fractures with a wide range of orientations
d. Reduced tectonic stress, extension fractures develop	−8.2	−13.5	**−24** *−29*	**−33** *−50*	**−24** *−33 **	**−28.5** *−43 **	MPa	Extension fractures develop per-pendicular to least com-pressive stress
e. Reduced tectonic stress, extension fractures in all orientations	N/A	N/A	**N/A** *N/A*	**N/A** *N/A*	**N/A** *N/A*	**N/A** *N/A*	MPa	Requires increase in fluid pressure
f. Increased tectonic stress, some reactivation of cohesionless	48	43	**0** (normal) **180** (thrusts) *0* (normal) *400* (thrusts)	**170** *370*	**145** *320*	**140** *315*	MPa	Reactivation of cohesionless fractures in shear
g. Increased tectonic stress, new shear fractures can develop	95	90	**370** *590*	**360** *580*	**275** *450*	**270** *440*	MPa	Creation of new shear fractures

5.2. Devonian and Carboniferous Greywackes and Slates

Critical values predicted by the modelling for the Devonian and Carboniferous greywackes and slates at depths of 2 km and 4.5 km are presented in Table 4. These depths are used because they represent the proposed depths for an initial research well and an exploitation well, respectively. The results of the modelling enable the following predictions to be made.

1. Very low geostatic stress ratios (e.g., 0.125) in the greywackes and slates are required to reactivate cohesionless fractures without fluid overpressure or applied tectonic stresses (base-case model; e.g., Figure 3a);
2. A fluid pressure of about 50 MPa would be needed to reactivate cohesionless fractures in both the greywackes and slates at a depth of 2 km if the geostatic stress ratio is high (e.g., 0.41; e.g., Figure 3b), with fluid pressures reaching lithostatic pressures;
3. Gently-dipping extension fractures will start to develop in the greywackes at a depth of 2 km if the fluid pressure is about 72 MPa, but may develop in the slates at slightly lower pressures (about 68 MPa). The models predict that, in the absence of cohesionless fractures, increasing fluid pressure will initially create gently-dipping extension fractures. This is because of the assumption of uniaxial strain (i.e., that the rocks are laterally confined). Higher fluid pressures will be required to generate steeply-dipping extension fractures, in the absence of horizontal tensile stresses, such as those related to tectonic forces or to cooling (Section 6);
4. Steeply-dipping extension fractures are predicted to develop in the greywackes at 2 km depth if there is a tensile tectonic stress between about -24 and -40 MPa, but are likely to develop in the slates at tensile tectonic stress between about -24 to -33 MPa (e.g., Figure 3d);
5. Extension fractures in all directions may develop in the greywackes at 2 km depth if the fluid pressure is between about 100 and 210 MPa, but would develop in the slates if the fluid pressure is between about 90 and 105 MPa (e.g., Figure 3e);
6. Shear (normal faulting) may begin on favourably-orientated cohesionless fractures in the greywackes at a depth of 2 km if there is a tensile tectonic stress between about 0 and -7.5 MPa, and in the slates if there is a tensile tectonic stress between about -2 and -6.5 MPa (e.g., Figure 3b);
7. Favourably-orientated cohesionless fractures may be reactivated as thrusts in the greywackes at 2 km depth if there is a compressive tectonic stress of about 170 MPa, and in the slates at about 140 MPa (e.g., Figure 3f);
8. New thrusts will begin to develop in the greywackes at a depth of 2 km if there is a compressive tectonic stress of about 360 MPa, but will develop in the slates if there is a compressive tectonic stress of about 270 MPa (e.g., Figure 3g).

The different critical values for the greywackes and slates predicted by the modelling are caused by the different physical properties that have been used for these rock types. The different geostatic stress ratios and failure criteria (especially cohesion and tensile strengths) are particularly important. The different geostatic stress ratios of different rock types may explain variations of fractures with lithology. This is commonly seen in greywacke/slate and sandstone/mudstone sequences (e.g., [96]). The tensile strengths of slates tend to be lower than that of greywackes, so extension fractures may develop in the slates at lower fluid pressures and lower tensile tectonic stresses than in the greywackes. Although this suggests that fractures will be created in slates before fracturing occurs in greywackes during stimulation (Section 6.1), it does not necessarily mean that there will be more, wider or better-connected fractures in the slates than in the greywackes. The model cannot predict the fluid flow properties of the slates or greywackes.

5.3. Possible Effects of Exhumation on the Bunter Sandstone

Here, we use simple modelling to determine at what depth joints are likely to have developed in the Bunter Sandstone of the Leinetal Graben, based on the changes in stresses and temperatures the rocks are likely to have experienced during exhumation. The Bunter

Sandstone is currently exposed on the flanks of the Leinetal Graben but at depths of between ≈300 m and ≈850 m in Borehole Sudhein II in the Graben [37]. Note that the top of the Bunter Sandstone has been faulted out and the base is mixed with Zechstein Salt in Borehole Sudhein II [37]. The values used in the modelling are shown in Table 5, with the Mohr diagrams shown in Figure 6. The vertical stress changes as the thickness and, therefore, weight of the overlying material changes during burial or exhumation, and this causes a change in the horizontal stresses that is proportional to the geostatic stress ratio (Figure 4b). A change in depth typically causes a change in temperature that is related to the geothermal gradient (e.g., [97]). A decrease in temperature will tend to cause contraction of rock, determined by the coefficient of thermal expansion, and this will tend to cause a reduction in compressional stresses in the rock (e.g., [98]).

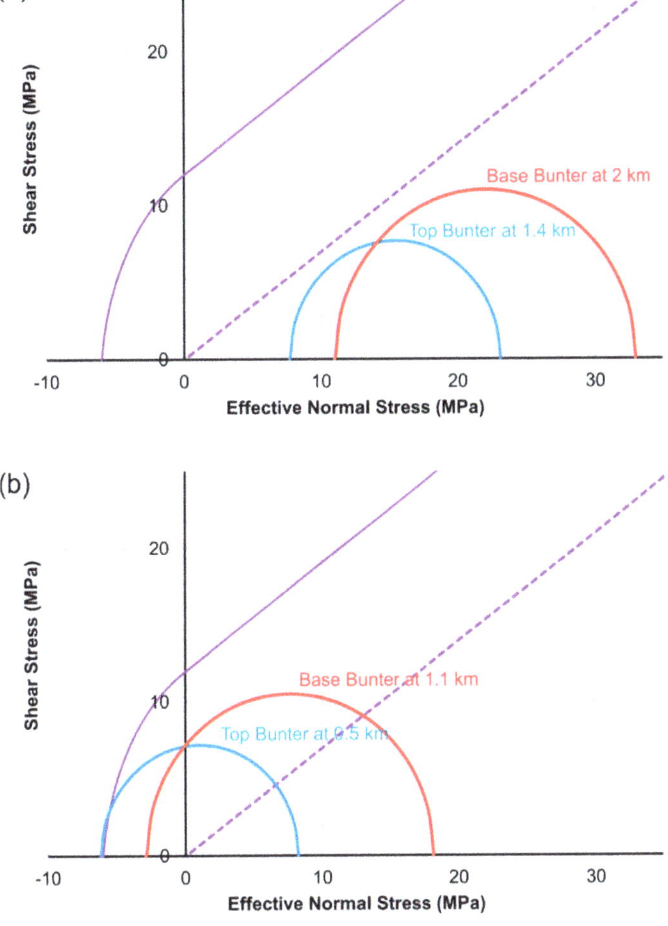

Figure 6. Mohr diagrams illustrating the possible effects of exhumation and cooling on the Bunter Sandstone. See Table 5 for parameters used in the model. (**a**) Model for the base of the Bunter Sandstone at a depth of 2 km and the top of the Bunter Sandstone at 1.4 km. The model predicts the rock is in a stable stress state. (**b**) Model for the Bunter Sandstone after 0.9 km exhumation. The top of the Bunter Sandstone is now at a depth of 0.5 km, and the stress state promotes the development of steeply-dipping extension fractures. Notice that the differential effecting stresses in (**a**,**b**) are similar, because cooling during exhumation tends to generate horizontal tensile stresses.

The model presented in Figure 6 includes reductions in horizontal stresses caused by decreases in overburden and temperature to predict at what depth extension fractures (e.g., joints) may develop in the Bunter Sandstone. We use a geothermal gradient of 30 °C per km, which is similar to that modelled by [99]. The model assumes, for simplicity, an <u>initial</u> 1.4 km depth for the top of the Bunter Sandstone, and an <u>initial</u> depth of 2 km for the base of the Bunter Sandstone (Figure 6a). Vertical and horizontal stresses reduce during exhumation, with the model predicting the development of joints beginning after 0.9 km exhumation, with the top of the Bunter at a depth of 0.5 km (Figure 6b). Note, however, that this base case model assumes hydrostatically-pressured pore fluids and no applied tectonic stresses. Joints would form at greater depths if the fluids are over-pressured, if there is a tensile tectonic stress, or if the geothermal gradient were higher.

Table 5. Initial values used for modelling the effects of exhumation and cooling on the Bunter Sandstone.

Parameter	Value	Unit
Density	2.68	g/cm^3
Porosity	10	%
Tensile strength	6	MPa
Cohesion	12	MPa
Poisson ratio	0.25	
Geostatic stress ratio	0.333	
Young's modulus	22	GPa
Coefficient of thermal expansion	11.25	$10^{-6}\,°C^{-1}$
Geothermal gradient	30	°C per km
Fluid pressure	Hydrostatic	
Tectonic stresses	Zero	
Initial top of unit	1.4	km
Initial base of unit	2	km

6. Discussion

6.1. Possible Effects of Stimulation on Devonian and Carboniferous Rocks at 2 km Depth

Here, we consider the possible effects of stimulation by increasing fluid pressure on the folded, cleaved and veined Devonian and Carboniferous greywackes and slates, based on the modelling discussed in Section 5 and using a simple schematic figure for these rocks (Figure 7a). The modelling suggests the following sequence of events as fluid pressure is gradually increased (Figure 7):

1. Favourably-orientated cohesionless fractures (joints) may be critically-stressed in greywackes with a very low geostatic stress ratio, so may undergo normal faulting even at hydrostatic fluid pressure (Figure 7a). Faulted joints are described by [100];
2. Shear can occur along favourably-orientated cohesionless fractures (e.g., joints) in the slates at fluid pressures of about 50 MPa, which is an overpressure of about 30 MPa (Figure 7b);
3. Gently-dipping extension fractures can be generated in the slates a fluid pressure of about 68 MPa, and gently-dipping cohesionless fractures in the greywackes may be reactivated as extension fractures (Figure 7c);
4. Gently-dipping extension fractures can be generated in the greywackes at a fluid pressure of about 72 MPa (Figure 7d);
5. Extension fractures with any orientation may develop in the slates at fluid pressure between about 90 and 105 MPa (Figure 7e);
6. Extension fractures with any orientation may develop in the greywackes if the fluid pressure is between about 100 and 210 MPa (Figure 6f).

6.2. Possible Effects of Stimulation on Devonian and Carboniferous Rocks at 4.5 km Depth

The critical values for deformation predicted by the modelling for the greywackes and slates at a depth of 4.5 km are shown in Table 4. These critical values are higher than predicted for a depth of 2 km, but suggest that stimulation would have similar effects as at a depth of 2 km. One notable difference is that higher differential stresses at depth would mean that stimulation is more likely to reactivate fractures in shear, or to create shear fractures, at greater depths. The modelling predicts that the slates will be particularly likely to develop shear fractures during stimulation.

Figure 8 highlights the effects of the different fluid pressures needed to stimulate reservoirs at different depths. The difference in fluid pressure between the top and bottom of a 2500 m column of water would be approximately 25 MPa. If fluid pressure is increased at a depth of 4.5 km such that it exceeds the horizontal stress, this may cause the fluid pressure at 2 km to exceed the vertical stress, if these different depths are hydraulically-connected (Figure 8). This suggests that care will be needed during stimulation if it is not intended to fracture rocks at shallower levels. We note that the Zechstein salts are likely to act as an effective top-seal (e.g., [101]) during stimulation.

6.3. Possibility of Open Fractures in the Devonian and Carboniferous Rocks below Göttingen

Although we have carried out simple modelling of joint development in the Bunter Sandstone, we have not performed so for the Devonian and Carboniferous rocks that are inferred to occur below about 1.5 km under Göttingen. This is because we consider there to be too many uncertainties for meaningful analysis of the effects of Late Cretaceous to Tertiary exhumation and cooling on these deeper and older rocks. There are, however, two arguments for the occurrence of open fractures in the Devonian and Carboniferous rocks. Firstly, joints will have developed in those rocks as they were exhumed at the end of the Variscan Orogeny, before deposition of the Permian and Mesozoic rocks unconformably above. It is possible, however, that these late- or post-Variscan joints will have subsequently been mineralised. Secondly, there is evidence that joints and other open fractures can occur at the depths of the proposed geothermal reservoir at Göttingen. For example, joints have been reported at depths of >2 km in tunnels (e.g., [102]) and in mines (e.g., [103,104]). Similarly, open fractures have been reported in fractured "basement" petroleum fields at depths of >4 km (e.g., [105]).

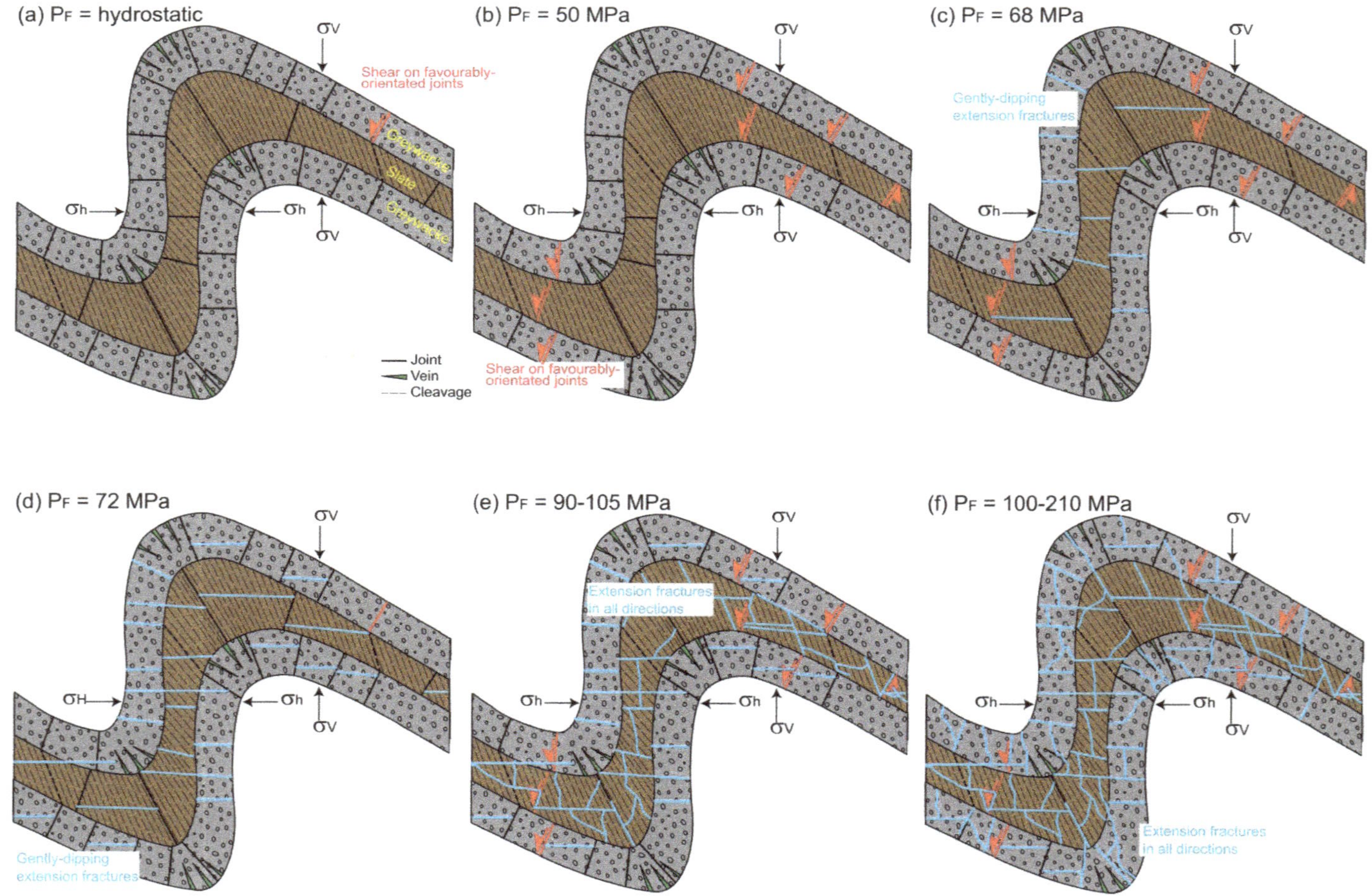

(a) P_F = hydrostatic
(b) P_F = 50 MPa
(c) P_F = 68 MPa
(d) P_F = 72 MPa
(e) P_F = 90-105 MPa
(f) P_F = 100-210 MPa
σv
σh
Cleavage
Slate
Cleavage
Joint
Vein
Cleavage
Shear on favourably-orientated joints
Shear on favourably-orientated joints
Gently-dipping extension fractures
Gently-dipping extension fractures
Extension fractures in all directions
Extension fractures in all directions

Figure 7

Figure 7. (previous page). Schematic figure for likely effects of stimulation on the folded and cleaved Devonian and Carboniferous greywackes and slates, based on the results presented in Table 4. The fold has an amplitude of a few metres to tens of metres. The model is for a depth of 2 km, at which hydrostatic fluid pressure would be 19.62 MPa. Veins are shown in the greywackes in the outer arcs of folds, and a fanning axial planar cleavage is shown. Joints are drawn approximately perpendicular to bedding. We predict the following sequence of events as fluid pressure is increased. (**a**) Favourably-orientated cohesionless fractures (joints) may be critically-stressed in greywackes with very low geostatic stress ratios, so may undergo normal faulting even at hydrostatic fluid pressure. (**b**) Shear can occur along favourably-orientated joints in the slates at fluid pressures of about 50 MPa. (**c**) Gently-dipping extension fractures can be generated in the slates at fluid pressures at about 68 MPa, and gently-dipping joints in the greywackes may be reactivated as extension fractures. (**d**) Extension fractures with any orientation may develop in the slates at fluid pressure between about 90 and 105 MPa. (**e**) Gently-dipping extension fractures can be generated in the greywackes at fluid pressures of about 72 MPa. (**f**) Extension fractures with any orientation may develop in the greywackes if the fluid pressure is between about 100 and 210 MPa.

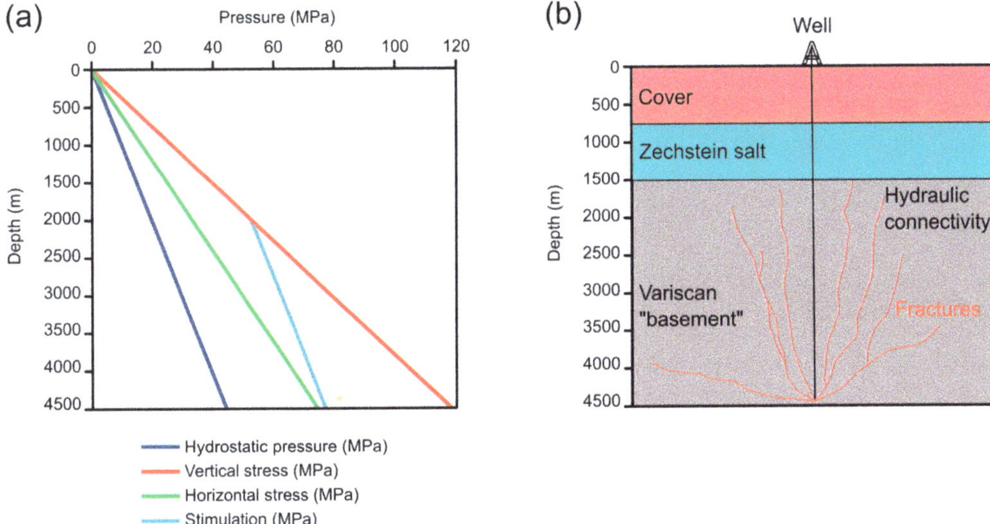

Figure 8. Possible effects of stimulation at a depth of 4.5 km on the rocks at a depth of 2 km, modelled for a greywackes and slates with a geostatic stress ratio of 0.41. (**a**) Graph of pressure against depth. (**b**) Schematic cross-section. At a depth of 4.5 km, hydrostatic fluid pressure = 44.15 MPa, σ_V = 118.31 MPa and σ_h = 74.55 MPa. At a depth of 2 km, hydrostatic fluid pressure = 19.62 MPa, σ_V = 52.58 MPa and σ_h = 33.14 MPa. If there is hydraulic connectivity through the greywackes and slates, increasing fluid pressure at 4.5 km to above 77 MPa, such that it exceeds σ_h, would increase fluid pressure at 2 km 52.58 MPa, at which point it will equal σ_V.

6.4. Potential Benefits, Problems, and Improvements

Although the modelling presented in this paper is simple, it has several uses when attempting to predict what is likely to occur in the sub-surface. Firstly, constructing the models requires consideration of the range of the conditions that are likely to occur, including the lithologies and their mechanical properties, stresses, fluid types and fluid pressures, and the presence of different types of fractures. Secondly, predictions can be made about whether natural open fractures (e.g., joints) may occur, and whether exhumation might have created joints at reservoir depths. Thirdly, it enables predictions to be made about the orientations and types of induced fractures that are likely to occur under a range of possible conditions (Figure 7). It may also enable predictions to be made about whether there will be induced seismicity during stimulation.

There are various potential problems with the simplicity of the approach presented here, but these suggest ways in which the models can be improved. For example:

1. The Mohr diagram models used give little direct information about potential fluid flow in the sub-surface. The approach could, however, be used in combination with

other modelling approaches. For example, it would be useful to compare predictions of critically stressed fractures from Mohr diagrams with distinct element analysis of fracture networks (e.g., [106,107]);

2. The values for rock properties used are based on triaxial tests, which are probably over-estimates because small, unfractured samples are generally used (e.g., [86]). More accurate methods for estimating the material properties of rock masses are available (e.g., [108]), and these methods could be used when more detailed information becomes available about the fracture patterns in the rock mass;

3. Similarly, we have used rock mechanical properties from the literature and have made various simplifying assumptions (e.g., no applied tectonic stresses, Biot coefficient = 1). The modelling can be improved and the assumptions properly tested as the input parameters become better constrained, for example as borehole data become available;

4. The anisotropy of the slates has not been modelled in a sophisticated way here, and this can be improved using more detailed information about the relationships between the angle between in situ stresses and cleavage (e.g., [109]);

5. The Mohr diagram analysis used here is two-dimensional, mainly because the magnitudes and orientations of the horizontal stresses are unknown. The analysis could be expanded to three dimensions when such information becomes available, for example from well data. Although predictions can be made about the stresses involved in the Variscan Orogeny and the formation of the Leinetal Graben, those predictions do not help with making predictions about the present-day stresses.

7. Conclusions

This paper shows how simple mechanical modelling, using Mohr diagrams and reasonable ranges of values for rock properties, stresses, and fluid pressures, can be used to predict fracturing in a potential geothermal reservoir. Inferences can be made about the range of structures likely to exist in the sub-surface and that may be generated during stimulation. Critical values of fluid pressure and applied tectonic stresses determine under what conditions different types and orientations of fractures are likely to occur or be generated during stimulation.

A model is presented for the development of shear and extension fractures in the Devonian and Carboniferous greywackes and slates as fluid pressure increases, as would occur during hydraulic stimulation (Figure 7). For a depth of 2 km, this involves: (a) shear on favourably-orientated cohesionless fractures in greywackes with a very low geostatic stress ratio under hydrostatic fluid pressure (Figure 7a); (b) shear along favourably-orientated joints in the slates at fluid pressures of about 50 MPa (Figure 7b); (c) generation of gently-dipping extension fractures in the slates at fluid pressures between about 68 MPa (Figure 7c); (d) development of gently-dipping extension fractures in the greywackes at fluid pressures of about 72 MPa (Figure 7d); (e) creation of extension fractures with any orientation in the slates at fluid pressure between about 90 and 105 MPa (Figure 7e); (f) Generation of fractures with any orientation in the greywackes at fluid pressures between about 100 and 210 MPa (Figure 7f). A similar sequence is predicted for a modelled depth of 4.5 km, although higher differential stresses at greater depths imply that hybrid or shear fractures are more likely to form than are extension fractures as fluid pressure increases.

The modelling addresses many key questions asked in Section 1. (1) The tensile strengths of slates tend to be lower than those of greywackes, suggesting lower fluid pressure or lower tensile tectonic stresses are needed to hydraulically fracture slates than greywackes. (2) Critical values of fluid pressure and tectonic stresses for fracturing can be predicted (Table 4). (3) Steeply-dipping cohesionless fractures (e.g., joints) are most likely to be reactivated first, as shear fractures. (4) Existing cohesionless fractures (e.g., joints) may initially reactivate in shear while cohesional fractures (e.g., veins) will tend to reactivate as extension fractures. (5) Shear fracturing generally requires the reactivation of cohesionless fractures as tensile tectonic stresses are applied, or the application of compressive tectonic stresses to generate new shear fractures. (6) Exhumation and related cooling may be responsible for the creation of joints in the Bunter Sandstone and the Variscan rocks beneath Göttingen. It is likely that joints developed in the Variscan rocks prior to deposition of the Permian rocks, but these may have subsequently been mineralised.

The relationships between stresses, fluid pressure and the types and orientations of fractures are important inputs for more detailed modelling techniques (e.g., DFN modelling [110]), and can be used to test the mechanical implications of those detailed modelling techniques. Not only does this modelling help geoscientists consider and model the ranges of mechanical properties of rock, stresses, fluid pressures and the resultant fractures that are likely to occur in the sub-surface, it encourages them to consider the ranges of key parameters and their effects.

Author Contributions: Conceptualization, D.C.P.P. and D.J.S.; methodology, D.C.P.P. and D.J.S.; software, D.J.S.; validation, D.C.P.P., D.J.S. and B.L.; resources, B.L.; data curation, B.L.; writing—original draft preparation, D.C.P.P.; writing—review and editing, D.J.S. and B.L.; supervision, B.L.; project administration, B.L.; funding acquisition, B.L. All authors have read and agreed to the published version of the manuscript.

Funding: This work has received funding from the European Union's Horizon 2020 research and innovation programme (grant agreement № 792037-MEET Project).

Data Availability Statement: No new data were created or analysed in this study. Data sharing is not applicable to this article.

Acknowledgments: Carl-Heinz Friedel is thanked for helpful advice and discussions. We are grateful to John Reinecker and three anonymous reviewers, whose comments greatly improved this manuscript.

Conflicts of Interest: The authors declare no conflict of interest.

References

1. Leiss, B.; Tanner, D.; Vollbrecht, A.; Wemmer, K. Tiefengeothermisches Potential in der Region Göttingen—Geologische Rahmenbedingungen. In *Neue Untersuchungen zur Geologie der Leinetalgrabenstruktur*; Leiss, B., Tanner, D., Vollbrecht, A., Arp, G., Eds.; Universitätsverlag Göttingen: Göttingen, Germany, 2011; pp. 163–170.
2. Leiss, B.; Romanov, D.; Wagner, B. Risks and challenges of the transition to an integrated geothermal concept for the Göttingen University Campus. In Proceedings of the European Association of Geoscientists and Engineers, Conference Proceedings, 1st Geoscience and Engineering in Energy Transition Conference, Strasbourg, France, 16–18 November 2020; pp. 1–5. [CrossRef]
3. Leiss, B.; Wagner, B.; Heinrichs, T.; Romanov, D.; Tanner, D.C.; Vollbrecht, A.; Wemmer, K. Integrating deep, medium and shallow geothermal energy into district heating and cooling system as an energy transition approach for the Göttingen University Campus. In Proceedings of the World Geothermal Congress 2021, Reykjavik, Iceland, 24–27 October 2021; pp. 1–9.
4. Arp, G.; Vollbrecht, A.; Tanner, D.C.; Leiss, B. Zur Geologie des Leintalgrabens—Ein kurzer Überblick. In *Neue Untersuchungen zur Geologie der Leinetalgrabenstruktur*; Leiss, B., Tanner, D., Vollbrecht, A., Arp, G., Eds.; Universitätsverlag Göttingen: Göttingen, Germany, 2011; pp. 1–7.
5. Trullenque, G.; Genter, A.; Leiss, B.; Wagner, B.; Bouchet, R.; Léoutre, E.; Malnar, B.; Bär, K.; Rajšl, I. Upscaling of EGS in different geological conditions: a European perspective. In Proceedings of the 43rd Workshop on Geothermal Reservoir Engineering, Stanford University, Stanford, CA, USA, 12–14 February 2018.
6. Dalmais, E.; Genter, A.; Trullenque, G.; Leoutre, E.; Leiss, B.; Wagner, B.; Mintsa, A.C.; Bär, K.; Rajsl, I. MEET Project: toward the spreading of EGS across Europe. In Proceedings of the European Geothermal Congress, Den Haag, The Netherlands, 11–14 June 2019; p. 8.
7. Leiss, B.; Wagner, B.; MEET Konsortium. EU-Projekt MEET: Neue Ansätze »Enhanced Geothermal Systems (EGS)«—Göttinger Unicampus als Demoprojekt. *Geotherm. Energ.* **2020**, *91*, 26–28.
8. Welsch, B. Technical, Environmental and Economic Assessment of Medium Deep Borehole Thermal Energy Storage Systems. Ph.D. Thesis, Technische Universität Darmstadt, Darmstadt, Germany, 2019; 192p.
9. Krzywiec, P. Triassic-Jurassic evolution of the Pomeranian segment of the Mid-Polish Trough—Basement tectonics and subsidence patterns. *Geol. Quart.* **2006**, *50*, 139–150.
10. Doornenbal, H.; Stevenson, A. (Eds.) *Petroleum Geological Atlas of the Southern Permian Basin Area*; EAGE: Houten, The Netherlands, 2010; 342p.
11. Arp, G.; Hoffmann, V.E.; Seppelt, S.; Riegel, W. Trias und Jura von Göttingen und Umgebung. In *Geobiologie 2—Jahrestagung der Paläontologischen Gesellschaft*; Reitner, J., Reich, M., Schmidt, G., Eds.; Universitätsverlag Göttingen: Göttingen, Germany, 2004; Volume 74, pp. 147–192.

12. Leiss, B.; Tanner, D.; Vollbrecht, A.; Arp, G. (Eds.) *Neue Untersuchungen zur Geologie der Leinetalgrabenstruktur*; Universitätsverlag Göttingen: Göttingen, Germany, 2011; 170p.

13. Kirnbauer, T. (Ed.) *Geologische Exkursionen in die Region um Göttingen*; Jahresberichte und Mitteilungen des Oberrheinischen geologischen Vereins; E. Schweizerbart'sche Verlagsbuchhandlung: Stuttgart, Germany, 2013; Volume 95, 319p.

14. Sanderson, D.J. Field-based structural studies as analogues to sub-surface reservoirs. In *The Value of Outcrop Studies in Reducing Subsurface Uncertainty and Risk in Hydrocarbon Exploration and Production*; Bowman, M., Smyth, H.R., Good, T.R., Passey, S.R., Hirst, J.P.P., Jordan, C.J., Eds.; Geological Society London, Special Publications: London, UK, 2015; Volume 436.

15. Martel, S.J. Progress in understanding sheeting joints over the past two centuries. *J. Struct. Geol.* **2017**, *94*, 68–86. [CrossRef]

16. Fairhurst, C. Fractures and fracturing: hydraulic fracturing in jointed rock. In *Effective and Sustainable Hydraulic Fracturing*; Jeffrey, R., Ed.; IntechOpen Limited: London, UK, 2013; pp. 47–79. ISBN 978-953-51-6341-1.

17. Ramsay, J.G. *The Folding and Fracturing of Rocks*; McGraw-Hill: New York, NY, USA, 1967; 568p.

18. Fossen, H. *Structural Geology*, 2nd ed.; Cambridge University Press: Cambridge, UK, 2016; 524p.

19. Mahmoodpour, S.; Singh, M.; Turan, A.; Bär, K.; Sass, I. Key parameters affecting the performance of fractured geothermal reservoirs: a sensitivity analysis by thermo-hydraulic-mechanical simulation. *Geophysics*. in review.

20. Friedel, C.H.; Huckriede, H.; Leiss, B.; Zweig, M. Large-scale Variscan shearing at the southeastern margin of the eastern Rhenohercynian belt: a reinterpretation of chaotic rock fabrics in the Harz Mountains, Germany. *Int. J. Earth Sci.* **2019**, *108*, 2295–2323. [CrossRef]

21. Fielitz, W. Variscan transpressive inversion in the northwestern central Rhenohercynian belt of western Germany. *J. Struct. Geol.* **1992**, *14*, 547–563. [CrossRef]

22. Franke, W. The mid-European segment of the Variscides: tectonostratigraphic units, terrane boundaries and plate tectonic evolution. In *Orogenic Processes: Quantification and Modelling in the Variscan Belt*; Franke, W., Haak, V., Oncken, O., Tanner, D., Eds.; Geological Society of London, Special Publications: London, UK, 2000; Volume 179, pp. 35–61.

23. Franke, W. Rheno-Hercynian belt of central Europe: review of recent findings and comparisons with south-west England. *Geosci. SW Eng.* **2007**, *11*, 263–272.

24. Friedel, C.H.; Hoffmann, C. Stopp 4: Schichtgebundene Deformationsstrukturen in der Kulmgrauwacke des Oberharzes (Bundesstraße 242). In *Harzgeologie 2016. 5. Workshop Harzgeologie Kurzfassungen und Exkursionsführer*; Friedel, C.H., Leiss, B., Eds.; Universitätsverlag Göttingen: Göttingen, Germany, 2016; pp. 85–90.

25. Zeuner, M. Conceptual 3D-Structure Model of the Variscan "Culm Fold Zone" in the Vicinity of the Oktertal dam (Harz Mountains, Germany) in Terms of Geothermal Reservoir Development. Master's Thesis, Georg-August-Universität Göttingen, Göttingen, Germany, 2018.

26. Wagner, B.; Leiss, B.; Tanner, D.C. Stopp 2: Gefalteter unterkarbonischer Kieselschiefer am Bielstein nördlich Lautenthal (Innerstetal). In *Harzgeologie 2016. 5. Workshop Harzgeologie Kurzfassungen und Exkursionsführer*; Friedel, C.H., Leiss, B., Eds.; Universitätsverlag Göttingen: Göttingen, Germany, 2016; pp. 69–78.

27. Franzke, H.J.; Hauschke, N.; Hellmund, M. Spätpleistozäne bis holozäne Tektonik an der Harznordrand-Störung bei Benzingerode (Sachsen-Anhalt). In *Harzgeologie 2016. 5. Workshop Harzgeologie Kurzfassungen und Exkursionsführer*; Friedel, C.H., Leiss, B., Eds.; Universitätsverlag Göttingen: Göttingen, Germany, 2016; pp. 13–17.

28. De Graaf, S.; Lüders, V.; Banks, D.A.; Sośnicka, M.; Reijmer, J.J.G.; Kaden, H.; Vonhof, H.B. Fluid evolution and ore deposition in the Harz Mountains revisited: isotope and crush-leach analyses of fluid inclusions. *Miner. Depos.* **2020**, *55*, 47–62. [CrossRef]

29. Ramsey, J.M.; Chester, F.M. Hybrid fracture and the transition from extension fracture to shear fracture. *Nature* **2004**, *428*, 63–66. [CrossRef]

30. Wagner, B.; Günther, S.; Ford, K.; Sosa, G.; Leiss, B. The "Hexagon concept": a fundamental approach for the geoscientific spatial data compilation and analysis at European scale to define the geothermal potential of Variscan and pre-Variscan low- to high-grade metamorphic and intrusive rocks. In Proceedings of the World Geothermal Congress 2021, Reykjavik, Iceland, 24–27 October 2021; pp. 1–10.

31. Arp, G.; Tanner, D.C.; Leiss, B. Struktur der Leinetalgraben-Randstörung bei Reiffenhausen (Autobahn 38 Heidkopftunnel-Westportal). In *Neue Untersuchungen zur Geologie der Leinetalgrabenstruktur*; Leiss, B., Tanner, D., Vollbrecht, A., Arp, G., Eds.; Universitätsverlag Göttingen: Göttingen, Germany, 2011; pp. 17–21.

32. Tanner, D.C. Exkursion nördliches Leinetal. In *Neue Untersuchungen zur Geologie der Leinetalgrabenstruktur*; Leiss, B., Tanner, D., Vollbrecht, A., Arp, G., Eds.; Universitätsverlag Göttingen: Göttingen, Germany, 2011; pp. 27–38.

33. Arp, G.; Bielert, F.; Hoffmann, V.E.; Löffler, T. Palaeoenvironmental significance of lacustrine stromatolites of the Arnstadt Formation ("Steinmergelkeuper", Upper Triassic, N-Germany). *Facies* **2005**, *51*, 419–441. [CrossRef]

34. Ritter, M.; Vollbrecht, A.; van den Kerkhof, A.; Wemmer, K. Sedimentgänge im Bausandstein der Solling-Folge NW' von Billingshausen. In *Neue Untersuchungen zur Geologie der Leinetalgrabenstruktur*; Leiss, B., Tanner, D., Vollbrecht, A., Arp, G., Eds.; Universitätsverlag Göttingen: Göttingen, Germany, 2011; pp. 89–106.

35. Reyer, D.; Bauer, J.F.; Philipp, S.L. Fracture systems in normal fault zones crosscutting sedimentary rocks, Northwest German Basin. *J. Struct. Geol.* **2012**, *45*, 38–51. [CrossRef]

36. Grupe, O. Über die Zechsteinformation und ihr Salzlager im Untergrunde des hannoverschen Eichsfeldes und angrenzenden Leinegebietes nach den neueren Bohrergebnissen. *Z. Fur Prakt. Geol.* **1909**, *17*, 185–205.

37. Tanner, D.C.; Musmann, P.; Wawerzinek, B.; Buness, H.; Krawczyk, C.M.; Thomas, R. Salt tectonics of the eastern border of the Leinetal Graben, Lower Saxony, Germany, as deduced from seismic reflection data. *Interpretation* **2015**, *3*, T169–T181. [CrossRef]

38. Tanner, D.C.; Leiss, B.; Vollbrecht, A. *Strukturgeologie des Leinetalgrabens (Exkursionen G1 und G2 am 4. und 5. April 2013)*; Jahresberichte und Mitteilungen des Oberrheinischen Geologischen Vereins Band; E. Schweizerbart'sche Verlagsbuchhandlung: Stuttgart, Germany, 2013; Volume 95, pp. 131–168.

39. Brink, H. The Variscan Deformation Front (VDF) in northwest Germany and its relation to a network of geological features including the ore-rich Harz Mountains and the European Alpine Belt. *Int. J. Geosci.* **2021**, *12*, 447–486. [CrossRef]

40. Vincent, C.J. *Porosity of the Bunter Sandstone in the Southern North Sea Basin Based on Selected Borehole Neutron Logs*; IR/05/074; British Geological Survey Internal Report: Keyworth, UK, 2005; 20p.

41. McNamara, D.D.; Faulkner, D.; McCarney, E. Rock properties of greywacke basement hosting geothermal reservoirs, New Zealand: preliminary results. In Proceedings of the Thirty-Ninth Workshop on Geothermal Reservoir Engineering Stanford University, Stanford, CA, USA, 24–26 February 2014.

42. Gholami, R.; Rasouli, V. Mechanical and elastic properties of transversely isotropic slate. *Rock Mech. Rock Eng.* **2013**, *47*. [CrossRef]

43. Menezes, F.F.; Lempp, C.; Svensson, K.; Neumann, A.; Pöllmann, H. Geomechanical behavior changes of Bunter Sandstone and borehole cement due to scCO$_2$ injection effects. In Proceedings of the IAEG/AEG Annual Meeting, San Francisco, CA, USA, 17–21 September 2018; Shakoor, A., Cato, K., Eds.; Springer Nature: Cham, Switzerland, 2019; Volume 1, pp. 111–118.

44. Agustawijaya, D.S. The uniaxial compressive strength of soft rock. *Civil Eng. Dimen.* **2007**, *9*, 9–14.

45. Menezes, F.F.; Lempp, C. On the structural anisotropy of physical and mechanical properties of a Bunter Sandstone. *J. Struct. Geol.* **2018**, *114*, 196–205. [CrossRef]

46. Havlíčková, D.; Závacký, M.; Krmíček, L. Anisotropy of mechanical properties of greywacke. *GeoSci. Eng.* **2019**, *65*, 46–52. [CrossRef]

47. Rodríguez-Sastre, M.A.; Gutiérrez-Claverol, M.; Torres-Alonso, M. Relationship between cleavage orientation, uniaxial compressive strength and Young's modulus for slates in NW Spain. *Bull. Eng. Geol. Environ.* **2008**, *67*, 181–186. [CrossRef]

48. Price, N.J. *Fault and Joint Development in Brittle and Semi-Brittle Rock*; Pergamon Press: Oxford, UK, 1966; 176p.

49. Martel, S.J. Effects of cohesive zones on small faults and implications for secondary fracturing and fault trace geometry. *J. Struct. Geol.* **1997**, *19*, 835–847. [CrossRef]

50. Alneasan, M.; Behnia, M.; Bagherpour, H. The effect of Poisson's ratio on the creation of tensile branches around dynamic faults. *J. Struct. Geol.* **2020**, *131*, 103950. [CrossRef]

51. Dobbs, M.R.; Cuss, R.J.; Ougier-Simonin, A.; Parkes, D.; Graham, C.C. Yield envelope assessment as a preliminary screening tool to determine carbon capture and storage viability in depleted southern north-sea hydrocarbon reservoirs. *Int. J. Rock Mech. Min. Sci.* **2018**, *102*, 15–27. [CrossRef]

52. Jefferies, G.M.; Crooks, J.H.A.; Becker, D.E.; Hill, P.R. Independence of geostatic stress from overconsolidation in some Beaufort Sea clays. *Can. Geotech. J.* **1987**, *24*, 342–356. [CrossRef]

53. Cai, J.; Du, G.; Ye, H.; Lei, T.; Xia, H.; Pan, H. A slate tunnel stability analysis considering the influence of anisotropic bedding properties. *Adv. Mater. Sci. Eng.* **2019**, *2019*, 4653401. [CrossRef]

54. Griffiths, L.; Heap, M.J.; Xu, T.; Chen, C.; Baud, P. The influence of pore geometry and orientation on the strength and stiffness of porous rock. *J. Struct. Geol.* **2017**, *96*, 149–160. [CrossRef]

55. Wang, H.; Mang, H.; Yuan, Y.; Pichler, B.L.A. Multiscale thermoelastic analysis of the thermal expansion coefficient and of microscopic thermal stresses of mature concrete. *Materials* **2019**, *12*, 2689. [CrossRef]

56. Pogacnik, J.; Elsworth, D.; O'Sullivan, M.; O'Sullivan, J. A damage mechanics approach to the simulation of hydraulic fracturing/shearing around a geothermal injection well. *Comput. Geotech.* **2016**, *71*, 338–351. [CrossRef]

57. Lee, C.; Park, J.W.; Park, C.; Park, E.S. Current status of research on thermal and mechanical properties of rock under high-temperature condition. *Tunn. Undergr. Space* **2015**, *25*, 1–23. [CrossRef]

58. Engelder, T.; Fischer, M.P. Influence of poroelastic behavior on the magnitude of minimum horizontal stress, S_h, in overpressured parts of sedimentary basins. *Geology* **1994**, *22*, 949–952. [CrossRef]

59. du Rouchet, J. Stress fields, a key to oil migration. *AAPG Bull.* **1981**, *65*, 74–85.

60. Mesri, G.; Hayat, T.M. The coefficient of earth pressure at rest. *Can. Geotech. J.* **1993**, *30*, 647–666. [CrossRef]

61. Gercek, H. Poisson's ratio values for rocks. *Int. J. Rock Mech. Min. Sci.* **2007**, *44*, 1–13. [CrossRef]

62. Jaeger, J.C. Mohr diagram. In *Structural Geology and Tectonics*; Encyclopaedia of Earth Science; Springer: Berlin, Germany, 1987.

63. Hoek, E.; Martin, C.D. Fracture initiation and propagation in intact rock—A review. *J. Rock Mech. Geotech. Eng.* **2014**, *6*, 287–300. [CrossRef]

64. Anderson, E.M. *The Dynamics of Faulting*; Oliver & Boyd: Edinburgh, UK, 1951.

65. Cook, N.G.W. The failure of rock. *Int. J. Rock Mech. Min. Sci. Geomech. Abs.* **1965**, *2*, 389–403. [CrossRef]

66. Ucar, R. Determination of shear failure envelope in rock masses. *J. Geot. Eng.* **1986**, *112*, 303–315. [CrossRef]

67. Patel, S.; Martin, C.D. Effect of stress path on the failure envelope of intact crystalline rock at low confining stress. *Minerals* **2020**, *10*, 1119. [CrossRef]

68. Sibson, R.H. Brittle failure mode plots for compressional and extensional tectonic regimes. *J. Struct. Geol.* **1998**, *20*, 655–660. [CrossRef]

69. Brace, W.F. An extension of the Griffith theory of fracture to rocks. *J. Geophys. Res.* **1960**, *65*, 3477–3480. [CrossRef]

70. McClintock, F.A.; Walsh, J.B. Friction on Griffith cracks in rocks under pressure. In Proceedings of the fourth U.S. National Congress of Applied Mechanics, Berkeley, CA, USA, 18–21 June 1962; Volume 2, pp. 1015–1021.

71. Jaeger, J.C.; Cook, N.G.W. *Fundamentals of Rock Mechanics*; Methuen: London, UK, 1969; 515p.

72. Chang, C.; Zoback, M.D.; Khaksar, A. Empirical relations between rock strength and physical properties in sedimentary rocks. *J. Pet. Sci. Eng.* **2006**, *51*, 223–237. [CrossRef]

73. Zoback, M.D. *Reservoir Geomechanics*; Cambridge University Press: Cambridge, UK, 2007; 449p.

74. Sibson, R.H. A note on fault reactivation. *J. Struct. Geol.* **1985**, *7*, 751–754. [CrossRef]

75. Scholz, C.H. *The Mechanics of Earthquakes and Faulting*, 2nd ed.; Cambridge University Press: Cambridge, UK, 2002; 504p.

76. Meissner, R.; Strehlau, J. Limits of stresses in continental crusts and their relation to the depth-frequency distribution of shallow earthquakes. *Tectonics* **1982**, *1*, 73–89. [CrossRef]

77. Malama, B.; Kulatilake, P.H.S.W. Models for normal fracture deformation under compressive loading. *Int. J. Rock Mech. Min. Sci.* **2003**, *40*, 893–901. [CrossRef]

78. Corcoran, D.V.; Doré, A.G. Depressurization of hydrocarbon-bearing reservoirs in exhumed basin settings: evidence from Atlantic margin and borderland basins. In *Exhumation of the North Atlantic Margin: Timing, Mechanisms and Implications for Hydrocarbon Exploration*; Doré, A.G., Cartwright, J.A., Stoker, M.S., Turner, J.P., White, N., Eds.; Geological Society of London, Special Publications: London, UK, 2002; Volume 196, pp. 457–483.

79. Terzaghi, K. *Erdbaumechanik auf Bodenphysikalischer Grundlage*; Franz Deuticke: Vienna, Austria, 1925.

80. Terzaghi, K. *Theoretical Soil Mechanics*; Wiley: New York, NY, USA, 1943.

81. Bredehoeft, J.D.; Wesley, J.B.; Fouch, T.D. Simulations of the origin of fluid pressure, fracture generation, and the movement of fluids in the Uinta Basin, Utah. *AAPG Bull.* **1994**, *78*, 1729–1747.

82. Biot, M.A.; Willis, D.G. The elastic coefficients of the theory of consolidation. *J. Appl. Mech.* **1957**, *24*, 594–601. [CrossRef]

83. Barnhoorn, A.; Cox, S.F.; Robinson, D.J.; Senden, T. Stress- and fluid-driven failure during fracture array growth: implications for coupled deformation and fluid flow in the crust. *Geology* **2010**, *38*, 779–782. [CrossRef]

84. Crawford, B.R.; Webb, D.W.; Searles, K.H. Plastic compaction and anisotropic permeability development in unconsolidated sands with implications for horizontal well performance. In Proceedings of the 42nd US Rock Mechanics Symposium and 2nd U.S.-Canada Rock Mechanics Symposium, San Francisco, CA, USA, 29 June–2 July 2008.

85. Ferrill, D.A.; Smart, K.J.; Cawood, A.J.; Morris, A.P. The fold-thrust belt stress cycle: superposition of normal, strike-slip, and thrust faulting deformation regimes. *J. Struct. Geol.* **2021**, *148*, 104362. [CrossRef]

86. Becker, D.E. Testing in geotechnical design. *Geotech. Eng. J. SEAGS AGSSEA* **2010**, *41*, 10.

87. Bär, K.; Arbarim, R.; Turan, A.; Schulz, K.; Mahmoodpour, S.; Leiss, B.; Wagner, B.; Sosa, G.; Ford, K.; Trullenque, G.; et al. *Database of Petrophysical and Fluid Properties and Recommendations for Model Parametrization of the Four Variscan Reservoir Types*; MEET Report, Deliverable D5.5; MEET Consortium European, June 2020; 110p, Available online: https://www.meet-h2020.com/project-results/deliverables/ (accessed on 23 September 2021).

88. Tingay, M.R.P.; Hillis, R.R.; Morley, C.K.; Swarbrick, R.E.; Okpere, E.C. Variation in vertical stress in the Baram Basin, Brunei: tectonic and geomechanical implications. *Mar. Pet. Geol.* **2003**, *20*, 1201–1212. [CrossRef]

89. Tan, C.P.; Willoughby, D.R.; Zhou, S.; Hillis, R.R. An analytical method for determining horizontal stress bounds from wellbore data. *Int. J. Rock Mech. Min. Sci. Geomech. Abs.* **1993**, *30*, 1103–1109. [CrossRef]

90. Lund, B.; Townend, J. Calculating horizontal stress orientations with full or partial knowledge of the tectonic stress tensor. *Geophys. J. Int.* **2007**, *170*, 1328–1335. [CrossRef]

91. Zhang, J.; Hüpers, A.; Kreiter, S.; Kopf, A.J. Pore pressure regime and fluid flow processes in the shallow Nankai Trough Subduction Zone based on experimental and modeling results from IODP Site C0023. *J. Geophys. Res.* **2021**, *126*, e20248. [CrossRef]

92. Gao, B.; Flemings, P.B.; Nikolinakou, M.A.; Saffer, D.M.; Heidari, M. Mechanics of fold-and-thrust belts based on geomechanical modelling. *J. Geophys. Res.* **2018**, *123*, 4454–4474. [CrossRef]

93. Heidbach, O.; Rajabi, M.; Reiter, K.; Ziegler, M.; WSM Team. World Stress Map Database Release 2016. V. 1.1. *GFZ Data Serv.* **2016**. [CrossRef]

94. Hobbs, B.E.; Means, W.D.; Williams, P.F. *An Outline of Structural Geology*; Wiley: New York, NY, USA, 1976; 671p.

95. Peacock, D.C.P.; Tavarnelli, E.; Anderson, M.W. Interplay between stress permutations and overpressure to cause strike-slip faulting during tectonic inversion. *Terra Nova* **2017**, *29*, 61–70. [CrossRef]

96. Laubach, S.E.; Olson, J.E.; Gross, M.R. Mechanical and fracture stratigraphy. *AAPG Bull.* **2009**, *93*, 1413–1426. [CrossRef]

97. Japsen, P.; Green, P.F.; Nielsen, L.H.; Rasmussen, E.S.; Bidstrup, T. Mesozoic-Cenozoic exhumation events in the eastern North Sea Basin: a multi-disciplinary study based on palaeothermal, palaeoburial, stratigraphic and seismic data. *Basin Res.* **2007**, *19*, 451–490. [CrossRef]

98. Nadan, B.J.; Engelder, T. Microcracks in New England granitoids: a record of thermoelastic relaxation during exhumation of intracontinental crust. *Geol. Soc. Am. Bull.* **2009**, *121*, 80–99. [CrossRef]

99. Albero, F.; Tanner, D.C. Modellierung des Temperaturfeldes um eine tiefe Förderbohrung am östlichen Rand des Leinetalgrabens—Abschätzung des geothermischen Potentials. In *Neue Untersuchungen zur Geologie der Leinetalgrabenstruktur*; Leiss, B., Tanner, D.C., Vollbrecht, A., Arp, G., Eds.; Universitätsverlag Göttingen: Göttingen, Germany, 2011; pp. 115–123.

100. Wilkins, S.J.; Gross, M.R.; Wacker, M.; Eyal, Y.; Engelder, T. Faulted joints: kinematics, displacement–length scaling relations and criteria for their identification. *J. Struct. Geol.* **2001**, *23*, 315–327. [CrossRef]
101. Karnin, W.D.; Idiz, E.; Merkel, D.; Ruprecht, E. The Zechstein Stassfurt Carbonate hydrocarbon system of the Thuringian Basin, Germany. *Pet. Geosci.* **1996**, *2*, 53–58. [CrossRef]
102. Bucher, K.; Stober, I.; Seelig, U. Water deep inside the mountains: unique water samples from the Gotthard rail base tunnel, Switzerland. *Chem. Geol.* **2012**, *334*, 240–253. [CrossRef]
103. Van Der Heever, P. The Influence of Geological Structure on Seismicity and Rockbursts in the Klerksdorp Goldfield. Master's Thesis, Rand Afrikaans University, Johannesburg, South Africa, 1982; 156p.
104. Gumede, H.; Stacey, T.R. Measurement of typical joint characteristics in South African gold mines and the use of these characteristics in the prediction of rock falls. *J. S. Afr. Inst. Min. Metall.* **2007**, *107*, 335–344.
105. Areshev, E.G.; Dong, T.L.; San, N.T.; Shnip, O.A. Reservoirs in fractured basement on the continental shelf of southern Vietnam. *J. Pet. Geol.* **1992**, *15*, 451–464. [CrossRef]
106. Zhang, X.; Sanderson, D.J. Numerical study of critical behaviour of deformation and permeability of fractured rock masses. *Mar. Pet. Geol.* **1998**, *15*, 535–548. [CrossRef]
107. Camac, B.A.; Hunt, S.P. Predicting the regional distribution of fracture networks using the distinct element numerical method. *AAPG Bull.* **2009**, *93*, 1571–1583. [CrossRef]
108. Cundall, P.A.; Pierce, M.E.; Mas Ivars, D. Quantifying the size effect of rock mass strength. In Proceedings of the First Southern Hemisphere International Rock Mechanics Symposium, Perth, Australia, 16–19 September 2008; Potvin, Y., Carter, J., Dyskin., A., Jeffrey, R., Eds.; Australian Centre for Geomechanics: Perth, Australia, 2008; pp. 3–15.
109. Donath, F.A. Experimental study of shear failure in anisotropic rocks. *Geol. Soc. Am. Bull.* **1961**, *72*, 985–989. [CrossRef]
110. Yao, C.; Shao, J.F.; Jiang, Q.H.; Zhou, C. A new discrete method for modeling hydraulic fracturing in cohesive porous materials. *J. Pet. Sci. Eng.* **2019**, *180*, 257–267. [CrossRef]

Article

Fracture Transmissivity in Prospective Host Rocks for Enhanced Geothermal Systems (EGS)

Johannes Herrmann [1,*], Valerian Schuster [2], Chaojie Cheng [2], Harald Milsch [2] and Erik Rybacki [2]

[1] Institute of Drilling Engineering and Fluid Mining, Technische Universität Bergakademie Freiberg, 09599 Freiberg, Germany

[2] GFZ German Research Centre for Geosciences, 14473 Potsdam, Germany; valerian.schuster@gfz-potsdam.de (V.S.); chaojie@gfz-potsdam.de (C.C.); milsch@gfz-potsdam.de (H.M.); uddi@gfz-potsdam.de (E.R.)

* Correspondence: johannes.herrmann1@tbt.tu-freiberg.de

Abstract: We experimentally determined the hydraulic properties of fractures within various rock types, focusing on a variety of Variscan rocks. Flow-through experiments were performed on slate, graywacke, quartzite, granite, natural fault gouge, and claystone samples containing an artificial fracture with a given roughness. For slate samples, the hydraulic transmissivity of the fractures was measured at confining pressures, p_c, at up to 50 MPa, temperatures, T, between 25 and 100 °C, and differential stress, σ, acting perpendicular to the fracture surface of up to 45 MPa. Fracture transmissivity decreases non-linearly and irreversibly by about an order of magnitude with increasing confining pressure and differential stress, with a slightly stronger influence of p_c than of σ. Increasing temperature reduces fracture transmissivity only at high confining pressures when the fracture aperture is already low. An increase in the fracture surface roughness by about three times yields an initial fracture transmissivity of almost one order of magnitude higher. Fractures with similar surface roughness display the highest initial transmissivity within slate, graywacke, quartzite and granite samples, whereas the transmissivity in claystone and granitic gouge material is up to several orders of magnitude lower. The reduction in transmissivity with increasing stress at room temperature varies with composition and uniaxial strength, where the deduction is lowest for rocks with a high fraction of strong minerals and associated high brittleness and strength. Microstructural investigations suggest that the reduction is induced by the compaction of the matrix and crushing of strong asperities. Our results suggest that for a given surface roughness, the fracture transmissivity of slate as an example of a target reservoir for unconventional EGS, is comparable to that of other hard rocks, e.g., granite, whereas highly altered and/or clay-bearing rocks display poor potential for extracting geothermal energy from discrete fractures.

Keywords: Enhanced Geothermal Systems (EGS); Variscan rocks; slate; quartzite; granite; claystone; graywacke; gouge; fracture transmissivity; effective stress

Citation: Herrmann, J.; Schuster, V.; Cheng, C.; Milsch, H.; Rybacki, E. Fracture Transmissivity in Prospective Host Rocks for Enhanced Geothermal Systems (EGS). *Geosciences* **2022**, *12*, 195. https://doi.org/10.3390/geosciences12050195

Academic Editors: Béatrice A. Ledésert, Ronan L. Hébert, Ghislain Trullenque, Albert Genter, Eléonore Dalmais, Jean Hérisson and Jesus Martinez-Frias

Received: 8 April 2022
Accepted: 27 April 2022
Published: 3 May 2022

Publisher's Note: MDPI stays neutral with regard to jurisdictional claims in published maps and institutional affiliations.

1. Introduction

Extracting geothermal energy from underground is of major interest in the transition from energy recovered from conventional resources such as coal or oil towards renewable energies [1]. Geothermal energy is expected to have a great potential to meet future energy demands. However, exploitation is highly dependent on the presence of accessible hot fluids within the reservoir formation to ensure sufficient energy extraction. In contrast to hydrothermal geothermal systems, where steam or hot water is extracted from the subsurface, hot (>150 °C) and deep (>3 km), but low permeable, reservoir rocks represent the largest geothermal energy resources [2–5]. To facilitate the extraction of geothermal energy from tight reservoirs, a network of highly conductive fractures is required that serves as heat exchanger [6–8] in Enhanced Geothermal Systems (EGS). An EGS typically relies on two wells, an injection and a production well, which are connected by a network of open (conductive to fluid flow) fractures. In order to create such a fracture network, artificial fractures are often generated by hydraulic fracturing (HF). These are believed to be connected to pre-existing natural fractures in the reservoir [9]. The efficiency and sustainability of an EGS are critically dependent on sufficient water flow through the fracture network and on the conductive properties of the separate fractures [1]. This prerequisite, in addition to a relatively high geothermal gradient, is necessary to successfully run an EGS over several years [10,11].

Fluid flow within rough natural fractures is a complex process, influenced by many parameters, for example thermodynamic boundary conditions, surface roughness and chemical processes, such as mineral precipitation [12]. In addition, the bedding and cleavage orientation of anisotropic rocks such as shale or slate can have a strong effect on the development and permeability of fractures [13–15]. The surface roughness of natural and artificially created fractures has a strong impact on the (hydraulic) aperture [12,16–18]. Fracture transmissivity can rapidly decrease with increasing confining pressure (p_c) [12,16] and increasing effective stress, σ, oriented perpendicular to the fracture surface (e.g., [19]). Increasing temperature, T, may (completely) seal fractures in sediments by thermal expansion [17]. Fracture sealing may be also induced by mineral precipitation processes, depending on fluid and rock composition [10,12,20]. Injecting fluid into a fluid-bearing fracture with a different composition may lead to dissolution-precipitation reactions due to local changes in the chemical equilibrium, which typically results in a decrease in fracture transmissivity [10,17,21,22]. The transmissivity may be also reduced by the clogging of flow channels due to the migration of fine particles, e.g., clay [1,13,23,24] or by the production of a fine-grained gouge layer resulting from shear displacement [19]. On the other hand, fracture transmissivity may be enhanced with low effective normal stress acting on the fracture due to shear-induced dilatancy, resulting in self-propping of the fracture surfaces [1,12,16,20,25]. The transmissivity of the propped fractures may be influenced by proppant embedment (e.g., [24]).

Many existing EGS are located in granitic rocks, which often exhibit a relatively large amount of natural, highly permeable, open fractures and have a high potential for successfully creating fracture networks with stimulation techniques [26,27]. However, metamorphic rocks may also be considered as potential host rocks for EGS. Within the European initiative 'Multidisciplinary and multi-context demonstration of Enhanced Geothermal Systems Exploration and Exploitation Techniques and potentials (MEET)', the suitability of Variscan rocks, such as granites [27], quartzites [28], slates and graywacke [29–31] were investigated to assess their potential for extracting geothermal energy. In this study, we focused on the influence of confining pressure, stress, temperature, and surface roughness on the fracture transmissivity of Wissenbach slate which is the expected target rock for a planned EGS in Göttingen, Germany, dedicated to heating the University Campus [31]. Since the productivity of fractured rocks may be influenced by their composition, as well as their metamorphic and alteration grade, we additionally performed flow through experiments at elevated stress on fractured graywacke, quartzite, fresh and altered (gouge) granite, and claystone, where the latter may be regarded as an end-member rock type of highly altered formations. This may help to assess the potential of different geological settings for future EGS.

2. Materials and Methods

2.1. Sample Material

Most of the experiments were performed on slates derived from the Middle Devonian (Eifelian age) Wissenbach Slate (WBS) Formation, one of the intended reservoir target horizons for the planned EGS in Göttingen. Because only a few existing wells in the Göttingen area penetrate the Paleozoic rock units below the Permian and Mesozoic sedimentary cover [29,31], rock samples were taken from an analogue site in the Western Harz Mountains. Samples were prepared from cores collected at a depth of 1134–1210 m from the scientific well 'Hahnenklee', which was drilled in the northwestern part of the Harz mountains in the early 1980's [32–36]. On macroscopic scale, WBS appears as black, homogeneous, argillaceous rock with sporadically distributed pyrite aggregates (<0.25 mm). The very few carbonate layers that occur are aligned parallel to the visible and characteristic cleavage, which is rarely interrupted by carbonate-filled veins (<0.5 mm). An X-Ray diffraction (XRD) analysis revealed that this metamorphic rock is mainly composed of quartz (Qtz), calcite (Ca), dolomite (Dol), muscovite (Ms), illite (Ill), chlorite (Chl), and feldspar (Fsp) with minor amounts of apatite (Ap), pyrite (Py), and organic matter (Om). The bulk composition was categorized into three main groups consisting of mechanically weak (Phyl = phyllosilicates), intermediate strong (Cb = carbonates), and strong (QFSO = Qtz + Fsp + Sulfides + Oxides) minerals (Table 1) and plotted in a ternary diagram (Figure 1).

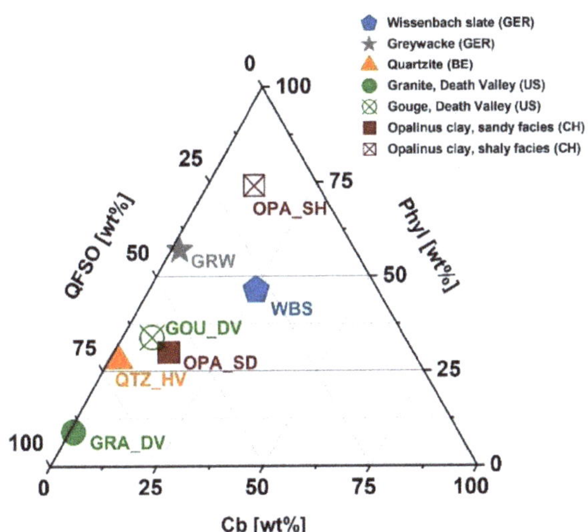

Figure 1. Ternary plot displaying mineral composition of investigated samples. Composition is separated into mechanically strong (QFSOP = Qtz + Fsp + Sulfides + Oxides), intermediate strong (Cb = Carbonates), and weak (Phyl = Phyllosilicates) fractions. Mineral data given in wt%., WBS = Wissenbach slate, GRW = Graywacke, QTZ_HV = Quartzite Havelange, GRA_DV = Granite Death Valley, GOU_DV = Gouge Death Valley, OPA_SD = Opalinus clay, sandy facies, OPA_SH = Opalinus clay, shaly facies.

Optical and electron microscopy revealed a complex (expressed by a large range of mineral types and grain sizes), fine-grained microstructure of the starting material (Figure 2a,b), with quartz, feldspar, chlorite and carbonates making up the largest mineral grains (<50 μm) that are dispersed in the phyllosilicate-rich matrix. Cleavage planes are characterized by phyllosilicates whose longest axes are oriented subparallel to each other. The bulk density of dried (110 °C for >48 h) WBS is in the range of around 2.8 g/cm^3 with a respective porosity measured with He-pycnometry (Micrometrics, AccuPyc 1340) of $\phi_{He} \approx 2$ vol% (including micro pores). Matrix permeability, k_{matrix}, is less than 10^{-19} m^2, which is the detection limit of the gas permeameter used at TU Darmstadt (K. Bär, personal communication).

Table 1. Petrophysical and mechanical properties.

Formation	Sample ID	Depth [m]	ρ [g/cm^3]	ρ_g [g/cm^3]	ϕ_{He} [%]	Phyl [wt%]	Cb [wt%]	QFSO [wt%]	S_{q_ini} [mm]	S_{q_def} [mm]	Mechanical Properties	Experimental Conditions	Fluid Medium
WBS		1134–1210	2.76	2.82	2.0	46	26	28	0.006 (low)		Perpendicular: $\sigma_{UCS} = 219 \pm 2$ MPa E = 27.4 $\pm$ 0.3 GPa	UCS	-
									0.019 (high)		Parallel: $\sigma_{UCS} = 124 \pm 1$ MPa E = 31.4 $\pm$ 0.3 GPa	UCS	-
											Perpendicular: $\sigma_{TCS} = 498 \pm 20$ MPa E = 69 $\pm$ 14 GPa	Triaxial: $p_c = 50$ MPa T = 100 °C $\dot{\varepsilon} = 5 \times 10^{-4}$ s^{-1}	-
	WBS$_T$									0.019		Hydrostatic: $p_c = 5$ MPa $p_P = 1$ MPa, T = 25–100 °C	H$_2$O
	WBS$_{pc}$									0.020		Hydrostatic: $p_c = 2$–25 MPa $p_P = 1$ MPa, T = 25 °C	H$_2$O
	WBS$_{pc_\sigma_T}$									0.021		Triaxial: $p_c = 5$–50 MPa $p_P = 1$ MPa, $\sigma = 0$–45 MPa T = 25–90 °C	H$_2$O
	WBS$_{\sigma AR}$									0.020		Triaxial: $p_c = 14$ MPa $p_P = 10$ MPa, $\sigma = 0$–45 MPa T = 25 °C	Ar
	WBS$_{pc_lowrough}$									0.006		Triaxial: $p_c = 1$–5 MPa $p_P = 0.5$–1 MPa, T = 25 °C	H$_2$O
GRW		843	2.67	2.69	0.7	57	2	41	0.016		$\sigma_{UCS} = 185 \pm 2$ MPa E = 35.2 $\pm$ 0.3 GPa	UCS	-
	GRW$_\sigma$									0.016			Triaxial: $p_c = 5$ MPa, $p_P = 1$ MPa, $\sigma = 0$–45 MPa T = 25 °C	H$_2$O
QTZ_HV		4732	2.69	2.7	0.2	20	3	77	0.014		$\sigma_{UCS} = 175 \pm 2$ MPa E = 35.6 $\pm$ 0.4 GPa	UCS	-
	QTZ_HV$_\sigma$									0.014			Triaxial: $p_c = 5$ MPa $p_P = 1$ MPa, $\sigma = 0$–45 MPa T = 25 °C	H$_2$O

Table 1. *Cont.*

Formation	Sample ID	Depth [m]	ρ [g/cm^3]	ρ_g [g/cm^3]	ϕ_{He} [%]	Phyl [wt%]	Cb [wt%]	QFSO [wt%]	S_{q_ini} [mm]	S_{q_def} [mm]	Mechanical Properties	Experimental Conditions	Fluid Medium
GRA_DV		OC	2.63	2.66	1.3	9	1	90	0.019		$\sigma_{UCS} = 160 \pm 2$ MPa $E = 31.7 \pm 0.3$ GPa	UCS	-
	GRA_DV$_\sigma$									0.019		Triaxial: $p_c = 5$ MPa $p_P = 1$ MPa, $\sigma = 0$–45 MPa $T = 25\,°C$	H$_2$O
GOU_DV		OC	2.24	2.56	12.6	34	7	59	0.247		$\sigma_{UCS} = 4.5 \pm 0.7$ MPa $E = 0.7 \pm 0.3$ GPa	UCS	-
	GOU_DV$_\sigma$									0.175		Triaxial: $p_c = 14$ MPa $p_P = 10$ MPa, $\sigma = 0$–20 MPa $T = 25\,°C$	Ar
OPA_SD		MT_URL	2.38	2.7	11.8	30	13	57	0.012		$\sigma_{UCS} = 49.6 \pm 0.5$ MPa $E = 5.3 \pm 0.1$ GPa	UCS	-
	OPA_SD$_\sigma$									0.012		Triaxial: $p_c = 14$ MPa $p_P = 10$ MPa, $\sigma = 0$–45 MPa $T = 25\,°C$	Ar
OPA_SH		MT_URL	2.39	2.77	13.8	74	11	15	0.016		$\sigma_{UCS} = 34.9 \pm 0.3$ MPa $E = 1.8 \pm 0.1$ GPa	UCS	-
	OPA_SH$_\sigma$									0.015		Triaxial: $p_c = 14$ MPa $p_P = 10$ MPa, $\sigma = 0$–30 MPa $T = 25\,°C$	Ar

WBS = Wissenbach slate, GRW = Graywacke, QTZ_HV = Quartzite Havelange, GRA_DV = Granite Death Valley, GOU_DV = Gouge Death Valley, OPA_SD = Opalinus clay sandy facies, OPA_SH = Opalinus clay shaly facies, MT_URL = 'Mont Terri' Underground research laboratory, OC = outcrop, σ_{TCS} = triaxial compressive strength, σ_{UCS} = uniaxial compressive strength, E = static Young's modulus, wt% = weight percent, ρ = bulk density, ρ_g = grain density, ϕ_{He} = porosity determined by Helium pycnometry, p_P = pore pressure, S_{q_ini} = initial fracture surface roughness, S_{q_def} = fracture surface roughness of the deformed sample (after testing), Phyl = phyllosilicates, Cb = carbonates, QFSO = quartz + feldspar + sulfides + oxides, p_c = confining pressure, T = temperature, σ = axial stress perpendicular to fracture surface, S_q = root mean square value of fracture surface roughness, H$_2$O = distilled water, Ar = Argon gas.

Another potential unit that has sufficient thickness for developing an EGS at the Göttingen site are Lower Carboniferous (Kulm facies) graywacke-successions [31]. Graywacke (GRW) samples also originate from an analogue site in the Western Harz Mountains and were taken at a depth of 843 m from the well 'Wulpke-2' that was drilled in the 1980's [32]. This grey to light-green marine psammite shows angular detrital quartz and feldspar grains (<200 μm) embedded in a fine-grained matrix made up of chlorite, feldspar and mica (Figure 2c,d). The samples were taken from a relatively homogeneous part without any obvious grading. Larger, planar aligned biotite grains (<400 μm) indicate the macroscopically visible bedding orientation. The main mineral components determined by the XRD analysis show dominantly quartz, feldspar and micas with minor amounts of carbonates. Porosity and bulk density of dried samples are 0.7 vol% and 2.67 g/cm^3, respectively.

A further demonstration site studied in the frame of the MEET-project is the Havelange deep borehole in Belgium (Wallonia). Here, Variscan quartzite formations are being considered for the development of potential EGS. Therefore, we also studied the fracture transmissivity of quartzite samples (QTZ_HV) obtained from a depth of z = 4732 m of the Havelange well, drilled in the Dinant Synclinorium in the early 1980's [28]. The grey to light-green samples are characterized by a granoblastic fabric of fine-grained (<150 μm) quartz with illite and sparsely appearing dolomite (Figure 2e,f). The Lower Devonian (Pragian) samples contain roughly 77 wt% quartz and 20 wt% clays and micas with minor amounts of dolomite. The porosity of the used samples was lowest off all the tested materials and ranges around $\phi_{He} \approx 0.2$ vol%.

To capture the influence of alteration on fracture transmissivity, tests were performed on fresh (GRA_DV), altered and sheared granite gouge (GOU_DV) samples recovered from an exposure located in the Noble Hills area in the southern part of the Death Valley (US) as an easily accessible analogue material to Variscan granites [37,38]. The light-reddish to pinkish, equigranular Noble Hill granite (Figure 2g,h) is mainly composed of medium to coarse grained (<3 mm) plagioclase (35 wt%), quartz (30 wt%), K-feldspar (30 wt%) and biotite (10 wt%) [38] with an initial porosity of 1.3 vol%. On the other hand, the light-yellowish to orange gouge (Figure 2i,j), which was highly altered due to multiple shearing events and weathering processes, is characterized by granite clasts (<5 mm) and brecciated quartz veins (<1 cm) embedded in a carbonate- and clay-rich matrix [37]. In contrast to the granite, the gouge displays a high porosity of 12.6 vol%.

In addition, we performed measurements on claystone samples to shed light on the influence of consolidation and metamorphic grades on fracture transmissivity in clay-rich formations, which may result from alteration processes in hard rocks. Claystone material was recovered from the Opalinus Clay Formation (OPA), whose mechanical properties have been well-studied in the context of nuclear waste disposal (e.g., [39–41]). Compared to WBS, these rocks exhibited a much lower burial depth and temperature [42]. We investigated samples from the sandy (OPA_SD) (Figure 2k,l) and shaly facies (OPA_SH) (Figure 2m,n), both collected from the Underground Research Laboratory (URL) 'Mont Terri' (Switzerland, St. Ursanne), gratefully provided by the Swiss Federal Office of Topography–swisstopo. The fine-grained OPA_SD is mainly composed of Qtz (48 wt%), carbonates (20 wt%), and Fsp (9 wt%), but contains a distinctly lower amount of weak sample constituents (30 wt%), such as clay and mica, compared to WBS samples (Figure 1). Note that these values may vary by about 10 wt% because of the compositional heterogeneity of this facies (e.g., [41]). The matrix permeability of OPA_SD is in the same range compared to WBS ($k_{matrix\_OPA\_SD} = 10^{-19}$–$10^{-21}$ m^{-2}) [43], whereas the porosity of OPA_SD ($\phi_{He_OPS_SD} = 11.8$ vol%) is significantly higher. In comparison to the sandy facies of OPA, the shaly facies contain less quartz (9 wt%) and more clay minerals (74 wt%). The main constituents of OPA_SH are clay minerals, quartz, carbonates, feldspar, pyrite, and organic matter. The permeability of OPA_SH is similar to that of OPA_SD ($k_{matrix\_OPA\_SH} = 10^{-19}$–$10^{-21}$ m^{-2}, [43]), but with a slightly higher porosity of 13.8 vol%.

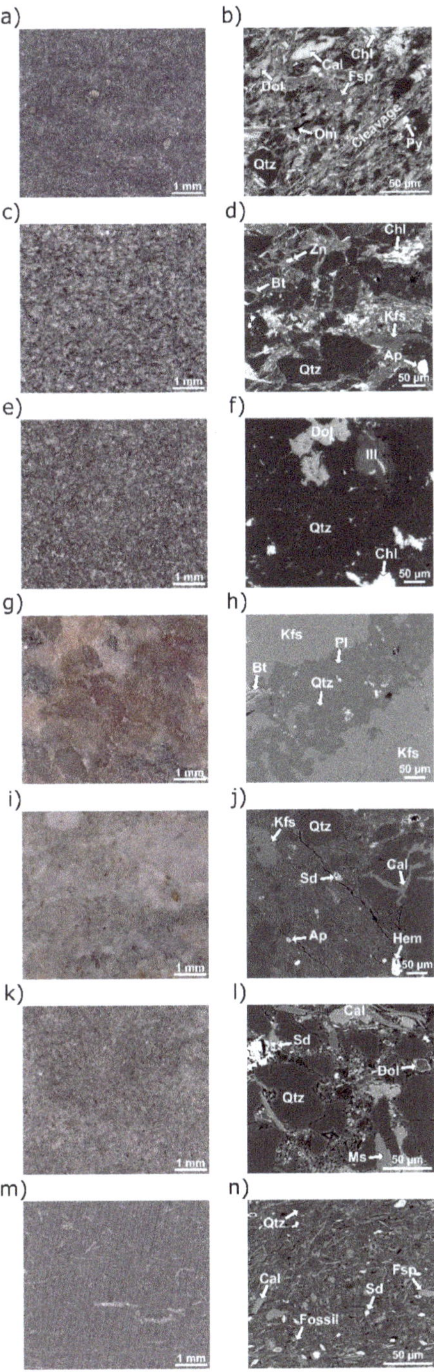

Figure 2. Optical micrographs (left column) and SEM-backscattered (BSE) images (right column) of investigated starting materials showing different composition and grain size (see text for details).

(**a,b**) Wissenbach slate, (**c,d**) graywacke, (**e,f**) quartzite, (**g,h**) granite, (**i,j**) altered granitic gouge, (**k,l**) Opalinus clay—sandy facies, and (**m,n**) Opalinus clay—shaly facies. Note different scales. Dol = Dolomite, Cal = Calcite, Chl = Chlorite, Qtz = Quartz, Om = Organic Matter, Fsp = Feldspar, Py = Pyrite, Bt = Biotite, Zn = Zinc, Kfs = Kalifeldspar, Ap = Apatite, Ill = Illite, Pl = Plagioclase, Hem = Hematite, Sd = Siderite.

2.2. Methods

Flow through experiments were performed on cm scale specimens at various p_c, T, σ-conditions using a modified MuSPIS (multiple sample production and injection simulator, [10]) apparatus for long-term petrophysical investigations. The sample assembly consists of two separate cylindrical objects with a thickness of t* = 10 mm (Figure 3a,b, Table A1), which were isolated by rubber (neoprene) jackets from the confining medium (hydraulic oil) inside the pressure vessel.

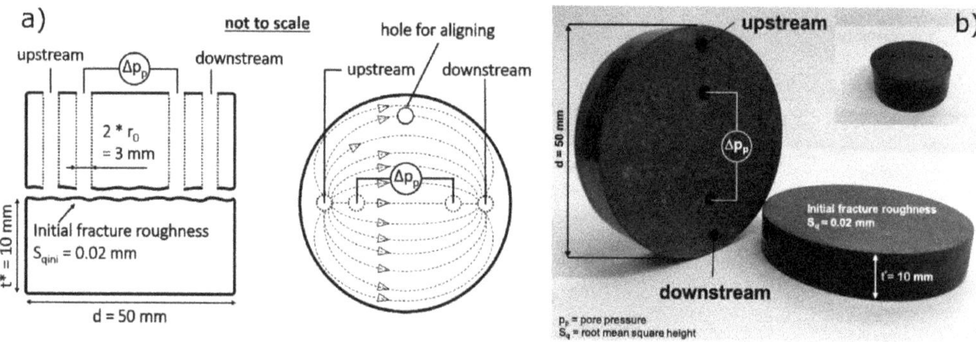

Figure 3. Schematic experimental setup of performed flow through experiments (**a**) and an example of a Wissenbach slate sample (**b**). Flow direction (dashed lines) is upstream borehole to downstream borehole. Differential pore pressure, Δp_p, for the determination of fracture transmissivity is measured using a highly accurate pressure transducer by utilizing two additional boreholes. d = diameter, t* = thickness.

The confining pressure is controlled by a syringe pump (Isco 65D, max p_c = 138 MPa). The maximum temperature is T = 200 °C, using a resistance-heating element (Thermocoax) mounted on a cylindrical stainless-steel tube placed inside the pressure vessel. To measure the transmissivity of the surface between the two disks, a continuous fluid flow at a pre-defined constant fluid pressure can be generated using two syringe pumps (Isco 100DM), which are connected to the surface via boreholes (up- and downstream boreholes in Figure 3a). The maximum pore (fluid) pressure is p_p = 70 MPa at flowrates of $Q = 2.5 \times 10^{-2} - 1 \times 10^{-6}$ L/min. The fluid pressure is measured separately across two additional boreholes with an additional pressure transducer (Siemens Sitrans P DS III), which is capable of measuring differential pressures of $\Delta p_p = \pm$ 10 mbar (1000 Pa). This was so that we could measure very low-pressure differences at low flow rates within the fracture. As our experiments were performed on low permeable rocks (matrix permeability $<10^{-17}$ m^2), we assumed the contribution of fluid flow through the matrix to be negligible. Deviatoric stresses up to σ = 100 MPa perpendicular to the interface can be applied using an axial piston driven by another syringe pump (Isco 65D). The axial load is measured using an internally mounted load cell (200 kN). The displacement of the axial piston is measured by Linear Variable Differential Transformers (LVDT) with an uncertainty of ± 1 μm, allowing to correlate fracture transmissivity with changing fracture aperture and sample deformation induced by variations in axial stress, confining pressure, and temperature.

For anisotropic rocks, the disks were prepared with the cylinder axis oriented perpendicular to bedding. The surface roughness of the artificial fracture, represented by the interface of the two sample disks, was controlled by grinding the flat saw-cut surfaces using

SiC-grains of a defined diameter (Kl60 $\approx$ 260 µm). The resulting initial fracture surface roughness, S_q (root mean square value), prior to testing was S_{q_ini} = 0.016 $\pm$ 0.003 mm, except for the granitic gouge that revealed a considerably higher initial roughness of S_{q_ini} = 0.247 mm (Table 1), which was measured with a surface scanner (white light profilometer, Keyence VR3200, accuracy $\approx$ 3µm). Since we also tested the influence of roughness on fracture transmissivity, an additional sample of Wissenbach slate was roughened with K600 SiC-grains (diameter $\approx$ 9 µm) resulting in lower S_{q_ini} of 0.006 mm.

Experiments on Wissenbach slate, Havelange quartzite, graywacke and the fresh Noble Hill granite from the Death Valley were performed using distilled water at a constant fluid pressure of p_p = 1 MPa (downstream borehole in Figure 3) as fluid flowing through the artificial fracture. Samples prepared from Opalinus Clay, and the gouge material from the Death Valley were tested using Argon gas to avoid sample disintegration during fluid flow. For comparison, one additional test on slate was also performed with Argon. Due to the high compressibility of the Argon gas, these experiments were performed at elevated p_p = 10 MPa to reduce the apparent permeability enhancement through the Klinkenberg effect (e.g., [44]). The effective confining pressure p_{eff} = p_c − p_p was kept constant (4 MPa) throughout most experiments so as to compare results gained with either distilled water or Argon gas. In a few cases, p_{eff} could not be held constant, for example, during tests with step-wise increasing confining pressure. Here, p_p was kept below p_c to avoid a 'blow-up' of the fracture.

Fracture transmissivity was calculated by assuming the fluid flow pattern within the fracture to be comparable to the flow of electric current through a resistive solid within an insulating medium (dashed lines in Figure 3a). Assuming that the up- and downstream borehole form the dipole of a magnetic field and comparing Ohm's law with Darcy's law for steady fluid flow to characterize the transmissivity of thin cracks in various rocks, Ohm's law may be rewritten as [19,45]:

$$k * t = \frac{Q * \eta * ln\left(\frac{2a}{r_0} - 1\right)}{\Delta p_p * B * \pi} \tag{1}$$

where the term k∗t is the transmissivity of the fracture, with k and t representing fracture permeability and fracture thickness, respectively. Q is the volumetric flow rate, η is the (temperature-dependent) dynamic viscosity of the fluid, 2a is the distance between up- and downstream borehole with radius r_0, Δp_p is the pressure difference within the fracture, and B (=0.75 for our geometry) is a geometry factor accounting for the aspect ratio of the fracture [19]. Here, we assumed single-phase fluid flow within the fracture, which is certainly a simplification compared to multi-phase fluid flow that may occur in natural fractures and the surrounding matrix [12].

A detailed microstructural analysis of the starting and the post-experimental samples was performed on mechanically as well as broad ion-beam polished, carbon-coated thin sections using a scanning electron microscope (SEM FEI Quanta 3D dual beam). SEM sections were prepared perpendicularly to the fracture surface. Broad ion-beam (BIB JEOL IB-19520CCP) polishing was used for the gouge and OPA samples to avoid preparation-induced damage of clay minerals. Due to its fragile character, the gouge was additionally solidified using epoxy resin prior to broad ion beam-milling.

In addition, we performed uniaxial compression (UCS) tests to determine the basic mechanical properties such as the uniaxial compressive strength (σ_{UCS}) and tangent Young's Modulus (E) of each sample material. The tests were performed according to the suggested ISRM method [46] using cylindrical specimens with a length-to-diameter ratio of 2:1. Results from UCS tests (Table 1) indicate that despite the high content of weak mineral phases, the strength and stiffness of WBS are comparable to granite, quartzite and graywacke samples that contain significantly more strong minerals. Furthermore, triaxial compressive strength (σ_{TCS}) and static secant Young's modulus of WBS were determined in constant strain rate tests of $\dot{\varepsilon}$ = 5 $\times$ 10^{-4} s^{-1} at simulated reservoir conditions with a confining pressure of p_c = 50 MPa and temperature of T = 100 °C using a Paterson-type deformation apparatus [47].

3. Results

In total, we performed five flow through experiments on Wissenbach slate to investigate the evolution of fracture transmissivity with changing confining pressure, differential stress oriented perpendicular to the fracture interface, temperature, and surface roughness. In addition, six tests under room temperature, constant confining pressure and stresses up to 45 MPa were conducted on the other rock types to evaluate the influence of composition on fracture transmissivity. The experimental duration for most of the experiments was between one and two weeks. Fracture transmissivity was determined after reaching steady-state flow conditions at the different thermodynamic boundary conditions.

Previously performed triaxial tests reveal σ_{TCS} = 498 ± 20 MPa and E = 69 ± 14 GPa (Table 1) that are comparable to the mechanical data of Westerly granite, Panzhihua gabbro or Novaculite [48] obtained at the same conditions.

3.1. Effect of Thermodynamic Boundary Conditions

Fracture fluid flow experiments at elevated p_c, T and σ-conditions were performed on Wissenbach slate samples with an average initial fracture surface roughness of S_q = 0.019 mm.

The influence of temperature on fracture transmissivity was evaluated at a confining pressure of p_c = 5 MPa and pore pressure of p_p = 1 MPa in the range of T = 20–100 °C (Table 2).

Table 2. Fracture transmissivity in Wissenbach slate at elevated temperatures (p_c = 5 MPa, σ = 0 MPa).

T [°C]	p_p [MPa]	$k*t$ [m³]
25	1	1.33×10^{-14}
40	1	1.02×10^{-14}
60	1	1.02×10^{-14}
80	1	1.07×10^{-14}
100	1	9.40×10^{-15}

T = temperature, p_p = pore pressure, $k*t$ = fracture transmissivity.

Except for a moderate decrease from T = 20 °C to T = 40 °C, fracture transmissivity does not change significantly at up to 100 °C (Figure 4), showing that the fracture transmissivity of Wissenbach slate is insensitive to changes in temperature at a low confining pressure.

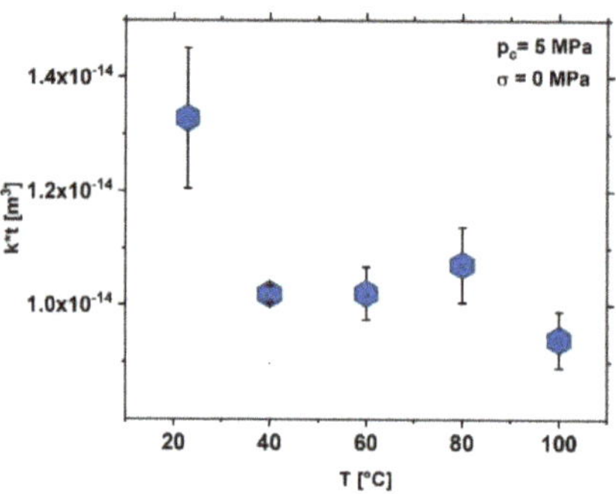

Figure 4. Effect of temperature, T, on fracture transmissivity, k*t, in Wissenbach slate. Fracture transmissivity decreases only slightly from T = 20 °C towards T = 40 °C and remains almost constant with increasing temperature up to 100 °C. Experimental conditions are indicated.

Hydrostatic tests with confining pressures ranging from p_c = 2–25 MPa and p_p = 1 MPa were performed at a constant temperature of T = 25 °C (Table 3).

Table 3. Fracture transmissivity in Wissenbach slate at elevated confining pressures (σ = 0 MPa, T = 25 °C).

p_c [MPa]	p_p [MPa]	$k*t$ [m^3]
2	1	1.46×10^{-14}
5	1	1.26×10^{-14}
10	1	3.01×10^{-15}
15	1	2.27×10^{-15}
20	1	1.24×10^{-15}
25	1	1.71×10^{-15}

p_c = confining pressure, p_p = pore pressure, $k*t$ = fracture transmissivity.

Fracture transmissivity decreased non-linearly by about one order of magnitude from $\approx 1.5 \times 10^{-14}$ m^3 to 1.7×10^{-15} m^3 with increasing confining pressures, approaching almost constant values at high p_c (Figure 5a).

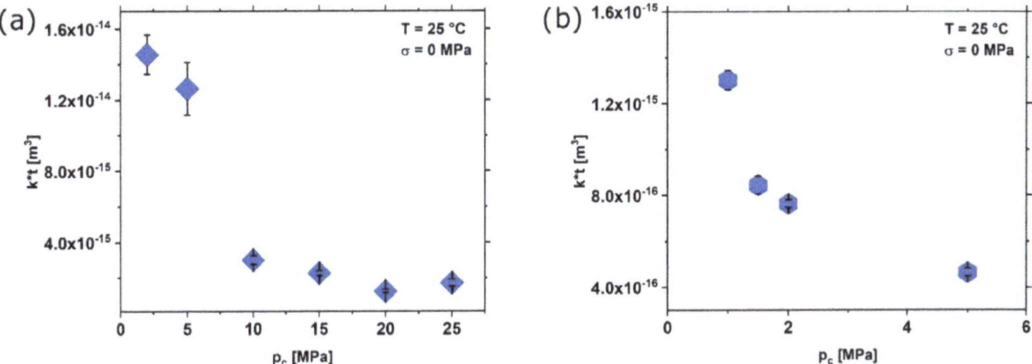

Figure 5. Effect of confining pressure, p_c, on fracture transmissivity, $k*t$, in Wissenbach slate with high (S_q = 0.0196 mm) (**a**) and low (S_q = 0.006 mm) initial roughness (**b**). Fracture transmissivity is decreasing with increasing confining pressure. At relatively low p_c ($\leq$5 MPa), fracture transmissivity is up to one order of magnitude lower for the sample with low initial roughness.

Experiments performed at p_c = 5 MPa, p_p = 1 MPa and T = 25 °C reveal a similar decrease in fracture transmissivity with increasing axial differential stress of σ = 0–45 MPa, but with a roughly 2-fold higher transmissivity than measured at similar p_c values (Table 4, Figure 6a).

Table 4. Fracture transmissivity in Wissenbach slate at elevated differential stress (p_c = 5 MPa, T = 25 °C) using water as flow medium.

σ [MPa]	p_p [MPa]	$k*t$ [m^3]
0.5	1	1.47×10^{-14}
5	1	1.01×10^{-14}
10	1	6.30×10^{-15}
15	1	4.83×10^{-15}
20	1	3.62×10^{-15}
25	1	2.67×10^{-15}
30	1	2.12×10^{-15}

Table 4. *Cont.*

σ [MPa]	p_p [MPa]	$k{*}t$ [m^3]
35	1	1.51×10^{-15}
40	1	1.22×10^{-15}
45	1	8.87×10^{-16}
40	1	9.11×10^{-16}
35	1	9.22×10^{-16}
30	1	1.01×10^{-15}
25	1	1.04×10^{-15}
20	1	1.16×10^{-15}
15	1	1.29×10^{-15}
10	1	1.44×10^{-15}
5	1	1.63×10^{-15}
0.5	1	2.03×10^{-15}

σ = differential stress, p_p = pore pressure, $k{*}t$ = fracture transmissivity.

Interestingly, upon unloading, the fracture transmissivity did not recover and remained considerably lower than the initial transmissivity. For comparison, we performed a second test on the WBS slate using Argon gas as fluid medium at p_c = 14 MPa, p_p = 10 MPa, i.e., a similar effective pressure of p_{c_eff} = p_c − p_p = 4 MPa, assuming that Terzaghi's principle [49] is valid. For this sample, the initial transmissivity was found to be distinctly lower than of the sample with water as the fluid medium (Table 5, small symbols in Figure 6a), but approaching similar values at high σ ($\geq$40 MPa) and showing similar residual transmissivity upon unloading.

Table 5. Fracture transmissivity in Wissenbach slate at elevated differential stress (p_c = 14 MPa, T = 25 °C) using Argon gas as flow medium.

σ [MPa]	p_p [MPa]	$k{*}t$ [m^3]
0	10	4.62×10^{-15}
5	10	3.95×10^{-15}
10	10	3.4×10^{-15}
15	10	2.83×10^{-15}
20	10	2.3×10^{-15}
25	10	1.8×10^{-15}
30	10	1.41×10^{-15}
35	10	1.08×10^{-15}
40	10	8.4×10^{-16}
45	10	6.58×10^{-16}
40	10	5.82×10^{-16}
35	10	6.41×10^{-16}
30	10	6.84×10^{-16}
25	10	7.44×10^{-16}
20	10	8.25×10^{-16}
15	10	9.3×10^{-16}
10	10	1.1×10^{-15}
5	10	1.37×10^{-15}
0	10	1.93×10^{-15}

σ = differential stress, p_p = pore pressure, $k{*}t$ = fracture transmissivity.

For the sample tested with water, after the axial loading (phase I) and unloading (phase II) steps, we subsequently increased p_c from 5 to 50 MPa (phase III), resulting in a fracture transmissivity slightly lower than that measured at σ = 45 MPa (Table 6, Figure 7), confirming the stronger influence of p_c on transmissivity reduction than of σ as described above.

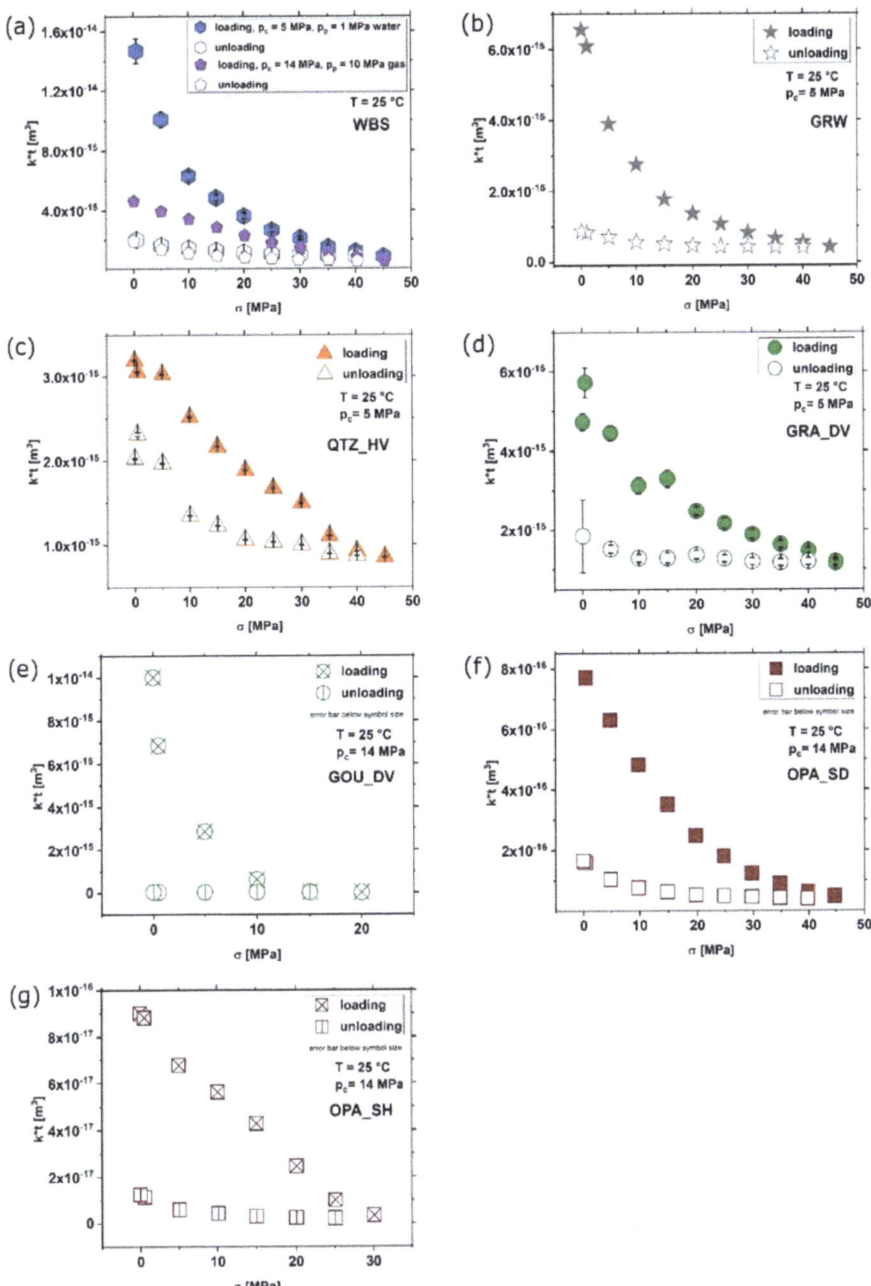

Figure 6. Effect of differential stress, σ, on fracture transmissivity, k∗t, in Wissenbach slate (**a**), Graywacke (**b**), Quartzite-Havelange (**c**), Granite-Death Valley (**d**), Gouge-Death Valley (**e**), Opalinus clay-sandy facies (**f**), and Opalinus clay-shaly facies (**g**). Experiments conducted at p_c = 14 MPa were performed using Argon gas as flow medium (gouge, OPA), all other with distilled water. The effective pressure ($p_c - p_p$) is 4 MPa in all cases. Experimental conditions are indicated. Note different scales.

Table 6. Fracture transmissivity in Wissenbach slate at elevated differential stress, confining pressures and temperatures.

Step	σ [MPa]	p_c [MPa]	T [°C]	p_p [MPa]	$k{*}t$ [m^3]
I	0.5	5	25	1	1.47×10^{-14}
	45	5	25	1	8.87×10^{-16}
II	0.5	5	25	1	2.03×10^{-15}
III	0	50	25	1	5.39×10^{-16}
IV	30	50	25	1	2.44×10^{-16}
V	30	50	90	1	1.88×10^{-17}

σ = differential stress, p_c = confining pressure, T = temperature, p_p = pore pressure, $k{*}t$ = fracture transmissivity.

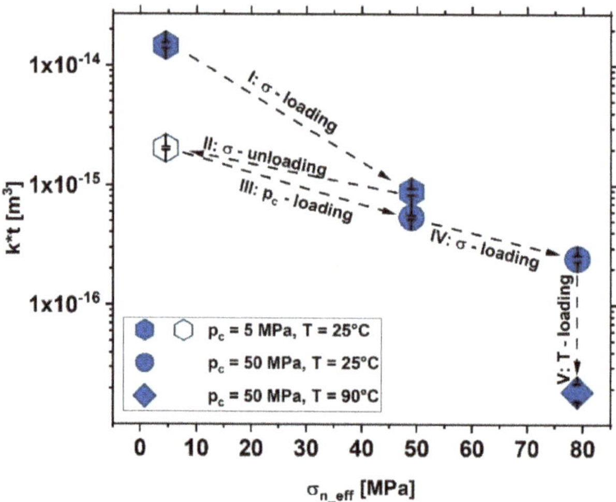

Figure 7. Combined effect of differential stress, σ, confining pressure, p_c, and temperature, T, on fracture transmissivity, $k{*}t$, plotted versus effective normal stress, σ_{n_eff}, acting on the fracture surface in Wissenbach slate. Fracture transmissivity decreases along σ-loading path (I) and recovers only partly during unloading (II). Increasing p_c (III), σ (IV) and T (V) further decreases fracture transmissivity.

Note that for comparison we plotted the transmissivity data which are presented in Figure 7 as a function of the effective normal stress $\sigma_{n\_eff} = \sigma + p_c - p_p$, acting normal to the fracture surface. In phase IV, axial stress was increased again to $\sigma = 30$ MPa at a constant p_c of 50 MPa ($\sigma_{n\_eff} = 79$ MPa), which yields a further transmissivity reduction, again with a lower gradient than observed in phase III (Figure 7). Finally, we increased temperature from 25 to 90 °C at constant σ and p_c (phase V), which induced a strong transmissivity reduction of almost one order of magnitude.

3.2. Effect of Fracture Surface Roughness

The influence of reduced fracture surface roughness on the change of fracture transmissivity with increasing confining pressure of up to 5 MPa was measured on the Wissenbach slate at T = 25 °C (Table 7).

The initial fracture surface roughness of this sample was $S_{q\_ini\_low} = 0.006$ mm, which is roughly three times smaller than the surface roughness of previously examined samples ($S_{q\_ini} = 0.019$ mm). Comparable to the rough surface, fracture transmissivity decreases non-linearly with increasing confining pressure (Figure 5b), but at a similar p_c the transmissivity values are about one order of magnitude lower (Table 3). At higher confining pressures ($p_c > 5$ MPa), we were not able to resolve transmissivity due to technical issues (below measurable range).

Table 7. Low-roughness-fracture transmissivity in Wissenbach slate at elevated confining pressures (σ = 0 MPa, T = 25 °C).

p_c [MPa]	p_p [MPa]	$k*t$ [m^3]
1	0.5	1.30×10^{-15}
1.5	0.5	8.44×10^{-16}
2	0.5	7.63×10^{-16}
5	1	4.65×10^{-16}

p_c = confining pressure, p_p = pore pressure, $k*t$ = fracture transmissivity.

3.3. Effect of Sample Composition

To determine the impact of composition on fracture transmissivity, we performed flow through experiments at T = 25 °C, with an effective pressure of p_{c_eff} = 4 MPa, and by increasing differential stress up to 45 MPa. Tests were conducted at p_c = 5 MPa, p_p= 1 MPa using water as fluid medium for samples WBS, GRW, QTZ_HV and GRA_DV. For the clay-rich samples GOU_DV, OPA_SD and OPA_SH, we used Argon at p_c =14 MPa and p_p = 10 MPa to avoid disintegration and swelling of the samples. As observed for WBS (Figure 6a), in all other rocks, fracture transmissivity decreased non-linearly with increasing σ and demonstrated limited recovery upon unloading (Figure 6, Table 8).

Table 8. Fracture transmissivity of and Graywacke (GRW), Quartzite-Havelange (QTZ_HV), Granite Death-Valley (GRA_DV), Gouge-Death Valley (GOU_DV), Opalinus clay-sandy facies (OPA_SD), Opalinus clay-shaly facies (OPA_SH) at elevated differential stress and T = 25 °C.

Formation	Fluid	σ [MPa]	p_c [MPa]	p_p [MPa]	$k*t$ [m^3]
GRW	H$_2$O	0	5	1	6.55×10^{-15}
		1			6.08×10^{-15}
		5			3.89×10^{-15}
		10			2.75×10^{-15}
		15			1.77×10^{-15}
		20			1.37×10^{-15}
		25			1.07×10^{-16}
		30			8.26×10^{-16}
		35			6.55×10^{-16}
		40			5.44×10^{-16}
		45			4.12×10^{-16}
		40			4.14×10^{-16}
		35			4.19×10^{-16}
		30			4.20×10^{-16}
		25			4.26×10^{-16}
		20			4.56×10^{-16}
		15			5.02×10^{-16}
		10			5.53×10^{-16}
		5			7.00×10^{-16}
		1			8.26×10^{-16}
		0			8.68×10^{-16}
QTZ_HV	H$_2$O	0	5	1	3.20×10^{-15}
		0.5			3.06×10^{-15}
		5			3.03×10^{-15}
		10			2.52×10^{-15}
		15			2.17×10^{-15}
		20			1.89×10^{-15}
		25			1.67×10^{-15}
		30			1.50×10^{-15}
		35			1.11×10^{-15}
		40			9.22×10^{-16}

Table 8. *Cont.*

Formation	Fluid	σ [MPa]	p_c [MPa]	p_p [MPa]	$k*t$ [m^3]
		45			8.50×10^{-16}
		40			8.68×10^{-16}
		35			8.90×10^{-16}
		30			9.98×10^{-16}
		25			1.03×10^{-15}
		20			1.06×10^{-15}
		15			1.23×10^{-15}
		10			1.35×10^{-15}
		5			1.97×10^{-15}
		0.5			2.32×10^{-15}
		0			2.03×10^{-15}
GRA_DV	H$_2$O	0	5	1	4.73×10^{-15}
		0.5			5.72×10^{-15}
		5			4.44×10^{-15}
		10			3.13×10^{-15}
		15			3.30×10^{-15}
		20			2.48×10^{-15}
		25			2.16×10^{-15}
		30			1.88×10^{-15}
		35			1.63×10^{-15}
		40			1.47×10^{-15}
		45			1.16×10^{-15}
		40			1.20×10^{-15}
		35			1.17×10^{-15}
		30			1.20×10^{-15}
		25			1.27×10^{-15}
		20			1.37×10^{-15}
		15			1.28×10^{-15}
		10			1.29×10^{-15}
		5			1.52×10^{-15}
		0			1.84×10^{-15}
GOU_DV	Argon	0	14	10	1.00×10^{-14}
		0.5			6.86×10^{-15}
		5			2.85×10^{-15}
		10			6.17×10^{-16}
		15			4.69×10^{-17}
		20			2.14×10^{-17}
		15			2.84×10^{-17}
		10			3.02×10^{-17}
		5			3.22×10^{-17}
		0.5			3.81×10^{-17}
		0			3.86×10^{-17}
OPA_SD	Argon	0.5	14	10	7.71×10^{-16}
		5			6.30×10^{-16}
		10			4.83×10^{-16}
		15			3.51×10^{-16}
		20			2.48×10^{-16}
		25			1.79×10^{-16}
		30			1.23×10^{-16}
		35			8.95×10^{-17}
		40			6.05×10^{-17}
		45			4.78×10^{-17}

Table 8. *Cont.*

Formation	Fluid	σ [MPa]	p_c [MPa]	p_p [MPa]	$k*t$ [m^3]
		40			3.91×10^{-17}
		35			4.22×10^{-17}
		30			4.52×10^{-17}
		25			4.88×10^{-17}
		20			5.20×10^{-17}
		15			6.23×10^{-17}
		10			7.58×10^{-17}
		5			1.04×10^{-16}
		0.5			1.59×10^{-16}
		0			1.65×10^{-16}
OPA_SH	Argon	0	14	10	8.99×10^{-17}
		0.5			8.82×10^{-17}
		5			6.76×10^{-17}
		10			5.61×10^{-17}
		15			4.27×10^{-17}
		20			2.44×10^{-17}
		25			9.86×10^{-18}
		30			1.85×10^{-18}
		25			2.06×10^{-18}
		20			2.30×10^{-18}
		15			2.98×10^{-18}
		10			4.20×10^{-18}
		5			5.81×10^{-18}
		0.5			1.13×10^{-17}
		0			1.24×10^{-17}

p_c = confining pressure, σ = differential stress, p_p = pore pressure, $k*t$ = fracture transmissivity.

The initial transmissivity of the graywacke sample (Figure 6b) is comparable to that of Havelange quartzite (Figure 6c) and Noble Hill granite (Figure 6d). However, the fracture transmissivity of the graywacke decreases at a higher rate upon differential loading, whereas the transmissivity of quartzite and granite decreases almost at the same point. The permanent reduction in transmissivity after unloading is lowest for the quartzite followed by the granite. On the other hand, the graywacke shows a reduction in transmissivity almost threefold that of the quartzite. Compared to WBS, the quartzite and granite were less sensitive to increasing stress, while the graywacke demonstrated a comparable sensitivity (Figure 6). The Death Valley gouge shows a very strong transmissivity reduction already at low stress and almost no recovery after unloading (Figure 6e), which is likely related to the relatively high preparation-related initial roughness, almost an order of magnitude higher than of the other rocks. For Opalinus Clay, transmissivity within the sandy facies (Figure 6f) is about one order of magnitude higher than for the shaly facies (Figure 6g). Interestingly, the non-linearity of transmissivity with increasing stress is least pronounced for OPA_SH. Note that the maximum stress on OPA_SH and GOU_DV samples was less than for the other rocks to avoid potential major creep deformation of these weak rocks.

For comparison, all data are shown in Figure 8 in a semi-logarithmic scale, revealing that hard rocks such as granite and quartzite display transmissivity of several orders of magnitude higher at high σ and less irreversible damage than relatively weak OPA and gouge. Moreover, slate and graywacke transmissivity is also relatively high (in the same range as of granite and quartzite transmissivity) at high σ but shows much less reversibility upon unloading. Note, however, that the initial roughness of the gouge sample is higher than of the other tested rocks, which may obscure the systematics to some extent.

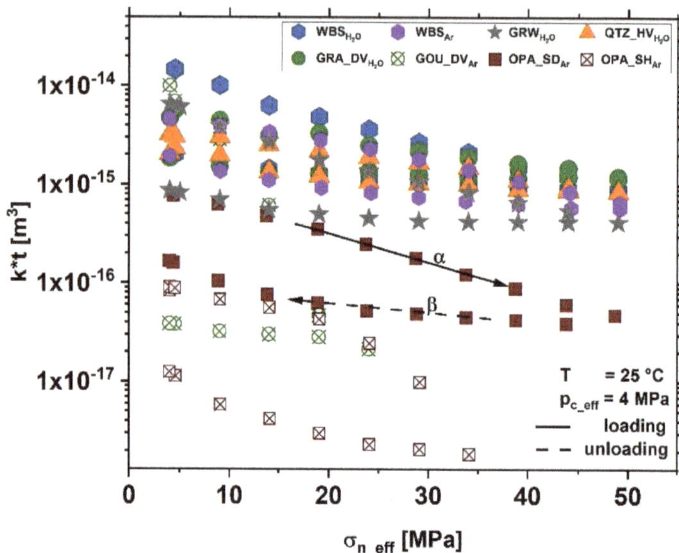

Figure 8. Comparison of the influence of effective normal stress, σ_{n_eff}, acting on the fracture surface on fracture transmissivity, k*t. Data are the same as shown in Figure 6, but plotted in semi-logarithmic scale. α and β indicate the linear slopes for fitting of the loading and unloading path to an exponential relationship (c.f., Equation (2)). Note that for unloading the slope is deviating from linearity at low stress.

3.4. Microstructures

After each experiment, the topography of the fracture surface of each sample was measured again with a white light profilometer in order to quantify changes due the applied experimental procedures. Thin sections of the tested specimens were then prepared for a subsequent SEM analysis.

In order to compare the surface topographies before and after the experiments, we show surface height distribution of the lower sample half (without injection/production holes, see Figure 3a) of each sample in Figures A1 and A2. In comparison to the initial surface, the height distribution of WBS samples (Figure A1) measured after testing at increasing p_c, σ or T became slightly wider but remained more or less unaffected. This is also reflected by the determined surface roughness measurements, which are relatively similar to the initial roughness values (Table 1). The same applies to the other samples, showing slight widening of height distributions after deformation (Figure A2). However, height distributions of the quartzite and granite samples show the least changes (Figure A2b,c), indicating no obvious change in the topography after loading. On the other hand, the distribution and fracture roughness of the gouge (Figure A2d, Table 1) display the highest grade of asperity degradation, although the maximum applied stress did not exceed 20 MPa. Independent of the facies, the surface of fractures within the Opalinus Clay (Figure A2e,f) exhibit a moderate alteration due to axial loading.

SEM micrographs of the Wissenbach slate and of the other rocks prepared perpendicular to the fracture surface are shown in Figures 9 and 10, respectively.

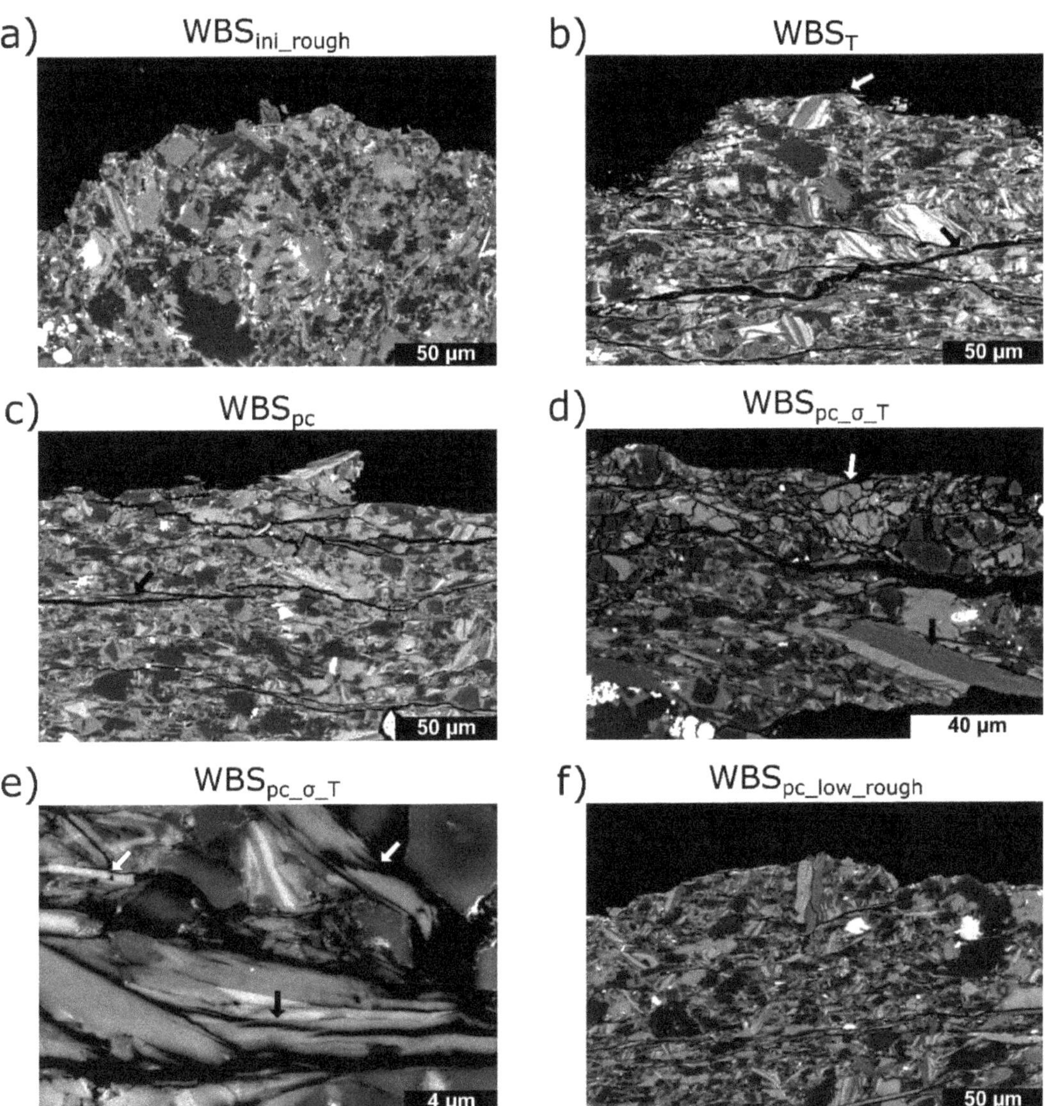

Figure 9. Scanning electron microscope-back scattered electron (SEM-BSE) photographs of sections prepared from Wissenbach slate (WBS) samples. All thin sections have been prepared perpendicular to the artificially prepared macro fracture located at the top of each image. Panel (**a**) shows the initial (ini_rough) fracture prior to testing. Subscripts 'T', 'pc', and 'pc_σ_T' refer to tests performed at increasing temperature (**b**), confining pressures (**c**), a combination of pressure, stress, and temperature (**d**,**e**). Panel (**f**) shows the post-experimental profile of the sample with initial low roughness. Arrows indicate microfracturing and intracrystalline plasticity (see text for details). Note the different scales.

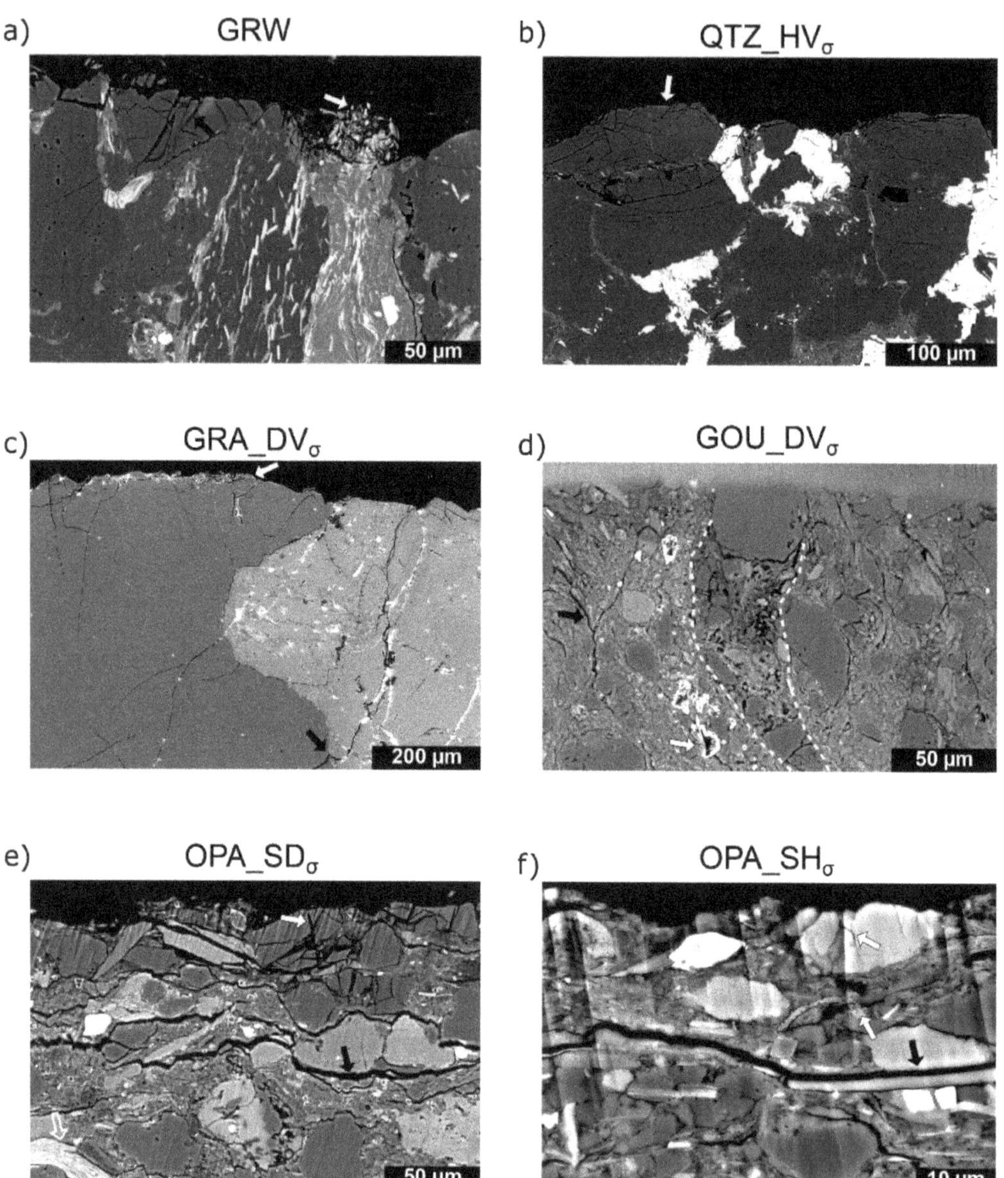

Figure 10. Scanning electron microscope-back scattered electron (SEM-BSE) photographs of sections prepared from samples of various formations after applying differential stress normal to the fracture surface. All thin sections have been prepared perpendicular to the artificially prepared macro fracture (located at the top of each figure). (**a**) Graywacke (GRW), (**b**) quartzite recovered from the Havelange borehole (QTZ_HV), (**c**) granite recovered from the Death Valley (GRA_DV), (**d**) gouge like material recovered from the Death Valley (GOU_DV), (**e**) Opalinus clay material recovered from the sandy facies (OPA_SD), (**f**) Opalinus Clay recovered from the shaly facies (OPA_SH). Arrows indicate testing-induced damage (see text for details). Note different scales.

For WBS, in comparison to the intact initial rough artificial fracture surface (Figure 9a), the matrix close to the surface of samples after testing is severely damaged (Figure 9b–e). Samples being step-wisely exposed to increasing temperature locally displayed a slightly flattened fracture surface profile (Figure 9b, white arrow). Sporadically, quartz, carbonate and dolomite grains are fractured, potentially induced by local stress concentration. On the other hand, small intergranular fractures (<5 µm opening; black arrow in Figure 9b) oriented subparallel to the macroscopic artificial fracture can be frequently observed, which may be partially induced by unloading and/or cooling and follows a cleavage orientation. Deformation microstructures of the sample deformed at varying confining pressure of up to 25 MPa are similar compared to the previously described sample. However, the density of intergranular, cleavage-parallel fractures is higher in comparison (Figure 9c). Damage of the sample subjected to combined pressure, axial stress, and temperature variations was highest when a large number of intra- and transgranular cracks in calcite, dolomite and quartz mineral grains (white arrow in Figure 9d) were detected. In addition, this sample shows several bent, kinked and delaminated micas (black arrows in Figure 9d,e), indicative for crystal plasticity. Furthermore, we observed boudinage micas suggesting shear motion at grain boundary surfaces (white arrows in Figure 9e). For all samples, the maximum damage zone is limited to <150 µm below the fracture surface profile line. As expected, the surface of the sample with a low initial roughness was considerably smoother compared to the other samples (Figure 9f). Except for the subparallel fractures with a small aperture, damage of the surface by the applied pressure was hardly visible, likely because of the relatively low applied confinement.

The damage near the fracture surface of the other rocks after stress stepping varies with composition. The graywacke sample shows some spalling microfractures in quartz and feldspar grains close to the fracture interface (white and black arrow in Figure 10a). Rarely, minor intergranular fractures (aperture < 5 µm) propagate into the rock. The damage zone is limited to <150 µm.

In the Havelange quartzite, microfractures in quartz are confined to single grains. The intragranular cracks form near the surface and lead to the spalling of grain fragments, indicative of small grain contact areas during differential loading (white arrow in Figure 10b). Despite the high number of fractures, the grain shape still remains relatively intact, resulting in a recognizable topography of asperities on the fracture surface. The damage zone below the fracture surface does not exceed 150 µm and is limited to the first layer of grains below the fracture boundary. The Noble Hill granite also exhibited spalling in quartz and feldspar grains at the contact to the fracture surface (white arrow in Figure 10c). In addition, we observed several intra- and trans-granular fractures opening subparallel to the loading direction (black arrow in Figure 10c) as well as the refracturing of previously healed, pre-existing fractures. Compared to the quartzite, the damage zone in the granite sample was found to be significantly larger. Single fractures extend up to 700 µm below the surface. Due to the high alteration and tectonic overprint [37], deformation structures that were generated by our experimental procedure were difficult to identify in the gouge from the Death Valley. Besides the cataclastic fabric and alteration of minerals (e.g., siderite, white arrow Figure 10d), we identified deformation features such as strongly bent phyllosilicates, the collapse of pore space and altered mineral grains (area between dashed white lines in Figure 10d). Clastic mineral grains (quartz, carbonates, feldspars) and mineral fragments show rare evidence of experimentally induced fractures. We observed several intergranular fractures that potentially opened during unloading (black arrow in Figure 10d). Note that the milling artefacts at the fracture surface are caused by the partly remaining resin, which was applied after testing to avoid preparation-induced damage. Opalinus Clay of the sandy (Figure 10e) and shaly (Figure 10f) facies typically displayed matrix deformation by the bending of phyllosilicates (open arrow in Figure 10e) and small fractures opening subparallel to the fracture surface (black arrows in Figure 10e,f). Clastic mineral grains in the sandy facies display several transgranular microfractures (white arrow in Figure 10e). On the other hand, the shaly facies show a significantly lower amount of fractured mineral grains compared to the sandy facies. Furthermore, these are limited to calcite grains and fossil shells, whereas quartz grains remained unfractured (white arrows in Figure 10f). In both facies, the "card house" structure of the clay matrix observed in the undeformed material collapsed with clay minerals being reoriented, with their longest axis parallel to the facture surface and perpendicular to the maximum principal stress direction. The damage zone extends up to 150 µm below the surface in the sandy facies and up to 300 µm in the shaly facies.

4. Discussion

Our set of flow through experiments demonstrated that at given fracture surface roughness increasing confining pressure, differential stress, and temperature at elevated pressure reduces the fracture transmissivity of slates (Figures 4–7). Furthermore, the initial transmissivity reduced considerably for smooth surfaces compared to rough surfaces (Figure 5). A strong influence of sample composition and associated mechanical properties on transmissivity is also evident (Figures 6 and 8). These effects will be discussed in the following paragraphs.

4.1. Influence of Thermodynamic Boundary Conditions (T, p_c, σ) on Fracture Transmissivity of Wissenbach Slate

Temperature has a strong effect on the interaction between the fluid transported through the fracture and the adjacent host rock [50–53]. The rate of dissolution of asperities is often enhanced at high temperatures and can reduce the mean fracture aperture due to mineral precipitation [52–55]. For WBS flushed with water, [22] observed a slight (few %), time-dependent fracture permeability decline in long-term flow through tests at p_c = 10 MPa, σ = 0 MPa and elevated temperatures up to 90 °C due to pressure solution and free face dissolution. However, using saline fluids, the authors measured an increase in permeability with increasing T, probably related to enhanced dissolution kinetics. In our tests with distilled water, we observed that at p_c = 5 MPa and σ = 0 MPa the fracture transmissivity of WBS was hardly affected by temperature between 25 and 100 °C (Figure 4), except a small decrease ($\approx$23%) between 25 and 40 °C. This confirms that temperature sensitive processes, like chemical dissolution-precipitation processes or stress corrosion at areas of high stress concentration were not very effective within the relatively short time span of our experiments. In contrast, transmissivity dropped considerably more ($\approx$93%) at p_c = 50 MPa and σ = 30 MPa in response to a temperature increase from 25 to 90 °C (Figure 7). This observation suggests that thermal dilation is more effective at high effective normal stress since the mean fracture aperture is lower than at low p_c, σ conditions (e.g., [52,56]). However, performing experiments over a longer time range may be necessary to reliably record chemical processes such as diffusion and mineral precipitation as reported by [22]. These authors performed similar flow-through experiments on samples prepared from the same slate material emphasizing the effect of fluid-rock interactions on the time-dependent transmissivity of fractures.

Concerning the influence of pressure and stress on transmissivity, we observed that the transmissivity is slightly (about two times) higher at enhanced p_c compared to a similar enhancement of σ alone, although the effective stress acting normal to the interface is the same (e.g., Figure 7). The difference may be explained by the increasing strength of the matrix adjacent to the fracture and the suppression of tensile fracturing with increasing pressure, whereas with increasing stress but constant (low) confining pressure the strength of the aggregate is lower than at high p_c. In both cases, the observed reduction in fracture transmissivity with increasing p_c or σ (Figures 5 and 6) is likely induced by indentation and damage of fracture surface asperities (Figures 9 and 10), resulting in a change of contact area and therefore mean aperture. A similar trend of decreasing fracture transmissivity induced by mechanical and/or pressure solution processes is reported for other rocks, e.g., shales, (tight) sandstones, granites, and granodiorite [1,16,19,50,52,53,57–59].

Interestingly, the transmissivity of WBS at low applied stress was distinctly lower for the sample tested with Argon at $p_c = 14$ MPa, $p_p = 10$ MPa than for the sample flushed with water at $p_c = 5$ MPa, $p_p = 1$ MPa, i.e., at the same effective pressure (Figure 6a). The difference may be attributed to the experimental protocol, where in the case of Argon the initial 14 MPa confining pressure was applied before applying the fluid pressure. This may induce more initial damage of the surface compared to the test with water at only 5 MPa confining pressure (cf., Figure 5a). In addition, it is not well established that Terzaghi's principle of effective stresses, i.e., $p_{c\_eff} = p_c - \delta\, p_p$ with $\delta = 1$, is valid here because at the microscale asperities are in contact, which reduces the Biot coefficient δ to a value < 1. This effect may result in a higher effective pressure in the experiment conducted with Argon gas compared to that with water, which would reduce the transmissivity difference at low stress. On the long term, chemical effects may also contribute to the transmissivity evolution due to different chemical fluid-rock interaction rates for different types of fluids.

In Figure 11, a comparison of our data with that measured by [19] on Westerly granite, Pennant sandstone and Bowland shale is presented, revealing that the influence of effective pressure on transmissivity disappears at high p_{c_eff}. The transmissivity of WBS appears to be comparable to that of Pennant sandstone, whereas the transmissivity of our granite appears to be considerably lower than that of Westerly granite. Bowland shale exhibits the lowest transmissivity values, which are even lower than our clay-rich samples. The difference may be attributed to the initial roughness of the samples used by [19], which was unfortunately not provided by the authors.

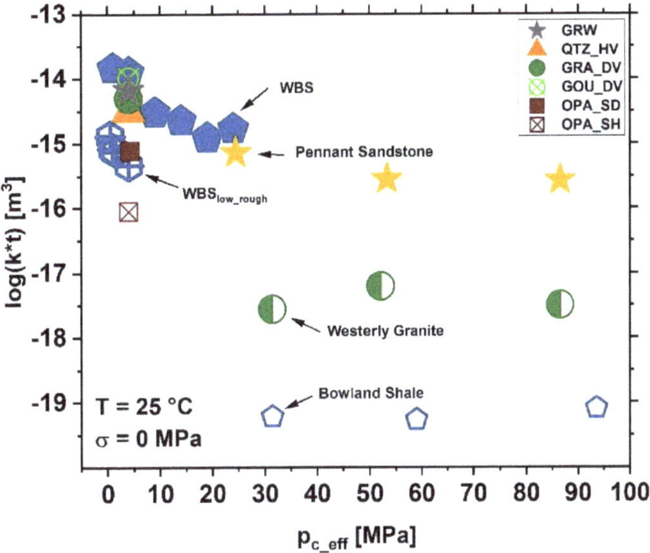

Figure 11. Effect of effective confining pressure, $p_{c\_eff} = p_c - p_p$, on fracture transmissivity, $k*t$, in Wissenbach slate (WBS) in comparison with other formations as reported by [19]. Data of the other formations investigated here (at $\sigma = 0$ MPa, c.f., Figure 6) are shown in addition.

The non-linear transmissivity reduction with increasing pressure and/or stress (Figures 5, 6 and 8) may be explained by the initial elastic deformation of the asperities that is gradually replaced by irreversible damage due to high stress concentrations. Since the effective contact area also increases with load (e.g., [60]), deformation or the breaking of further asperities or grains in the vicinity of the fracture surface become increasingly hampered and additional deformation is increasingly promoted by matrix deformation. As a consequence, the fluid flow pattern likely changes towards a higher tortuosity of the flow path and probably towards more localized flow through channels at high pressures, as also observed for other rocks (e.g., [50]). A number of theoretical attempts were made to relate fracture transmissivity to surface roughness and the applied normal stress (e.g., [61–66]). From that and from empirical correlations based on experiments, the change in fracture permeability with stress was described by a hyperbolic relationship [67], power law [68], or an exponential law [17,61]. Here, we used an exponential law of the following form:

$$k * t = c * e^{x * \sigma_{n\_eff}} \tag{2}$$

where c is a constant and x is the slope in a plot of $\ln(k*t)$ vs. σ_{n_eff}, with $x = \alpha$ for the loading path and $x = \beta$ for the unloading path (c.f., Figure 8). For WBS, β is about 3 times smaller than α (Table 9), showing that irreversible damage changes the stress-sensitivity of transmissivity considerably.

Table 9. Influence of host rock composition on fracture transmissivity.

Sample	α	β	B_{compo}
WBS	-0.061 ± 0.002	-0.018 ± 0.002	0.32
WBS$_{Ar}$	-0.044 ± 0.001	-0.025 ± 0.002	0.32
GRW	-0.061 ± 0.003	-0.019 ± 0.003	0.41
QTZ_HV	-0.030 ± 0.001	-0.022 ± 0.003	0.71
GRA_DV	-0.032 ± 0.001	-0.008 ± 0.002	0.90
GOU_DV	-0.319 ± 0.023	-0.026 ± 0.003	0.61
OPA_SD	-0.065 ± 0.001	-0.030 ± 0.005	0.61
OPA_SH	-0.110 ± 0.014	-0.064 ± 0.008	0.16

α = slope of loading branch, β = slope of the unloading branch, B_{compo} = brittleness index based on sample composition. WBS = Wissenbach slate, WBS$_{Ar}$ = Wissenbach slate tested with Argon, GRW = Graywacke, QTZ_HV = Quartzite Havelange, GRA_DV = Granite Death Valley, GOU_DV = Gouge Death Valley, OPA_SD = Opalinus clay sandy facies, OPA_SH = Opalinus clay shaly facies.

4.2. Sample Composition and Mechanical Properties

To better quantify the influence of composition and mechanical properties on fracture transmissivity, we applied the exponential law on the transmissivity–stress data of all other rocks, yielding values in the range of -0.03 and -0.32 for α, and -0.008 and -0.064 for β (Table 9).

Lowest (absolute) α-values were fitted for granite and quartzite samples, whereas absolute α-values of the graywacke and slate are about twice as high and comparable to the sandy facies of OPA. Lowest (absolute) β-values were found in the granite and graywacke. In order to relate these values to composition, we superimposed them in ternary diagrams (Figure 12).

Low (absolute) α-values, indicative of a weak reduction in transmissivity at elevated stress appear to be correlated with a high fraction of strong components (QFSO) and a low amount of weak constituents (Phyl), whereas carbonates appear to have no influence (Figure 12a). For the unloading sensitivity β, which is high for strong fracture transmissivity recovery, we observed a trend of enhanced unloading recovery with an increasing quantity of weak minerals and decreasing amount of QFSO (Figure 12b). Note that other parameters, which likely influence α and β, for example, porosity and cementation, are not captured in the diagrams.

In addition, we compared the σ-sensitivity of transmissivity for the different rock types with their brittleness defined by composition. Here, we used the empirical brittleness definition suggested by [69]:

$$B_{compo} = \frac{w_{QFSO} * f_{QFSO}}{w_{QFSO} * f_{QFSO} + w_{Cb} * f_{Cb} + w_{Phyl} * f_{Phyl}} \tag{3}$$

where f_{xx} is the fraction of minerals xx given in wt% and w_{QFSO}, w_{Cb}, and w_{Phyl} are the weighting factors ranging from 0 to 1. We set $w_{QFSO} = w_{Phyl} = 1$ and $w_{Cb} = 0.5$ as suggested by [69]. Calculated brittleness values are given in Table 9, ranging from 0.16 to 0.90. Based on the simplified assumption that in the absence of strong (QFSO) minerals the mechanical behavior is dominantly ductile, B values can vary between 0 and 1, indicating ductile and brittle deformation behavior, respectively. For rocks with high brittleness, the effect of increasing pressure or axial stress on fracture transmissivity is expected to be lower than for those with low B values. With the exception of the gouge, absolute values of α decrease with increasing B_{compo} (Figure 13a) in line with the common assumption that the deformation of brittle rocks is less stress-sensitive than that of ductile rocks. The low α value of the gauge-like material recovered from the Death Valley is most likely due induced by the high initial roughness and a high porosity of about 13 vol% (Table 1). Similarly, the (absolute) unloading sensitivity β seems to slightly decrease with increasing brittleness (Figure 13b), which may be explained by the crushing of strong and brittle minerals at the fracture surface at elevated stresses, where the produced fines prevent the recovery of fracture transmissivity during unloading.

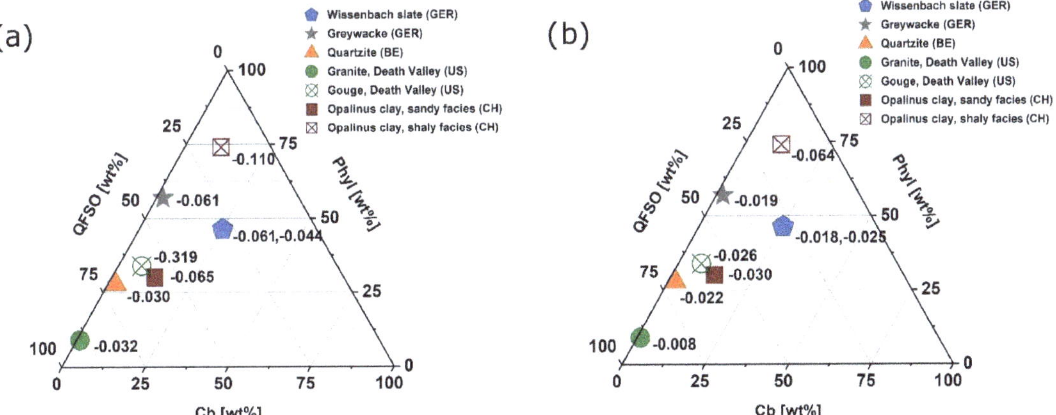

Figure 12. Influence of sample composition on the (effective) stress-sensitivity of fracture transmissivity indicated by superimposed α-values for the loading path (**a**) and by β-values for the unloading path (**b**). The (absolute) sensitivity generally decreases for a high fraction of strong (QFSO) minerals and low amount of weak (Phyl) constituents.

Another parameter that likely explains the stress-sensitivity of the fracture transmissivity of the different investigated rocks are their mechanical properties. The measured unconfined uniaxial strength, σ_{UCS}, and the static Young's moduli, E, vary over wide range and appear to be linearly correlated (Table 1, Figure A3), as has been observed frequently (e.g., [48,70]). Compared to composition-based brittleness, both (absolute) α and β values decrease with increasing σ_{UCS} and E (c.f., Figure 13c–f, respectively), revealing that rocks with a high strength and/or elastic stiffness are less sensitivity to stress-induced damage of the fracture surface and associated transmissivity reduction.

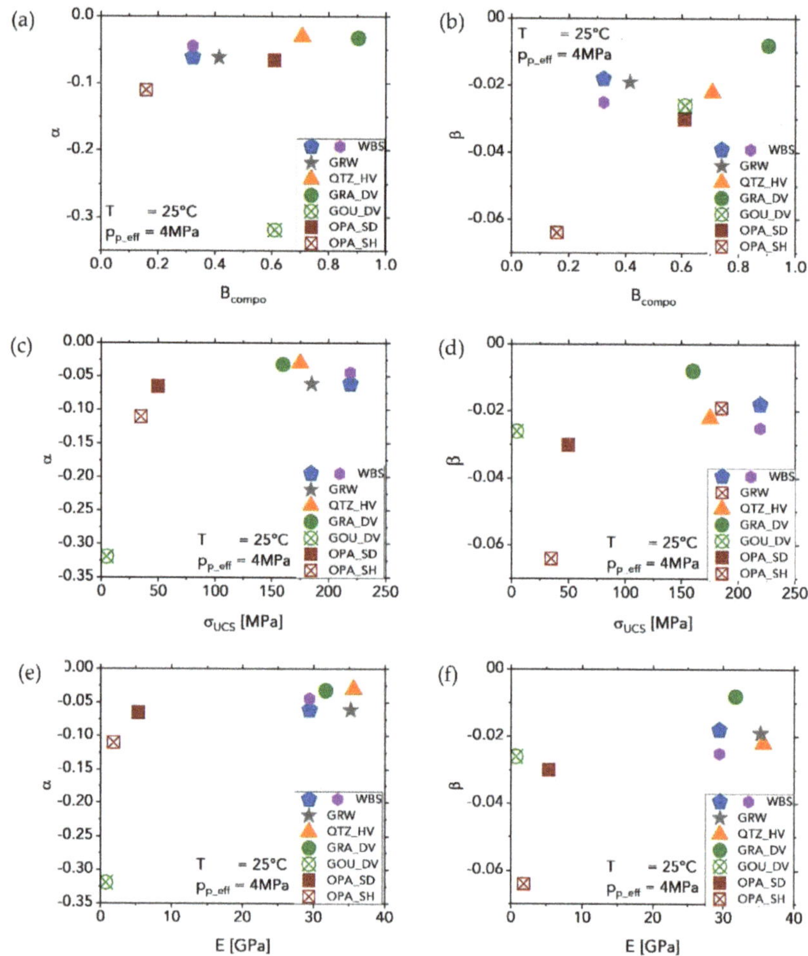

Figure 13. Influence of brittleness based on composition, B_{compo}, uniaxial compressive strength, σ_{UCS}, and static Young's modulus, E, on the (effective) stress-sensitivity of fracture transmissivity, α, for the loading path (**a,c,e**) and for the unloading path β (**b,d,f**), respectively. The (absolute) sensitivity decreases with increasing brittleness, strength, and elastic stiffness.

4.3. Implications for EGS in Different Host Rocks

An assessment of the performance and sustainability of fractured rock systems for EGS requires knowledge of the response of fractures to the acting effective stress, temperature, and fluid flow. The effective normal stress on fractures changes during and after hydraulic stimulation of reservoirs. As may be expected, our results of flow through experiments on various rock types (i.e., slate, quartzite, graywacke, granite, claystone, and gouge material) demonstrate that the key parameters that determine the stress resistance of fracture transmissivity are fracture surface roughness and the mechanical properties of the respective rock material. Fractures with similar surface roughness display the highest transmissivity in slate, quartzite, graywacke, and fresh granite, all of which contain a high amount of strong minerals and display high brittleness, high strength, and high elastic stiffness. On the other hand, in weak and porous rocks such as claystone and gouge material, transmissivity is reduced by several orders of magnitude by deformation of the soft matrix and limited crushing of contact asperities. Therefore, highly altered, clay-bearing formations are not favorable candidates for EGS. For hard rocks, grain size may also play a role if fracture propagation is predominantly intergranular. For example, the grain size of slate and graywacke is much smaller than that of quartzite and granite (Figure 2), so that at the microscale the real contact area of fine-grained rocks is higher, which reduces the stress acting on contact asperities, thereby maintaining a higher transmissivity.

Our results also imply that beside fracture surface roughness and composition, the degree of consolidation and metamorphic grade affects the potential of reservoir rocks for EGS. For example, the shaly facies of Opalinus Clay, graywacke and slate are relatively rich in phyllosilicates (Figure 1), but the latter two experienced a higher degree of metamorphosis with a strong cementation of grains, resulting in a much higher strength and elastic stiffness (Figure A3). Accordingly, fractures in slates are much more resistant against stress-induced fracture closure than in OPA (Figure 8).

We also noticed that (for slate samples) an increase in the fracture surface roughness (S_q) by about three times yields a fracture transmissivity of almost one order of magnitude higher at a low effective pressure (<5 MPa), which vanishes at high pressure (up to 25 MPa in our experiments), possibly due to the fact that the mechanical strength of slate at the microscale is not sufficient to resist high effective stresses. During hydraulic stimulation in EGS, tensile fractures are created that may connect to a preexisting fracture network. The surface roughness of these fractures is likely higher than the roughness of the artificially prepared surfaces in our study. For example, we measured a significantly larger (up to 20×) and more heterogeneously distributed roughness of tensile fractures created in WBS by Brazilian Disk (BD) experiments in divider, short-transverse, and arrester configuration following the ISRM suggested method [71]. However, even fractures with high initial roughness, e.g., created during hydraulic fracturing, may show a strong transmissivity decline at high stress conditions as long as they are not self-propped by shearing or artificially kept open by the addition of proppants (e.g., [72]). For the latter, high stress may crush the proppants and embed them into the matrix, which leads again to a transmissivity reduction (e.g., [73]).

Other parameters, which were not investigated in this study, may also affect the efficiency and sustainability of an EGS in different formations, for example self-propping due to a shear deformation event or fine production and clogging of flow path by ongoing deformation. Additionally, chemical effects due to long-term fluid–rock interactions or scaling may change the fracture transmissivity, depending on fluid and rock composition and thermodynamic boundary conditions.

5. Conclusions

Fracture transmissivity decreases asymptotically with increasing confining pressure and stress due to the damage caused to surface asperities and the matrix deformation of weak rocks, which is largely irreversible. A lower initial roughness also reduces transmissivity, while a temperature increase only significantly reduces the transmissivity by thermal dilation if the fracture aperture is already low. As may be expected, the fracture transmissivity of hard brittle rocks with a high fraction of strong minerals is less sensitive to stress-induced fracture closure and exhibits less irreversible damage. However, since the transmissivity of slates is much higher than of claystone, as well as the transmissivity of granite compared to a highly altered and sheared granitic gouge, the grade of consolidation, metamorphosis, and alteration of rocks is also important for fluid flow within discrete fractures under in situ conditions. Our results suggest that Variscan metamorphic rocks such as slate, graywacke and quartzite can be considered as host rocks for unconventional Enhanced Geothermal systems, with resistance to fracture transmissivity against thermodynamic boundary conditions that is as good as that of granite or quartzite for similar initial fracture roughness.

Author Contributions: Conceptualization, J.H.; methodology, J.H. and V.S.; software, J.H., V.S. and E.R.; validation, J.H., V.S., C.C., H.M. and E.R.; formal analysis, J.H.; investigation, J.H. and V.S.; resources, J.H., V.S. and C.C.; data curation, J.H. and V.S.; writing—original draft preparation, J.H.; writing—review and editing, J.H., V.S., C.C., H.M. and E.R.; visualization, J.H., V.S., C.C. and E.R. All authors have read and agreed to the published version of the manuscript.

Funding: This research was funded by the "Multidisciplinary and multi-context demonstration of Enhanced Geothermal Systems Exploration and Exploitation Techniques and potentials (MEET)" research project funded by the European Union's Horizon 2020 research and innovation programme under Grant Agreement No. 792037.

Data Availability Statement: Not applicable.

Acknowledgments: We thank Michael Naumann for the construction of the modified MuSPIS apparatus and for their assistance with the uniaxial and triaxial deformation as well as fracture transmissivity experiments, Stefan Gehrmann for sample and thin section preparation, Anja Schreiber for carbon coating of SEM samples, Bianca Wagner and Bernd Leiss from the University of Göttingen for XRD results and providing Wissenbach slate and graywacke samples, Yves Vanbrabant from Institut Royal des Sciences Naturelles de Belgique for providing Havelange quartzite samples, Ghislain Trullenque and Johanne Klee from Institut Polytechnique LaSalle Beauvais for providing granite and gouge samples, and Kristian Bär and Aysegül Turan from the Technical University of Darmstadt for providing some mechanical and petrophysical data.

Conflicts of Interest: The authors declare no conflict of interest.

Appendix A

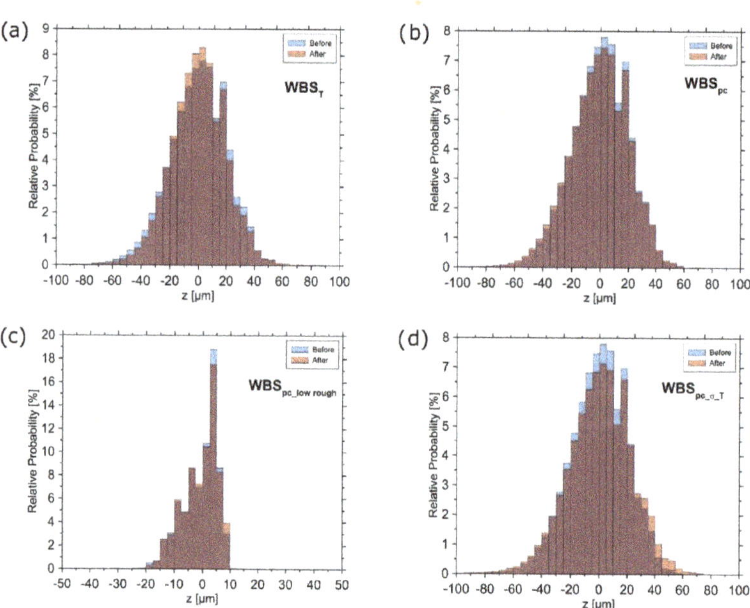

Figure A1. Fracture surface height distribution before and after experiments for Wissenbach slate tested at different temperatures (**a**), confining pressures (**b,c**), and a combination of different pressures, stresses and temperatures (**d**). Initial sample roughness was $S_q = 0.019$ mm in (**a,b,d**) and $S_q = 0.006$ mm in (**c**).

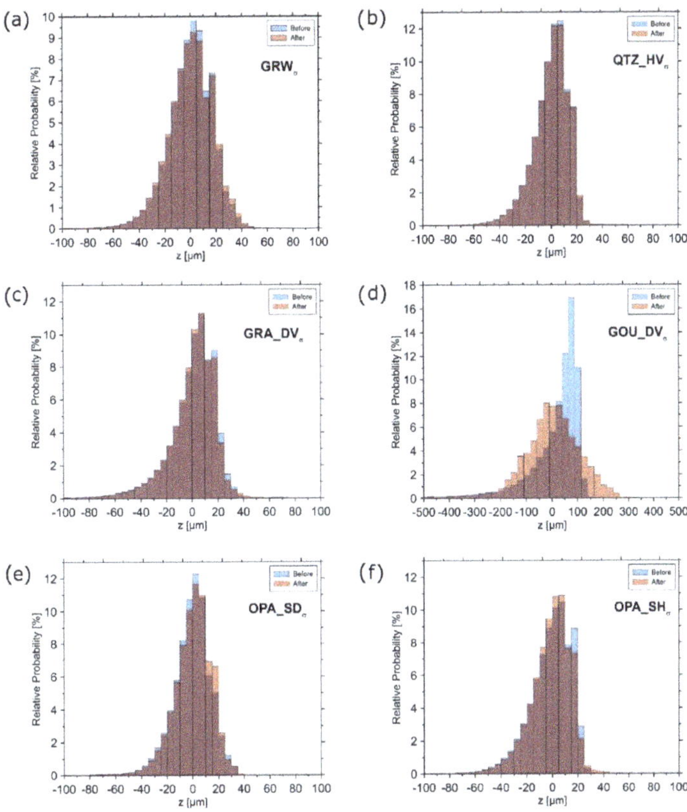

Figure A2. Fracture surface height distribution before and after testing at elevated stresses for graywacke (**a**), quartzite (**b**), fresh granite (**c**), altered granitic gouge (**d**), Opalinus clay—sandy facies (**e**), and Opalinus clay—shaly facies (**f**). Wissenbach slate tested at different temperatures (**a**), confining pressures (**b**,**c**), and a combination of different pressures, stresses and temperatures (**d**).

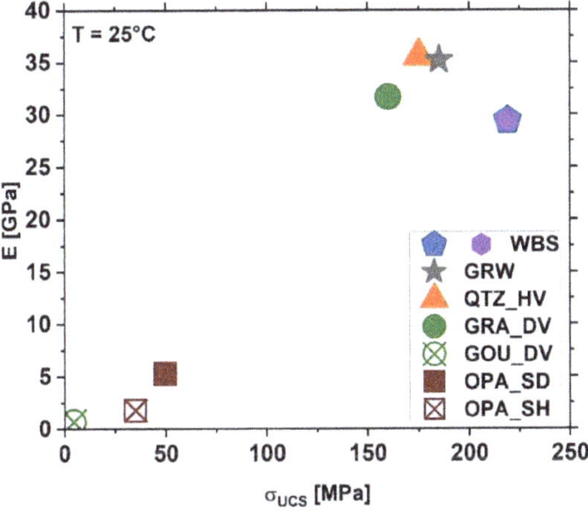

Figure A3. Static Young's modulus E vs. uniaxial compressive strength σ_{UCS} of all tested rock types.

Table A1. List of symbols.

Symbol	Description	Unit
ρ	Bulk density	g/cm^3
ρ_g	Grain density	g/cm^3
ϕ_{He}	Sample Porosity (using Helium pycnometry)	vol%
k_{matrix}	Matrix permeability	m^2
σ_{UCS}	Uniaxial compressive strength	MPa
σ_{TCS}	Triaxial compressive strength	MPa
E	Static Young's modulus	GPa
p_c	Confining pressure	MPa
p_p	Fluid pressure	Pa
T	Temperature	°C
σ	Axial deviatoric stress	MPa
S_q	Fracture surface roughness (root mean square)	mm
k	Fracture permeability	m^2
t	Fracture thickness/hydraulic aperture	mm
Q	Volumetric flow rate	ml/min
η	Dynamic viscosity	Pa∗s
2a	Distance between up- and downstream borehole	mm
r_0	Radius of up- and downstream borehole	mm
Δp_p	Differential pore pressure within the fracture	Pa
Δx	Distance within the fracture over which is Δp_p measured	mm
B	Geometry factor	/
B_{compo}	Brittleness based on composition	/
w_{xx}	Mineral weighting factor	/
f_{xx}	Mineral fraction	wt%

References

1. Vogler, D.; Amann, F.; Bayer, P.; Elsworth, D. Permeability evolution in natural fractures subject to cyclic loading and gouge formation. *Rock Mech. Rock Eng.* **2016**, *49*, 3463–3479. [CrossRef]
2. Lu, S.M. A global review of enhanced geothermal system (EGS). *Renew. Sustain. Energy Rev.* **2018**, *81*, 2902–2921. [CrossRef]
3. Favier, A.; Lardeaux, J.M.; Corsini, M.; Verati, C.; Navelot, V.; Géraud, Y.; Diraison, M.; Ventalon, S.; Voitus, E. Characterization of an exhumed high-temperature hydrothermal system and its application for deep geothermal exploration: An example from Terre-de-Haut Island (Guadeloupe archipelago, Lesser Antilles volcanic arc). *J. Volcanol. Geotherm. Res.* **2021**, *418*, 107256. [CrossRef]
4. Roche, V.; Bouchot, V.; Beccaletto, L.; Jolivet, L.; Guillou-Frottier, L.; Tuduri, J.; Bozkurt, E.; Oguz, K.; Tokay, B. Structural, lithological, and geodynamic controls on geothermal activity in the Menderes geothermal Province (Western Anatolia, Turkey). *Int. J. Earth Sci.* **2019**, *108*, 301–328. [CrossRef]
5. Beauchamps, G.; Ledésert, B.; Hébert, R.; Navelot, V.; Favier, A. The characterisation of an exhumed high-temperature paleo-geothermal system on Terre-de-Haut Island (the Les Saintes archipelago, Guadeloupe) in terms of clay minerals and petrophysics. *Geotherm. Energy* **2019**, *7*, 6. [CrossRef]
6. Gringarten, A.C.; Witherspoon, P.A.; Ohnishi, Y. Theory of heat extraction from fractured hot dry rock. *J. Geophys. Res.* **1975**, *80*, 1120–1124. [CrossRef]
7. Liotta, D.; Brogi, A.; Ruggieri, G.; Zucchi, M. Fossil vs. active geothermal systems: A field and laboratory method to disclose the relationships between geothermal fluid flow and geological structures at depth. *Energies* **2021**, *14*, 993. [CrossRef]
8. Brogi, A.; Alçiçek, M.C.; Liotta, D.; Capezzuoli, E.; Zucchi, M.; Matera, P.F. Step-over fault zones controlling geothermal fluid-flow and travertine formation (Denizli Basin, Turkey). *Geothermics* **2021**, *89*, 101941. [CrossRef]
9. Li, Q.; Xing, H.; Liu, J.; Liu, X. A review on hydraulic fracturing of unconventional reservoir. *Petroleum* **2015**, *1*, 8–15. [CrossRef]
10. Milsch, H.H.; Spangenberg, E.; Kulenkampff, J.; Meyhöfer, S. A new apparatus for long-term petrophysical investigations on geothermal reservoir rocks at simulated in-situ conditions. *Transp. Porous Media* **2008**, *74*, 73–85. [CrossRef]
11. Voltolini, M.; Ajo-Franklin, J. Evolution of propped fractures in shales: The microscale controlling factors as revealed by in situ X-Ray microtomography. *J. Pet. Sci. Eng.* **2020**, *188*, 106861. [CrossRef]
12. Crandall, D.; Moore, J.; Gill, M.; Stadelman, M. CT scanning and flow measurements of shale fractures after multiple shearing events. *Int. J. Rock Mech. Min. Sci.* **2017**, *100*, 177–187. [CrossRef]
13. Carey, J.W.; Lei, Z.; Rougier, E.; Mori, H.; Viswanathan, H. Fracture-permeability behavior of shale. *J. Unconv. Oil Gas Resour.* **2015**, *11*, 27–43. [CrossRef]
14. Leung, C.T.O.; Zimmerman, R.W. Estimating the Hydraulic Conductivity of Two-Dimensional Fracture Networks Using Network Geometric Properties. *Transp. Porous Media* **2012**, *93*, 777–797. [CrossRef]

15. Stober, I.; Bucher, K. Hydraulic conductivity of fractured upper crust: Insights from hydraulic tests in boreholes and fluid-rock interaction in crystalline basement rocks. *Geofluids* **2015**, *15*, 161–178. [CrossRef]
16. Cho, Y.; Apaydin, O.G.; Ozkan, E. Pressure-dependent natural-fracture permeability in shale and its effect on shale-gas well production. *SPE Reserv. Eval. Eng.* **2013**, *16*, 216–228. [CrossRef]
17. Gutierrez, M.; Øino, L.E.; Nygård, R. Stress-dependent permeability of a de-mineralised fracture in shale. *Mar. Pet. Geol.* **2000**, *17*, 895–907. [CrossRef]
18. Liu, E. Effects of fracture aperture and roughness on hydraulic and mechanical properties of rocks: Implication of seismic characterization of fractured reservoirs. *J. Geophys. Eng.* **2005**, *2*, 38–47. [CrossRef]
19. Rutter, E.H.; Mecklenburgh, J. Influence of normal and shear stress on the hydraulic transmissivity of thin cracks in a tight quartz sandstone, a granite, and a shale. *J. Geophys. Res. Solid Earth* **2018**, *123*, 1262–1285. [CrossRef]
20. Blöcher, G.; Kluge, C.; Milsch, H.; Cacace, M.; Jacquey, A.B.; Schmittbuhl, J. Permeability of matrix-fracture systems under mechanical loading—Constraints from laboratory experiments and 3-D numerical modelling. *Adv. Geosci.* **2019**, *49*, 95–104. [CrossRef]
21. Cheng, C.; Milsch, H. Permeability variations in illite-bearing sandstone: Effects of temperature and NaCl fluid salinity. *J. Geophys. Res. Solid Earth* **2020**, *125*, 9. [CrossRef]
22. Cheng, C.; Herrmann, J.; Wagner, B.; Leiss, B.; Stammeier, J.A.; Rybacki, E.; Milsch, H. Long-Term Evolution of Fracture Permeability in Slate: An Experimental Study with Implications for Enhanced Geothermal Systems (EGS). *Geosciences* **2021**, *11*, 443. [CrossRef]
23. Walsh, S.D.C.; Smith, M.; Carroll, S.A.; Crandall, D. Non-invasive measurement of proppant pack deformation. *Int. J. Rock Mech. Min. Sci.* **2016**, *87*, 39–47. [CrossRef]
24. Zhang, J.; Ouyang, L.; Zhu, D.; Hill, A.D. Experimental and numerical studies of reduced fracture conductivity due to proppant embedment in the shale reservoir. *J. Pet. Sci. Eng.* **2015**, *130*, 37–45. [CrossRef]
25. Watanabe, N.; Ishibashi, T.; Ohsaki, Y.; Tsuchiya, Y.; Tamagawa, T.; Hirano, N.; Okabe, H.; Tsuchiya, N. X-ray CT based numerical analysis of fracture flow for core samples under various confining pressures. *Eng. Geol.* **2011**, *123*, 338–346. [CrossRef]
26. Schill, E.; Genter, A.; Cuenot, N.; Kohl, T. Hydraulic performance history at the Soultz EGS reservoirs from stimulation and long-term circulation tests. *Geothermics* **2017**, *70*, 110–124. [CrossRef]
27. Vidal, J.; Genter, A. Overview of naturally permeable fractured reservoirs in the central and southern Upper Rhine Graben: Insights from geothermal wells. *Geothermics* **2018**, *74*, 57–73. [CrossRef]
28. Vanbrabant, Y.; Stenmans, V.; Burlet, C.; Petitclerc, E.; Meyvis, B.; Stasi, G.; Abarim, R.; Bar, K.; Goovaerts, T. Havelange deep borehole (Belgium): A study case for the evaluation of metasedimentary formations as potential geothermal reservoir—H2020 MEET project. In Proceedings of the EGU General Assembly Conference 2020, Online, 4–8 May 2020; p. 10943.
29. Leiss, B.; Tanner, D.; Vollbrecht, A.; Wemmer, K. *Neue Untersuchungen zur Geologie der Leinetalgrabenstruktur*; Universitätsverlag Göttingen: Göttingen, Germany, 2011.
30. Leiss, B.; Tanner, D.; Vollbrecht, A.; Wemmer, K. Tiefengeothermisches Potential in der Region Göttingen—Geologische Rahmenbedingungen. In *Neue Untersuchungen zur Geologie der Leinetalgrabenstruktur*; Leiss, B., Tanner, D., Vollbrecht, A., Arp, G., Eds.; Universitätsverlag Göttingen: Göttingen, Germany, 2011; pp. 163–170. [CrossRef]
31. Leiss, B.; Wagner, B.; Heinrichs, T.; Romanov, D.; Tanner, D.C.; Vollbrecht, A.; Wemmer, K. Integrating deep, medium and shallow geothermal energy into district heating and cooling system as an energy transition approach for the Göttingen University Campus. In Proceedings of the World Geothermal Congress 2021, Reykjavik, Iceland, 24–27 October 2021; pp. 1–9.
32. Brinckmann, J.; Brüning, U. (Eds.) *Das Bundesbohrprogramm im West-Harz, Paläogeographische Ergebnisse (The Federal Drilling Program in the Western Harz Mountains: Paleogeographic Results and Five Additional Contributions to the Geology of the Western Harz Mountains)*; Geologisches Jahrbuch Reihe D: Hanover, Germany, 1986; Band D.
33. Gerling, P.; Kockel, F.; Krull, P. *The HC Potential of Pre-Westphalian Sediments in the North German Basin—A Synthesis*; DGMK-Research Report 43; 1999. Available online: https://www.osti.gov/etdeweb/biblio/695510 (accessed on 7 April 2022).
34. Littke, R.; Krooss, B.; Uffmann, A.K.; Schulz, H.M.; Horsfield, B. Unconventional gas resources in the Paleozoic of Central Europe. *Oil Gas Sci. Technol.* **2011**, *66*, 953–977. [CrossRef]
35. Schubert, M. *Die Dysaerobe Biofazies der Wissenbach Schiefer (Rheinisches Schiefergebirge, Harz, Devon)*; Institute der Georg-August-Universit: Göttingen, Germany, 1996.
36. Sáez, R.; Moreno, C.; González, F.; Almodóvar, G.R. Black shales and massive sulfide deposits: Causal or casual relationships? Insights from Rammelsberg, Tharsis, and Draa Sfar. *Miner. Depos.* **2011**, *46*, 585–614. [CrossRef]
37. Klee, J.; Potel, S.; Ledésert, B.; Hébert, R.; Chabani, A.; Barrier, P.; Trullenque, G. Fluid-rock interactions in a Paleo-geothermal reservoir (Noble Hills Granite, California, USA). Part 1: Granite pervasive alteration processes away from fracture zones. *Geosciences* **2021**, *11*, 325. [CrossRef]
38. Klee, J.; Chabani, A.; Ledésert, B.A.; Potel, S.; Hébert, R.L.; Trullenque, G. Fluid-rock interactions in a paleo-geothermal reservoir (Noble hills granite, california, usa). part 2: The influence of fracturing on granite alteration processes and fluid circulation at low to moderate regional strain. *Geosciences* **2021**, *11*, 433. [CrossRef]
39. Amann, F.; Wild, K.M.; Loew, S.; Yong, S.; Thoeny, R.; Frank, E. Geomechanical behaviour of Opalinus Clay at multiple scales: Results from Mont Terri rock laboratory (Switzerland). *Swiss J. Geosci.* **2017**, *110*, 151–171. [CrossRef]

40. Bossart, P.; Bernier, F.; Birkholzer, J.; Bruggeman, C.; Connolly, P.; Dewonck, S.; Fukaya, M.; Hexfort, M.; Jensen, M.; Matray, J.-M.; et al. Mont Terri rock laboratory, 20 years of research: Introduction, site characteristics and overview of experiments. *Swiss J. Geosci.* **2017**, *110*, 3–22. [CrossRef]

41. Schuster, V.; Rybacki, E.; Bonnelye, A.; Herrmann, J.; Schleicher, A.; Dresen, G. Experimental deformation of Opalinus Clay at elevated temperature and pressure conditions: Mechanical properties and the influence of rock fabric. *Rock Mech. Rock Eng.* **2021**, *54*, 4009–4039. [CrossRef]

42. Nussbaum, C.; Kloppenburg, A.; Caër, T.; Bossart, P. Tectonic evolution around the Mont Terri rock laboratory, northwestern Swiss Jura: Constraints from kinematic forward modelling. *Swiss J. Geosci.* **2017**, *110*, 39–66. [CrossRef]

43. Yu, C.; Matray, J.M.; Gonçalvès, J.; Jaeggi, D.; Gräsle, W.; Wieczorek, K.; Vogt, T.; Skyes, E. Comparative study of methods to estimate hydraulic parameters in the hydraulically undisturbed Opalinus Clay (Switzerland). *Swiss J. Geosci.* **2017**, *110*, 85–104. [CrossRef]

44. McKernan, R.; Mecklenburgh, J.; Rutter, E.; Taylor, K. Microstructural controls on the pressure-dependent permeability of Whitby mudstone. *Geol. Soc.* **2017**, *454*, 39–66. [CrossRef]

45. Ji, Y.; Hofmann, H.; Rutter, E.H.; Xiao, F.; Yang, L. Revisiting the Evaluation of Hydraulic Transmissivity of Elliptical Rock Fractures in Triaxial Shear-Flow Experiments. *Rock Mech. Rock Eng.* **2022**, 1–9. [CrossRef]

46. Bieniawski, Z.T.; Bernede, M.J. Suggested methods for determining the uniaxial compressive strength and deformability of rock materials: Part 1. Suggested method for determining deformability of rock materials in uniaxial compression. *Int. J. Rock Mech. Min. Sci. Geomech. Abstr.* **1979**, *16*, 138–140. [CrossRef]

47. Paterson, M.S. A high-pressure, high-temperature apparatus for rock deformation. *Int. J. Rock Mech. Min. Sci.* **1970**, *7*, 517–526. [CrossRef]

48. Rybacki, E.; Reinicke, A.; Meier, T.; Makasi, M.; Dresen, G. What controls the mechanical properties of shale rocks?—Part I: Strength and Young's modulus. *J. Pet. Sci. Eng.* **2015**, *135*, 702–722. [CrossRef]

49. von Terzaghi, K. Die Berechnung der Durchas- sigkeitsziffer des Tones aus dem Veriauf der hydrodynamischen Spannungser- scheinungen, Sitzungsber. *Akad. Wiss. Wien Math Naturwiss.* **1923**, *132*, 105.

50. Cheng, C.; Milsch, H. Evolution of fracture aperture in quartz sandstone under hydrothermal conditions: Mechanical and chemical effects. *Minerals* **2020**, *10*, 657. [CrossRef]

51. Li, F.B.; Sheng, J.C.; Zhan, M.L.; Xu, L.M.; Wu, Q.; Jia, C.L. Evolution of limestone fracture permeability under coupled thermal, hydrological, mechanical, and chemical conditions. *J. Hydrodyn.* **2014**, *26*, 234–241. [CrossRef]

52. Lima, M.G.; Vogler, D.; Querci, L.; Madonna, C.; Hattendorf, B.; Saar, M.O.; Kong, X.Z. Thermally driven fracture aperture variation in naturally fractured granites. *Geotherm. Energy* **2019**, *7*, 23. [CrossRef]

53. Yasuhara, H.; Kinoshita, N.; Ohfuji, H.; Lee, D.S.; Nakashima, S.; Kishida, K. Temporal alteration of fracture permeability in granite under hydrothermal conditions and its interpretation by coupled chemo-mechanical model. *Appl. Geochem.* **2011**, *26*, 2074–2088. [CrossRef]

54. Kamali-Asl, A.; Ghazanfari, E.; Perdrial, N.; Bredice, N. Experimental study of fracture response in granite specimens subjected to hydrothermal conditions relevant for enhanced geothermal systems. *Geothermics* **2018**, *72*, 205–224. [CrossRef]

55. Yasuhara, H.; Elsworth, D. Compaction of a rock fracture moderated by competing roles of stress corrosion and pressure solution. *Pure Appl. Geophys.* **2008**, *165*, 1289–1306. [CrossRef]

56. Rutqvist, J. Fractured rock stress-permeability relationships from in situ data and effects of temperature and chemical-mechanical couplings. *Geofluids* **2015**, *15*, 48–66. [CrossRef]

57. Cao, N.; Lei, G.; Dong, P.; Li, H.; Wu, Z.; Li, Y. Stress-dependent permeability of fractures in tight reservoirs. *Energies* **2019**, *12*, 117. [CrossRef]

58. Rutter, E.H.; Mecklenburgh, J. Hydraulic conductivity of bedding-parallel cracks in shale as a function of shear and normal stress. *Geol. Soc.* **2017**, *454*, 67. [CrossRef]

59. Yasuhara, H.; Elsworth, D. Evolution of permeability in a natural fracture: Significant role of pressure solution. *J. Geophys. Res.* **2004**, *109*, 03204. [CrossRef]

60. Bowden, F.P.; Tabor, D. *The Friction and Lubrication of Solids*; Clarendon Press: Oxford, UK, 1964; Volume 2, No. 2.

61. Huo, D.; Benson, S.M. An Experimental Investigation of Stress-Dependent Permeability and Permeability Hysteresis Behavior in Rock Fractures. Dynamics of Fluids and Transport in Complex Fractured-Porous Systems. *Geophys. Monogr.* **2015**, *210*, 99–114.

62. Gangi, A.F. Variation of whole and fractured porous rock permeability with confining pressure. *Int. J. Rock Mech. Min. Sci.* **1978**, *15*, 249–257. [CrossRef]

63. Tsang, Y.W.; Witherspoon, P.A. The dependence of fracture mechanical and fluid flow properties on fracture roughness and sample size. *J. Geophys. Res.* **1983**, *88*, 2359–2366. [CrossRef]

64. Walsh, J.B. Effect of pore pressure and confining pressure on fracture permeability. *Int. J. Rock Mech. Min. Sci. Geomech. Abstr.* **1981**, *18*, 429–435. [CrossRef]

65. Witherspoon, P.A.; Wang, J.S.Y.; Iwai, K.; Gale, J.E. Validity of cubic law for fluid-flow in a deformable rock fracture. *Water Resour. Res.* **1980**, *16*, 1016–1024. [CrossRef]

66. Zimmerman, R.W.; Chen, D.W.; Cook, N.G.W. The effect of contact area on the permeability of fractures. *J. Hydrol.* **1992**, *139*, 79–96. [CrossRef]

67. Barton, N.; Bandis, S.; Bakhtar, K. Strength, Deformation and Conductivity Coupling of Rock Joints. *Int. J. Rock Mech. Min. Sci. Geomech. Abstr.* **1985**, *22*, 121–140. [CrossRef]
68. Gale, J.E. The effects of fracture type induced versus natural on the stress-fracture closure-fracture permeability relationships. In Proceedings of the 23rd U.S. Symposium on Rock Mechanics, University of California, Berkeley, CA, USA, 25–27 August 1982; pp. 290–298.
69. Rybacki, E.; Meier, T.; Dresen, G. What controls the mechanical properties of shale rocks?—Part II: Brittleness. *J. Pet. Sci. Eng.* **2016**, *144*, 39–58. [CrossRef]
70. Herrmann, J.; Rybacki, E.; Sone, H.; Dresen, G. Deformation experiments on Bowland and Posidonia shale—Part I: Strength and Young's modulus at ambient and in situ pc-T conditions. *Rock Mech. Rock Eng.* **2018**, *51*, 3645–3666. [CrossRef]
71. Ulusay, R. *The ISRM Suggested Methods for Rock Characterization, Testing and Monitoring 2007–2014*; Springer: Cham, Switzerland, 2015. [CrossRef]
72. Durham, W.B.; Bonner, B.P. Self-propping and fluid flow in slightly offset joints at high effective pressures. *J. Geophys. Res.* **1994**, *99*, 9391–9399. [CrossRef]
73. Reinicke, A.; Rybacki, E.; Stanchits, S.; Huenges, E.; Dresen, G. Hydraulic fracturing stimulation techniques and formation damage mechanisms—Implications from laboratory testing of tight sandstone–proppant systems. *Geochemistry* **2010**, *70* (Suppl. S3), 107–117. [CrossRef]

 *geosciences*

Article

Impact of Well Placement in the Fractured Geothermal Reservoirs Based on Available Discrete Fractured System

Saeed Mahmoodpour [1,*], Mrityunjay Singh [1,*], Kristian Bär [1] and Ingo Sass [1,2]

[1] Group of Geothermal Science and Technology, Institute of Applied Geosciences, Technische Universität Darmstadt, 64287 Darmstadt, Germany; baer@geo.tu-darmstadt.de (K.B.); sass@geo.tu-darmstadt.de (I.S.)
[2] Darmstadt Graduate School of Excellence Energy Science and Engineering, Technische Universität Darmstadt, 64287 Darmstadt, Germany
* Correspondence: saeed.mahmoodpour@tu-darmstadt.de (S.M.); mrityunjay.singh@tu-darmstadt.de (M.S.)

Abstract: Well placement in a given geological setting for a fractured geothermal reservoir is necessary for enhanced geothermal operations. High computational cost associated with the framework of fully coupled thermo-hydraulic-mechanical (THM) processes in a fractured reservoir simulation makes the well positioning a missing point in developing a field-scale investigation. To enhance the knowledge of well placement for different working fluids, we present the importance of this topic by examining different injection-production well (doublet) positions in a given fracture network using coupled THM numerical simulations. Results of this study are examined through the thermal breakthrough time, mass flux, and the energy extraction potential to assess the impact of well position in a two-dimensional reservoir framework. Almost ten times the difference between the final amount of heat extraction is observed for different well positions but with the same well spacing and geological characteristics. Furthermore, the stress field is a strong function of well position that is important concerning the possibility of high-stress development. The objective of this work is to exemplify the importance of fracture connectivity and density near the wellbores, and from the simulated cases, it is sufficient to understand this for both the working fluids. Based on the result, the production well position search in the future will be reduced to the high-density fracture area, and it will make the optimization process according to the THM mechanism computationally efficient and economical.

Keywords: well placement; CO_2-EGS; water-EGS; discrete fracture networks; THM modeling

Citation: Mahmoodpour, S.; Singh, M.; Bär, K.; Sass, I. Impact of Well Placement in the Fractured Geothermal Reservoirs Based on Available Discrete Fractured System. *Geosciences* **2022**, *12*, 19. https://doi.org/10.3390/geosciences12010019

Academic Editors: Ronan L. Hébert, Ghislain Trullenque, Albert Genter, Eléonore Dalmais, Jean Hérisson and Jesus Martinez-Frias

Received: 23 September 2021
Accepted: 30 December 2021
Published: 4 January 2022

Publisher's Note: MDPI stays neutral with regard to jurisdictional claims in published maps and institutional affiliations.

1. Introduction

Geothermal field development and management is a complex process. Engineering a geothermal system requires appropriate well placement and fracture connectivity to ensure well connectivity and least fluid loss [1,2]. Placement of injection and production wells or a doublet system in a given geological framework to achieve maximum geothermal energy extraction is one of the most complicated and expensive procedures. The location of the injection well concerning production well decides the production mass flux [3,4]. Practically, there is an infinite number of sites where an injection well can be placed in designing an enhanced geothermal system (EGS). Well placement in association with fracture network requires two critical aspects to ensure high heat extraction potential. First, the fractures must be connected sufficiently, and they must provide a high fluid flow rate at a low-pressure difference, and secondly, fluid residence time in the fractures should be increased to allow sufficient heat exchange. Longer residence time enhances the heat extraction capacity and reduces the chances of short-circuiting [5,6]. Figure 1 shows a subsurface fracture network where red is the high-temperature region. Hot water is produced through the red color well from the reservoir, and after passing it through the heat exchanger, it is reinjected to the reservoir with the blue color well. Fractures are the main paths for fluid flow that allow for heat extraction from the various MEET geothermal sites, including Soultz sous Forêts, United Down, Göttingen, and Havelange. Discrete Fracture Network (DFN) characterization is an essential step toward the simulation of reservoir performance. However, the total number of fractures resulting from DFN characterization is a high number (in the order of millions of fractures) which is not feasible for performing numerical simulation (due to computational costs) while considering all of them discretely through the thermal-hydraulic (TH) or thermal-hydraulic-mechanical (THM) simulator. Recently, Lepillier et al. [7] combined TH behavior with the steady-state solid mechanics process to examine the well positioning impact with four doublet scenarios. However, transient temperature and pressure changes will affect the stress field, and four scenarios are not sufficient to accurately demonstrate the impact of well position.

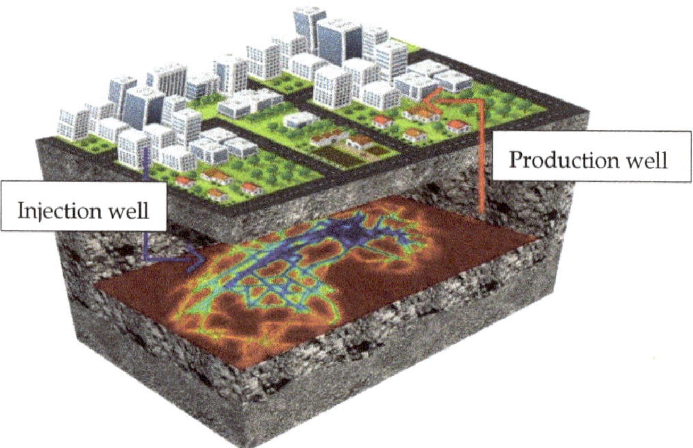

Figure 1. An enhanced geothermal system. The subsurface plan shows an intricate network of fractures. For optimized power generation, appropriate placement of the injection and production wells is necessary.

At the same time, the fracture network alignment also contributes to the thermal drawdown, mass flux, and extracted energy. Therefore, it is essential to estimate the well locations a priori for better connectivity and maximum energy extraction. For example, a second well was designed at Rittershoffen (Upper Rhine Graben, France) in the damage zone of the Rittershoffen fault after the drilling of the first one and an additional geophysical survey [8].

This paper considers a two-dimensional fractured reservoir for a potential enhanced geothermal system. Fully coupled thermo-hydro-mechanical (THM) processes are simulated on the fractured reservoir to estimate maximum geothermal energy extraction potential.

Several optimization techniques are available for determining well placement in a reservoir [9]. Some of these methods are gradient-free methods, including genetic algorithms [10], particle swarm optimization algorithm [11], fast marching method [12], and simultaneous perturbation stochastic approximation [13,14], and gradient-based optimization methods, including adjoint methods [15–17]. These models lack geological uncertainty, e.g., fracture network connectivity, while considering well placement optimization [18]. Few thermo-hydraulic compositional reservoir simulation-based models on well spacing optimization [19–26]. Based on a coupled thermo-hydraulic model, Akin et al. [27] developed artificial neural networks (ANN) and a search algorithm to optimize an injection well for a geothermal reservoir whereas, Samin et al. [28] developed a hybrid approach integrating a multi-objective genetic algorithm with finite element modeling of thermo-hydraulic processes. EGS involves complex THM processes. Gudmundsdottir and Horne [29] developed an ANN model to characterize fractured geothermal reservoirs for a coupled TH model. Training data necessary for creating a robust ANN model based on a coupled thermo-hydro-mechanical process requires many numerical reservoir simulations. A recent study using an ANN model for a coupled thermo-hydraulic approach supports this idea for a fracture in a hot geothermal reservoir [30] and supercritical geothermal reservoirs [31].

While performing a parametric investigation for a multi-well reservoir, Chen and Jiang [19] found that production well configuration concerning injection well affects the heat mining potential. Chen et al. [32] used a multivariate adaptive regression spline technique coupled with hydrothermal numerical simulation to optimize the well placement under the given fault size and permeability for a prospective geothermal site near Superstition Mountain in Southern California USA. They found that for the maximum net profit over fifty years, the optimal well spacing is 473 m at 30.7 kg/s. For 45° angle between fracture orientation and inlet-outlet connection in a given fracture network, Zhang et al. [33] observed optimized geothermal energy extraction performance. They obtained a stable heat mining rate at reduced efficiency for higher orientation angles. Zhang et al. [2] found that the presence of many fractures in the vicinity of the production well increases the working fluid residence time, and heat recovery efficiency significantly improves. They suggested that thermally-induced fractures near the production well assist in greater power generation than when the fracture density is high in the vicinity of the injection well. They also observed that placing the production well in the high permeability region increases heat production. Gao et al. [34] used a coupled thermo-hydraulic model for a discrete fracture network in a fractured geothermal reservoir to investigate heat extraction performance. They used multilateral well orientations with a varying number of branch wells and well orientation. They found that production temperature decreases with an increase in the well and fracture intersections, whereas injection pressure increases. Aliyu et al. [35] and Aliyu and Chen [36] used COMSOL Multiphysics to develop a model depicting THM and TH processes in a geothermal reservoir for two fractures and single fractures, respectively. They estimated the impact of well spacing on thermal energy extraction performance.

The MEET project framework considers water as the working fluid or heat-carrying fluid from a geothermal system. However, this study finds CO_2 as an alternative to water because the loss of CO_2 as the heat-carrying fluid is environment friendly [37,38]. Furthermore, the use of supercritical CO_2 may assist in the formation of an interconnected fracture network of multiple channels at a lower pressure than water due to smaller fluid density and viscosity [38]. Due to the lower reactivity of CO_2 in comparison to water, the possible silica dissolution and precipitation at high temperatures and pressure decreases [37–39]. Additionally, Bongole et al. [40] observed that the reservoir deformation is more minor when CO_2 is the working fluid compared to water due to the lower heat capacity of CO_2. The lower freezing point of CO_2 than water helps heat rejection at a much lower temperature. Therefore, its geothermal systems may work even for cold climatic conditions where water is unusable [41]. The above literature shows no available THM model for determining the well placement in a geothermal reservoir concerning a given fracture map to maximize mass flux, energy extraction, and the thermal drawdown duration. In this work, a fully coupled THM model is developed and used to demonstrate the importance of well position by characterizing the fracture network connectivity and density. This study will build a basis for future well placement optimization considering THM processes for a given fracture network. The present study is organized in the following manner. First, a mathematical and numerical model is presented for a coupled THM process followed by results and discussion on optimizing well positions in a two-dimensional fracture network based on thermal drawdown, mass flux, and energy extraction potential followed by conclusions.

2. Methodology

This study uses a fracture network based on outcrop fractures mapped from Otsego County in New York state [42] for THM modeling. The total number of fractures in this outcrop map is 440. As depicted in Figure 1, this study considers a two-dimensional geometry at subsurface conditions. The reservoir geometry is a two-dimensional planar model (1000 m × 600 m), and the injection-production wells are placed 500 m apart. An initial case is considered where the injection well is present at point 0 and the production well at point 180, as shown in Figure 2. Considering this axis as diameter, a circular zone is assumed, and the perimeter is divided into 36 equal intervals. These 36 intervals are considered for placing the injection and production wells, and they are arranged at α angle from the base case. All fractures are assumed as interior boundaries, and the displacement is constrained in all normal directions. The side boundaries are assumed as no flow for both heat and mass exchange.

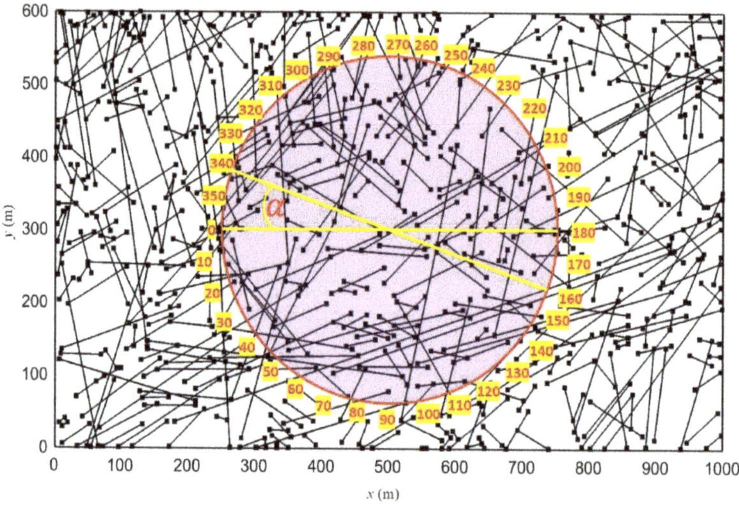

Figure 2. The geometry of the reservoir. Injection and production wells are placed 500 m apart. Total 36 cases or 36 values of α are considered for simulations. Here, when the injection well is present at 0 and production well is present at 180 and they are 500 m placed apart, the value of α is 0. When the injection well is present at 340 and the production well is present at 160, the value of α is 340.

Conservation equation of mass when coupled with pore volume and fluid temperature alteration for a porous medium is [43]:

$$\rho_1(\phi_m S_1 + (1 - \phi_m)S_m)\frac{\partial p}{\partial t} - \rho_1(\alpha_m(\phi_m \beta_1 + (1 - \phi_m)\beta_m))\frac{\partial T}{\partial t} + \rho_1 \alpha_m \frac{\partial \varepsilon_V}{\partial t} = \nabla \cdot \left(\frac{\rho_1 k_m}{\mu}\nabla p\right) \tag{1}$$

All the parameters are listed in Appendix A. Water and supercritical CO_2 are considered heat-transmitting fluids in this study. The equation that governs fluid flow along the internal fractures is:

$$\rho_1\left(\phi_f S_1 + \left(1 - \phi_f\right)S_{mf}\right)e_h\frac{\partial p}{\partial t} - \rho_1\left(\alpha_f\left(\phi_f \beta_1 + \left(1 - \phi_f\right)\beta_f\right)\right)e_h\frac{\partial T}{\partial t} + \rho_1 \alpha_f e_h \frac{\partial \varepsilon_V}{\partial t} = \nabla_T \cdot \left(\frac{e_h \rho_1 k_f}{\mu}\nabla_T p\right) + n \cdot Q_m \tag{2}$$

In Equation (2), fluid flow along the fracture width is ignored because fracture aperture is much smaller than the fracture length. Fractures and the rock matrix are assumed at thermodynamic inequilibrium. In other words, the local thermal non-equilibrium model is implemented in this investigation.

$$(1 - \phi_m)\rho_m C_{p,m}\frac{\partial T_m}{\partial t} = \nabla \cdot ((1 - \phi_m)\lambda_m \nabla T_m) + q_{ml}(T_l - T_m) \tag{3}$$

$$\left(1 - \phi_f\right)e_h\rho_f C_{p,f}\frac{\partial T_m}{\partial t} = \nabla_T \cdot \left(\left(1 - \phi_f\right)e_h\lambda_f \nabla_T T_m\right) + e_h q_{fl}(T_l - T_m) + n \cdot (-(1 - \phi_m)\lambda_m \nabla T_m) \tag{4}$$

The energy balance equation for the rock matrix and fractures are shown by Equations (3) and (4), respectively. The energy balance equation for either water or CO_2 is:

$$\phi_m \rho_l C_{p,l} \frac{\partial T_l}{\partial t} + \phi_m \rho_l C_{p,l} \left(-\frac{k_m \nabla p}{\mu} \right) \cdot \nabla T_l = \nabla.(\phi_m \lambda_l \nabla T_l) + q_{ml}(T_m - T_l) \tag{5}$$

The following equation can write heat exchange between a rock and the fracture matrix:

$$\phi_f e_h \rho_l C_{p,l} \frac{\partial T_l}{\partial t} + \phi_f e_h \rho_l C_{p,l} \left(-\frac{k_f \nabla_T p}{\mu} \right) \cdot \nabla_T T_l = \nabla_T \cdot \left(\phi_f e_h \lambda_l \nabla_T T_l \right) + e_h q_{fl}(T_m - T_l) + n \cdot (-\phi_l \lambda_l \nabla T_l) \tag{6}$$

In Equation (6), the Darcy flux in the fractures is $u_f = -\frac{k_f \nabla_T p}{\mu}$ and heat flux is $n.q_l = n \cdot (-\phi_l \lambda_l \nabla T_l)$.

A fully coupled thermo-hydro-mechanical model is developed in this study. If effective stress is $\sigma_{eff}^{ij} = \sigma_{ij} + \alpha_p p \delta_{ij}$ and the volumetric expansion coefficient of porous media is $\beta_T = \phi_l \beta_l + (1 - \phi_m)\beta_m$, then the stress-strain relationship considering fully coupled thermoelastic and poroelastic stress can be written as:

$$\sigma_{ij} = 2G\varepsilon_{ij} + \lambda tr\varepsilon \delta_{ij} - \alpha_p p \delta_{ij} - K'\beta_T T \delta_{ij} \tag{7}$$

The reservoir deformation equation can be written as:

$$G u_{i,jj} + (G + \lambda)u_{j,ji} - \alpha_p p_{,i} - K'\beta_T T_{,i} + f_i = 0 \tag{8}$$

The opening and closure of the thermo-poroelastic stress-dependent fracture aperture are modeled using the Barton and Bandis model [44,45] as follow:

$$\Delta e_n = \frac{e_0}{1 + 9\frac{\sigma_{eff}^n}{\sigma_{nref}}} \tag{9}$$

In Equation (9), Δe_n is the fracture aperture change under in-situ stress conditions.

Thermodynamic properties of water and CO_2 are represented by dynamic viscosity (Equations (10) and (11)), specific heat capacity (Equations (12) and (13)), density (Equations (14) and (15)), and thermal conductivity (Equations (16) and (17)) [42], and they are implemented in Equations (1)–(6).

$$\mu_w = 1.38 - 2.12 \times 10^{-2} \times T^1 + 1.36 \times 10^{-4} \times T^2 - 4.65 \times 10^{-7} \times T^3 + 8.90 \times 10^{-10} \times T^4$$
$$-9.08 \times 10^{-13} \times T^5 + 3.85 \times 10^{-16} \times T^6 \quad (273.15 - 413.15 \ K) \tag{10}$$

$$\mu_{CO_2} = -1.49 \times 10^{-6} - 6.47 \times 10^{-8} \times T^1 - 3.66 \times 10^{-11} \times T^2 + 1.25 \times 10^{-14} \times T^3 \quad (220 - 600 \ K) \tag{11}$$

$$C_{p,w} = 1.20 \times 10^4 - 8.04 \times 10^1 \times T^1 + 3.10 \times 10^{-1} \times T^2 - 5.38 \times 10^{-4} \times T^3 + 3.63 \times 10^{-7} \times T^4 \tag{12}$$

$$C_{p,CO_2} = 459.91 + 1.86 \times T^1 - 2.13 \times 10^{-3} \times T^2 + 1.22 \times 10^{-6} \times T^3 \quad (220 - 600 \ K) \tag{13}$$

$$\rho_w = 1.03 \times 10^{-5} \times T^3 - 1.34 \times 10^{-2} \times T^2 + 4.97 \times T + 4.32 \times 10^2 \tag{14}$$

$$\rho_{CO_2} = pA \times 0.04401 / RT \tag{15}$$

$$\kappa_w = -8.69 \times 10^{-1} + 8.95 \times 10^{-3} \times T^1 - 1.58 \times 10^{-5} \times T^2 + 7.98 \times 10^{-9} \times T^3 \tag{16}$$

$$\kappa_{CO_2} = -1.32 \times 10^{-3} + 4.14 \times 10^{-5} \times T^1 + 6.71 \times 10^{-8} \times T^2 - 2.11 \times 10^{-11} \times T^3 \tag{17}$$

In Equation (15), pA is the absolute pressure, and R is the molar gas constant. Coefficients in Equations (10)–(17) are constants and obtained from various correlations [42].

COMSOL Multiphysics version 5.5 [43] is used to perform numerical modeling of THM processes. It uses a finite element method to solve general purpose partial differential equations. The full mesh contains 112,818 domain elements and 13,071 boundary elements. This free triangular mesh is generated by using the maximum element size of 37 m, and minimum element size of 0.125 m, maximum element growth of 1.25 with the curvature factor 0.25, and the resolution of the narrow regions is 1. For the numerical modeling purpose, we have used a scaled absolute tolerance of magnitude 10^{-8} and automatic time step constraint. We assumed Backward Differentiation Formula (BDF) for time-stepping with maximum BDF order as two and minimum BDF order as 1. Further, we have validated our model with a soil thermal consolidation model as Bai [46] demonstrated in Mahmoodpour et al. [47].

3. Results and Discussions

Numerical simulation results from coupled THM mechanisms associated with a geothermal energy extraction process from a fractured reservoir are presented in this section. First, we performed a sensitivity analysis for three different mesh elements. For the following stage, the adopted sequence of presentation is:

(a)   coupled THM mechanisms for heat mining using water as heat-carrying fluid,
(b)   coupled THM processes when CO_2 is the heat-carrying fluid, and
(c)   predicting a suitable doublet well position for a given fracture network to obtain highest mass flux from the production well and maximize the heat production.

The results are presented for two working fluids: water and CO_2. Reservoir permeability of 2 mD and 5 mD are considered. These values are chosen in a way that sweep efficiency with the different working fluids to be similar at the same time.

Furthermore, permeability values are kept constant to understand the working fluid effect by running another case 180. Other parameters are listed in Table 1 that are not site specific and selected to represent a generic geothermal system, and the fracture map is the same for all scenarios. Constant injection and production pressures are considered at both the wellbores. A two-dimensional horizontal cross-sectional reservoir is considered for all the simulations.

We performed a mesh sensitivity analysis with case 180 for CO_2 (it has the highest velocity variation, and the simulations convergence is the most complex among all the cases). Here, the mesh sensitivity analysis is attached for the simulations with (a) 92,655 domain elements and 12,103 boundary elements, (b) 112,818 domain elements and 13,071 boundary elements and (c) 181,410 domain elements and 15,687 boundary elements. Convergence was not achieved with the mesh size of the 71,089 domain elements and 10,357 boundary elements. The maximum element size for the standard case is (a) 67 m, (b) 37 m, and (c) 18 m, whereas the minimum element size is (a) 0.3 m, (b) 0.125 m, and (c) 0.075 m. The maximum element growth rate is (a) 1.3, (b) 1.25, and (c) 1.2, the curvature factor is (a) 0.3, (b) 0.25, and (c) 0.25, and the resolution of the narrow regions is 1 for all three cases. Free triangular meshes are used for discretizing this domain.

Based on this description, our results are insensitive to the mesh refinements (see Figure 3).

Figure 4 shows the time evolution of reservoir temperature distribution during the heat extraction operation using water and CO_2 for case 180. Here, the injection well is present at 180, and the production well is present at 0 as shown in Figure 2. The reservoir permeability for the left (Figure 4(a1–d1)) and right (Figure 4(a3–d3)) columns is 5 mD whereas the middle column has 2 mD permeability (Figure 4(a2–d2)). Higher reservoir permeability in the case of the left column causes faster cold fluid propagation through the fracture network. Additionally, water propagation through the fractures becomes less dominant, and it starts flowing through the rock matrix at higher permeability as shown in Figure 4(c1,d1). We adopted smaller reservoir permeability for well placement when water is the working fluid to account for both these factors. The cold-water propagation is aligned along the dominant fracture rather than the horizontal axis between the doublet. The reason behind selecting different permeability values for water and CO_2 is to reach a similar sweep efficiency with different fluids; however, we provided the quantitative comparison between simulation of water at 5 mD, 2 mD and CO_2 at 5 mD for case 180 as shown in Figure 5. Figure 5a–c shows that CO_2 is a better-working fluid concerning the breakthrough time, mass flux, and cumulatively extracted energy, respectively over 30 years.

Table 1. Numerical simulation parameters (see the range of database in [48]).

Parameter	Magnitude for Water-Based Simulations	Magnitude for CO$_2$ Based Simulations
Young's modulus	40 GPa	40 GPa
Poisson's ratio	0.25	0.25
Rock density	$2500\ \frac{kg}{m^3}$	$2500\ \frac{kg}{m^3}$
Horizontal stress	50 MPa	50 MPa
Vertical stress	50 MPa	50 MPa
Initial pressure	30 MPa	30 MPa
Injection pressure	50 MPa	50 MPa
Rock porosity	0.2	0.2
Rock permeability	2 mD	5 mD
Fracture zone porosity	0.5	0.5
Fracture roughness	1	1
Fracture aperture	0.2 mm	0.2 mm
Closure stress	150 MPa	150 MPa
Wellbore radius	0.2 m	0.2 m
Rock thermal conductivity	$3\ \frac{W}{m \times K}$	$3\ \frac{W}{m \times K}$
Fracture zone thermal conductivity	$2.5\ \frac{W}{m \times K}$	$2.5\ \frac{W}{m \times K}$
Rock specific heat capacity	$800\ \frac{J}{kg \times K}$	$800\ \frac{J}{kg \times K}$
Fracture zone specific heat capacity	$800\ \frac{J}{kg \times K}$	$800\ \frac{J}{kg \times K}$
Initial temperature	200 °C	200 °C
Biot coefficient	0.7	0.7
Thermal expansion coefficient	$10^{-5}\ \frac{1}{K}$	$10^{-5}\ \frac{1}{K}$
Injection temperature	70 °C	70 °C

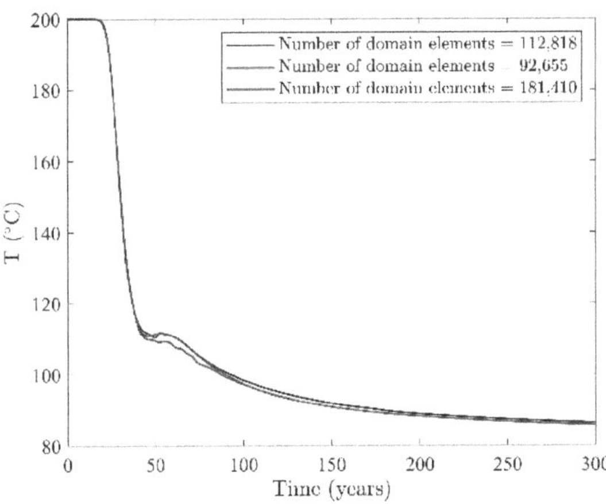

Figure 3. Mesh sensitivity results for three different number of mesh elements.

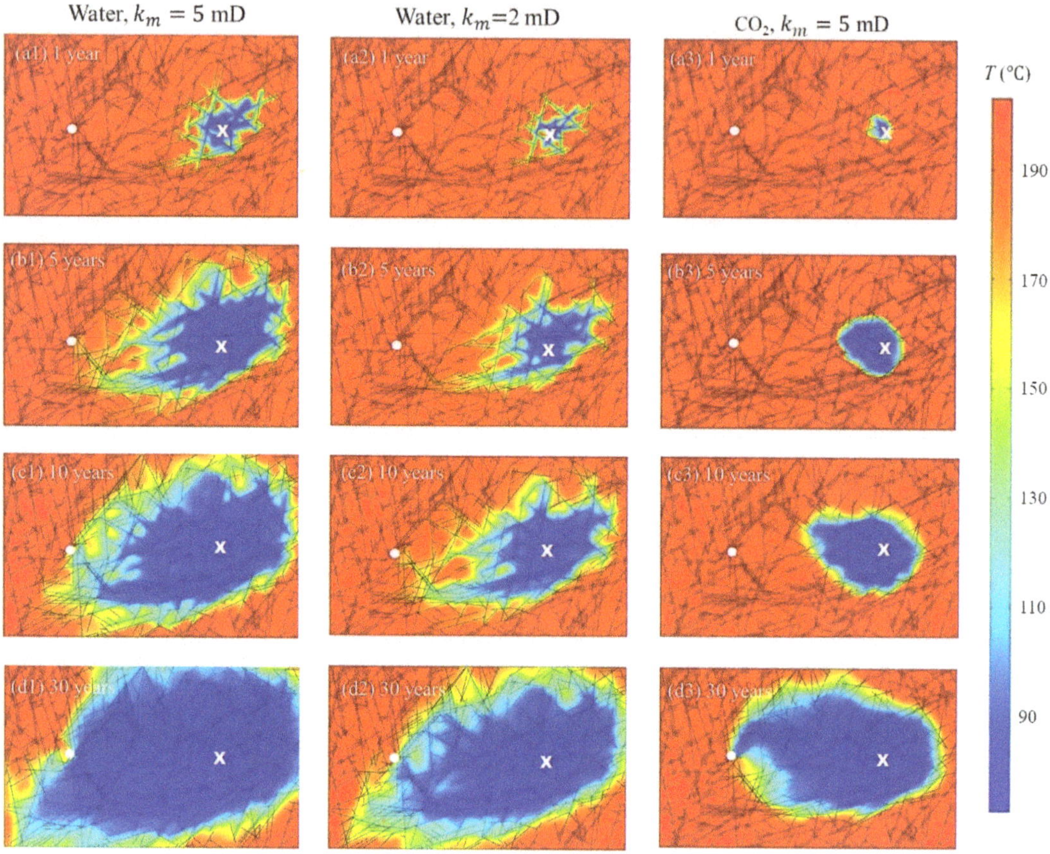

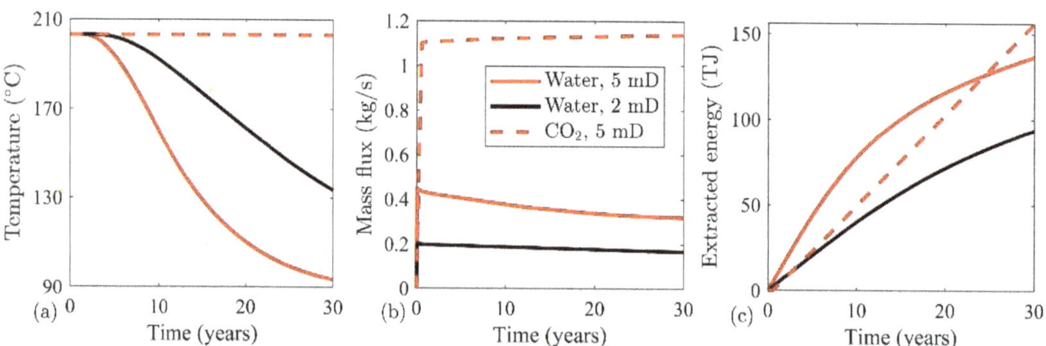

Figure 4. Reservoir temperature distribution at time 1 year (**a1–a3**), 5 years (**b1–b3**), 10 years (**c1–c3**) and 30 years (**d1–d3**) when the injection well is present at 180 and the production well is placed at 0. Results from water for reservoir permeability 5 mD and 2 mD are shown by (**a1–d1**) and (**a2–d2**) respectively. CO_2 as working fluid results are displayed in (**a3–d3**). The reservoir permeability for CO_2 simulations is 5 mD. The injection wellbore position is shown by cross symbol whereas a circle indicates production wellbore position.

Figure 5. Comparison of working fluid effects, (**a**) temperature at the production well, (**b**) mass flux of different fluid, and (**c**) extracted energy.

The viscosity of supercritical CO_2 at injection conditions is approximately half compared to water. Higher reservoir permeability and lower viscosity indicate CO_2 propagation through the fractures as well as through the matrix rather than flowing through fractures only as seen in Figure 4(a2,b2). Figure 4(a3–d3) shows that the cold fluid plume spread is much diffusive compared to water, and flow is primarily occurring through the matrix. However, Figure 4(d3) shows that CO_2 flows through the dominant fractures near the production well. Therefore, fluid propagation through the fractures is the principal mechanism between the doublet, which is assisted by flow through permeable rock matrix.

Furthermore, convective heat transfer inside the rock matrix and the fracture is the primary heat transfer mechanism. Therefore, the fractures control the heat transfer in a fractured reservoir. To show the relative importance of convective to conductive heat transfer, we calculated Peclet number (*Pe*), a nondimensional number which indicates the ratio of convective to conductive heat transfer [49] and it can be written as $Pe = \frac{uL\rho_i C_{p,i}}{k_i}$, where u is the fluid velocity, L is the characteristic length (here it is 500 m, the distance between the two wells), the subscript i indicates the fluid either water or CO_2, ρ_i is the fluid density, $C_{p,i}$ is the specific heat capacity of fluid and k_i is the fluid thermal conductivity. Figure 6a shows the *Pe* value after one year for the entire reservoir for case 180, where CO_2 is the working fluid. To elaborate on the relative impact of convective heat transfer inside the fracture, one fracture is selected as shown in Figure 6b, and the corresponding *Pe* number is shown in Figure 6c. *Pe* number is estimated for five different times, and for all the cases, *Pe* number is significantly larger than 1, indicating more vigorous convection than conduction.

Results obtained from the 36 reservoir simulations on the well positioning are shown in Figure 7. For water simulations, reservoir permeability is 2 mD, whereas, for cases with CO_2 as working fluid, permeability is 5 mD. Water-based models demonstrate faster thermal breakthrough (Figure 7a) due to higher specific heat capacity than CO_2. Therefore, water simulations are presented for 30 years whereas CO_2 results are plotted for 300 years. We magnified the results for the CO_2 over 30 years in Figure 7b,d,f to compare it with water (see Figure 7a,c,e). Figure 7a shows thermal drawdown at the production well when water is the working fluid. The fastest thermal drawdown was observed for case 220 (see Figure 8(c1)) whereas the slowest thermal drawdown occurred for case 130 (see Figure 8(a1)). From Figure 2, it is clear that case 220 has a well position along a dominant fracture supported by minor intersecting fractures, whereas case 130 wells are aligned approximately orthogonal to this prevalent fracture. Figure 8(a2,a3) show thermoelastic stress along the horizontal and vertical directions, respectively, indicating stress localization spans across the cold fluid plume region. Greater concentration of connected fractures in the area away from the doublet axis causes prolonged thermal breakthrough time. Therefore, in 30 years, the temperature drop is approximately 40 °C for case 130, whereas case 220 shows a 75 °C temperature drop at the production well.

In comparison to water, CO_2 has approximately seven times smaller thermal conductivity at the injection conditions. Due to this, thermal depletion time is prolonged when CO_2 is the working fluid compared to water as the heat-carrying fluid. Figure 7b shows the thermal drawdown at the production well when CO_2 is the operating fluid. It shows that the thermal drawdown curves depend significantly on the fracture network connectivity than water. The slowest thermal drawdown is demonstrated by case 40, where production well temperature drops by approximately 20 °C in 300 years. In contrast, the fastest thermal drawdown is displayed by case 180, where around 90 °C temperature drop is estimated. Figure 9(a1,a2) show reservoir temperature distribution for case 40 and case 180, respectively. In Figure 9(a1), the cold fluid spread is extremely slow since a high fracture density is present near point 40, as shown in Figure 2 that is present away from the doublet axis. This leads to a reduced amount of cold fluid injection and restricted heat exchange between the fluid–fractures and fluid–matrix in the reservoir, decreasing the horizontal and vertical thermoelastic stress as shown in Figure 9(a2,a3), respectively. A detailed sensitivity analysis of dependent parameters is performed for water [47] and

CO_2 [50] based geothermal systems for the same fractured reservoir as mentioned in this paper. Our numerical simulations consider poroelastic stress, but we have not shown here that they contribute little due to the fluid injection and production, as shown in our previous findings [47,50]. For case 180, Figure 9(b1) shows reservoir temperature distribution after 300 years. It indicates that the hot fluid has been completely extracted between the doublet, and the heat replenishment is too slow to recharge this depleting heat content. Figure 9(b2,b3) approves this reasoning that due to favorable fracture density along the doublet axis, higher fluid flux reinjection results in higher thermoelastic stress evolution.

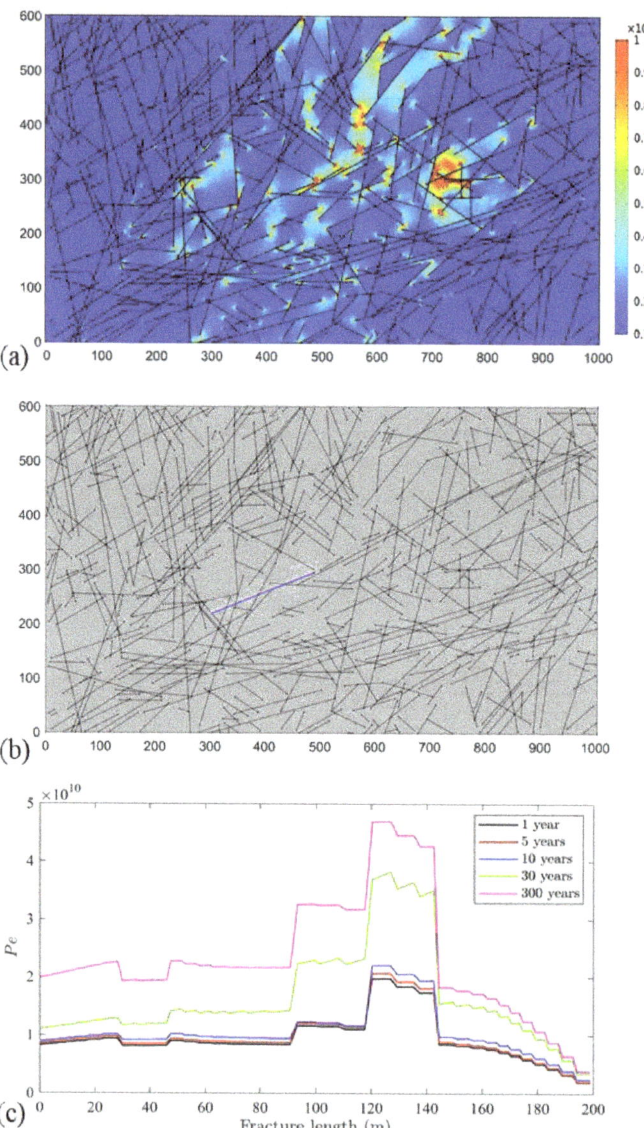

Figure 6. For case 180 and CO_2 as the working fluid after one year, (**a**) Peclet number distribution across the reservoir, (**b**) fracture considered for estimating fracture length, and (**c**) Peclet number along with the fracture length.

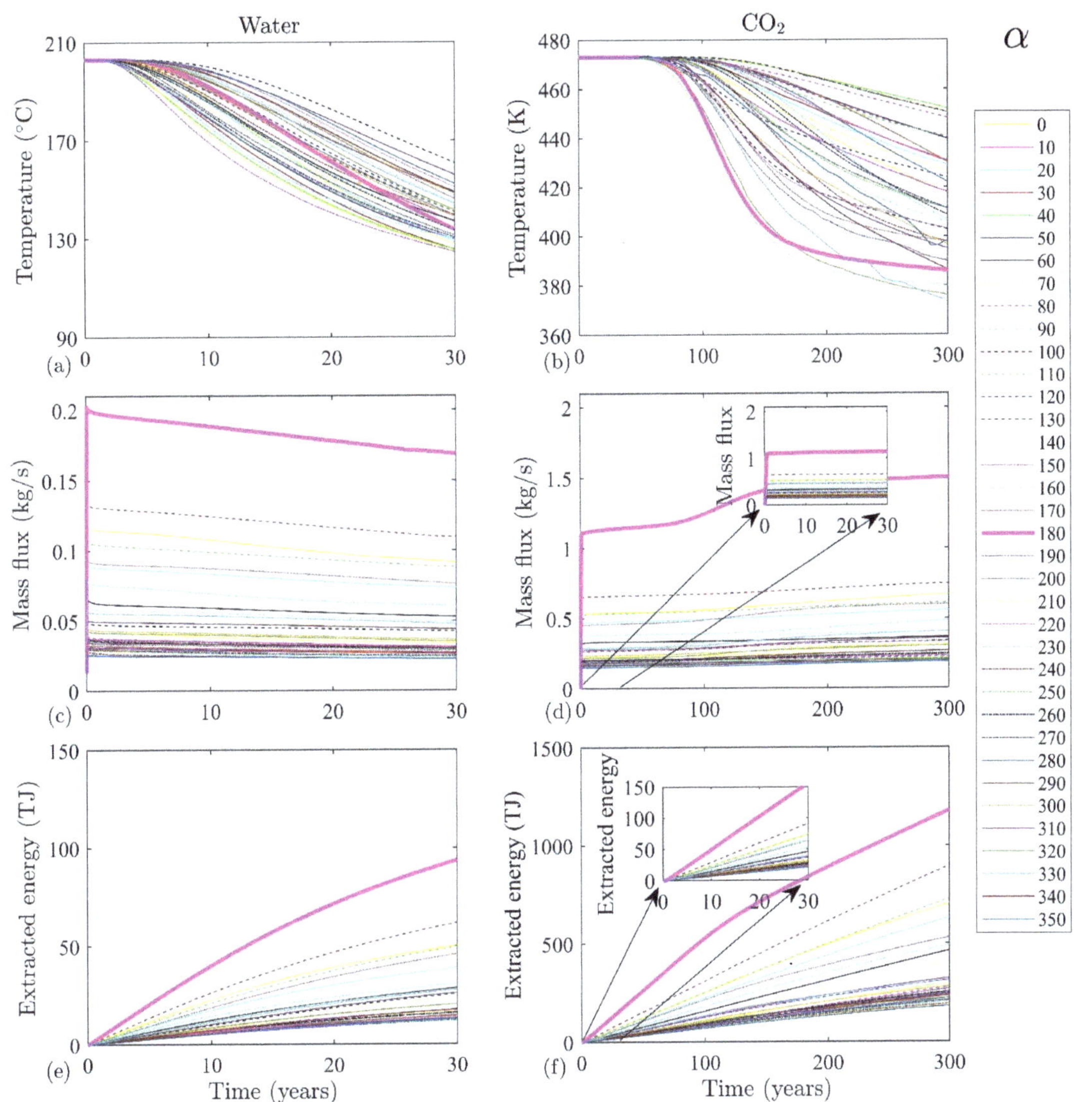

Figure 7. The temperature at the production well for (**a**) water and (**b**) CO$_2$, mass flux at the production well for (**c**) water and (**d**) CO$_2$, and cumulative energy extraction using (**e**) water and (**f**) CO$_2$ as heat-carrying fluid. Results from 36 simulation cases are plotted as shown by the legend, and case 180 is indicated by bold magenta color. Here Case 180 means injection well is present at 180, and production well is present at 0 (see Figure 2).

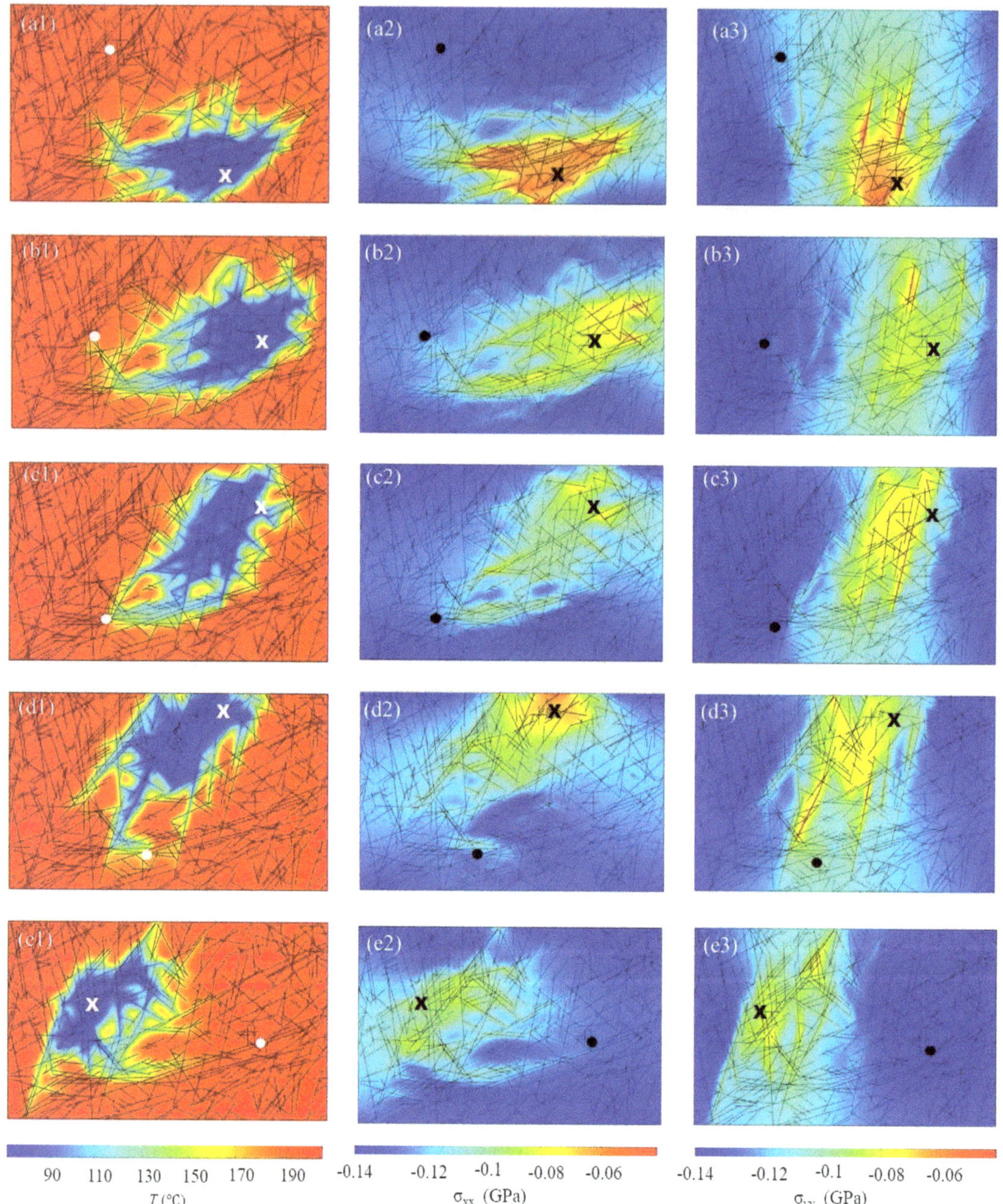

Figure 8. Distribution of (**a1**–**e1**) reservoir temperature, (**a2**–**e2**) horizontal thermoelastic stress and (**a3**–**e3**) vertical thermoelastic stress when water is used as the working fluid. Here Case 130 is displayed by (**a1**–**a3**), Case 180 is displayed by (**b1**–**b3**), Case 220 is displayed by (**c1**–**c3**), Case 250 is displayed by (**d1**–**d3**), and Case 350 is displayed by (**e1**–**e3**). All contours are plotted at ten years. The injection wellbore position is shown by cross symbol whereas a circle indicates production wellbore position.

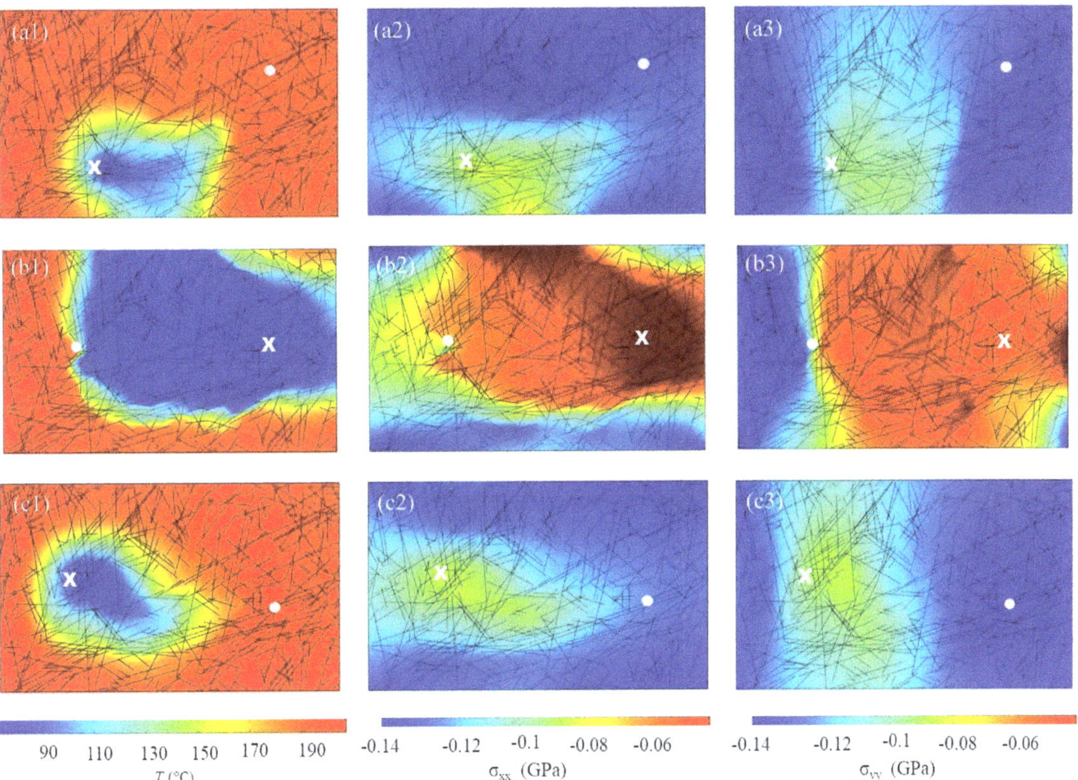

Figure 9. Distribution of (**a1–c1**) reservoir temperature, (**a2–c2**) horizontal thermoelastic stress and (**a3–c3**) vertical thermoelastic stress when CO_2 is used as the working fluid. Here Case 40 is displayed by (**a1–a3**), Case 180 is displayed by (**b1–b3**), and Case 350 is displayed by (**c1–c3**). All contours are plotted at 100 years. The injection wellbore position is shown by cross symbol whereas a circle indicates production wellbore position.

Figure 7c,d show mass flux at the production well for 36 cases when the working fluid in the reservoir is water and CO_2, respectively. The mass flux for CO_2 is approximately five times higher than water to compensate for smaller viscosity and higher permeability by maintaining the reservoir injection pressure. For both the fluids highest mass flux is observed for case 180, and the smallest mass flux is marked for case 350. Even though these two doublet arrangements have approximately the same axis (endpoints of a single line connecting injection and production wells), the fracture density near the production well plays a crucial role in mass flux; a similar observation was made by Zhang et al. [2]. For case 180, fractures are well connected near the production well, which assists in higher fluid production, whereas in the case of 350, fractures are not connected in a wide area leading to smaller fluid production. The temperature front in Figure 8(e1) shows the weak convective flow for case 350. This can be easily seen from the stress distribution plots in Figure 8(e2,e3) for case 350 when water is the working fluid and in Figure 9(c2,c3) for case 350 when CO_2 is the working fluid. The decrease in mass flux for all the cases with time is due to a in water viscosity with increased fluid temperature. However, we observe that the mass flux increases with time if CO_2 is the working fluid. This increase is approximately <30% between the period when CO_2 production starts till 300 years of numerical simulation. This increase is pronounced for the case 180 where we observe that the mass flux increases from 1.15 to 1.5 kg/s and the increase starts after approximately 50 years from the beginning of the operation. This discrepancy is observed due to limitations in the equation of state used in modeling using COMSOL Multiphysics. Since viscosity is a function of temperature only, the mass flux increase is observed after the breakthrough time (see Equation (11)).

The energy extraction potential from the reservoir for both the fluids are approximately the same (see Figure 7e,f) since in the case of water simulations, reservoir permeability is 2.5 times more negligible compared to CO_2, and the mass flux of CO_2 is approximately 3–5 times greater than water. In contrast, the specific heat capacity of water is around three times higher compared to CO_2 with three times higher viscosity. Due to higher mass flux and delayed thermal drawdown operation, total energy extraction potential is significantly higher when CO_2 is the working fluid. Case 180 for both the fluids shows the highest energy extraction potential due to the maximum mass flow rate in Figure 8(b1–b3). However, due to higher mass flux, thermal depletion is also fastest, and therefore, doublet placed for case 180 may not show a longer operation when water is the working fluid. Case 250 for water and case 350 for CO_2 show the least energy extraction potential over 30 and 300 years, respectively. Figure 8(d1) shows the reservoir temperature distribution for case 250 when water is used for heat transmission. It is visible from Figure 8d1 that a more incredible amount of cold fluid is present near the injection well, and there is only one large fracture along the doublet axis. This limits the fluid transmission at a higher rate only through a narrow region causing limited energy extraction.

Furthermore, Figure 8(d2,d3) shows the corresponding stress distribution for the horizontal and vertical directions. On the other hand, Figure 9(c1) shows reservoir temperature distribution for case 350 and CO_2 is used as working fluid and localization of cold fluid near the injection well in the absence of any dominating fracture system. The passage of fluid is limited through the fractures toward the production well. The stress field in Figure 8(c2,c3) shows the horizontal and vertical stress are well aligned with the temperature propagation. However, since thermal breakthrough is slower when CO_2 is used for heat transmission, energy extraction potential may enhance if EGS operation is performed beyond 300 years.

4. Conclusions

Geothermal energy extraction from deep fractured reservoirs can support high energy demand for a long duration. Water and CO_2 are two fluids that can extract energy from the subsurface. A fractured reservoir shows a complex network of fractures, and fracture conductivity controls the primary fluid passage for heat extraction longevity of the operation. Well placement for a given fracture network should consider the fracture density and orientation. Keeping all the parameters constant except the injection–production doublet axis orientation, we observe a difference of approximately ten times of energy extracted among the studied cases. High fracture density in the vicinity of the production well is the reason behind this increased energy extraction. The doublet axis orientation affects the injectivity (poroelastic stress) and temperature propagation (thermoelastic stress). It has a great impact on the stress field development during heat extraction.

Fluid type plays a significant role in determining the THM behavior of the EGS. The viscosity of fluid determines the temperature propagation through the fractures, as well as through the rock matrix. CO_2 with lower viscosity can penetrate easily inside the matrix zone. This effect, combined with the lower specific heat capacity of CO_2, eventuates the cold front of fluid propagation through the matrix and fracture. While water with high viscosity and specific heat capacity mainly transmits heat alongside the fracture and results in early breakthrough time. Different cases with water have a small range of breakthrough time compared to CO_2. While CO_2 shows a higher flow rate, resulting from the lower viscosity, this behavior is compensated by the higher heat capacity of water. Therefore, the overall heat extraction is comparable for both fluids.

Author Contributions: Conceptualization, S.M. and M.S.; methodology, S.M. and M.S.; software, S.M. and M.S.; validation, S.M. and M.S. writing—original draft preparation, S.M. and M.S.; writing—review and editing, K.B.; visualization, S.M. and M.S.; supervision, K.B. and I.S.; project administration, K.B.; funding acquisition, K.B. All authors have read and agreed to the published version of the manuscript.

Funding: The work is conducted as part of the MEET project that has received funding from the European Union's Horizon 2020 research and innovation programme under grant agreement No 792037.

Data Availability Statement: Data used in this paper are synthetic and can be reproduced from the Table 1.

Acknowledgments: Group of Geothermal Science and Technology, Institute of Applied Geosciences, Technische Universität Darmstadt has provided institutional support to authors. Authors would like to acknowledge anonymous reviewers for their valuable comments and suggestions.

Conflicts of Interest: The authors declare no conflict of interest.

Appendix A

Table A1. List of parameters.

Symbol	Parameter
p	Fluid pressure
T	Fluid Temperature
ε_V	Pore volumetric strain
α_m	Biot's coefficient of porous media
α_f	Biot's coefficient of the fracture
ϕ_m	Reservoir porosity
ϕ_f	Fracture zone porosity
S_m	Storage coefficients of fluid
S_1	Storage coefficients of rock matrix
S_f	Storage coefficients of fracture
β_1	Thermal expansion coefficients of fluid
β_m	Thermal expansion coefficients of rock matrix
β_f	Thermal expansion coefficient of fracture
$\rho \ \& \ \rho_1$	Fluid density
k_m	pressure-dependent rock matrix permeability
k_f	stress-dependent fracture permeability
e_h	hydraulic aperture between two fracture surfaces
nQ_m	$n.\left(-\frac{\rho k_m}{\mu \nabla p}\right)$, mass flux exchange between porous media and the fracture
∇_T	Gradient operator restricted to the fracture's tangential plane
T_m	Rock matrix temperature
T_l	Fluid temperature
ρ_m	Rock density
$C_{p,m}$	Specific heat capacity of the rock matrix
λ_m	Heat conductivity of the rock matrix

Table A1. *Cont.*

Symbol	Parameter
q_{ml}	Rock matrix-pore fluid interface heat transfer coefficient
ρ_f	density of the fracture zone
$C_{p,f}$	Specific heat capacity of the fracture
λ_f	Heat conductivity of the fracture
q_{fl}	Rock fracture-fluid interface heat transfer coefficient
C_p & $C_{p,l}$	Heat capacity of the fluid at a constant pressure
λ_l	Heat conductivity of the fluid
σ_{ij}	Total stress
G & λ	Lame's constants
tr	Trace operator
K'	$\frac{2G(1+\nu)}{3(1-2\nu)}$, bulk modulus of the drained porous media
β_T	Volumetric thermal expansion coefficient of porous media
δ_{ij}	Dirac dealt function
α_p	Biot's coefficient
σ_{eff}^{ij}	Effective stress
f_i	External body force
Δe_n	Change in the initial aperture of the fracture under in-situ stresses
e_0	Initial aperture of the fracture
σ_{eff}^n	Effective normal stress acting on the fracture surface
σ_{nref}	Effective normal stress required to cause 90% reduction in fracture aperture
μ	CO_2 dynamic viscosity
κ	CO_2 thermal conductivity

References

1. Singh, M.; Tangirala, S.K.; Chaudhuri, A. Potential of CO_2 based geothermal energy extraction from hot sedimentary and dry rock reservoirs, and enabling carbon geo-sequestration. *Geomech. Geophys. Geo-Energy Geo-Resour.* **2020**, *6*, 16. [CrossRef]
2. Zhang, H.; Huang, Z.; Zhang, S.; Yang, Z.; Mclennan, J.D. Improving heat extraction performance of an enhanced geothermal system utilizing cryogenic fracturing. *Geothermics* **2020**, *85*, 101816. [CrossRef]
3. Stefansson, V. Geothermal reinjection experience. *Geothermics* **1997**, *26*, 99–139. [CrossRef]
4. Mahmoodpour, S.; Singh, M.; Turan, A.; Bär, K.; Sass, I. Hydro-Thermal Modeling for Geothermal Energy Extraction from Soultz-sous-Forêts, France. *Geosciences* **2021**, *11*, 464. [CrossRef]
5. Hofmann, H.; Babadagli, T.; Zimmermann, G. Hot water generation for oil sands processing from enhanced geothermal systems: Process simulation for different hydraulic fracturing scenarios. *Appl. Energy* **2014**, *113*, 524–547. [CrossRef]
6. Hofmann, H.; Babadagli, T.; Yoon, J.S.; Blöcher, G.; Zimmermann, G. A hybrid discrete/finite element modeling study of complex hydraulic fracture development for enhanced geothermal systems (EGS) in granitic basements. *Geothermics* **2016**, *64*, 362–381. [CrossRef]
7. Lepillier, B.; Daniilidis, A.; Gholizadeh, N.D.; Bruna, P.-O.; Kummerow, J.; Bruhn, D. A fracture flow permeability and stress dependency simulation applied to multi-reservoirs, multi-production scenarios analysis. *Geotherm. Energy* **2019**, *7*, 24. [CrossRef]
8. Baujard, C.; Genter, A.; Dalmais, E.; Maurer, V.; Hehna, R.; Rosillettea, R.; Vidal, J.; Schmittbuhl, J. Hydrothermal characterization of wells GRT-1 and GRT-2 in Rittershoffen, France: Implications on the understanding of natural flow systems in the rhine graben. *Geothermics* **2017**, *65*, 255–268. [CrossRef]
9. Islam, J.; Vasant, P.M.; Negash, B.M.; Laruccia, M.B.; Myint, M. Watada A holistic review on artificial intelligence techniques for well placement optimization problem. *Adv. Eng. Softw.* **2020**, *141*, 102767. [CrossRef]
10. Zhang, L.; Deng, Z.; Zhang, K.; Long, T.; Desbordes, J.K.; Sun, H.; Yang, Y. Well-placement optimization in an enhanced geothermal system based on the fracture continuum method and 0–1 programming. *Energies* **2019**, *12*, 709. [CrossRef]

11. Ansari, E.; Hughes, R.; White, C.D. Well placement optimization for maximum energy recovery from hot saline aquifers. In Proceedings of the Thirty-Ninth Workshop on Geothermal Reservoir Engineering, Stanford University, Stanford, CA, USA, 24–26 February 2014.

12. Yousefzadeh, R.; Sharifi, M.; Rafiei, Y. An efficient method for injection well location optimization using Fast Marching Method. *J. Pet. Sci. Eng.* **2021**, *204*, 108620. [CrossRef]

13. Bangerth, W.; Klie, H.; Wheeler, M.; Stoffa, P.; Sen, M. On optimization algorithms for the reservoir oil well placement problem. *Comput. Geosci.* **2006**, *10*, 303–319. [CrossRef]

14. Onwunalu, J.E.; Durlofsky, L.J. Application of a particle swarm optimization algorithm for determining optimum well location and type. *Comput. Geosci.* **2010**, *14*, 183–198. [CrossRef]

15. Forouzanfar, F.; Li, G.; Reynolds, A.C. A two-stage well placement optimization method based on adjoint gradient. In Proceedings of the SPE Annual Technical Conference and Exhibition, Florence, Italy, 20–22 September 2010.

16. Sarma, P.; Durlofsky, L.J.; Aziz, K. Computational techniques for closed–loop reservoir modeling with application to a realistic reservoir. *Petrol. Sci. Technol.* **2008**, *26*, 1120–1140. [CrossRef]

17. Zandvliet, M.; Handels, M.; Essen, G.; Brouwer, R.; Jansen, J. Adjoint-based well-placement optimization under production constraints. *SPE J.* **2008**, *13*, 392–399. [CrossRef]

18. Vaseghi, F.; Ahmadi, M.; Sharifi, M.; Vanhoucke, M. Generalized multi-scale stochastic reservoir opportunity index for enhanced well placement optimization under uncertainty in green and brownfields. *Oil Gas Sci. Technol.—Rev. IFP Energ. Nouv.* **2021**, *76*, 41. [CrossRef]

19. Chen, J.; Jiang, F. Designing multi-well layout for enhanced geothermal system to better 489 exploit hot dry rock geothermal energy. *Renew. Energy* **2015**, *74*, 37–48. [CrossRef]

20. Kong, Y.; Pang, Z.; Shao, H.; Kolditz, O. Optimization of well-doublet placement in geothermal reservoirs using numerical simulation and economic analysis. *Environ. Earth Sci.* **2017**, *76*, 118. [CrossRef]

21. Liang, X.; Xu, T.; Feng, B.; Jiang, Z. Optimization of heat extraction strategies in fault-controlled hydro-geothermal reservoirs. *Energy* **2018**, *164*, 853–870. [CrossRef]

22. Babaei, M.; Nick, H.M. Performance of low-enthalpy geothermal systems: Interplay of spatially correlated heterogeneity and well-doublet spacings. *Appl. Energy* **2019**, *253*, 113569. [CrossRef]

23. Jiang, Z.; Xu, T.; Wang, Y. Enhancing heat production by managing heat and water flow in confined geothermal aquifers. *Renew. Energy* **2019**, *142*, 684–694. [CrossRef]

24. Blank, L.; Rioseco, E.M.; Caiazzo, A.; Wilbrandt, U. Modeling, simulation, and optimization of geothermal energy production from hot sedimentary aquifers. *Comput. Geosci.* **2021**, *25*, 67–104. [CrossRef]

25. Chen, S.-Y.; Hsieh, B.-Z.; Hsu, K.-C.; Chang, Y.-F.; Liu, J.-W.; Fan, K.-C.; Chiang, L.-W.; Han, Y.-L. Well spacing of the doublet at the Huangtsuishan geothermal site, Taiwan. *Geothermics* **2021**, *89*, 101968. [CrossRef]

26. Zhang, S.; Jiang, Z.; Zhang, S.; Zhang, Q.; Feng, G. Well placement optimization for large-scale geothermal energy exploitation considering nature hydro-thermal processes in the Gonghe Basin, China. *J. Clean. Prod.* **2021**, *317*, 128391. [CrossRef]

27. Akin, S.; Kok, M.V.; Uraz, I. Optimization of well placement geothermal reservoirs using artificial intelligence. *Comput. Geosci.* **2010**, *36*, 776–785. [CrossRef]

28. Samin, M.Y.; Faramarzi, A.; Jefferson, I.; Harireche, O. A hybrid optimization approach to improve long-term performance of enhanced geothermal system (EGS) reservoirs. *Renew. Energy* **2019**, *134*, 379–389. [CrossRef]

29. Gudmundsdottir, H.; Horne, R.N. Inferring interwell connectivity in fractured geothermal reservoirs using neural networks. In Proceedings of the World Geothermal Congress 2020+1, Reykjavik, Iceland, 24–27 October 2021.

30. Pandey, S.N.; Singh, M. Artificial neural network to predict the thermal drawdown of enhanced geothermal system. *J. Energy Resour. Technol.* **2021**, *143*, 010901. [CrossRef]

31. Ishitsuka, K.; Kobayashi, Y.; Watanabe, N.; Yamaya, Y.; Bjarkson, E.; Suzuki, A.; Mogi, T.; Asanuma, H.; Kajiwara, T.; Sugimoto, T.; et al. Bayesian and neural network approached to estimate deep temperature distribution for assessing a supercritical geothermal system: Evaluation using a numerical model. *Nat. Resour. Res.* **2021**, *30*, 3289–3314. [CrossRef]

32. Chen, M.; Tompson, A.F.B.; Mellors, R.J. Abdalla An efficient optimization of well placement and control for geothermal prospect under geological uncertainty. *Appl. Energy* **2015**, *137*, 352–363. [CrossRef]

33. Zhang, W.; Qu, Z.; Guo, T.; Wang, Z. Study of the enhanced geothermal system (EGS) heat mining from variably fractured hot dry rock under thermal stress. *Renew. Energy* **2019**, *143*, 855–871. [CrossRef]

34. Gao, X.; Zhang, Y.; Huang, Y.; Ma, Y.; Zhao, Y.; Liu, Q. Study on heat extraction considering the number and orientation of multilateral wells in a complex fractured geothermal reservoir. *Renew. Energy* **2021**, *177*, 833–852. [CrossRef]

35. Aliyu, M.D.; Chen, H.-P.; Harireche, O. Finite element modeling for productivity of geothermal reservoirs via extraction well. In Proceedings of the 24th UK conference of the Association for Computational Mechanics in Engineering, Cardiff University, Cardiff, UK, 31 March–1 April 2016; pp. 331–334.

36. Aliyu, M.D.; Chen, H.-P. Optimum control parameters and long-term productivity of geothermal reservoirs using coupled thermo-hydraulic process modeling. *Renew. Energy* **2017**, *112*, 151–165. [CrossRef]

37. Brown, D. A hot dry rock geothermal energy concept utilizing supercritical CO_2 instead of water. In Proceedings of the 25th Workshop on Geothermal Reservoir Engineering, Stanford University, Stanford, CA, USA, 24–26 January 2000; p. SGP-TR-165.

38. Isaka, B.A.; Ranjith, P.; Rathnaweera, T. The use of supercritical carbon dioxide as the working fluid in enhanced geothermal systems (EGSs): A review study. *Sustain. Energy Technol. Assess* **2019**, *36*, 100547.

39. Atrens, A.D.; Gurgenci, H.; Rudolph, V. Electricity generation using a carbon-dioxide thermosiphon. *Geothermics* **2010**, *39*, 161–169. [CrossRef]

40. Bongole, K.; Sun, Z.; Yao, J.; Mehmood, A.; Yueying, W.; Mboje, J.; Xin, Y. Multifracture response to supercritical CO2- EGS and water-EGS based on thermo-hydro-mechanical coupling method. *Int. J. Energy Res.* **2019**, *43*, 7173–7196.

41. Atrens, A.D.; Gurgenci, H.; Rudolph, V. CO_2 thermosiphon for competitive geothermal power generation. *Energy Fuels* **2009**, *23*, 553–557. [CrossRef]

42. Tran, N.H.; Chen, Z.; Rahman, S.S. Object-based global optimization in modeling discrete-fracture network map: A case study. In Proceedings of the SPE Annual Technical Conference and Exhibition, Denver, CO, USA, 5–8 October 2003; p. 84456.

43. COMSOL Multiphysics®v. 5.5. COMSOL AB, Stockholm, Sweden. Available online: www.comsol.com (accessed on 11 November 2020).

44. Barton, N.; Bandis, S.; Bakhtar, K. Strength, deformation and conductivity coupling of rock joints. *Int. J. Rock Mech. Min. Sci. Geomech. Abstr.* **1985**, *22*, 121–140. [CrossRef]

45. Bandis, S.C.; Lumsden, A.C.; Barton, N.R. Fundamentals of rock joint deformation. *Int. J. Rock Mech. Min. Sci. Geomech. Abstr.* **1983**, *20*, 249–268. [CrossRef]

46. Bai, B. One-dimensional thermal consolidation characteristics of geotechnical media under non-isothermal condition. *Eng. Mech.* **2005**, *22*, 186–191.

47. Mahmoodpour, S.; Singh, M.; Turan, A.; Bär, K.; Sass, I. Key parameters affecting the performance of fractured geothermal reservoirs: A sensitivity analysis by thermo-hydro-mechanical simulation. *arXiv* **2021**, arXiv:2107.02277.

48. Bär, K.; Reinsch, T.; Bott, J. P^3-PetroPhysical Property Database. V. 1.0. GFZ Data Services. 2019. Available online: https://doi.org/10.5880/GFZ.4.8.2019.P3 (accessed on 1 April 2021).

49. Bergman, T.L.; Lavine, A.S.; Incropera, F.P. *Fundamentals of Heat and Mass Transfer*, 7th ed.; John Wiley & Sons, Incorporated: Hoboken, NJ, USA, 2011; ISBN 9781118137253.

50. Mahmoodpour, S.; Singh, M.; Bär, K.; Sass, I. Thermo-hydro-mechanical modeling of an Enhanced geothermal system in a fractured reservoir using CO_2 as heat transmission fluid—A sensitivity investigation. *arXiv* **2021**, arXiv:2108.05243.

geosciences

Article

Study of Corrosion Resistance Properties of Heat Exchanger Metals in Two Different Geothermal Environments

Svava Davíðsdóttir [1], Baldur Geir Gunnarsson [1], Kjartan Björgvin Kristjánsson [1], Béatrice A. Ledésert [2] and Dagur Ingi Ólafsson [1,*]

[1] Tæknisetur ehf, Árleynir 2-8, 112 Reykjavík, Iceland; svava@taeknisetur.is (S.D.); baldur@taeknisetur.is (B.G.G.); kjartan@taeknisetur.is (K.B.K.)
[2] GEC Laboratory, CY Cergy Paris Université, 1 Rue Descartes, 95000 Neuville Sur Oise, France; beatrice.ledesert@cyu.fr
* Correspondence: dagur@taeknisetur.is

Abstract: Geothermal fluids harnessed for electricity production are generally corrosive because of their interaction with the underground. To ensure the longevity and sustainability of geothermal Organic Rankine Cycle (ORC) powerplants, the choice of heat exchanger material is essential. The performance of heat exchangers is affected by corrosion and scaling due to the geothermal fluids, causing regular cleaning, part replacement, and in the worst cases, extensive repair work. The properties of geothermal fluids vary between geothermal settings and even within geothermal sites. Differences in exposure conditions require different material selection considerations, where factors such as cost, and material efficiency are important to consider. This work studies in-situ geothermal exposure testing of four metals at two geothermal locations, in different geological settings. Four corrosion-resistant materials were exposed for one month at Reykjanes powerplant in Iceland and four months at Chaunoy oil field in France as material candidates for heat exchangers. The tested alloys were analysed for corrosion with macro- and microscopic techniques using optical and electron microscopes, which give an indication of the different frequencies of repairs and replacement. Inconel 625 showed no effects at Reykjanes and cracks at Chaunoy. The others (316L, 254SMO, and titanium grade 2) showed either corrosion or erosion traces at both sites.

Keywords: geothermal; oil; heat exchanger; corrosion; scaling; geology

Citation: Davíðsdóttir, S.; Gunnarsson, B.G.; Kristjánsson, K.B.; Ledésert, B.A.; Ólafsson, D.I. Study of Corrosion Resistance Properties of Heat Exchanger Metals in Two Different Geothermal Environments. *Geosciences* **2021**, *11*, 498. https://doi.org/10.3390/geosciences11120498

Academic Editors: Romain Augier, Thomas Hermans and Jesus Martinez-Frias

Received: 8 October 2021
Accepted: 2 December 2021
Published: 7 December 2021

Publisher's Note: MDPI stays neutral with regard to jurisdictional claims in published maps and institutional affiliations.

1. Introduction

Material selection for geothermal equipment is a crucial parameter to consider when constructing an Organic Rankine Cycle (ORC) geothermal power plant. The heat exchanger transfers heat energy from the geothermal power source to the organic fluid within the closed loop of the ORC, boiling it. The organic fluid in its vapour form is then used to transfer the energy to the turbines of the ORC plant, generating electricity, before being condensed to repeat the continuous cycle. Geothermal environments differ between locations and the material selection for the heat exchanger needs to be tailored for each environment. The geothermal fluids can be challenging for heat exchanger materials due to high velocity, abrasive particles, reactive gasses such as H_2S, and in some cases salinity [1,2].

The focus of the current study is to investigate heat exchange plate candidates, metals, and metal alloys which are more prone to local corrosion rather than uniform corrosion. Local corrosion is often associated with metals and metal alloys forming natural passivation layers, which protect the surface of the metal. The performance of the metals and alloys is often dependent on its natural initial passivation layer and its ability to re-passivate if it gets attacked during operation. Alloying elements such as Cr and Mo form protective oxides at the metal surface [3]. Furthermore, the amount of Cr and Mo within the alloy is often used for ranking the alloys with pitting resistance equivalent number (PRE) [4].

Oxygen is rarely present within the geothermal fluid of heat exchange systems and if it is, special precautions need to be taken [5]. Gasses such as CO_2 and H_2S are common in geothermal environments [2,6].

The surface area and thermal conductivity of the heat exchanger plates are crucial for optimising the efficiency of the heat transfer. Scaling and corrosion often result in insulating layers forming on the surface of the plates, which reduce the heat transfer efficiency from the geothermal fluid. Scaling and corrosion are therefore important to avoid or minimize.

There are three main failure mechanisms of a heat exchanger: leakages, blocking due to deposits, and material thinning [7]. The main reasons for material thinning are erosion and corrosion. Erosion is caused by mechanical abrasion which results in mechanical removal of the surface material. Erosion depends on fluid velocity and the presence of abrasive particles in the fluid [7]. There is a correlation between material hardness and erosion resistance [8]. To minimize maintenance costs, production downtime, and part replacement in geothermal power plants, it is important to study the corrosion, erosion, and scaling issues affecting the various equipment in contact with the geothermal brine and to acquire a better understanding of how to ensure the sustainability of the plants. In the case of ORC power plants, the main equipment in contact with the geothermal brine are heat exchangers.

The majority of geothermal power plants, and especially those that run on the ORC principle, operate at low or medium temperatures, below 200 °C. This is reflected in the demonstration sites of the MEET H2020 project (Multidisciplinary and multi-context demonstration of Enhanced geothermal systems exploration and Exploitation Techniques and potentials) to which this study is dedicated, such as Soultz-sous-Forêts, Grásteinn, Cazaux, Chaunoy, and Krauma. Various corrosion and scaling studies have been conducted at Soultz-sous-Forêts where the geothermal fluid is quite saline, corrosive, and tends to form scales which accumulate naturally occurring radioactive materials [9–12]. Ledésert et al. [13] studied the scaling formation at Soultz-sous-Forêts when the reinjection temperature of the geothermal fluid was reduced from 70 °C to 40 °C. They concluded that the scaling formation observed was not influenced by the alloy of the heat exchanger material.

In recent years, several material experiments have been conducted at supercritical temperatures [14,15] where the conditions vary significantly from standard geothermal plants. For geothermal power plants operating at a low or medium temperature a few studies have been published [16–18] but no review paper is available. For a tailored material selection, a common evaluation method is to expose material candidates in the actual geothermal environment. It is recommended for material evaluation that the samples be tested for as long as possible although commonly it is based on the convenience of the plant operators [19]. Frequently, test durations range from a month and upwards. The exposure time is short compared to the lifetime of the material but gives an indication of the performance of the material in the actual environment. The most common method to evaluate exposed materials for geothermal applications is with weight loss measurements using standard ASTM G1-03, which estimates uniform corrosion. However, the materials evaluated in this study are more prone to local corrosion.

Metal and metal alloys forming natural passivation layers are more often affected by local corrosion than uniform corrosion. The passivation layer is locally broken, resulting in the direct electrical contact of the metal with the fluid. The fluid can either be in liquid or vapour form depending on the temperature and pressure. Geothermal fluids usually contain ions which increase their conductivity which can accelerate corrosion [2]. Furthermore, Cl ions in connection with oxygen and water have been reported to assist with the breakdown of the passivation layer of stainless steel resulting in corrosion [20]. For example, 316L stainless steel has been reported to have low performance in humid and oxygen rich environments; however, it has been successfully used in oxygen free CO_2 brines at temperatures as high as 150 °C [21].

The 316L and 304L are some of the most used and studied materials in the geothermal industry and are widely used in heat exchangers [5]. Nickel alloys have shown promising results in the S rich environment and Ti and its alloys in an oxygen rich environment [5,22]. However, due to the lower hardness of pure Ti, its alloys are often preferred for the geothermal environment [23,24]. Furthermore, from the oil and gas industry it is reported that the main failure mechanism of Ni-based alloy is from hydrogen embrittlement [25]. It has been reported by Karlsdóttir et al. [14] that the most corrosive resistant material tested for geothermal heat exchange application in superheated geothermal steam was Inconel 625 and Ti-alloy (Ti gr. 7). Furthermore, the group reported that 254SMO had a corrosion rate of 0.001 mm/year measured according to ASTM G1-03 [14], which is in line with results from the same group's measuring in simulated superheated geothermal conditions' corrosion rate of 0.001 mm/year [26]. Therefore, 254SMO has been recommended to be used by the Icelandic geothermal industry [10]. However, due to its cost it is not as commonly used in geothermal environments [27] as other cheaper materials such as 316L, for example.

The pH of the fluid is an important factor for determining the lifetime of components and 316L is, for example, not recommended to be used at a pH lower than 3.2 [28] where superaustenitic steels, such as 254SMO, can be a good option [29]. For geothermal applications, stainless steels and Ni-alloys are recommended to have a minimum of 3% Mo [30]. However, high levels of Mo could result in more risk for inducing segregation, the formation of hard and brittle phases within the material, which could initiate cracks and voids during exposure [31]. The grain size of steel can be crucial as smaller grain size can reduce interdenritic segregation as well as the chances of high temperature cracking [32]. Inclusions are common within stainless steels and have been reported to be found in 254SMO [31,32] and 316 [33]. The inclusions can be a preferential place for crack development at the subsurface of the material [33]. Inclusions can create geometric discontinuities where stress and strain can accumulate [34]. Moreover, for surfaces with a passive layer, an inclusion might act as an initiation site for pitting [35].

To identify suitable materials for heat exchangers in low-temperature volcanic and sedimentary ORC power plants, four metal samples were exposed in two different geothermal environments, namely at high salinity locations in Reykjanes, Iceland and Chaunoy, France. The effects on the coupons were then examined macro- and microscopically. The selected materials consisted of stainless steels 316L (EN 1.4404) and 254SMO (EN 1.4547), nickel alloy Inconel 625 (EN 2.4856), and titanium grade 2 (EN 3.7035). All these materials are reported to be used in the geothermal industry [29]. In general, failure during the plate heat exchange operation accrues in the weakest points of the material. Additionally, the materials studied here are more prone to local corrosion than uniform corrosion. Therefore, the main interest of this study is to investigate the most affected areas of the exposed materials and report on the observed local corrosion. The testing period was one and four months, respectively, for Reykjanes and Chaunoy, while the expected lifetime of a heat exchanger is around 20–30 years. All samples tested were affected by the exposure, including 254SMO and Inconel 625. Further investigation is needed to determine the development of the defects found after the exposure.

1.1. Geological Context

1.1.1. Reykjanes

Iceland began to form in Tertiary (16 Ma) [36] and is considered to be the product of anomalously high volcanism related to the interaction of the divergent plate boundaries on the Mid-Atlantic Ridge (MAR) and the Iceland mantle plume [37]. The MAR cuts across Iceland in a roughly SW-NE trending direction and at the crest of it is where the most vigorous magmatic activity occurs. Thus, numerous active volcanic systems can be found along the ridge axis [38]. The Reykjanes Peninsula, in SW Iceland, is a continuation of the median fault zone of the MAR and continues as Reykjanes Ridge offshore SW of Iceland. The Peninsula is formed of hyaloclastite ridges and basaltic lavas [39]. The oldest lavas on the surface are from the Late Pleistocene age while the youngest are from the Holocene [40]. Furthermore, the area is intensely fractured with normal faults and extension fractures, within a narrow belt, along with eruptive fissures lying perpendicular to those extensional rifts [39,41]. There are four distinct volcanic systems on the Reykjanes Peninsula, from west to east: Reykjanes, Krísuvík, Brennisteinsfjöll, and Hengill fissure swarms [42,43], see Figure 1.

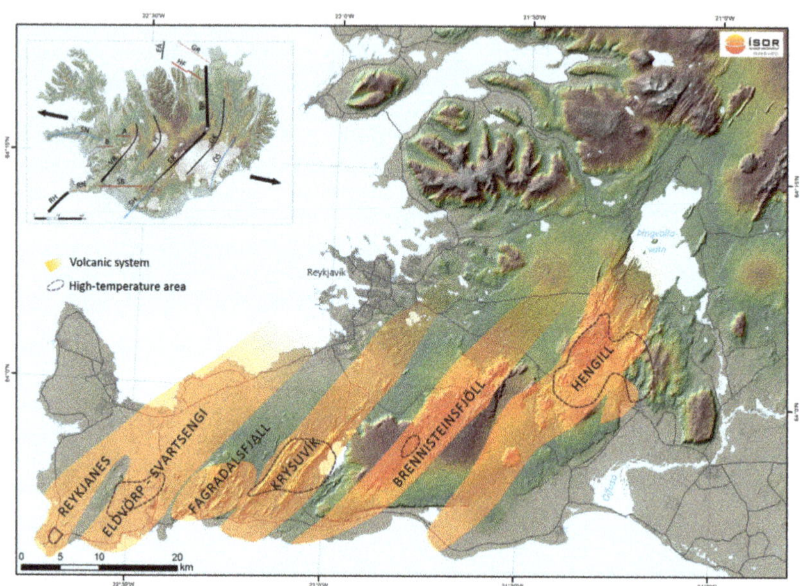

Figure 1. Geological map showing the location of the Reykjanes Peninsula and the fissure swarms in the area (image courtesy of Iceland GeoSurvey (ÍSOR)).

The Reykjanes fissure swarm, which the Reykjanes geothermal field is a part of, is at least 45 km in length, where 30 km are on land [44]. The Reykjanes geothermal field is at the SW tip of the Peninsula and is one of the smallest geothermal areas in Iceland. It is about 1–1.5 km^2, based on the geothermal features displayed on the surface [39]. The heat source of the Reykjanes geothermal field is thought to be dykes, thin sills, and/or a sheeted dyke complex. At a 5.5–6 km depth, a brittle-ductile transition marks the bottom of potential permeability, hence, the base of the hydrothermal system [45]. The chemical composition of the high-temperature hydrothermal fluid derives from the interaction of seawater with basaltic rocks made up mostly by hyaloclastite, volcanic breccias, and tuffaceous units to a depth of 1000 m [46]. Observations from the 2054 m deep well, RN-10, shows that at increased depths the stratigraphy consists predominantly of pillow basalts and formations exhibit relatively high porosity and low permeability with aquifers that are related to fractures along sub-vertical dyke intrusions [47]. The chemical composition of the liquid phase of the geothermal brine in well RN-29 at Reykjanes geothermal powerplant was measured by Iceland GeoSurvey (ÍSOR) in 2014 [48] and is shown in Table 1.

1.1.2. Chaunoy

The Keuper Triassic sediments were deposited during an extensional phase that induced the subsidence of the Paris basin [49]. The sediments were deposited in a continental environment, ranging from alluvial (west of the basin) to evaporitic (east of the basin) [50]. The maximum subsidence occurred east of Chaunoy while, to the west, the basin overlapped gently with the armorican variscan massif [51]. The Chaunoy reservoir is part of the Chaunoy sandstone, a Carnian-Norian lithostratigraphic formation due to a second-order Scythian-Carnian cycle of tectonic origin [52,53]. It lies conformably on reddish dolomitic shales deposited in a coastal plain [51]. The Chaunoy oil field (Figure 2A), located 50 km south-east of Paris, is the largest oil field of the Paris basin [54]. It is a broad north-south anticline structure 15 km long and 5 km wide, with a faulted eastern flank [51,55]. The average thickness of the reservoir is approximately 66 m [51,55]. It developed in the distal part of alluvial deposits. Small ribbon channel deposits are interbedded with a flood plain and lacustrine deposits. The channel amalgamation occurred during periods of low accommodation, producing highly heterogeneous sand sheets [51]. During higher accommodation periods, channels became progressively isolated within flood plain mudstones. Finally, lacustrine mudstones were deposited, creating a vertical permeability barrier. Then, the decrease in accommodation induced a strong pedogenetic alteration responsible for dolocrete and groundwater dolomite. The amalgamation rate varied with cyclic lake-level variations, which directly controlled the reservoir geometry [51]. Due to this history, the reservoir is strongly heterogeneous. An upper siliciclastic/dolomitic member can be divided into two units with porous conglomeratic channels interfingered with cemented lagoonal dolomites. A lower siliciclastic member shows four heterogeneous sand sheets (7 m thick), which have been correlated across the field. Each of them is made up of stacked single channel sequences. The sand sheets are separated by extensive lacustrine and flood plain mudstone layers acting as permeability barriers. Bourquin et al. [56] presented a high-resolution sequence stratigraphy of the Chaunoy reservoir. The maximum net oil pay is 25 m with a 11 m average [51], in which the thickness of each reservoir unit ranges from 1 m to 5 m. In this field, the well spacing, 600 m in average, is larger than the channel width [51,54] which does not allow for a complete understanding either of the structure or of fluid flows as well-to-well correlations are hypothetical. The Liassic shales located directly above the Triassic reservoir constitute the oil source rock [51].

Table 1. Chemical composition of the liquid phase of well RN-29 at Reykjanes geothermal powerplant.

Analysis Year: 2014		
Chemical	**Value**	**Unit**
CO_2	45.5	mg/kg
H_2S	1.44	mg/kg
NH_3	2.06	mg/kg
B	13.1	mg/kg
SiO_2	1088	mg/kg
Na	14630	mg/kg
K	2520	mg/kg
Mg	2.86	mg/kg
Ca	1950	mg/kg
F	0.27	mg/kg
Cl	28600	mg/kg
SO_4	16.3	mg/kg
Al	0.022	mg/kg
Fe	3.82	mg/kg
Acidity	5.15	pH

The timing of the in-situ material exposure performed at Chaunoy oil field coincided with an ORC powerplant demonstration performed within the framework of the MEET project. Exhaust water coming out of the CNY40 oil well (Figure 2B) was connected to the inlet of an ORC powerplant. The samples tested in this work at Chaunoy were placed within the exhaust water flowline (Figure 2C) connected to the inlet of the small-scale ORC unit (Figure 2D). The cooled water coming out of the ORC was reinjected with the whole exhaust flow. The oil field showed a high water cut with approximately 96 L of water for every 1 L of oil produced (Vermilion, personal communication).

Figure 2. The small-scale mobile Organic Rankine Cycle (ORC) unit installed at Chaunoy oil field. (**A**): one of the oil wellheads, (**B**): head of the CNY40 well from which the oil and hot water are extracted, (**C**): the connection of the hot water pipe to the ORC unit, (**D**): the mobile ORC unit connected to the water pipe. The samples tested at Chaunoy in this work were placed within the inlet of the ORC unit during a 4-month demonstration.

2. Materials and Methods

The corrosion experiments were based on the Standard Guide for Conducting Corrosion Tests in Field Applications, ASTM-G4-1 [4]. The corrosion experiments were conducted by exposing the four potential heat exchanger materials in the form of metal coupons at the two locations: Reykjanes and Chaunoy. The coupons were placed on a sample holder, with ceramic isolation rings to avoid the electrical contact between the holder and the coupon samples. The sample holder was placed in a pressure vessel for Reykjanes and in a flow-line pipe for Chaunoy. The 254SMO was bought from Outokumpu, Inconel 625 was bought from ThyssenKrupp, 316L was bought from Outokumpu and Stálnaust, and Ti gr. 2 was acquired from The Welding Institute (TWI). The dimensions of the coupons were in all cases $50 \times 25 \times 3$ mm except for Ti gr. 2 which was 4 mm in thickness. The elemental compositions of the materials tested are shown in Table 2.

2.1. Reykjanes

The coupon samples were exposed inside a pressure vessel at Reykjanes, seen in Figure 3, for 30 days. The temperature and pressure were 200 °C and 18 bar, respectively. The fluid in the pressure vessel was saturated with geothermal steam which had already passed through a separator. The gas concentrations of the steam present in the pressure vessel, which is the same steam as is fed to the turbines of the Reykjanes power plant, can be seen in Table 3.

Table 2. Elemental composition of the materials tested in this work.

Element (% wt.)	Material			
	316L	Inconel 625	254SMO	Ti gr. 2
C	0.02	-	-	0.08
Si	0.5	0.2	0.4	-
Mn	1.0	0.1	0.7	-
Fe	Balance	4.5	Balance	0.3
Ni	10.1	Balance	17.8	-
Mo	2.0	6.7	6.1	-
Cr	17.1	21.4	19.8	-
Cu	-	-	0.3	-
N	0.04	-	0.3	0.03
Ti	-	0.2	-	Balance
Nb	-	3.4	-	-
H	-	-	-	0.015
O	-	-	-	0.025

Figure 3. The coupons being removed from the pressure vessel at Reykjanes power plant where they were exposed.

Table 3. Gas concentrations in the pressure vessel at Reykjanes power plant [57].

Chemical	Value	Unit
CO_2	6600	mg/kg
H_2S	220	mg/kg
H_2	1.9	mg/kg
N_2	66	mg/kg
CH_4	0.9	mg/kg

2.2. Chaunoy

The Chaunoy oil field is in the Paris basin, in northern France, where the salinity is around 6.5 wt.%. The fluid from the tested oil well is mostly hot geothermal water, with a small amount of oil. The test was conducted by placing the coupons on a sample holder rack, which was placed inside a flowline at the inlet of an ORC power production unit. The sample rack before and after testing can be seen in Figure 4.

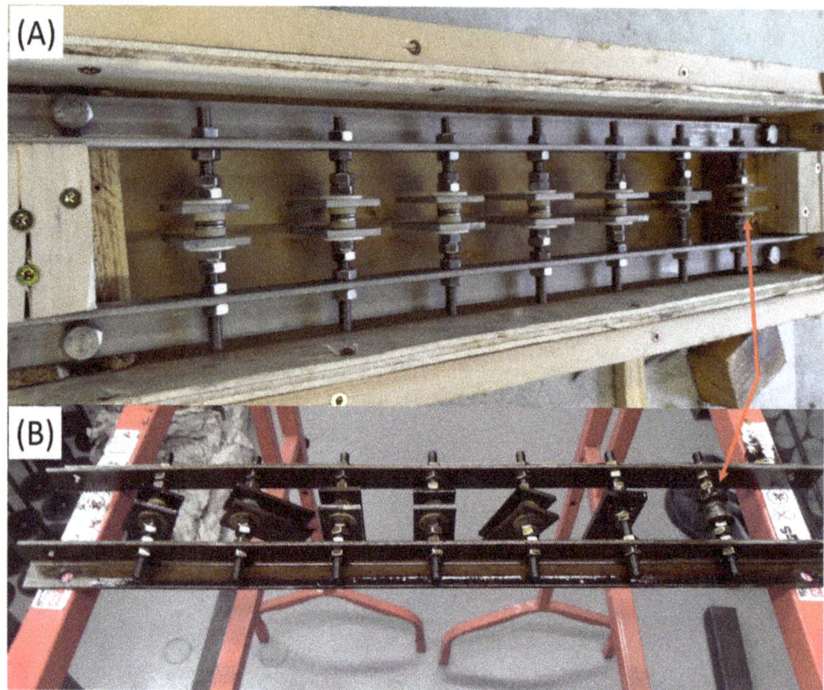

Figure 4. (A): The samples on a rack which was inserted into an oil flowline at Chaunoy. **(B)**: The rack and samples after testing had been conducted. It is interesting to note that two carbon steel coupons, which are not investigated in this work, had corroded severely, and fallen off of the threaded rod during exposure as marked with red arrows.

The Chaunoy exposure was conducted in well CNY40 for four months. The temperature at the wellhead was measured at 94 °C, the oil ratio of the fluid was 1.85%, and the flow rate was 6.25 l/s. The pressure was on average 9.5 bar. A chemical analysis of the fluid can be seen in Table 4.

Table 4. Chemical analysis of the fluid in well CNY40.

Analysis Year: 2009		
Chemical	**Value**	**Unit**
Ba	2.82	mg/L
Ca	3986	mg/L
Fe	1.6	mg/L
Mg	572.7	mg/L
Mn	0.29	mg/L
Na	22910	mg/L
Sr	297.8	mg/L
K	572.8	mg/L
Cl^-	41967	mg/L
S	399	mg/L
$HCO_3{}^-$	1696.1	mg/L
Acidity	7.15	pH

2.3. Material Assessment

The samples were analysed both topographically and cross-sectionally using an optical microscope, AXIO from Zeiss, and a Scanning Electron Microscope (SEM), Supra 25. A chemical analysis was conducted using an Energy-Dispersive X-ray spectroscopy (EDX) detector, Oxford Instrument X-max. For simplicity, the analysis was only given for elements over 0.5 wt.%.

The samples were dismounted from the racks and photocopied with Canon MG5450 before assessment. The Backscatter Electron (BSE) detector of the SEM was used for the topography imaging and a secondary detector was used for cross-section imaging. The cross-sections were prepared by cutting the samples with a diamond cutter and baking in conductive phenolic epoxy for 3.5 min at 150 °C. The samples were polished in steps until reaching 1 μm roughness. The macro and micro-structure images were taken at the most affected areas with a good distance from the edges to avoid artifacts from surface preparation. For the EDX analysis, the cross-section might contain C and O contamination from sample preparation and handling as well as from the vacuum chamber of the SEM.

3. Results and Discussion

3.1. Material Evaluation

Visual appearance is the first indicator of material performance after exposure and the photocopied samples can be seen in Figure 5. Where the coupons were fastened to the rack, marked with a red circle, there was an area which was masked, sometimes partially, from the geothermal environment. Utilising masked areas as a reference in corrosion exposure has been reported [58]. The difference in surface appearance between the exposed areas compared to the masked areas was evident for 316L, 254SMO, and Ti gr. 2 tested at Reykjanes. It is important to note, due to the testing set-up, that only the area, marked with a blue box, of Ti gr. 2 was exposed. Inconel 625 did not show visible discoloration after being exposed at Reykjanes. The discoloration formed at the surface of the 316L, 254SMO, and Ti samples tested at Reykjanes was well adhering. The samples exposed at Chaunoy had an indication of fluid deposits where layers had loosely formed on the surfaces of the samples with low adherence.

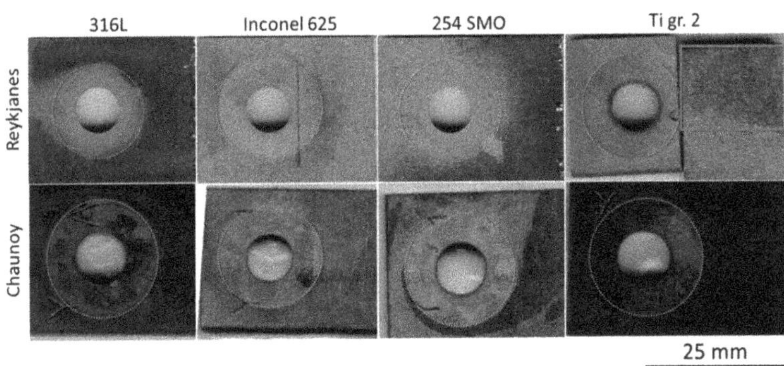

Figure 5. Coupons after exposure, the sample size is 25 mm in the vertical direction. The red circles indicate the covered areas, unexposed, due to ceramic shielding of the fastener. The blue square on the Ti gr. 2 sample exposed in Reykjanes marks the main affected area of the sample. The samples exposed in Reykjanes showed discoloration on all samples except the Inconel 625. The samples exposed in Chaunoy showed an indication of deposits from the fluid for all samples.

The macroscopic topography evaluation of the tested samples gave an indication of the general surface texture of the samples which can be seen in Figure 6, where high roughness is evident for 254SMO and Inconel 625. There was a surface texture difference within 316L exposed at different locations which might partly be due to a different provider.

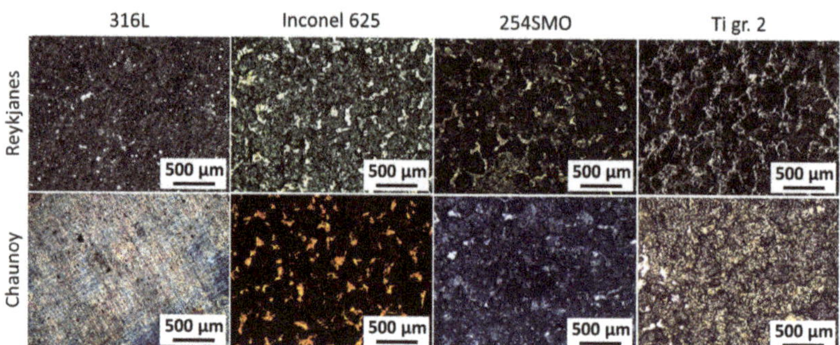

Figure 6. Optical topography images where the above line has the samples tested in Reykjanes and below are the samples tested in Chaunoy.

Topography images taken with the BSE detector are shown in Figure 7, where the 316L had microscopic defects at the surface both when tested in Reykjanes and Chaunoy. Inconel 625 and 254SMO had high surface roughness. Furthermore, there was a pit present at the surface of 254SMO exposed at Reykjanes. The topography microstructure indicated erosion in the Ti gr. 2 both tested at Reykjanes and Chaunoy where valleys are present. Furthermore, there were Fe-S rich deposits at the surface. Similar deposits have been reported for Ti in supercritical geothermal environments [59]. The microstructure analyses of the surfaces demonstrated point defects on 316L, micro-roughness on Inconel 625, defect on 254SMO, and Ti gr. 2 showed an indication of erosion.

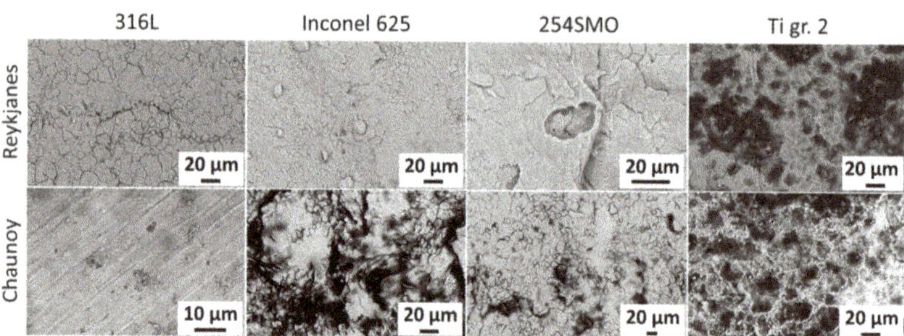

Figure 7. Topography images taken with Backscatter Electron (BSE) detector of the surfaces of the exposed coupons.

Similar to the topography images, seen in Figure 7, the microscopic cross-sectional images, seen in Figure 8, contained micro roughness for 254SMO and Inconel 625. The 316L tested at Chaunoy had a smooth surface with a defect of around 5 μm into the subsurface containing corrosion deposits. The Reykjanes-exposed 316L sample had unhomogenized corrosion layers covering the surface. The 254SMO had subsurface cracks after exposure in Reykjanes and Chaunoy while Inconel 625 had subsurface cracks only for samples tested in Chaunoy. Both 254SMO and Inconel 625 contained high amounts of Ni and Mo. Furthermore, there was the presence of inclusions within both materials contributing to weak areas.

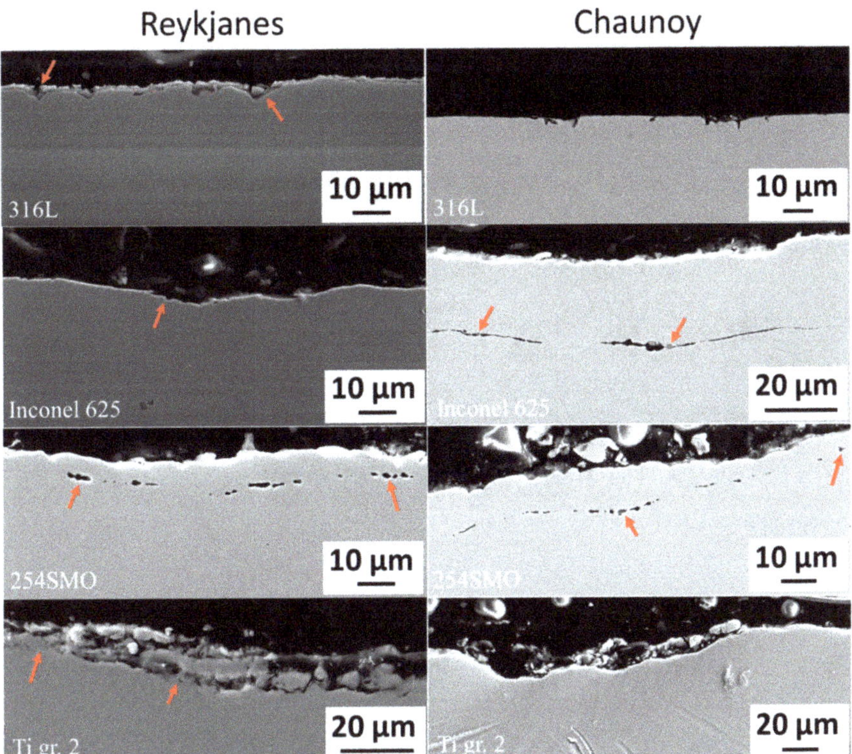

Figure 8. Scanning Electron Microscope (SEM) cross-section images taken of the most affected areas detected after exposure in both locations. The arrows indicate areas of interest: corrosion layer for 316L (Reykjanes), subsurface cracks for Inconel 625 (Chaunoy) and 254SMO (Chaunoy and Reykjanes), and abrasion of Ti gr. 2 surfaces (Chaunoy and Reykjanes).

3.2. Material Characterisation

The results from the microstructural analysis are detailed in the section below. The chemical composition of the bulk material for reference is included in the material characterisation.

3.2.1. Austenitic Stainless Steel—316L

The 316L is one of the most commonly used materials in the geothermal environment [29]. Figure 9 shows the cross-section of 316L where two corrosion layers are present along the profile. The estimated pH level of the environment was 5 at Reykjanes, which is an unstable state for Cr. The inner corrosion layer marked 1, 2, and 5 in Figure 9 consisted of high Cr/Fe and Mo/Fe ratios of 0.51–0.76 and 0.05–0.07 respectively. The measured reference ratio of Cr/Fe and Mo/Fe for bulk material was 0.26 and 0.03, respectively. The outer corrosion layer marked 3, and 4 consisted mostly of Fe and O products. The elemental analysis of the locations in the figure is detailed in Table 5. This finding is in line with 316L tested in an H_2S environment where the inner layer is Cr and Mo rich and the outer layer has Fe segregation measured at pH 4 [60].

Unlike after exposure in Reykjanes, the 316L after exposure in Chaunoy, which had a pH level of 7, did not show the same corrosion layer formation. However, there were local defects scattered over the surface as seen in the topography image in Figure 10A. Furthermore, defects were seen in the cross-section images, Figure 10B. The elemental analysis of locations shown in Figure 10A is detailed in Table 6. There were traces of S and Si from the exposure found within corrosion deposits at the defects. The deposits mainly consisted of Fe, O, and Cr. Traces of Al were found, which can be a typical inclusion material in stainless steels [61].

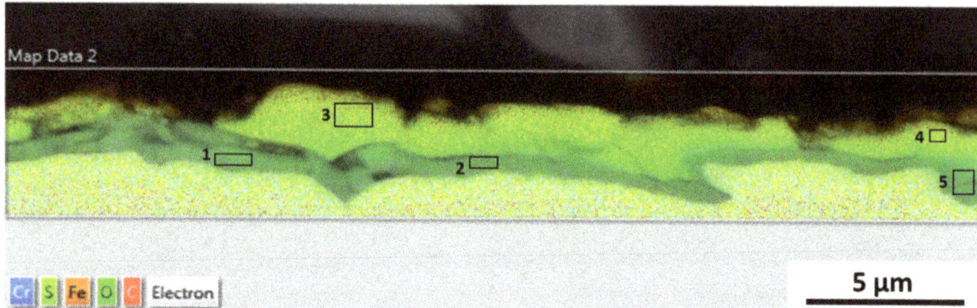

Figure 9. Elemental analysis and mapping of 316L after being exposed in the pressure vessel at Reykjanes at 200 °C, showing locations where Energy-Dispersive X-ray spectroscopy (EDX) elemental analysis was performed. There are two semi-homogenous layers over all the cross-sections analysed. The layer in locations 1, 2 and 5 is an inner corrosion layer while the top layer, locations 3 and 4, is an outer corrosion layer mostly composed of Fe and O. The colour map shows the relative composition of Cr, S, Fe, O and C in the material.

Table 5. EDX elemental analysis of locations shown in Figure 9. For comparison, an analysis of the bulk material after Reykjanes exposure is included.

	Location					
Element (wt. %)	1	2	3	4	5	Bulk
C	20.5	8.8	27.2	27.6	6.8	5.0
O	22.6	19.0	26.6	23.9	19.5	-
Si	0.7	0.7	0.5	0.3	0.8	0.60
S	2.3	3.5	0.6	2.1	3.9	-
Cr	18.2	18.7	1.4	3.2	20.9	16.8
Fe	23.8	37.0	43.2	38.5	34.8	65.8
Ni	6.1	8.2	-	2.6	8.3	9.6
Mo	1.7	1.8	-	0.9	2.2	2.1
Ba	2.5	2.2	-	0.4	2.5	-

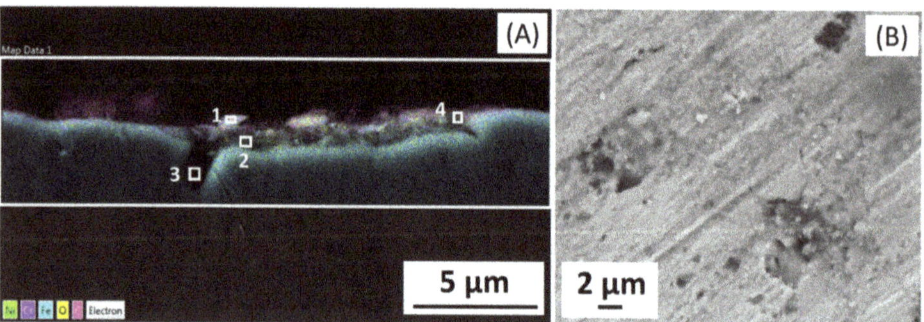

Figure 10. EDX elemental analysis of 316L after exposure in Chaunoy. (**A**): Cross-sectional elemental mapping of a surface defect, where locations marked for EDX elemental analysis are shown. Location 3 is inside the small surface defect, while locations 1, 2 and 4 are corrosion deposits around the defect. The colour map shows the relative elemental composition of Ni, Cr, Fe, O and C in the image. (**B**): Topography image of the surface defect by SEM.

Table 6. EDX elemental analysis of locations shown in Figure 10A.

Element (wt. %)	Location				
	1	**2**	**3**	**4**	**Bulk**
C	69.2	39.5	6.0	49.8	4.7
O	11.1	13.0	-	16.8	-
Al	0.7	1.0	-	0.9	-
Si	1.1	1.7	-	1.3	0.4
S	0.8	2.0	-	1.9	-
Ca	1.6	0.8	0.2	0.8	-
Cr	2.9	7.3	16.2	5.7	16.8
Fe	10.4	29.6	65.6	18.7	66.7
Ni	1.4	3.9	9.7	3.3	9.9
Mo	-	-	1.5	-	1.6

3.2.2. Nickel Alloy—Inconel 625

Inconel 625 is a Ni-alloy used within the geothermal sector and is recommended for sulphur rich environments [62]. Figure 11 shows locations of the chemical and microstructural analysis of Inconel 625 after exposure in Reykjanes. Table 7 details the chemical analysis of the locations analysed in the figure. A horizontal crack was detected within the material after exposure. The crack contained O and a high ratio of Cr/Ni was measured at 1.4–1.8 while the base material had a ratio of 0.3. Furthermore, the Mo/Ni ratio was slightly higher in the crack. The results indicate that when the crack is formed, Cr and Mo will diffuse into the crack and together with oxygen seal it. A trace of S was also found within the crack. The surface seemed to be unaffected by the exposure which is reflected in Figure 5 where the exposed sample contains a similar appearance to the masked area.

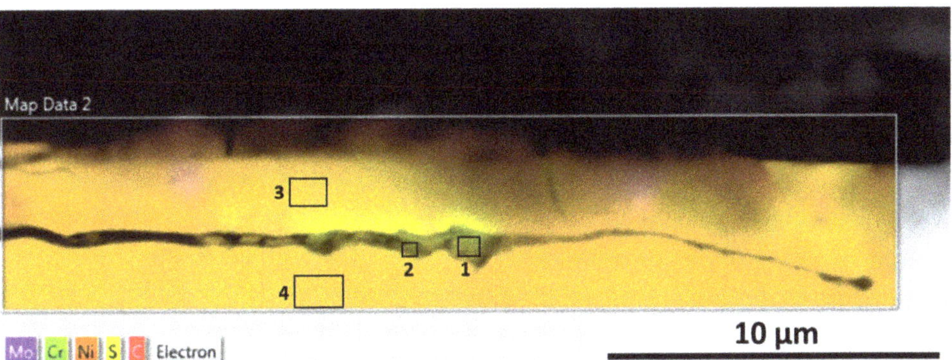

Figure 11. Elemental mapping and EDX analysis of Inconel 625 exposed in Reykjanes, which showed crack propagation that might have initiated from inclusions as a high amount of Nb is present in location 1. Nb carbides are commonly found inclusions within Inconel 625 [63]. The measured area within the crack shows traces of S in locations 1 and 2. The colour map shows the relative composition of Mo, Cr, Ni, S and C in the material.

The Inconel 625 exposed in Chaunoy contained subsurface cracks parallel to the surface which crossed the inclusions as seen in Figure 12. Due to geometric discontinuity, the inclusions often act as weak points in materials [34] and can be prone to cracking and crack propagation.

Table 7. EDX elemental analysis of locations shown in Figure 11. For comparison, an analysis of the bulk material after exposure at Reykjanes is included.

Element (wt. %)	Location				
	1	**2**	**3**	**4**	**Bulk**
C	9.1	7.0	12.8	8.7	8.6
F	-	-	4.0	3.3	-
O	20.7	15.2	-	-	-
S	2.3	4.4	-	-	-
Cr	36.6	38.0	16.4	16.1	16.4
Fe	2.3	2.2	3.7	3.9	4.1
Ni	20.1	26.6	53.3	58.6	61.3
Nb	5.6	2.5	2.5	2.2	-
Mo	2.4	3.2	6.9	6.9	7.1

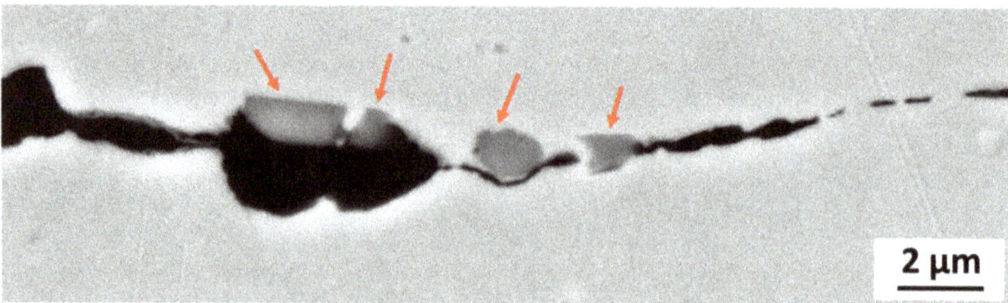

Figure 12. SEM cross-sectional image of the Inconel 625 after exposure in Chaunoy where inclusions are seen within crack propagation, marked with arrows.

Microstructural analysis of another crack propagation within the Chaunoy-exposed Inconel 625 can be seen in Figure 13. The Cr/Ni and Mo/Ni ratios were 0.26–0.3 and 0.1–0.12, respectively, in bulk material and within cracks where the inclusions were present as seen in Table 8. Similar ratio was found within the bulk material of the Reykjanes-exposed samples. The pH level of the environment was 7 which indicates a passive state of Cr and Mo, resulting in poor mobility for the elements to diffuse into the crack and seal it.

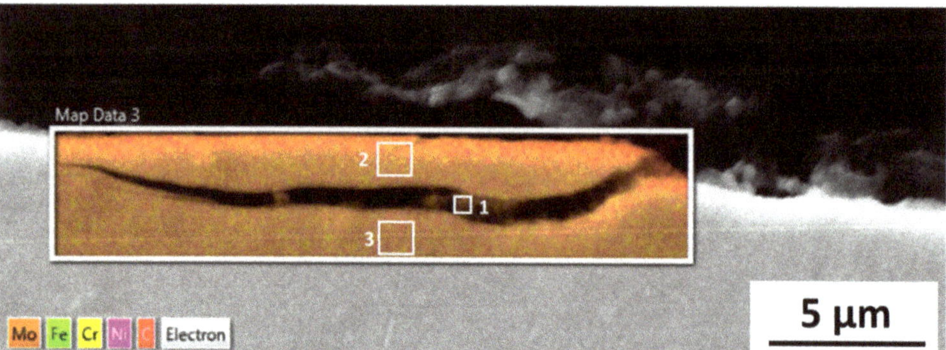

Figure 13. Microstructural analysis of Chaunoy-exposed Inconel 625 taken with SEM, where inclusions are seen within crack propagation in location 1. The colour map shows the relative composition of Mo, Fe, Cr, Ni and C in the material.

Table 8. EDX elemental analysis of locations shown in Figure 13. For comparison, an analysis of the bulk material after exposure at Chaunoy is included.

	Location			
Element (wt. %)	1	2	3	Bulk
C	13.4	28.8	8.6	7.5
O	3.9	2.6	0.8	-
Cr	14.6	12.4	16.3	18.2
Fe	3.8	3.0	4.1	4.0
Ni	55.8	46.0	60.1	60.5
Nb	1.5	1.3	2.1	2.2
Mo	5.6	4.7	7.0	7.4

The crack propagation from the weak inclusion areas could be due to the brittleness of the alloy when exposed to rapid temperature changes. From Figure 5, and the microscopic analysis, there is little to no evidence of corrosion taking place at the surface.

3.2.3. Super Austenitic Stainless Steel 254SMO

The 254SMO is a super austenitic stainless steel, which contains high amounts of alloying elements Cr, Ni, and Mo. Furthermore, 254SMO has shown good performance in acidic [33] and sea water environments [21]. The cross-section of 254SMO and topography after exposure in Reykjanes can be seen in Figure 14. The topography image shows the typical surface texture found over the surface of the sample after being exposed at Reykjanes. Furthermore, contamination was present at the surface. The cross-section shows voids at the subsurface in close proximity to grooves at the external surface. These voids contained O and S as can be seen from Table 9.

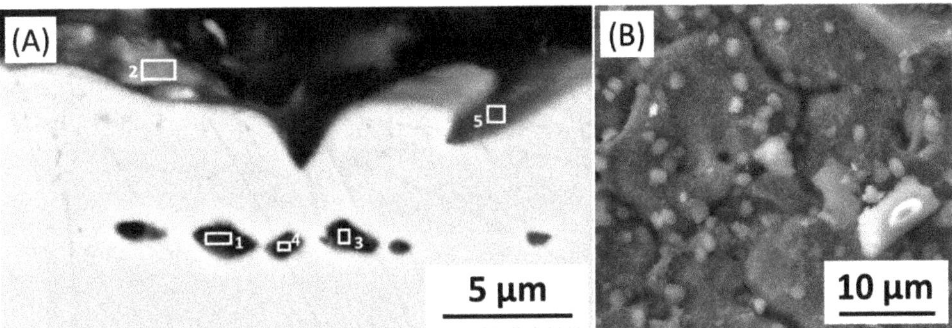

Figure 14. The 254SMO after exposure in Reykjanes taken with BSE detector (**A**): Cross-sectional SEM of the surface of the sample showing locations elementally analysed. (**B**): Topography SEM image of the 254SMO after exposure at Reykjanes.

A pit was found on the surface of the material after testing in Reykjanes as seen in Figure 15A. Due to the naturally formed passivation layer on the 254SMO surface, pits grow under the surface layer [64]. The cross-section, in Figure 15B, shows a crack propagation at the subsurface which led to the breaking of the material causing direct exposure of the metal which could result in pit formation as seen in Figure 15A.

Table 9. EDX elemental analysis of locations shown in Figure 14. For comparison, an analysis of the bulk material after exposure at Reykjanes is included.

	Location					
Element (wt. %)	1	2	3	4	5	Bulk
C	13.3	49.1	36.9	32.1	46.0	6.1
O	5.7	-	4.4	3.5	10.1	-
Al	-	3.6	-	-	-	-
S	1.1	0.8	0.6	-	0.9	-
Ca	4.6	0.6	3.3	2.5	0.7	-
Cr	18.1	9.0	12.7	13.5	7.9	19.5
Fe	44.5	28.8	31.6	35.1	23.6	50.4
Ni	12.0	7.5	8.5	9.8	7.5	17.0
Cu	-	0.6	-	-	-	0.8
Mo	-	-	1.5	2.3	1.4	5.6

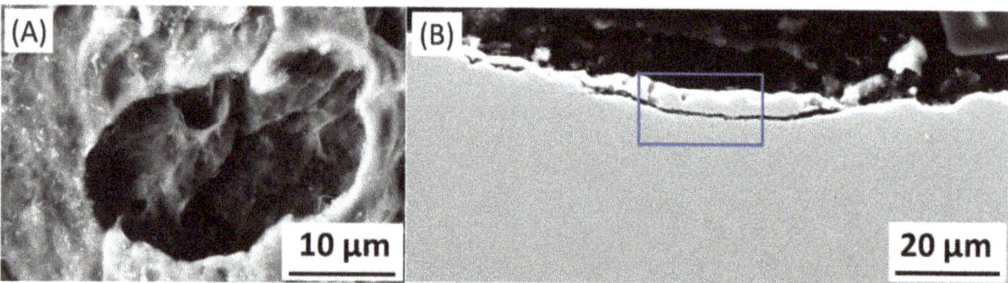

Figure 15. 254SMO microstructural analysis using secondary detector (SEM) from Reykjanes exposed sample (**A**): Pit found at the surfaces of the 254SMO. (**B**): Cross-section where a sub-crack is present. The area marked in blue is further analysed in Figure 16.

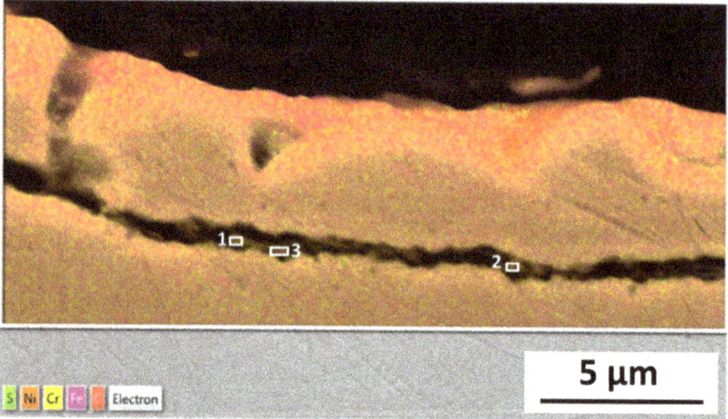

Figure 16. Elemental mapping of within the subsurface crack propagation of 254SMO after Reykjanes exposure, showing the locations of EDX elemental analysis inside the crack. The colour map details the relative elemental composition of S, Ni, Cr, Fe and C in the material.

Higher magnification of the crack in the area marked with blue in Figure 15B, is shown in Figure 16 and an elemental analysis of locations in the crack is presented in Table 10. Within the crack was a high ratio of Cr/Fe, or 0.6–0.8, compared with the bulk material and voids at the subsurface with around 0.4 Cr/Fe ratio. Like Inconel 625, the Cr could have diffused from the bulk material forming the thin layer where the pH level was 5. There was a high wt.% ratio of O/Fe within the crack of 0.4–0.6. The Cr rich layer in the crack seemed to be brittle, and part of the material was breaking off at places. The 254SMO might contain weak areas due to inclusion and increased brittle areas due to high wt.% of Mo [31]. The practical environment could promote crack propagation from weak areas resulting in material thinning.

Table 10. EDX elemental analysis of locations shown in Figure 16. For comparison, an analysis of the bulk material after exposure at Reykjanes is included.

	Location			
Element (wt. %)	1	2	3	Bulk
C	13.0	10.0	12.1	6.2
O	17.5	14.1	15.1	-
Al	0.9	1.1	1.2	-
S	0.9	1.1	1.2	-
Cr	23.4	21.4	22.5	19.5
Mn	2.7	3.1	2.6	-
Fe	29.8	35.2	33.6	50.4
Ni	8.5	10.7	8.3	17.0
Cu	-	-	0.6	0.8
Mo	3.0	4.1	3.2	5.6

The sample exposed in Chaunoy, seen in Figure 17, had a crack forming parallel to the surface of 254SMO. Between the detachment layer and the 254SMO there were both inclusion elements such as Al and Si as well as elements from the fluid such as Cl, Ca, Na, and S as seen in Table 11.

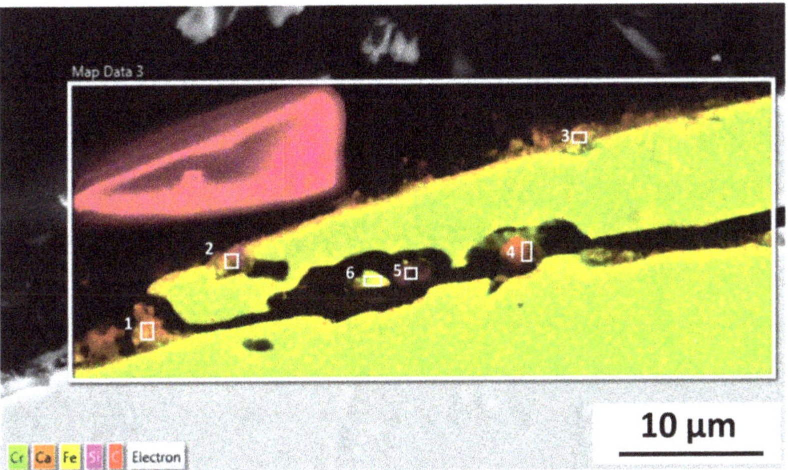

Figure 17. Element mapping and analysis of 254SMO after being exposed in Chaunoy. Within the crack there is visible inclusion of Al and Si, number 5. Furthermore, there is corrosion deposit in locations number 1, 2, and 3, containing S and elements from the base material, Cr, Fe, and Ni. The colour map shows the relative chemical composition of Cr, Ca, Fe, Si and C in the material.

3.2.4. Titanium Grade 2

Ti gr. 2 is commercially pure, and is often preferred where high specific strength is needed [65]. The Ti gr. 2 naturally forms a TiO_2 passivation layer of a few nm which protects the bulk material. However, due to the low thickness of the TiO_2 passivation layer it can easily erode during exposure from the high velocity fluid which can contain abrasive particles. Figure 18 shows the sample after exposure where the affected area is evident. There were parallel subsurface cracks where the metal was adhering in a few contact points.

The EDX revealed Cu deposits, which can be found at Reykjanes [66]. The cross-section clearly supports the observation of the topography, from Figure 7, where erosion has taken place at the surface as can be seen in Figure 18. The EDX analysis performed in between the loosely adhering Ti contained high wt.% of O and C, presented in Table 12. Due to the position of the point evaluation, it cannot be excluded that phenolic epoxy, the sample mounting material, contributed to the signal from C and O. The Ti gr. 2 has been reported to be successfully used in an oxygen rich environment [67–69]. However, in geothermal environments, where free O is limited, a Ti alloy might be better suitable.

Table 11. EDX elemental analysis of locations shown in Figure 17. For comparison, an analysis of the bulk material after exposure at Chaunoy is included.

	Location						
Element (wt. %)	1	2	3	4	5	6	Bulk
C	54.0	35.3	59.0	81.6	29.6	35.9	6.8
O	10.1	15.5	5.7	3.0	28.2	3.9	-
Na	2.5	1.6	0.8	3.2	5.0	3.7	-
Al	0.3	1.6	0.1	-	3.8	0.2	-
Si	0.6	3.5	0.2	0.1	10.5	0.5	0.5
P	0.5	0.2	0.6	-	-	0.1	-
S	6.6	7.8	6.3	0.4	1.5	0.8	-
Cl	2.7	0.5	0.7	2.5	1.4	3.1	-
Ca	1.9	1.1	1.8	0.3	1.0	0.6	-
Cr	2.2	3.0	2.8	2.0	4.2	10.3	19.4
Fe	15.9	25.0	18.4	5.5	12.4	29.1	50.1
Ni	1.5	2.6	2.3	1.2	2.3	9.1	16.7
Cu	0.2	-	-	-	-	-	0.7
Mo	-	0.9	-	-	-	2.2	5.8
Ti	-	-	0.7	-	-	-	-

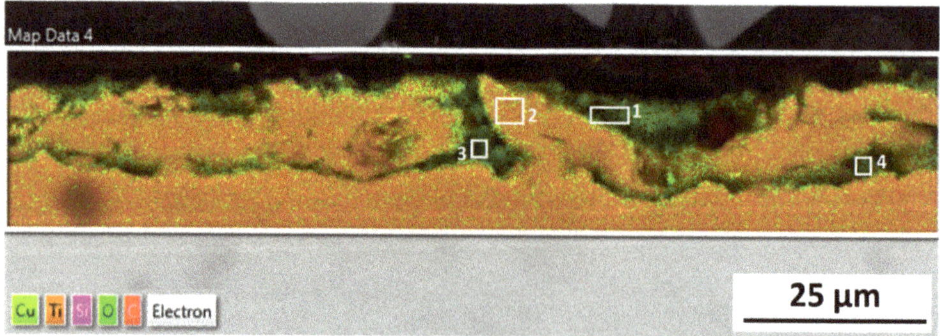

Figure 18. Ti gr. 2 after exposure in Reykjanes where around 10 µm of material is loosely attached to the base material. Four locations analysed with EDX are shown. Locations 1,3 and 4 show high amounts of Cu deposits while location 2 is nearly pure Ti. The profile shows an indication of erosion. The colour map shows the relative composition of Cu, Ti, Si, O and C in the material.

The Ti gr. 2 tested in Chaunoy, seen in Figure 6, showed an indication of erosion at the surface which is further supported with topography images. Figure 19 shows the affected area where Ti was eroding and a 10 µm metal part was loosely adhering to the surface along with S-Fe rich corrosion products. Table 13 details the elemental analysis of locations shown in the figure.

Table 12. EDX elemental analysis of locations shown in Figure 18. For comparison, an analysis of the bulk material after exposure at Reykjanes is included.

Element (wt. %)	Location				
	1	**2**	**3**	**4**	**Bulk**
C	49.1	4.0	45.6	39.3	2.5
O	21.4	-	19.8	15.3	2.5
Ca	0.3	-	1.0	0.7	-
Ti	5.1	95.0	7.0	24.7	94.9
Fe	1.2	0.2	0.7	0.9	0.1
Cu	21.2	0.3	24.8	18.5	-

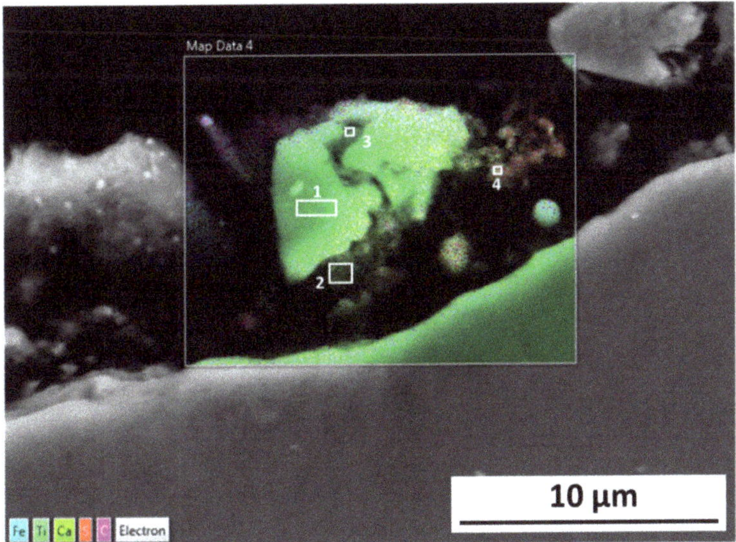

Figure 19. Ti gr. 2 after exposure at Chaunoy where there is visible detachment of Ti subsurface. The detached material is seen in location number 1. Furthermore, locations 2 and 4 contain Ti with high amounts of Fe and S. The colour map shows the relative composition of Fe, Ti, Ca, S and C in the material.

Table 13. EDX elemental analysis of locations shown in Figure 19. For comparison, an analysis of the bulk material after exposure at Chaunoy is included.

Element (wt. %)	Location				
	1	**2**	**3**	**4**	**Bulk**
C	4.3	21.3	28.4	38.2	2.2
O	6.8	11.6	10.5	8.3	3.9
Si	-	0.7	0.5	0.3	0.1
S	0.2	11.4	1.5	16.3	-
Ca	0.4	1.2	0.9	1.7	0.1
Ti	87.9	34.7	52.6	5.9	93.6
Cr	-	-	4.6	0.5	-
Fe	0.4	18.7	-	28.5	-

4. Conclusions

The most accurate method for evaluating the material performance for a geothermal application is with direct exposure of the material in the geothermal environment. Such experiments give an important indication of the viability of engineering materials in the particular geothermal setting in order to ensure the sustainability of geothermal projects. The material exposure is dependent on fluid parameters such as the ratio between steam and liquid form, corrosive ions and gases, temperature, and pressure. Furthermore, the abrasiveness of particles found within the fluid coming from the underground or high fluid velocity can also affect the material.

While the two geothermal locations differ widely in geological setting, they have some similarities when it comes to geochemistry. Both geothermal fluids, at the wellhead, have high salinity and calcium content. However, the fluid in Reykjanes is quite a bit more acidic when compared to the Chaunoy fluid, and while Reykjanes holds a high amount of SiO_2, the fluid at Chaunoy is a mixture of geothermal brine and oil. The experiment at Reykjanes, a volcanic geothermal area, exposed the samples to a separated vapour state 200 °C and 18 bar geothermal fluid, resulting in a visually homogeneous appearance after exposure. The Chaunoy experiment, performed in a sedimentary basin, on the other hand, exposed the materials to a 94 °C and 9.5 bar liquid state geothermal fluid.

The duration of the exposure was 1 and 4 months for Reykjanes and Chaunoy, respectively. The Reykjanes geothermal exposure conditions are substantially more severe, resulting in similar or more affected samples compared to Chaunoy.

The microstructure analysis of 316L at Reykjanes showed the formation of two inhomogeneous layers where the inner layer is Cr and Mo rich, and the outer layer contains Fe segregation. The 316L tested in Chaunoy showed local corrosion of around 4 μm in diameter and 2.5 μm in depth. The 254SMO tested in both locations showed subsurface elongated cracks of approximately 100 μm and longer.

The cracks in the 254SMO tested at Reykjanes showed a Cr rich layer forming within them as well as areas where material is breaking off. Corrosion was not observed on the surface of the Inconel 625 after Reykjanes testing. After both exposures, the material showed subsurface cracking. The Ti gr. 2 samples tested both in Reykjanes and Chaunoy showed an indication of erosion. The results are both supported by topography and cross-section images.

Even though all tested materials showed defects after exposures at both locations, no defect was observed to reach deeper than 50 μm into the coupon samples. Further testing is required to accurately predict the long-term sustainability of the tested materials for usage in heat exchanger plates at the tested locations and to investigate how the different defects evolve over time.

Author Contributions: Conceptualisation, S.D., B.G.G., K.B.K., B.A.L. and D.I.Ó.; Formal analysis, S.D.; Investigation, S.D.; Methodology, S.D. and D.I.Ó.; Project administration, D.I.Ó.; Supervision, D.I.Ó.; Writing—original draft, S.D.; Writing—geological context, K.B.K., B.G.G., D.I.Ó. and B.A.L.; Writing—review and editing, S.D., B.G.G., K.B.K., B.A.L. and D.I.Ó. All authors have read and agreed to the published version of the manuscript.

Funding: This research was funded by the European Union's Horizon 2020 research and innovation program and was part of the H2020 project MEET: Multidisciplinary and Multi-context Demonstration of EGS Exploration and Exploitation Techniques and Potentials. Call H2020-LCE-2017-RES-IA, grant number 792037.

Data Availability Statement: Not applicable.

Acknowledgments: We would like to thank Geir Þórólfsson in assisting with testing at Reykjanes as well as André-Charles Mintsa do Ango at Enogia and Eric Leoutre at Vermilion in assisting with testing at Chaunoy. We would also like to thank Helen Ósk Haraldsdóttir and Kolbrún Ragna Ragnarsdóttir for their contribution to the MEET project.

Conflicts of Interest: The authors declare no conflict of interest.

References

1. Sundén, B.; Manglik, R.M. *Plate Heat Exchangers: Design, Applications and Performance*, 11th ed.; Wit Press: Southampton, UK, 2007.

2. Nogara, J.; Zarrouk, S.J. Corrosion in geothermal environment: Part 1: Fluids and their impact. *Renew. Sustain. Energy Rev.* **2018**, *82*, 1333–1346. [CrossRef]

3. Wang, Z.; Paschalidou, E.M.; Seyeux, A.; Zanna, S.; Maurice, V.; Marcus, P. Mechanisms of Cr and Mo Enrichments in the Passive Oxide Film on 316L Austenitic Stainless Steel. *Front. Mater.* **2019**, *6*, 232. [CrossRef]

4. Cabrini, M.; Lorenzi, S.; Pastore, T.; Favilla, M.; Perini, R.; Tarquini, B. Materials selection for dew-point corrosion in geothermal fluids containing acid chloride. *Geothermics* **2017**, *69*, 139–144. [CrossRef]

5. Rafferty, K. Chapter 11: Heat Exchangers. In *Geothermal Direct Use: Engineering and Design Guidebook*; Oregon Institute of Technology: Klamath Falls, OR, USA, 1998; pp. 1–32.

6. Csaki, I.; Ragnasdottir, K.R.; Buzaianu, A.; Leosson, K.; Motoiu, V.; Guðlaugsson, S.; Lungu, M.V.; Haraldsdottir, H.O.; Karlsdottir, S.N. Nickel based coatings used for erosion-corrosion protection in a geothermal environment. *Surf. Coat. Technol.* **2018**, *350*, 531–541. [CrossRef]

7. Addepalli, S.; Eiroa, D.; Lieotrakool, S.; François, A.L.; Guisset, J.; Sanjaime, D.; Kazarian, M.; Duda, J.; Roy, R.; Phillips, P. Degradation Study of Heat Exchangers. *Procedia CIRP* **2015**, *38*, 137–142. [CrossRef]

8. Hattori, S.; Kitagawa, T. Analysis of cavitation erosion resistance of cast iron and nonferrous metals based on database and comparison with carbon steel data. *Wear* **2010**, *269*, 443–448. [CrossRef]

9. Mundhenk, N.; Scheiber, J.; Zorn, R.; Huttenloch, P.; Genter, A.; Kohl, T. Corrosion and Scaling in the Geothermal Cycle of Soultz-sous-Forêts (France). In Proceedings of the Corrosion 2014, San Antonio, TX, USA, 9–13 March 2014.

10. Ravier, G.; Seibel, O.; Pratiwi, A.S.; Mouchot, J.; Genter, A.; Ragnarsdottir, K.R.; Sengelen, X. Towards an optimized operation of the EGS Soultz-sous-Forêts power plant (Upper Rhine Graben, France). In Proceedings of the European Geothermal Congress 2019, Den Haag, The Netherlands, 11–14 June 2014.

11. Francesca Baticci, R.Z.; Genter, A.; Huttenloch, P. Corrosion and Scaling Detection in the Soultz EGS Power Plant, Upper Rhine Graben, France. In Proceedings of the World Geothermal Congress, Bali, Indonesia, 25–29 April 2010.

12. Sanjuan, B.; Millot, R.; Dezayes, C.; Brach, M. Main characteristics of the deep geothermal brine (5 km) at Soultz-sous-Forêts (France) determined using geochemical and tracer test data. *C. R. Geosci.* **2010**, *342*, 546–559. [CrossRef]

13. Ledésert, B.A.; Hébert, R.L.; Mouchot, J.; Bosia, C.; Ravier, G.; Seibel, O.; Dalmais, E.; Ledésert, M.; Trullenque, G.; Sengelen, X. Scaling in a Geothermal Heat Exchanger at Soultz-Sous-Forêts (Upper Rhine Graben, France): A XRD and SEM-EDS Characterization of Sulfide Precipitates. *Geosciences* **2021**, *11*, 271. [CrossRef]

14. Karlsdottir, S.N.; Ragnarsdottir, K.R.; Thorbjornsson, I.O.; Einarsson, A. Corrosion testing in superheated geothermal steam in Iceland. *Geothermics* **2015**, *53*, 281–290. [CrossRef]

15. Sarrade, S.; Féron, D.; Rouillard, F.; Perrin, S.; Robin, R.; Ruiz, J.C.; Turc, H.A. Overview on corrosion in supercritical fluids. *J. Supercrit. Fluids* **2017**, *120*, 335–344. [CrossRef]

16. Batis, G.; Kouloumbi, N.; Kotsakou, K. Corrosion and protection of carbon steel in low enthalpy geothermal fluids. The case of Sousaki in Greece. *Geothermics* **1997**, *26*, 65–82. [CrossRef]

17. Ungemach, P. Handling of corrosion and scaling shortcommings in low enthalpy geothermal environments. In Proceedings of the European Summer School on Geothermal Energy Applications, Oradea, Romania, 26 April–5 May 2001; pp. 113–127.

18. Stănășel, I.; Stănășel, O.; Gilău, L.; Sebeșan, M.; Gavriș, G. Control of Corrosion and Scaling in Selected Geothermal Wells from Romania. In Proceedings of the Proceedings World Geothermal Congress 2010, Bali, Indonesia, 25–29 April 2010.

19. ASTM G4-01. *Standard Guide for Conducting Corrosion Tests in Field Applications*; ASTM International: West Conshohocken, PA, USA, 2001; Volume 100.

20. Qurashi, M.S.; Cui, Y.; Wang, J.; Dong, N.; Bai, J.; Han, P. Erosion and passivation of borated 254 SMO stainless steel in simulated flue gas desulfurization solution. *Int. J. Electrochem. Sci.* **2020**, *15*, 2987–3002. [CrossRef]

21. Geary, E.A. *A Review of Performance Limits of Stainless Steels for the Offshore Industry*; Report 44; Health and Safety Laboratory: Derbyshire, UK, 2011.

22. Kritzer, P. Corrosion in high-temperature and supercritical water and aqueous solutions: A review. *J. Supercrit. Fluids* **2004**, *29*, 1–29. [CrossRef]

23. Neville, A.; Mcdougall, B.A.B. Erosion– and cavitation–corrosion of titanium and its alloys. *Wear* **2001**, *250*, 726–735. [CrossRef]

24. Khayatan, N.; Ghasemi, H.M.; Abedini, M. Study of Erosion–Corrosion and Corrosion Behavior of Commercially Pure-Ti During Slurry Erosion. *J. Tribol.* **2018**, *140*, 061609. [CrossRef]

25. Martelo, D.; Sampath, D.; Monici, A.; Morana, R.; Akid, R. Correlative analysis of digital imaging, acoustic emission, and fracture surface topography on hydrogen assisted cracking in Ni-alloy 625+. *Eng. Fract. Mech.* **2019**, *221*, 106678. [CrossRef]

26. Thorbjornsson, I.O.; Stefansson, A.; Kovalov, D.; Karlsdottir, S.N. Corrosion testing of materials in simulated superheated geothermal environment. *Corros. Sci.* **2020**, *168*, 108584. [CrossRef]

27. Karlsdottir, S. *Corrosion, Scaling and Material Selection in Geothermal Power Production*; Elsevier: Amsterdam, The Netherlands, 2012; pp. 241–259.

28. Nogara, J.; Zarrouk, S.J.; Seastres, J. Surface analysis of metal alloys exposed to geothermal fluids with high non-condensable gas content. *Geothermics* **2018**, *72*, 372–399. [CrossRef]

29. Nogara, J.; Zarrouk, S.J. Corrosion in geothermal environment Part 2: Metals and alloys. *Renew. Sustain. Energy Rev.* **2018**, *82*, 1347–1363. [CrossRef]

30. Sanada, N.; Lichtf, K.A. Prospects for the evaluation and development of materials under IEA research collaboration program on deep geothermal resources. In Proceedings of the NEDO International Geothermal Symposium, Sendai, Japan, 11–12 March 1997; pp. 192–199.

31. Wang, Q.; Wang, L.; Zhang, W.; Li, J.; Chou, K. Effect of Cerium on the Austenitic Nucleation and Growth of High-Mo Austenitic Stainless Steel. *Metall. Mater. Trans. B Process Metall. Mater. Process. Sci.* **2020**, *51*, 1773–1783. [CrossRef]

32. Van Der Eijk, C.; Walmsley, J.; Klevan, O.S.; Refinement, G.; Metals, R.E. Grain refinement of fully austenitic stainless STEELS using a Fe-Cr-Si-Ce master alloy. In Proceedings of the 59th Electric Furnace Conference, Phoenix, AZ, USA, 11–14 November 2001.

33. Al-Subai, S.G. Corrosion Resistance of Austenitic Stainless Steel in Acetic Acid Solution Containing Bromide Ions. Ph.D. Thesis, University of Manchester, Manchester, UK, 2011.

34. Giordani, E.J.; Guimarães, V.A.; Pinto, T.B.; Ferreira, I. Effect of precipitates on the corrosion-fatigue crack initiation of ISO 5832-9 stainless steel biomaterial. *Int. J. Fatigue* **2004**, *26*, 1129–1136. [CrossRef]

35. Wranglen, G. Pitting and sulphide inclusions in steel. *Corros. Sci.* **1974**, *14*, 331–349. [CrossRef]

36. Moorbath, S.; Sigurdson, H.; Goodwin, R. K-Ar ages of oldest exposed rocks in Iceland. *Earth Planet. Sci. Lett.* **1968**, *4*, 197–205. [CrossRef]

37. Vink, G.E. A hotspot model for Iceland an the Vøring Plateau. *J. Geophys. Res.* **1984**, *89*, 9949–9959. [CrossRef]

38. Thordarson, T.; Larsen, G. Volcanism in Iceland in historical time: Volcano types, eruption styles and eruptive history. *J. Geodyn.* **2007**, *43*, 118–152. [CrossRef]

39. Björnsson, S.; Arnórsson, S.; Tómasson, J. Exploration of the Reykjanes thermal brine area. *Geothermics* **1970**, *2*, 1640–1650. [CrossRef]

40. Sæmundsson, K.; Jóhannesson, H.; Hjartarson, Á.; Kristinsson, S.; Sigurgeirsson, M. *Geological map of SW-Iceland, 1:100 0000*; Iceland GeoSurvey: Reykjavík, Iceland, 2010.

41. Clifton, A.E.; Schlische, R.W. Fracture populations on the Reykjanes Peninsula, Iceland: Comparison with experimental clay models of oblique rifting. *J. Geophys. Res. Space Phys.* **2003**, *108*, 2074. [CrossRef]

42. Jakobsson, S.P. Chemistry and distribution pattern of recent basaltic rocks in Iceland. *Lithos* **1972**, *5*, 365–386. [CrossRef]

43. Pálmason, G.; Sæmundsson, K. Iceland in relation to the Mid-Atlantic Ridge. *Annu. Rev. Earth Planet. Sci.* **1974**, *2*, 25–50. [CrossRef]

44. Sæmundsson, K.; Sigurgeirsson, M.Á.; Friðleifsson, G.Ó. Geology and structure of the Reykjanes volcanic system, Iceland. *J. Volcanol. Geotherm. Res.* **2020**, *391*, 106501. [CrossRef]

45. Gudnason, E.A.; Agustsson, K.; Gunnarsson, K.; Flovenz, O.G. *Seismic Activity on Reykjanes December 2014–November 2015*; Report ÍSOR-2015/068; Iceland GeoSurvey: Reykjavík, Iceland, 2015; 31p.

46. Tómasson, J.; Kristmannsdóttir, H. High temperature alteration minerals and thermal brines, Reykjanes, Iceland. *Contrib. Miner. Petrol.* **1972**, *36*, 123–134. [CrossRef]

47. Franzon, H.; Thordardson, S.; Bjornsson, G.; Gudlaugsson, S.T.; Richter, B.; Fridleifsson, G.O.; Thorhallsson, S. Reykjanes High-temperature field SW-Iceland. Geology and hydrothermal alteration of well RN-10. In Proceedings of the 27th Workshop on Geothermal Reservoir Engineering, Stanford University, Stanford, CA, USA, 28–30 January 2002; pp. 233–240.

48. Berehannu, M.M. *Geochemical Interpretation of Discharge from Reykjanes Wells Rn-29 and Rn-32, Sw-Iceland*; United Nations University: Reykjavik, Iceland, 2014.

49. Brunet, M.-F.; Le Pichon, X. Subsidence of the Paris Basin. *J. Geophys. Res. Space Phys.* **1982**, *87*, 8547–8560. [CrossRef]

50. Courel, L.; Durand, M.; Maget, P.; Maiaux, C.; Menillet, F.; Pareyn, C. Trias. In *Synthèse Géologique Du Bassin Paris, Mémoire*; Mégnien, C., Ed.; BRGM: Orleans, France, 1980; Volume 101, pp. 37–74.

51. Eschard, R.; Lemouzy, P.; Bacchiana, C.; Desaubliaux, G.; Parpant, J.; Smart, B. Combining sequence stratigraphy, geostatistical simulations, and production data for modeling a fluvial reservoir in the Chaunoy field (Triassic, France). *Am. Assoc. Pet. Geol. Bull.* **1998**, *82*, 545–568.

52. Bourquin, S.; Friedenberg, R.; Guillochea, F. Depositional sequences in the Triassic series of the Paris Basin: Geodynamic implications. *J. Iber. Geol. Int. Publ. Earth Sci.* **1995**, *19*, 337–362.

53. Bourquin, S.; Guillocheau, F. Depositional sequences geometry of Keuper age (Ladinian to Rhaetian) of the Paris Basin—Geodynamic implications. *C. R. L Acad. Sci.* **1993**, *317*, 1341–1348.

54. Eschard, R.; Desaubliaux, G.; Lemouzy, P.; Bacchiana, C.; Parpant, J.; Chautru, J. Geostatistical simulations of alluvial sandbodies in the Triassic series of the Chaunoy field, France. In Proceedings of the American Association of Petroleum Geologists (AAPG) Mid-Continent Section Meeting, Amarillo, TX, USA, 10–12 October 1993. AAPG Bulletin.

55. Bacchiana, C.; Parpant, J.; Smart, B. Management of Chaunoy Oil Field Multilayered Reservoir. In *Hydrocarbon and Petroleum Geology of France*; Mascle, A., Ed.; Springer: Berlin/Heidelberg, Germany, 1994; pp. 147–153. [CrossRef]

56. Bourquin, S.; Rigollet, C.; Bourges, P. High-resolution sequence stratigraphy of an alluvial fan–fan delta environment: Stratigraphic and geodynamic implications—An example from the Keuper Chaunoy Sandstones, Paris Basin. *Sediment. Geol.* **1998**, *121*, 207–237. [CrossRef]

57. Karlsdottir, S.N.; Csaki, I.; Antoniac, I.V.; Manea, C.A.; Stefanoiu, R.; Magnus, F.; Miculescu, F. Corrosion behavior of AlCrFeNiMn high entropy alloy in a geothermal environment. *Geothermics* **2019**, *81*, 32–38. [CrossRef]

58. Mundhenk, N.; Knauss, K.; Bandaru, S.R.S.; Wonneberger, R.; Devine, T.M. Corrosion of carbon steel and the passivating properties of corrosion films formed under high-PT geothermal conditions. *Sci. Total Environ.* **2019**, *677*, 307–314. [CrossRef] [PubMed]
59. Ragnarsdóttir, K.R. Corrosion Experiments in Dry Superheated Steam From IDDP-1. Master's Thesis, University of Iceland, Reykjavik, Iceland, 2013.
60. Wang, Z.; Feng, Z.; Zhang, L. Effect of high temperature on the corrosion behavior and passive film composition of 316 L stainless steel in high H2S-containing environments. *Corros. Sci.* **2020**, *174*, 108844. [CrossRef]
61. Da Costa E Silva, A.L.V. The effects of non-metallic inclusions on properties relevant to the performance of steel in structural and mechanical applications. *J. Mater. Res. Technol.* **2019**, *8*, 2408–2422. [CrossRef]
62. Ganesan, P.; Renteria, C.M.; Crum, J.R. Versatile Corrosion Resistance of INCONEL Alloy 625 in Various Aqueous and Chemical Processing Environments. *Superalloys* **1991**, *718*, 663–680. [CrossRef]
63. Sundararaman, M.; Mukhopadhyay, P.; Banerjee, S. Carbide Precipitation in Nickel Base Superalloys 718 and 625 and Their Effect on Mechanical Properties. *Superalloys* **1997**, *718*, 625–706. [CrossRef]
64. Abd El Meguid, E.A.; Abd El Latif, A.A. Critical pitting temperature for Type 254 SMO stainless steel in chloride solutions. *Corros. Sci.* **2007**, *49*, 263–275. [CrossRef]
65. Ji, X.; Qing, Q.; Ji, C.; Cheng, J.; Zhang, Y. Slurry Erosion Wear Resistance and Impact-Induced Phase Transformation of Titanium Alloys. *Tribol. Lett.* **2018**, *66*, 64. [CrossRef]
66. Uz-Zaman, S. *Chemical Assessment of Icelandic Geothermal Fluids for Direct Applications*; United Nations Geothermal Training Programme; United Nations University: Reykjavik, Iceland, 2013.
67. Thomas, R. Titanium in the geothermal industry. *Geothermics* **2003**, *32*, 679–687. [CrossRef]
68. De, C. Use of Titanium and its Alloys in Sea-Water Service. *High Temp. Mater. Process.* **1993**, *11*, 61–96. [CrossRef]
69. Banker, J.G. *Titanium for Secondary Marine Structures*; International Titanium Association: Northglenn, CO, USA, 1996.

Article

Analysis of Enhanced Geothermal System Development Scenarios for District Heating and Cooling of the Göttingen University Campus

Dmitry Romanov [1,2,3,*] and **Bernd Leiss** [1,2]

1 Geoscience Centre, Department of Structural Geology and Geodynamics, Georg-August-Universität Göttingen, Goldschmidtstraße 3, 37077 Göttingen, Germany; bleiss1@gwdg.de
2 Universitätsenergie Göttingen GmbH, Hospitalstraße 3, 37073 Göttingen, Germany
3 HAWK Hildesheim/Holzminden/Faculty of Resource Management, Göttingen University of Applied Sciences and Arts, Rudolf-Diesel-Straße 12, 37075 Göttingen, Germany
* Correspondence: dromano@gwdg.de or dmitry.romanov2@hawk.de

Abstract: The huge energy potential of Enhanced Geothermal Systems (EGS) makes them perspective sources of non-intermittent renewable energy for the future. This paper focuses on potential scenarios of EGS development in a locally and in regard to geothermal exploration, poorly known geological setting—the Variscan fold-and-thrust belt —for district heating and cooling of the Göttingen University campus. On average, the considered single EGS doublet might cover about 20% of the heat demand and 6% of the cooling demand of the campus. The levelized cost of heat (LCOH), net present value (NPV) and CO_2 abatement cost were evaluated with the help of a spreadsheet-based model. As a result, the majority of scenarios of the reference case are currently not profitable. Based on the analysis, EGS heat output should be at least 11 MW_{th} (with the brine flow rate being 40 l/s and wellhead temperature being 140 °C) for a potentially profitable project. These parameters can be a target for subsurface investigation, reservoir modeling and hydraulic stimulation at a later stage. However, sensitivity analysis presented some conditions that yield better results. Among the most influential parameters on the outcome are subsidies for research wells, proximity to the campus, temperature drawdown and drilling costs. If realized, the EGS project in Göttingen might save up to 18,100 t CO_2 (34%) annually.

Keywords: deep geothermal energy; EGS; Variscan fold-and-thrust belt; district heating and cooling; economic indicators; CO_2 abatement cost; sensitivity analysis

Citation: Romanov, D.; Leiss, B. Analysis of Enhanced Geothermal System Development Scenarios for District Heating and Cooling of the Göttingen University Campus. *Geosciences* **2021**, *11*, 349. https://doi.org/10.3390/geosciences11080349

Academic Editors: Béatrice A. Ledésert, Ronan L. Hébert, Ghislain Trullenque, Albert Genter, Eléonore Dalmais, Jean Hérisson and Jesús Martínez-Frías

Received: 7 June 2021
Accepted: 16 August 2021
Published: 19 August 2021

Publisher's Note: MDPI stays neutral with regard to jurisdictional claims in published maps and institutional affiliations.

1. Introduction

According to the report by the German Federal Ministry for Economic Affairs and Energy (BMWi) [1], the share of geothermal energy in renewable-based electricity generation in Germany in 2019 was just 0.1%. The analogous value for heat generation is 8.9%. While 8.2% are related to shallow geothermal energy, which is usually used for local, decentralized low temperature applications in urban areas [2], deep geothermal energy accounts for only 0.7%. At the same time, deep geothermal energy is potentially an enormous source of renewable energy of a non-intermittent nature that has low land and water requirements and significant CO_2 sequestration potential [3,4]. Other positive and negative sustainability issues of geothermal energy are reviewed in Ref. [5].

An Enhanced (or engineered) Geothermal System (EGS), which is also referred to as Hot Dry Rock (HDR) in some works, is a technology implying artificial enhancements of rock permeabilities, e.g., by the creation of new fractures in rocks or by opening and/or widening preexisting ones to extract geothermal energy from depths of 3–5 km where sufficient natural permeability is low or absent [6]. In paper [7], the authors developed a subsurface model for evaluating maximum electric output from an EGS in dependence on brine flow rate and the distance between the wells. The authors estimated that 13,450 EGS plants can be built in crystalline areas in Germany providing 474 GW_{el} (4155 TWh_{el}). At the same time, the technical potential of EGS in Europe was assessed at 6560 GW_{el} [8], which is a significant amount of renewable energy. However, the risks and uncertainties of developing EGS are still high. This is why various research groups currently focus on the development and exploitation of EGS and overcoming related geological, technical, economic, ecological and social issues and risks [9–14].

As of now, the technology is not mature enough, and there are just a few successful R&D or commercial EGS projects, e.g., [15,16]. Most of them have been realized in igneous and sedimentary rocks and have reservoir temperatures less than 165 °C and flow rates less than 40 l/s [17]. However, there are some exceptions, e.g., the geothermal heat plant in Rittershoffen producing more than 70 l/s of brine with temperature of 170 °C [18].

Many research works related to EGS focus on electricity generation or combined generation of heat and power via Organic Rankine Cycle (ORC) or Kalina cycle [19–21]. In [22], multiple-criteria decision making (MCDM) for EGS and a decision-making tool were presented, and the authors used the tool to calculate the levelized cost of electricity (LCOE) and perform a sensitivity analysis. Levelized costs of electricity generated from EGS were acquired and analyzed in other works as well [23–26]. The values of LCOE for solar-geothermal plants in Northern Chile were estimated in Ref. [27]. In the work [28], Monte Carlo simulations were used to assess the LCOE of a double-flash geothermal plant, and the values were compared with gas prices. The investments in the plant are more attractive if natural gas prices are higher. Two software packages, EURONAT and GEOPHIRES, were used in Ref. [29] for economic studies, and the authors concluded that EGS facilities are not likely to be competitive with either renewable on non-renewable energy sources by 2030. Nevertheless, the latest studies and reports by different organizations [30–33] have shown that the LCOE of renewables, including geothermal energy, is already competitive or even lower than the LCOE of fossil fuel-based alternatives. The findings of work [34] show that 4600 GW_{el} of EGS with an LCOE less than 50 €/MWh_{el} can be installed worldwide by 2050.

While renewable energy sources met 42.1% of Germany's gross electricity consumption in 2019, the share of renewables in the final energy consumption in the heating/cooling sector was just 14.7%. In addition, final energy consumption was 576 TWh_{el} and 1218 TWh_{th}, respectively [1]. It can be observed that Germany's energy transition (Energiewende) focuses much more on the electricity sector than on the heating and cooling one. Additionally, electricity (e.g., from distant wind farms) can be transmitted on long distances easier than heat. This is why this work considers EGS as a locally available energy source to cover the base load for heat and cold supply rather than for electricity supply. The latter can be met by various other renewable energy sources like solar, wind energy or biomass.

In work [35], the authors also used the tool GEOPHIRES to estimate LCOE and levelized cost of heat (LCOH) for different technology readiness levels of EGS. The estimated value of LCOH for today's technology is about 42 €/MWh_{th}. In the study [36], LCOH for a doublet in the West Netherlands Basin with a production rate of 200 m^3/h was estimated at around 30 €/MWh_{th}. The cost of geothermal heat for oil sands extraction in Northern Alberta (Canada) was estimated at up to 38 €/MWh_{th} [37]. An economic analysis made for a university in the USA showed that a low-potential geothermal reservoir at a 3-km depth assisted by a heat pump can supply the university's district heating system, the LCOH being about 20 €/MWh_{th} [38]. The perspective development of CO_2 storage technologies is CO_2-EGS, which utilizes CO_2 as the circulating heat exchange fluid or the working fluid. For potential cogeneration of CO_2-EGS in Central Poland, the calculations showed that LCOH varies from 25 to 45 €/MWh_{th} [39].

The existing literature gives quite a good overview of economic indicators of electricity and heat generation from geothermal systems. Meanwhile, the majority of LCOH values found in different works show a quite promising and optimistic economic picture when considering that the current average costs of heat from oil and gas boilers in German households are 65–75 €/MWh_{th} [40] and from natural gas CHP—74 €/MWh_{th} [41]. This means that energy transition from fossil fuels to geothermal energy in Germany might be quite attractive. However, this work focuses on the EGS exploration of metasedimentary sequences of the Variscan fold-and-thrust belt in the Göttingen region, which has been poorly investigated as of yet, especially with regard to geothermal exploration and exploitation. This is why one of the goals of this work is to perform an economic and ecological analysis of different potential scenarios on the preliminary stage of EGS development for district heating and cooling of the Göttingen University campus. Such analysis is necessary because of many geological uncertainties of EGS exploration in a Variscan geotectonic setting, and it is supposed to show the minimum required brine parameters of a successful EGS project, which will be a target for subsurface investigation, reservoir modeling and hydraulic stimulation at a later stage. Another goal is to show potential investors and stakeholders the range of possible outcomes of EGS development and define which factors are the most important for the outcome and need of increased attention when developing EGS systems.

2. Background

The campus of the Georg August University (UGOE) and University Medical Centre (UMG) (both UGOE and UMG referred to as "the university" further on) takes up a large area in the center and in the north of the city of Göttingen. Being a demo site of the EU Horizon 2020 project MEET (Multidisciplinary and multi-context demonstration of EGS exploration and Exploitation Techniques and potentials [42,43]), the university has a good chance to commit itself to renewable energy utilization, and particularly to geothermal energy by developing an EGS concept in deformed metasedimentary rocks. In the best-case scenario, the Göttingen demo site can become a real laboratory for exploring and expanding the knowledge about EGS in Variscan fold-and-thrust belt and serving as a representative case study for other places with similar geotectonic settings in Europe.

2.1. Geological Setting

The geological setting in Göttingen and its vicinity is quite poorly investigated since there are only a few exploration wells with a maximum depth of 1500 m in the surrounding area. This indicates that the exploration of geothermal energy potential in Göttingen is currently at a very early stage. However, some progress was made in 2015 when a seismic campaign with two profiles crossing the campus area at an exploration depth of 1500 m validated that the upper several thousand meters of the subsurface of Göttingen are built up of three main units [44]:

- The lowermost unit (below 1500 m) represents low-grade metamorphic basement mainly consisting of Devonian and Carboniferous metasedimentary and metavolcanic successions (greywackes, slates, quartzites, cherts, diabase) that have been folded and thrusted during the Variscan Orogeny in the late Carboniferous;
- A Permian sedimentary sequence (several hundred meters of thickness) on top of the basement unit. It starts with either no or only locally deposited metavolcanics or sandstones of the Rotliegend as well as sequences of rock salt, potash salt, anhydrite, dolomite and clay-dominated layers of the Zechstein age;
- The uppermost major unit comprises the sedimentary cover (500 to 800 m of thickness) made up mainly of sandstones, clay rocks and limestones of the Triassic age (Buntsandstein, Muschelkalk and Keuper).

The whole sequence is overprinted tectonically by the north–south striking Leinetal Graben structure which developed during Mesozoic to Cenozoic times. It is still not clear whether the faults continue into the Variscan rock sequences or if they are decoupled mechanically by the Zechstein successions and possibly located elsewhere. Within the Leinetal Graben structure, Quaternary alluvial and wind-carried sediments form an additional unit of minor thickness but of importance regarding the utilization of shallow geothermal systems.

Considering the very low natural permeability of the basement, either EGS or a deep closed loop system with horizontal multilateral wellbores can be applied [45,46]. In this study, only EGS is considered, which means that permeability is to be increased by hydraulic stimulation. However, for the Variscan geotectonic setting underneath Göttingen, the fluid flow rate after the stimulation is not expected to be higher than 50 l/s.

2.2. Current Energy System of the Campus

The campus' main energy supplier is a combined heat and power (CHP) plant which includes a gas turbine and several steam and hot water boilers. The existing high-temperature district heating (HTDH) network (13 km of pipelines) delivers heat from the plant to the consumers of the campus. Apart from electricity and heat for district heating, the UMG also needs steam and cold. The latter is produced by both absorption cooling (base load) and vapor compression machines (peak load). In 2018, total natural gas consumption of the CHP plant and boilers for producing electricity (partly for cold), hot water (district heating) and steam (partly for cold) for the campus was about 358 GWh/a [47], and corresponding CO_2 emissions were about 72,000 t/a. Additional indirect emissions resulted from an external power grid and external district heating for the campus (about 22,000 and 5000 t/a, respectively).

The complete renewal of most of the UMG buildings within the next 15–20 years and the soon-expected end of the gas turbine lifetime (produced in 1997) put a question to the UGOE and the UMG on what their energy supply system should consist of in the future. The plans of the university involve the construction of not only new buildings but also a low-temperature district heating (LTDH) network. Although it is not exactly clear at the moment at which level of temperatures the planned LTDH network will operate, design supply and return temperatures in the network for this work were assumed to be 70 °C and 40 °C, respectively. This level of temperatures correlates with Refs. [48,49]. Additionally, steam absorption cooling machines are planned to be replaced with low-temperature ones supplied by hot water with the minimum temperature of 70 °C. Although these measures are good prerequisites for utilization of geothermal energy at different depths and for energy transition at the campus, other additional measures can be also considered [50], including an integrated energy concept, energy efficient construction of new buildings and refurbishment of old buildings of the campus. The latter aspect is one of the key elements of a successful energy transition and CO_2 neutrality until 2050 [51].

2.3. Initial Data and Scenarios

Since there are no geological and geophysical well data and no reliable numerical reservoir models yet, several probable scenarios for brine flow rate and wellhead temperature were considered. The values varied from 10 to 50 l/s with a step of 10 l/s and from 90 to 140 °C with a step of 10 °C, respectively. Higher flow rates and temperatures can hardly be expected in this Variscan geological setting. The density and specific heat capacity of the brine were derived from dependencies provided in Ref. [52]. Average geothermal gradient was considered to be a standard value for Europe, which is 30 °C/km [53]. There are no accessible or reliable data on the geothermal gradient for a better specification for the Variscan basement. Several other uncertain parameters were composed in the reference case and two cases for sensitivity analysis: unfavorable and favorable deviations, which are presented in Table 1. The values of CO_2 tax equal to 55–65 €/t correspond to the ones set by the German Government and coming into effect from 2026 [54]. Since CO_2 taxes in Sweden and Switzerland are already much higher [55], 100 €/t was also considered as additional favorable scenario in this work. Parameter "Subsidy for production well" included 50% or 80% (both favorable deviations) subsidy for drilling and stimulation of the research well (to be transformed in the production well later on).

Table 1. Parameters for the reference case and two cases with parameters for sensitivity analysis.

Case	L [km]	C_{drill} [%]	$OPEX$ [%]	d_n [%]	S_{gov} [%]		C_{carb} [€/t]		β [%]	τ [years]	T_{draw} [%/years]	T_{inj} [°C]
Unfavorable deviation	10	130	130	9.1	—		—		15	8	2	70
Reference case	5	100	100	7.0	0		55		10	6	1	60
Favorable deviation	0.5	85	85	6.0	50	80	65	100	5	4	0.5	55

L—distance to the campus; C_{drill}—cost of drilling; $OPEX$—operational expenditures; d_n—nominal discount rate; S_{gov}—subsidy for research (production) well; C_{carb}—CO$_2$ tax; β—brine salinity; τ—construction time; T_{draw}—temperature drawdown; T_{inj}—injection temperature.

The EGS system in Göttingen is supposed to represent a doublet (one production and one injection well). The research (production) well depth was assumed to be 5000 m since it is not known at what depth the most suitable conditions can be found. Only after a first research well is drilled, it will be clear at what depth an EGS could be developed with the highest efficiency. Thus, the depth of the second well (injection) is a variable, and it is defined by the considered temperature scenarios and the geothermal gradient.

For different values of the parameter "Distance to the campus", heat losses in the pipelines and specific electricity consumption for pumping the heat carrier from the site to the campus were assumed based on Ref. [56] and are shown in Table 2.

Table 2. Heat losses and specific electricity consumption for different distances to the campus [56].

L [km]	q_{hl} [%]	e_{spec} [kW$_{el}$/MW$_{th}$]
0.5	5	5
5	10	7.5
10	15	10

L—distance to the campus; q_{hl}—heat losses in the pipelines; e_{spec}—specific electricity consumption for pumping.

Usually, it takes from 5 to 7 years to develop a deep geothermal project from the early stages of exploration to operating facilities [57,58]. In order to make planning of the geothermal system construction clear and coherent with the planning of the reconstruction of the campus' buildings, the Gantt chart for potential EGS development in Göttingen is proposed in Table 3. There are also the reference case, unfavorable deviation and favorable deviation in the chart. Parameter "Construction time" from Table 1 correlates with the parameter "Start of operation" from Table 3.

For investors, important milestones in the chart are marked with darker shades, which means moments when investments for the project are needed. The investments can be split up into three parts: the research well (on average 45%), the second well (on average 37%), and surface infrastructure (on average 18%). The decisive part of the whole project is the first one, when the research well drilling is financed, the research work is done and it will be clear what outcome can be achieved. In case of geologically unsuitable and unpromising conditions leading to a failure of the project, the first part of the investments is lost and the other two parts make no further sense for investors. This shows that EGS projects can be quite risky and not very attractive for investors. However, there are initiatives and projects aiming at establishing financial instruments for insurance of deep geothermal projects [59] which might be able to attract investors.

Table 3. Gantt chart for potential development of EGS for Göttingen demo site. Squares with green color—favorable deviation; squares with yellow color—reference case; squares with red color—unfavorable deviation; darker shades—moments for investments.

Activity	Year								
	0	1	2	3	4	5	6	7	8
Feasibility study	green								
Permitting and public survey	green								
Research well financing (~45%)	dark green								
Research well drilling		green							
Research work		green	green	yellow					
Stimulation tests			green	yellow					
Injection well financing (~37%)			dark green	yellow	brown				
Injection well drilling				green	yellow	red			
Transformation of the well: research → production				green	yellow	red			
Stimulation tests				green	yellow	red			
Surface infrastructure financing (~18%)				dark green		yellow	brown		
Surface infrastructure construction				green		yellow		red	
Start-up and commissioning				green		yellow		red	
Start of operation					green		yellow		red

3. Materials and Methods

Based on the test reference year data from the German weather service [60] and internal documents from the university, an expected heat load profile of the campus after its reconstruction, which includes the heat demand for the new (to-be-built) buildings, for the existing buildings (a part of the existing buildings is planned to be deconstructed, and only remaining buildings are meant here) and for the absorption cooling machines was compiled. The assumption was made that a potential EGS supplies the new buildings as the first priority, then the remaining existing buildings, and, in the last turn, the absorption chillers. The calculations of the total EGS heat generation were performed considering the limitation that the injection temperature is at least 5 °C higher than the return temperature from a consumer and not lower than noted in Table 1.

The main focus of the methodology is the calculation of LCOH, net present value (NPV), and CO_2 abatement cost for the campus for different scenarios described in Section 2.3. These parameters are one of the main indicators for potential investors and stakeholders to make a decision with regard to an EGS project. A spreadsheet-based (Microsoft Excel 2019) model, which is explained below, was developed for evaluating those parameters.

Capital expenditures (CAPEX) include the following components:

$$\text{CAPEX} = C_{drill} + C_{stimul} + C_{pipes} + C_{land} + C_{manage} + C_{equip} \tag{1}$$

where:

C_{drill}—cost of drilling. Dependencies of drilling costs from depth were acquired from several sources [61–65], and the average values are plotted in Figure 1 and were taken for the calculations.

C_{stimul}—cost of hydraulic stimulation; assumed to be 2 M€/well.

C_{pipes}—cost of the main pipelines from the site to the campus (distribution pipelines are already a part of the existing HTDH network and not included here, and the cost of the planned distribution LTDH network is also not included); derived from work [66].

C_{land}—cost of land; specific value was assumed to be 37.5 €/m^2 [64]. Land requirement is 5000 m^2/MW [67]. For the scenario "0.5 km from the campus", this cost is zero since the university already owns the land.

C_{manage}—cost of project management, cost of campaigning for public acceptance and other costs [64,68].

C_{equip}—cost of equipment (pumps, heat exchangers, piping valves, auxiliaries), which can be calculated as follows:

$$C_{equip} = C_{sub\_pumps} + C_{pumps} + C_{HEX} + C_{valves} \tag{2}$$

C_{sub_pumps}—cost of submersible pumps; derived from Ref. [69] using Producer Price Index (PPI) equal to 1.6 [70] (exchange rate in November 2020: 1 USD = 0.85 EUR). The pumps are to be replaced every 5 years.

C_{pumps}—cost of circulation pumps for district heating network. The values were obtained from price lists of manufacturers.

C_{HEX}—cost of surface heat exchangers; average specific cost is 0.009 M€/MW$_{th}$ [61].

C_{valves}—cost of piping valves and auxiliaries; assumed to be 25% of the equipment cost.

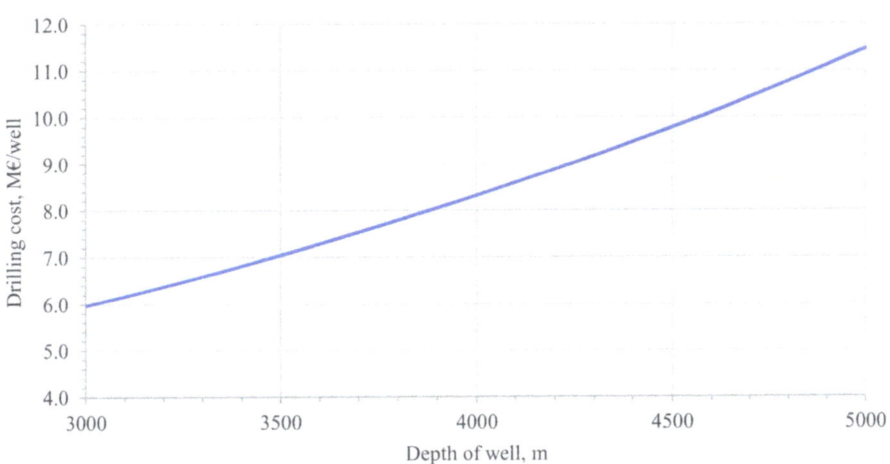

Figure 1. Cost of drilling assumed for the calculations.

Operational expenditures (OPEX) include the following components:

$$OPEX = C_{el\_pump} + C_{labor} + C_{mainten} + C_{insur} \tag{3}$$

where:

C_{el_pump}—annual cost of electricity for pumping. Submersible pumps' electricity consumption was assumed to be 10% of the total heat production from EGS [67,71]. For the circulation pumps in the district heating network, specific values from Table 2 were used. The electricity price for non-households in Germany in 2020 was 178 €/MWh$_{el}$ [72].

C_{labor}—annual cost of labor [64].

$C_{mainten}$—annual cost of maintenance and repair [64].

C_{insur}—annual cost of insurance and legal assistance [64].

LCOH was calculated according to Ref. [73]:

$$\text{LCOH} = \frac{\sum_{j=0}^{L}\left(CAPEX_j + OPEX_j\right)\cdot(1+d_n)^{-j}}{\sum_{j=S}^{L} Q_{egsj}\cdot(1+d_n)^{-j}} \tag{4}$$

where $CAPEX_j$—capital expenditures from year 0 to S; $OPEX_j$—annual operational expenditures from year S to L; S—year of the operation start; L—total project lifetime; Q_{egsj}—annual amount of heat delivered from the EGS to the campus (from year S to L) considering heat losses derived from Table 2; d_n—nominal discount rate [74], which can be calculated by Equation (5):

$$d_n = (1+d_r)\cdot(1+e) - 1 \tag{5}$$

where d_r—real discount rate; e = 0.015—annual average inflation rate [72].

Operational lifetime was defined by temperature drawdown: the operation ends when the wellhead temperature reaches the value of 10 °C higher than the injection temperature. Otherwise, operational lifetime was considered to be 30 years.

NPV was calculated according to Equation (6):

$$\text{NPV} = \sum_{j=0}^{L} \frac{-CAPEX_j - OPEX_j + Q_{egsj}\cdot C_{heat}}{(1+d_n)^{j}} \tag{6}$$

where C_{heat} = 89 €/MWh$_{th}$—prognosed heat tariff (price) for the university from fossil-fuel based system taking into account the CO_2 tax in Germany equal to 55 €/t from the year of 2026 [54]. For the favorable deviations of CO_2 tax in Table 1, the heat tariff is 91 and 100 €/MWh$_{th}$, respectively.

The CO_2 abatement cost was calculated according to Equation (7):

$$AC = \frac{-\text{NPV}}{\sum_{j=S}^{L} CO2_j^{sav}} \tag{7}$$

where $CO2_j^{sav}$—annual CO_2 savings during operation from year S to L considering electricity mix and natural gas emission factors equal to 397 t CO_2/GWh$_{el}$ and 202 t CO_2/GWh, respectively [75,76]. Positive values of AC show how much money is required to avoid one ton of CO_2 emission, while negative values show that the process is economically profitable.

4. Results

4.1. Heat Demand of the Campus and Potential Heat Supply from EGS

Figure 2 illustrates the expected heat load of the campus when its reconstruction is finished. The design heat load of the to-be-built buildings and the remaining existing buildings is estimated at 32.6 and 23.2 MW$_{th}$, respectively, while the heat load of absorption chillers can reach up to 10.7 MW$_{th}$ in summer. Thus, the influence of the to-be-built buildings on the future heat and cold supply of the campus might be quite high. The heat demand of the campus in summer is, in many scenarios, lower than potential heat production. That is why practically achievable EGS heat generation is lower than the potential one. The red dashed line in Figure 2 displays the estimated maximum geothermal output, and in this case, the load factor is 88%, while other values of the output lead to higher load factors. The average distribution of geothermal energy between the heat demand of the to-be-built buildings, remaining existing buildings, and absorption cooling machines is 91%, 7% and 2%, respectively.

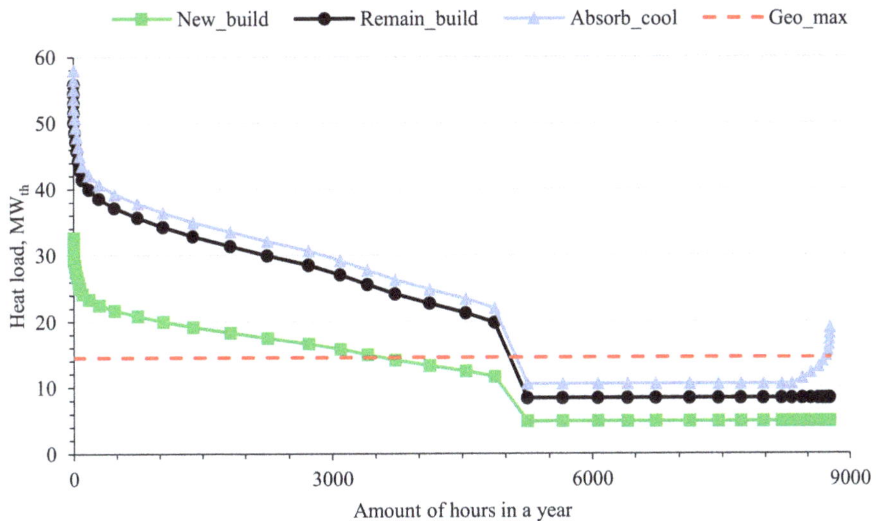

Figure 2. Expected heat load duration curve after reconstruction of the campus (including to-be-built buildings, remaining existing buildings, and heat for absorption cooling) in comparison with the estimated maximum geothermal output. Note: absorption cooling is assumed to be constant throughout the year, except for summer peak.

In Figure 3, lifetime-average EGS heat generation (without heat losses in the network) and design EGS heat output for the reference case are presented depending on brine flow rate and wellhead temperature. The heat generation can vary from 7.6 to 86.0 GWh_{th}/a, while the heat output—from 1.1 to 14.5 MW_{th}. Table 4 displays how much the heat demand of different consumers of the campus can be potentially covered by an EGS. On average, the values amount to 30.8%, 5.0% and 6.1% of the heat demand of new buildings, remaining existing buildings and absorption chillers, respectively. Existing heat and cold supply sources of the campus, which are supposed to cover the remaining demand, and back-up options were not considered in this study. Additionally, CO_2 emissions from the fossil-fuel based heating and cooling system of the campus (the specific CO_2 emissions are 252.7 g/kWh_{th}) are shown in Table 4. Thus, from 3.6 to 41.1% of CO_2 emissions could be theoretically saved if the EGS caused no greenhouse gas emissions during operation.

Table 4. Lifetime-average potential heat supply from EGS as a share of heat demand of different consumers and CO_2 emissions of a future fossil-fuel based heating and cooling system.

Application	Type of Heat Consumer	Expected Heat Demand (100%) [GWh_{th}/a]	CO_2 Emissions from Fossil-Fuel System [t/a]	Potential Heat Supply from EGS		
				Minimum [%]	Average [%]	Maximum [%]
	New buildings	110.8	27,991	6.8	30.8	60.6
Heating	Remaining buildings	78.8	19,920	0.0	5.0	15.3
	Buildings total	189.6	47,910	4.0	20.1	41.8
Cooling	Absorption chillers	19.9	5029	0.0	6.1	34.0
Heating & Cooling	Total	209.5	52,939	3.6	18.8	41.1

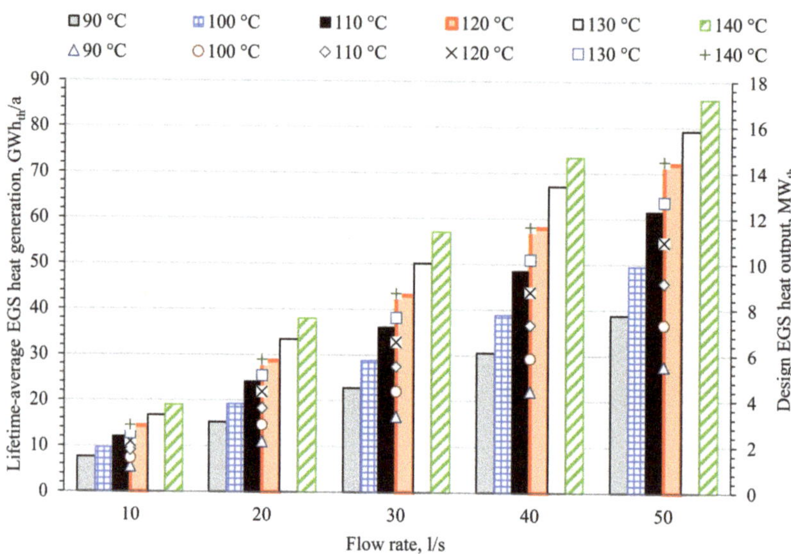

Figure 3. Lifetime-average EGS heat generation (columns) and design EGS heat output (symbols) for the reference case.

4.2. Economic and Ecological Results

Capital expenditures for the reference case conditions along with 0.5 km, 5 km and 10 km scenarios are shown in Figure 4. Costs of drilling and stimulation of two wells represent from 56 to 86% of CAPEX, and the next costly items are the costs of pipelines and the cost of project management. It is worth noting that components "Land", "Pipes_5 km" and "Pipes_10 km" are applicable only for 5 km and 10 km scenarios. Since drilling-related costs form the biggest part of the CAPEX, the components of CAPEX are presented in the chart only in dependence to the wellhead temperature, which is directly related to the drilling depth, and the influence of fluid flow rate on the total CAPEX is relatively small. In the same chart, specific CAPEX for different flow rates in the reference case are shown. The smallest values correspond to the highest considered flow rate (50 l/s); they vary from 5.8 to 2.9 k€/kW$_{th}$ for the temperatures from 90 to 140 °C. At the same time, the values for 10 l/s vary from 26.1 to 12.3 k€/kW$_{th}$.

In Figure 5, temperature- and flow rate-average structure of operational costs and average annualized capital costs for the reference case are compared with each other. Annualized capital costs were obtained with the help of an annuity factor [77], but such an approach was used only in this part of the work to allow for the comparison in Figure 5. It can be noted that electricity costs for pumping represent the biggest part of the OPEX.

The results of LCOH and NPV calculations for the reference case are demonstrated in Figure 6. For illustrative comparison, the prognosed heat tariff for the campus (89 €/MWh$_{th}$) from a fossil-fuel based system under CO_2 tax equal to 55 €/t and a hypothetic heat tariff (100 €/MWh$_{th}$ if CO_2 tax is 100 €/t) are also plotted in the chart. LCOH varies from 80 to 525 €/MWh$_{th}$ for the highest and lowest parameters of brine, respectively. It is clear that the majority of the scenarios of the reference case have worse LCOH than the prognosed fossil fuel-based heat tariff. The exceptions are the few scenarios with brine temperatures of 120 °C or higher and brine flow rates equal to 50 l/s, as well as the scenario with parameters 140 °C and 40 l/s. Although the scenarios with high temperatures and flow rates are considered in this work, their practical accomplishment in a Variscan geological setting is doubtful, especially regarding quite high flow rates such as 40–50 l/s. A slightly more realistic flow rate is 30 l/s, for which high temperature scenarios result in an LCOH of about 111 and 104 €/MWh$_{th}$, i.e., relatively close to the prognosed heat tariff and might be even profitable under some favorable conditions. Scenarios with flow rates less than 30 l/s are far from being economically feasible.

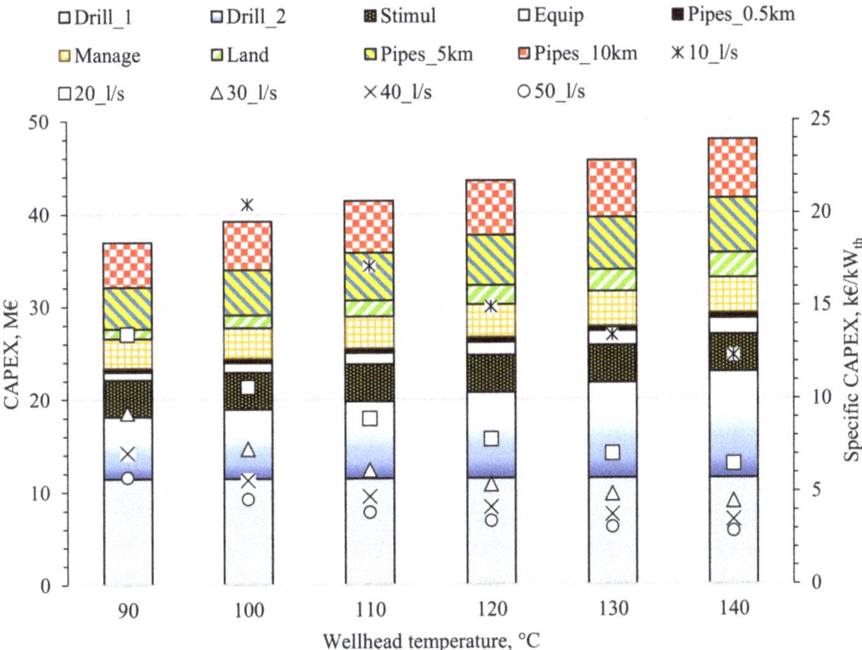

Figure 4. Components of CAPEX for different wellhead temperatures and different distances to the campus under the reference case conditions (columns) and specific CAPEX for different flow rates under the reference case conditions (symbols). Note: the value of specific CAPEX for 10 l/s under 90 °C is 26.1 k€/kW$_{th}$.

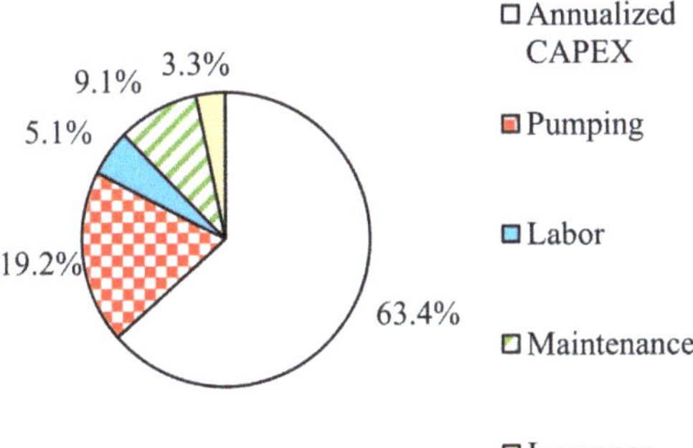

Figure 5. Average structure of operational costs and average annualized capital costs for the reference case.

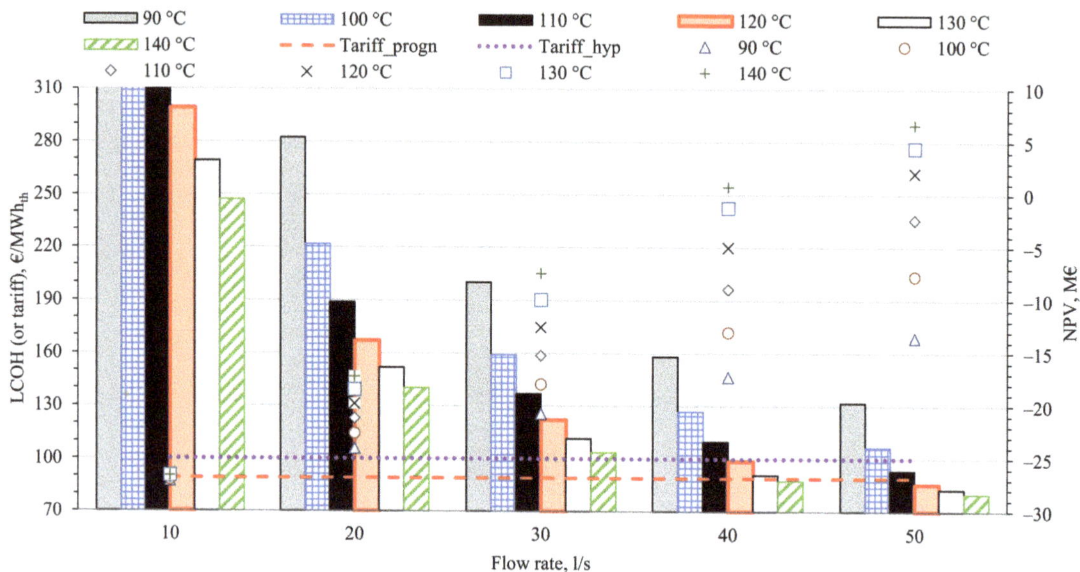

Figure 6. LCOH (columns), prognosed heat tariff for the campus (dashed line), hypothetic heat tariff for the campus if CO_2 tax is 100 €/t (dotted line), and NPV (symbols) for the reference case. Note: the values of LCOH for 90 °C, 100 °C and 110 °C under 10 l/s are 525, 406 and 342 €/MWh$_{th}$, respectively.

As for the NPV values, they vary from −27.1 to 6.6 M€ for the lowest and highest parameters of brine, respectively. Most of the calculated values are below zero, except for the same high temperature and high flow rate scenarios as in the LCOH part. In general, the values of NPV show similar trends to LCOH. When looking at NPV values, the scenarios with the flow rate of 30 l/s and high temperatures seem to be not so optimistic since the highest achievable NPV for such scenarios is −7.5 M€; however, these can be improved under favorable conditions.

Even though many scenarios are not economically beneficial, their ecological effect is an important factor to consider. CO_2 abatement cost and lifetime-average operational CO_2 savings for the reference case are presented in Figure 7. The CO_2 savings vary from 1.6 to 18.1 kt/a (3–34% of the emissions from the fossil-fuel based heat supply system of the remaining existing buildings, new buildings and absorption chillers) and the CO_2 abatement costs—from −12 to 655 €/t. The highest parameters of the brine yield the best results, as it was demonstrated for LCOH and NPV. The high-temperature scenarios with brine flow rates of 30 l/s present quite acceptable CO_2 abatement costs, which range from 21 to 31 €/t.

The average specific value of CO_2 emissions resulting from the operation of a potential geothermal plant in Göttingen is 42.4 g/kWh$_{th}$. In work [16], life cycle assessment for currently operating direct-use geothermal plant Rittershoffen in France was performed, which indicated that specific CO_2 emissions are 7.0–9.2 g/kWh$_{th}$. However, it should be noted that the nuclear power-dominated French electricity mix is nine times less carbon-intensive than the German one (44 g/kWh$_{th}$ vs. 397 g/kWh$_{th}$) [75], which explains the difference.

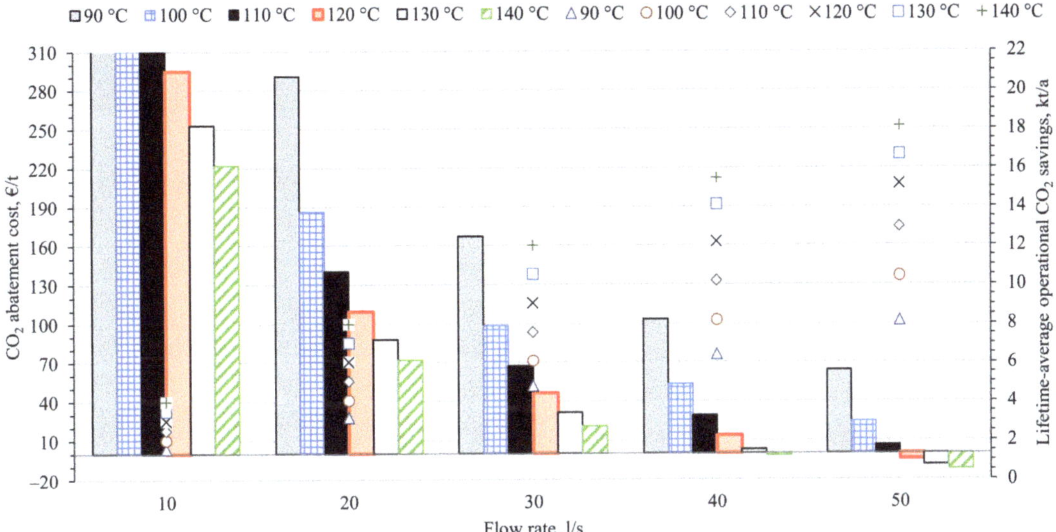

Figure 7. CO_2 abatement cost (columns) and lifetime-average CO_2 operational savings (symbols) for the reference case. Note: the values of CO_2 abatement cost for 90, 100 and 110 °C under 10 l/s are 655, 445 and 355 €/t, respectively.

4.3. Sensitivity Analysis

In order to cope with the uncertainty of the parameters and to get a better understanding of potential deviations from the obtained results of the reference case, a sensitivity analysis was carried out. The temperature-averaged results of LCOH sensitivity analysis for the parameters from Table 1 (unfavorable and favorable deviations) are displayed in Figure 8.

Increasing temperature drawdown from 1%/year to 2%/year leads to the biggest increase of LCOH (24–18% for the flow rates from 10 to 50 l/s, respectively). The other significant factor leading to increase of LCOH by 16–20% is a 30% increase of the nominal discount rate. Additional parameters worth noting are a 10 °C increase of injection temperature, 10 km-distance from the field to the campus and a 30% increase of drilling costs contributing 13–21%, 17–18% and 13–18% to the increase of LCOH, respectively. Less important parameters are OPEX, construction time (or the year of operation start) and brine salinity.

The most important parameter leading to decrease of LCOH is the research well subsidy of 50 or 80%, which allows to achieve 13–18% or 21–29% of LCOH reduction, respectively. Decreasing the distance from 5 km to 0.5 km has also a large effect (15–18%) on LCOH decrease. The remaining considered parameters are of less importance and contribute not more than 10% to LCOH decrease each.

It can be noted that, in general, factors influencing LCOH have less effect on higher flow rate scenarios which can be explained by larger amount of produced heat by EGS, thus smoothing the fluctuations. The exceptions are "Distance" and "OPEX" parameters showing the opposite trend since they are both proportionally and strongly related to the amount of produced heat.

The temperature-averaged results of NPV sensitivity analysis for the parameters from Table 1 (unfavorable and favorable deviations) are presented in Figure 9.

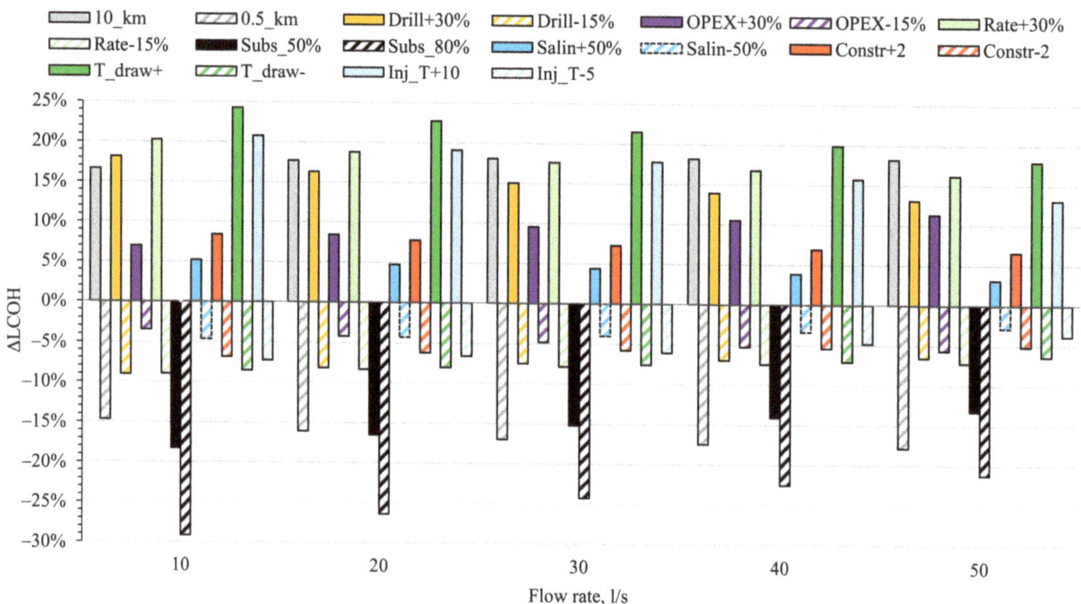

Figure 8. LCOH sensitivity analysis for the parameters from Table 1.

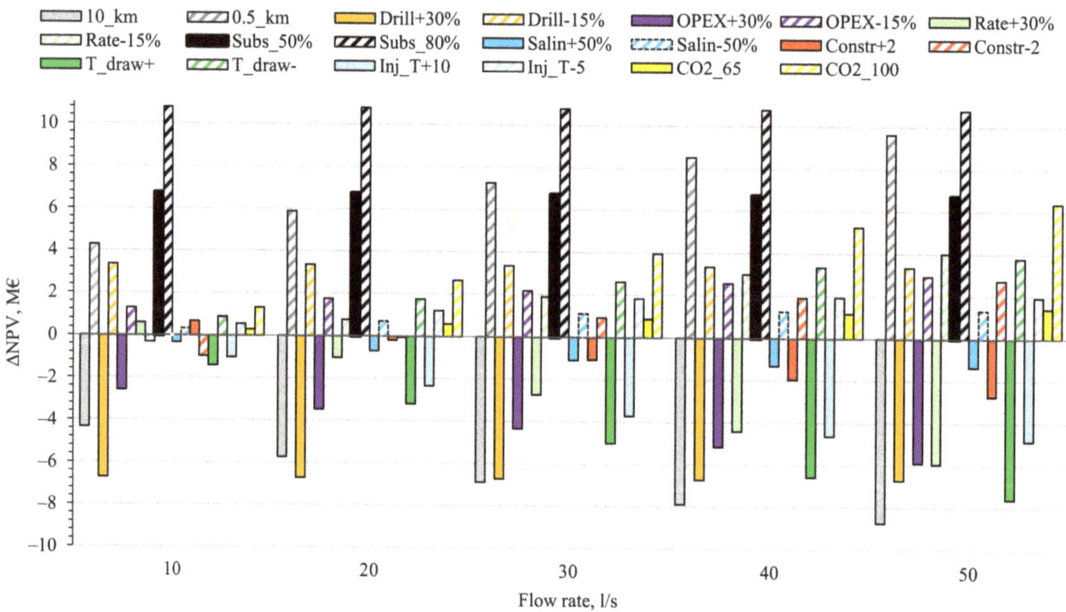

Figure 9. NPV sensitivity analysis for the parameters from Table 1.

Parameters "Distance" and "Drilling cost" are very influential on unfavorable deviations of NPV. If the distance from the field to the campus is increased from 5 to 10 km, NPV decreases by 4.3–8.7 M€ for the scenarios with flow rates from 10 to 50 l/s. Increasing drilling cost by 30% leads to NPV decrease by 6.7 M€. Temperature drawdown is also a parameter worth noting, especially for high flow rates, since it can worsen NPV value by up to 7.6 M€. Other parameters—OPEX, nominal discount rate and injection temperature—have some impact on NPV, and the remaining parameters have a relatively minor one.

As for favorable deviation of the parameters, 50 and 80% subsidy will lead to NPV increase by 6.7 and 10.8 M€, respectively. The possibility to build the geothermal plant just 0.5 km from the campus will result in 4.3–9.6 M€ increase of NPV for the scenarios with flow rates from 10 to 50 l/s. Reducing drilling cost by 15% can improve NPV by 3.4 M€. A potential increase of CO_2 tax up to 100 €/t can lead to 1.3–6.4 M€ increase of NPV. While the influence of temperature drawdown, nominal discount rate and OPEX becomes significant under high flow rates, the other remaining parameters have much less effect.

As seen in Figure 9, the influence of most of the parameters intensifies for higher flow rates. And for low flow rates, e.g., 10 l/s, some parameters barely lead to any change in NPV. Parameters "Construction time" and "Nominal discount rate" show shifting behavior when the flow rate increases. For small flow rates, they lead to NPV increase, while NPV decreases for large flow rates. It can be explained by the fact that longer construction time leads to a shift in the schedule for investments (Table 3) which is "beneficial" for low flow rate scenarios because of later discounting of those CAPEX. It simply means that the scenarios are not profitable anyway, but the losses are a bit reduced because the investments were made later. For high flow rates, the situation is the opposite since longer construction time delays obtaining relatively high revenues from the plant operation. Being discounted in later years, those revenues obtain less value and influence on the overall result leading to decrease of NPV with regard to the reference case. Moreover, a similar behavior is true for the nominal discount rate.

5. Discussion

Although the results were acquired for the Göttingen demo site, the considered parameters for the calculations and analysis were quite typical. This is why the results can be also applied to an early-stage development analysis of other EGS projects in poorly known geological settings for district heating and cooling. However, it should be noted that electricity prices in Germany are among the highest in Europe [72], and the district heating prices are above the average level [78]. Therefore, the results will be economically better for the countries with lower electricity prices and higher heat prices.

Interconnection and interdependence between different parameters were not considered during the sensitivity analysis. For example, large extraction of heat from the geothermal reservoir (high brine flow rate and low injection temperature) might lead to a bigger and faster temperature drawdown, and consequently, to a much smaller operation lifetime of the project, which puts forward a question of sustainable energy generation from EGS. Additionally, brine salinity and injection temperature are also correlated parameters since injection temperature might be limited by scaling and corrosion issues in the case of a high-salinity brine. Thus, some parameters from the sensitivity analysis can depend on each other and/or aggregate, leading to more complex effects on the final result.

The development of the EGS system in the Variscan basement in Göttingen is associated with many uncertainties and risks. Although the risks are not explicitly considered in this study, they are addressed in a follow-up work [79]. One of the biggest uncertainties and risks for any EGS is induced seismicity risk since it affects public acceptance of EGS projects [80] and can completely freeze any further works and lead to the cancellation of a project [81]. Cost-benefit analysis was applied to quantify the trade-off between seismicity risks and proximity to district heating and heat consumers in Ref. [82]. The authors concluded that remote EGS is less favorable even if the seismicity risk is considerably decreased or close to zero. The results of the sensitivity analysis of this work have also demonstrated that proximity of the geothermal plant to the campus significantly improves the economic indicators of the project. Nevertheless, it might be quite challenging to drill the wells, conduct hydraulic stimulation tests and build the plant close to the campus without public acceptance. That is why public acceptance is one of the prerequisites of future successful realization of the EGS project in Göttingen.

One of the obstacles of the project is to correlate and synchronize the university's plans of the campus reconstruction and the development of the EGS plant, which was partially addressed in this study by proposing the Gantt diagram of the EGS development. Nevertheless, effective communication and cooperation between different stakeholders within the university and outside of it is also one of the key prerequisites to launch the project.

The performed analysis has demonstrated that the reference case might not currently be competitive with the existing fossil fuel alternatives. Moreover, such long-term projects are usually not very attractive for private investors. Therefore, the government's support is another necessary prerequisite for the project.

Even if the economic part of the project might happen to be not very promising after the research drilling and hydraulic stimulation tests are carried out, the importance of its ecological effect is undoubtful. Economic and political measures for CO_2 emission reduction are likely to become stricter in the future which will pave the way for initially commercially unfeasible projects to be supported and implemented. The hypothetic value of the heat tariff depicted in Figure 6 and the values in Figure 9 show that CO_2 tax can be a powerful tool of the government to impact the profitability of EGS projects. On the other hand, other renewable options, which also benefit from high CO_2 tax and can yield better results than EGS due to more mature technological level and undercut the heat tariff, were not considered in this work.

The next important and essential step of the project development is to get research well funding. An additional opportunity of EGS projects in sites with poorly known geological settings, which can be provided by a deep research well, is investigation of shallow and medium layers and their further exploitation, e.g., for underground seasonal thermal energy storages. Such an integrated approach helps not only to increase the overall contribution of geothermal energy, i.e., renewable energy, but also to maximize the public subsidies for a research well, since the research focus is on both deep and medium deep target zones. In Germany, this is crucial when applying for public subsidies because financial support for drilling is preferentially given to the drilling sections defined as not yet investigated target rock units.

6. Conclusions

The geological setting in Göttingen—Variscan fold-and-thrust belt—is relatively poorly investigated for Enhanced Geothermal Systems exploitation. Nevertheless, there are expectations that geothermal energy will be able to partially meet the heat and cold demands of the Göttingen University campus (a demo site of the MEET project). This is why various scenarios of potential development of EGS for the campus were considered and analyzed in order to deal with different geological, technical and economic uncertainties of the project. On average, the considered single EGS doublet might cover about 20% of the expected heat demand and 6% of the cooling demand of the campus after its reconstruction. For different wellhead temperatures (90–140 °C) and flow rates (10–50 l/s) of brine, the obtained values of LCOH, NPV and CO_2 abatement cost vary from 80 to 525 €/MWh$_{th}$, from −27.1 to 6.6 M€ and from −12 to 655 €/t, respectively.

The most influential parameters on LCOH and NPV, which were identified during the sensitivity analysis, are subsidies for research wells, distance from the field to the campus, temperature drawdown and drilling costs. These parameters should be primarily dealt with when considering EGS development. CO_2 tax, operational expenditures, injection temperature and nominal discount rate can be also quite influential, especially under high brine flow rates. Other analyzed parameters, namely construction time and brine salinity, have significantly less effect on the economic indicators.

For the considered initial conditions in the reference case, it can be concluded that EGS heat output should be at least 11.0 MW$_{th}$ (corresponds to brine flow rate and wellhead temperature of 40 l/s and 140 °C, respectively) in order to have a more or less economically justified EGS project. If the distance between the field and the campus is 0.5 km, minimum EGS heat output can be 7.2 MW$_{th}$ (30 l/s and 125 °C). In case of 50 and 80% subsidy, the minimum heat output is 8.1 MW$_{th}$ (30 l/s and 135 °C) and 7.2 MW$_{th}$, respectively. The abovementioned parameters of brine can serve as a benchmark for geologists, engineers, managers, investors and other involved stakeholders to evaluate the success rate of the project, especially with regard to subsurface investigation, reservoir modeling and hydraulic stimulation.

The support of the government, public acceptance and effective cooperation between all stakeholders are the key prerequisites for launching the EGS project in Göttingen, which might save 1600–18,100 t CO_2 annually (3–34% of the emissions from the fossil fuel-based heat supply system of the remaining existing buildings, new buildings and absorption chillers).

Author Contributions: Conceptualization, D.R. and B.L.; methodology, D.R.; investigation, D.R.; writing—original draft preparation, D.R.; writing—review and editing, D.R. and B.L.; supervision, B.L.; funding acquisition, B.L. All authors have read and agreed to the published version of the manuscript.

Funding: This work has received funding from the European Union's Horizon 2020 research and innovation program (grant agreement № 792037—MEET Project).

Acknowledgments: The authors are grateful to Bianca Wagner for having contributed to the funding strategy for this study as well as for her discussions. Inga Moeck's helpful comments and remarks on the manuscript are appreciated. We thank Real Estate and Facilities Management (Gebäudemanagement) departments of UGOE and UMG for providing data and discussions.

Conflicts of Interest: The authors declare no conflict of interest.

References

1. BMWi. *Renewable Energy Sources in Figures. National and International Development 2019*; BMWi: Berlin, Germany, 2020; p. 88.
2. Bayer, P.; Attard, G.; Blum, P.; Menberg, K. The geothermal potential of cities. *Renew. Sustain. Energy Rev.* **2019**, *106*, 17–30. [CrossRef]
3. Lu, S.M. A global review of enhanced geothermal system (EGS). *Renew. Sustain. Energy Rev.* **2018**, *81*, 2902–2921. [CrossRef]
4. Anderson, A.; Rezaie, B. Geothermal technology: Trends and potential role in a sustainable future. *Appl. Energy* **2019**, *248*, 18–34. [CrossRef]
5. Shortall, R.; Davidsdottir, B.; Axelsson, G. Geothermal energy for sustainable development: A review of sustainability impacts and assessment frameworks. *Renew. Sustain. Energy Rev.* **2015**, *44*, 391–406. [CrossRef]
6. Olasolo, P.; Juárez, M.C.; Morales, M.P.; Damico, S.; Liarte, I.A. Enhanced geothermal systems (EGS): A review. *Renew. Sustain. Energy Rev.* **2016**, *50*, 133–144. [CrossRef]
7. Jain, C.; Vogt, C.; Clauser, C. Maximum potential for geothermal power in Germany based on engineered geothermal systems. *Geotherm. Energy* **2015**, *3:15*, 1–20. [CrossRef]
8. Chamorro, C.R.; García-Cuesta, J.L.; Mondéjar, M.E.; Pérez-Madrazo, A. Enhanced geothermal systems in Europe: An estimation and comparison of the technical and sustainable potentials. *Energy* **2014**, *65*, 250–263. [CrossRef]
9. Garapati, N.; Alonge, O.B.; Hall, L.; Irr, V.J.; Zhang, Y.; Smith, J.D.; Jeanne, P.; Doughty, C. Feasibility of development of geothermal deep direct-use district heating and cooling system at west Virginia university campus-Morgantown, WV. In Proceedings of the 44th Workshop on Geothermal Reservoir Engineering Stanford University, Stanford, CA, USA, 11–13 February 2019; p. 12.
10. Gustafson, J.O.; Smith, J.D.; Beyers, S.M.; Aswad, J.A.; Jordan, T.E.; Tester, J.W.; Ming Khan, T. Risk reduction in geothermal deep direct-use development for district heating: A cornell university case study. In Proceedings of the 44th Workshop on Geothermal Reservoir Engineering Stanford University, Stanford, CA, USA, 11–13 February 2019; p. 23.
11. Dalmais, E.; Genter, A.; Trullenque, G.; Leoutre, E.; Leiss, B.; Wagner, B.; Mintsa, A.C.; Bär, K.; Rajsl, I. MEET Project: Toward the spreading of EGS across Europe. In Proceedings of the European Geothermal Congress, Den Haag, The Netherlands, 11–14 June 2019; p. 8.
12. Menberg, K.; Pfister, S.; Blum, P.; Bayer, P. A matter of meters: State of the art in the life cycle assessment of enhanced geothermal systems. *Energy Environ. Sci.* **2016**, *9*, 2720–2743. [CrossRef]
13. Chavot, P.; Heimlich, C.; Masseran, A.; Serrano, Y.; Zoungrana, J.; Bodin, C. Social shaping of deep geothermal projects in Alsace: Politics, stakeholder attitudes and local democracy. *Geotherm. Energy* **2018**, *6*, 1–21. [CrossRef]

14. Ragnarsson, Á.; Óladóttir, A.A.; Hauksdóttir, S.; Maury, J.; Maurel, C.; Manzella, A.; Ámannsson, H.; Drouin, V.; Haraldsdóttir, S.H.; Guðgeirsdóttir, G.; et al. Report on Environmental Concerns. Overall State of the Art on Deep Geothermal Environmental Data; Geoenvi. 2021. Available online: https://www.geoenvi.eu/publications/report-on-environmental-concerns-overall-state-of-the-art-on-deep-geothermal-environmental-data/ (accessed on 7 June 2021).

15. Genter, A.; Evans, K.; Cuenot, N.; Fritsch, D.; Sanjuan, B. Contribution of the exploration of deep crystalline fractured reservoir of Soultz to the knowledge of enhanced geothermal systems (EGS). *Comptes Rendus. Geosci.* **2010**, *342*, 502–516. [CrossRef]

16. Pratiwi, A.; Ravier, G.; Genter, A. Life-cycle climate-change impact assessment of enhanced geothermal system plants in the Upper Rhine Valley. *Geothermics* **2018**, *75*, 26–39. [CrossRef]

17. Breede, K.; Dzebisashvili, K.; Liu, X.; Falcone, G. A systematic review of enhanced (or engineered) geothermal systems: Past, present and future. *Geotherm. Energy* **2013**, *1*, 4. [CrossRef]

18. Mouchot, J.; Ravier, G.; Seibel, O.; Pratiwi, A. Deep geothermal plants operation in upper Rhine graben: Lessons learned. In Proceedings of the European Geothermal Congress 2019, Den Haag, The Netherlands, 11–14 June 2019; pp. 1–8.

19. Meng, N.; Li, T.; Jia, Y.; Qin, H.; Liu, Q.; Zhao, W.; Lei, G. Techno-economic performance comparison of enhanced geothermal system with typical cycle configurations for combined heating and power. *Energy Convers. Manag.* **2020**, *205*, 112409. [CrossRef]

20. Moya, D.; Aldás, C.; Kaparaju, P. Geothermal energy: Power plant technology and direct heat applications. *Renew. Sustain. Energy Rev.* **2018**, *94*, 889–901. [CrossRef]

21. Wei, G.; Meng, J.; Du, X.; Yang, Y. Performance analysis on a hot dry rock geothermal resource power generation system based on Kalina cycle. *Energy Procedia* **2015**, *75*, 937–945. [CrossRef]

22. Raos, S.; Ilak, P.; Rajšl, I.; Bilić, T.; Trullenque, G. Multiple-criteria decision-making for assessing the enhanced geothermal systems. *Energies* **2019**, *12*, 1597. [CrossRef]

23. Sanyal, S.K.; Morrow, J.W.; Butler, S.J.; Robertson-Tait, A. Is EGS commercially feasible? In Proceedings of the Annual Meeting of the Geothermal Resources Council 2007: Renewable Baseload Energy: Geothermal Heat Pumps to Engineered Reservoirs, Reno, NV, USA, 30 September–3 October 2007; p. 10.

24. Sanyal, S.K.; Morrow, J.W.; Butler, S.J.; Robertson-Tait, A. Cost of electricity from enhanced geothermal systems. In Proceedings of the 32nd Workshop on Geothermal Reservoir Engineering, Stanford, CA, USA, 22–24 January 2007; p. 11.

25. Van Erdeweghe, S.; Van Bael, J.; Laenen, B.; D'haeseleer, W. Feasibility study of a low-temperature geothermal power plant for multiple economic scenarios. *Energy* **2018**, *155*, 1004–1012. [CrossRef]

26. Heidinger, P. Integral modeling and financial impact of the geothermal situation and power plant at Soultz-sous-Forêts. *Comptes Rendus Geosci.* **2010**, *342*, 626–635. [CrossRef]

27. Cardemil, J.M.; Cortés, F.; Díaz, A.; Escobar, R. Thermodynamic evaluation of solar-geothermal hybrid power plants in northern Chile. *Energy Convers. Manag.* **2016**, *123*, 348–361. [CrossRef]

28. Sener, A.C.; Van Dorp, J.R.; Keith, J.D. Perspectives on the economics of geothermal power. In Proceedings of the Transactions—Geothermal Resources Council, Reno, NV, USA, 4–7 October 2009; p. 8.

29. Olasolo, P.; Juárez, M.C.; Olasolo, J.; Morales, M.P.; Valdani, D. Economic analysis of Enhanced Geothermal Systems (EGS). A review of software packages for estimating and simulating costs. *Appl. Therm. Eng.* **2016**, *104*, 647–658. [CrossRef]

30. *Renewable Power Generation Costs in 2019*; International Renewable Energy Agency: Abu Dhabi, United Arab Emirates, 2020; 144p, Available online: https://www.irena.org/-/media/Files/IRENA/Agency/Publication/2020/Jun/IRENA_Power_Generation_Costs_2019.pdf (accessed on 7 June 2021).

31. Kost, C.; Shammugam, S.; Jülch, V.; Nguyen, H.-T.; Schlegl, T. *Stromgestehungskosten Erneuerbare Energien*; Fraunhofer-Institut für Solare Energiesysteme ISE: Freiburg, Germany, 2018; 44p.

32. Lazard. *Lazard's levelized cost of energy analysis—Version 13.0*; Lazard: Hamilton, Bermuda, 2019; 20p.

33. *Levelized Cost and Levelized Avoided Cost of New Generation Resources in the Annual Energy Outlook 2020*; U.S. Energy Information Administration: Washington, DC, USA, 2020; 22p.

34. Aghahosseini, A.; Breyer, C. From hot rock to useful energy: A global estimate of enhanced geothermal systems potential. *Appl. Energy* **2020**, *279*, 115769. [CrossRef]

35. Beckers, K.F.; Lukawski, M.Z.; Anderson, B.J.; Moore, M.C.; Tester, J.W. Levelized costs of electricity and direct-use heat from Enhanced Geothermal Systems. *J. Renew. Sustain. Energy* **2014**, *6*, 013141. [CrossRef]

36. Willems, C.J.L.; Nick, H.M. Towards optimisation of geothermal heat recovery: An example from the West Netherlands Basin. *Appl. Energy* **2019**, *247*, 582–593. [CrossRef]

37. Pathak, V.; Babadagli, T.; Majorowicz, J.A.; Unsworth, M.J. Evaluation of engineered geothermal systems as a heat source for oil sands production in Northern Alberta. *Nat. Resour. Res.* **2014**, *23*, 247–265. [CrossRef]

38. Galantino, C.R.; Beyers, S.; Lindsay Anderson, C.; Tester, J.W. Optimizing Cornell's future geothermal district heating performance through systems engineering and simulation. *Energy Build.* **2021**, *230*, 110529. [CrossRef]

39. Gładysz, P.; Sowiżdżał, A.; Miecznik, M.; Pająk, L. Carbon dioxide-enhanced geothermal systems for heat and electricity production Energy and economic analyses for central Poland. *Energy Convers. Manag.* **2020**, *220*, 113142. [CrossRef]

40. IEA. *Germany 2020—Energy Policy Review*; International Energy Agency: Paris, France, 2020; 229p.

41. Rubczyński, A.; Rączka, J.; Macuk, R. *Good heating practices from Denmark and Germany. Conclusions for Poland*; Forum Energii: Warsaw, Poland, 2018; 84p.

42. Trullenque, G.; Genter, A.; Leiss, B.; Wagner, B.; Bouchet, R.; Léoutre, E.; Malnar, B.; Bär, K.; Rajšl, I. Upscaling of EGS in Different Geological Conditions: A European Perspective. In Proceedings of the 43rd Workshop on Geothermal Reservoir Engineering Stanford University, Stanford, CA, USA, 12–14 February 2018; pp. 1–10.

43. Leiss, B.; Wagner, B. EU-Projekt MEET: Neue Ansätze "Enhanced Geothermal Systems (EGS)"—Göttinger Unicampus als Demoprojekt. *Geotherm. Energie* **2019**, *91*, 26–28. [CrossRef]

44. Leiss, B.; Vollbrecht, A.; Tanner, D.; Wemmer, K. Tiefengeothermisches Potential in der Region Göttingen—geologische Rahmenbedingungen. In *Neue Untersuchungen zur Geologie der Leinetalgrabenstruktur—Bausteine zur Erkundung des geothermischen Nutzungspotentials in der Region Göttingen*; Leiss, B., Tanner, D., Vollbrecht, A., Arp, G., Eds.; Universitätsverlag Göttingen: Göttingen, Germany, 2011; pp. 163–170.

45. Wang, G.; Song, X.; Shi, Y.; Yang, R.; Yulong, F.; Zheng, R.; Li, J. Heat extraction analysis of a novel multilateral-well coaxial closed-loop geothermal system. *Renew. Energy* **2021**, *163*, 974–986. [CrossRef]

46. Lea, E. *A Step Towards A Lower Carbon Future: Integrating Closed Loop Geothermal Technology in District Cooling Applications*; University of Calgary: Calgary, AB, Canada, 2020; 79p.

47. Leiss, B.; Romanov, D.; Wagner, B. Risks and Challenges of the Transition to an Integrated Geothermal Concept for the Göttingen University Campus. In Proceedings of the 1st Geoscience & Engineering in Energy Transition Conference, Virtual Conference, 16–18 November 2020; pp. 1–4. Available online: https://www.earthdoc.org/content/papers/10.3997/2214-4609.202021044 (accessed on 7 June 2021).

48. Lund, H.; Werner, S.; Wiltshire, R.; Svendsen, S.; Thorsen, J.E.; Hvelplund, F.; Mathiesen, B.V. Generation district heating (4GDH). integrating smart thermal grids into future sustainable energy systems. *Energy* **2014**, *68*, 1–11. [CrossRef]

49. Kaarup Olsen, P.; Holm Christiansen, C.; Hofmeister, M.; Svend Svendsen, J.-E.T. Guidelines for Low-Temperature District Heating. *EUDP 2010-II Full-Scale Demonstr. Low-Temp. Dist. Heat. Exist. Build.* **2014**, *68*, 1–43.

50. Leiss, B.; Wagner, B.; Heinrichs, T.; Romanov, D.; Tanner, D.; Vollbrecht, A.; Wemmer, K. Integrating deep, medium and shallow geothermal energy into district heating and cooling system as an energy transition approach for the Göttingen University Campus. In Proceedings of the World Geothermal Congress 2021, Reykjavik, Iceland, 24–27 October 2021; pp. 1–9.

51. Agemar, T.; Suchi, E.; Moeck, I. *Die Rolle der Tiefen Geothermie bei der Wärmewende—Wie Deutschland 60% Erneuerbare Wärme bis 2050 Schaffen KöNnte*; LIAG: Hannover, Germany, 2018; 14p.

52. Lukosevicius, V. *Thermal Energy Prodcution from Low Temperature Geothermal Brine—Technological Aspects and Energy Efficiency*; United Nations University: Tokyo, Japan, 1993; p. 42.

53. Bauer, M.; Freeden, W.; Jacobi, H.; Neu, T. *Handbuch Tiefe Geothermie. Prospektion, Exploration, Realisierung, Nutzung*; Springer: Berlin/Heidelberg, Germany, 2014; 867p.

54. CO_2-Bepreisung. Available online: https://www.bundesregierung.de/breg-de/themen/klimaschutz/co2-bepreisung-1673008 (accessed on 12 December 2020).

55. Carbon Taxes in Europe. Available online: https://taxfoundation.org/carbon-taxes-in-europe-2020/ (accessed on 16 April 2021).

56. International Energy Agency District Heating. IEA-ETSAP Techonology Brief E16. *ETSAP Energy Technol. Syst. Anal. Program.* **2013**, *7*. Available online: https://iea-etsap.org/E-TechDS/PDF/E16_DistrHeat_EA_Final_Jan2013_GSOK.pdf (accessed on 7 June 2021).

57. Pan, S.Y.; Gao, M.; Shah, K.J.; Zheng, J.; Pei, S.L.; Chiang, P.C. Establishment of enhanced geothermal energy utilization plans: Barriers and strategies. *Renew. Energy* **2019**, *132*, 19–32. [CrossRef]

58. Stefánsson, V. Investment cost for geothermal power plants. *Geothermics* **2002**, *31*, 263–272. [CrossRef]

59. GEORISK: Providing financial de-risking schemes for geothermal. Available online: https://www.georisk-project.eu/wp-content/uploads/2020/08/D6.4-GEORISK_Brochure_smaller.pdf (accessed on 11 December 2020).

60. Deutscher Wetterdienst. Available online: https://www.dwd.de/EN/Home/home_node.html (accessed on 11 December 2020).

61. Bertani, R. Deep geothermal energy for heating and cooling. In *Renewable Heating and Cooling: Technologies and Applications*; Woodhead Publishing: Sawston, UK, 2016; pp. 67–88. ISBN 9781782422181.

62. Beckers, K.F.; McCabe, K. GEOPHIRES v2.0: Updated geothermal techno-economic simulation tool. *Geotherm. Energy* **2019**, *7*. [CrossRef]

63. Eyerer, S.S.; Hofbauer, S.C.; Wieland, C.; Zosseder, K.; Bauer, W.; Baumann, T.; Heberle, F.; Hackl, C.; Irl, M.; Spliethoff, H. *Potential der Hydrothermalen Geothermie zur Stromerzeugung in Deutschland*; Geothermie-Allianz Bayern: Munich, Germany, 2017; 53p, Available online: https://www.mse.tum.de/fileadmin/w00bvc/www/gab/Potential_der_hydrothermalen_Geothermie_zur_Stromerzeugung_in_Deutschland.pdf (accessed on 7 June 2021).

64. Weinand, J.M.; McKenna, R.; Kleinebrahm, M.; Mainzer, K. Assessing the contribution of simultaneous heat and power generation from geothermal plants in off-grid municipalities. *Appl. Energy* **2019**, *255*, 113824. [CrossRef]

65. Yost, K.; Valentin, A.; Einstein, H.H. Estimating cost and time of wellbore drilling for Engineered Geothermal Systems (EGS)—Considering uncertainties. *Geothermics* **2015**, *53*, 85–99. [CrossRef]

66. Vivian, J.; Emmi, G.; Zarrella, A.; Jobard, X.; Pietruschka, D.; De Carli, M. Evaluating the cost of heat for end users in ultra low temperature district heating networks with booster heat pumps. *Energy* **2018**, *153*, 788–800. [CrossRef]

67. Danish Energy Agency. *Technology Data for Generation of Electricity and District Heating*; DEA: Copenhagen, Denmark, 2020; 414p.

68. Schlagermann, P. *Exergoökonomische Analyse geothermischer Strombereitstellung am Beispiel des Oberrheingrabens*; Dr. Hut: München, Germany, 2014.

69. Mines, G. *GETEM User Manual*; INL: Idaho Falls, ID, USA, 2016; 194p.
70. Federal reserve bank of, St. Luis Producer Price Index by Commodity: Machinery and Equipment: Pumps, Compressors, and Equipment. Available online: https://fred.stlouisfed.org/series/WPU1141 (accessed on 11 December 2020).
71. Majorowicz, J.; Grasby, S.E. Deep geothermal energy in Canadian sedimentary basins vs. fossils based energy we try to replace—Exergy [KJ/KG] compared. *Renew. Energy* **2019**, *141*, 259–277. [CrossRef]
72. The Statistical Office of the European Union. Available online: https://ec.europa.eu/eurostat (accessed on 12 December 2020).
73. IEA. *Projected Costs of Generating Electricity 2015*; IAE: Paris, France, 2015.
74. Short, W.; Packey, D.; Holt, T. *A Manual for the Economic Evaluation of Energy Efficiency and Renewable Energy Technologies*; NREL: Golden, CO, USA, 1995; 121p.
75. Buck, M.; Redl, C.; Hein, F.; Jones, D. *The European Power Sector in 2019*; Agora Energiewende and Sandbag: Berlin, Germany, 2020; 48p.
76. Juhrich, K. *CO$_2$ Emission Factors for Fossil Fuels*; Umweltbundesamt: Dessau-Roßlau, Germany, 2016; 48p.
77. Eyerer, S.; Schifflechner, C.; Hofbauer, S.; Bauer, W.; Wieland, C.; Spliethoff, H. Combined heat and power from hydrothermal geothermal resources in Germany: An assessment of the potential. *Renew. Sustain. Energy Rev.* **2020**, *120*, 109661. [CrossRef]
78. Werner, S. *European District Heating Price Series*; Energiforsk: Stockholm, Sweden, 2016; 58p.
79. Romanov, D.; Leiss, B. *Summary of Göttingen Site Optimization for Integration of Geothermal Resources in the Energy Supply*; MEET Report, Deliverable D6.12; Universitätsenergie Göttingen: Göttingen, Germany, 2021; 22p.
80. Kunze, C.; Hertel, M. Contested deep geothermal energy in Germany—The emergence of an environmental protest movement. *Energy Res. Soc. Sci.* **2017**, *27*, 174–180. [CrossRef]
81. Deichmann, N.; Giardini, D. Earthquakes induced by the stimulation of an enhanced geothermal system below Basel (Switzerland). *Seismol. Res. Lett.* **2009**, *80*, 784–798. [CrossRef]
82. Knoblauch, T.A.K.; Trutnevyte, E. Siting enhanced geothermal systems (EGS): Heat benefits versus induced seismicity risks from an investor and societal perspective. *Energy* **2018**, *164*, 1311–1325. [CrossRef]

geosciences

Article

Two-Stage Geothermal Well Clustering for Oil-to-Water Conversion on Mature Oil Fields

Josipa Hranić [1,*], Sara Raos [1], Eric Leoutre [2] and Ivan Rajšl [1]

1 Faculty of Electrical Engineering and Computing, University of Zagreb, Unska 3, 10000 Zagreb, Croatia; sara.raos@fer.hr (S.R.); ivan.rajsl@fer.hr (I.R.)

2 Vermilion REP SAS, 1752, Route de Pontenx, 40160 Parentis-en-Born, France; eleoutre@vermilionenergy.com

* Correspondence: josipa.hranic@fer.hr

Abstract: There are numerous oil fields that are approaching the end of their lifetime and that have great geothermal potential considering temperature and water cut. On the other hand, the oil industry is facing challenges due to increasingly stringent environmental regulations. An example of this is the case of France where oil extraction will be forbidden starting from the year 2035. Therefore, some oil companies are considering switching from the oil business to investing in geothermal projects conducted on existing oil wells. The proposed methodology and developed conversions present the evaluation of existing geothermal potentials for each oil field in terms of water temperature and flow rate. An additional important aspect is also the spatial distribution of existing oil wells related to the specific oil field. This paper proposes a two-stage clustering approach for grouping similar wells in terms of their temperature properties. Once grouped on a temperature basis, these clusters should be clustered once more with respect to their spatial arrangement in order to optimize the location of production facilities. The outputs regarding production quantities and economic and environmental aspects will provide insight into the optimal scenario for oil-to-water conversion. The scenarios differ in terms of produced energy and technology used. A case study has been developed where the comparison of overall fields and clustered fields is shown, together with the formed scenarios that can further determine the possible conversion of petroleum assets to a geothermal assets.

Keywords: geothermal; conversion; clustering; upscaling; heat; electricity; scenarios; LCOE; LCOH; NPV; CO_2 emissions

Citation: Hranić, J.; Raos, S.; Leoutre, E.; Rajšl, I. Two-Stage Geothermal Well Clustering for Oil-to-Water Conversion on Mature Oil Fields. *Geosciences* **2021**, *11*, 470. https://doi.org/10.3390/geosciences11110470

Academic Editors: Suzanne Golding and Jesús Martínez-Frías

Received: 26 October 2021
Accepted: 12 November 2021
Published: 16 November 2021

Publisher's Note: MDPI stays neutral with regard to jurisdictional claims in published maps and institutional affiliations.

1. Introduction

Geothermal heat has been traditionally extracted at locations characterized by hydro-geological anomalies, but recent advances in engineering have enabled the development of alternative approaches such as enhanced geothermal systems (EGS) and borehole heat exchangers (BHE) [1–3]. Both technologies can enable harvesting Earth's heat without any (or little) location constraints. EGS systems are used to produce energy by enhancing in situ permeability and harvesting heat from hot rock reservoirs [4]. The connection between production and injection wells in EGS is engineered by various stimulation techniques. The viability of an EGS project is mostly influenced by brine flow rate and production temperature, where higher flow rates and temperatures support electricity generation and lower values support direct usage of hot water, i.e., heating power production. Regarding fluid flow rates, the increase in low rates could be achieved by applying reservoir stimulation, whereas temperatures can be increased only by drilling deeper wells [5]. BHEs harvest geothermal energy without direct interaction of flowing fluid with the soil or rock. Different from the EGS, the efficiency of deep BHEs strongly depends on heat exchanger configurations and the host rock thermal properties [6]. The economic viability of both technologies, especially considering high depths (>3 km), depends on emerging technologies, drilling technologies, reservoir technologies, etc.

In order to bypass exploration and drilling risks, mature and abandoned oil wells could be used. There are thousands of onshore wells in Europe, and most of them are mature oil provinces where it is expected that the existing wells are now producing much more water than oil, with an average water to oil ratio higher than 90%; thus, the cost of wastewater disposal increases. The oil reservoir's depth ranges between few hundreds to few thousand meters; therefore, fluid temperature at the surface can reach up to 90 $^\circ$C and more, thus enabling the production of electricity, heat, or both. In most cases, hot water is reinjected into the reservoir to increase production through pressure support and sweep; hence, the calorific energy of water is wasted. This is the coupling point between oil industry and geothermal energy production. Namely, the possibility of using these high temperature fluids to produce geothermal energy during the final stage of the life of an oil field and converting the field into a geothermal one is an emerging and interesting option for energy strategy. Numerous studies have been conducted on mature oil fields where geothermal potential has been proven with simulations or with actual exploitation [7–16]. In reference [17], the authors revised mature oil and gas fields across the world where waste heat from geothermal water has already been recovered or its potential has been determined. In order to ensure profitable waste heat recovery, a list of criterions formed on reservoir, geological, production, and economic characteristics was suggested. The criteria were used as a guideline in the assessment of geothermal energy utilization and were tested on the Villafortuna-Trecate field in Italy. The results showed that roughly 25 GWh of electricity could be produced with installed capacity of 500 kW from a single well in the period of 10 years. Another case of retrofitting the hydrocarbon wells into a geothermal ones was introduced in [18], where the method used for exploiting geothermal energy took into account economic and environmentally friendly solutions for the efficient production of electricity by considering mathematical and 3D numerical models of heat extraction. The model resulted in viable and efficient electricity and heat generation over the lifetime of the reservoir. The conducted sensitivity analysis of main parameters controlling the outlet fluid temperature implied that abandoned gas wells are applicable sources of geothermal energy. In reference [19], the authors evaluated the abandoned petroleum wells in Hungary, which are suitable for potential applications of enhanced geothermal systems. The database of 168 wells defined with moderate to high heat flow (75–100 mW/m^2) proved the feasibility of using abandoned wells for direct uses, all using either hydrothermal or EGS with identified influencing factors such as well geometries, geothermal gradient, pipe diameter, etc. The authors in [20] investigated the possible production of geothermal energy from inactive wells in the Arun Field, and their study confirmed the feasibility of extracting geothermal energy for electricity and heat generation and stated that, with 2.56 kg/s of mass flow and 170 $^\circ$C, it is possible to produce 2900 kW of electricity and satisfy the heating and cooling demands of various industry objects. Such positive retrofitting project outcomes have significant contributions to meeting rising global energy demands with renewable energy use without necessitating additional land usage and costs such as exploration, drilling, casing, surface pipeline, and decommissioning costs.

However, it is important to determine the optimal applicable exploitation technology with respect to the site and potentially close end users for the heating power production case. Given its promising future, plenty of studies on geothermal energy extractions from abandoned oil wells have been carried out and appraised [21–25]. The focus of the mentioned studies was on retrofitting an abandoned oil well for feasible technical and economic exploitation of geothermal energy, performance during the operational phase, decision on open-loop or closed-loop geothermal extraction choice between borehole heat exchanger (double pipe or U-tube), and heat transfer improvements. The fundamental parameters such as the working fluid characteristics, well geometry, and operational parameters that concern working fluid flow rate, inlet temperature, operating pressure, etc., were likewise examined [26–28].

Moreover, the majority of work that has been performed on retrofitting abandoned petroleum wells as a source for geothermal energy has been focused on open loop systems that repurpose the petroleum reservoir as a geothermal reservoir [29]. There are multiple countries that have sponsored research and/or investigations specific to adapting an open loop design for abandoned wells, including the following: Albania, China, Croatia, Hungary, Israel, New Zealand, Poland, and the US. Additionally, Vermilion Energy is recovering heat from two producing oil fields in sedimentary basins in France. On the Parentis oil field in SW France, 60 wells producing a total of 400 m^3/h water at 60 °C water have been used to heat up 8 ha of tomato greenhouses since 2008, creating more than 100 jobs. In La Teste in SW France, two producing wells yielded 40 m^3/h at 70 °C, which is enough to cover 80% of the heat needed for 450 new flats. These two projects demonstrate that recovering heat from produced water creates value and jobs at any scale (small or large oil fields). Based on these successfully conducted projects, the idea of shifting the paradigm from investing in geothermal projects from the beginning, starting with exploration and drilling activities, to start where geology is already known through existing wells in the oil industry emerged. Therefore, the end-of-life oil well conversion methodology is part of the Horizon 2020 project: Multidisciplinary and multi-context demonstration of EGS exploration and Exploitation Techniques and potentials (MEET, GA No 792037).

End-of-life oil well conversion methodology towards the geothermal wells defines the roadmap for further conversion of oil wells into geothermal production wells, thereby enabling a certain niche for geothermal energy penetration into the market. Namely, notable potentials for conversion to geothermal wells include abandoned, mature, or high water-cut wells since they are almost instantly available, i.e., there is no need for drilling, and available and thorough logging of production data facilitate well performance assessment which results in diminishing risks and enhancing cost estimation [7,29]. Furthermore, petroleum infrastructure and facilities available on the field can be converted to enable geothermal exploitation; in doing so, major costs related to drilling a new geothermal well and power plant are economized [29,30]. Retrofitting petroleum wells into geothermal wells also prospers from reducing or even excluding the cost of decommissioning of the oil well, thus maintaining the economic viability of the well.

The methodology conducted in this study and corresponding support tool for an economic evaluation of end-of-life conversion will enable pre-technical economic feasibility studies for converting an oil field to a geothermal field at the end of its economic "petroleum" life, including geological, technological, financial, and environmental aspects of an oil field and the technology used. The clustering feature, where wells can be clustered based on production temperature and spatial distribution, enables including wells at a specific oil field in the calculations that are best suited for a certain option—only heating power production, electricity generation, or both (combined heat and electricity production, CHP). This two-level clustering method facilitates the decision process regarding the possible usage of produced heat. The first step starts with temperature clustering, which is based on sorting the oil wells into different groups based on the temperature ranges from modified Lindal diagram [31]. Additionally, spatial clustering, which is based on the grouping a certain number of wells into one group according to their mutual distances, enables the best allocation of power plant installation and piping connection system between the selected wells. The output results of the methodology are based on economic metrics (net present value (NPV), levelized cost of electricity (LCOE), and levelized cost of heat (LCOH)) and production metrics (yearly/monthly production values, avoided CO_2 emissions) that are used in the decision-making process with respect investing in a specific project or not.

2. Background

The mentioned conversion is based on input data from the oil field, default values about the heating demand, energy prices, emission factors that can be changed by the user, proxy values of pump power consumption, thermal efficiency of Organic Rankine Cycle power plant, etc. Based on the mentioned data, five scenarios of geothermal energy production with different production technologies are developed with the main goal of comparing different options for heat and/or electricity production and to choose the optimal one. One of the main features of the conversion is temperature and spatial clustering, which clusters the wells according to the geothermal fluid temperature into a different end-use group and, once again, clusters the wells into spatial groups according to the distance between each well. Spatial clustering enables the user to include all wells on the field with high water cut in the conversion in order to upscale production quantities and to decrease piping connection cost. Additionally, three submodules are developed to calculate the power consumption of the production pump, injection pump, and deep borehole heat exchanger pump. After entering the input data for each scenario, the conversion tool will calculate four main outputs: produced energy quantities, LCOE or LCOH, NPV, and avoided CO_2 emissions. Based on these results, the user can decide which conversion option is optimum for a given petroleum asset.

2.1. Developed Scenarios

The methodology for an economic evaluation of end-of-field life conversion is a decision-making framework that uses different input data in which the main goal is to compare different options for heat and/or electricity production and to choose the most suitable option. The main purpose of the methodology is to offer the optimal scenario for converting the petroleum asset to a geothermal one. Based on the input data of mature or abandoned petroleum fields, economic or environmental parameters, and technological features, five scenarios are modelled, and the result is output data. The output data, based on the extensive and thorough calculations, provide insight into the economic and environmental aspects of the geothermal project for each scenario.

One of the key benefits of the proposed work is the avoidance of decommissioning the cost of wells and surface facilities and generating income through electricity and heat production by repurposing the mature oil field into a geothermal asset. One of the main contributions of the methodology is two-stage clustering that enables the temperature and spatial arrangement of the wells and, among the oil wells, also includes the wells from the field that were previously flooded and were not producing oil or newly drilled wells in terms of upscaling geothermal energy production. Two-stage clustering is an optimization process because it clusters the wells according to the temperature of the end-use and according to the spatial distribution so that the position of the geothermal plant can be determined along with the inclusion of the wells in the gathering system corresponding to the shortest distance from the geothermal plant.

The developed methodology should serve as a pre-feasibility study of converting a petroleum field to a geothermal one. The methodology provides guidelines in terms of retrofitting mature or abandoned petroleum fields to geothermal energy exploitation and user-friendly environment for which its outputs could encourage possible users to invest in geothermal projects. In the following bulleted list, the developed scenarios are described.

- Scenario 1—"Do nothing"

This scenario refers to plugging and dismantling all the wells and surface facilities and can represent hundreds of thousands of Euros of abandonment cost per well required by mining law. The operating life of an oil field has a certain limitation, and when reaching the end of its viable life, the next step is strategy planning for plugging and abandonment operations.

This is dependent on factors such as well location and depth, type of the surface and subsurface facility, number and weight of structures needed to be removed, removal method, transportation, and disposal options, etc [32].

- Scenario 2—"Heat doublets"

The developed scenario concerns heat production from production wells and the injection of geothermal fluid into the reservoir by using the injection wells. The main challenge that concerns the geothermal industry is associated with capital-intensive costs of drilling geothermal wells; hence, the utilization of abandoned petroleum wells is encouraged. The aforementioned wells can potentially be harnessed for geothermal energy for direct usage depending on the temperature of geothermal water [3,22,33]. This scenario consists of two sub-scenarios: *Temperature range* sub-scenario and *Heat needs* sub-scenario. The *Temperature range* sub-scenario is the scenario where heat production is based on utilizing the temperature range of geothermal fluid (production temperature and fixed injection temperature). The latter scenario, *Heat needs* sub-scenario, is based on satisfying the heat demand of the end-user. The heat demand is set as the user's input, or it is calculated based on the heat demand of three different type of buildings.

- Scenario 3—"Heat via BHE"

The modelled scenario regards heat production using one well, i.e., the borehole heat exchanger. Borehole heat exchangers are used to extract heat without producing geothermal fluid from wells, i.e., with running circulation fluid through the wellbore. The usage of abandoned wells in such a manner can decrease gas emissions with respect to the atmosphere and the energy needed for reinjection. The circulating fluid is injected through annular space and produced at the wellhead through production tubing or vice versa [1,23,34]. This scenario consists of two sub-scenarios, *Temperature range* sub-scenario and *Heat needs* sub-scenario, which is the same as described in Scenario 2.

- Scenario 4—"ORC power production"

This scenario represents electric power generation using the Organic Rankine Cycle (ORC). Electricity can be produced by using production and injection wells or using a deep borehole heat exchanger. The power capacity is determined primarily by the production rate, temperature of produced water, ambient temperature, water salinity, conversion efficiency of the geothermal power plant, heat transfer efficiency between the reservoir rocks and circulating fluid, etc. [3,8,17,21,35,36].

- Scenario 5—"Combined power and heat"

The developed scenario refers to combined heat and power production (CHP) with parallel configuration modes [37]. The total geothermal fluid flow is divided into two branches as follows: Primarily, heat demand is satisfied, and electric power is then produced with the residual flow. Two sub-scenarios are developed: the first one with the production and injection wells and second one with BHE. The well for BHE is the well with the highest temperature according to the wells clustered by the "electricity" end-use [3,8,21–23,34].

2.2. Input Data

The main input data used in the methodology for calculations and clustering process are shown in Table 1.

Even at very high water cut, an oil field often displays mixed flow, meaning that a given geological layer produces both oil and water. It is, therefore, expected to produce both water and oil after conversion. Since the oil cut is very low, it is expected that gravity separation in water tanks will take place. Yearly water-cut increment is a linear percentage value of the annual average water-cut increase, based on historical data. Yearly thermal dropdown is defined as the annual average temperature decline rate for petrothermal reservoirs, as the reservoir is expected to be cooled down by colder fluid injection. Additionally, at the beginning of calculations, it should be determined if the production pump is already installed and running or not. If the pump is already installed, additional input regarding the pump power is required, which is used afterwards to calculate pump consumption power, i.e., parasitic load. In both cases, if the pump is already installed or not yet installed, the user should proceed with the calculation related to the electric submersible pump (ESP) design in order to either design the required new pump that should be installed or to estimate pump consumption for the already installed pump. Temperature loss along the wellbore is also stated as the user's input, and it is automatically subtracted from the reservoir temperature to calculate the wellhead's temperature, i.e., production temperature.

The rest of the input data follows the developed scenario's data and will be set as the default or calculated with the possibility of user's input.

Table 1. Input data required for conducting clustering methods and further calculations.

Input Data
Well name
Longitude
Latitude
Well temperature
Temperature loss through gathering lines
Oil production
Water production
Bottomhole pressure
Density of oil
Density of geothermal fluid
Specific heat capacity of oil
Specific heat capacity of geothermal fluid
Well depth
Yearly thermal drawdown
Yearly thermal water-cut increment
Water-cut
Production pump installed
Temperature loss along the wellbore
Reservoir temperature at the well depth
Downtime of the plant
Outlet temperature of the plant

3. Materials and Methods

At the end of its economic life, a certain spatial footprint of oil field exists. Based on the development history of the oil fields, well pads are made of several wells drilled from the same surface location and are connected to the main facilities by flowlines [38]. Each of the wells on the oil field has different surface flow rates and temperatures. When converting the oil field to geothermal usage, the wells on the field are optimized and the wells that deliver the most suitable flow rate and temperature are kept. The example of temperature and spatial clustering was shown in [39], where the author used the Cluster and Outlier Analysis tool for spatial and temperature well clustering for deep borehole heat exchanger (DBHE) geothermal systems, which solves for the Anselin Local Moran's „I" statistic of spatial association. The statistic was used to identify the aggregation of wells with high bottomhole temperatures. Temperature data of 42,601 wells were collected, and areas with significant densities of oil and gas wells with the accompanying high temperatures were outlined. The described approach could result in an increase in system efficiency and economic viability of the geothermal projects, which are based on the already built subsurface infrastructure of oil and gas fields. The main advantage of clustering methods is the possibility of selecting clusters and/or wells that are already connected to built surface piping infrastructure or are near existing power distribution infrastructure. Moreover, the ability to connect new wells that so far have not produced any oil and gas and have high water cut to a gathering system would result in upscaling the overall capacity of geothermal energy production.

The basis of the developed methodology and the supporting tool is the clustering of the wells, both in terms of temperature and spatial clustering. For both clustering layers, the Python programming language is used with integrated pre-made libraries.

In the first layer of clustering, i.e., temperature clustering, the production wells are sorted into the temperature groups according to their well temperature, and each well is sorted into groups for one or more end-uses. The well that has more than one end-use is used in calculations for more than one scenario.

The second layer of clustering is spatial clustering where the used method of clustering sorts the wells in certain number of clusters based on their distances between each other. Spatial clustering enables the inclusion of unused wells on the field in further calculations that have a high water cut that is suitable for geothermal energy production; the wells that were not previously included in oil production; and newly drilled wells that have a high water cut and are drilled for geothermal purposes on the mature or abandoned oil field. Spatial clustering also defines the data point (centroid, most commonly an imaginary point), which is in the middle of the cluster and the well (existing data point that is nearest to the centroid) upon which the new thermal or power plant should be built with minimum cost of a new pipeline system.

3.1. Temperature Clustering

The temperature clustering layer is based on Lindal's diagram [31] with minor modifications. Minor modifications of Lindal's diagram and the possible applications of geothermal energy made for the purpose of the methodology concern the expansion of temperature ranges for end-uses. The main modification is the expansion of temperature range for electricity production using Organic Rankine Cycle (ORC) smart mobile units, which is one on the main goals of the MEET project, i.e., enhancing heat-to-power conversion at low temperature (60 °C–90 °C). The temperature ranges for different end-uses are shown in Figure 1. The temperature spans from 0 °C to 200 °C with the heat pump, heat generation, and electricity generation as the end-uses. Electricity generation end-use covers electricity production by using smart mobile ORC units and electricity production in binary systems (ORC). The temperature range for heat pump is stated here as the informational data, and it does not proceed to further calculations for the purposes of methodology and tools.

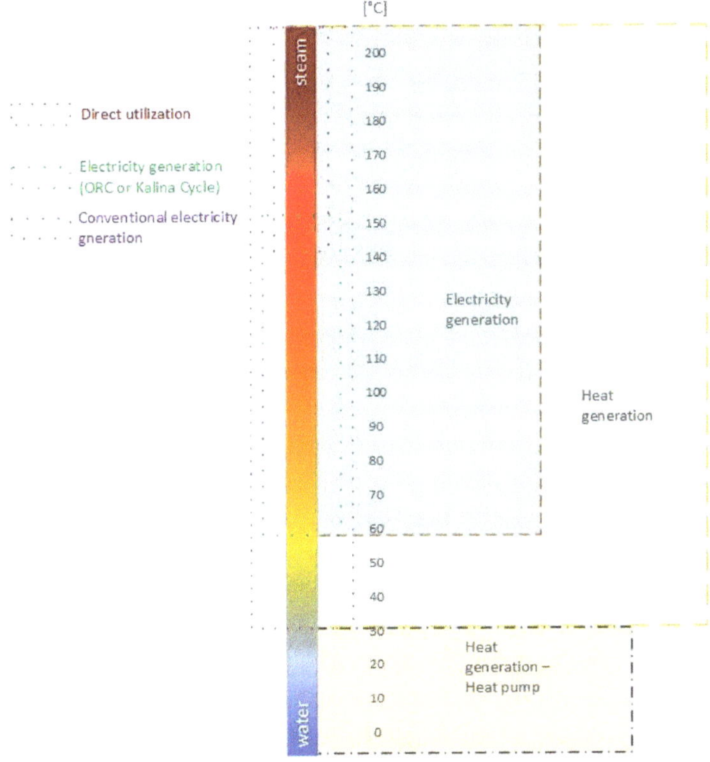

Figure 1. Modified Lindal's diagram with different end-uses.

3.2. Spatial Clustering

The method used in the developed methodology and case study is the Density-Based Spatial Clustering of Applications with Noise or DBSCAN, which is an unsupervised machine learning algorithm. Unsupervised machine learning algorithms are used to allocate unlabelled data. DBSCAN method examines the clusters as high-density regions separated by low-density regions; therefore, the clusters found by DBSCAN method can be of any shape [40].

For the mentioned clustering method, a few important parameters need to be predetermined:

- Epsilon, as the maximum distance between two data points for one to be considered as in the neighbourhood;
- Minimum samples, as the number of data points (or total weight) in a neighbourhood for a point to be recognized as a core point and includes the point itself;
- Metric, as the metric used when calculating the distance between instances in the array.

DBSCAN creates a circle of *Epsilon* radius around every data point and classifies them into core point, border point, and noise point. A data point is a core point if the circle around it contains at least a number of *Minimum Samples* points. If the number of points is less than the *Minimum Samples* number, then it is classified as a border point, and if there are no other data points around any data point within the epsilon radius, then it is treated as a noise point [41]. The *Epsilon* value can be calculated as the average distance between each point in the data set and its *Minimum samples* number of nearest neighbours. The average distance is then plotted by ascending value where the sorted values produce an elbow plot that indicates the maximum curvature on the point, which is the *Epsilon* value.

The main advantage of the considered method is the determination of outlier points and the selection of clusters according to the different shapes of the data set. The main weakness of the method is that it does not work well with the data set that has different densities (different distances between the points), and due to the fact that *Epsilon* is a fixed value, it will characterize the points with different densities as outlier points.

3.3. Outputs

When two-layer clustering is conducted, the number of clusters in the specific field is obtained, and the filtering option is enabled. Namely, various filtering options are possible: filtering of the individual wells, filtering the number of end-uses for each well, filtering desired end-uses to be included in further calculations, filtering regarding the wellhead temperature, and filtering according to the number of the cluster in which the well is located. This type of listing and filtering later enables the calculation of results for each scenario, including both heat and electricity generation, and provides the user with an option to include or exclude a particular well or cluster from further calculations and scenario development.

3.3.1. Energy Production Quantities

Regarding energy production, the quantities of produced electricity and/or heat are calculated for each modelled scenario. For both electricity and heat generation scenarios, when using the production–injection wells, the temperature of the mixed fluids, density of the mixed fluids, and specific heat capacity of the mixed fluids are all computed from all wells from the field, which are filtered after the clustering process. The temperature of filtered wells is calculated by using Richmann's rule of mixing [42], shown in Equation (1):

$$T = \frac{\sum_{i=1}^{n} \dot{m}_i \cdot c_i \cdot T_i}{\sum_{i=1}^{n} \dot{m}_i \cdot c_i}, \tag{1}$$

where the T represents the fluid's temperature (°C), $\dot{m}$ is the mass flow (kg/s), c is the specific heat capacity (J/kg°C) of the geothermal fluid from each well, and n is the number of wells. Moreover, the density of the mixed fluid [43] from the geothermal water from all wells used in the methodology is calculated by using Equation (2):

$$\rho_f = \frac{\sum_{i=1}^{n} \rho_i \cdot \dot{m}_i}{\sum_{i=1}^{n} \dot{m}_i},$$ (2)

where ρ_f represents the density of the geothermal water (kg/m^3) from a specific well, and $\dot{m}$ is the mass flow (kg/s) from the well. The specific heat capacity of the mixed fluid is determined by using Equation (3) [44]:

$$c_p = \frac{\sum_{i=1}^{n} \dot{m}_i \cdot c_i}{\sum_{i=1}^{n} \dot{m}_i},$$ (3)

where the c_p refers to the specific heat capacity of geothermal fluid (J/kg°C), and $\dot{m}$ is the mass flow (kg/s).

For the scenarios with heat production, thermal energy production concerns the exploitation of a fixed temperature range between geothermal fluid production temperature and fixed outlet temperature from the plant where heat can be delivered to multiple end-users during the entire year or serves as the base load thermal power plant. The installed capacity (Q_{th}) is a direct function of specific heat capacity of geothermal fluid, density of geothermal fluid, fluid flow, and the temperature difference between the temperature inlet and outlet in the thermal power plant, as shown in Equation (4):

$$Q_{th} = c_p \cdot \rho_f \cdot q \cdot (T_i - T_o),$$ (4)

where c_p is the specific heat capacity of the geothermal fluid (J/kg°C), ρ_f represents the density of the geothermal water (kg/m^3), q is the fluid flow (m^3/s), T_i is the wellhead temperature (°C), i.e., the temperature at the inlet of the thermal power plant, and T_o is the temperature (°C) at the outlet of the thermal power plant. The produced heat is calculated by using Equation (5):

$$E_{th} = c_p \cdot \rho_f \cdot q \cdot (T_i - T_o) \cdot t \cdot \eta_{HE},$$ (5)

where the t is time (hours) in which the thermal power plant is operating, and η_{HE} is the efficiency (%) of the heat exchanger between the geothermal (circulating) fluid and the working fluid in the secondary loop (end user side).

For Scenario 3, i.e., heat production using the borehole heat exchanger, heat transfer between the reservoir rocks and the circulating fluid is quantified by using the temperature ratio (X_{TR}). Temperature ratio is the number that represents heat transfer correlation between the reservoir rock and circulating fluid, including the heat transfer through cement, casing, tubing, and tubing isolation, i.e., the ratio of the temperature outlet from the deep borehole heat exchanger and bottomhole temperature. It is assumed that the reservoir temperature is the same as the temperature of the reservoir (geothermal) fluid and that the changes in reservoir porosity and thermal conductivity do not change significantly in the reservoir. The theoretical lower limit of X_{TR} is zero, which means that there is no heat transfer between the fluid and the rock, and the theoretical upper limit is one, which means that heat transfer from the reservoir rock to circulating fluid happened completely. The temperature ratio is derived from the database of several real and simulated cases of deep borehole heat exchanger performances provided in Appendix A. The temperature ratio should imply how much heat is lost in the transfer process by using circulating fluid rather than geothermal fluid. The maximum ratio is 0.864, which means that more than 86% of heat from geothermal reservoir is transferred on the circulating fluid. The minimum ratio is 0.240, which means that only 24% of heat from reservoir rock is transferred to the circulating fluid. The generated temperature ratio enables the simplified estimation of heat transfer without the need for simulation or measurements on the field. General knowledge about favorable technical and geological parameters and configuration is of key importance. The explained factor, X_{TR}, is included in Equations (6) and (7) for the calculation of the installed capacity and heat production as the factors, which are multiplied with reservoir temperature:

$$Q_{th} = c_p \cdot \rho_f \cdot q \cdot (X_{TR} \cdot T_r - T_o) \tag{6}$$

$$E_{th} = c_p \cdot \rho_f \cdot q \cdot (X_{TR} \cdot T_r - T_o) \cdot t \cdot \eta_{HE} \tag{7}$$

where X_{TR} is the temperature ratio used to describe the heat transfer between the reservoir and the circulating fluid (-), T_r is the reservoir temperature, and T_o is the injection temperature ($^\circ$C) which is the temperature at the outlet of the thermal power plant.

For the scenarios with electricity production using Organic Rankine Cycle (ORC) technology to assess heat exchange performances of the used binary power plant, thermal efficiency is analysed and calculated by using the following equations. For the wellhead temperatures higher than 120 $^\circ$C, the method proposed by the Massachusetts Institute of Technology [3] is used. Namely, regression Equation (8), based on the data from fourteen ORC power plant, is used to calculate thermal efficiency (%):

$$\eta_{ORC} = 0.0005 \cdot T_{Inlet}^2 - 0.0577 \cdot T_{Inlet} + 8.2897, \tag{8}$$

where T_{inlet} ($^\circ$C) represents the production temperature in the scenario with the production and injection wells, and in the BHE scenarios, it represents the product of temperature ratio and the reservoir temperature at a certain depth.

The installed power is calculated by using Equation (9) [3].

$$Q_{el} = c_p \cdot \rho_f \cdot q \cdot (T_{Inlet} - T_o), \tag{9}$$

The produced energy is a direct function of installed power, thermal efficiency, and operating time, as shown in Equation (10).

$$E_{el} = c_p \cdot \rho_f \cdot q \cdot (T_{inlet} - T_o) \cdot t \cdot \eta_{ORC} \tag{10}$$

Additionally, for wellhead temperatures lower than 120 $^\circ$C, the approach from the Deliverable D7.1, based on the data provided from ENOGIA for the purposes of the EU Horizon 2020 project MEET [45], was applied in order to evaluate the ORC power plant production. The following parameters should be considered:

- DT—temperature difference on primary side of dedicated heat exchanger;
- η_{ORC}—net ORC power plant efficiency as function of geothermal brine extraction temperature (circulating fluid temperature in case of BHE) and DT.

As observed, both η_{ORC} are a function of two variables. In addition, there was a limited number of ORC operating points available from ENOGIA. For that reason, the "MATLAB Curve Fitting Tool" was used to approximate these three-dimensional relationships. Polynomial approximation including third degree was performed.

Equation (11) represents the functional relationship between net ORC power plant efficiency (z), geothermal water extraction temperature, or circulating fluid temperature in case of BHE (y) and DT (x).

$$\begin{aligned} z(x,y) = \quad & -0.06849 - 0.001452 * x + 0.002209 * y - 1.017e^{-5} * x^2 \\ & +1.639e^{-5} * x * y - 1.096e^{-5} * y^2 + 3.241e^{-8} * x^2 * y \\ & -4.203e^{-8} * x * y^2 + 1.866e^{-8} * y^3, \end{aligned} \tag{11}$$

It should be noted that relationship from Equation (11) between these variables is best suited for brine extraction temperature values in the range from 80 $^\circ$C to 120 $^\circ$C and for DT values in the range from 0 $^\circ$C to 40 $^\circ$C. In cases when Equation (11) is used for values outside of the suggested ranges, slightly less accurate results can be expected.

Finally, installed power and produced electricity are calculated according to Equations (9) and (10) with corresponding power plant efficiencies.

3.3.2. Levelized Cost of Energy

The levelized cost of electricity or heat (LCOE or LCOH) is defined as the total discounted lifetime costs of an energy project divided by the total discounted amount of energy it either produces or saves in its lifetime [46].

The approach used in this methodology is based on a discounted cash flow (DCF) analysis. Additionally, it must be emphasized that the LCOE/LCOH metric should be considered rather as an informing measure for investment decisions than an absolute decision metric. Actual system and project planning should also consider reliability issues and other factors. Namely, the availability factor of the power plant, i.e., the time that the plant is available for running influences the produced amount of electricity in a specific period.

The LCOE/LCOH is calculated according to Equations (12) and (13).

$$LCOE = \frac{\sum_{t=1}^{TPL} \frac{I_t - S_t}{(1+r)^t} + \sum_{t=1}^{TPL} \frac{OM_t \cdot (1-TR)}{(1+r)^t}}{\sum_{t=1}^{TPL} \frac{EE_t}{(1+r)^t}} \tag{12}$$

$$LCOH = \frac{\sum_{t=1}^{TPL} \frac{I_t - S_t}{(1+r)^t} + \sum_{t=1}^{TPL} \frac{OM_t \cdot (1-TR)}{(1+r)^t}}{\sum_{t=1}^{TPL} \frac{EH_t}{(1+r)^t}} \tag{13}$$

In Equations (12) and (13), TPL represents the total lifetime of the project [years], r represents the nominal discount rate (%/100), I_t represents investment costs in year t, S_t represents incentives or subsidies in year t, OM_t represents operation and maintenance costs in year t, TR represents effective tax rate, EE_t represents generated electricity in year t, and EH_t represents produced heating energy in year t. Total investment costs I_t for specific year t in Equations (12) and (13) are calculated as shown in Equation (14):

$$I_t = I_t^{exp,est} + I_t^{prod,inje} + I_t^{ppinst} + I_t^{admi,man} + I_t^{other}, \tag{14}$$

where $I_t^{exp,est}$ represents yearly exploration and establishment costs (summarizes the cost of concession or lease acquisition of oil field, permissions, environmental studies, civil work, support facilities, surface exploring, shallow drilling, make-up well deepening, and pre-feasibility and feasibility studies), $I_t^{prod,inje}$ represents yearly production and injection wells and system costs (includes mobilization, drilling, logging, testing, production piping, separators, water tanks, injection piping, production and injection pumps, and corrosion inhibitor systems), I_t^{ppinst} represents yearly power plant installation costs (it includes power plant design and engineering, procurement procedures and complete phase of construction, testing and controlling, grid connection, and transmission), $I_t^{admi,man}$ represents yearly administration and management costs (it includes project management, project and company administration, insurance costs, and different financing fees), and $I_{other,t}$ represents yearly other investment costs not included in any of the aforementioned categories. Additionally, operation and maintenance costs OM_t in year t are calculated according to Equation (15):

$$OM_t = FO\&M_t + O\&M_t^{production\ pump} + O\&M_t^{injection\ pump} + O\&M_t^{other}, \tag{15}$$

where $FO\&M_t$ represents yearly fixed O&M (including labor costs, maintenance of field and/or wells and/or power plant) in Euros, $O\&M_t^{production\ pump}$ (€) represents yearly production pump variable costs that depend on the installed power of the pump, working hours and electricity price, $O\&M_t^{injection\ pump}$ (€) represents yearly injection pump variable costs that depend on the installed power of the pump, working hours, and electricity price, and $O\&M_t^{other}$ (€) represents yearly variable costs that were not covered by other defined categories.

The nominal discount rate, r, is calculated from the real discount rate, r_r, and inflation rate, i, according to Equation (16).

$$r = (1 + r_r) \cdot (1 + i) - 1 \tag{16}$$

For combined heat and power (CHP) applications, more complex equations are used, dependent on what the main product is. Namely, the LCOE is used if the decision maker chooses the main product of interest as electricity; consequently, when calculating the LCOE for CHP plant, revenues from heat sales must be deduced, and if the main product is heat, when calculating the LCOH for CHP plant, revenues from electricity sales must be deduced (Equations (17) and (18)):

$$LCOE(chp) = \frac{\sum_{t=1}^{TPL} \frac{I_t - S_t}{(1+r)^t} + \sum_{t=1}^{TPL} \frac{OM_t \cdot (1-TR)}{(1+r)^t} - \sum_{t=1}^{TS} \frac{RHS_t \cdot (1-TR)}{(1+r)^t} - \sum_{t=TS+1}^{TPL} \frac{RHM_t \cdot (1-TR)}{(1+r)^t}}{\sum_{t=1}^{TPL} \frac{EH_t}{(1+r)^t}}, \tag{17}$$

$$LCOH(chp) = \frac{\sum_{t=1}^{TPL} \frac{I_t - S_t}{(1+r)^t} + \sum_{t=1}^{TPL} \frac{OM_t \cdot (1-TR)}{(1+r)^t} - \sum_{t=1}^{TS} \frac{RES_t \cdot (1-TR)}{(1+r)^t} - \sum_{t=TS+1}^{TPL} \frac{REM_t \cdot (1-TR)}{(1+r)^t}}{\sum_{t=1}^{TPL} \frac{EE_t}{(1+r)^t}}, \tag{18}$$

where RHS_t represents revenues from subsidized heating power sales in year t, RHM_t represents revenues from the market heating power sales in year t, RES_t represents revenues from subsidized electricity sales in year t, REM_t represents revenues from the market electricity sales in year t, and TS represents the duration of subsidized price of electricity or heating power.

3.3.3. Net Present Value

The NPV metric is in this methodology is calculated as shown in Equation (19):

$$NPV = \sum_{t=0}^{TPL} a_t \cdot S_t = \frac{S_0}{(1+r)^0} + \frac{S_1}{(1+r)^1} + \ldots + \frac{S_T}{(1+r)^T}, \tag{19}$$

where S_t is the balance of cash flow (inflows minus outflows) at time t, a_t is the financial discount factor chosen for discount at time t, and r is the nominal discount factor. The nominal discount factor is calculated according to Equation (16). Inflows include revenues obtained from electricity and/or heat sells. Outflows include investment costs, which are calculated according to Equation (14) and operating costs, which are further calculated according to Equation (15) but also include yearly tax payments.

3.3.4. Avoided CO$_2$ Emissions

In order to assess the environmental impact of such conversion projects and, consequently, to approximate the money savings based on this indicator, calculation of avoided CO$_2$ emissions during operational phase of the plant is proposed and calculated in this methodology. The avoided emissions during operational phase are calculated based on the comparison with the production of the same services with the reference electricity mix and reference heat mix, respectively. The reference mixes are country specific and represent business-as-usual developments until 2019 for each country.

For scenarios with only electricity generation, the amount of avoided CO$_2$ emissions (tons) is calculated as stated in Equation (20):

$$E_{CO_2} = \sum_{p=1}^{top} \left(\dot{E}_p \cdot e_{CO_2,elemix} \right), \tag{20}$$

where t_{op} represents the duration of the operational phase of the plant, $\dot{E}_p$ is the net electricity production by system at the operating conditions of period p (MWh$_e$), and $e_{CO_2,elemix}$

is the specific CO_2 emissions of electricity production from the reference electricity mix ($kgCO_2/MWh_e$).

For scenarios with only heating power production, the amount of avoided CO_2 emissions (tons) is calculated as stated in Equation (21):

$$E_{CO_2} = \sum_{p=1}^{t_{op}} \left(\dot{Q}_p \cdot e_{CO_2,heatmix} \right), \tag{21}$$

where t_{op} represents the duration of the operational phase of the plant, $\dot{Q}_p$ is the produced heat energy to cover heating requirement during period p (MWh_{th}), and $e_{CO_2,heatmix}$ is the specific CO_2 emissions of heating production from a heat mix ($kgCO_2/MWh_{th}$).

In case of CHP scenario, Equations (20) and (21) are combined into Equation (22).

$$E_{CO_2} = \sum_{p=1}^{t_{op}} \left(\dot{E}_p \cdot e_{CO_2,elemix} + \dot{Q}_p \cdot e_{CO_2,heatmix} \right), \tag{22}$$

4. Case Study

A case study was formed, i.e., mature oil field with high water-cut production served as the basis for forming the case and conduction of two-layer clustering. The oil field formed for the case study was slightly altered from the existing oil field for the purposes of retaining realistic parameters needed for the conduction of further calculations. The remaining required input data were modelled in such manner as to replicate geothermal systems that could be found in reality regarding technology, modelling of the developed scenarios, environmental and economic data such as the market price of electricity, emission factors, share of fossil fuels in total energy mix for each country, weather data for each country, etc. The rest of the input data for the purposes of the heat demand calculation and variable operational cost of production pumps, injection pumps, and BHE pumps are based on proxy values and can be replaced with the user's input.

In order for the outputs of methodology to be comparable with the outputs from other scenarios, it is desirable for the input data to be similar, referring to data such as heat needs, temperature difference, downtime, etc.

Regarding the heat production, the heat produced by exploiting the default temperature range will be shown. The calculation of heat demand is based on the building's heating system, i.e., heating curve [47]. The operational cost of production pumps is based on [48], where inserting data is required with respect to the well's geometrics for each well, well fluids parameters, productivity, and the associated pressures such as dynamic pressure, differential pressure, hydrostatic pressure, pump intake pressure, etc. For the wells that are newly included in the production of geothermal water, at the start of the calculation there is a short check up to verify if there is a need for production pump installation; if the outcome is positive, the well with its parameters enters the above-mentioned calculation. For the selection of the production pump, an optimization process of selection based on the Schlumberger catalogue [49] is developed where the pump with highest efficiency at the corresponding flow is chosen while satisfying the minimum velocity check-up and operating range check-up. The operational cost of injection pump is based on the performance curves of injection pumps installed at the facilities for geothermal exploitation. The operational cost of BHE pump is estimated by calculating the pump's head loss and Darcy–Weisbach friction factor by using Colebrook's equation [50]. For operational costs, it is assumed that the electricity from the grid is used at the market price and electricity generated from the power plant is sold at the subsidized price.

For the purposes of the case study, the oil asset will be called "Reservoir 1," and it is determined to be in France in the Aquitaine basin. The temperatures in the basin are mainly between 65 °C and 90 °C. The reservoir is characterized with tidal and fluvial sandstones interbedded with clays with thermal conductivity of 3 W/m/K. Reservoir 1 for the conversion to geothermal field consists of 26 production wells and 10 injection wells. The choice for performing deep borehole technology can be any well from the field with suitable production temperature of circulating fluid. The well chosen for Scenario 3 will be the well with the maximum temperature of the geothermal fluid at the wellhead. All injection wells are considered to be of suitable properties for the injection of overall fluid flow. In Figure 2, the spatial distribution of production wells is shown.

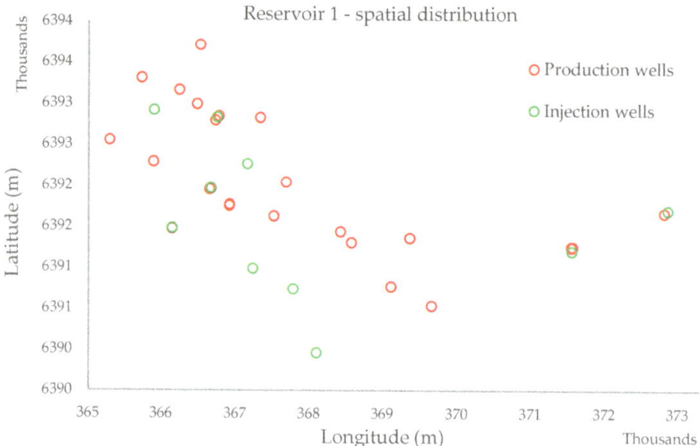

Figure 2. Spatial distribution of production and injection wells at Reservoir 1.

The main input parameters with respect to Reservoir 1 are shown in the Table 2. Most of the data are shown as the average of all wells taken into calculation. The wells on the field were all oil production wells with no newly added wells, and the data used for further clustering and calculations are more detailed and well specific. Depending on well depth, wells on the field range from 22 °C to 97 °C with respect to the temperature of geothermal water at the wellhead. Depending on the distance from the gathering station, pipe material, and insulation, temperature loss from gathering lines from the wellhead also varies from 0.2 °C to 2 °C overall. The yearly thermal dropdown explains the annual decline rate of reservoir temperature, and the yearly water-cut increment represents annual linear water-cut increase. For the simplification of the calculations, two mentioned values are taken as constant during the duration of project.

Table 2. General data about the Reservoir-1.

Input Data	Value	Unit
Overall fluid flow	0.042332	m^3/s
Average wellhead temperature of considered wells	55.67	°C
Average temperature loss through gathering lines	1.23	°C
Reservoir pressure gradient	0.0874	bar/m
Density of produced oil	850	kg/m^3
Density of geothermal water	1014	kg/m^3
Specific heat capacity of geothermal water	3914	J/kg °C
Yearly thermal dropdown	0.5	%
Average reservoir water-cut production	84.22	%
Yearly water-cut increment	0.15	%
Minimum number of well required for the spatial clustering	4	-

4.1. Modelling of the Heat Production Scenarios

The input data for modelling heat production scenarios for all wells on the field are presented in Table 3. Scenario 1, which is about decommissioning oil assets, has no input parameters regarding the production, and it will not be shown in this subchapter or the following one. The downtime presents the yearly percentage of time when the well, plant, and other surface facilities were not operating. It could be due to disruption in production, maintenance, or similar reasons. The outlet temperature is the temperature of the fluid at the outlet of the plant. In Scenario 5, a parallel configuration model was applied where heat demand is calculated based on input data stated in Table 3 needed for heating curve performance computation and, consequently, the building's heat demand. The pipeline temperature coefficient corresponds to temperature loss caused by transmission pipelines from the plant to the end-user. It ranges from 0 to 1, where 1 means all the heat is transferred through the pipeline, and 0 corresponds to total temperature loss. The presented value is dependent on the pipeline material and geometry.

Table 3. Input data for the heat production scenarios.

Scenario 2			Scenario 3					
Input Data	**Value**	**Unit**	**Input Data**	**Value**	**Unit**	**Input Data**	**Value**	**Unit**
Downtime	10	%	Downtime	10	%	Circulating fluid flow	0.004	m^3/s
Outlet temperature	70	°C	Outlet temperature	70	°C	Efficiency of surface heat exchanger	100	%
Efficiency of heat exchanger	100	%	Well depth	3500	m	Temperature ratio	0.718	-
			Specific heat capacity of circulating fluid	4187	J/kg °C	Density of circulating fluid	1000	kg/m^3
Temperature loss from the gathering system to the plant	1	°C	Yearly thermal dropdown of the wellbore	0.5	%	Temperature loss along the wellbore	4	°C
						Geothermal gradient of the well	0.033	°C/m

Scenario 5								
Input Data	**Value**	**Unit**	**Input Data**	**Value**	**Unit**	**Input Data**	**Value**	**Unit**
Type of building	Public building	-	Temperature loss from the gathering system to the plant	1	°C	Pipeline temperature coefficient	0.94	-
Required inside temperature	19	°C	Building surface	12,000	m^2	Thermal pinch-point in heat exchanger	1.5	°C
Outdoor non-heating temperature of the pivot point	20	°C	Minimum water temperature of the pivot point	20	°C	Specific heat capacity of the cold loop fluid	4180	J/kg °C
Outdoor non-heating temperature	17	°C	Maximum flow in the cold loop	30	m^3/h	Density of the cold loop fluid	1000	kg/m^3
Minimum water temperature	35	°C	Minimum flow in the cold loop	10	m^3/h			

4.2. Modelling of the Electricity Production Scenarios

The input data for modelling electricity production scenarios for all wells on the field are presented in Table 4. The outlet temperature is the temperature of the fluid at the outlet of the ORC power plant.

Table 4. Input data for the electricity production scenarios.

Scenario 4			Scenario 5		
Input Data	**Value**	**Unit**	**Input Data**	**Value**	**Unit**
Downtime	10	%	Downtime	10	%
Outlet temperature	70	°C	Outlet temperature	70	°C
Temperature loss from the gathering system to the power plant	1	°C	Temperature loss from the gathering system to the power plant	1	°C
			Pipeline temperature coefficient	0.94	-

4.3. Environmental and Economic Input Parameters

For further calculations of the outputs, it is important to define environmental and economic parameters that will be used for calculating avoided CO_2 emissions, levelized cost of electricity or heat, and net present value. The stated outputs are dependent on production quantities and will have the same input parameters except the specific costs that are related to the installed power.

Economic input parameters used in calculations are shown in Table 5. For the market price of electricity, the ARIMA model developed in MATLAB was used to predict the market price of electricity for the time of the project duration based on the historical values [51].

Table 5. Economic parameters used in calculations.

Input Data	Value				Unit
	Scenario 2	Scenario 3	Scenario 4	Scenario 5	
Effective tax rate			30		%
Inflation rate			2.3		%
Discount rate			6.5		%
Electricity market price (average)	-	-	0.03628	0.03628	€/kWh
Electricity selling price	-	-	0.065	0.065	€/kWh
Heat selling price	0.045	0.045	-	-	€/kWh
Lifetime of the project			20		years

The input values regarding the environmental aspects are stated in the Table 6. The share of each fossil fuel in total fossil fuel electricity or heat generation is taken here as the default value and is based on the data from [52] for a chosen country. The emission factors of each fossil fuel for each energy type, i.e., electricity or heat, are obtained from [53] and will not be publicly shown due to legal reasons.

Table 6. Environmental parameters used in calculation.

Input Data	Value	Unit
Share of coal in total fossil fuel electricity generation	23	%
Share of oil in total fossil fuel electricity generation	12	%
Share of natural gas in total fossil fuel electricity generation	65	%
Share of coal in total fossil fuel heat generation	7	%
Share of oil in total fossil fuel heat generation	11	%
Share of natural gas in total fossil fuel heat generation	82	%

5. Results

After conducting two-stage clustering, the results of temperature and spatial clustering are shown in the Table 7. The column "Number of end-uses" represents how many end-uses are possible for the conversion of each well based on the wellhead temperature, respectively. The first eight wells have low temperature for district heating and electricity generation but are adequate for the installation of heat pump systems; as such, they are automatically excluded from further calculations. The remaining wells from the field, i.e., eighteen wells in total, were chosen for further calculations of methodology outputs. The first subcase "Whole field" includes all eighteen wells. Furthermore, after applying the DBSCAN method for spatial clustering, the wells were sorted into two clusters, "Cluster 1" and "Cluster 2," without outlier wells. Namely, eighteen wells in total were chosen for further calculations of methodology outputs for sub-cases as follows:

- Whole field—18 production wells;
- Cluster 1—16 production wells;
- Cluster 2—2 production wells.

Table 7. Results of temperature and spatial clustering.

Well Name	Wellhead Temperature (°C)	Number of End-Uses	Cluster Number
Well 1	22	1	1
Well 2	23	1	1
Well 3	23	1	2
Well 4	25.3	1	2
Well 5	27	1	1
Well 6	27	1	1
Well 7	29	1	1
Well 8	32	1	1
Well 9	39	1	1
Well 10	47.6	1	1
Well 11	49.3	1	1
Well 12	52.8	1	1
Well 13	54.5	1	1
Well 14	55	1	1
Well 15	60	2	1
Well 16	61.15	2	1
Well 17	68.8	2	1
Well 18	74	2	1
Well 19	74.95	2	2
Well 20	75	2	2
Well 21	78	2	1
Well 22	81.5	2	1
Well 23	83.6	2	1
Well 24	92	2	1
Well 25	95	2	1
Well 26	97	2	1

The input data for the subcases Whole field and Cluster 1 remain the same, as described in Sections 4.1 and 4.2. For Cluster 2, the data are changed in order to have realistic scenarios and meaningful production and are shown in Table 8. The spatial representation of conducted clustering of production wells is shown in Figure 3.

Table 8. The changed input values for the sub-case "Cluster 2".

Input Data	Value	Unit
Outlet temperature from the plant in Scenario 2	50	°C
Outlet temperature from the plant in Scenario 4	47	°C
Outlet temperature from the plant in Scenario 5	47	°C

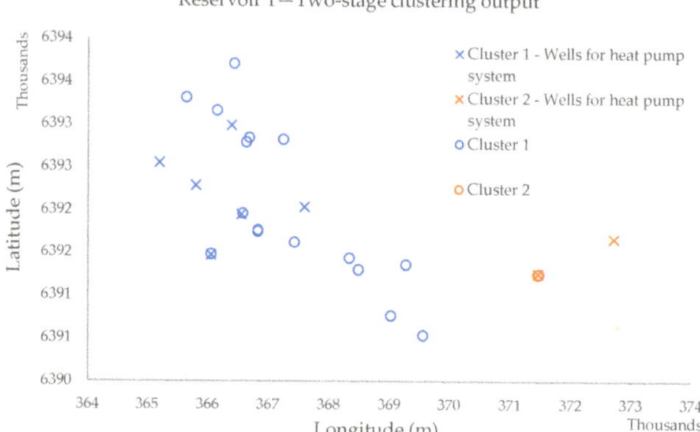

Figure 3. Graphical results of the two-stage clustering.

5.1. Calculated Input Values

After the two-stage clustering, the values described in Section 3.3.1. are calculated and shown for each scenario, i.e., subcase in Tables 9–12: calculated input values for Scenario 5. The calculated values, together with the required input data, are substituted into further calculation of methodology outputs.

Table 9. Calculated input values for Scenario 2.

Calculated Value	Unit	Whole Field	Cluster 1	Cluster 2
Wellhead temperature	°C	81.16	82.10	72.93
Specific heat capacity of geothermal water	J/kg °C	3904.49	3906.12	3890.20
Density of geothermal water	kg/m^3	1012.35	1011.38	1020.79
Total geothermal water flow	m^3/s	0.0406	0.0364	0.0042

Table 10. Calculated input values for Scenario 3.

Calculated Value	Unit	Whole Field	Cluster 1	Cluster 2
Wellhead temperature	°C	78.93	78.93	82.44

Table 11. Calculated input values for Scenario 4.

Calculated Value	Unit	Whole Field	Cluster 1	Cluster 2
Wellhead temperature	°C	84.48	85.96	72.93
Specific heat capacity of geothermal water	J/kg °C	3900.69	3902.03	3890.20
Density of geothermal water	kg/m^3	1012.30	1011.22	1020.79
Total geothermal water flow	m^3/s	0.0366	0.0325	0.0042
Thermal efficiency of the ORC plant	%	4.40	4.44	2.35

Table 12. Calculated input values for Scenario 5.

Calculated Value	Unit	Whole Field	Cluster 1	Cluster 2
Wellhead temperature	°C	84.48	85.96	72.93
Specific heat capacity of geothermal water	J/kg °C	3900.69	3902.03	3890.20
Density of geothermal water	kg/m^3	1012.30	1011.22	1020.79
Total geothermal water flow	m^3/s	0.0366	0.0325	0.0042
Thermal efficiency of the ORC plant	%	4.40	4.44	2.35
Available fluid for the electricity generation	m^3/s	0.0307	0.0271	0.0006

Contrary to greenfield geothermal projects, end-of-life oil wells conversion into geothermal ones enables omitting more than a half of the costs related to drilling and stimulation. The values for CAPEX and OPEX are for the purpose of this study estimated based on real data collected by the authors. CAPEX is represented with specific investment costs in Euro per kilowatt and consists of costs included in Equation (14), which depend on the analysed scenario. For each scenario, OPEX is calculated according to Equation (15). Additionally, tax rates are country specific, the discount rate was calculated according to Equation (16) where annual inflation rate for France at the moment of the analysis was 2.3% [54], and the discount rate was considered to be 6.5% [55,56].

CAPEX and OPEX for each scenario and "Whole field" case and additional subcases "Cluster 1" and "Cluster 2" are shown in Table 13.

Table 13. Calculated CAPEX and OPEX for each case and each scenario.

Input Data			Value				Unit
			Scenario 2	Scenario 3	Scenario 4	Scenario 5	
Whole field	CAPEX	ORC	-	-	5667.48	5566.87	€/kW
		Heat	449.26	2340.47	-	385.14	
	OPEX (average)		0.0906	0.0121	1.0217	0.3147	€/kWh
Cluster 1	CAPEX	ORC	-	-	6333.58	5469.57	€/kW
		Heat	440.83	2340.47	-	380.72	
	OPEX (average)		0.0733	0.8337	0.0121	0.2713	€/kWh
Cluster 2	CAPEX	ORC	-	-	6710.21	5714.34	€/kW
		Heat	392.69	1511.67	-	367.37	
	OPEX (average)		0.0396	0.0086	1.2715	0.0780	€/kWh

5.2. Methodology Outputs

The graphical results for each subcase and its outputs for each scenario are shown for the first year of operation. The production quantities of heat and electricity scenarios are shown in Figures 4 and 5.

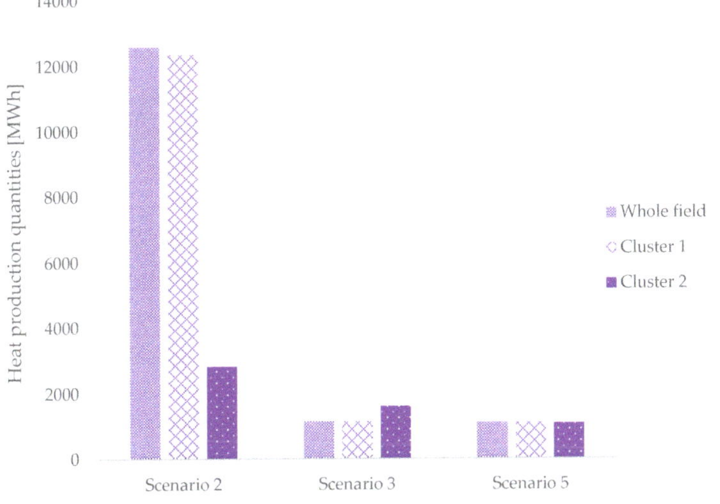

Figure 4. Heat production quantities for each of three sub-cases.

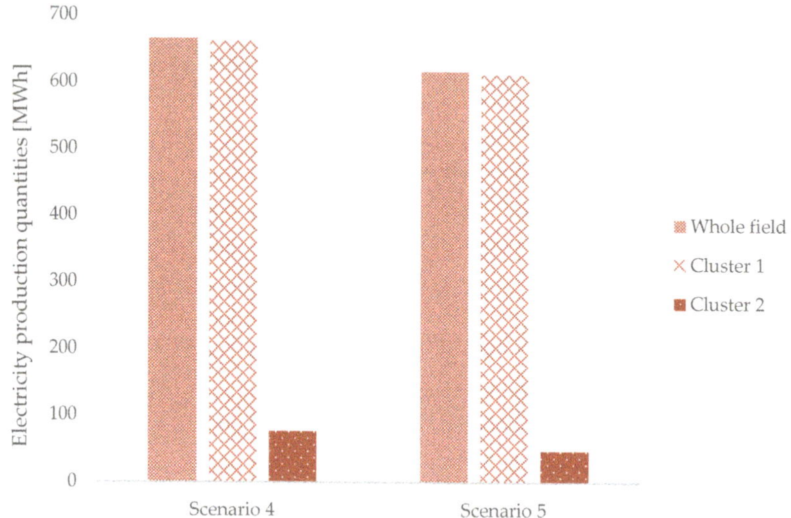

Figure 5. Electricity production quantities for each of three subcases.

The differences between the production quantities are due to number of wells included in each subcase; thus, fluid flow and the temperature varies. In Scenario 3, the production quantities between each subcase are directly dependent of the fluid's temperature, since the remaining input data are the same; hence, heat production is the greatest for Cluster 2. In Scenario 5, the heat production quantities are similar for all three sub-cases since it is required in order to satisfy the heat demand first. The electricity production temperatures are directly dependent on fluid flow and thermal efficiency of the ORC turbine, which is conditioned by the geothermal fluid temperature at the inlet of the power plant and the temperature difference between the mentioned temperature and the outlet temperature from the power plant. The levelized costs of heat and levelized costs of electricity are shown in Figures 6 and 7.

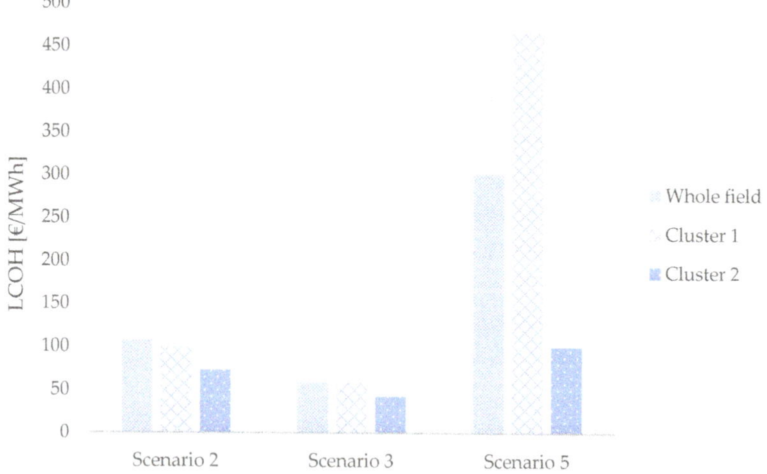

Figure 6. Levelized cost of heat for each of three subcases.

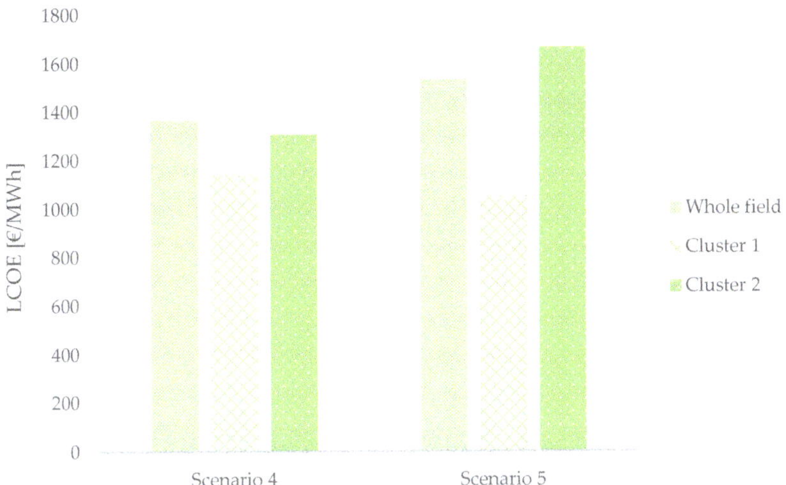

Figure 7. Levelized cost of electricity for each of three subcases.

The levelized cost of heat is greater that the levelized cost of electricity since the production of electricity is significantly lower than the production of heat, according to the set case study. The net present value is generally negative since it is required in investing in the conversion to geothermal assets, and due to the great investments in the first year of the operation period, the expenses exceed revenues. The net present value for each scenario of three subcases is shown in Figure 8.

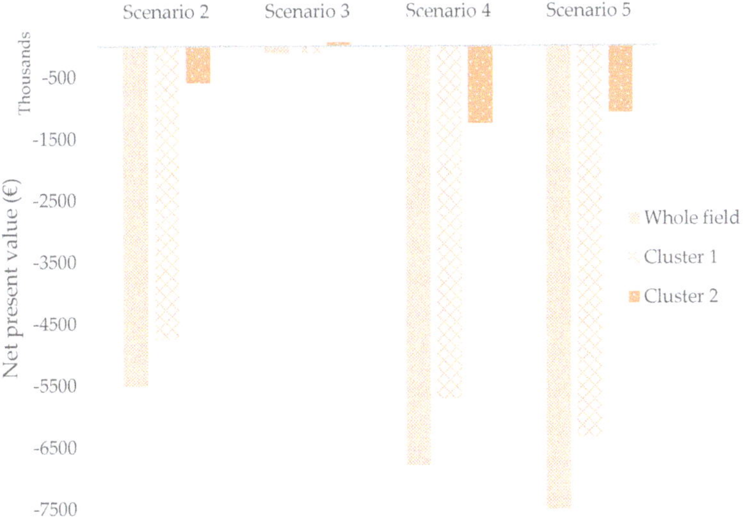

Figure 8. Net present value for each scenario of three subcases.

The CO_2 emissions that are avoided in the production of geothermal energy are directly dependent of the energy production quantities, since the emission factors and the share of each fossil fuel in the fossil fuel mix are the same and are, as said, country specific. The avoided CO_2 emissions for each scenario of the three sub-cases are shown in Figure 9.

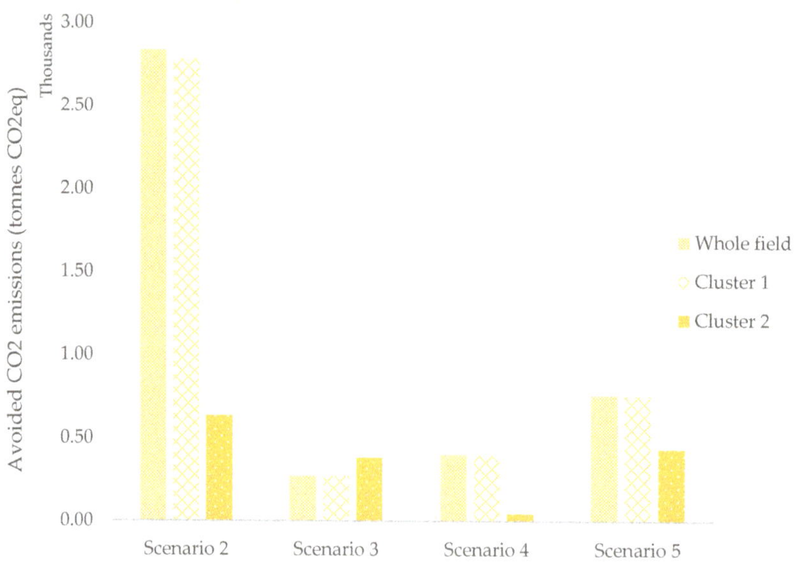

Figure 9. Avoided CO_2 emissions for each scenario of three subcases.

6. Discussion

When comparing the production of energy in each scenario and for each subcase, the production directly depends on fluid flow and the inlet temperature of the geothermal fluid or circulating fluid. When producing heat, in Scenario 2, for the Whole field and Cluster 1 subcases, the quantities are nearly the same; the greater inlet temperature in Cluster 1 compensates for the part of the lower flow that is caused by the lower number of production wells compared to Whole field. In Scenario 3, heat production using the borehole heat exchanger produces the most heat in Cluster 2 since the wellhead temperature is higher by more than 3 °C compared to rest of the subcases. In Scenario 5, subcases Whole field and Cluster 1 with similar inlet temperatures and flow of geothermal fluid managed to satisfy the heat needs where the remaining available flow was directed to the electricity production facility. Regarding Cluster 2, the changed input parameter of the outlet temperature, i.e., the greater exploitable temperature range, delivered enough heat to satisfy head demand, and more than 60% of the available flow was directed to the electricity production facility. Regarding electricity production, subcase Whole field produced more electricity than Cluster 1, where greater fluid flow in Whole field compensated for greater wellhead temperature and thermal efficiency of ORC turbine in Cluster 1. Cluster 2, with its two production wells, the wellhead temperature of 72.93 °C, and a low thermal efficiency of the ORC turbine (2.35%), produced about 90% less electricity than Whole field and Cluster 1, even with decreased outlet temperature from the ORC power plant. As stated before, in line with the objectives of MEET 2020 for enhancing heat-to-power conversion, the modelled case study uses low temperature (60 °C—90 °C) at the inlet of the power plant where smart mobile Organic Rankine Cycle units can be used for electricity production. Using mobile ORC greatly enlarges the potential sites that could be exploited together with the use of abandoned oil wells. Such usage of low temperature sources can result in uneconomic scenarios with respect to lower energy production quantities, but existing oil wells can minimize capital investments and increase cost competitiveness.

6.1. Economic Results

The levelized cost of electricity or heat and net present values are the main indicators for investment. For the modelled case study and associated subcases, the mentioned outputs quite differ depending on each subcase and scenario.

For the Whole field subcase, LCOH varies from 59.41 €/MWh in Scenario 3 to 301.36 €/MWh in Scenario 5 for heat generation, and the LCOE varies from 1365.82 €/MWh in Scenario 4 up to 1532.94 €/MWh in Scenario 5 for electricity production. Such variations can be explained by the different costs of capital investment, operational costs, and produced energy. It can be observed in the scenarios of electricity generation that lower thermal efficiency greatly affects production where revenue from selling electricity cannot exceed the cost of capital investment and high operational cost of running the production pumps. The operational cost of running the production pumps corresponds to the changes in electricity market price since the electricity from the grid is used to power pumps.

For the Cluster 1 subcase, LCOH varies from 59.41 €/MWh in Scenario 3 up to 465.76 €/MWh in Scenario 5, and as for LCOE, it varies from a minimum of 1052.14 €/MWh in Scenario 5 up to 1143.25 €/MWh in Scenario 4. Different values between two mentioned subcases can be explained with the difference in the number of production wells where there are 18 production wells in Whole field and 16 production wells in Cluster 1. For Scenario 3, a similar value of LCOH results from the fact that the same well was chosen to be converted into the borehole heat exchanger.

For subcase Cluster 2, there are two production wells, and LCOH ranges from 42.76 €/MWh in Scenario 3 to 100.21 €/MWh in Scenario 5, and LCOE ranges from 1307.08 €/MWh in Scenario 4 to 1664.96 €/MWh in Scenario 5. The lower values of LCOH in this subcase can be explained by lower pump operation costs and higher inlet temperatures in Scenario 3 for the borehole heat exchanger.

As for the entire case study, Cluster 2 has the lowest values of LCOH. In Scenario 3, Whole field and Cluster 1 have the same LCOH, but they slightly differ in Scenario 2. The largest difference is in Scenario 5 where the LCOH of Cluster 1 exceeds the LCOH of Whole field since the calculation of LCOH counts for the overall investment cost of combined heat and power, and the revenue from the electricity generation is subtracted. The LCOE values are the lowest for Cluster 1 in both scenarios and for Scenario 4; for the subcases Whole field and Cluster 2, the LCOE is slightly lower in the latter subcase due to lower thermal efficiency. In Scenario 5, Cluster 2 has the highest LCOE due to lower electricity production quantities and the result of subtracting revenues from selling heat.

As for the net present value of the case study, the values range from −7.616 M€ for Scenario 5 up to −0.116 M€ for Scenario 3 in the Whole field subcase, from −6.351 M€ for Scenario 5 up to −0.116 M€ in Scenario 3 in the Cluster 1 subcase, and from −1.263 M€ in Scenario 4 to 0.063 M€ in Scenario 3 in sub-case Cluster 2. The negative values of the net present value are the result of the investment cost of the plant at the beginning of the project and the replacement cost of the production pumps in year 15 of the project duration. Another reason for the negative NPV values is the high operating costs of production pumps and injection pumps that depend upon the electricity market price and the running time of the facility. In general, the lowest net present value for all scenarios is the Whole field followed by Cluster 1 where the differences are manifested from the production pump investment and operational cost of the pumps, among other associated costs. Cluster 2 has higher net present values since it consists of only two production wells and one injection well. Scenario 3 for all subcases has the highest net present value since the investment consists of plant and well configuration costs. For subcase Cluster 2, Scenario 3 has a positive net present value since the revenue from the heat produced exceeds the investment costs.

In general, the heat production scenarios are more economically feasible than the electricity generation scenarios due to low production quantities of electricity. The greater heat production quantities cover the initial investment cost of the oil-to-water conversion, but pump replacement lowers the cumulative cash flow of each scenario, along with the operational cost of each pump.

6.2. Environmental Results

Although almost all the scenarios of subcases are not economically feasible, it is important to elaborate about their environmental footprint regarding the potentially produced CO_2 emissions. The CO_2 savings during the operational period are shown as the avoided CO_2 emission, which is substituted with the geothermal exploitation. The quantity of the avoided emissions is directly correlated with energy production, emission factors of each fossil fuel, and the share of fossil fuel type in the total share of fossil fuels. The latter two are the same for all scenarios performed; thus, the only influencing value is the energy produced. The avoided CO_2 emissions range from 49.94 to 2837.73 t of CO_2 eq/year for Whole field, from 49.64 to 2783.77 t of CO_2 eq/year for Cluster 1, and it ranges from 29.54 to 637.01 t of CO_2 eq/year for Cluster 2. It can be concluded that the highest production generates the greatest CO_2 savings, and it cannot stand alone as the output based on which decisions will be made.

7. Conclusions

The presented methodology and the demonstrated case study offer solutions for the conversion of mature or abandoned oil fields to a geothermal asset and, for this reason, extend the production life of the reservoir. The comprehensive methodology takes into account production technology, economic and environmental parameters, and, together with the presented two-stage clustering, provides various options for further converting petroleum to a geothermal facility while regulatory and policy aspects of such action are left with the knowledge of the user or potential investor, since this is highly country and project specific.

One of the main features of the conversion is bi-level clustering, which facilitates firstly clustering of the wells according to geothermal fluid temperature into a different end-use group, and secondly, clustering of the wells into spatial groups according to the distance between each well. This approach allows optimal conversion and usage of the cumulative production flow from the production wells, simultaneously minimizing the costs for piping infrastructure and power plant spatial positioning and avoiding the decommissioning and abandonment costs of an oil field. An extensive review of input parameters and calculated values such as wellhead temperature, geothermal water flow, specific heat capacity, and density of geothermal fluid produced a thorough background for creating the different scenarios for conversion.

The methodology was applied to the modelled Reservoir 1, which replicates the petroleum reservoir that could be found in reality, in order to evaluate the best conversion scenario for the Whole field or the given clusters for the modelled case study. The outputs indicate that the best scenarios for the oil-to-water conversion were heat production scenarios due to highest production and avoided CO_2 emission quantities, which are directly related. The calculated economic parameters, LCOE, LCOH, and NPV, indicate that the optimal scenario for conversion was Scenario 3 for performing the deep borehole heat exchanger in all three subcases due to its lowest investment and operating costs, followed by Scenario 2 where production and injection wells are used to generate heat. Temperature clustering enabled considering a greater number of wells in heat production calculation rather than in electricity generation scenarios that had influence on the cumulative flow and the temperature of the geothermal fluid. The clustered wells showed different outputs in each cluster, which considered pipeline costs due to spatial clustering, and since there were no newly added wells, the pipeline cost was reduced to a minimum.

The mentioned scenarios resulted in the different main outputs such as production quantities, levelized cost of electricity, levelized cost of heat, and net present value which served as the peculiar roadmap towards the optimal oil-to-water conversion.

Author Contributions: Conceptualization, J.H. and S.R.; methodology, J.H.; validation, J.H., S.R. and E.L.; formal analysis, J.H. and S.R.; data curation, J.H., S.R. and I.R.; writing—original draft preparation, J.H. and S.R.; writing—review and editing, J.H., S.R. and I.R.; visualization, J.H. and S.R. All authors have read and agreed to the published version of the manuscript.

Funding: This work has received funding from the European Union's Horizon 2020 research and innovation program (Grant agreement № 792037—MEET Project).

Data Availability Statement: All data related to this study can be obtained from the corresponding author upon request.

Acknowledgments: The authors are grateful to Guillaume Ravier for contributing to cost estimation for corresponding outputs and for his helpful comments.

Conflicts of Interest: The authors declare no conflict of interest.

Appendix A

Table A1. Temperature ratio database.

Source	Case	Working Fluid	Depth (m)	Bottomhole Temperature (°C)	Outlet Temperature from BHE (°C)	Temperature Ratio
[57]	Simulation	R–C318	5950	165.00	100.38	0.608
[58]	Real	Isobutane	1050	154.70	75.95	0.491
	Real	Isobutane	1050	154.70	76.37	0.494
	Real	Isobutane	1050	154.70	74.51	0.482
	Real	Isobutane	1050	154.70	71.21	0.460
	Real	Propane	1050	154.70	77.75	0.503
	Real	Propane	1050	154.70	76.10	0.492
	Real	Propane	1050	154.70	73.61	0.476
	Real	Isopentane	1050	154.70	81.97	0.530
	Real	Isopentane	1050	154.70	81.72	0.528
	Real	Isopentane	1050	154.70	80.71	0.522
	Real	Butane	1050	154.70	78.51	0.507
	Real	Butane	1050	154.70	77.54	0.501
	Real	Butane	1050	154.70	74.48	0.481
[59]	Stimulation	Water	5593	350.00	84.00	0.240
[24]	Stimulation	Decafluoro-Butene	1909	295.50	150.00	0.508
[60]	Real	Water	6800	211.48	130.00	0.615
	Real	Water	6000	186.60	130.00	0.697
	Real	Water	4900	152.39	130.00	0.853
[61]	Real	Water	2295	73.00	43.00	0.589
[62]	Stimulation	Water	3950	105.70	68.00	0.643
	Stimulation	Water	3950	105.70	86.60	0.816
	Stimulation	Water	3950	105.70	53.00	0.501
[63]	Stimulation	Water	2340	73.18	19.90	0.272
[64]	Real	Water	1000	185.00	128.00	0.692
[65]	Stimulation	Water	4423	159.80	138.00	0.864
[28]	Stimulation	CO_2	1800	54.00	24.19	0.448
	Stimulation	Water	1800	54.00	18.43	0.341
	Stimulation	R134a	1800	54.00	27.30	0.506
	Stimulation	R152a	1800	54.00	27.69	0.513
	Stimulation	R227ea	1800	54.00	27.65	0.512
	Stimulation	R245fa	1800	54.00	26.48	0.490
	Stimulation	R1234ze	1800	54.00	27.85	0.516
	Stimulation	R600a	1800	54.00	28.92	0.536
	Stimulation	Pentane	1800	54.00	28.09	0.520
[66]	Stimulation	Water	4000	180.00	129.88	0.722
	Stimulation	Water	4000	180.00	129.28	0.718
	Stimulation	Water	4000	180.00	128.93	0.716
	Stimulation	Water	4000	180.00	128.96	0.716
	Stimulation	Water	4000	180.00	128.50	0.714
	Stimulation	Water	4000	180.00	128.35	0.713
	Stimulation	Water	4000	180.00	128.22	0.712
	Stimulation	Water	4000	180.00	128.11	0.712
	Stimulation	Water	4000	180.00	128.01	0.711
	Stimulation	Water	4000	180.00	127.92	0.711

References

1. Caulk, R.A.; Tomac, I. Reuse of abandoned oil and gas wells for geothermal energy production. *Renew. Energy* **2017**, *112*, 388–397. [CrossRef]
2. Alimonti, C.; Soldo, E. Study of geothermal power generation from a very deep pil weel with a wellbore heat exchanger. *Renew. Sustain. Energy Rev.* **2016**, *86*, 292–301. [CrossRef]
3. Alimonti, C.; Falcone, G.; Liu, X. Potential for Harnessing the Heat from a Mature High-Pressure-High- Temperature Oil Field in Italy Case Study: The Villafortuna—Trecate Oil Field. In Proceedings of the SPE Annual Technical Conference and Exhibition, Amsterdam, The Netherlands, 27–29 October 2014.
4. Abuaisha, M.S. *Enhanced Geothermal Systems: Permeability Enhancement through Hydraulic Fracturing in a Poro-Thermoelastic Framework*; University of Grenoble: Grenoble, France, 2014.
5. Karvounis, D.C. Simulations of enhanced geothermal systems with an adaptive hierarchical fracture representation. Ph.D. Thesis, ETH Zurich, Zürich, Switzerland, 2013.
6. Le Lous, M.; Larroque, F.; Dupuy, A.; Moignard, A. Thermal performance of a deep borehole heat exchanger: Insights from a synthetic coupled heat and flow model. *Geothermics* **2015**, *57*, 157–172. [CrossRef]
7. Kurnia, J.C.; Shatri, M.S.; Putra, Z.A.; Zaini, J.; Caesarendra, W.; Sasmito, A.P. Geothermal energy extraction using abandoned oil and gas wells: Techno-economic and policy review. *Int. J. Energy Res.* **2021**. [CrossRef]
8. Singh, H.; Falcone, G.; Volle, A.; Guillon, L. Harnessing geothermal energy from mature onshore oil fields, the Wytch Farm case study. *Work. Geotherm. Reserv. Eng.* **2017**, *17*, 13–15.
9. Akhmadullin, I. Utilization of co-produced water from oil production, energy generation case. In Proceedings of the SPE Health, Safety, Security, Environment & Social Responsibility Conference, New Orleans, LA, USA, 18–20 April 2017.
10. Xin, S.; Liang, H.; Hu, B.; Li, K. Electrical power generation from low temperature co-produced geothermal resources at Huabei oilfield. In Proceedings of the Thirty-Seventh Workshop on Geothermal Reservoir Engineering, Stanford, CA, USA, 30 January–1 February 2012.
11. Westphal, D.; Ruud, W. Economic appraisal and scoping of geothermal energy extraction projects using depleted hydrocarbon wells. *Energy Strateg. Rev.* **2018**, *22*, 348–364. [CrossRef]
12. Davis, A.P.; Michaelides, E.E. Geothermal power production from abandoned oil wells. *Energy* **2009**, *34*, 866–872. [CrossRef]
13. Wang, K.; Wu, X. Extension of oil well economic life by simultaneous production of oil and electricity. In Proceedings of the Society of Petroleum Engineers—SPE Oklahoma City Oil and Gas Symposium 2019, OKOG 2019, Oklahoma City, OK, USA, 9–10 April 2019.
14. Noorollahi, Y.; Pourarshad, M.; Jalilinasabady, S.; Yousefi, H. Numerical simulation of power production from abandoned oil wellsin Ahwaz oil field in southern Iran. *Geothermics* **2015**, *55*, 16–23. [CrossRef]
15. Macenić, M.; Kurevija, T. Revitalization of abandoned oil and gas wells for a geothermal heat exploitation by means of closed circulation: Case study of the deep dry well Pčelić-1. *Interpretation* **2018**, *6*, SB1–SB9. [CrossRef]
16. Ziabakhsh-Ganji, Z.; Nick, H.M.; Bruhn, D.F. Investigation of the synergy potential of oil and geothermal energy froma fluvial oil reservior. *J. Pet. Sci. Eng.* **2019**, *181*, 106–195. [CrossRef]
17. Liu, X.; Falcone, G.; Alimonti, C. A systematic study of harnessing low-temperature geothermal energy. *Energy* **2018**, *142*, 346–355. [CrossRef]
18. Mehmood, A.; Yao, J.; Fan, D.; Bongole, K.; Liu, J.; Zhang, X. Potential for heat production by retrofitting abandoned gas wells into geothermal wells. *PLoS ONE* **2019**, *14*, e0220128. [CrossRef]
19. Toth, A.N.; Szucs, P.; Pap, J.; Nyikos, A.; Fenerty, D.K. Converting Abandoned Hungarian Oil and Gas wells into Geothermal Sources. In Proceedings of the 43rd Workoshop on Geothermal Reservoir Engineering, Stanford, CA, USA, 12–14 February 2018.
20. Syarifudin, M.; Octavius, F.; Maurice, K. Feasibility of Geothermal Energy Extraction from non-activated petroleum wells in Arun field. In Proceedings of the 5th ITB Internation Geothermal Workshop, Bandung, Indonesia, 28 March–2 April 2016.
21. Al-Mahrouqi, J.; Falcone, G. An Expanded Matrix to Scope the Technical and Economic Feasibility of Waste Heat Recovery from Mature Hydrocarbon Fields. In Proceedings of the PROCEEDINGS Geothermal Reservoir Engineering Stanford University, Stanford, CA, USA, 10–12 February 2020; Volume 3, pp. 1–16.
22. Soldo, E.; Alimonti, C. From an Oilfield to a Geothermal One: Use of a Selection Matrix to Choose Between Two Extraction Technologies. In Proceedings of the World Geothermal Congress, Kyushu-Tohoku, Japan, 28 May–10 June 2000; pp. 19–25.
23. Kaplanoglu, M.A.; Baba, A.; Akkurt, G.G. Use of abandoned oil wells in geothermal systems in Turkey. *Geomech. Geophys. Geo-Energ. Geo-Resour.* **2020**, *6*, 10. [CrossRef]
24. Alimonti, C.; Soldo, E.; Berardi, D.; Bocchetti, D. A matrix method to select the more suitable extraction technology for the Campi Flegri geothermal area (Italy). In Proceedings of the European Geothermal Congress 2016, Strasbourg, France, 19–24 September 2016; p. 10.
25. Alimonti, C.; Soldo, E.; Berardi, D.; Bocchetti, D. A comparison between energy conversion systems for a power plant in Campi Flegrei geothermal district based on a WellBore Heat eXchanger. In Proceedings of the European Geothermal Congress 2016, Strasbourg, France, 19–24 September 2016; pp. 19–24.
26. Franco, A.; Villani, M. Optimal design of binary cycle power plants for water-dominated, medium-temperature geothermal fields. *Geothermics* **2009**, *38*, 379–391. [CrossRef]

27. Kharseh, M.; Al-Khawaja, M.J.; Hassani, F. Utilization of oil wells for electricity generation: Performance and economics. *Energy* **2015**, *90*, 1–7. [CrossRef]
28. Zhang, Y.; Yu, C.; Li, G.; Guo, X.; Wang, G.; Shi, Y.; Peng, C. Performance analysis of a downhole coaxial heat exchanger geothermal system with various working fluids. *Appl. Therm. Eng.* **2019**, *163*, 13. [CrossRef]
29. Templeton, J.D.; Ghoreishi-Madiseh, S.A.; Hassani, F.; Al-Khawaja, M.J. Abandoned petroleum wells as sustainable sources of geothermal energy. *Energy* **2014**, *70*, 366–373. [CrossRef]
30. Soldo, E.; Alimonti, C.; Scrocca, D. Geothermal Repurposing of Depleted Oil and Gas Wells in Italy. *Multidiscip. Digit. Publ. Inst. Proc.* **2020**, *58*, 9. [CrossRef]
31. Gudmundsson, J.S.; Freeston, D.H.; Lienau, P.J. Lindal Diagram. *Trans. Geotherm. Resour. Counc.* **1985**, *9*, 15–19.
32. Jusoh, N.A. *Decommiccioning Cost Estimation Study*; Universiti Teknologi Petronas: Perak, Malaysia, 2014.
33. Reyes, A.G. *Abondoned Oil and Gas Wells: A Reconnaissance Study of an Unconventional Geothermal Resource*; GNS Science: Lower Hutt, New Zealand, 2007.
34. Alimonti, C.; Soldo, E.; Bocceti, D.; Berardi, D. The Wellbore Heat Exchangers: A Technical Review. *Renew. Energy* **2018**, *123*, 353–381. [CrossRef]
35. Sanyal, S.K.; Butler, S.J. Geothermal Power Capacity from Petroleum Wells—Some Case Histories of Assessment. In Proceedings of the World Geothermal Congress 2010, Bali, Indonesia, 25–29 April 2010.
36. Johnson, L.A.; Walker, E.D. Oil production waste stream, a source of electrical power. In Proceedings of the Thirty-Fifht Workshop on Geothermal Reservoir Engineering, Stanford, CA, USA, 1–3 February 2010.
37. Van Erdeweghe, S.; Van Bael, J.; Laenen, B.; William, D. Comparison of series/parallel configuration for a low-T geothermal CHP plant, coupled to thermal networks. *Renew. Energy* **2017**, *111*, 494–505. [CrossRef]
38. Abramov, A. Optimization of well pad design and drilling—Well clustering. *Pet. Explor. Dev.* **2019**, *46*, 614–620. [CrossRef]
39. Jones, M.C. *Implications of Geothermal Energy Production via Geopressured Gas Wells in Texas: Merging Conceptual Understanding of Hydrocarbon Production and Geothermal Systems*; University of Texas at Austin: Austin, TX, USA, 2016.
40. Maklin, C. DBSCAN Python Example: The Optimal Value for Epsilon (EPS). Available online: https://towardsdatascience.com/machine-learning-clustering-dbscan-determine-the-optimal-value-for-epsilon-eps-python-example-3100091cfbc (accessed on 10 February 2021).
41. Sharma, A. How Does DBSCAN Clustering Work? *DBSCAN Clustering for ML*. Available online: https://www.analyticsvidhya.com/blog/2020/09/how-dbscan-clustering-works/ (accessed on 10 February 2021).
42. tec-science.com. Final Temperature of Mixtures (Richmann's Law). Available online: https://www.tec-science.com/thermodynamics/temperature/richmanns-law-of-final-temperature-of-mixtures-mixing-fluids/ (accessed on 10 February 2021).
43. Mixing of Fluids Having Different Densities—Pipelines, Piping and Fluid Mechanics Engineering. Available online: https://www.eng-tips.com/threadminder.cfm?pid=378 (accessed on 10 February 2021).
44. Rule of Mixtures Calculator for Specific Heat Capacity. Available online: https://thermtest.com/thermal-resources/rule-of-mixtures (accessed on 10 February 2021).
45. Trullenque, G.; Genter, A.; Leiss, B.; Wagner, B.; Bouchet, R.; Léoutre, E.; Malnar, B.; Bär, K.; Rajšl, I. Upscaling of EGS in Different Geological Conditions: A European Perspective. In Proceedings of the 43rd Workshop on Geothermal Reservoir Engineering Stanford University, Stanford, CA, USA, 12–14 February 2018; pp. 1–10.
46. Short, W.; Packey, D.; Holt, T. A manual for the economic evaluation of energy efficiency and renewable energy technologies. *Renew. Energy* **1995**, *95*, 73–81. [CrossRef]
47. Pumps, N.R. How to Read a Pump Curve. Available online: https://www.northridgepumps.com/ (accessed on 16 February 2021).
48. Fetoui, I. ESP Design—Hand Calculations. Available online: https://production-technology.org/esp-design-hand-calculations/ (accessed on 26 February 2021).
49. Schlumberger REDA Electric Submersible Pump Systems Technology Catalog 2017, 360. Available online: https://www.slb.com/-/media/files/al/catalog/artificial-lift-esp-technology-catalog.ashx (accessed on 29 February 2021).
50. The Engineering Toolbox Darcy-Weisbach Pressure and Major Head Loss Equation. Available online: https://www.engineeringtoolbox.com/darcy-weisbach-equation-d_646.html (accessed on 29 February 2021).
51. ENTSO-E Data View—Transmission. Available online: http://dataview.ofsted.gov.uk/#/Tab/?percentageType=1&remit=3&deprivation=0&providerType=7&judgement=1&provisionType=0&year=2013-08-31&areaType=1®ionId=0&similarDate=2013-08-31®ionOne=0®ionTwo=0&eightRegions=false&tabName=LocalAuthorityFocus&_=13 (accessed on 1 March 2021).
52. Eurostat Database—Eurostat. Available online: http://ec.europa.eu/eurostat/data/database (accessed on 1 March 2021).
53. International Energy Agency Data Overview—IEA. Available online: https://www.iea.org/data-and-statistics (accessed on 1 March 2021).
54. The Statistical Office of European Union. Available online: https://ec.europa.eu/eurostat/ (accessed on 1 March 2021).
55. Sigfússon, B.; Uihlein, A. *2015 JRC Geothermal Energy Status Report*; EUR 27623; Publications Office of the European Union: Luxembourg, 2015.
56. Sigfusson, B.; Uihlein, A. *2014 JRC Geothermal Energy Status Report*; EUR 26985; Publications Office of the European Union: Luxembourg, 2015.
57. Alimonti, C.; Berardi, D.; Bocchetti, D.; Soldo, E. Coupling of energy conversion systems and wellbore het exchanger in a depleted oil well. *Geotherm. Energy* **2016**, *4*, 17. [CrossRef]

58. Immanuel, L.G.; Almas, G.S.F.U.; Dimas, T.M. Experimental Desing of Wellbore Heat Exchanger in Binary Optimization for Low—Medium Enthalpy to Utillize Non-Sefl Discharge Wells in Indonesia. In Proceedings of the 43rd Workshop om Geothermal Reservoir Engineering, Stanford, CA, USA, 12–14 February 2018; p. 8.
59. Nalla, G.; Shook, G.M.; Mines, G.L.; Bloomfield, K.K. Parametric sensitivity study of operating and design variables in wellbore heat exchangers. In Proceedings of the Twenty-Ninth Workshop on Geothermal Reservoir Engineering, Stanford, CA, USA, 26–28 January 2004.
60. Wright, N.M.; Bennett, N.S. Geothermal energy from abandoned oil and gas wells using water in combination with a closed wellbore. *Appl. Therm. Eng.* **2015**, *89*, 908–915. [CrossRef]
61. Kohl, T.; Brenni, R.; Eugster, W. System performance of a deep borehole heat exchanger. *Geothermics* **2002**, *31*, 687–708. [CrossRef]
62. Kujawa, T.; Nowak, W.; Atachel, A.A. Utilization of existing deep geological wells for acquisitions of geothermal energy. *Energy* **2006**, *31*, 650–664. [CrossRef]
63. Sliwa, T.; Gonet, A.; Sapinska-Sliwa, A.; Knez, D.; Juzuit, Z. Applicability of Borehole R-1 as BHE for Heating of a Gas Well. In Proceedings of the Proceedings World Geothermal Congress 2015, Melbourne, Australia, 19–25 April 2015; p. 10.
64. Alimonti, C.; Soldo, E.; Moroni, E. Evaluation of geothermal energy production using a WellBore Heat eXchanger in the reservoirs of Vampi Flegri and Ischia Island. In Proceedings of the European Geothermal Congress 2016, Strasbourg, France, 19–24 September 2016; p. 6.
65. Noorollahi, Y.; Yousefi, H.; Pourarshad, M. Three Dimensional Modeling of Heat Extraction from Abandoned Oil Well for Applicatioin in Sugarcane Industry in Ahvaz—Southern Iran. In Proceedings of the World Geothermal Congress 2015, Melbourne, Australia, 19–25 April 2015; p. 11.
66. Bu, X.; Ma, W.; Li, H. Geothermal energy production utilizing abandoned oil and gas wells. *Renew. Energy* **2012**, *41*, 80–85. [CrossRef]

MDPI

St. Alban-Anlage 66

4052 Basel

Switzerland

Tel. +41 61 683 77 34

Fax +41 61 302 89 18

www.mdpi.com

Geosciences Editorial Office

E-mail: geosciences@mdpi.com

www.mdpi.com/journal/geosciences